MANUAL OF PHYTOGEOGRAPHY

En souvenir de mon Père, Victor Croizat
Chambéry (Savoie) 1856 — Torino (Piemonte) 1915

LÉON CROIZAT

MANUAL OF
PHYTOGEOGRAPHY

OR

AN ACCOUNT OF PLANT-DISPERSAL
THROUGHOUT THE WORLD

Springer-Science+Business Media, B.V.
1952

ISBN 978-94-017-5744-7   ISBN 978-94-017-6113-0 (eBook)
DOI 10.1007/978-94-017-6113-0

by N.V. DRUKKERIJ HOOIBERG, EPE

CHAPTER INDEX

NOTICE

Prefaces being themselves often redundant, I should not bring to these pages an additional Notice, and thus sin twice perhaps against the unwritten code of authors. To account for this breach of the rules, I should inform the reader that, written amid the royal splendor of the bibliographic means at one time available to me as a member of the Arnold Arboretum of Harvard University, this Manual reached my hands in proof-stage in the city of Caracas, capital of the Republic of Venezuela, South America. In consideration of the somewhat less than abundant literature here available and of a heavy schedule of field-duty between the Andes of Mérida and the sources of the Orinoco, I have definitely not been in the position of lavishing upon proofreading a wealth of means and a store of quiet times. In particular, great many titles cited in the coming pages could no longer be had, and thus had to be left unchecked. This may explain why errors and omissions easily avoidable otherwise prove on the contrary unavoidable under the conditions now ruling.

In the three years elapsed following my departure from the United States I have put together a second work, larger than the present, in which the dispersal of animals (birds, mammals, fishes, reptilians, insects, etc.) is worked out along the pattern of investigation outlined in the coming pages. It is likely that this second work will soon be released from the press.

Planning, and even less putting through this undertaking, would have been impossible outright had I not acquired, while busy on the present Manual, some understanding of the fundamentals of distribution. Thus, I had a welcome opportunity of checking the main conclusions upon which this Manual rests. On the whole, its fundamentals proved serviceable, which induces me to believe they are at least to some extent true.

In sum, were I to rewrite the book here offered for the attention of the reader, I should stand by the very same principles and methods. Of course, the dispersal of animals differs to an extent from that of plants. Problems like an exact definition of what is "antarctic" dispersal, for example, can hardly be worked out clean by making reference only to plant-life because of the limitations inherent in the subject. The same is true of certain phases of the distribution currently understood as "holarctic". Issues of the kind can only then be seen lucidly when flowering and non-flowering plants, birds and mammals, fishes and reptilians are dealt with as one. Having worked out my subjects one by one, in order to test the same fundamentals over and over again, I am not unmindful of considerable accretions in my knowledge as the result of years of attention given to the same general matters. However, I have no reason to lament that the Manual here offered for attention should leave issues less than well closed which I could, perhaps, bring nearer to a solution today. It is desirable that the limitations of a subject should be accepted for what they are, and that reason be never strained beyond the limits of the known. A day is definitely to come, when the whole of dispersal, animals' as well as plants', shall be seen as one. When this day dawns, all of us will understand the two much better as part of a single whole, not only, but do so with a precise understanding of the peculiarities and technical limitations of either at the price of having sailed along with time and available means.

Venezuela, a country unparalleled in natural wealth and a paradise for the naturalist throughout, has brought to my attention many problems more or less immediately related to the subject dealt with in these pages. Although present means are here still below requirements, still something may be done. Time is not yet for me to put on paper what I have observed in Venezuela but I have some confidence that whatever I have learned is to fit in without difficulty with the principles first stated and applied in the pages that follow.

Leon Croizat.

April 1st, 1950—April 1st, 1952.

Museo de Ciencias Naturales
Caracas, Venezuela, South America
Universidad de Los Andes
Mérida, Venezuela, South America

VIII

PREFACE

The notes on which this manual rests were put together to answer my own questions and to satisfy my own curiosity. The problem of origins in time and space was constantly before me for the best part of ten years devoted to taxonomy and general botany.

Naming plants is but a means to an end which is deeper knowledge of life in all its aspects. I was struck speechless one day seeing that one of the alpine spurges of the Western Alps, *Euphorbia Valliniana,* could hardly be told apart from *E. Potaninii* in far away Mongolia. Soon I learned to take events of the kind for granted because there was no end of them in and outside the Euphorbiaceae. As I inquired about the reasons, why and how dispersal constantly defies time and space, I became familiar with an extensive literature.

Much of this literature was instructive and interesting, no doubt, but it failed to satisfy me. I could find little merit in efforts directed to explain migration as the byproduct of carriage of seeds by winds, birds, etc., for it was patent that a plant like *Euphorbia Valliniana* had no known „Means of dispersal." Moreover, supposing that these „Means" had been at work, the question remained whether Mongolia had fed the Western Alps or the other way around, or whether, perhaps, a third center had independently fed these two regions, and when. In short, here was something before me apalling in terms of mileage and time. To account for it something to match its scope seemed to me absolutely necessary. I was impressed that hardly anything of this nature was to be found in the literature known to me.

In the years that followed I undertook to break the riddle by the simplest of all procedures. One after the other, I reviewed the distribution of hundreds of different groups and families. In the end, I became thoroughly satisfied that all these groups and families spoke the same language and that the least of my difficulties was the lack of facts. Indeed, I had too many of them. Certain concepts occurred to me readily, but it proved difficult to refine them, and years were actually spent laboring over issues which in the end turned out to be quite simple. I should have known forthwith what it took me a long time to learn, which is the reason why I still find it difficult to believe that the human mind can easily be relieved of toil, and science freed of much useless discussion.

It is clear before me that this manual represents a beginning. The record of distribution available is still largely insufficient, and an enormous amount of field-work and exploration remains to be done before certain of the issues brought to focus in these pages are seen with the required lucidity. I believe on the other hand that the principles advanced in the coming pages, and the examples dealt with, will be found useful in the work ahead. The only justification that I can give myself and others for this work is that *it answers practical needs*. Claims to full adequacy or complete accuracy in details are not in my intentions. I have used many records, perfect and imperfect alike, and to the mistakes of others I have undoubtedly added some of my own. This manual will justify, and in part condemn itself, only with time. For this reason, it is not my intention to enter into controversy, but to allow the record to speak for itself.

Were I asked to state in all candor what is in my opinion the most relevant contribution from these pages to the study of life, I would answer without hesitation. *This contribution is in presenting a global picture of plant-dispersal in time and space.* The record cannot be brought farther back than the Jurassic for the present, but the lapse of time it covers after this age is adequate in most respects.

Geophysicists, geologists, and biogeographers generally, must work together, for all of them have something vital to contribute to the success of the work ahead. I may speak only for the phytogeographer, but speaking for him, I feel to be right in stating that to this very day plant-dispersal has only been handled piecemeal. Should the geophysicist and the geologist turn to the phytogeographer, and pointedly ask that he come forward with a rational and fairly complete account of angiospermous dispersal in the first place, the phytogeographer would have nothing to offer. Indeed, he could submit theories, generalizations about plant-geography being "descriptive" before "interpretative", tales of sundry mysterious "disconnections" to be bridged by the help of "Extraordinary capacities for transoceanic dispersal," etc. He could refer to "polyploidy," "species-senescence," quote presumed authorities who insist that only "perfect" taxonomy is to make "perfect" phytogeography, and expatiate perhaps for long hours on these and similar arguments. Pressed, however, to furnish *something concrete*, at least a tentative chronology of dispersal, an outline of the major channels of migration, a rationalization of the interrelation of evolution and migration, the phytogeographer would now be pressed in vain.

Having nothing concrete to offer, the phytogeographer can ask nothing concrete in return. He may not go to the geophysicist and geologist to inquire of them what is their explanation of the fact that great depths of the ocean fill today the space which dispersal shows was given over to lands in former epochs.

He cannot press them with questions, how they happen to believe that only the coasts of the Western Pacific crumbled, when vegetable distribution intimates that the same fate also befell those of the Eastern Pacific. He cannot, in short, *trade knowledge*, and for this reason he, not only, but those who ought to rely upon him are made all the poorer.

This fateful bar to knowledge has been paramount in my mind, and I have bent every effort in order to lessen its weight. Whatever be the result of my work, a platform of concrete discussion has at least been made available in its pages on the strength of elements that are essentially factual. Were not a single one of my conclusions to stand whole, still this work would have performed the not indifferent service of forcing the replacement of these conclusions with others, and better ones, of the same *general scope*. We do not need facts as much as we need to organize the facts we have within a rational pattern of interpretation. Botany has lost sight of this homely truth, and is in the danger of sinking for this very reason into a morass of unrelated petty specialties, extolling description as the supreme goal of scientific inquiry. This may be a comfortable belief, but is essentially a false one.

Knowing that geophysicists, geologists and biogeographers are definitely bound to work together, I have been careful throughout to present a picture of vegetable dispersal which stands on its own merits and demerits. Nothing could be easier than to pore over the literature of sciences unrelated to botany, and to yield to the temptation of seeking corroboration for my own opinions in the geological maps of Africa, the New World, etc. This temptation I have resisted as fundamentally contrary to the purpose for which this manual was written. Naturally, I have some reason to believe that my opinions on the subject of vegetable dispersal do not fundamentally conflict with the facts in the records of other sciences. There is some measure at least of guarantee to this effect, because my own conclusions are based upon the record of plant-distribution in the first place. This record cannot be fictitious, and what legions of plants bespeak cannot be a dream. These plants behave in an orderly fashion, indeed, monotonously, and the monotony of their behavior gives at least hope that the underlying reasons of their migrations may be known. In short, vegetable life speaks for itself in the first place, and it should be a mistake, I believe, to handle it at first with constant reference to other non-botanical sciences. A conciliation can and will no doubt be made between phytogeography and these sciences, but coming collaboration and discussion would gain nothing in lucidity if different wines were poured together into one skin in the beginning. It is up to each department to tabulate its own items first, later to merge the whole within a single ledger. *If we are to strive for something, we must have a cleancut idea of what we seek in the first place, and of the means most appropriate to secure its accomplishment.*

No doubt, botanists will find in this manual answers to some of the questions which they have long sought, and the suggestion of many more problems the solution of which lies in the future. Whatever be their immediate field of endeavor, theoretical or practical, they will be able at least to set their minds at rest as to the approaches of the fundamental problems in dispersal, and to glance upon the whole of it at one stroke. They will easily pass from the Caribbeans to Malaysia following well worn tracks of migration.

Likewise, geophysicists and geologists who do not speak the language of systematic and taxonomic botany will be offered in this manual something tangible in their own fields. They will be hopefully told what vegetable dispersal demands of the maps of the past, and be invited thereby to collaborate with the phytogeographer much more closely than they ever did in a time when the lack of suitable texts made collaboration practically impossible.

This collaboration is imperative, and rapid steps toward better knowledge can be secured if collaboration is achieved. The earth and its vegetable mantle are so intimately bound together that, paradoxical as the statement may sound, this mantle is in the ultimate reality a layer of the earth itself. It is in a sense *a geological feature*, one which in some "modern" aspects easily goes back to the earliest Cretaceous at least.

No doubt, animal life obeys the same laws as do plants, though in a different degree, and under limitations that are peculiarly its own. It is not for me to identify this degree and to work out these limitations, but it is clear to me that biogeography in general will go nowhere if it persists in assuming "Extraordinary capacities for transoceanic dispersal" without cloaking these and similar slogans with a precise meaning as to what "Extraordinary capacities" are, or may be. Words do not make for thoughts.

Brief notes are not amiss on certain minor aspects of this undertaking.

I am definitely set against the opinion that a large bibliography, and lavish compilations are of the essence of a scientific undertaking. Bibliography answers two definite needs, and two needs only. It acquaints the reader with the authority for certain statements, which is but fair to him. It guides the reader to sources which he may like to consult for further instruction or pleasure, which is the duty of the author to furnish. Considerable latitude of judgement must be exercised in meeting both these needs, and I am far from sure that my judgement in choosing references has constantly been happy.

At any rate, I have made no effort to bring forward literature that has purely historical value or deals with theories that do not demand serious discussion. Compilation is space-consuming, and print can as a rule be given to more useful undertakings. To omit this literature seemed to me all the more desirable, because a wealth of it is contained in recent publications. Three well known works of E. V. Wulff are useful as regards references of this kind. Two of them are in the Russian language, "Vvedenie v istoriceskuio Geografiio Rastenii (English subtitle: Introduction to the Historical Geography of Plants), published as "52nd Supplement of the Bulletin of Applied Botany, Genetics and Plant-Breeding", Leningrad, 1932. (Prilozenie 52 k Trudam po Prikladnoi Botanike Genetiki i Selektzii), and Istoriceskaia Geografia Rastenii (Historical Geography of Plants) released in full book-form, Leningrad, 1944. The former of these works has recently been translated in a somewhat abridged form into English as, An Introduction of Historical Plant Geography (Transl.

E. BRISSENDEN) by the Chronica Botanica Co., of Waltham Mass., U.S.A., 1943. Although the Russian text is not easily accessible to Western readers, the bibliography is given in the original languages and can readily be consulted. This bibliography covers the field adequately enough to meet the needs of practically every botanist who is keen to learn about the historical aspects of phytogeography, and to assemble reference on special subjects. *)

The taxonomic records which I have used are often referred to their sources. To give full references proved at times impracticable for the simple reason that the work of several authors were consulted to assemble the data relating to a single group. In some cases I have implemented these data with material taken directly from the herbarium or card-indexes. Standard taxonomic sources, such as the Index Kewensis and the Gray Index were often fingered, and access was had throughout to a vast array of standard floras. Concerning these, we are fortunate in the possession of an excellent source-book, Geographical Guide to Floras of the World, Part I, by S. F. BLAKE and A. C. ATWOOD, published by the U. S. Department of Agriculture, Miscellaneous Publication No. 401, Washington, 1942.

Little need be said about the use made of certain geographical names. Malaysia has become the currently used designation of the islands large and small contained, roughly speaking, within the triangle Sumatra-Philippines-New Guinea, and I have employed it throughout in this sense. The designation Malacca is manifestly improper as I have maintained it. Malacca is but part of the peninsula which almost entire, though not absolutely whole, is occupied by the Federate Malay States. This peninsula is often referred to as Malaya or Malayan Peninsula, but to avoid confusion I have kept to the designation Malacca, thus following the precedent set in the Flora of British India by J. D. HOOKER.

I have further standardized reference to various quarters of the same land as Northern, Southern and the like, except in very few cases when awkwardness would result; Northern America has hardly currency against North America. I have chosen the longer ending against the shorter for the simple reason that, for instance, South Australia is genuine political division of Australia, while Southern Australia includes not only South Australia but the southern section of Western Australia and New South Wales with the whole of Victoria.

---

*) Ecology does not concern this manual, but a brief discussion of its limits toward phytogeography will be found in a separate chapter. The literature of ecology is enormous, and can have no place in these pages. Handy bibliography which illustrates the main aspects of standard ecological work, and the use of the current terminology of this science, is represented by, (a) WOOD, J. C. — The Vegetation of South Australia; pp. 1—164, map. Adelaide, 1937; (b) VAUGHAN, R. E. & WIEHE, P. O. — Studies on the Vegetation of Mauritius, I. II; in Journal of Ecology 25: 289-343, map 1937, op. cit. 27: 263-281. 1939; (c) CASTELLANOS, A. & PEREZ MOREAU, R. A. — Los Tipos de Vegetación de República Argentina; Publicación No. 3, Universidad de Buenos Aires, Facultad de Ciencias Exactas, Físicas y Naturales, pp. 1-154, map. 1945 (1944). The first of these works contains a condensed but interesting bibliography and notes on the interrelation of soils and vegetable life; the second is illustrative of generalities of current methods, and highlights suggestive issues of evolution and morphology; the third is invaluable on account of its giving some of the main technical terms in Spanish, English, German and French, and a very extensive bibliography. These comments are intended to direct the curious reader, not to judge of the works in question, or those of other authors whom brevity forbids to mention.

The islands of Hispaniola in the West Indies is in two political divisions, Haiti and Santo Domingo. No attempt has been made to discriminate in detail as between these countries whenever the record failed to advise the contrary. Occasionally, special reference, has been made to certain part of a country as against certain others, witness repeated mentions of Eastern Cuba or, Cuba (Oriente). Eastern Cuba has significant position in dispersal. India, of course, is a very broad designation, and for this reason I have more often than not spoken of the Himalayas, Bengal, Eastern India, Punjab and the Kashmir. The Caucasus is rich in subdivisions, some of them significant, but the nature of this manual is such that detailed accounts of them could not be entertained. So far as possible the Mexican, Venezuelan, Peruvian, etc., states have been mentioned in detail.

The difficulty of suitably illustrating a manual of this nature is truly tremendous. On this account the maps in its pages easily lend themselves to misunderstandings which it is desirable to forestall.

Like a figure a map answers a definite purpose, and only then is seen in proper light when the purpose is known for which it was made *).

In current botanical practice, a map showing plant-distribution is prepared, and used, with the prior tacit understanding that the records shown are as nearly accurate in every detail as the map's author can make them. So viewed, a map does more than to integrate the text, for it may indeed replace it as to the essentials.

The maps that illustrate this manual answer *an entirely different purpose*. They do not strive to picture the full record of actual distribution in the first place, and they cannot be used without reference to the notes and elucidations in the main text.

These maps are *essentially diagrammatic* and aim to present what I take to be essential, and useful, ideas how dispersal effects itself. It is obvious that for this very reason static records of ranges in conventional outline are not satisfactory to my purpose. This manual need not have been written to bring records of this kind to the fore. These records are available in other works and easily compiled out of the standard indices and monographs of taxonomic botany, not only, but authorities for my data are currently mentioned in the text.

In order that my thought on the score may be entirely clear I refer for elucidation to the map, for example, which illustrates the relations which Malaysia bears to the main arteries of angiospermous distribution. This map shows (Fig. 31) an overwhelming tangle of tracks, and it is physically impossi-

---

*) Much advanced biogeographic work in the German language especially makes no use of actual maps, and relies for the graphic presentation of its subjects upon straight symbols, in which reference is made to the name of detailed or generalized localities connected by lines or arrows oftentimes orientated to show the trends of migration. This method is indeed excellent, and readers having intimate knowledge of world geography demand hardly any other. After mature thought, I have decided not to follow it, however, preferring to offer the readers of these pages outlines of land and sea which are beyond doubt familiar to all.

ble to work out this tangle into details unless by using whole series of maps on a large scale.

Were such maps made out, and the whole pictured accurately in detail, with possible alternatives, references to specific instances of migrations and the like, the reader would forthwith lose the impression — which he secures at a glance *by one map* — how mistaken are the beliefs of authors who suppose that Malaysian dispersal can be handled as a local Malaysian problem. The reader would know everything in detail, perhaps, but he would not be offered graphic global material which in my opinion is beyond comparison more useful at this stage of work than a graphic, pedantic knowledge of details. This map consequently is *true and accurate as to its main purpose,* and this main purpose is singled out pointedly. I wish it were possible to work out the details in addition referring to great many maps, but this is manifestly outside the province of such a manual as this. The reader will have to read the text, follow the discussions and notes patiently, and reach the same goal by means other than being offered here a complete sets of extensive records of distribution on maps in an easily readable scale.

Another instance worth mentioning is that of the map (Fig. 45) which illustrates the dispersal of the Ericaceae Vaccinioideae. This map gives a seemingly disproportionate amount of symbols to a petty outlier of two species of *Vaccinium* in South Africa, and accordingly appears to slight the enormous massing of species in this genus and its immediate allies endemic to the Far East. The truth is, however, that the insignificant outlier in South Africa (insignificant, that is, from the standpoint of taxonomy) is vital to a proper appreciation of the origin of this large group.

The map illustrating the main trends of dispersal of the Ericaceae Gaultherieae (Fig. 44 A) stresses the lone species of genus *Wittsteinia* in cleancut fashion, and handles but superficially, therefore in a sense less than accurately, the massive component of this group which is in the Americas. The reason for this is that *Wittsteinia* — and its lone species — are cardinal if we are to appreciate the genetic ties between Gaultherieae and Vaccinioideae. Hundreds of other species in these two groups have no value to compare with this single one.

In certain cases, a map cannot be made to contain the distribution of a single group of genera, or family, in a satisfactory manner, for the group or family is widely dispersed, for instance, in Africa, the Americas, Australia and the northern hemisphere generally which demands the use of a map on a big scale. Such a map cannot be made to contain details, and the tracks it presents, consequently, are bound to be quite generalized.

In conclusion, all the maps in this work convey *ideas* which — in my opinion — are relevant and surely necessary. These maps do not attempt to tabulate *facts in detail.* Naturally, these maps could not be correctly used, and even less justly understood, if these primary factors were ignored or neglected. Dispersal cannot be made the subject of a written work which exempts the reader *from continuous access to an atlas and sundry taxonomic data.* The reader must know, or be willing to learn, geography in the first place, and to have of general botany some knowledge beforehand.

It could be suggested that maps which portray *ideas,* therefore carry their own moral on the sleeve, as it were, are propaganda. This suggestion need not be taken seriously, and will most likely not be taken seriously by the great majority of the readers. Ideas are essential to science, because facts about which no one has any ideas for the time being are hardly existent as regards knowledge.

It is to be anticipated that some of the maps in these pages will shock readers accustomed to conventional discussion as bold beyond words. I feel confident, nevertheless, that when these readers are through with the entire evidence, and thoroughly well apprised of the fact that distribution cannot be studied as a provincial issue interesting the Mediterranean, North America, Malaysia or Argentina, they will most likely reach different conclusions. If there is something seemingly bold and novel in these maps, the readers to whom they may seem so may well ask themselves the question, "What could I do to replace these maps with better ones?"

This is not a challenge, nor an expression of undue faith in my conclusions. It is, rather, the candid result of my wish that, whenever and wherever I may be off the path of reason, somebody steps in who is able to contribute constructive changes. A subject that has the earth for its theater during uncounted ages, and involves as one a multitude of specialized sciences, has tremendous limitations and is bound to be mastered by degrees after the pioneer work meant to clear its approaches has been done at least to an extent.

Leon Croizat

TO THE READER

Every science tends to speak a language of its own, and a textbook of mathematics contains, for example, abundance of symbols and formulae. Brought together in certain ways, these symbols and formulae readily satisfy a reader familiar with their use that a conclusion follows as against another. A reader unacquainted with this language will, on the contrary, see nothing whatever. Clearly, the author of this handbook may not be quicker of wits than the reader who is unable to follow him. It may be that the former knows conventions and postulates of which the latter has as yet no knowledge.

Like every other science, phytogeography (or, the study of plant dispersal) speaks a language of its own, which consists in the last resort of plant-names and records of localities. Let us repeat that a sound grasp of geography; visualizing the map as something better than a diagram to consult reluctantly; an appreciation of the fundamentals of botany throughout are prerequisite to a study in dispersal.

Doubtless, the reader is familiar with the last of these requirements. As regards the first two, he may test himself on the spot, seeing for instance whether these names bring something definite to his mind;

(1) Kerguelen Islands; (2) Mascarene Islands (or Mascarenes); (3) Madagascar; (4) Ceylon; (5) Indochina; (6) Java; (7) Celebes; (8) Philippines; (9) Formosa; (10) Queensland; (11) New Guinea; (12) Fiji; (13) Rapa Island; (14) Juan Fernandez Islands.

If most, when not all, these names bring to him the image of certain points of the map connected in one direction or other by a common thread, the reader is ready to follow a discussion in dispersal, whether plants, fishes, birds or snails are involved. So readied, he will understand why the author of this discussion compares various patterns of distribution, and drives home the ultimate conclusion that the facts warrant certain inferences.

A reader who cannot follow the matter while seeing the world before him all the time, is bound to have reference to the map. With the map before him (Fig. 1) he will easily locate these points, not only, but form a concept of *track*, thread or channel of migration, connecting various regions wherein allied plants appear.

Here is a difficulty. Many localities from all over the world, and many tracks must be introduced to a study of dispersal, some of which are so small, or so short, that the maps condensed to illustrate a textbook are not made to show them. Frequent are the cases in which a track that interests two or more continents highlights somewhere an island or mountain which is less than a speck against the background of the whole.

This difficult can be laid to rest in part by suitable illustrations, but cannot be eliminated altogether *). The author of a work on dispersal is bound to assume that the reader is familiar with the map or conversely willing to satisfy himself as to the precise location of a region new to him.'

The reader, no doubt, will be surprised at first seeing how spectacular is the pageant of vegetable dispersal, and how easy it proves for lowly herbs to travel at ease over the five continents. He will thus readily come the conclusion that there can not be phytogeography without geography, and that a subject so interesting is well worth initial efforts.

It is possible, however, that the reader does not care to wade through many detailed records of dispersal, and abundant notes on what these records appear to suggest, as the price to pay for reaching in the end conclusions the significance of which may not be clear to him from the start.

In consideration of this feeling, and as a compromise between it and the inescapable requirements of our subject, we venture to suggest that the Chapter XII of this manual contains a digest of our conclusions with the addition of brief considerations. The reader will also find in this chapter an outline of the thirteen major channels of vegetable dispersal.

If, then, the reader turns to that chapter first, he will forthwith learn what our conclusions are, and have something tangible in his hands to work out some of his own problems, to see whether what we suggest is workable, and worth reading. It is not impossible that the reader will decide in the end to peruse the chapters that precede the last, for only in these chapters will he find authority, and documents, for the ultimate conclusions and observations with which he will begin.

---

*) See the concluding part of the Preface for an account of some of these difficulties and limitations.

CHAPTER I: PRELIMINARIES

Whatever concerns plant-distribution belongs of necessity to sciences outside of botany. Vegetables are fed upon by animals, and travel together with animals over well worn paths which jointly interest all naturalists. This being the case, we may not enter our own investigation of dispersal without first looking around, and taking brief notice of certain works in the field of general biogeography.

It is unfortunate that, while excellent in most regards, many of these works are essentially theoretical or deal extensively with branches of lore alien to botany. As botanists, we may not be qualified to handle the intricacies of the classification of birds, fishes and the like, and it would be presumption on our part to share here in controversies to which this classification gives rise *). We believe that we can best serve the study of vegetable dispersal at this stage of work, if we steer clear of subjects which we are as yet not thoroughly ready to discuss, and do not invite controversy *as to details*. Once we are on record with cleancut generalities of method and an adequate supply of facts in our own field, time will do its work for and against us, without it being necessary that we compile diligently, but perhaps not always successfully, from the contributions of naturalists busy outside of botany.

If we see no reason for arguing and compiling at length in proud anticipation that we may seem to be well informed inside and outside botany, still we may not ignore altogether some of the tendencies now current in biogeography. As it is well known, many of the zoologists of a former generation saw no harm in accounting for dispersal through landbridges, and took for granted, for instance, that a community of mammals or birds between Africa and South America could come about, and was indeed established, only because these continents were once connected by an intervening landmass. These authors believed in orderly massive migrations such as could be accomplished only overland, and were willing to visualize landbridges fit to suit their belief.

Today, the trend in certain branches of zoology is wholly to the contrary, and seemingly good authorities dismiss authors who trust landbridges as naive and wanting in seriousness. Dispersal — these authorities maintain — takes place by means other than overland connections, and were we to hold

---

*) These controversies are as a rule of no consequence. We will prove it in a coming work on animal dispersal.

to the contrary we might be dismissed as reactionaries living in the past of the biological sciences.

Naturally, we could ignore these strictures in the knowledge beforehand that current fads have a great deal to do with them, and that a day is to come when landbridges now derided by certain American ornithologists, for example, are again *à la mode*, patiently waiting for the pendulum to take another swing. It will repay our time, nevertheless, if, without waiting, we briefly review the utterances of authors who do not believe in landbridges. The issue that faces us here is important, and it is clearly desirable that we know what to think of its main aspects.

Dealing with the origin of the bird-fauna of Hawaii, an ornithologist (1; p. 47) writes as follows, "There is not a single serious modern student (I use the term *serious* advisedly) who believes in the former existence of land bridges between America and Hawaii, or between Polynesia and Hawaii." Three years later, this author (2; p. 12, 36) submits a diagram intended to elucidate the components of the bird-world of the Americas (Fig. 2), and explains that favorable conditions were lacking for tropical life to pass freely between Asia and America in the latitude of the Aleutians. This explanation is followed by a statement to the effect that, "The close relationship between the Old and New World members of the Pantropical element, whose ranges are now widely discontinuous, proves that such a faunal exchange must have taken place, and this places the zoogeographer in a real quandary. The customary solution for the problem is to ignore it . . . . In view of the improbability of a North Atlantic land connection, various attempts have been made to find new routes for the transpacific migration. I shall refrain from a discussion of the various proposed transpacific land bridges. They are faunistically possible, but find no geological support. There is, however, some evidence for considerable recent tectonic activity in and south of the Aleutian island region, as well as for a pronounced lowering of the floor of the Pacific as a whole."

It is not our intention, in quoting these statements and bringing forth this map, to imply that the author in question is glaringly inconsistent. Consistency — as it it has been shrewdly remarked — may indeed be a mere synonym of obstinacy.

It is patent, nevertheless, that the author of the statements quoted was by no means averse — when penning the latter if not the former — to entertaining the hypothesis that there once was some sort of "landbridge" south of the Aleutians, and stood in fact ready to accept a pronounced lowering of the floor of the Pacific Ocean *as a whole* as one of the causes, perhaps the main cause, why this "landbridge" is no longer to be seen on our maps. It seemed in the end wiser to this author to believe in a transpacific connection south of the Aleutians rather than to remain in what we himself defines "A real quandary," or to dismiss fundamental problems in dispersal as something which need be allowed to sleep lest it becomes prurient. In short, there is no doubt from the texts quoted that this author was willing to handle these problems both cherishing and ridiculing in turn landbridges and similar connections.

If, then, this is the case, we cannot accept as valid the logic of the author quoted. This author, we believe, does not think lucidly in the first place, and what he may know of detail is not served adequately in the higher sphere of pure thought. It is also less than correct to aver that a landbridge in the approximate modern latitude of Hawaii finds no support in geology. This very same bridge, moreover, is shown to have been in existence by an overwhelming evidence from plant-distribution, much of it reviewed later in these pages.

One of the busy minds of science, Darwin, eloquently pleads in the pages of his classic (3) that dispersal is reduced, in the end, to chance dissemination, and introduces painstaking experiment to prove that particles of mud attached to the feet of certain birds may convey germs of life across thousands of miles. Darwin, nevertheless, is well aware that even this does not account for everything in plant-dispersal *) and introduces to the work just cited the following passage, "The facts seem to indicate that distinct species belonging to the same genera have migrated in radiating lines from a common center, and I am inclined to look in the southern, as in the northern hemisphere, to a former and warmer period, before the commencement of the last Glacial period, when Antarctic lands, now covered with ice, supported a highly peculiar and isolated flora. It may be suspected that before this flora was exterminated during the last Glacial epoch, a few forms had already widely dispersed to various points of the southern hemisphere by occasional means of transport, and by the aid as halting-place of now sunken islands. Thus the southern shore of America, Australia and New Zealand may have become slightly tinted by the same peculiar forms of life." Darwin further states, "New Zealand, South America and the other southern lands have been stocked in part from a nearly intermediate though distant point, namely from the antarctic islands, when they were clothed with vegetation, during a warmer tertiary period, before the commencement of the last Glacial period." This is clear enough, and the insight of the author of these statements as to the main potential of distribution, when not all its details, is all the more striking as he could not know of all the facts we have today before us.

The work of Darwin's rival, Wallace, is replete with explanations which may be introduced as evidence that Wallace disbelieved landbridges, and for this reason he is often quoted by such among the modern zoologists who wish landbridges to be ludicrous. However, these zoologists are certainly not familiar with Wallace's interpretation of the similarities and dissimilarities in the floras of Australia and New Zeland. Wallace unequivocally admits (5; p. 435, 459) that these regions were once connected by a continuous landbridge, and only through reference to this landbridge he rationalizes the interchanges in those floras. Wallace, indeed, produces a map to support his

---

*) It is doubtful whether Guppy's (4) aversion to landbridges does justice to Darwin's thought, and represents the substances of the teaching to which Guppy professes himself mainly indebted.

belief showing that the Pacific's floor is comparatively shallow in the ranges where he places the landbridge in question.

Were it necessary, other authors could be found who write in the manner of those just quoted. This, however, seems no longer necessary, if only we stop to consider that naturalists who profess to see nothing good or constructive in landbridges are sooner or later enmeshed within "Real quandaries," and are then willing to intimate that landbridges indeed existed. In brief, it is useless in our considerate opinion to take serious account of works in which a deep knowledge of details, perhaps, goes dissociated in the end from the essentially scientific power of cogent, lucid thinking. *So many are the facts, that unless rigorous premises are constantly being driven home with reference to pertinent facts, the facts themselves cease as a whole to have a meaning.*

The Pacific is the biggest ocean on earth, and the one which, to judge by current maps, is the most difficult to bridge. Had dispersal taken place over lands and seas even faintly resembling those of the present, we could certainly not find continuous evidence of a community of vegetable and animal life between Malaysia, for instance, and Central America. Dispersal would have been forced to follow the Pacific's continental shores in the main, and reached under the most favorable circumstances only certain of the islands nearest this shore.

It is a fact, on the contrary, that dispersal did not behave at all as modern maps would have it, and authors who argue against landbridges, whatever their nature, cannot hope to retain the attention of readers trained in precise thinking. Let us repeat, "Real quandaries" are not passed under silence, or explained by devious and contradictory statements, denying as ludicrous the existence of ancient landconnections around Hawaii, only to have recourse in the next breath to a pronounced lowering of the Pacific's floor *as a whole* to rationalize distribution in some other corner in this ocean. The question is here no longer with facts and honest differences of interpretation. The question is with arguments which are either only part sincere, for they withold some of the evidence, or conducted with seemingly scanty regard for elementary requirements of logic.

A reader trained in the use of common sense, moreover, is hard put to account for the current fad against landbridges when there is not the slightest possibility of dispensing with them. Palaeobotanists tell us that the Eocene floras of Europe and North America were remarkably different, and modern ornithologists inform us that certain birds, Trogonidae, for instance, were in France and Central America in the opening stages of the Tertiary.

This being the record, it is *a fact* that the floras of Europe and North America had been originally laid down as different prior to the Eocene, which is to say in the Cretaceous. It is also *a fact* that the Trogonidae originated as a modern family in the Cretaceous, not only, but had so widely migrated prior to the Eocene as to have reached France and Central America by then.

If, then, informed as *to these facts*, we take in hand an elementary textbook

of geology, we forthwith learn that the geography of the earth was in the Cretaceous not at all the same as it is today. There were once seas where lands now
stand, and the other way around. Whether we believe that the continents
"floated", or were sunk by pieces does not make a particle of difference, for
it is *a fact* that immense changes in geography took place after the Cretaceous.

Seeing that birds like the Trogonidae existed as modern in the Late Cretaceous at least, and that floras living in comparable conditions already
differed in the Eocene, we are bound to conclude by common sense that the
dispersal of these birds and floras was conditioned in the first place by a type
of geography that has no counterpart in the geography of the immediate
present. These birds, these plants came into being, moved right and left,
south and north, *upon a world that was, not upon the world that is, or but a
slightly altered version of it.*

It must come as incredible to a man of common sense, not to mention a
trained scientist, that these migrations can seriously be approached and interpreted in a manner that fits the facts on the strength of maps other than those
of the Cretaceous when indeed not of the Jurassic. It has been openly acknow-
ledged by a great foe of landbridges (6) that about 70 % of the living families
of seedplants are distributed in a manner that does not agree in the least with
modern maps. The question follows, on what maps, then, were these plants
first distributed? The answer must be, once more, that they were distributed
at least some times during the Cretaceous, and over the maps of the Cretaceous. Considering next, that land stood in these maps where only the ocean
is to be seen today, it is correct to infer that points in the modern world separated by waters at this hour once were connected overland, which forthwith
settles the question, whether "landbridges" could exist or not, in the affirma-
tive. Moreover, if the maps of the present epoch and of the Mid- and Late
Tertiary do not agree with dispersal, which is pacific, the conclusion further
follows, as two and two make four, *that modern dispersal is older than these
maps.* It finally follows that, dispersal being the older *it is dispersal that gives
a clue to the maps that are relevant in the case, not the other way around.*
Geography, then, ancient and modern, not theories which disregard geography and modern distribution alike are of the essence of our study.

If naiveté is found anywhere, we may freely look for it in the efforts of
authors who replace pertinent maps with assumptions of "Extraordinary
capacities for transoceanic dispersal," and believe they can rationalize dispersal by means that they cannot fathom. If these authors cannot rationalize
their own explanations, and have recourse instead to haphazard and marvellous events, they have nobody to blame but themselves if they fail to find
an audience among readers for whom rigorous thinking *in essentials* is an
absolute prerequisite of scientific discussion.

It is not yet our intention to press inquiries into fields that lie outside
botany, and to enter into technical details which may obscure the fundamentals which we seek as the proper introduction to our subject. We are not

interested in proving true this or that learned "theory," but in laying on the contrary solid foundations for the massive work of detail which lies ahead in botany throughout. This work must be freed from loose thinking *in fundamentals*, and this is what we have in mind to achieve above all.

We cannot refrain from mentioning, on the other hand, that merely by fingering the records of the natural sciences we find abundant evidence of types of dispersal that, calling into play various animals, are exactly matched by plants, witness, for example, the distribution (Fig. 3) of the Heliornithidae. Fossil or subfossil remnants of birds in Chatham Islands by New Zealand point to affinities toward the Mascarenes and South America which are of the commonest among plants, it being remarkable indeed that petty islands like Chatham and the Mascarenes constantly recur in the annals of animal and vegetable distribution. This fact has never been noticed, and even less accounted for, by authors to whom "Extraordinary means of transoceanic dispersal" are the sesame that opens the "mysteries" of migration.

It is claimed that the Iguanidae to which the well known lizards *Iguana* belong are massively American to the exception of three genera, one in Fiji, two in Madagascar, which again matches standard aspects of plant-migration, the tie America-Madagascar (when not continental Africa) being endlessly repeated in vegetables. We should not be surprised in the least if the authors who impart these data are contradicted by other authors. We should not be taken aback if specialists told us that the authorities upon whom we rely are wrong throughout, and that, in reality, the three supposed genera of the Iguanidae alien to the New World form three or six distinct families of their own. Discrepancies of nomenclature, and differences of opinion as to the degree of consanguineity of related forms are the rule in all sciences that cater to classification, and it is not to be anticipated that herpetologists are free from the troubles that beset taxonomic botany. These troubles, on the other hand, need not be exaggerated, and even less converted into standards of truth. Affinities become neither looser not closer when names are changed, and it is ample for our purpose that a certain affinity in, or near, the Iguanidae is massive in the New World but elsewhere known only from Madagascar and Fiji. In short, it is odd that these three regions should be brought together by lizards quite as much as they are by plants and this coincidence is proof against changes in nomenclature in the classification of lizards and that of vegetables alike.

Biogeographic work which pays scanty heed to the requirements of every day's logic ends by taking inconsistencies for granted. This work easily credits certain birds, Sungrebes for example, with the proud possession of "Extraordinary means of transoceanic dispersal," whatever these means may be. These birds — this work assumes — could fly the Pacific, Atlantic and Indian Oceans at ease (Figs. 2, 3) and perform wondrous feats of migration. These birds knew well enough — we are asked to believe — to sail in the teeth of certain auspicious gales, and to rule their travels with so sensitive a compass as to secure a safe landing with mates of the opposite sex only at certain spots, half one island in preference to the other half, and the like. Meantime, this work also tells us, fully 70 % of the birds of New Guinea were prevented from migrating to the adjacent island of New Britain by a paltry stretch of the sea hardly 45 miles wide. At one hand, then, we span oceans, *in imagination*, without making the slightest use of landconnections; at the other hand, we should visualize a lowering of the Pacific's floor as a whole in order to explain why certain landconnections — most desirable ones indeed — are no longer to be seen; in another direction, we ought to take for granted that a ferryride caused the birds native to New Guinea to be different from those of New Britain, while those of Britain are certainly not much different from those of France.

The issue involved is no longer here with facts, we may repeat, and tenable differences of opinion. Profound misconceptions as to the proper use of evidence from the standpoint of pure logics; radical failures in methods and approaches underlie the work of authors who attempt to rationalize migration through irrational means, postulating migrations *in vacuo* in the teeth of winds, waves and the like.

Obviously, we cannot follow these authors in their dismissal of landbridges as something which no longer is *à la mode*, and constantly make reference to their conclusions while introducing ours. It is not facts we need, quite as much as a rational approach to the evidence, for without this we cannot understand the facts at all.

In this knowledge, let us dismiss from our thoughts the notion that compilation is desirable, and that no work is complete that fails to record most dutifully, and to discuss, the private opinions of authors A, B, C. If phytogeography has so far indulged in these compilations, it is perhaps because phytogeography has failed to find a way to more constructive undertakings. The distinction sometimes attempted between descriptive and interpretative plantgeography is wide of the mark, because description and interpretation go necessarily hand in hand. Description unaccompanied by interpretation is cataloguing at the best, propaganda at the worst, and the fundamentals of a science cannot be mastered either one way or the other.

To enter our subject in a spirit which strikes us as factual, therefore as fitting, let us glance at a map (Fig. 4) which illustrates the broad outlines of the dispersal of a large family of seed-plants, Bignoniaceae, known to all inside and outside botany as the source of showy ornamentals both in the tropical and temperate lands of the two hemispheres. Full lines identify on the face of this map the regions where the Bignoniaceae are today in existence, and dotted connections the probable paths of migrations, which, at various times during epochs past, made it possible to different members of this family to reach their present abodes. It is obvious that these paths of migrations are charted on the strength of affinities. If two or more allied forms are in the Far East and the United States, it is consistent with reason to assume a migrational flow running between these two centers, it being at first immaterial whether this flow has streamed west to east or the other way around.

We do not know — nor do we need to know — whether the author of this map (7) is a sound taxonomist, or whether the Bignoniaceae consist of a hundred or a hundred and fifty genera. This map charts dispersal, and can be read without making reference to specific concepts, generic names, and sundry amenities. As we glance at it, we are told, (a) The Bignoniaceae are essentially tropical, because the 40° Lat. is their approximate limit of distribution in both hemispheres; (b) These plants are seemingly intolerant on the whole of scanty precipitation, for none occur in the drier districts of Africa, and none thrive in Arabia and lands adjacent; (c) Two lines of presumed migration lead from the southern Pacific northward, entering Tasmania, respectively, and southern South America; (d) Contrariwise, no line of migration leads from the northern polar regions southward; (e) Definite ties obtain between Eastern Asia and Northern America; (f) The Pacific Ocean is ringed throughout by routes of migration, these routes, or tracks as we may call them, being subject to local deviations and disconnections on account of factors which the map suggests to be due to climate and precipitation; (g) The same is true of the Indian Ocean, in which two tracks, one pointedly identified (Madagascar-Southern India), the other presumed (East Africa-India) freely cross the water.

This map, then, speaks for itself with the eloquence of a statistical graph, and whether the conclusions it presents are wholly, or only in part, right or wrong as to taxonomy is not of the essence of the matter which should retain our attention at this juncture. It is thoroughly relevant to this matter, on the contrary, that the distribution of the Bignoniaceae cannot be judged otherwise than on the strength of elements of world-wide significance, and with a fitting understanding of the chronology it involves. We cannot hope to understand the facts, and to interpret them, if we forthwith proceed to enmesh ourselves within definitions of our own choice, academic discussions of "landbridges," "bipolarism," and the like, or if we decide that we would like to judge of the facts only as students of the flora of the United States, Malaysia or Madagascar. Even less may we state at first that this dispersal was effected in the Eocene rather than in the mid-Cretaceous. As the evidence stands before us, we must perforce enlarge our horizons at the very first, and be ready to take in the world, for the Bignoniaceae belong to the flora of the world. This is the first and foremost requirement which speaks in this evidence, and it is not for us to argue about it but to accept it *as it stands*.

Some authors believe (8) that we cannot know the direction of the main flows of migration, for some appear to run one way, others the opposite. A palaeobotanist (9; p. 35, 36) goes even further, and is thoroughly skeptical of the prospects of phytogeography, for he submits a map of the distribution of a family of seed-plants, Monimiaceae (Fig. 5), and argues as follows with reference to the genera of this aggregate, "If *Laurelia* came from Antarctica or if it crossed Antarctica, why did not some of the genera also, such as *Molle-nedia* and *Siparuna?* Why did not *Laurelia* reach Australia, inasmuch as *Hedycarya* is common to Australia and Africa? .... As it will be seen, we arrive to no conclusion, and it seems to me that we cannot hope to solve problems of this sort until we have sufficient facts, no matter how clever we are in speculation and prophesy."

It will be learned in due course that these questions can be answered in a manner acceptable to reason, and that the answer is to come neither by clever-ness or prophecy but by the ordinary process of painstaking and critical comparison of the data already in our hands. Indeed, there are as many facts available to us as are the plants of the earth, and it is odd that we should crave more evidence when, perhaps, we may not be able to put to good use what we already have. If, dismissing these queries as irrelevant, and this pessimism as entirely unjustified, we compare the dispersal of the Monimiaceae with that of the Bignoniaceae, we readily perceive that these two families have something at least in common from the standpoint of dispersal because, (a) They both are essentially tropical and both have migrated most widely; (b) They both are in West and East Africa, and both are strung along the arc Madagascar-Ceylon-Malaysia; (c) Both cling to the coasts of the Western Pacific, and both are in Tropical America.

On these grounds, we may affirm as a preliminary that the Bignoniaceae and the Monimiaceae are probably in the same category as dispersal goes, and that both have migrated taking advantage of the same ancient land-connec-tions. It is inconceivable that one exacts more cleverness in speculation and prophecy than the other, and the correct assumption is that a critical, per-sistent comparison of dispersals will in the end tell us why families patently unrelated from the phylogenetic standpoint, such as are these two, migrate along comparable lines of dispersal. If we cannot handle this evidence to a telling purpose, the fault is certainly with us rather than with a bountiful nature which has provided some threehundred families of seed-plants for us to study as one. True, nature will not yield its answers without some gentle nudging from our side, and the facts will not automatically arrange themselves before our own eyes in harmonious patterns if we choose to be blind to their message.

To abide in the line of thought suggested by a rough comparison of the dispersal of the Bignoniaceae and Monimiaceae, let us next consider the map (Fig. 6; 8) which illustrates the distribution of the scrophulariaceous genus *Euphrasia*, and bring together this map and the one illustrating the dispersal of the Bignoniaceae. The composite picture thus secured (Fig. 7) reveals that the Bignoniaceae and *Euphrasia* share almost the same abode in Australia and Malaysia. As to the balance of the distribution, the Bignoniaceae dwell in warm and moist ranges, while *Euphrasia* tends to avoid them, and to thrive in cooler regions. It is clear that as to *potential of migration* — in other words, ability to negotiate enormous distances at ease — the Bignoniaceae and

Euphrasia are in the same class. Were the climatic and edaphic preferences of these groups reversed, we should find today the Bignoniaceae in the place of *Euphrasia* and the other way around. Mileage and modern maps are patently irrelevant to these plants alike, for they play fast and loose with oceans and enormous distances.

Would we serve any useful purpose, if facing these facts we were to indulge as an introduction in academic discussions about „landbridges", or the like? Plainly not, because the evidence before us has to be approached in a spirit fitting its scope. Something most ancient; something orderly and so vast in space and time as almost to defy reason cannot be understood in the light of preconceptions. The issue is here between ourselves as inquirers and the bosom of nature, and it avails us nothing, that we theorize and compile if we cannot tackle its marrows straightforwardly.

How did *Euphrasia* disperse? This genus is thoroughly disconnected in the Americas for about 70° Lat. north to south, or south to north, and it is manifest that this disconnection must be bridged in the first place in a manner consistent with the whole of the distribution. To do this within reason we need a starting point first and foremost. As the map reads, it would seem that *Euphrasia* closed upon the New World in the manner of a huge pincer, simultaneously biting into boreal North America and southern South America from a westerly direction. If this be correct, the center of dispersal, and the probable place of origin of *Euphrasia* may lie somewhere in a base of distribution located in the south of the Pacific Ocean, and that from this base the dispersal could be effected as follows, (1) From the base itself, to reach in the east South America, in the west New Zealand, Australia and Malaysia; (2) From Malaysia toward the Far East (Japan and Kamchatka in particular) next crossing the Kuriles and the Aleutians to reach Alaska, Canada, Greenland and the northern United States, stopping perhaps at the Azores; (3) From Japan speeding westward and northward ultimately to attain the whole of extra-tropical Asia, Europe, Northern Africa and possibly points beyond, always in the west.

If, to verify what the map seems to advise, we turn next to the classification of *Euphrasia*, we learn (10, 11) something of interest. One of the species of South America (*E. formosissima* of the Juan Fernandez Islands) is unrelated to the populations of the New World, but allied to those of New Zealand, and, though more distantly, also those of New Guinea and Borneo. The species of Borneo and New Guinea, for their part, are akin to those of Australia and New Zealand. There occurs in New Zealand (11) a monotypic plant, *Siphonidium longiflorum* which is quite close to *Euphrasia*. It would then seem that *Euphrasia* had indeed a well marked center of evolution located in the Southern Pacific, and that the effects of this center are felt across an immensity of modern waters as far as the approaches of the South American coast, where the Juan Fernandez Islands rate as one of the most interesting domains of general phytogeography. The author, who imparts the bulk of the taxonomic data used here, tells us in addition that the Australian and American ends of the dispersal of *Euphrasia* were beyond doubt connected *via* an antarctic continent, not only, but that the peculiar connecticns with forms of New Zealand of a species of *Euphrasia* belonging by geography to the New World, are repeated in other genera, witness *Nothofagus*, discussed elsewhere in this handbook.

Mysterious as they may seem to us at first, the transpacific migrations of *Euphrasia*, *Nothofagus*, Monimiaceae and Bignoniaceae, are commonplace. At least 20% of the living families of seed-plants (12) exhibit affinities of the closest between the Australian and American World. More diluted forms of kinship, involving no longer an identity of species but groups of higher order both below and above the genus, are in evidence in almost all large families of Angiosperms, to omit mention of other groups. All in all, there is no risk in affirming that about 40%, when not more, of the major angiospermous families reveal wellmarked "antarctic" migrations in their ranks. This given, we readily understand why Darwin and Wallace forsook any instrument of dispersal but straight landbridges when facing the problem of migration in the southern hemisphere, and why Darwin freely advocated radiations of vegetable progenitors from Antarctica. The traffic taking place in this hemisphere is so massive, and involves so many groups of all sorts, that it would be nonsense to insist that, after all, landconnections have nothing to do with it, but certain mysterious, unexplained and unexplainable agencies blowing seeds right and left. If botanists have argued how the austral landconnections were used, and when, none of their persuasion ever came forward to deny them. None came forward to this effect, of course, because none could.

The evidence, consequently, fully justifies us in reading the map of the dispersal of *Euphrasia* from south moving north. Under this approach, we see *Euphrasia* emerging together with *Siphonidium* somewhere in the Southern Pacific, and proceed from a base-line of distribution primarily stretching between New Zealand and southern South America in two directions. Eastward, the migration flows to Magellania and Chile, and here stops. Westward, on the contrary, the stream of the progenitors hits Australia and New Guinea, then proceeds in the manner we have already described.

We are ready at this juncture to put on record several observations, all of which will come under stringent discussion, directly or by immediate implication, in the coming pages, as follows, (a) *Euphrasia* is conventionally "bipolar" in the New World because it there suffers a huge disconnection. As we have intimated, this disconnection is a byproduct of the total dispersal, not necessarily an "American" phenomenon. From the center of emergence in Southern Pacific a huge pincer was clamped upon the Americas with a short and long arm respectively going to South and North America. It is confirmed, accordingly, that dispersal cannot be interpreted piecemeal, but must be read whole; (b) A disconnection existing in the range between Southeastern Australia and New Guinea may mean that *Euphrasia* died out in Northeastern

Australia (Queensland, particularly) while migrating northward. This disconnection, however, may also be understood as the byproduct of a radiation westward from a center of origin in the Pacific, diverging streams from this center reaching independently Southern Australia, New Zealand and New Guinea; (c) Malaysia is entered by *Euphrasia* from the east, and it is erroneous to assume that plants common to Polynesia and Malaysia are necessarily "malaysian" immigrants into Polynesia. It is altogether possible, on the contrary, that Polynesia furnished plants to Malaysia, and such terms as "malaysian", "polynesian", etc. etc., are meaningless so long as they are not assumed with critical regard of the whole of migration; (d) As Malaysia is entered by *Euphrasia* from the east, it is not to be excluded that Malaysia may be entered by other groups from the west, and it is altogether probable that Malaysia is a crossroads of dispersal on account of this possibility, some of its plants coming in from the west, other from the east. So much, of course, does not rule out the probability that Malaysia originated within its limits numerous species and genera out of primitive streams of immigration; (e) *Euphrasia* occurs in the Philippines, so far as known, only in the island of Luzon, and is here endemic to a peculiar plant-world, the so called "Bontoc flora." This flora is commonly supposed to consist of "holarctic" elements, and to have entered the Philippines *via* Formosa. This is not necessarily correct, and authors who maintain the contrary are liable to grave errors; (f) A track of migration exists which, beginning in the Pacific somewhere in the immediate vicinity of the Solomon Islands, leads into Malaysia along two lines of penetration, one going directly through New Guinea to Celebes, the other following from New Guinea to Luzon and veering ultimately southwestward to enter Northeastern Borneo; (g) A group which originates in the Southern Pacific like *Euphrasia* may reach in the end the shores of the Atlantic anywhere between Norway and Morocco, colonizing meantime the whole of Eurasia; (h) Considering that *Euphrasia* is in Europe from the east, and boreal North America from the west, we may not say whether Greenland received the genus all from Europe, all from America, or part from Europe or America. It is possible, as the map reads, that the American end of the migration crossed Greenland throughout moving eastward, or that the European stream worked itself into Greenland over the opposite path, going westward from a starting point located anywhere between Norway, and the shores of the Arctic Ocean; (i) By the same token, the Azores might have been reached either from Europe or America.

These observations are consistent with the facts in the map, and free of theoretical bias. They follow from the assumption of a starting point in the south which has in its favor a considerable weight of evidence, and yields a simple, global interpretation of the dispersal of *Euphrasia*, not only, but of groups that migrated like *Euphrasia*. If we consider the distribution of three species of *Carex*, *C. Lachenalii*, *C. pyrenaica* and *C. Gaudichaudiana* as something entirely unrelated, we have little before us but disconnections of the widest in every single case. If, contrariwise, we bring these three distributions together upon a single map (Fig. 8), we secure a composite picture with all the essential characters of *Euphrasia*, minus the South American outlier *). Likewise, knowing that a base-line of distribution rings Antarctica we have little difficulty in rationalizing the dispersal of such plants (Fig. 9) as *Azorella Selago* or *Taraxacum magellanicum*, for we face in all these types of distribution the same main order of causes.

The reader may, or may not, believe everyone of our statements and subscribe to all our assumptions, but will be inclined to agree, perhaps, that reading dispersal as dispersal is manifestly effected, *globally*, we stand at least a fair chance of securing data of a positive nature, and such as to promote fruitful discussion. It is not probable, to say the least, that the seeds of *Carex*, *Euphrasia*, Bignoniaceae, *Taraxacum* and the like were casually distributed by birds, winds or waves over maps even faintly resembling those of the modern earth. The whole is too big to be handled by means of the kind, and it is patent that radical changes took place in the geography of the earth following the early days when these plants began to migrate. It should be futile to press geophysicists and geologists for proofs they themselves are not always ready or able to furnish. It is we, as botanists, who — for the present at least — have the better part of the bargain, for it is we who can line up something like 40% of the families of the seed-plants, proving that these vegetables dispersed in the southern hemisphere in a manner which does not agree at all with the maps of the present.

Once more, then, we are brought to face the fact that modern floras were put together in epochs wholly bygone, and that instead of reading and interpreting dispersal in the light of the outlines of sea and land now current we must read and interpret it, on the contrary, on the maps of the *Cretaceous* at the latest. Considering that we have of these maps but rough outlines, it is consistent with reason that we read dispersal at first in the manner we have chosen to follow, that is to say, *globally*, reconstructing in the beginning only the main outlines of migrations, and ancient lands, at the same time. We court disaster if we choose to drown ourselves in theories at the start, or to handle the evidence piecemeal.

Euphrasia has revealed to us a facet of migration, and *Evolvulus*, a genus of the Convolvulaceae, may well now show us another. *Evolvulus* consists of about a hundred species (13) only two of which occur outside of the Americas. This genus is massive in Eastern and South-eastern Brazil, where it has

*) This outlier would readily be furnished if we were to add to the range of the three species of *Carex* above mentioned the range of *C. magellanica* as given by the author here cited (see (8), p. 218, fig. 3). We omit this range in order not to complicate the figure with the addition of further symbols, *also to stress once again the point that whenever the mainsprings of distribution are correctly grasped details readily take care of themselves.* Let us notice, finally, that the range of *C. magellanica* is homologous of that of *Primula*, elsewhere discussed in this handbook. Peculiarly, the range of *Primula* is accepted by many authors as „proof" that this genus is of „holarctic" origin throughout. These authors reach this „proof" by (1) Dismissing the austral American distribution of *Primula* without adequate discussion of its meaning; (2) Disregarding comparable distribution in *Carex*, *Euphrasia*, Lardizabalaceae, etc., etc. We regret that we cannot consider the conclusions of these authors as weighty.

a center of evolution rooted in lands which are known to have remained in state of emersion lasting aeons. To avoid cluttering these pages with tedious records, we will tabulate only part of one of the sections of *Evolvulus*.

Evolvulus Sect. *Alsinodei* Subsect. *Pedunculati* is distributed in this manner.

(1) E. *alsinoides*

Var. *typicus* — Madagascar, Tropical East Africa, Maldives, Ceylon, India, Nicobars, Indochina, Southern China, Philippines.

Var. *hirsutus* — Ceylon, India, Burma, Sumatra, Java, Philippines.

Var. *philippinensis* — Philippines (Luzon).

Var. *rotundifolius* — Formosa.

Var. *thymoides* — Madagascar, Aldabra Isld.

Var. *adscendens* — Texas, Mexico, Central America.

Var. *Grisebachianus* — Florida, Bahamas, West Indies, Central America, Guyana, Brazil (Amazonas).

Var. *debilis* — Colombia, Central America, Mexico.

Var. *acapulcensis* — Southwestern United States, Mexico.

Var. *linifolius* — Cabo Verde Islds., Senegal, Nubia, Ethiopia, Angola, Congo, Transvaal, Madagascar, Comores, Socotra, Yemen.

Var. *glaber* — Cameroon, Belgian Congo, Natal.

Var. *Wallichii* — India.

Var. *decumbens* — Indochina, South-eastern China, Sumatra, Timor, Moluccas, Philippines, Boeroe, Amboina, Kei Islds., New Guinea, West Australia, North Australia, Queensland, New Caledonia, Fiji.

Var. *javanicus* — Java, Moluccas.

Var. *villosicalyx* — Western and Central Australia.

Var. *sericeus* — Queensland.

(2) E. *tenuis* — Brazil, Paraguay, Bolivia, Columbia, Venezuela, West Indies, Mexico (Yucatan).

(3) E. *linarioides* — Brazil (Minas Geraës), Paraguay.

(4) E. *saxifragus* — Brazil (Bahia, Minas Geraës, Piauhy).

(5) E. *filipes* — Brazil, Paraguay, Peru, Ecuador, Colombia, Venezuela, British Guyana (Mt. Roraima), West Indies, Mexico.

(6) E. *serpylloides* — Brazil (Minas Geraës, São Paulo).

(7) E. *vimineus* — Brazil (Minas Geraës).

(8) E. *arizonicus* — North America (Arizona, Sonora), South America (Argentina: Cordoba, Mendoza, Tucuman, Salta, Santiago del Estero, La Rioja, Catamarca, San Luis, Jujuy).

(9) E. *corumbaensis* — Brazil (Matto Grosso), Paraguay, Bolivia.

(10) E. *Herrerae* — Peru (Cuzco).

(11) E. *Fieldii* — Peru (Tumbez, Piura).

(12) E. *magnus* — Peru (Cajamarca).

(13) E. *argyreus* — Bolivia, Peru (Ayacucho, Huànuco, Cajamarca), Ecuador, Colombia (Huila), West Indies (Curaçao, Bonaire, Aruba, Martinica, Antigua).

(14) *E. piurensis* — Peru (Piura).

(15) *E. boliviensis* — Bolivia (Cochabamba).

(16) *E. helianthemoides* — Peru (Lima, Arequipa, Piura).

(17) *E. villosus* — Peru (Lima, Arequipa).

(18) *E. anagalloides* — Brazil (Piahuy, Ceará).

(19) *E. pusillus* — Brazil (Rio Janeiro, São Paulo, Paranà, Santa Catharina).

(20) *E. bogotensis* — Colombia (Cundinamarca).

(21) *E. Grisebachii* — Western Cuba, Isle of Pines.

(22) *E. incanus* — Peru (Huànuco).

(23) *E. flexuosus* — Brazil (Bahia, Espiritu Santo).

(24) *E. stellariifolius* — Brazil (Minas Geraës).

(25) *E. glaber* — Brazil (Matto Grosso), Paraguay (Chaco), Bolivia (Santa Cruz), Galápagos Islds., Peru (Tumbez), Ecuador, Colombia (Magdalena, Bolívar), Venezuela, Lesser Antilles, Portorico, Jamaica, Hispaniola, Cuba, Bahamas, Mexico (Oaxaca), United States (Florida, Louisiana).

The casual reader of a tabulation such as this sees in it nothing clearer than meaningless localities scattered over the better part of four continents. There seems to be neither rhyme nor reason in a dispersal which overruns large tracts of Africa and Asia, and penetrates Polynesia, with the varieties of a single species that further occurs in Brazil and Texas. This and similar types of distribution are seldom approached in phytogeography because they are believed to be too "pantropic" to yield concrete indications of tracks and center of origin.

In reality, the migrations of this group are simple. As we know, a base-line of distribution in the Southern Pacific is operative in numerous angiospermous families, and this base-line further extends all around Antarctica. The taxonomist who furnished us the tabulation given above tells us that *E. alsinoides* var. *villosicalyx* (Western and Central Australia) suggests in many of its characters African forms of *E. alsinoides* var. *linifolius*. These affinities are less than surprising, for they come to the fore along that segment of the southern base-line of distribution which runs between Southeastern Africa and West Australia.

If, then, doing for *Evolvulus* what we have done for *Euphrasia* we begin to read the map from the south on, and take a starting point in South Africa, we have before us the following picture: *Evolvulus* migrates east and west dispatching at first an arm of dispersal toward Western Australia (*E. alsinoides* var. *villosicalyx*), which ranges to Central and Eastern Australia (Queensland), and here comes to a dead end. Always from South Africa, the dispersal proceeds northward from Transvaal to Ethiopia and Arabia (*E. alsinoides* var. *linifolius*). A track to parallel this one in Africa (*E. alsinoides* var. *typicus*) runs from Madagascar and East Africa generally across the Indian Ocean (Maldives) to India (Ceylon), further extending to Indochina, China and the Philippines. Asia, Malaysia and the Far East are the seat of peculiar forms (*E. alsinoides* var. *hirsutus, Wallichii, philippinensis, javanicus, rotundifolius*) with a parallel development in Australia (*E. alsinoides* var. *sericeus*), and

Madagascar (*E. alsinoides* var. *thymoides*). In the eastern quarter of this dispersal there further occurs a widespread aggregate (*E. alsinoides* var. *decumbens*) which has its seat in the triangle Fiji, West Australia and Indochina.

Returning now to South Africa to explore the western end of the migration, we have before us a line of migration running from Angola to the Senegal and Cabo Verde (*E. alsinoides* var. *linifolius*), and different forms in the New World (*E. alsinoides* var. *Grisebachianus, debilis, adscendens, acapulcensis*) which turn up in Brazil, and from Brazil through Colombia and the Guianas extend to the West Indies, Central America, Mexico and the United States.

So much for *E. alsinoides* and its varieties. The balance of the group consisting of twentyfour species repeats with variants of detail the dispersal of the American end of *E. alsinoides*. The track runs between Eastern and Southeastern Brazil, Paraguay, Northwestern Argentina, Bolivia, Peru, Ecuador, Colombia, the Guianas, the West Indies, Mexico and the United States.

Considering that *E. alsinoides* var. *decumbens* is connected by transitional forms in South-eastern Asia with *E. alsinoides* var. *typicus*, we may readily plot this migration on the map (Fig. 10), with the result that Africa is before us as the center of the dispersal. From the region Angola-Transvaal various lines radiate to the New World, Arabia and the Senegal, to the Far East, Australia and Western Polynesia. The base-line of distribution still lies in the southern hemisphere, but, quite unlike *Euphrasia*, *Evolvulus* is not rooted in the Pacific.

We do not ask the reader to take for granted that we are certainly and definitely in the right, but feel free to comment that a type of dispersal seemingly meaningless on its very face, readily yields to rationalization once it is read from a southern base-line of distribution. This group of *Evolvulus* falls into three centers, American, African and Eastern Old World, Australia being possibly entered both from the north-west (Queensland) and the south-west (West Australia).

Let us now carefully observe certain peculiarities of the dispersal of *Evolvulus* in regard of that of *Euphrasia*. In Polynesia, *Evolvulus* does not range beyond the line Bismarck Archipelago-Fiji, and turns up in South America precisely in the regions which *Euphrasia* leaves untouched (i. e. South America north of Chile). The migrations of these two genera do not intimately come together except within the triangle New Guinea, Philippines and Timor. It might be suggested that these plants are incompatible as to climate and ecology, which may be true to some extent, but cannot be true altogether, for one species at least of *Evolvulus* (*E. pilosus* of Sect. *Alsinodei* Subsect. *Epedunculati*) makes its home in a range which is anything but tropical (Montana and North Dakota to Arizona, New Mexico, Texas, Arkansas and Tennessee). The reason why *Evolvulus* and *Euphrasia* do not come together except at very few points is in the fact that these genera started their now visible migrations at different points *of a world that was*, and dispersed using basically different land-connections. *Evolvulus* took advantage of land-connections between Africa and America, Africa and the Far East in the first place, while *Euphrasia* was served, always in the first place, by land connections between Tasmania, New Zealand and Polynesia generally with South America and Japan. If we return now to the map illustrating the distribution of the Sun-grebes (Fig. 3) we readily understand that these birds also took advantage of the land-connections offered *Evolvulus*, and that some birds of the „Pantropic element" common to the Far East and America (see Fig. 2) utilized perhaps bridges not unknown to *Euphrasia* altogether. We may not enter the subject beyond these cursory generalities, for space forbids it, but it would seem that correct interpretations of what plants did throw at least a beginning of light upon the doings of birds that put the zoogeographer in a „Real quandary."

Euphrasia, as we know, entered the „Bontoc flora" of Luzon from the east, that is to say, from the direction of the Pacific. *Evolvulus* also is in the region of Bontoc, but this time it is in from the west, that is to say from the direction of the Indian Ocean. Thus is verified as a beginning the surmise that Malaysia was fed the archetypes of its plant-world *from two main directions*, and some explanation is forthcoming for the fact that authors who persist in interpreting the broad outlines of the phytogeography of Malaysia as something which is „Malaysian" go nowhere.

It is not yet time for us to enter into details of dispersal which, though most interesting, would cause us to indulge in digressions, and lose sight in consequence of the rudiments which we strive for at this juncture. We may observe, nevertheless, that Java possesses at least two varieties of *Evolvulus alsinoides* (namely, *hirsutus, javanicus*), but appears to lack one of them (var. *decumbens*) which is widespread in Sumatra, Timor, the Moluccas, Kei Islands and New Guinea. Strange to say, Borneo seems to have no *Evolvulus* at all.

In the Americas, we have before us aspects of migration not less puzzling. *Evolvulus arizonicus*, the lone species of Subsect. *Pedunculati* to inhabit Argentina, is massive in Northwestern Argentina, where the genus is otherwise very nearly or wholly unknown. Departing from this center by apparently mysterious means, this species fares over a huge disconnection to northwestern North America (Sonora and Arizona). Other species in the same affinity, on the contrary, stream from Eastern Brazil to Paraguay and Bolivia, and follow the Andes northward without ever being lost of sight.

It stands to reason that we may never hope to fathom these puzzles if we assume „Enormous capacities for transoceanic dispersal," without making a preliminary effort at casting before us a concrete outline of the main channels of migrations active throughout the earth. The problems that crop up in Malaysia and South America are but local byproducts of the distribution as a whole, and the moment we subscribe to the tenet that distribution is by casual agencies operative upon the maps of the present, or closely resembling those of the present, we are lost in figuring out why *Evolvulus* is not in Borneo, or disappears between Argentine and Mexico. We may imagine, of course, that these oddities of dispersal are the direct result of unsuitable ecologic conditions, or the like, but we still may not give ourselves a rational account how these conditions work, for there is no present reason why *Evolvulus* should occur in Sumatra but not Borneo. These islands are intimately related in many respects, and *Evolvulus* which fared through oceans should find it easy to reach Borneo in massive numbers. Dispersal in general, and dispersal in detail, thus interlock most closely, and if one goes unaccounted for, the other does not make sense either.

So pressing is this problem that students of dispersal faced with it are oftentimes inclined to charge disconnections or peculiarities which they cannot fathom to missing or erroneous records. A puzzling record of the Lauraceae, for example, is dismissed by a palaeobotanist (9) as a „Taxonomic faux-pas." In the opinion of this worker, the record in question has no place where it occurs, but being there, nevertheless, must be ruled out as a mistake. It is anything but a mistake, of course, but, alas, the opinion of the author in question is apt to receive attention, while the taxonomic and phytogeographic aspects of the matter are ignored altogether.

It must be admitted that taxonomists lay themselves open to suspicion by frequent changes of names, and that some of them bring to phytogeography so narrow a view of what is a legitimate record of distribution as to discredit that branch of botany which they profess. A legitimate question is, nevertheless, how we may ever hope to work with records that are in a state of continuous flux, and with tabulations of genera that can never be complete.

This question is apt to disturb only those among the students of biogeography who are not familiar with taxonomy and its methods, and confuse classification with dispersal. The two are not necessarily related to the point of dependency, and it is a mistake to maintain that nomenclatural amenities, the latest in classification, 'etc., etc., are indispensable props of sound phytogeography. Much has been written which may foster this notion, but it does not follow that the opinions so voiced are correct.

We have used a classification of *Evolvulus* which credits *Evolvulus alsinoides* with numerous varieties. Of a certainty, this classification may prove objectionable to taxonomists other than the one whom we followed, and many among them will be found to maintain that these varieties are in reality „good" species. We may safely anticipate, however, that hardly two of these dissenters will agree how many of the varieties in question are „good" species, for some will claim that three are so, others ten, and so on without end. To us, these controversies are trifling, for the record does not vary on the map, whether we speak of „species" or „varieties." To us, taxonomic labels are largely symbolic, and we would have the track of *Evolvulus* before us on the map if we agreed to designate *Evolvulus* with letter A and its subordinate entities with letters a, b, c, etc. Let us observe that we have interpreted the dispersal of *Euphrasia* and the Bignoniaceae *without even knowing how many were the genera and species involved under these groups.*

Phytogeography is primarily interested in affinities and localities, and while it is true that affinities are usually expressed in formal taxonomy by appropriate subordinations, placing for instance all the species of a certain group under a section or subgenus rather than another, finer points of opinion as to whether the group in question should be a section or a subgenus, and how it should be named, are immaterial to us. As we shall duly learn taxonomists seemingly at hopeless odds in their interpretations may on the contrary be wholly agreed as to essentials. A perfect record, moreover, is out of this world, and it is certain that the tabulations we use today as the latest product of the best available taxonomic talents will look, perhaps, less than excellent to our grandsons. In short, it is the manner in which we use the evidence in our hands; the manner in which we think that counts first and foremost. Every age has its problems, and we should be humorless, were we to claim that only „perfect" taxonomy must be our fare. Taxonomy of the sort appeals only to those who fancy they can write it *).

As we have emphasized in the preceding pages the antiquity of the dispersal of living plants, we may now turn our attention to fossil tracks. It stands to reason that if dispersal was laid down in the Cretaceous, when not earlier, the tracks of Cretaceous and Tertiary fossils must to some extent agree with those of living entities.

To test this assumption, we may cast upon the map of the dispersal of *Euphrasia* the fossil and living records of the coniferous genus *Libocedrus*. The map (Fig. 11) suggests that a track runs between Western America and the western coast of Greenland. *Euphrasia* follows this route to this day, and *Libocedrus* followed it in the remote past, the only difference being that *Euphrasia* is still living at both ends of the track, while *Libocedrus* failed to survive in Greenland, where we have record of its former presence as a fossil. The same is true in the Eurasian range, but here *Euphrasia* is living throughout, while *Libocedrus* is extinct almost throughout surviving only in South-western China and Formosa.

A joint consideration of the tracks of these two genera tells us, then, that *Euphrasia* and *Libocedrus* once travelled together, and both reached the same stations in Eurasia, Greenland and boreal North America. Later on, *Libocedrus* succumbed to climatic rigors which *Euphrasia* proved able to endure, but the tracks, *as the tracks themselves read*, are substantially the same for both, the fossil records being complementary to the living. It might be pointed out that there is no proof that *Libocedrus* and *Euphrasia* actually travelled together,

---

*) Not all taxonomists are agreed that classification and phytogeography ought to be married in pedantry. Precise taxonomy is resilient enough in the eyes of all its most thoughtful students to allow for opinions, and minor discrepancies. Between the notions of FERNALD (see, for example, Quart. Rev. Biol. 1: 218. 1926) who constantly harps upon „taxonomic" errors to condemn this and that in phytogeography, and the equilibrate notes of BALDWIN (see, for instance, Jour. Hered. 38: 54. 1947) who entertains the possibility that *Hevea* may consist of anywhere between one and twenty species, of though thinking that nine is best in his opinion, there is indeed a great deal of difference. BALDWIN is perfectly aware, in addition, that palaeogeography is material to an interpretation of present speciation. His work takes place among the taxonomic contributions which a phytogeographer — regardless of minor differences in viewpoints — is glad to consider as relevant.

because the former is a very ancient conifer, and the latter a supposedly „derivative" angiospermous plant. This objection might be serious if it did not happen that *Libocedrus* and *Euphrasia* both occur today in South America and Australia and its geographic dependencies. This being the case, these genera fared together when all these regions were still connected by landbridges in the distant past, and it remains to be proved why and how plants which patently used the same channels of migrations in the southern hemisphere could not do the same in the northern. We may be in error taking this view but if we are, the burden of proof rests upon shoulders other than our own.

In the course of a discussion of the distribution of the Droseraceae in North America, a taxonomist refers (14) to *Drosera filiformis*, saying that this species has closer kinship with forms of Asia and Australia than with its congeners in the boreal New World and adds that the range of *Dionaea* belongs to the „Cretaceous (hardly Tertiary or Quaternary) Coastal Plain of the Carolinas." In the belief of this author, „The antiquity of the insectivorous *Droseraceae*, therefore, can scarcely be doubted."

The antiquity of the insectivorous plants in general is further affirmed by another author (15, p. 17) who refers to the range of *Sarracenia Sledgei* as follows, „One or two colonies are reported to survive in the Cumberland Plateau of Tennessee, presumably in its ancestral home before the rise of the peneplain of the Cretaceous." Both these statements rhyme to the same, namely, that *Sarracenia*, *Dionaea* and *Drosera* were essentially modern in the early Cretaceous, and by then had colonized the modern Atlantic coast of the United States. Turning this into figures, we learn that these genera are about 100.000.000 years old, and had anyone of them been buried in earlier ages amid conditions favoring fossilization, we would speak of it today as an arrant „Cretaceous" fossil.

In this knowledge, let us next turn to the fossils of this very same region. In Maryland and Virginia are located the so called Potomac Beds (16) which contain three distinct fossil horizons of approximately Early Cretaceous age. The oldest of these horizons, the Patuxent Bed, yielded a flora of some hundred species in which Conifers dominate (40%), next followed by Ferns (35%) and Cycads (25%). Above Patuxent are other strata, rich this time in relics of angiospermous plant of a modern type.

The Potomac Beds contain, consequently, the record of transition effected toward the beginning of the Cretaceous between a plant-world of conventional Jurassic facies and the Angiosperms. This transition took place in a period of active emersions and submersions, and it appears probable that both the ancient Jurassic and the novel angiospermous flora lived for some time, perhaps millions of years, side by side, until the former was supplanted in the end by the latter. We have no fossil records to show of *Drosera*, *Dionaea* and *Sarracenia*, which is regrettable, but we have on the other hand the evidence of modern dispersal to tell us that these genera stood in the general vicinity of the Patuxent Bed within 10.000.000 years at the most of these fateful events.

This granted, it would seem that the tracks of these *living genera* are material for a fuller understanding of the routes taken by the angiospermous hosts embedded in the strata above Patuxent, and that, though living, these genera cast considerable light upon the fossils in the region. In other words: phytogeography can view fossil and live plants as equivalent for its own purpose of investigation, and is not at all disposed to accept as final the conclusions of palaeobotanists who tend to emphasize the former against the latter.

There exists in Venezuela a chain of sandstone heights known as the Pacaraima Mountains. Geological opinion is not agreed as to their age, but a conservative estimate is that they are not younger than the early Cretaceous (17), and may represent riverine deposits laid down in a former broad delta.

These mountains are inhabited by a peculiar flora, and exceedingly rich in endemisms of their own. It belongs to the geological record that the Pacaraima was uplifted and eroded, but left free otherwise from catastrophic disturbances. There is sound reason to believe, consequently that the flora presently on these heights stems from early angiospermous prototypes which first colonized them at least during the Cretaceous *). This flora contains several species of the insectivorous genus *Heliamphora*, which is related to *Sarracenia* of the Eastern United States and *Darlingtonia* of California and Oregon. It would seem, then, that the archetypes of the Sarraceniaceae in a state of advanced segregation, and beyond doubt modern as to genera, were in the immediate region of the Pacaraima in the early Cretaceous as they were by then in the Appalachian range of North America. The fossils of the Potomac Beds, then, cannot be more interesting to a phytogeographer than the living *Heliamphora* and *Sarracenia*, and claims voiced by specialists in petrifacts that dispersal cannot be understood at all if we are short of fossils are patently wide of the mark. These claims would not be heard, indeed, if their authors had better knowledge of the global evidence.

Let us ponder the facts carefully. We find in Cretaceous beds of Alaska and Greenland fossils of *Platanus*, *Acer*, etc., which we have no difficulty in identifying, for they resemble their living descendants in every essential. The species may themselves no longer exist, but the genera are modern. This means that *Platanus* and *Acer* endured for something like 80.000.000 or 120.000.000 years to this day, which prompts us next to inquire, how long did it take these genera to become modern out of the primaeval angiosper-

---

*) It will later be shown (see p. 400) that plant-migrations (the Angiosperms particularly) had period of intense activity from toward the middle Jurassic to the Early Cretaceous, and again in, or about, the Mid Cretaceous, Referring to northern Central Venezuela's past, a geologist (KAMEN-KAY in Bol. Soc. Venezol. Cienc. Nat. 10: 221. 1946) visualizes prior to the Cretaceous „A gigantic delta" reminiscent of the Orinoco's in the immediate present. During the Cretaceous there was active peneplanation, followed by a marine transgression which was „Part of a world-wide epoch of submergence." The Sarraceniaceae, Cyrillaceae, Ochnaceae, etc., were evidently in the region which is now Venezuela when this region was mostly occupied by the gigantic delta in question. The visualizations of the geologist referred to are accordingly in full agreement with the data of dispersal, though it remains to be seen whether the Pacaraima dates from the Mid-Cretaceous, or is earlier. By settling the issue of chronology, we will merely decide, then, whether the Sarraceniaceae, Cyrillaceae, Ochnaceae, etc. colonized the Pacaraima in the first or second era of maximum migration, which is comparatively minor.

mous matrix? The question is speculative, of course, and so must be the answer. It is within sober reason, however, that *Platanus* and *Acer* began to differentiate long aeons of geological time before the Cretaceous, and the most conservative estimate is bound to place their ultimate origin in the Permian, indeed, most likely in the Carboniferous. Another question then follows: What were *Platanus* and *Acer* doing in the millions of years prior to the Cretaceous? This question, too, is speculative, but a rational answer is that they were in active evolution and migration throughout all this time, and that, as a matter of fact, *they never ceased migrating and evolving*.

If this is what common sense dictates, whether we like to set our reckonings at some 20.000.000 years earlier or later does not matter, how may we venture to build theories of angiospermous „origin" on the strength of Cretaceous fossils? How can we insist that the angiosperms originated in „Holarctis" or some such mythic land because certain *modern* genera of their number are fossils in Alaska and in Greenland? Where did the Angiosperms arise which colonized in the very early Cretaceous the regions by the Potomac and the Pacaraima? Where had these plants lived and reproduced before then?

Proof that theories of angiospermous origin built upon foundations so unsound are worthless, readily greets us in the work of authors who believe in a nebulous „Holarctis," and locate therein the angiospermous cradle. Faced by records of distribution which they dislike, these authors, as we have seen, rule them out as taxonomic *faux pas*, and boldly go on record announcing that the study of dispersal must wait endlessly for the perfect score to turn out, or, as an alternative, rely upon cleverness and prophecy. It does not occur to these authors, it seems, that plants are millions, and that some piece missing in the puzzle somewhere can readily be replaced by a piece drawn from elsewhere. Even less do these authors suspect that the source of the trouble is in their own assumptions.

Faulty thinking in fundamentals is beyond doubt responsible for ninety per cent of the difficulties in modern biogeography, for we have right now more facts than we can handle. The root of the evil is in our coining definitions or entering uncautious theories, and finding in the end that we cannot break out of the ring which we have forged.

The Pacaraima mountain-system is notoriously older than the Andes. We are not sure about the age of the latter, and it may be granted that they began to stir long before they reached their present majestic heights. It would seem, nevertheless, that the greatest measure of uplift dates from the Tertiary, while the Pacaraima was laid down, if not as a mountain-system, at least as a broad alluvial plain long before then.

This being the case, it is peculiar that the Pacaraima should be credited with the possession of „Andean" elements, while a regard for elementary chronology indicates that it is the Andes, on the contrary, which are tinted with „Pacaraiman" life.

An ornithologist (18) contributes the following tabulation of the affinities of the birds of the Pacaraima,

(1) Andean .... 38 species
(2) Southeastern Brazilian .... 2 species
(3) Guatemalan .... 2 species
(4) „Tropical" generally .... 22 species
(5) Unknown .... 21 species;

and next concludes that the close relationships existing between the birds of the Andes and the Pacaraima are most mysterious. These relationships may, perhaps, be accounted for by later secondary migrations taking place between the Andes and the Pacaraima during the „Ice Age."

It seems to us that there is nothing mysterious in the case. As we see it, there lived in the *pre-andean* lands of northern South America the progenitors of some 95 species of modern birds. As the Andes began to rise, the progenitors of 38 of these species were caught in the uplift, and made „Andean." As, on the other hand, these progenitors had alliances in the Pacaraima region *before the uplift*, we find today an „Andean" element in this region. The birds themselves did not budge; the earth, on the contrary, shifted under them. By playing with labels such as „Andean" and the like we scramble chronology and phylogeny at the same time, and are left as a result with a „mystery" in our lap.

The fundamental error is here the usual one. All too many are the authors who try to interpret migration on *modern maps* with scanty understanding of the basic fact that we deal, on the contrary, with migration effected *on ancient maps*. It matters not that we have ten or ten thousand facts at our disposal if we muster them out in the light of false logic. True, it is possible that some migrations actually took place between the Andes and the Pacaraima in the Tertiary, but the first stop necessary to elucidate all migrations in this region is to have of the entire evidence a lucid perception. It may be affirmed as one of the most fundamental generalities of biogeographical studies in South America that the Andes are late comers, and that, if we stress them as an original feature of the map of this continent, we can understand next to nothing in the end.

In the preceding pages, Darwin was quoted who spoke of „Sunken islands" acting as „Halting-places" in the process of global migration. It is interesting now to learn how these „Sunken islands" and „Halting-places" performed in this regard.

We have a record (19; p. 34) of the geological events affecting Alaska for scores of millions years, which summarily reads as follows,

(a) In early Permian or Pennsylvanian a widespread marine submergence caused Alaska te be covered with thick limestone deposits.

(b) At the end of the Permian volcanic outbursts took place, and a land emerged which lasted throughout the Lower and Middle Triassic.

(c) In the Upper Triassic a new marine submergence carried the sea into the areas of the present major mountains of the Alaskan peninsula.

(d) At the end of the Triassic the sea withdrew from the whole of Alaska.

(e) During the Jurassic a new marine transgression of a more or less gradual nature established a permanent shore-line in, or near the present position of the Alaska Range.

(f) In the Lower Cretaceous the sea swept again over most of Alaska, even covering the greatest part of the Yukon Valley which had been above water ever after the Palaeozoic. Volcanic activity appears to have been absent during this era.

(g) In the beginning of the Upper Cretaceous the sea receded from Alaska, and sedimentation began of a new type, in areas different from those involved in sedimentation during the very early Cretaceous.

(h) At some time during the latter half of the Cretaceous the sea invaded once more part of the southern coast of Alaska.

(i) As the Cretaceous came to a close, a new period of active mountain-building began which caused violent volcanism. When this period ended, the sea had withdrawn, and folding and erosion of the Cretaceous rocks started.

This record may freely be discounted to allow for discrepancies of opinion among geologists, imperfect chronology, outright errors and omissions and present lack of certain necessary data. This done, enough remains to tell us that the Alaska of the modern map had a violent past, during which it stood in turn above and below oceanic waters, endured volcanism for ages, saw mountains coming and going. Most, when not all, of the land which is now Alaska was at some time or other *both above and below water*.

During the Mid-Cretaceous, Alaska nurtured a comparatively rich flora (19), consisting of a quantity of ferns, well diversified conifers including *Gingko* and angiospermous plants of a thoroughly modern type, among others *Platanus*, *Quercus*, *Magnolia* and *Lindera*. Palaeobotanists visualize in this flora supposedly tropical and subtropical forms, *Credneria*, *Sterculia*, *Grewiopsis*, *Persea*, etc., which does not satisfy us altogether, but the matter need not be discussed at this juncture, considering that there can be no doubt as to *Platanus*, *Quercus*, *Magnolia* and *Lindera* at least having been in this plant-world. As we shall see, Alaska is by no means the only landbridge that stood in the Northern Pacific, but this phase of the evidence, too, is to be discussed later.

Taking the record as it reads, we see in Alaska during the Middle Cretaceous plants which are now living along the North American Atlantic coast. The evidence, then, is that while Alaska and the lands in its vicinity were being convulsed by geological turmoil, and in turn sunk under water and uplifted, the Angiosperms continued to filter in, reaching in the end the Atlantic clear across the continent. How many were the „Sunken islands" and the „Halting-places" inside and outside Alaska of our maps which favored their progress we do not know. Even less do we know whether Alaska and the regions nearby always consisted of large islands throughout, of small islands with some larger continental mass inbetween, of lowlands or high mountains. This we do not know, but what we definitely know, and is material to us, is that Western North America behaved, *as to the ultimate effects*, as a landbridge between the Old and the New World throughout. In other words: *A landbridge is judged primarily by its effects upon dispersal*, very much in the manner in which ashes bespeak the presence of a former fire. It does not make a particle of a difference to a phytogeographer what this landbridge was *in details* at intervals of ten million years more or less, for that which counts for him is the fact that plants came through. Naturally, we would like to know everyone, of the details, and have perfect continuous records, but in the absence of these details and records we are by no means left in the dark. If the plants came through they did so *because they could pass*.

How, then, could they pass? To answer this question in the light of sound probabilities, we may visualize a land fully stocked with vegetation, and a land newly emerged nearby, which has as yet no plant-cover. Beyond doubt, the latter will in due course be colonized by the former, and should the former in due course sink, the latter may again colonize newly emerged lands, and so on without end. In sum, a continuous flow of organized life will take place from the land which has this life toward the land which is without this life but has conditions suitable to its survival and reproduction.

In the light of the example we have just introduced, it is altogether sound to postulate that *the colonization of vacant lands* is effected, as claimed by Darwin and others, by „casual means of dispersal," winds, waves, birds and the like. It is furthermore possible to visualize passages of flora between contiguous regions effecting themselves over chains of ancient or newly uplifted mountains. The rise of certain mountains may also modify the climate of a region, and open it to the inroads of new forms of life. The same is true of gentle changes in level bringing a former plain to an height suitable to subalpine or alpine floras. In sum, we may imagine anything and everything we like as a combination of these factors, constantly using „casual means of dispersal" and continuous land-connections, both of them, in different degrees of blending to further migration.

It stands to reason that all these means will be operative *only insofar land is available upon which they can operate*. As we know, massive exchanges of flora took place between Australia and southern South America, for example, and their very massiveness is proof that the „casual means of dispersal" active between these regions worked under the most favorable possible conditions. In other words, there stood at various epochs in the past, chains of islands, sundry landmasses and the like which connected Australia and southern South America, and allowed a constant flow of angiospermous creation to move toward them both, and ultimately to colonize them both. By the same token, Africa and South America were connected in an homologous manner, and so were India and East Africa. How do we know, then? We know for the simple reason that there is a community of flora between these regions, a community of flora which is massive and persistent enough to assure us that „casual means of dispersal" worked at the greatest possible advantage *in certain directions rather than in certain others*. It is also for this reason that vegetation moves along certain fairly well defined tracks, and that plants of unlike ancestry all follow the same routes.

It is not easy to understand why an abstract „Enormous capacity for trans-oceanic dispersal", or words to the same effect, should be introduced into scientific thinking, when, were this putative capacity entirely independent from the presence of definite landconnections, this very same capacity would eventually work itself out radially and uniformly from a center instead of following precise directions, and we should have no explanation whatever of the fact that arborescent Cactaceae are massive in the New World, where succulent tree-like Euphorbias are unknown, while the latter teem in Africa, a continent that the Cactaceae hardly touch. *The plain truth is that capacity for dispersal, and lands upon which this capacity can spend itself to a purpose, are synonymous. If we had this capacity, great as we may like to fancy it, and no suitable lands this capacity would be wholly sterile; conversely, if we had lands open to colonization, but no capacity for dispersal, migrations could not take place either.*

Darwin, then, is in the right, as we see it, when he visualizes a certain type of flora dispersing from a focus of origin, and using „Sunken islands" as „Halting-places" to reach the lands in the modern maps. He is also right, as we believe, in assuming that dispersal utilizes birds, waves, winds, in order to proceed from a point well stocked with life toward another point which is without life as yet. Naturally, all this presupposes the existence of „land-bridges." The evidence is here global.

Authors who believe in „casual means of dispersal" as the reason for dispersal all too frequently forget that life flows well only between points stocked with life and points bereft of life, but hardly flows at all between points which both are stocked with life. True, occasional exchanges may take place all over, but a seed which is wafted by the winds and deposited in a region already in the firm grip of a flora has very little chance of starting a new progeny. What ecologists describe as „closed associations," meaning societies of plants in full possession of a suitable ground, are very nearly impervious to immigrants, and the most pronounced „Capacity for transoceanic dispersal" is stopped dead in its tracks by such associations.

The certain sterility attending the explanations of authors who profess to work dispersal as it were *in vacuo* is well attested in the literature. A phytogeographer who has excellent knowledge of Malaysia (20) writes as follows, „If we consider the means of dispersal of the temperate mountain flora of the Lesser Soenda Islands as a whole . . . . it appears that from 81 species 21 can be assumed to be dispersed by *animals* (endozoically or epizoically) and 18 by *wind*, but 42 species have *no special means* of dispersal . . . . Why have

Distylium, Ranunculus, Euphorbia, etc. crossed Bali Straits but *Gentiana* and numerous species of *Carex* not? Why has *Valeriana* with its very effective dispersal method not been able to reach Bali from Mt. Idjen *(in Java)*? I believe that no one can answer these intricate questions; I have in vain tried to trace a relation between distribution and dispersal method and I have not the least idea for explanation. That dispersal alone is responsible cannot be accepted .... Two endemics are wind-dispersed, viz. *Gerbera peregrina* and *Senecio lombokensis.* The first is known from Bali only and occurs at a distance of 3000 kilometers from its nearest Asiatic congeners. The second is endemic in Lombok and occurs at about the same range from its allies in SE. Australia. Why are these wind-dispersed plants endemic in small spots? *On the whole I cannot trace any relation between* (means of distribution) *and what is known of dispersal.*"

These questions can be answered by quoting another botanist (21) also active in the tropical Far East. Discussing the flora endemic to the summit of Mt. Gunong Tapis in Malacca, this botanist writes, „If we consider the floristic composition of the mountain top we find that there are several species .... that are of typical Australian genera or, at least, closely allied to Australian forms .... Mr. Ridley has drawn attention to this phenomen and in explanation postulates the theory that these plants are relics of an Australian flora that covered the „cool dry tableland with Mt. Kinabalu *(of Borneo)* on one side and Gunong Kerbau and Mt. Ophir *(of Malacca)* on the other" in Tertiary times. Some such hypothesis appears essential if we are to explain these relics. Perhaps, we may go further and say that not only are the species relics, but also the plants communities to which they belong."

We may, on our part, go further still, and state that, in most cases when not all, not only the plant-communities are relics but *the very land on which they grow is a fragment from the geologic past.* Such is the Pacaraima. Of course, the Pacaraima is no longer of loose sand, for the sand hardened into the modern sandstone. However, the whole plant-world endemic to the Pacaraima summits is still very much the same as it was prior to the hardening of these sands. As a matter of fact, we can see loose sand being compacted by various agencies into a magma under our own eyes. This takes place in the Malaysian „padangs" (22), and it is by no means surprising that Ericaceae and Droseraceae, to mention examples, are characteristic of the „padangs" of Malaysia and abundant in the Pacaraima vegetation.

When these harbors of ancient flora are carefully searched, forms turn up showing extraordinary disconnections. The reason is simple, of course, for the ranges intervening between these forms and their next of kin have been blotted out by geological changes. Accordingly, we find the Sarraceniaceae represented at the Pacaraima by a genus, *Heliamphora;* in California and Oregon by a second, *Darlingtonia;* in the Eastern United States by a third, *Sarracenia.* As we shall see, there once was a continuous range between the Guianas and the Eastern United States, and this range, marked by extra-

ordinary disconnection in the Sarraceniaceae, is still fairly closely held together by the distribution of *Utricularia* and the Cyrillaceae. Little wonder,
then, that the Sarraceniaceae still appear at various points of this range. Their
dispersal is often given up as a mystery (23), but we do not share this opinion,
for no mystery at all is involved in this and homologous distributions. We do
not understand the facts merely *because we look at them under wrong premises*.
Patently, it is not more facts that we need, rather methods and approaches
fit to piece the facts together within a common-sense whole.

Mount Ophir is a comparatively isolated peak in Malacca proper, and its
flora produced two striking disconnections at least. *Linaria alpina* native to
the European Pyrenees, Alps and Transsylvania, was collected on this peak
under conditions vouching for its being certainly endemic at the spot. Its
occurrence on the Malaccan heights is dismissed with the laconic comment
(20), „I cannot account for this extraordinary disconnection."

Always on the summit of Mt. Ophir, was located a species of the barringtoniaceous genus *Foetidia*, *F. ophirensis*, which is otherwise restricted to Madagascar and the Mascarenes, but has a form, *F. mauritiana* in the Mascarenean island of Mauritius, with a variety *elongata* in Eastern Bengal. Although less startling, the presence of *Foetidia* on Mt. Ophir is quite as remarkable as that of *Linaria alpina*. There can be no question here of birds
having carried the seed of these plants, or the like. These are relic-ranges,
and as such they must be dealt with.

In Java was once collected on the slopes of Mt. Dieng *Alisma natans* (20)
otherwise known from Europe. This record is also commented upon as follows, „I cannot account for this enormous discontinuity." There is no doubt
as to this being, once more, a relic-range.

We will learn in the coming pages that these relic-ranges are precisely
what we have the right to expect on account of the operation of certain channels of migration streaming to Malaysia from the direction of Africa. Considering that these channels are standard, and their operations can easily
be identified, *Linaria*, *Foetidia* and *Alisma* are in Malacca and Java by ordinary
means, not by a miracle. These plants survived at certain isolated stations
while they died out elsewhere, and if they survived at these stations, this is
because the land there stood still, as it were. Mount Ophir, Mt. Dieng, the
Pacaraima are all in one and the same phytogeographic class, and once we
multiply „miracles" to take care of the Sarraceniaceae, Scrophulariaceae,
Barringtoniaceae and Alismataceae in South and North America and the Far
East it is time for us to think that, perhaps, there is nothing extraordinary in
the whole, and we had better change tune, and stop wondering.

Naturally, if we insist upon „occasional means of dispersal," and reject
former landconnections as something no longer *à la mode*, we may keep up
these „miracles" to the end, and we may find so many of them, as a matter of
fact, that we can easily turn the whole of a science, phytogeography, into a
collection of mystery-tales.

Why, then, are the „wind-dispersed" *Gerbera peregrina* and *Senecio lombokensis* endemic to small spots?

The answer pointedly is that the wind has nothing to do with it any longer. These species came to these small spots, either as the modern species or in another form, *when these spots were not quite as small as they are today.* In other words, we merely delude ourselves when we pin faith in „means of dispersal" that are academic.

May we hope to reconstruct the shattered ranges of these and similar waifs? By all means, yes. This we can do, if abandoning a wholly unsuitable approach to phytogeography, and desisting from theorizing, we look at the facts in the light of common sense. The first thing to do is to compare the migrations of many unrelated plants; to study their flow, and to follow them from one continent to the other if need arises. We cannot handle dispersal piecemeal. The evidence has its own requirement, and it is not academic habits or preconceptions that have a saying in the matter. Either we take things as they are, and approach them suitably, or go nowhere. If we have gone nowhere so far, and multiplied dark corners around us rather than doing away with them, the fault is ours, not of things in nature.

The tracks of *Euphrasia* and certain species of *Carex* prove most conclusively that a continuous channel of migration connects Australia with the Mediterranean. Much has been written by various authors (for instance, 24) versed in palaeobotany to affirm, or deny the presence of putative Tertiary fossils in Europe having „Australian" affinities. Although we are not palaeobotanists, we do not scruple going on record with the statement that it is altogether possible that the ancient floras of the Mediterranean and Europe contained authentic „Australian" plants. Some of these plants are definitely identified (25), and the presence of these „Australian" migrants in Europe is commonplace. The flora of that continent, as a matter of fact, contains cleancut „Australian" elements to this day, witness *Carex pyrenaica* and *C. Lachenalii*, not only, but an „Australian" element typified by the former is in Greenland. This would sound preposterous if the maps of dispersal did not vouch for it. *Euphrasia*, too, is an „Australian" plant in Europe, Greenland and the Moroccan mountains of the Atlas. In sum, when we deal with labels and definitions, everything becomes possible, which is reason why we should not deal with labels and definitions at all, merely stand by the facts, and *understand them before we try to catalogue them.* The last thing we will do, of course, is to follow palaeobotanists who argue fossils without taking strict account of living plants, and how these plants migrate. As we have stated, only artificial issues are injected in the discussion when we draw sharp lines between the past and the present, and deal with botany, *in any of its phases,* without simultaneously heeding every one of these phases. Let us be certain in our mind that what we need are not facts, of which we have perhaps too many at hand but correct methods of evaluating the evidence not as palaeobotanists, morphologists, ecologists, etc., etc., but simply as botanists, that is, students of plants.

Christmas Island is a speck of land about 43 square miles big which lies in the Indian Ocean 190 miles south of Java and about 900 miles north-west of West Australia. This island has a fauna and a flora rich of various endemisms and with affinities in different directions. As seen by a geologist (26), this outpost is somewhat less than conventional for a conventional „oceanic" island, because its consists of limestones of Tertiary age overlaying a deep submarine platform. There is good evidence that at some time during the Tertiary this island consisted in reality of a galaxy of atolls or islets, upon which certain flying creatures deposited large amounts of guano, the source of modern phosphates. What this island was prior to the Tertiary we do not well know.

Reviewing the origin of the fauna and flora of Christmas Island, a naturalist (26; p. 302) affirms that, indeed, this flora and fauna came in on account of „Occasional means of dispersal," rafts, etc., etc. This naturalist agrees, however, that „For an extremely long time no bird has become a permanent denizen of these islands," which might have come from Java or elsewhere, and further states that the ancestors of the endemic birds arrived when „Different metereological conditions prevailed" from those which now obtain. Of course, we have no objection against accepting colonization by „Occasional means of dispersal" effected an extremely long time ago, indeed when conditions of all sorts prevailed at this point which have no counterpart in those of the present. Mount Gunong Tapis also was reached by the „Australian" flora under similar conditions of time, and, probably, of climate, and once we are fairly safe as to chronology, and do not hurry nature in its work, we may allow honest discrepancies of opinion. If five seemingly related species of *Balanophora* have the following distribution (27),

(1) *B. fungosa* — Society, Caroline and Loyalty Islands, New Hebrides, New Caledonia, Eastern Australia.

(2) *B. Hildebrandtii* — Tahiti, Comoro Islands.

(3) *B. insularis* — Christmas and Pulau Aor Islands.

(4) *B. Micholitzii* — Tanimbar Laoet Islands.

(5) *B. Zollingeri* — Salayer Island;

we may be certain that these species did not migrate yesterday, and that the „Occasional means of dispersal" which served them to wander had abundant facilities in the geography of the remote past for doing their work. The map is eloquent (Fig. 12), and should it be so, that these species are less intimately related than supposed, or occur elsewhere as yet unreported by botanical explorers, still we could say that it took some time for *B. Hildebrandtii* „sensu latissimo" or „sensu strictissimo" to effect a journey between the Comoros and Tahiti. We could add, moreover, that this journey took place when these islands stood in contact in a manner that seems to us mysterious, indeed, but we must leave no stone unturned to fathom through and through. Clearly, it is the function of science to do away with mysteries replacing them with reason, and where a mystery stands, there we must go forthwith. As we are to learn, *Balanophora* migrates more or less as other plants, *Korthalsella* or

Nesogenes do, for instance, and critical, insistent comparisons of dispersal are our best weapons.

We do not believe that there is basic conflict between an orderly approach to the problem of dispersal made in the light of geological factors, and the conclusions of an author (28), who, dealing *with the flora of Norway,* remarks that some thirty local species give every indication of having been introduced by birds or similar agencies. Unquestionably, this author has sound evidence as regards the origin of several of these plants, and one of his authorities may be correct, who maintains that the center of distribution of the above-named species is along a regular line of migration followed by birds of passage from Denmark to Norway. There is a question in our mind, however, how the statements of this author can be reconciled with the diametrically opposite declarations of the previously quoted botanist who avers that the whole of the plant-communities on the Malaccan Gunong Tapis are „Australian," and the agnosticism of still another author who, as we have heard, is sure that „means of dispersal" and actual distribution bear one to the other no rational relation. These sources all disagree in some essential or detail, which is merely proof that all of them have seen an aspect of the problem, but not the whole, and that, accordingly, they cannot be reconciled *except by seeing the whole in the first place.*

The author on Norway so writes, „It is no longer possible to maintain the old dogma that the entire plant community migrated step by step, like a regiment of soldiers, and took possession of the country under climatic conditions that were favorable to the various species, while the previous vegetation was decimated and only survived in especially favorable localities; for *vegetable species generally immigrate singly and independently of one another,"* only to conclude, „All things considered, I am inclined to believe that in trying to explain the distribution of vegetable species and the paths they have followed, we shall arrive at better results by studying the ways in which they spread at the present time than by setting up hypotheses of tremenduous convulsions of nature . . . . which can neither be proved nor disproved, as they lie beyond the spheres in which our present knowledge has a firm foundation on which to stand."

This author, evidently, is averse to „Cleverness and prophesy," but we would like to have him explain, merely as an example, the fact that two „Norwegian endemics" which he mentions, *Alnus glutinosa* and *Prunus avium,* occur *wholly disconnected* in the Southwestern Cape of South Africa, and the Kulu District of Eastern Punjab. This author speaks of an „*Altai Flora*" described as „A northeastern and eastern vegetation, which came into Europe from Siberia after the Glacial Period." This is an error outright, for the Altai node, as we shall see, was feeding Eurasia and the high north *long before* the Glacial Period was ever heard of.

There might be some difference between some thirty or sixty or ninety species migrating into Norway in a manner that suggests the action of birds, and the like, and truly formidable masses of vegetation spreading all over the world in the wake of thoroughly modern early Cretaceous ancestors. Norway, to begin with, is a land which was de-glaciated only in the very recent geological past, and may therefore still be open to readjustments in its floristic composition. It is possible to conceive of exchanges taking place only between it and lands to the south, east and north having, on the whole, comparable climate and ecology. Indeed, the world has in common a large population of weeds, and partial interchanges of plant-life constantly take place between neighbouring lands, the very same district harboring in epochs of heavy rains certain associations which seem to disappear as if by magic in years of marked drought.

However the chief weakness of arguments such as have been quoted is that they lump fundamental problems, and some of their very minor aspects, as something which is liable to rationalization of details in exactly the same measure. To authors who believe in these arguments, the massive exchanges of flora which, taking advantage of a type of geography now entirely lost, migrated between Africa and the Americas, Asia and the New World are wholly different from the dispersal of thirty species from Danemark to Norway in the historical present. Authors of this persuasion may be excellent *local* ecologists, but they have doubtless no clear perception of the issues that face the phytogeographer *at the same time all over the earth.* To these authors, *Alnus* and *Prunus* are „Norwegian," or at the most „Eurasian" genera, which is false from the ground up. These authors fail to see that the de-glaciation of Norway in a petty geological past opened lands to vegetable colonization very much in the same general manner as the submersion and emersion of parts of the world in the dim Jurassic and Cretaceous made it possible to other, and much earlier plants, to spread on to fresh pastures. These authors, obviously, fail to acquire the proper perspective as between „means of dispersal" and their manner of operation *in time and space.* They are not well aware of the principle that plants evolve as they migrate and that, accordingly, the first step necessary to pass sweeping opinions on dispersal with knowledge is in reaching of time and space a fitting concept. Naturally, these authors are flatly contradicted by other authors who dwell on other sides of the evidence, and their notions merely contribute to the babel of uncritical statements that blights the study of dispersal to its core. For each statement in the record we have another contradictory statement, which is reason why we intend to move first toward generalities of method in order that we may judge as between this welter of opposites. Were we to regret anything of the record most keenly, we would, no doubt, bemoan the fact that uncritical assumptions present themselves as scientific prudence, and that a negative attitude toward the duty of thought is enforced on the ground that, to be truly imbued with scientific spirit, we must count the species which might have been carried to Norway by birds of Danemark, and forget all the while the evidence from the rest of the world. The truth is not this way, which is to conclude that science is certainly not there, where these authors see its shadow.

BIBLIOGRAPHY

Chapter I

Depending upon the subject, the page-number may refer, (a) To the first page of the work cited; (b) To the page other than the first, specifically involved in the discussion. Page-numbers are omitted when general reference is had to works in book-form.

(1) MAYR, E. — The zoogeographic Position of the Hawaii Islands; The Condor 45: 45. 1943.

(2) MAYR, E. — History of the North American Bird Fauna; The Wilson Bulletin 58: 3. 1946.

(3) DARWIN, C. — The origin of species by means of natural selection. 1859.

(4) GUPPY, H. B. — Plants, Seeds, and Currents in the West Indies and Azores. 1917.

(5) WALLACE, A. R. — Island Life or, the Phenomena and Causes of Insular Faunas and Floras. 1881.

(6) GUPPY, H. B. — Plant-distribution from the standpoint of an idealist; Jour. Linn. Soc. (London), Bot., 44: 439. 1919.

(7) VAN STEENIS, C. G. G. J. — Malayan Bignoniaceae, their Taxonomy, Origin and Geographical Distribution; Rec. Trav. Bot. Néerl. 34: 787. 1927.

(8) DU RIETZ, G. E. — Problems of Bipolar Plant Distribution; Acta Phytog. Suec. 13: 215. 1940.

(9) BERRY, E. W. — Tertiary Flora from the Rio Pichileufu, Argentina; Geol. Soc. America, Spec. Papers No. 12. 1938.

(10) SKOTTSBERG, C. (editor) — The Natural History of Juan Fernandez and Easter Islands, No. 7 (SKOTTSBERG, C., author) Phanerogams of the Juan Fernandez. 1921.

(11) DU RIETZ, G. E. — Two new species of Euphrasia from the Philippines and their phytogeographical significance; Svensk. Bot. Tidskr. 25: 500. 1931.

(12) SKOTTSBERG, C. — Notes on the Relations between the floras of Subantarctic America and New Zealand; The Plant World 18: 129. 1915.

(13) VAN OOSTROOM, S. J. — A monograph of the genus Evolvulus; Mededeel. Bot. Mus. Herb. Univ. Utrecht, No. 14, 1. 1934.

(14) FERNALD, M. L. — Specific Segregations and Identities in some

44

floras of Eastern North America and the Old World; Rhodora 33: 25. 1931.

(15) LLOYD, F. E. — The Carnivorous Plants. 1942.

(16) DARRAH, W. C. — Textbook of Paleobotany. 1939.

(17) LIDDLE, R. A. — The Geology of Venezuela and Trinidad, 1928.

(18) CHAPMAN, F. M. — Problems of the Roraima-Duida region as presented by bird-life; Geographical Review 21: 363. 1931.

(19) HOLLICK, A. (& MARTIN, G. C.) — The Upper Cretaceous Floras of Alaska; U. S. Dept. Inter., Geol. Survey, Profess. Paper 159. 1930.

(20) VAN STEENIS, C. G. G. J. — On the origin of the Malaysian Mountain Flora, Pt. III; Bull. Jard. Bot. Buitenzorg, III, 14: 56. 1936.

(21) SYMINGTON, D. F. — The Flora of Gunong Tapis in Pahang; Jour. Malayan Branch Roy. Asiatic Soc. 14: 333. 1936.

(22) VAN STEENIS, C. G. G. J. — Botanical results of a trip to the Anambas and Natoena Islands with notes on the vegetation of Djemada by M. R. Henderson; Bull. Jard. Bot. Buitenzorg, III, 12: 179. 1932.

(23) GLEASON, H. A. — Botanical Results of the Tyler-Duida Expedition: *Sarraceniaceae;* Bull. Torrey Bot. Club 58: 366. 1931.

(24) DEANE, H. — Observations on the Tertiary Flora of Australia, with special reference to Ettingshausen's Theory of the Tertiary Cosmopolitan Flora, Pt. II; Proceed. Linn. Soc. New South Wales 25: 581. 1901.

(25) BANDULSKA, H. — On the cuticles of some fossil and recent *Lauraceae* Jour. Linn. Soc. (London), Bot., 47: 383. 1926; On the cuticle of some Recent and Fossil Myrtaceae, ibid. 48: 657. 1931.

(26) ANDREWS, C. W. (*et. al.*) — A monograph of Christmas Island (Indian Ocean). 1900.

(27) RIDLEY, H. N. — An expedition to Christmas Island; Jour. Straits Branch Roy. Asiatic Soc. No. 45: 137. 1905.

(28) WILLE, N. — The flora of Norway and its immigration; Ann. Missouri Bot. Gard. 2: 59. 1915.

CHAPTER II

THE CONCEPT OF *GENORHEITRON*

The group of *Evolvulus* reviewed in the preceding chapter falls into two groups, Africa being central to the dispersal. One of these groups, represented by *E. alsinoides*, overruns the map to an extent beyond comparison greater than the other.

In the opinion of the taxonomist whom we have followed, *E. alsinoides* consists of several populations with the rank, or status, of varieties. As we have pointed out, this opinion may not be agreable to other taxonomists, some of whom, doubtless, will prefer to identify one or more of these populations as „good" species.

Divergences of the kind do not interest us, for only localities and affinities are of our immediate concern. We mention them, nevertheless, to stress the fact that *E. alsinoides* is the heart of a galaxy of minor forms lacking formal characters of the kind which taxonomists regard as „strong." If some of these forms are „species," still none, or at the best few of them, may claim the distinction of being „strong species." It follows that the stream of migration repeating its origin from *E. alsinoides*, though wide-ranging in three continents, and more important from the standpoint of pure geography than all the rest of the alliance, was not laden in origin with great potentials of evolution. As it migrated overruning litterally millions of square miles, this stream failed to unburden itself of local entities which we might uncontroversially identify as species. It released out of its bosom, on the contrary, nothing better than comparatively weak offshoots. Some of these offshoots are interesting for a phytogeographer, but none of them is major from the standpoint of taxonomy.

Quite to the contrary, *Evolvulus* swept upon the New World with a flow of migration bursting with evolutive power, and in this direction brought forward numerous distinct species. Once again, some of these species will not seem „good" to all taxonomists alike, but there is no mistaking the fact that *Evolvulus* dispatched to America ancestors competent to do in evolution what the ancestors directed to Asia and Australia could not achieve.

A behavior of this sort is by no means peculiar to *Evolvulus*, witness, for instance, the dispersal of two closely allied families, Turneraceae and Malesherbiaceae (1, 2), as follows,

A) Turneraceae

(1) *Wormskioldia* — 8 species in Tropical Africa.

(2) *Streptopetalum* — 2 species in Tropical East Africa.

(3) *Piriqueta* — 15 species (Sect. *Eupiriqueta*) in Brazil, Paraguay, Bolivia, Guianas, Venezuela, West Indies, United States; 4 species (Sect. *Ehrblichia*) in South Africa (here 1 species), Madagascar (2 species), Central America (1 species).

(4) *Mathurinia* — 1 species in the Mascarenes (Rodriguez).

(5) *Hyalocalyx* — 1 species in Madagascar.

(6) *Turnera* — 57 species in Argentina, Brazil, Paraguay, Bolivia, Peru, Guiana, Venezuela, West Indies, Mexico; 1 species (*T. ulmifolia*), ranging from Argentina to Mexico and the West Indies, further extending with a variety, or form, from the Mascarenes and Seychelles to the tropical Far East and Malaysia.

(7) *Loewia* — 2 species in tropical Eastern Africa.

B) Malesherbiaceae

(1) *Malesherbia* — 23 species in Argentina, Central and Northern Chile to Southern Peru.

It is apparent (Fig. 13) that this dispersal matches in essentials that of *Evolvulus*, though it leaves West Australia out of its reach. *Piriqueta* splits into two sections unequally divided between Africa and America. *Turnera* does the same, if to a different degree, for it is all in the New World to the exception of a form or variety that fares to the Far East through the Mascarenes and Seychelles. *Malesherbia* is distributed as is *Evolvulus arizonicus* in South America. While *Piriqueta* freely streams from Brazil to Bolivia across Paraguay, *Malesherbia* keeps a somewhat parallel course to the south reaching Argentina in the first place.

The data we use may be imperfect, for no record is ever final in taxonomy, but mean as they read, (a) *Evolvulus*, Turneraceae and Malesherbiaceae followed the same tracks to the New World, therefore took advantage of the same land-connections; (b) The stream of migration which the closely allied Turneraceae and Malesherbiaceae dispatched to the Americas was laden with potentials of evolution beyond comparison greater than the stream that went to Africa and the Far East, at least as to species.

Groups that migrate along lines comparable to those we have just illustrated are numerous, both under the Dicotyledons and the Monocotyledons, witness, for example, the family Rapateaceae (3) of the latter, as follows (Fig. 14),

(1) *Maschalocephalus* — 1 species in Africa (Liberia).

(2) *Rapatea* — 12 species in Northern and Central Brazil, Eastern Peru, Guianas and Venezuela.

(3) *Stegolepis* — 4 species in Venezuela (Pacaraima Mts.)

(4) *Saxofridericia* — 3 species in the Guianas.

(5) *Cephalostemon* — 5 species in Eastern, Central and Northern Brazil.

(6) *Monotrema* — 3 species in Venezuela.

(7) *Spathanthus* — 1 species in Northern Brazil and Guiana.

(8) *Winsdorina* — 1 species in Guiana.

(9) *Amphiphyllum* — 1 species in Venezuela.

It will readily be noticed that all these groups come to the New World with streams of dispersal beyond comparison stronger than those they dispatch to the Old, and that all of them repeat their origin from a region that corresponds to the vicinity of West Africa on our maps.

The question arises, then, whether these streams actually consisted of the modern genera when the migrations first began, or merely contained, as yet unevolved, genetic powers that could evolve these genera in a later time. This question is important, for its answer is material to an understanding of phylogeny.

A glance at the dispersal of all the groups tabulated readily shows that certain species or genera were beyond doubt modern in a very early stage of the migrations. *Piriqueta,* for instance, must have existed very much as *Piriqueta* is today prior to the final breaking apart of America and Africa, for it is represented both in Madagascar, South Africa and the United States. We may indeed go further, and assume within reason that Sect. *Eupiriqueta* existed before the connections were severed which took the archetypes of the Sarraceniaceae to the North American Atlantic coast, also the Guianas and Venezuela. Knowing that dispersal was effected upon ancient maps, and that conventional „means of dispersal" had little to do with it, we can justifiably read the evidence in distribution for what it says, and conclude in consequence.

Less clear is the case of other groups. We no longer know whether *Wormskioldia* and *Streptopetalum* are quite as old as *Piriqueta,* for both these genera are in Africa, and it is accordingly difficult for us to test them against geographical standards such as the severance of Africa from South America or the like.

However, if we are not privileged in every case to decide whether a certain genus or species is older than another, and are unable to settle its position, whether primitive or derivative, we may have no doubt that dispersal was effected by broad streams of life in motion. These stream released at some point or other of their tracks taxonomic entities, whether families, genera or species, fit to take advantage of peculiarly favorable conditions. The Rapateaceae, evidently, found these conditions in northern South America, and it is in this region that they are most numerous and best developed. The Rapateaceae, too, had begun their process of genus-making before Africa and America separated, for one of their genera at least, *Mascalocephalus*, is in West Africa.

Streams of plant-life in motion laden with evolutive potential are known (4) as *genorheitra* *) This potential makes itself felt along the track delivering varieties, species, genera, etc., as the case may be at appropriate points. In follows that the concept of *genorheitron* and orderly migration are inseparable, for it is only when we follow the migration in correct sequence, from start to termination, that we reach a sound understanding of the *genorheitron's* operation on the map.

It is easy to illustrate by concrete examples what this means. Let us take *Turnera* for the purpose. This genus is massive in warmer America, but one of its species *T. ulmifolia* also extends to the Mascarenes, Seychelles, the tropical Far East and Malaysia.

We have so far read migration from a starting point in the south going north, and in every case read it easily. It has been possible for us, as a matter of fact, to put order in the seemingly confuse distribution of *Evolvulus* by so doing, which is sound evidence that we have followed the *genorheitron* correctly in its wanderings on the map.

If we accept for *Turnera* a starting point in the south once more, we see the genus breaking before us in two masses, one by far larger than the other. The larger mass goes to the New World ultimately reaching the West Indies and Mexico. In this stream of migration is included also part of *T. ulmifolia*.

The remaining segment of *T. ulmifolia*, whether it be the type-species itself, or a variety, follows on the contrary a different path, going to Malaysia and the Far East.

The whole, genus, species, and variety, comes together south of Africa. *Evolvulus* also does the same. We have, then, identical patterns of behavior in genetically unrelated entities, which is a first indication that our starting point is correctly chosen. We know that there existed lands in the southern hemisphere through which close to 40 % of the angiospermous families could migrate between Australia and South America, and we have no reason to reject as misleading the thought that the very same lands mark the starting point of the migrations of *Turnera*, generally speaking.

The *genorheitron* of this genus, then, had reached an advanced state of evolution even before it entered lands which still stand on our maps. Before leaving their apparent cradle in the south, the Turneraceae had marched far enough to feed part of one species to the Americas, sending the balance of it to the tropical Far East.

The distribution of the Malesherbiaceae further tells us that we cannot be far from correct in interpreting the evidence before us as we do. This family belongs to the *genorheitron* of the Turneraceae, and follows the same tracks to the New World. Malesherbiaceae, Turneraceae, part of *Turnera* all flow in

---

*) The author here cited prefers the spelling *Genorheithrum*. In using a slightly simplified form of the term we do not intend to imply that the original spelling is objectionable from the standpoint of etymology.

the same direction from the same point, and so does the larger segment of *Evolvulus*. The balance of *Turnera* and *Evolvulus* streams in another direction to the Far East, but the whole is orderly, and keys together to a common point south of Africa.

Indeed, we may add the Rapateaceae to the Malesherbiaceae Turneraceae and *Evolvulus*, always with the same result. The *genorheitra* uniformly come back to a region meridional to the Dark Continent, and depart from this region northward, eastward and westward like rays out of a single focus.

Were we to try to read these migration from other points of the compass, we could imagine that the Turneraceae, Malesherbiaceae, Rapateaceae originated in northern South America, but we should be hard put indeed to account for half *Turnera ulmifolia* running to Malaysia without turning up in Africa meantime. We could not understand the strange preeminency that certain petty islands, Mascarenes and Seychelles, seem to possess in dispersal. Even less could we hope to view the migrations of *Evolvulus* as something orderly and precise.

It is clear, then, that reading migration sequentially we readily understand *genorheitra* and their operations. Reading migrations correctly, as a matter of fact, makes things so simple and easy that we may wonder why all this has not been see long time ago, and why we have indulged in theoretical abstractions instead of going back to the map, and some elementary principle for guidance.

Much indeed has been written which would never have made print if the concept of *genorheitron* had been familiar to certain writers. Some of them insisted that the Aceraceae, for example, originated in „Holarctis" because *Acer* is fossil in the Cretaceous of Greenland and Alaska. It is true that *Acer* is fossil in those regions, which, as we have noticed, does not mean as yet that *Acer* originated there. Granting, however, that the testimony from fossil relicts is quite as relevant as these writers claim, the fact still remains that the Aceraceae are part and parcel of the *genorheitron* which gave birth to the Sapindaceae. If it were true that *Acer* originated in Alaska or Greenland, it is in Alaska and Greenland that we should also seek theorigin of most Sapindaceae, for one *genorheitron* is common to these groups alike. It may indeed be easy to contend that *Acer* came out of Alaska, and streaming from Alaska in two directions reached in due course the United States Mexico and Central America, the Far East to Java and the whole of Eurasia. It is no longer easy to deal with the Sapindaceae themselves from „Holarctis", however, for in this family are groups so formed (5),

A) Tribe Cossignieae

(1) *Cossignia* — Sect. *Eucossignia*: 2 species in the Mascarenes; Sect. *Melicopsidium*: 1 species in New Caledonia.

(2) *Llagunoa* — 1 species in Central Chile; 1 species in Bolivia, Peru, Ecuador and Colombia.

Obviously, the Cossignieae are not straight from „Holarctis," and it so happens that certain petty islands off the east coast of Africa are once more called into play by their dispersal as they are by that of *Turnera*. What is the meaning of these outposts?

The peculiar floras of the Cape of South Africa and West Australia may illustrate for us other aspects of the *genorheitron*-concept, and the manner in which this concept necessarily permeates phytogeographic thought. These floras cannot be understood unless migration and phylogeny are considered as one.

The Euphorbiaceae are represented in Australia by two different groups, Platylobeae and Stenolobeae, the former having a wide range of affinities outside of Australia, the latter, on the contrary, restricted to this domain and New Zealand to the exception of one or two species native to New Caledonia.

The Stenolobeae consist of nine genera and close to ninety species which have place under none of the conventional categories of systematic botany. These entities cannot be usefully segregated out as a family of their own, because their kinship with the Euphorbiaceae is by far too close. They cannot be treated as a subfamily or tribe because they repeat their origin from unrelated phyla stemming from both subfamilies of the Euphorbiaceae, Phyllanthoideae and Crotonoideae. They stand out as a local group of genera having diverse phylogenetic origin, but held together by a peculiar habit, sclerophyllous foliage in the main, and narrow cotyledons.

The sclerophyllous foliage frequently found in Australian endemic genera is credited by some authors (6) to the soil deficient in nitrates, and thoroughly leached in the course of geological time. Whether ultimately correct or not, this explanation receives some support from the occurrence in West Australia of approximately fifty species of *Drosera*, representing half this genus. *Drosera* is not sclerophyllous, but its large speciation in West Australia suggests that a measure of interrelation between the nature of the soil and the evolution of peculiar endemisms, whether sclerophyllous or not, is probable.

This granted, the Stenolobeae might be understood as a highly specialized group of the Euphorbiaceae that developed narrow cotyledons as a derivative character under the influence of the peculiar Australian soil. It might be argued that these forms came to West Australia in a recent past, and peculiarly developed in a state of isolation.

This explanation receives scanty support, however, from the fact that while the Stenolobeae occur in West Australia with eight genera, the remaining Euphorbiaceae, Platylobeae, have there fourteen, three of which are typically Australian (*Petalostigma*, *Adriana* and *Calycopeplus*), and beyond doubt endemic wholly or in part to West Australia. *Calycopeplus* moreover, is an archetype of *Euphorbia*. Is is clear, consequently, that the Stenolobeae are not necessarily late migrants to Australia, and that the Euphorbiaceae as a whole have been in this domain from ages, not only, but mere part of them „acquired" the „stenolobeous" habit there.

Conventional methods and explanations thus leave the riddle of the „narrow-cotyledon" group of Euphorbiaceae unsolved, but this riddle can easily be disposed of by other, and more appropriate, means. Outside of the sclerophyllous Stenolobeae are two genera, *Dysopsis* (Juan Fernandez Islands, Chile through the Andes to Central America) and *Seidelia* (region of the Cape, South Africa) which have unusually narrow cotyledons. Is has been argued (17) that these cotyledons are not narrow enough to warrant the inclusion of these two genera under the Stenolobeae, and that, at any rate, neither *Dysopsis* nor *Seidelia* and its ally *Leidesia* agree with them in habit. This is correct from the standpoint of formal morphology, but not satisfactory in the light of phylogeny. It seems clear, on the contrary, that „narrow-cotyledon" Euphorbiaceae, whether such as to deserve the technical appellation Stenolobeae or not hardly matters, are endemic to, (a) South Africa; (b) Australia, West Australia particularly; (c) Magellania.

These three points belong to the base-line of distribution which is circumpolar to the south, and are typic of „antarctic" phyla. Considering that the affinity centering around *Dysopsis* and *Seidelia* directly involves some of the largest euphorbiaceous groups (e.g., *Acalypha, Mallotus, Mercurialis, Cleidion*), and that certain Stenolobeae (e.g., *Beyeria*) strongly suggest African Platylobeae (e.g., *Clutia*), it is doubtless correct to conclude as follows, (a) The euphorbiaceous *genorheitron* had secured an advanced stage of evolution prior to its reaching South Africa, Australia and South America; (b) These regions were entered from the same center of origin; (c) This center of origin is necessarily central to the regions in question, and falls for this reason in the south; (d) Following the invasion of Africa, Australia and South America, the Euphorbiaceae migrated northward in the main, reaching in the end nearly the whole of the northern hemisphere; (e) The „narrow-cotyledon" group is not derivative, and might indeed belong to the very earliest *genorheitron*. This group has possibly survived more fully in a center which, to all appearances, offered conditions not unlike those under which this *genorheitron* first came into being.

These conclusions can be translated into tracks (Fig. 15) and it will be observed that the trends of migrations thus revealed are not in conflict with other trends from groups previously studied *). The base-line of distribution

---

*) Studying Fig. 15, and comparing this figure with the map showing the dispersal of *Euphrasia*, for example (Fig. 6), the reader may easily reach the conclusion that the range of the Stenolobeae and *Dysopsis* stems from genorheitral streams of migration departing from a base-line *in the Pacific*, and charge Fig. 15 with failure to take graphic account of this possibility. In reality, as Fig. 9 well indicates, it is downright impossible to identify at a glance the detailed base-line of *genorheitra* which feed as one South America, South Africa and Australia, New Zealand. In the case at hand, we may consider a number of alternatives, for example, as follows, (1) The Stenolobeae and *Dysopsis* stem from a base-line in the Pacific, *Seidelia* and *Leidesia* from an outlier of this same base, either in the east or the west; (2) The Stenolobeae and *Dysopsis*, on the contrary, originate from an enlarged base south of Africa; etc. All these alternatives leave unchanged the fundamental fact that the base of migration always lies in the deep south of our maps, and is antarctic in the accepted meaning of the term. Indeed, these alternatives imply that the base of migration is meridional to South America, South Africa, Australia and New Zealand alike. Once the reader is familiar with the principles of method stressed in this hand-

is the same, and from this base-line streams of migration issue which are laden with evolutive potential in every case. In short, migration is orderly, and evolution by no means an haphazard process. *Genorheitron* and channels of dispersal work together in harmonious fashion.

Naturally, it proves impossible to claim that the Euphorbiaceae originated in „Holarctis" or some such mythic land. This cannot be claimed for the elementary reason that the claim in question immediately destroys orderly migration and *genorheitral* operations. There is no rational — *therefore scientific* — way of driving the Stenolobeae to Australia from the north, because these forms are also anchored at points other than Australia, namely, the Cape and Magellania. These three points, Australia, the Cape and Magellania, must be reached at the same time, or very nearly at the same time, and with a *genorheitron* still plastic, if migration is to make sense.

It has been suggested (8; p. 121) that the peculiar endemic flora of the Cape of South Africa is of „oceanic" type, and this suggestion is beyond doubt valid also as regards the flora of West Australia. A plant-world is stranded in these two centers which, despite further partial migrations northward (e.g., *Ricinocarpus* to New Caledonia), is markedly autochtonous. Were the Cape mountains, and the westernmost tip of Australia to be separated from the mainland of Africa and Australia by the narrowest channel, the „islands" so resulting would be quite as definitely „oceanic" as Saint-Helena. Patently, the origins of the two floras in question cannot be sought in „Holarctis," nor can any of the *genorheitra* endemic to them be easily driven from „Holarctis," or such other putative heaven of angiospermy. These floras spell the doom of attempts made at explaining away dispersal with nebulous migrations, casual operations of agencies of dissemination and the like. They cannot be dealt with unless we understand that a concept of *genorheitron* and orderly migration are fundamental to scientific phytogeography.

The affinities and dispersal of the Euphorbiaceae tell us that Australia, South Africa and southern South America were all entered at approximately the same time by a stil plastic *genorheitron* which originated in "Antarctica" and later evolved all over the earth. The floras of the Cape and West Australia reveal in addition characters justifying the conventional appellation „oceanic." Enough is here to make it interesting for us to work around these facts and concepts awhile, in order that we may perfect our grasp upon dispersal in general, and learn without delay of the problems that face us in our chosen field. It is obvious that these problems are not always seen in their true light, and for this reason much is written about dispersal which obscures rather than illuminates its fundamentals.

The endemic flora of West Australia includes (9) about 300 different species of phyllodine Acacias. As it is well known, these plants bear either cortical

---

book, he readily will see these alternatives, and their implications, in the light which is theirs, and understand that the seemingly omission of an obvious possibility of migration of the *genorheitra* of the Stenolobeae and *Dysopsis* does not at all weaken the discussion in this, and similar, cases. See further Fig. 44 and 44 A.

expansions, or dilated petioles (indifferently termed *phyllodes* in taxonomy, and, alas, morphology) which function as organs of assimilation, and often replace conventional leguminous leaflets. These 300 species fall under eight different sectional groups, in itself proof that this huge speciation represents in reality evolution from a few archetypes.

Phyllodine Acacias also occur in Asia, Malaysia and Polynesia (Java, Moluccas, Philippines, Formosa, New Guinea, New Caledonia, New Hebrides, Fiji) to the number of about ten or twelve species. One of their description, *A. heterophylla*, is native to the Mascarenes, the lone representative of the group in geographical Africa.

The occurrence of *A. heterophylla* in the Mascarenes has nothing in itself extraordinary. Interesting, rather, are the affinities that this species exhibits in various quarters.

Acacia heterophylla is so close to *A. Koa* of Hawaii that, were the ranges contiguous, one would beyond doubt be treated more often than not as a variety of the other. *Acacia Koa*, moreover, is surrounded by a galaxy of lesser forms, all endemic to Hawaii, one of which, *A. Koaia*, has affinities in the direction of the Formosan and Philippine *A. confusa*.

To rationalize this dispersal we are offered the choice of at least two tracks, as the map (Fig. 16) indicates. One of these tracks directly streams from the Mascarenes to Hawaii via Malaysia, the Philippines and Formosa; the other, on the contrary, proceeds from the Western Pacific to Hawaii. It will readily be observed that these tracks are not mutually exclusive, and that, regardless of their course northward and eastward, both take their start from a base-line drawn between the Mascarenes and Australia. Working from this base-line northward, the *genorheitron* migrated and speciated at the same time, and its operations fall in the standard pattern of *Evolvulus*, *Turnera*, the Euphorbiaceae etc.

Let us, then, check both these tracks against the migrations of other groups.

Two closely related species of the primulaceous genus *Lysimachia* (10) are distributed in this manner,

(1) *L. mauritiana* — Mascarenes, Southern India, Southeastern China, Corea, Japan, Philippines (Batan Islands), Formosa, Riu-kiu, Bonin Islands, Mariannes, Hawaii, New Caledonia.

(2) *L. glaucophylla* — Western Mexico (Nayarit) *).

These two species drive straight to Hawaii and Mexico from the Mascarenes, and New Caledonia is seemingly reached by a track in the open Pacific of our maps, east of the Philippines.

It has been remarked (11) that a pencilled-in note suggests that *L. glaucophylla*'s type-specimen was perhaps collected in the Bonin Islands. This note may be well taken or not, we do not know, but means nothing relevant either way. Dispersal readily effects itself between Hawaii and the New World at

---

*) Nayarit, or Tepic, is a Mexican state, between Jalisco and Sinaloa on the Pacific Ocean.

any rate, witness the distribution of the celastraceous genus *Perrottetia*,

(1)   1 species in Australia (Queensland).

(2)   6 species in New Guinea.

(3)   1 species in the Moluccas.

(4)   1 species in Borneo, Java, Sumatra, Malacca.

(5)   1 species in the Philippines.

(6)   1 species in Formosa.

(7)   1 species in Central China.

(8)   1 *species in Hawaii*.

(9)   2 species in Mexico (Oaxaca, Chiapas, Veracruz) and Guatemala.

(10) 1 species in Costarica.

(11) 1 or 2 species in Colombia and Venezuela.

The apocynaceous genus *Vallesia* is allied with forms in the Old World, but distributed thus,

(1)   1 *species in Hawaii*.

(2)   1 species in the United States (Florida) and the West Indies (Cuba, Bahamas).

(3)   1 species in Western Mexico, Guatemala, Southwestern Colombia, Galápagos Islands, Ecuador, Northern and Southern Peru, Southern Bolivia, Northwestern Argentina.

It is manifest that *Lysimachia* would experience no difficulty in crossing over from Hawaii to Mexico. Hawaii, as a matter of fact, is suggested as a secondary center of evolution of the *genorheitron* in the affinity of *Vallesia*.

However, Hawaii need not be in the picture to have perfect transpacific migration, witness the dispersal of the hamamelidaceous genus *Distylium* (12), as follows,

(1)   1 species in Celebes, Flores, Java, Sumatra, Malacca.

(2)   1 species in Formosa.

(3)   1 species in Southern China.

(4)   1 species in the Eastern Himalayas.

(5)   1 species in Southeastern China and Japan.

(6)   1 species in the Bonin Islands.

(7)   1 species in Guatemala and Honduras.

All these tracks (Fig. 17, Fig. 18) are comparable in the end, though everyone of them has individual peculiarities. The Pacific is uniformly crossed in the latitude of Hawaii, whether this archipelago is immediately invested or not does not matter. The crossing may be obvious throughout as with *Perrottetia*, or disconnected around Hawaii as with *Distylium*. It may be effected by identically the same species, or, as with *Vallesia*, by a *genorheitron* which came to maturity only in Hawaii and the New World, leaving behind in the Old affinities which taxonomists tag with generic names other than *Vallesia*. It may, lastly, be revealed by cryptic affinities whispered about as mysteries of nature, and taxonomic conundrums of the first water which leave everybody dealing with classification guessing as to what generic name is best.

The *genorheitral* ties of *Vallesia* with the Far East are, after all, by no means unusual, witness the distribution of *Lysimachia* subg. *Lysimachiopsis* (11),

(1) *L. alpestris* — Southeastern China (Hongkong).

(2) *L. Hillebrandii* — Hawaii.

(3) *L. Lydgatei* — Hawaii.

(4)· *L. Remyi* — Hawaii.

(5) *L. rotundifolia* — Hawaii.

(6) *L. daphnoides* — Hawaii.

Patently, there is nothing mysterious in all this, and we may then conclude this part of our study as follows, (a) The affinities between *Acacia heterophylla* of the Mascarenes and *A. Koa* of Hawaii are commonplace. It is a fact that other genera run a track between the Mascarenes and Hawaii, witness *Lysimachia*. These genera, as *Lysimachia* proves, may run this track in two ways (1) With the same species, e.g., *L. mauritiana*; (2) With an undifferentiated *genorheitron* which releases species or genera at definite points in the track, as shown by *Vallesia* and *Lysimachia* subg. *Lysimachiopsis*; (b) The presence of phyllodine Acacias in the Philippines and Formosa is a normal event, for to this point came either one or both of two tracks, (1) Directly from the Mascarenes; (2) Directly from the Southwestern Pacific.

Lengthy comments may now be dispensed with as regards the following conclusions, (a) „Holarctic" origins have nothing to do with these migrations; (b) A concept of *genorheitron* and orderly dispersal is fundamental to scientific phytogeography; (c) The notions that dispersal is effected by haphazard occasional means of transportation, etc., etc., operating on maps resembling those of the present has no claim to scientific attention; (d) Dispersal was effected in origin over maps entirely unlike those of the modern earth, consequently cannot be later than the Cretaceous at the latest; (e) Dispersal is approached in science only through critical comparisons of tracks, and with a definite understanding of phylogeny. Theories, academic preconceptions, etc. have no bearing upon it; (f) Considering that *Lysimachia mauritiana* cannot be labelled „Mascarenean" in China, „Chinese" in New Caledonia, „New Caledonian" in Hawaii and the like, we may not hope to write scientific phytogeography by seriously entertaining definitions of the sort. These definitions are radically mistaken on several grounds (1) They refer only to part of the track, not to the whole track which alone is most material to a critical phytogeographer; (2) They convey false indications of origin; (3) They lead to aimless discussion among biogeographers having local interests; (4) They refer to modern lands, therefore ignore the elementary fact that these lands might not have existed at all in their present form at the time when the migrations were first effected.

Concerning now the possibility that the phyllodine Acacias might reach Hawaii from Australia, we can do nothing better than to glance over the distribution of the rhamnaceous genus *Alphitonia* (13), as follows,

A) Sect. *Glabratae*

(1) *A. erubescens* — New Caledonia.
(2) *A. xerocarpa* — New Caledonia.

B) Sect. *Tomentosae*

(3) *A. Vieillardii* — New Caledonia.
(4) *A. neocaledonica* — New Caledonia.
(5) *A. franguloides* — Fiji.
(6) *A. zizyphoides* — Fiji, Tahiti, Cook Islands.
(7) *A. Whitei* — Queensland.
(8) *A. excelsa* — Queensland.
(9) *A. Petriei* — South Australia, Queensland, Northern Australia.
(10) *A. obtusifolia* — Queensland, Northern Australia.
(11) *A. moluccana* — Northern Queensland, New Guinea, Aroe Islands, Celebes, Amboina, Moluccas, Timorlaut.
(12) *A. philippinensis* — Philippines (Luzon, Mindoro, Palawan, Samar, Leyte, Panay, Negros, Guimaras, Mindanao).
(13) *A. ponderosa* — Hawaii.

It is clear from this record that Hawaii can be reached from Queensland or Southeastern Polynesia alike, which is mere detail, and that the phyllodine Acacias could come to that archipelago along either one, possibly both, the tracks we have first outlined (see Fig. 16 a, b). It is also manifest that all these tracks, and their possible variants (see, for example, the course of *Perrotetia, Distylium, Lysimachia* within Malaysia) hark back to a base-line of dispersal in the south.

Concerning this base-line, we may observe that it would seem to operate preferably in certain regions. We have seen numerous genera emerging from it around South Africa; some coming out in southern South America (e.g., *Llagunoa*); others in Australia (e.g., phyllodine Acacia); or in Western Polynesia (e.g., *Alphitonia*), but, as we shall see, the regions where this base-line releases the greatest number of *genorheitra* northward are actually three, (a) Africa; (b) Western Polynesia; (c) South America, Magellania in particular.

We propose to refer to these regions as *gates of angiospermy*, speaking, in the coming pages, for example, of an „African gate of angiospermy," to indicate that a certain *genorheitron* issues from a point, or region, located in the proximity of South Africa by modern maps.

The reason why gates of angiospermy exist is of the simplest. The cradle of the seed-plants is ultimately to be sought in southern quarters, as we have already seen and will much better see as this study proceeds. This center made contact with the modern continents at specific regions, and these regions were used as a matter of course by the angiospermous *genorheitra* to fare all over. It is altogether possible that certain portions of the modern continents do actually represents fragments of an ancient Antarctica as, Fuegia, the Cape region of South Africa, or the westernmost tip of South Australia, for instance.

If there is something speculative in asserting that these regions may be fragment of geologic Antarctica — and we freely grant this much — there would seem to be nothing speculative about the existence of former connections between Antarctica and the modern continental masses. One of these connections, as a matter of fact, is still outlined by a chain of islands stretching between Graham Land, an antarctic outpost, and the tip of Fuegia. The indication in question is beyond doubt interesting from the standpoint of geography and geophysics, but very tenuous from that of phytogeography. We mean, in other words, that the islands between Fuegia and Graham Land are not necessarily the ones which secured free passage to about 40% of the seedplants between Australia and South America. To secure this passage far more intimate connections between these continents were necessary than through the modern islands in question.

If we compare the present outlines of sea and land in the quadrangle New Zealand—Balleny Island—Graham Land—Fuegia, we are readily struck by the fact that while a land-connection is still well marked between Graham Land and Fuegia, nothing like it is to be seen between Balleny Island and New Zealand. Between these two outposts the sea is deep, and lands absent.

Yet, it is quite clear that there existed between Antarctica and New Zealand geologic connections beyond comparison more intimate than those now binding Graham Land and Fuegia. Without these intimate connections it should never be possible to have migrations of an „antarctic" character in close to half the families of seed-plants. We refer to about half, or 40% of these plants, because it is not in our intentions to trace these migrations to the end here, but we are quite sure that the percent in question could be greatly increased by a critical study of *genorheitra* which would reveal that not 40% of these *genorheitra* migrated along conventional „antarctic" channels, but a much larger proportion.

We do not believe it is necessary to discuss contentions that Antarctica never was an important channel of migration, or that „antarctic" migrations took place by whimsical winds, waves, birds and the like. The facts belie these assumptions, and strong evidence is lacking that the authors of these assumptions had of the necessary facts adequate command.

Considering that the gates of angiospermy are the probable abutments to the modern continents of ancient southern landconnections, and that not every *genorheitron* necessarily used the whole of each gate, we need not be surprised in seeing that the limits of the gates in question vary somewhat. Certain groups occur at one end of the gate rather than at the entire gate; some use two gates instead of one, which all is thoroughly understandable once we are aware of the prime movers. Differences in the operation of the gates readily bear being illustrated by concrete examples, as follows,

A) *Phylica* (14)

(1) About 150 species in the immediate Cape region of South Africa.

(2) *P. polifolia* — Saint-Helena.

(3) *P. arborea* — Tristan da Cunha, Inaccessible Gough and Nightingale Islands, Mauritius, Amsterdam Island.

(4) *P. mauritiana* — Madagascar, Mascarenes (Mauritius, Réunion).

(5) *P. emirnensis* — Madagascar; var. *nyasae*: Nyasaland.

B) *Sparrmannia* (15)

(1) *S. africana* — Eastern Cape, Eastern Madagascar.

(2) *S. ricinocarpa* — Natal to Ethiopia; 2 outliers in Angola and the Belgian Congo.

C) *Astelia* (16)

(1) Subg. *Euastelia* — 8 species in the Mascarenes, Tasmania, Southeastern Australia, New Zealand, New Guinea.

(2) Subg. *Asteliopsis* — 10 species in New Caledonia, New Zealand, Hawaii, Rapa, Magellania.

(3) Subg. *Tricella* — 5 species in Southeastern Australia, New Zealand.

It will be observed: (a) *Phylica* (Rhamnaceae) uses the African gate throughout from Saint-Helena and the Tristan da Cunha group of islands to the Cape and Amsterdam Island. *Phylica emirnensis* illustrates a direct penetration of continental Africa from Madagascar; (b) *Sparrmannia* (Tiliaceae) uses only the eastern end of this gate (Eastern Cape to Madagascar), immediately invading the continent with *S. ricinocarpa*; (c) *Astelia* (Liliaceae) takes advantage of three gates, African, Western Polynesian (New Zealand, New Caledonia, New Guinea) and American. One of its subgenera, *Euastelia*, uses two (African, Western Polynesian) simultaneously.

The gates, strictly speaking, overlap at times, witness,

(1) *Empetrum rubrum* (Empetraceae) — Tristan da Cunha, Falkland Islands, Fuegia through Chile to Juan Fernandez.

(2) *Caryophyllaceae Pychnophylleae* — *Lyallia*: 1 species in Kerguelen Islands; *Hectorella*: 1 species in New Zealand; *Pycnophyllum*: 16 species in Northern Chile, Northwestern Argentina, Bolivia and Peru.

It will be seen, (a) At Tristan da Cunha the tracks part, some going to America (*Empetrum rubrum*) others to Africa (*Phylica*); (b) The same occurs between the islands of Amsterdam and Kerguelen; channels of dispersal originate in this immediate vicinity, which depart to Africa (*Phylica*), or New Zealand and Andean America (Caryophyllaceae Pychnophylleae). It is to their nearness to important gates of angiospermy that certain petty islands, the Mascarenes or Juan Fernandez, for example, owe their preeminence in the annals of dispersal. It is ease of migration within „Antarctica", and outside the range of modern landmasses that made it possible for the same genus or affinity to take advantage of two or more gates of angiospermy. A map (Fig. 19) will show this clearly.

Further to test the concept of *genorheitron* and orderly migration, we may next turn to the work of a cytologist (17), who believes that the Compositae Crepidinae originated in „South-central Asia." This opinion is worth scrustiny because cytology, properly used, may prove most serviceable to phytogeography.

The work in question essentially rests upon a painstaking investigation of four genera, *Crepis*, *Hieracium*, *Lactuca* and *Prenanthes*. It is doubtful, whether these groups are the only ones to deserve the name Crepidinae, for the author whom we follow makes at least passing reference to *Chondrilla*, *Taraxacum*, *Sonchus* and *Launea*, and other authorities (18) bring about twenty genera under this subtribe. It may be thought, consequently, that conclusions as to the Crepidinae established by a study of about twenty, or even forty per cent, of their number are not necessarily final. These conclusions would be better grounded, doubtless, if more genera had been sampled, and the *genorheitron* of the affinity in question duly accounted for. It is our firm opinion that specialized investigations, deep and diligent as they may happen to be, are seldom productive of general conclusions, and that a broad survey of the field of work (19) is an absolute prerequisite of truly constructive endeavor.

Discussing *Crepis* subg. *Catonia* in particular, the author in question (17; p. 35) states, „Only three 10-chromosome species have thus far been obtained, but two of them, *sibirica* and *pontana*, are the most primitive types in the whole genus. The first is distributed from central Europe to Mongolia and southward to Transcaucasia, and the second occurs at higher altitudes from France to Rumania. The third, *aurea*, is much less primitive and occurs in various forms in the mountains of southern Europe and northern Asia Minor. The 8-chromosome species fall into two groups, one with four, the other with six species. Of the first group, *blattarioides* and *alpestris* are montane species of southern Europe; the latter also occurs in Asia Minor, which supplements the evidence from gross and chromosome morphology that it had a common origin with *C. hypochaeridea* of South Africa. Furthermore there are central African species which are certainly related to *hypochaeridea*, and one of these, *C. Newii*, has just been found to have a genom very similar to that of *C. hypochaeridea*. The fourth species, *Hookeriana*, a Moroccan plant of the Grand Atlas Mountains, is less closely related to the other three."

Remarking that *Crepis* is completely absent from South America, but *Hieracium* is there widely distributed, this author adds (17; p. 21), „The wide distribution of *Hieracium* in South America and the complete absence of *Crepis* from that continent presents a distributional problem calling for more detailed study. A suggestion worth considering is that distributional relations such as these will eventually be found to have a bearing on the concept of continental displacement . . . . However, if both genera (*Crepis* and *Hieracium*) originated in south-central Asia and migrated into North America by way of a land bridge, even though *Crepis* is the more primitive, a difference in time of migration and the special adaptation of *Hieracium* species for rapid dissemination would explain the present distributional relations of the two genera."

We definitely know (see *Gerbera peregrina* and *Senecio lombokensis*; p. 37) that „Special adaptation for rapid dissemination" explains nothing in dispersal, and that the interpretation of the dispersal of *Crepis* and *Hieracium* just quoted cannot be satisfactory for the very reason that it relies upon adaptations of this fictitious nature. It is obvious, on the other hand, that the peculiar limitations of the distribution of *Crepis*, which is unknown in South America, are a cardinal factor of the entire distribution of the Crepidinae, and that vague appeals to „Continental displacement" do not bring us a fraction of an inch nearer the solution we seek, not only in the case of the Crepidinae but of every other plant as well. Appeals of this sort are typical of work which eventually runs into basic difficulties and proves unable to meet them, even less to master them. Actual distribution may not tell us everything we need to know about „Continental displacement," but will certainly reveal something to us of its prime movers, for *migration is an orderly process in time and space*. In brief, the reason why *Crepis* is absent from South America, while *Hieracium* there abunds is the one we must seek first of all, and this reason, obviously, has very little to do at bottom with chromosome counts in certain species or groups of species. This reason is, we may say, of cosmic order, fundamental in phytogeography and phylogeny alike, and as such must be sought out in the first place.

The affinities affirmed on the basis of chromosome-count in the group of *Crepis* mentioned above, tally as follows.

A) 10-chromosome group

(1) *C. sibirica* — Central Europe to Mongolia.
(2) *C. pontana* — France to Rumania.
(3) *C. aurea* — Southern Europe to Asia Minor.

B) 8-chromosome group

(1) *C. hypochaeridea* — South Africa.
(2) *C. Newii* — Central Africa.

(3) *C. Hookeriana* — Morocco.
(4) *C. alpestris* — Southern Europe.
(5) *C. blattarioides* — Southern Europe.

In our viewpoint this gives a perfect track between South Africa and Central Europe, with the Moroccan *C. Hookeriana* in a somewhat isolated position, which is all as it should be. If there is a difference of opinion between the author whom we cite and ourselves, this difference is in the fact that he read the track north to south, while we read it south to north. We are agreed in everything else, insofar at least the track itself is concerned. We do not set up ourselves as judges of what is „primitive" and „derivative" under the cytologist's microscope, but may remark that even in the supposedly primitive 10-chromosomes group one species (*C. aurea*) would seem to be more „derivative" than the other two. In short, it is not certain, as we see it, that a count of 10 chromosomes is final evidence of „primitiveness" for species with 10 chromosomes exhibit different degrees of „primitiveness" or „derivativeness."

We see no difficulty, of course, in granting that „cold" affinities, such as *Crepis* might freely travel between Africa and the mountains of Central Europe and Mongolia, witness the following tracks, (10)

A) *Primula* sect. *Floribundae*

(1) *P. floribunda* — Western Himalayas (Kashmir), Afghanistan.
(2) *P. Lacei* — Beluchistan.
(3) *P. Aucheri* — Arabia (Oman).
(4) *P. verticillata* — Arabia (Yemen).
(5) *P. Boveana* — Arabia (Sinai).
(6) *P. simensis* — Ethiopia (Semien).

B) Sapindaceae Koelreuterieae (5)

(1) *Erythrophysa* — 1 species in South Africa; 1 species in Madagascar.
(2) *Stocksia* — 1 species in Persia, Afghanistan, Beluchistan.
(3) *Koelreuteria* — 5 species in China (Yunnan and Kwantung to Jehol and Shensi); 2 species in Formosa; 1 species in Corea.

We may work these tracks any way we care, south to north, or north to south, without for this reaching the conclusion that groups of „cold" nature (*Primula; Koelreuteria*) could not migrate between South Africa and the Himalayas. It is true that the tracks we introduce are disconnected between Ethiopia and Madagascar, but were we to fill this gap, we could without the slightest difficulty find appropriate material for the purpose, witness *Cerastium africanum*, which ranges from Natal to Ethiopia along the mountains of East

Africa. This species of *Cerastium* belongs to Sect. *Orthodon* which is said
(20) to stem from the northern hemisphere. We do not at all vouch for this
opinion, but observe that a continuous track between South Africa and Afghanistan is a fact, as it is a fact that this track is run by groups (*Primula*
and *Cerastium*, when not *Koelreuteria*) that are anything but tropical in their
climatic preferences.

In the light of this evidence, we conclude, (a) It is less than certain that the
species of *Crepis* with 10-chromosomes count are „primitive" either for
Crepis or the Crepidinae; (b) Assuming, however, that possession of 10
chromosomes is absolute evidence of primitiveness, it does not follow that
the species having this count could not reach Central Europe or Central
Asia from the general direction of South Africa. A track appropriate to plants
with „cold" affinities is operative between South Africa and the Himalayas.

The statement that the Crepidinae „originated" in Central Asia, or anywhere in the northern continent for that matter, must be corrected, therefore,
to read that *Crepis* has in the mountains of Europe and Central Asia species
with a count of 10-chromosomes, and that these species give indication of
„primitiveness." This is what the evidence warrants, no more.

A correction is all the more needed in that the opinion that the Crepidinae
originated in the northern hemisphere causes those who so believe to indulge
in vague assumptions to account for the South American component of
Hieracium. As regards this component we will observe, (a) At least one species
of *Taraxacum*, *T. magellanicum* (see Fig. 9) has absolute circumpolar distribution in the south; (b) There exists in New Zealand (21) a rare plant, variously classified under *Crepis*, *Hieracium*, *Sonchus* with the specific epithet
novae-zelandiae; (c) *Sonchus grandifolius* is „A handsome and distinct
species without close allies" (21) strictly endemic to Chatham Island south of
New Zealand. Always in this genus, *S. megalocarpum* is endemic to Tasmania
and Australia, and the nearly cosmopolitan *S. asper* has a form, var. *littoralis*
localized in New Zealand, Chatham and Stewart Island; (d) A critical population variously treated as *Crepidiastrum*, *Crepis*, *Prenanthes*, *Lactuca*, *Chondrilla* is endemic to Formosa, Corea and Quelpaert Island, further ranging
to the Riu-kiu and Bonin archipelagoes; (e) Two peculiar genera, *Thamno-
seris*, restricted to the Desventuradas Islands; and *Dendroseris*, native to Juan
Fernandez, both off the Chilean coast, suggest at least (22) certain of the
characters of the Crepidinae; (f) *Launea* ranges from the West Indies to
India through the Canary Islands, the Mediterranean and Socotra, and has an
outlier in South Africa. *Reichardia*, another genus of the Crepidinae extends
from the Canaries through the Mediterranean to Western India, and one of
its species, *R. picroides*, occurs in disjunction in the Mediterranean and South
Australia, where it is certainly endemic.

We may further remark that the Hawaiian endemics Compositae *Argyroxiphium* and *Wilkesia* (23) are suggestive of the Hawaiian *Dubautia* and
Raillardia, not only, but of *Robinsonia* of Juan Fernandez, *Brachionostylum*

of New Guinea and *Bedfordia* of Australia. The evidence is such that, in our
deliberate opinion, an author is in the right (23; p. 11) who concludes that
the affinities of the peculiar Hawaiian Compositae „Are to be found to the
south or southwest in the Pacific." The Pacific is ringed throughout, as a
matter of fact, by Compositae that it proves exceedingly difficult to classify,
and forms of the same controversial nature are reported from Saint-Helena
in addition.

It seems certain to us that the Compositae did not originate in Central
Asia or in the northern hemisphere for that matter. The Crepidinae, moreover, are reported from regions, New Zealand, South Africa, Magellania,
which ill agree with the hypothesis that they had their birth in the mountains
of Southeastern Asia. In our opinion, the migrations of *Crepis* and *Hieracium*
are fully taken care of in a map (Fig. 20) *) which interprets their dispersal
in the light of the dispersal of *Evolvulus* or *Turnera*, for example. This map
accounts without effort for the presence of *Hieracium* in South America,
South Europe and Central Asia, Africa, and if need be in New Zealand.
It is interesting to remark that the indications of „primitiveness" given by
certain species of *Crepis* of Europe and Central Asia must be offset in a
critical approach to the problem of the origin of this group by the puzzling
tangle of affinities apparent in *Crepis* (or *Sonchus*, or *Hieracium*) *novae-
zelandiae*. Here is a species well worth close study.

In conclusion, it appears most probable that cytology, though in most
respects an excellent tool of botanical investigation, and the standard by which
knotty questions of hybridism may be settled, has nothing to contribute *as
such* to the problem of origin of the Angiosperms which is outstanding.
Genorheitra came from the angiospermous home in the south to the Himalayas and Altais as to a secondary center of dispersal. Most of them fared across
the tropics along subalpine and alpine channels now vanished on account of
the disruption and final destruction of the continental mass that bound
Africa and Asia as one prior to the Mid-Cretaceous. Being thus fred of the
necessity of adaptation to tropical environment, these forms retained archaic
characters throughout, which were lost in their allies later adapted on the
contrary to this environment. This accounts for the fact that certain Hima-

---

*) It will be noticed that this map strictly accounts only for the origin of *Crepis* and *Hieracium*.
Crepis originates at the African Gate of Angiospermy, and migrates from this Gate northward and eastward in standard fashion. *Hieracium*, on the contrary, originates at the eastern end of the African Gate
of Angiospermy, as does *Crepis*, and additionally at the Magellanian Gate, running from these regions
its further course on modern maps in a way which is usual. This map does not account for the affinities
underscored in the text which involve the shores of the Pacific. These affinities take their start from a
base-line similar to that of *Euphrasia*, for example (see Fig. 6), and ring in time all the Pacific shores.
Thus, the affinities discussed in the text, including the Crepidinae, actually have *antarctic genorheitral
distribution*, and use all the Gates of Angiospermy (Magellanian, African, Western Polynesian). This is
normal, for the Compositae, like every other major angiospermous family, use all the Gates to achieve
in the end world-wide distribution. *Crepis* and *Hieracium* are manifestly but a mite in a distributional
maelstrom that reaches all the five continents alike, and it must be clear that partial cytogenetic studies
shorn of a basic understanding of distribution in time and space can yield but indifferent results. See
also the perfect circumpolar distribution of *Taraxacum magellanicum* (Fig. 9), which in itself tells us
where is to be sought the ultimate origin of the Compositae.

layan groups are, or seem to be, more „primitive" than their relatives in Tropical Africa or the warmer Far East.

The Crepidinae are lost in this massive stream of dispersal as a very insignificant part of the whole, and no more originated in the Himalayas or Central Asia than did *Alnus, Macropodium, Carex, Juncus, Cobresia* and uncounted other forms.

A wealth of literature conclusively shows that dispersal cannot be approached constructively in any other manner but with a precise understanding of its magnitude in time and space. An author, for instance, rejects the suggestion (24; p. 73) that *Crepis* originated in south Central Asia only to replace it with the surmise that this genus had its inception in Southwestern Asia. As we shall duly see, both these regions are reached by the very same tracks, so that nothing is gained or lost by shifting the supposed „Center of origin" of *Crepis* a few hundred miles. Another author categorically affirms (25) that *Juncus* stems from the mountains of Central Asia, but this author is disputed by a third (26), who writes as follows, „I am fully aware that the oldest history of this group can not be reconstructed from the present distribution. Nor can the original home of the family be found in that way. As has been mentioned above, Buchenau located the origin to high mountains of Eurasia and Irmscher to tropical regions from where the species were thought to have migrated towards the north and south into temperate latitudes. The present writer has not been able to find any weighty reasons for these assumptions. On the contrary many instances argue in favour of the hypothesis that the southern hemisphere has played an important role in the evolution of the family (Note the occurrence in the Cape of *Juncus lomatophyllus, J. viridifolius* and *Prionium*, in certain features intermediate between *Juncus* and *Luzula*, and in S. America — New Zealand of the small genera endemic there.) The family may have arisen in the Antarctic continent, the importance of which has formerly been much neglected."

We do not believe that the original home of the Juncaceae cannot be reconstructed from present distribution, for, to omit a detailed analysis of the *genorheitral* affinities among *Prionium, Juncus* and *Luzula*, of which the author quoted makes only brief mention, *Juncus effusus* and *J. inflexus* clearly exhibit (Fig. 21) a perfect pattern of migration from the African gate. This pattern is standard of literally hundreds of families of the Angiosperms and is primary, in reality, of the Angiosperms as a whole.

It is obvious that conflicting opinions such as we have summarily reported cannot be critically judged on the strength of the dispersal of *Crepis, Juncus* and the like. These opinions can be assayed at their true value only by broad comparisons of tracks and *genorheitra* in absolute disregard of preconceptions and academic habits of thought. Cytology, it may be repeated, is sovereign in certain domains, but contributes to others only data of potentially indifferent value. These data need be interpreted in the light of correctly assumed generalities, for something that is immediately from the heart of a cell does

not automatically promise to be foolproof. Cytology can beyond doubt press
for itself better claims than the power of reversing the course of standard
tracks of dispersal. Being by its very nature laborious and painstaking, cytology is precisely one among the branches of botany that demand a sound
grasp of the generalities of this science throughout. The most disturbing
aspect of modern investigation is that an appetite for generalities not only is
lost, but decried as redundant „philosophy." Armed with this belief, we bite
ever deeper into this or that „specialty" only to end so deep that we no longer
can correlate our own findings in the manner that makes them fruitful.
Nothing is more pressing than a reversal of this pernicious trend.

BIBLIOGRAPHY

Chapter II

(1) GILG, E. — *Turneraceae*; E. & P. Nat. Pflanzenf., ed. 2, 21: 459. 1925.

(2) HARMS, H. — *Malesherbiaceae*; E. & P. Nat. Pflanzenf., ed. 2, 21: 467. 1925.

(3) PILGER, R. — *Rapateaceae*; E. & P. Nat. Pflanzenf., ed. 2, 15a: 125. 1930.

(4) LAM, H. J. — Studies in Phylogeny. I. On the relation of Taxonomy, Phylogeny and Biogeography; Blumea 3: 114. 1938.

(5) RADLKOFER, L. — *Sapindaceae*; E. & P. Pflanzenr. 98 (iv. 165); 1328, 1337. 1933.

(6) WOOD, J. G. — The Vegetation of South Australia. 1937.

(7) GRUENING, G. — *Euphorbiaceae-Porantheroideae* et *Ricinocarpoideae*; E. & P. Pflanzenf. 58 (iv. 147). 1913.

(8) WEIMARCK, H. — Phytogeographical groups Centres and Intervals within the Cape Flora; Lunds Univ. Arsskr., N. F., Avd. 2, 37 (No. 5): I. 1941.

(9) GARDNER, C. A. — Enumeratio Plantarum Australiae Occidentalis, 1930-1931.

(10) PAX, F. & KNUTH, R. — *Primulaceae*; E. & P. Pflanzenr. 22 (iv. 237) 1905.

(11) HANDEL-MAZZETTI, H. — A revision of the Chinese species of Lysimachia, with a new system of the whole genus; Notes Roy. Bot. Gard. Edinburgh 16: 51. 1928.

(12) WALKER, E. H. — A revision of *Distylium* and *Sycopsis* (Hamamelidaceae); Jour. Arnold Arb. 25: 319. 1944.

(13) BRAID, K. W. — Revision of the genus *Alphitonia*; Kew Bull. 1925: 168. 1925.

(14) PILLAND, N. S. — The genus *Phylica*; Jour. South Afr. Bot. 8: I. 1942.

(15) WEIMARCK, H. — Die Verbreitung einiger Afrikanisch-Montanen Pflanzengruppen, I. Die Gattung Sparmannia; Svensk. Bot. Tidskr. 27: 400. 1933.

(16) SKOTTSBERG, C. — Astelia, an antarctic-pacific genus of Liliaceae; Proceed. Fifth Pacific Sc. Congr. 4: 3317. 1934.

(17) BABCOCK, E. B. — The origin of Crepis and related genera with

particular reference to distribution and chromosome relationships; Essays Geobot. in honor of W. A. Setchell (edit. T. H. Goodspead), Univ. California, 9. 1936.

(18) HOFFMANN, O. — *Cichorieae Crepidinae*; E. & P. Nat. Pflanzenf. 4 (5): 366. 1893.

(19) CROIZAT, L. — Trochodendron, Tetracentron, and their meaning in phylogeny; Bull. Torrey Bot. Club. 74: 60. 1947.

(20) WEIMARCK, H. — Die Verbreitung einiger Afrikanisch-Montanen Pflanzengruppen, II. Die *Cerastium Africanum*-Gruppe; Svensk. Bot. Tidskr. 27: 413. 1933.

(21) CHEESEMAN, T. F. — Manual of the New Zealand Flora, ed. 2 (W. R. B. Oliver). 1925.

(22) SKOTTSBERG, C. — Die Flora der Desventuradas-Inseln (San Felix und San Ambrosio); Göteborgs Vetensk. Vitterh. Sammhall. Handl. 5 Fold. Ser. B, 5 (No. 6): 1. 1937.

(23) KECK, D. D. — The Hawaiian Silverswords. Systematics, affinities and phytogeographic problems of the genus Argyroxiphium; Bishop Mus. Occ. Pap. II (No. 9): 1. 1936.

(24) STEBBINS, G. L. Jr. — Studies in the Chicorieae Dubyaea and Soroseris, Endemics of the Sino-Himalayan region; Mem. Torrey Bot. Cl. 19. (3): 1. 1940.

(25) BUCHENAU, F. — *Juncaceae*; E. & P. Pflanzenr. 25 (iv. 36). 1906.

(26) WEIMARCK, H. — Studies in Juncaceae with special reference to the species in Ethiopia and the Cape; Svensk. Bot. Tidskr. 40: 141. 1946.

CHAPTER III

INTERCONTINENTAL DISPERSAL I.

We propose to analyze in this chapter patterns of distribution that interest the five continents. In order to conduct our investigation against the proper background of distance, we will follow from beginning to end the tracks that make these patterns out, briefly commenting upon their course as necessity requires.

We have introduced several concepts in the preceding pages, such as *genorheitron*, gate of angiospermy, orderly migration, and begun to test these concepts in the solution of concrete problems of distribution. There is no reason to plead for these concepts in a theoretical vein, because, if adequate and correct, these concepts will yield results that speak for themselves. A track of migration may be assimilated to geometrical patterns. Mathematicians know how to bring these patterns to book by definite rules, and it remains to be proved that phytogeographers cannot do the same by the application of standard concepts to fit specific contingencies of migration.

Mileage is, as such, of no account in phytogeographical work, which we do not affirm as an opinion, but as a demonstrable fact. Dispersal was effected over lands in the past that hardly agree with the boundaries on the maps of our days. If, as it has already been shown by factual data, *Foetidia* occurs in Madagascar and India, and further turns up endemic to the summit of Mount Ophir in Malacca; if the Sarraceniaceae are known alike from Venezuela, the Eastern and Western United States; if *Linaria alpina* and *Alisma natans* are native to Mount Ophir once more, Java and Europe, we see no reason for standing by and theorizing, in the helpless belief that these events must forever defy reason. Whether the continents "pendulated," or rather "floated," is none of our immediate business. Six figures on a blackboard can always be compounded to effect an addition, and it remains to be seen why *Foetidia*, *Sarracenia*, *Heliamphora*, *Darlingtonia*, *Linaria*, *Alisma* and all their vagaries cannot be made to yield something satisfying to men of ordinary common sense, not to mention scientists. There stands at our disposal the dispersal of uncounted plants now living throughout the earth, and we can know the ranges of these plants, when not perfectly, at least with sufficient accuracy to work them out according to reason.

As a necessary compromise between immediate possibilities and the ultimate requirements of our subject, we will at first only rough out, as it were, the work before us, sketching in bold outline the major aspects of dispersal. In later chapters, we will deal with these aspects at greater length. The reader, we hope, will in the end agree that there is practically no limit to the questions plants may answer, and that a study of dispersal, given time and means, can be refined to a point of accuracy approaching pure mathematical sciences.

Geological and geophysical evidence of the soundest (1) informs us that a shore once ran in the Pacific much to the east of the present coasts of Australia and Malaysia. This shore (Fig. 22) began approximately at Chatham Island, extended northward to the general neighborhood of Samoa, then over a northwesterly path reached the Western Carolines. Opinions differ as to how this shore continued its course northward but some geophysicists think that it reached eventually the Mariannes and Japan.

It so happens that the boundary thus drawn on the maps of the Pacific by geologists and geophysicists closely agrees with the limits which a botanist assigns (Fig. 22; 2) to a fairly rich Orchid flora. If this is a coincidence, it must rate as a peculiar one. As for ourselves, we have already seen migrations of a type (*Acacia, Lysimachia, Perrotetia, Alphitonia*, etc.) which hardly bear being rationalized unless we believe that the geophysicists and botanists in question are correct as to their main premises.

Obviously, it is not our intention to claim that everything was solid land forever westward from this shore. It is sufficient for us to know that lands, either large islands or sizeable archipelagoes, once held quarters that are now either an emptiness of ocean, or are at the best studded by petty insular outposts in the present. We once more refer the reader to the thoughts suggested (see p. 33) by the geologic history of Alaska, which region many believed to be a classic "landbridge" between the Old and New World *).

The evidence which is satisfactory to us, and leads us to assume as most probable certain conclusions, is rejected on the contrary by two zoologists (3) who write, for example, as follows, "There is nothing in the bird fauna (of Polynesia) that would speak in favor of a former continental connection." Referring specifically to two genera of frogs widely dispersed from the Fiji to Asia, the other of the authors under discussion claims the following, "The fact that both genera (*Cornufer* and *Platymantis*) are in Fiji merely means that they have been successful in crossing over the water . . . . I assume this to mean that they arrived by flotsam-jetsam methods or perhaps were transported by native boats." **)

---

*) In later pages (p. 78), we point out that modern Alaska is not necessarily the classic landbridge between the Far East and North America which many take for granted. We show that Alaska was possibly reached by secondary migrations only, moving northward from a region more or less corresponding with today's California. Notice that, whatever the case, Alaska was in the end permeated by angiospermous plant-life, so that this plant-life passed through, whether it came primarily from the east or the south does not matter.

**) For a critical review of similar statements see also HARRISON, L., in Austral. Zoolog. 3 : 247. 1924; in Amer. Nat. 60 : 374. 1926. We should not care to have it understood that all ornithologists and herpetologists are of the same mind, and that the views that some authors introduce as authoritative are so accepted by all.

These statements demand a measure of attention, for if we overlook them they may be later introduced as proof that what we take for granted is merely our own opinion.

It is a fact (3; p. 204) that students of birds admit that Fiji and Samoa do not agree with the rest of Polynesia. These students are on record as follows, "Their fauna differs by several characters from those of the other subdivisions of Polynesia. First of all, it is a comparatively rich fauna, considering the distance from the nearest continent, and secondly, there is very little "pollution" by foreign elements, such as found in Micronesia (Palearctic and Philippine elements) . . . . These are the old islands of central Polynesia and the only ones which have a notably endemic fauna . . . . Elements that have developed in that region have subsequently spread into Micronesia, eastern Polynesia, and southern Melanesia." Following a review of the faunistic connections between Fiji and Samoa, the conclusion is advanced (3; p. 207), "Both Fiji and Samoa are oceanic islands that were populated by the same elements which possessed unusual colonizing abilities . . . . Both Fiji and Samoa are zoogeographically oceanic islands which derived their faunas across the sea."

It is immediately apparent by these very statements that, (a) Fiji and Samoa are a center of dispersal in their own right, for life originating from this center streams out to Micronesia, Eastern Polynesia and Central Melanesia; (b) Their fauna is comparatively rich and mostly autochthonous; (c) In order to believe that these islands were populated across the ocean from the nearest continent it proves necessary to assume that the animals that did so possessed "Unusual colonizing abilities."

In our opinion, the very face of the statements quoted proves, then, that these islands were not populated from across the ocean. We must emphatically reject the possibility of "Unusual colonizing abilities" effective only *in two islands of Polynesia,* for once we accept this possibility we are again faced with the exceptional and the unusual, which have no place in constructive thinking.

Fiji is well known to consist of an ancient thoroughly weathered core (4) modified by much later volcanism. The belief (3; p. 209) that New Caledonia "Apparently emerged from the sea during the Oligocene" is not certainly correct, for a fossil wood (*Cedroxylon;* 5) antedating by far the Cretaceous is known from this island. In short, we see no justification whatever for rejecting valuable evidence from geology and geophysics to the effect that the ancient continental shelf of Australia and Malaysia reached much deeper into the Pacific that it does now. The evidence from plant-distribution is formal to the effect that this shore stood much as geologists and geophysicists draw it, not only, but that south and east of Chatham Island this shore ran eastward to contact South America.

Naturally, we see no difficulty whatever in following tracks east of Australia and Malaysia that lead across the ocean. We would follow these tracks as a matter of course, even if we had no geological evidence in our favor *). The statement that this or that Polynesian islands contains "Philippine" elements does not mean that these elements are "Philippine" in the first place. Adjectives of the kind are misleading, as we have emphasized (see p. 55), and it stands to reason that if the same form of life occurs on islands A and B we have no justification for hurriedly concluding that island A fed island B, for the contrary may well be true.

I. Coriariaceae

The Coriariaceae consist of the lone genus *Coriaria* which has obscure affinities and isolated position. It does not seem probable that this genus is related with the Anacardiaceae as suggested by the classification of Engler & Prantl. In our opinion, this genus is an offshoot from the ranunculoid *genorheitron*.

The taxonomy of *Coriaria* is thoroughly unsettled because it is difficult to decide what is a "good species" in this genus. Some forms stand out with greater sharpness than others, but the whole is to too hazy for taxonomic comfort. It proves fortunate in the end that we may dispense with labels, specific and varietal names, etc., for the range of *Coriaria* breaks down into five disconnected segments, and it means little whether these segments each include one or ten "species." **)

These ranges tabulate as follows,

(1) *Polynesia* — Society Islands, Samoa, Fiji, New Hebrides; New Zealand, Kermadec, Stewart and Chatham Islands.

(2) *Malaysia and the insular Far East* — Southeastern New Guinea, Philippines (Luzon: Bontoc, Benguet on Mt. Data, Mt. Pauai, Mt. Santo Tomaso), Formosa, Riu-kiu, Japan to Yezo.

(3) *Eastern Asia* — China (Kwangsi, Kweichow, Yunnan, Hunan, Hupeh, Szechuan, Kansu, Shensi), Northern Burma, India (Sikkim).

---

*) The most unsatisfactory status of biogeography in these days stems well over ninety per cent from a general neglect of straight geography, and an unreasonable repugnance to take the facts in the record for what they are and mean. We strongly urge upon the reader to plot out dispersal on the map, factually and simply, without previous thought where dispersal is to lead him in the end. The reader will think out what seems to him most reasonable *after he has the facts in hand, not before*, which is of the essence of a scientific approach.

**) Many are the botanists who still believe that formal nomenclature can express the basic factors of plant-life, and cling to the opinion that a world of meaning attaches to the fact, as such, whether a plant is tagged out as "species" rather than "variety." The author of this manual is convinced, on the contrary, that formal nomenclature is becoming ever more incompetent to express the realities of plant-life, and that more than ever it proves necessary to understand a plant well before giving it a name and a rank. In this convinction the author no longer stands isolated (see, for instance, BALDWIN in Jour. Hered. 38: 55. 1947). It should be clearly understood that a healthy skepticism as regards the scope of formal nomenclature in biology does not necessarily imply that nomenclature is to be relegated to the background of botanical studies. On the contrary, formally correct nomenclature is essential to sound botany. Naturalists need to know exactly what they speak about, and precise language is impossible without correct nomenclature.

(4) *Europe* — Italy (Emilia to Liguria), Mediterranean shore of France, Corsica, Baleares, Spain (the northwestern provinces seemingly excluded).

(5) *America* — The Andes throughout from Southern Chile to Northern Guatemala, with a disconnection in northermost Chile and an outlier to Venezuela.

Although disfigured by wide disconnections (Fig. 23), the dispersal is easily plotted on the map, and shows that the tracks of *Coriaria* and *Euphrasia* (see Fig. 6) are homologous as to essentials. The base-line of distribution of both genera is in the Southern Pacific, and the dispersal streams out of this base-line to the Far East and the New World alike.

The wide disconnection noticeable between India and the Mediterranean is easily bridged, because an Oligocene fossil of *Coriaria* from Europe is close to a form still living in the Himalayas. This confirms that the Himalayas were once in intimate contact with Europe through the Caucasus, or Asia Minor, and renders it probable that *Coriaria* died out in the intervening range only because of local factors. It is also clear that *Coriaria* was in active migration before the Oligocene and that by then the genus was thoroughly modern.

The track over which *Coriaria* travelled to Europe is much older than the Oligocene, as a matter of fact. Refined palaeobotanical investigations reveal (6) that *Rhodomyrtus*, *Tristania* and *Cinnamomum* are fossil in the Eocene of England. All these genera are now known only from Australia or the Far East. *Tristania*, for example, is one the "Australian" elements endemic to the summit of Mt. Gunong Tapis in Malacca (see p. 37) and *Rhodomyntus* still ranges in New Caledonia, Eastern Australia, New Guinea, Philippines, Borneo, Java, Sumatra, Malacca, China, Southern India, Ceylon. A presumed fossil of *Tristania* is also on record from the Australian Miocene. Doubtless, the Oligocene fossil of *Coriaria* is anything but truly old, and this genus in all probability reached Europe in the Cretaceous, or earlier, together with *Rhodomyrtus*, *Tristania*, *Cinnamomum*. It must mean next to nothing to a phytogeographer that these genera died out, while *Coriaria* survived *).

The dispersal of *Coriaria* is significant on two counts, (a) It fared over a crumbled range in what is now ocean between New Guinea and Luzon; (b) It entered Formosa and Luzon *from the Pacific*. It is therefore well established that Malaysia and continental Asia were also open to invasion *from the east*.

It has been written (7) that *Coriaria* reached Luzon by a "secondary invasion." Nothing supports this opinion. The same species, *C. intermedia*, is native to Formosa and the Philippines, and numerous are the genera and species common to these islands. Moreover, the term, "Secondary invasion" is meaningless when dissociated from critical considerations of the main

---

*) The cult rendered petrifacts is oftentimes pathetic. A mind open to a sense of humor cannot fail, but to think that many credit fossils with powers to match those which ancient Chinese bestowed upon them out of outright faith in magics and "signatures." Many lightly dismiss living records of plant-life by the thousands to pore — often going wrong in the process — over a dubious fossil of recent geological age (e.g., the Eocene). *The basic rule is that from the earliest Cretaceous onward, living dispersal elucidates fossil, not the contrary.* See note p. 261, 387 on fossils of *Cochlospermum* and *Humiria.*

tracks. An immediate simultaneous invasion of Luzon and Formosa *from the Pacific* is indicated in the dispersal of *Coriaria*.

II. Lardizabalaceae

This family primarily belongs to the ranunculoid *genorheitron* and is, in part at least, correctly treated near the Berberidaceae. It is distributed in this manner,

A) Boquila

1 species in Central Chile.

B) Lardizabala

1 species in Central Chile.

C) Stauntonia

a) Subg. *Parvatia*

5 species in Southern and Western China (Kwantung, Kwangsi, Yunnan), and India (Assam, Eastern Bengal).

b) Subg. *Eustantonia*

11 species in Corea, Japan, Riu-kiu, Formosa, Southern China generally (Hainan, Kwantung, Hongkong, Fukien, Chekiang, Kiangsi, Hupeh, Hunan, Kweichow, Kwangsi), Indochina (Laos).

D) Holboellia

8 species in China (Kwantung, Hongkong, Fukien, Anhwei, Hupeh, Hunan, Kweichow, Szechuan, Yunnan), India (Sikkim, Kumaon), Indochina (Tonkin).

E) Decaisnea

2 species in China (Hupeh, Shensi, Hunan, Kweichow, Szechuan, Yunnan), India (Eastern Himalayas).

F) Sinofranchetia

ı species in China (Hupeh, Szechuan).

G) Akebia

4 species in Corea (Quelpaert Island), Japan (Honshu), Formosa, Southern China throughout.

H) Sargentodoxa

ı species in Southern China throughout, Indochina (Laos).

The dispersal of this family is seemingly homologous of that of the Coriariaceae. The American element is represented by two monotypic genera (*Boquila* and *Lardizabala*), the Asiatic by six genera with a variable number of species. The differences between these two families are (a) *Coriaria* immediately occurs in the Pacific, where the Lardizabalaceae are unknown; (b) *Coriaria* is endemic to New Guinea and Luzon, where the Lardizabalaceae are unrecorded; (c) *Coriaria* ranges to Europe which the Lardizabalaceae do not reach. The dispersal is otherwise in common in Japan, China and the Himalayas, South America.

It should be noticed that peculiar monotypic forms in the broad affinity of the ranunculoid *genorheitron* are known from Fiji (Degeneriaceae), Queensland (Austrobaileyaceae). The evidence warrants the tentative inference, consequently that the Lardizabalaceae originated in the Western Pacific and radiated from this point to South America and the Far East. The track enters Asia at a point which current dispersal suggests to lie north of Formosa.

"Holarctic" origins could easily — if mistakenly — be assumed for this group if it did not happen that two of its genera are definitely endemic to Central Chile.

III. Berberidaceae

The Berberidaceae are next of kin of the Lardizabalaceae. They include two large or fairly large genera, *Berberis* and *Mahonia*, of easy identification. The balance of the family is a loose aggregate of genera, some of which, witness *Ranzania* (8), are definitely controversial.

Certain of the lesser genera of this family are entirely restricted to the Old World, others evenly divided between the Old and New, as,

A) Leontice

(1) *L. Albertii* — Central Asia (Tianshan Mts.)

(2) *L. darwasica* — Central Asia (Pamir, Alatau Mts.)

(3) *L. altaica* — Central Asia (Altai Mts., region of the river Irtysh).

(4) *L. incerta* — Central Asia (Dzungaria, Western Sinkiang, Syr-Daria, regions of the Aral and Caspian Seas).

(5) *L. minor* — Southwestern Central Asia (Amu-darya, Iran (Persia), Caucasus).

(6) *L. Ewersmannii* — Southwestern Asia (regions of the Aral and Caspian Seas, Iran (Persia), Armenia).

(7) *L. Smirnowii* — Caucasus.

(8) *L. odessana* — Northern shores of the Black Sea, Crimea to Bulgaria (Dobruja).

B) Jeffersonia

(1) *J. dubia* — China, Japan, Manchuria, Eastern Siberia (Ussuri).

(2) *J. diphylla* — North America (Ontario and Wisconsin to Alabama and Maryland).

Mahonia is endemic both to the Old and New World. In China (Hupeh to Kansu southward) this genus has about twentytwo species. Three are in Formosa, one of which at least (*M. japonica*) ranges northward to Japan. Three species are in the tropical Far East and Malaysia, as follows,

(1) *M. philippinensis* — Philippines (Northern Luzon).

(2) *M. annamica* — Indochina (Annam).

(3) *M. nepalensis* — China (Kiangsu, Kansu, Szechuan), India (Eastern Himalayas), Northern Sumatra (Atjeh), Western and Central Java.

In the New World, *Mahonia* turns up in Mexico with a massive concentration of close to thirty species, some of which extend to the United States, witness,

(1) *M. Fremontii* — Mexico (Baja California, Sonora), United States (California to Utah and New Mexico); or range southward to Central America in this manner,

(2) *M. fascicularis* — Mexico (Guanajuato, Michoacán, Oaxaca, Veracruz), Guatemala, United States (California).

Two species at least are endemic to Central America,

(3) *M. Johnstonii* — Guatemala.

(4) *M. paniculata* — Costarica.

Mahonia and *Berberis* join hand on the slopes of the volcano Irazú in Costarica. To these slopes is endemic *Berberis nigricans* which has affinities in the direction of the Peruvian *B. latifolia*.

The majority of the species in Mexico are native to the state of Hidalgo, where not less than eight (9) would seem to be narrowly localized.

Mahonia is also well represented in the United States, California (about 8 species) and Oregon (5 species); further reaching northward to British Columbia (*M. nervosa, M. Aquifolium*), but not to Alaska, where neither *Mahonia* nor *Berberis* are recorded. Eastward, *Mahonia* reaches its farthest penetration along a line running from Alberta to Western Texas across Nebraska and Kansas. Its range, consequently, is clearly western in the United States. Both this genus and *Berberis* would seem to be entirely unknown in the Caribbeans.

Comparing the patterns of distributions of *Mahonia* and *Jeffersonia*, we may all too readily conclude that *Mahonia* crossed the Pacific in the approximate latitude of Hawaii, while *Jeffersonia* followed a track in the high north, leading from the Ussuri to Ontario in the first place. We should guard against jumping to conclusions, however, because — as it will be better seen in coming chapters — a single main track may be responsible for the seemingly opposite behaviors of *Jeffersonia* and *Mahonia*. This track comes in from the Far East across the Pacific, hits California and Northern Mexico first, then glides to one of the major — when not the major — North American phytogeographic centers located in the Ozark-Appalachian range. Out of this center, the track veers northward, and in many cases reaches Ontario and Wisconsin, possibly beyond, from Alabama, not the other way around. Ontario and Wisconsin, the former especially, are well known as a matter of fact as end-points of migrations running northward from the Ozark-Appalachian center *).

Berberis is a large genus. Close to a hundred of its species are endemic to South America, some of which range northward to Southeastern Brazil and Venezuela, one at least, as we have seen, reaching Costarica. The massing of these species is most marked in Chile and Patagonia, where *Berberis* occur in association with a flora of strong "antarctic" flavor (10), further extending to Juan Fernandez. Northward from Chile, *Berberis* follows the Andes through Bolivia (over 10 species), Peru (over 15 species), Ecuador (about 20 species) and Colombia (about 15 species). Against this wealth in the south, *Berberis* has very little to show in the north, for it does not occur in Mexico at all, and is represented in the United States only by two species B. *Fendleri* (Colorado and New Mexico), and B. *canadensis* (Virginia, North Carolina, Tennessee, Missouri). *Berberis*, unlike *Mahonia*, has thus an eastern range in the United States.

The affinities of the North American *Berberis* would seem to lie in the direction of the Far East, as the following indicates (11),

---

*) In a critical appraisal of migrations of the kind we need not overlook the circumstance that many are the groups living in the Eastern United States which once held sway also in the Western, but later died out. *Tilia* is a classic instance, for it ranged in sections of the Western United States (cf. CHANEY, for instance, in Carnegie Inst. Washington Publ. No. 553: 311. 1944) from which it has now disappeared. The living range of *Tilia* in the United States is strongly rooted in the Ozarkian-Appalachian center, but reaches from here as far north as Ontario and Wisconsin, not only, but Manitoba (see *T. americana*).

A) Sect. *Sinensis*

a) Subsect. *Eusinenses*

(1) *B. elegans* — China (Yunnan).
(2) *B. Thunbergii* — Japan (Kyushu, Honshu).
(3) *B. sinensis* — Northern China, Eastern Siberia and Mongolia.
(4) *B. iberica* — Caucasus, Northern Asia Minor.
(5) *B. Fendleri* — Colorado, New Mexico.
(6) *B. canadensis* — Virginia, North Carolina, Tennessee, Missouri (not in Canada).

b) Subsect. *Creticae*

(7) *B. crataegina* — Afghanistan, Asia Minor.
(8) *B. libanotica* — Syria (Mt. Lebanon).
(9) *B. cretica* — Aegean Islands, Greece.
(10) *B. australis* — Southern Spain, Algeria.
(11) *B. Garciae* — Spain (Aragon).

c) Subsect. *Sieboldianae*

(12) *B. Sieboldii* — Japan.
(13) *B. Rehderiana* — Japan(?).

If this classification be correct as to subsections — which is probable —
we must look to the Far East for the origin of the North American *Berberis*,
because *B. Fendleri* and *B. canadensis* are both closely allied with three Far
Eastern species (*B. elegans, B. Thunbergii, B. sinensis*). We will discuss in a
later part of our work the circumstance that these North American and Far
Eastern species are further related to plants from Afghanistan, the Caucasus
and the Mediterranean generally. Let us notice, meantime, that *B. Fendleri*
and *B. canadensis* range from New Mexico to Tennessee, and are accordingly
located along a track coming in from the direction of California toward the
Southeastern United States. Let us further remark that Colorado — the
northermost sector of the range of *B. Fendleri* — happens to be, on the contrary, the southernmost point in the dispersal of *Rhododendron albiflorum*
(Colorado to Montana, Washington, Alberta and British Columbia). As
R. albiflorum is allied with the Japanese *R. barbatum*, we may reasonably visualize
a track reaching the United States from China or Japan; driving straight to
the Ozarkian-Appalachian center; and branching from here by two paths,
one leading through the Eastern United States to New Brunswick and Eastern
Canada generally, the other following in disconnection to New Mexico and
Colorado westward as far as British Columbia. We will indeed have occasion

to investigate further migrations to, and from, the Ozarkian-Appalachian center. Above all, we need not rush to the conclusion that Alaska was *the* "Landbridge" between the Far East and the Northern New World.

Berberis in the New World is massive in the southern hemisphere, very scantily represented in the northern. In the Old World the situation is exactly reversed. It is difficult to estimate the number of species endemic to China, or extending from China to the Himalayas, but a total of about 125 may not be far from correct. Approximately 30 species are in the Far East exclusive of China and Europa as far as Madera.

In Malaysia *Berberis* is an alpine element (12) which is distributed thus,

(1) *B. Barandana* — Philippines (Luzon: Benguet).

(2) *B. Wallichiana* — China (Kwantung, Shensi), Himalayas, Indochina (Tonkin), Sumatra, Java, Lombok.

These two species are close, and the former hardly differs if indeed at all from the Formosan *B. Kawakamii*. They belong to Sect. *Wallichianae* which ranges to Eastern China (Kansu), the Central Himalayas (Nepal), and is massive in Southwestern China (Yunnan and Szechuan) where it is represented by not less than 25 species.

The Arabian and African *Berberis* can be correlated without doubt or difficulty to an Himalayan secondary center of dispersal, witness the following,

A) Sect. Tinctoriae

a) Subsect. A

(1) *B. Petitiana* — Tanganyka (Usambara), Kenya, Ethiopia.
(2) *B. Forskaliana* — Arabia.
(3) *B. afghanica* — Afghanistan.
(4) *B. gahrwalensis* — Western Himalayas.
(5) *B. umbellata* — Nepal.
(6) *B. Thomsoniana* — Eastern Himalayas (Sikkim).

b) Subsect. B

(7) *B. aristata* — Tanganyka, Southern India, Western Himalayas, Nepal.
(8) *B. ceylanica* — Ceylon, Southern India.
(9) *B. Wightiana* — Southern India.
(10) *B. tinctoria* — Ceylon, Southern India.
(11) *B. Huegeliana* — Western Himalayas (Kashmir).

It will be observed, (a) The track may have two branches, Tanganyka-Ethiopia-Arabia-Afghanistan-Himalayas; Tanganyka-Ceylon-Southern India (Nilghiri Mts.)-Himalayas; a cleancut tie between East Africa and the Himalayas being secured by either, or both *); (b) The region between Ethiopia and Tanganyka (Usambara in particular) stands as the southernmost (or northermost, depending on the track's direction, whether primary or secondary) limit of numerous forms (*Berberis, Primula, Populus, Sambucus,* etc.) common to Africa and the Himalayas. This region also marks a center of deflection westward (i.e., toward Western Africa) of *genorheitra* (e.g., Ericaceae Ericoideae) which fail to reach the Himalayas, wherein conditions would have been ideal for their survival and evolution. Very strong suggestions exist, consequently, to the effect that the range Tanganyka-Ethiopia-India has fundamental significance in dispersal. We are later to learn why this happens to be so.

The Malaysian *Berberis* is strewn along a track running the course, Southern China-Indochina-Formosa-Luzon-Sumatra-Java-Lombok. This track is one which we will carefully study in time, for it has unusual importance. For the present, and with regard of the evidence already before us, we will consider only two alternatives as to the trends in the general distribution of *Berberis*. These alternatives are illustrated in the map attached (Fig. 24); should the reader be struck at a glance by their seemingly boldness he ought to wait without rushing to conclusions on the spot. What may seem to him overbold right here will soon strike him as commonplace.

The alternatives in play are the following, (a) *The base-line of dispersal lies in the Pacific.* In this assumption the migrations of *Berberis* are in the main homologous of those of *Coriaria*. The Arabian and African outlier (*Berberis* sect. *Tinctoriae*) is fed by an eastern and southern track running out of the Himalayas; (b) *The base-line of dispersal is located in the Indian Ocean.* In this case the *genorheitron* streams from a point roughly located south of Africa both east (East Africa, Arabia and Asia generally), and west toward the New World (Magellania). It will be noticed that under either alternative *Mahonia* is distributed to the New World from the Far East.

These alternatives are not exclusive; cf Fig. 44, 44 A. They may be complementary, as a matter of fact, because the *genorheitral* ties and migrations in play are of the widest. Several factors speak for an origin of the Berberidaceae in the Pacific. The Lardizabalaceae, for example, are seemingly keyed to a base-line in this ocean, and their tracks also are widely disconnected between South America and the Far East. *Mahonia* is undoubtedly a transpacific migrant. The Magnoliaceae, *sensu lato,* originate in the Pacific, and stray monotypic families in their general affinity (Degeneriaceae, Austrobaileyaceae, Himantandraceae) are narrowly endemic to Fiji and Queensland. It is not certain that the *genorheitra* of the Magnoliaceae and Berberidaceae are closely related, but it is altogether probable that there exists between them some measure of

---

*) See the same branches stressed by Van Steenis, Fig. 4, in the dispersal of the Bignoniaceae.

affinity, and that the origin of the one may be near that of the other.

Against these considerations, arguing in favor of a Pacific origin of the Berberidaceae, stand other which suggest the Indian Ocean as the starting point of the berberidaceous tracks. *Berberis* is beyond doubt endemic to ancient lands in Eastern Africa, and the African gate of angiospermy feeds without difficulty the approaches of South America. The entire Asiatic massing of the Berberidaceae could easily be delivered from the Indian Ocean, and *Mahonia* next follow to the New World very much in the manner of countless genera that cross the Northern Pacific *).

We may not wish to press the evidence, as phytogeographers, beyond the limits advised by prudence, and it is immaterial to us in the end, where the berberidoid *genorheitron* first turned up ready to migrate northward, whether in the Pacific or Indian Ocean. Alternatives such as we have offered are satisfactory, whenever present knowledge or immediate necessity does not justify cleancut opinions one way or the other. Long years in the field of cytology, morphology and systematics will elapse before we are suitable informed as to the *genorheitron* of the Berberidaceae and its ties. Recent work (13) has merely confused the issue further by patent misapplication of the term "Ranales" (see for criticism, 14) in the direction of families such as the Trochodendraceae and Tetracentraceae. We will, then, decide between the alternatives, where and how originated the Berberidaceae, as soon as facts are at hand to support one or the other. For the present, it is sufficient to state the problem in a lucid manner.

Considering that it is commonly believed, perhaps with reason, that the *genorheitron* of the Berberidaceae is essentially "ranunculaceous," and that this family actually includes an element (the Podophyllaceae of certain systematists) of manifest "ranunculoid" affinity, we may do well if we review here the dispersal of some genuine ranunculoid plant, *Anemone*, for instance, which has two subgeneric groups ranging in this manner (15);

---

*) We do not intend in so early a stage of our study to draw the reader to conclude for alternatives which he may not be prepared as yet to indorse. We feel confident, on the other hand, that as the result of having perused the pages of this manual throughout, the reader will in the end believe, as we do, that the Berberidaceae originated in the modern Indian Ocean, and specifically, by the *Afroantarctic Triangle* (see Figs. 67, 102). This we believe because, (a) *Berberis* turns up as if a sudden in Tanganyka and Kenya, next following northward (see *Populus*, p. 217); (b) The path taking *Berberis* to the Magellanian Gate is strictly "antarctic"; (c) The *Berberis Wallichiana* aggregate is basically distributed along the lines of the *Schoenoxiphium-Cobresia* affinity (see p. 237), which is happily complementary of the African dispersal of the genus; (d) *Mahonia*, too, reiterates the main aspects of the migrations of the *Berberis Wallichiana* group; (e) *Leontice, Jeffersonia, Podophyllum* and their allies are mostly distributed from the *Altai Node*, which they reach by a standard major track (see track (4), p. 549; Fig. 102); (f) No Berberidaceae occur in Australia, Tasmania, New Zealand, Polynesia; (g) The South America massing of *Berberis*, and all its migrations in the New World, are consistent with the interpretation of the total dispersal here espoused; (h) The presence of possibly "berberidoid" *genorheitra* in the Pacific can easily be accounted for by early migrations along paths such as exemplified by Restionaceae, Centrolepidaceae (see p. 359), etc. These eight reasons clinch in our opinion the evidence in favor of the modern Berberidaceae having distributed by a crumbled center in the southwestern Indian Ocean of our maps.

A) Sect. *Rivularidium*

(1) *A. crassifolia* — Tasmania.

(2) *A. hepaticifolia* — Chile.

(3) *A. rigida* — Chile.

(4) *A. antucensis* — Chile.

(5) *A. Sellowii* — Brazil (Rio Janeiro).

(6) *A. Glazioviana* — Brazil (Rio Janeiro).

(7) *A. helleborifolia* — Peru (Cuzco to Cajamarca).

(8) *A. Leveilleii* — China (Kweichow).

(9) *A. rivularis* — Ceylon, Southern India (Nilghiri Mts.), Western Himalayas (Kashmir), Mongolia, China (Yunnan, Szechuan, Kansu, Shensi).

(10) *A. Richardsonii* — Eastermost Siberia, Alaska to Hudson Bay and Greenland, southward in boreal America to 55° Lat.

(11) *A. mexicana* — Mexico (Chiapas, Oaxaca, San Luis Potosi, Veracruz).

(12) *A. Hemsleyi* — Mexico (Veracruz) *).

B) Sect. *Pulsatilloides* subsect. *Longystylae*

(13) *A. capensis* — South Africa (Cape).

(14) *A. caffra* — South Africa (Cape).

(15) *A. Fanninii* — South Africa (Natal).

(16) *A. glaucifolia* — China (Yunnan) **).

The distribution of these two groups pointedly illustrates the alternatives in the dispersal of the Berberidaceae and their *genorheitron*. The range of sect. *Rivularidium* covers well indeed the compass of the distribution of *Berberis* and *Mahonia*, not only, but brings into focus a significant outlier in Tasmania stressing the Western Pacific. The Tasmanian species, *A. crassifolia*, is said to be closely allied with the boreal *A. Richardsonii*, which is not surprising in the light of the distribution (see p. 244) of certain species of *Carex*. The Subsection *Longistylae* illustrates clearly indeed a channel of migration leading from South Africa to Western China. We may then conclude that *Anemone*'s dispersal primarily rests upon a base-line in the deep south of modern maps (see lines connecting the gates of angiospermy, 1, 2, 3, in Fig. 25), and that *Anemone* issues northward from three gates *simultaneously*. If there is a

---

*) The reader may compare this type of dispersal with that of *Euphrasia* (see p. 19), noticing that the essentials in play are the very same.

**) The reader will duly learn in the progress of our study that the tracks of this Subsection are homologous of those of *Schoenoxiphium* (p. 237), therefore of the *Berberis Wallichiana* group. Yunnan, Laos (see range of the lardizabalaceous genus *Sargentodoxa*; p. 74), Northern Sumatra (Gajolands), Java, Luzon, Northeast Borneo all are within reach of the same basic channel of migration. See further notes, p. 116, on the dispersal of *Tilia* and its allies.

difference between *Anemone* and the Berberidaceae in phytogeography, this merely is, that we can identify the gates used by *Anemone* with fair accuracy, while we are still left guessing here as to those used by *Berberis* and *Mahonia* in the genorheitral stage. We would hardly stress this as a relevant difference. The reader is further to see in the pages of this manual numerous patterns of dispersal like *Anemone's* and the Berberidaceae's, and — to repeat — what may still seem to him bold, and less than certainly well founded, will soon impress him as commonplace. It might be noticed that the Lardizabalaceae, which migrate stressing the Pacific first and foremost, are related with the Berberidaceae, and go back for their origin to the same antarctic center. Groups that leave the African gate of angiospermy directly faring to the Pacific are well known (see *Uncinia* for example; p. 249), so that it proves inexpedient to take for granted that genera using different gates are necessarily unrelated from the standpoint of phylogeny. The concept of gate of angiospermy is essentially *geographic*, for the gates themselves are merely the regions where the tracks begin to show. Approached from the standpoint of *phylogeny*, one of their number, the African, is readily seen to be most important, for it is to it that most tracks go back in the end.

Primula is distributed very much along the lines of *Berberis* (see p. 76 and Fig. 40). Unlike *Berberis*, however, *Primula* is but thinly represented in South America by two or three species endemic to Chile and Argentina. These strayers belong to Sect. *Farinosae* (16), which, apart from its South American members, is distributed throughout the Far East, Europe and North America in the "bipolar" pattern of *Euphrasia*.

Commenting upon the dispersal of *Primula* in Malaysia an author primarily interested in the phytogeography of that region writes (12; p. 146) "It is clear that the center (birth place) of *Primula* is found in the SE. parts of the Asiatic continent, that the crossing of the tropics in S. America must be due or at least understood as a later migration and that the same must be held for the crossing of the equator in the palaeotropics . . . . The undeniable conclusion must be drawn that the genus *Primula* in some former period must have distributed itself to Malaysia (Sumatra, Java) by a species (*P. prolifera*) which originated within the generic population in Asia and within the sectional population restricted to some parts of the Himalaya and W. China. This migration must have been effected by natural means of dispersal (by wind, birds or otherwise). I believe that this evidence is so conclusive that no other explanation is possible." The same author further states (12; p. 238), "Primula . . . represented in Chile and Tierra del Fuego by a variety of the widely distributed *P. farinosa* which doubtless has migrated formerly along the andine bridge crossing the tropics."

These statements need be pointedly taken apart, for, were they correct, the *genorheitral* center of the Berberidaceae ought to be looked for in the Himalayas, too, with attending stultification of everything we have learned so far by immediate observations of the facts of dispersal.

Let us remark, as a preliminary, that the author whom we have just quoted to the effect that "Natural means of dispersal" were the necessary instruments of the distribution of *Primula prolifera* to Sumatra and Java, is also author of the following remarkably honest statement (17; p. 64), "I have in vain tried to trace a relation between distribution and dispersal methods and I have not the least idea for explanation. That dispersal alone is responsible cannot be accepted and about the effect of dispersal during former periods we can hardly guess. We have apparently to accept distribution as given in nature at present . . *On the whole I cannot trace any relation between [means of distribution, as currently understood] and what is known of dispersal.*"

As this author has no idea how dispersal was effected, and further discredits conventional means of distribution by wind, birds, etc., as something which has no cogency of action, it proves difficult to accept his explanation to the effect that *P. prolifera* reached Sumatra and Java by "Natural means of dispersal." Clearly, this author has no idea on fundamentals, and we cannot follow him, if we wished to do so on account of his praiseworthy candor. In reality *we believe that dispersal is primarily the result of ancient conditions of land and sea, and that conventional means of dispersal were active only insofar they could operate as regards these conditions.* We need not prove this here, for it is shown to be true throughout the body of this manual.

The occurrence of *Primula* in Sumatra and Java is part and parcel of a broad issue, how a flora of "Gajolandian" type reached Malaysia, which we discuss elsewhere (see p. 336). This issue involves continental masses of the past, not casual dissemination by birds, winds, and the like. Obviously, this issue makes no sense when unsuitably approached, as always great issues do when "elucidated" by petty explanations. Likewise, the occurrence of *Primula* in Magellania is part and parcel of a basic issue regarding the origin, and further dispersal, of an „Antarctic" flora having connection in the north, therefore "bipolar". This issue, too, is very broad, and certainly not to be dealt with properly (see discussion of *Juania*, p. 455, and *Carex incurva*, p. 463) by making reference to "migrations along the Andes," when the Andes did as yet not exist.

To the extent of the statements quoted, we can not agree with the author under discussion at all, for he himself is not sure in his own mind as to what he alike suggests and affirms. We might agree with him to the extent, that, perhaps, *Primula* reached Sumatra and Java from the Himalayas. That *Primula* so reached these islands is a possibility, by no means a certainty (see p. 239, discussion of the *Schoenoxiphium-Cobresia* affinity) Above all, whether sure or only possible, the "Himalayan" origin of *Primula* in Sumatra and Java is a trifling byproduct of the course of certain main channels of migrations upon ancient continental masses. To magnify it as a primary question we must subvert the normal order of phytogeographic thought. Naturally, we sharply disagree with the author under discussion, and see no reason whatsoever to alter our stand as regards the Berberidaceae and every other plant. It seems clear to us that something profoundly mistaken; basic inadequacies of method and thought stand in the way of interpretations of dispersal which thrive upon statements of the sort quoted. Failure is eventually admitted one way or the other, either by substantial contradictions or an open acknowledgement that the facts on record make no sense. We read, for example, declarations of this kind (18; p. 136) by a would-be phytogeographer mostly interested in palaeo-botany, "We arrive at no conclusion, and it seems to me that we cannot hope to solve problems of this sort until we have sufficient facts, no matter how clever we are in speculation or prophesy."

As we discuss this statement in a coming chapter (see p. 283) with reference to the very same dispersal that prompted it, we need not take issue with it here. Clearly, the facts will not answer our queries, if we do not know *how to use them*. In order that we may gain this knowledge, we need bow to the primordial necessity of thinking out fundamentals, not decry this thinking as a "theoretical" deviation from "science."

The Primulaceae are a large family, and one of their number, *Samolus repens* is conventionally "antarctic" (19) because it ranges to South America, New Zealand, Tasmania, Australia and New Caledonia. *Lysimachia*, moreover, is certainly anything but "Holarctic," as we have seen (refer to p. 53, Fig. 17). It is patent that if we arbitrarily decide that *Primula* is "Holarctic", we must quite as arbitrarily „de-antarcticize" *Samolus* and *Lysimachia*, lest the Primulaceae are torn to shreds, and their *genorheitron* ceases making sense.

There is another aspect of the matter which we may not overlook if we intend to stay within the limits of a critical scientific study. The limits of what we might describe as the "primulaceous" *genorheitron* reach far beyond *Primula, Samolus* and *Lysimachia*, and we certainly do knowledge a disservice, if we insist upon studying this *genorheitron* piecemeal. It is well known that the Primulaceae bear being separated from the Myrsinaceae only by tenuous, artificial characters. Systematicists call one group Primulaceae, the other Myrsinaceae, primarily because it is convenient to handle the two in classification as independent units, but phylogenists can accept this separation only with qualifications. As a matter of fact, other groups are in the same case as these two, witness Araliaceae and Umbelliferae, Ericaceae and Epacridaceae, Apocynaceae and Asclepiadaceae, etc. All these families, like Primulaceae and Myrsinaceae, are kept apart because of convenience; we often know the form which is intermediate between any two of them (see, for instance, the arali-aceous-umbelliferous *Myodocarpus;* BAUMANN, in Berichte Schweiz. Bot. Ge-sell. 56: 13. 1946), without for this reaching the conclusion that convenience is best served by merging them both as one. Convenience, however, is one thing, phylogeny another. Either one of them has a legitimate claim upon attention, but it is certain that, as phytogeographers, we need not be bound by convenience of a kind that merely serves the utilitarian ends of systematic botany. While studying the Primulaceae we may not dismiss the Myrsinaceae, and whatever we may legitimately conclude about the origin and dispersal of the former must take sharp account of the origin and dispersal of the latter.

Indeed, there is a strong probability that the Frankeniaceae also belong to primulaceous *genorheitron*, for it is not without reason that the Andean frankeniaceous genus *Anthobryum* is credited by some authors to the Primulaceae. Let us, however, study the dispersal of the Myrsinaceae in the first place.

IV. Myrsinaceae

This family has used two angiospermous gates in Africa and the Western Pacific, but, being absent from the third in Magellania, is practically unknown in Chile. By using only two gates, the Myrsinaceae have nevertheless managed to secure astounding dispersal, finally indulging in vagaries of migration that would stand beyond rationalization if we had no inkling of the starting points of the dispersal.

The monograph available for this family (20) is slightly over forty years old. Naturally, we would be glad to implement and to correct it, referring to much work which has been written of late concerning various myrsinaceous groups. We feel, however, that it is wiser to allow matters to stand as they are rather than to run the risk of making them worse by ill-fitting compilations.

Investigations in dispersal rely quite as much upon affinities as they do upon records of locality, and the former — *sectional and tribal affinities in particular* — are more important than the latter. Were we to choose in our work between reference to an absolutely up-to-date taxonomic text short of sectional and tribal classification, and a work of the same kind giving but half the known species, with, however, good subgeneric and subfamiliar groupings, we would not hesitate in preferring the latter. We are always eager to take the fullest possible advantage of the conclusions of a botanist who has seen a large group or family *as a whole*, and wary lest casual additions from indices or other readily available sources disturb the groups which this botanist provides. Experience has taught us, moreover, that the latest is not always the best, and that manipulations of nomenclature and changes are not necessarily the gate to the truth.

We need not enter into details here, but we could take classifications seemingly at great odds, and reach the conclusion in the end that the authors of these efforts cover each a part of the evidence in particular, though both have seen it whole, even if dimly, to begin with. Examples of seemingly hopeless conflicts between taxonomists are given in these pages (see *Phrygilanthus* p. 146; *Geranium* p. 112; *Juania* p. 455) leading to the ultimate conclusion that these conflicts are nugatory, and amount to plays of the sheerest opinion.

This being the case, we are not always inclined to doctor the classification we find in the naive anticipation we can render it perfect. Obviously, we do our best to verify and implement the records we use when important issues of range are involved, but not otherwise. For example, the myrsinaceous genus *Oncostemon* numbered forty years ago 57 species, of which 50 were in Madagascar, 5 in the Comoros, 2 in the Mascarenes. This total is probably larger now, and may include additions from continental Africa, but none of these additions can change the position of this genus in dispersal. On the contrary, dealing with *Suttonia* which is significantly distributed, we did our best to verify the score and bring it up to date. In short, we see no necessity for monographing on our own everyone of the groups we use, and constantly pouring new wine into old skins. Taxonomy labors under limitations of which we are keenly and constantly aware, and the reader is duly to appreciate in the end.

Maesa has an isolated position under the Myrsinaceae, and, peculiarly, is related to *Samolus*, the "antarctic" primulaceous genus. Some among the hundred species of *Maesa* range as follows,

(1) *M. rufescens* — Natal, Transvaal, Tanganyka, Cameroon.

(2) *M. lanceolata* — Madagascar, Angola, Cameroon, Fernando Po, Tanganyka, Ethiopia, Somaliland, Arabia.

These two species are very close, and represent in reality vicariant forms of a single broad entity. Their distribution is such as to tell us forthwith that the African Gate is in active operation, very much in the manner of *Evolvulus*. We may not be surprised, consequently, in finding a third species distributed thus,

(3) *M. ramentacea* — India, Andaman and Nicobar Islands, Himalayas, Southwestern China, Malacca, Sumatra, Java, Borneo.

Patently, three species out of a hundred are enough to tell us where we stand, and we need not pad the record with the latest in taxonomy to affirm that *Maesa* originates from the vicinity of South Africa.

Maesa is exceedingly well represented in Java (over 15 species), and by no means unknown in Malacca and Sumatra. Its track, moreover, sends a long finger into Malaysia in the direction of Timor (at least 2 species), Celebes (at least 4 species), and further still to the Moluccas (1 species at least in Amboina), Queensland (at least 2 species), New Guinea (at least 15 species), Fiji (at least 1 species), New Caledonia (at least 1 species), New Hebrides (at least 4 species), Tonga and Samoa (at least 2 species).

The track also invests Borneo (about 5 species), the Philippines (about 15 species) and over the Indochinese Peninsula reaches Southeastern China, Formosa, the Riu-kius (about 2 species).

Maesa, then, behaves conventionally for a genus of its origin. Scores of others behave in the same manner. The question arises, of course, whether it is *only* the African gate which has been active sending species to Samoa immediately from Madagascar, or whether the Western Pacific gate has been active instead, feeding Samoa and all the islands of the Pacific, as far at least as the Philippines, from the opposite direction.

This question may seem capital, forbidding indeed, because it involves the origin and direction of flow of the migration, and it may be thought that, once these factors are doubtful, we no longer have bearings by which to proceed in our quest. In reality, this question calls into play nothing capital. As we defend no theory, and freely admit that the same group of plants may stream from one or more gates of angiospermy in particular (see also p. 80; Berberidaceae), we cannot gain or loose anything by "proving" right here that Samoa received its species of *Maesa* from one or the other gate. When present knowledge is inadequate, and we do not exactly understand the affinities within a large group, we cannot do better than to outline alternatives which present knowledge justifies, leaving it to the future to settle whichever one be true, wholly or in part. The sum total of the distribution of the genus *Astelia* (see p. 58) would not change a bit if certain of its sectional and subgeneric units had used only one gate instead of two, and our understanding of the migrations of *Maesa* would not collapse only because we cannot definitely state at this hour, and in this very page, whether this genus is in Samoa mainly from the west or the east. That which we know is that *Maesa* massively streams out the African angiospermous gate, but this knowledge does not rule out the possibility that other gates were active in the dispersal of this genus. Let us not forget that all gates become confluent within a geologic continent of „Antarctica" in the end, and that an understanding of *genorheitra*, orderly tracks of migrations, affinities in general always give us the means of reaching tenable alternatives when certainty is still out of our immediate reach.

Under Subf. Myrsinoideae, by far the larger group of the Myrsinaceae, occurs *Discocalyx*, a sizeable genus in the Philippines (about 40 species) which ranges outside the Philippines as follows,

(1) *D. divaricata* — Fiji.

(2) *D. fusca* — Fiji.

(3) *D. multiflora* — Fiji.

(4) *D. Listeri* — Tonga.

(5) *D. palauensis* — Palau.

(6) *D. macrophylla* — Palau.

(7) *D. ladronica* — Mariannes.

(8) *D. megacarpa* — Mariannes.

Typical ranges in the Philippines and vicinity are,

(9) *D. cybianthoides* — Luzon nearly throughout, Mindoro, Leyte, Masbate, Borneo.

(10) *D. micrantha* — Luzon (Ilocos Norte), Mindanao (Surigao).

(11) *D. palawanensis* — Palawan, Mindoro, Bancalan, Banguey.

Discocalyx is allied with *Cybianthus*, which tabulated to 35 species nearly half a century ago. How many more have since swelled the record we do not exactly know, but the range of *Cybianthus* is covered by the following selection,

(1) *C. Cruegeri* — Trinidad.

(2) *C. Fendleri* — Venezuela (Aragua).

(3) *C. comatus* — French and British Guianas.

(4) *C. penduliflorus* — Brazil (Parà).

(5)   C. detergens — Brazil (Cearà, Rio Janeiro, São Paulo).

(6)   C. glaber — Brazil (Rio Janeiro: Serra dos Orgãos).

(7)   C. fuscus — Brazil (São Paulo, Matto Grosso: Cuyabà).

(8)   C.ʹresinosus — Peru (Loreto).

(9)   C. psychotriifolius — Brazil (Matto Grosso), Bolivia (La Paz).

Possibly less closely allied with *Discocalyx*, but still in this affinity is *Grammadenia*, as follows,

A) Subg. *Cybianthopsis*

(1)   G. Sintenisii — Portorico.

B) Subg. *Eugrammadenia*

(2)   G. parasitica — Guadeloupe, Dominica.

(3)   G. costaricensis — Costarica.

(4)   G. lineata — British Guiana (Mt. Roraima).

(5)   G. alpina — Venezuela.

(6)   G. Lehmannii — Colombia.

(7)   G. marginata — Colombia (Cauca).

(8)   G. pastensis — Colombia (Nariño).

(9)   G. magna — Colombia (Norte de Santander).

(10) G. nitida — Peru (Huànuco).

The dispersal of *Discocalyx*, *Cybianthus* and *Grammadenia* cannot be successfully studied in the light of methods now accepted, but to us it suggests certain *standard* alternatives, on the contrary, as follows, (a) From a point near modern South Africa the *genorheitron* directly fared to the Pacific, ultimately to come to emersion on our maps by the Western Polynesian gate near Fiji. Once in Fiji, the *genorheitron* streamed on, following in the wake of *Alphitonia* and *Perrottetia* (see p. 54) as far as the Far East and the New World; (b) Always from a point near modern South Africa *) the *genorheitron* first invaded South America, while dispatching an outlier to the Western Polynesian gate.

These alternatives are *standard*, we like to repeat, and numerous examples of them will come under our attention in these pages. They are not at all exclusive. *Grammadenia* — which occurs in the West Indies, Guianas, Colombia, Venezuela, Peru, but, so far as known, is not abundant, possibly un-

---

*) This point might in reality better agree with a center immediately in the Indian Ocean (see Fig. 26, R) connected with the approaches to the modern Caribbeans through the body of modern Central Africa. We regret we cannot always outline all possible alternatives in the course of the tracks, lest the attention of the reader be distracted from essentials, and the maps made too confuse. We feel sure, nevertheless, that the reader will soon be able to figure out alternatives for himself using the principles and methods constantly stressed and exemplified in this manual.

known, in Brazil — might have reached the Americas in the manner of *Perrotte-tia*, that is to say, crossing the Pacific in the north. *Cybianthus* on the contrary — which is well represented *in Brazil* — might have fared straight to the Americas in the manner of *Evolvulus* (see p. 24) or *Gnetum* (see p. 271). The key-point of all this whirlpool is, located in the end approximately south of Africa, precisely in a crumbled range which we have previously stressed. We are quite aware that dispersal of this type may seem bewildering to most, and too complicated to be encompassed at once. This is not the case for, as we shall see, dispersal of this type is *absolutely standard*, and we will have numerous examples of it before us in coming pages. Moreover, the map of myrsinaceous distribution given at the end of this discussion will easily be seen for what it is, very simple indeed at bottom.

Suttonia is distributed as follows,

A) Subg. *Eusuttonia*

1 species in Chatham Island; 1 species in Norfolk Island; 4 species in New Zealand.

B) Subg. *Rapaneopsis*

12 species in Hawaii *).

This is a very instructive type of dispersal matched in families, some discussed elsewhere in the pages of this manual, which are not at all allied to the Myrsinaceae. Though dispersal of this kind is commonly believed to be "antarctic," the reader will see that it might have been effected, on the contrary, by paths that had nothing to do with Antarctica.

Ardisia, one of the largest genera of this family is often dismembered into numerous lesser segregates, which is none of the phytogeographer's interest. Treated in a broad sense, it contains the following subgenera, for instance,

A) Subg. *Pimelandra*

1 species in Malacca; 1 species in Bangka; 3 species in Sumatra, Java, Borneo, Soembawa, Celebes; 1 species in Java, Timor; 1 species in Celebes; 1 species in Ternate; 3 species in the Philippines; 4 species in New Guinea.

---

*) The reader will duly learn in time that the distribution of *Suttonia* is a fair match of that of the violaceous genera *Melicytus* and *Hymenanthera* (see p. 181), and of the Oleaceae (see p. 355) endemic to the vicinity of the Coral Sea. He will then understand that the Myrsinaceae travelled the Pacific by channels such as those used by Violaceae and Oleaceae.

B) Subg. *Icacorea*

11 species in Mexico (Sinaloa, Tamaulipas, Veracruz to Oaxaca); 2 species in Eastern Cuba; 1 species in Haiti; 10 species in Central America; 2 species in the Amazonas; 2 species in Southeastern Brazil; 2 species wide-ranging (*A. compressa*: Mexico in Sinaloa, Veracruz, Oaxaca; Venezuela, Trinidad — *A. guianensis*: Trinidad, Tobago, Venezuela to Northern Peru and the Western Amazonas) *).

So far as known, *Ardisia* is unreported from Africa. Considering that a center near the Dark Continent is fundamental in the history of the dispersal of the Myrsinaceae, and all Angiosperms as well, we have two suggestions to offer, (a) The *immediate genorheitron* of this genus fared from this center directly to the New World, without leaving traces in Africa. This, as we know (see p. 389) is altogether possible; (b) The same *genorheitron*, on the contrary, left this center streaming south of modern Africa and America (or, conversely, proceeding south of the Indian Ocean) to the Western Polynesian gate, next to act in the manner of *Perrottetia* (see p. 54). Examples of distributions so effected are scattered throughout the pages of this manual.

The Myrsinaceae contain other examples of seemingly striking dispersal, witness these two small genera,

A) *Heberdenia*

(1) *H. penduliflora* — Mexico (Puebla, Veracruz, Oaxaca).
(2) *H. excelsa* — Madera, Canary Islands.

B) *Myrsine*

(1) *M. africana* — South Africa, Angola, Azores; Ethiopia, Socotra, Arabia, Afghanistan, Beluchistan, Himalayas (Kashmir), China **).
(2) *M. Mocquerysii* — Madagascar.
(3) *M. semiserrata* — Northwestern Himalayas to Burma and China (Yunnan to Hupeh).
(4) *M. marginata* — Southeastern China.

Dispersal of this type find such phytogeography as flourishes in our midst completely helpless. Not only there is nothing in the printed record to account even for its grossest aspects, but problems of this nature are not supposed to

---

*) As it will later appear (see p. 369, Fig. 70), these two species follow one of the most important channels of dispersal in the New World. The rate, consequently, as instances of standard distribution.

**) We may not enter a long digression here, but suggest without further that the dispersal of this species westward seemingly agrees with the migrations we have elsewhere postulated for the Cyrillaceae, *Ilex*, etc. (see p. 343), and that part of *Linum* (see note, p. 202) which reached the Caribbeans. *Myrsine africana*, however, went to the Azores rather than to modern America.

be phytogeography's legitimate concern. Right peculiarly, phytogeography's concern is supposed to lie with issues such as "Species senescence," which, were it a legitimate problem, could never be approached with hope of success in the absence of an adequate, critically elaborated, supply of facts culled out of the distributional record.

Dispersal of this type, on the other hand, is of the commonest. As we shall better learn in the coming chapters, it was effected precisely over a shore connecting Mexico with the Macaronesian domain (Azores, Canary Islands, Madera), and further extending eastward. This shore had in addition intimate contacts with other lands in the south. In brief, *dispersal may be taken at face-value, in the understanding that facts cannot be nonsense.* Without this shore, and without a free recognitions of the tracks involved therein *as something real,* phytogeography makes no sense. A few examples of distribution are next added to clinch the evidence, as follows,

A) *Thamnosma* (**Rutaceae**)

a) **Subg.** *Euthamnosma*

(1)   *T. montanum* — United States (South California, Nevada, Arizona, Utah), Northern Mexico.

(2)   *T. texanum* — United States (Texas, Arizona, Colorado), Northern Mexico.

(3)   *T. trifoliatum* — Mexico (Baja California).

(4)   *T. purpureum* — United States (Texas, Arizona, New Mexico).

b) **Subg.** *Palaeothamnosma*

(5)   *T. africanum* — Southern Rhodesia, Transvaal, Damaraland.

(6)   *T. socotranum* — Island of Socotra.

It will be noticed, (a) The Southwestern United States and adjacent Mexico are a very ancient harbor of angiospermy, which accounts for the comparatively heavy speciation in this hoary genus; (b) This distribution has the characters of that of the oleaceous *Menodora* (see p. 358) in the main, though differing in detail, because *Thamnosma* is not in South America, and *Menodora* is not in Socotra; (c) Eastern South Africa is a very ancient harbor of angiospermy, and so is the island of Socotra. It is interesting that these two centers are connected with equally ancient centers in the New World; (d) The genorheitral migrations are from the African gate of angiospermy (i.e., a center located in the water of the modern Indian Ocean). They reached Socotra, simultaneously streaming to the New World by either one of two channels, (i) Across Central Africa, or, (ii) Across Northern Africa. Dispersal of this nature is wholly standard, so that the seemingly "miraculously" distributed *Thamnosma* is in reality a phytogeographic common-place.

B) *Oligomeris* (Resedaceae)

(1) Sect. *Holopetalum*: 6 species in South Africa generally (Cape, South-west Africa, Namaland).

(2) Sect. *Resedella*: 1 species, *O. linifolia* distributed as follows: California to Texas and Mexico; Canary Islands, Northern Africa to Nubia; Arabia, Iran (Persia), Afghanistan, Beluchistan, India; 1 species, *O. Dregeana*, with range in South Africa.

The tie between the Old and New World is manifestly brought about by a line south to north in the Indian Ocean of our maps, crossed T-wise by another line running between California and India. The gate involved is the African one. This migration may be compared with that of *Cercis*, for instance, (see p. 137), of which it has certain basic elements.

C) *Littorella* (Plantaginaceae)

(1) *L. australis* — Falkland Islands, Magellania.

(2) *L. americana* — Newfoundland to Maine, Vermont, Ontario and Minnesota.

(3) *L. uniflora* — Azores, Western to Central Europe.

D) Malvaceae Tribe *Malopeae*

(1) *Palava* — 5 species in Chile, Peru, Mexico.

(2) *Malope* — 3 species in the Mediterranean.

(3) *Kitaibelia* — 1 species in Hungary and the Balkans.

E) Cneoraceae

(1) *Cneorum* — Subg. *Eucneorum*: 1 species (*C. trimerum*) in Eastern Cuba; 1 species (*C. tricoccum*) in Southern Spain, Baleares, Southern France, Italy (Tuscany). Subg. *Neochamaelea*: 1 species (*C. pulverulentum*) in the Canary Islands.

F) Labiatae (in part)

(1) *Minthostachys* — 1 species in Argentina (Catamarca, Tucumàn, Còrdoba); 6 species in Bolivia; 6 species in Peru; 3 species in Ecuador and Colom-

bia (Nariño, Cauca, Santander); 1 species in Peru (Huánuco, Lima, Junin), Colombia (Cauca, Huilla, Cundinamarca, Santander), Venezuela (Mérida).

(2) *Bystropogon* — 3 species in Madera, 6 species in the Canary Islands.

G) *Cardamine* sect. *Eucardamine* (in part)

(1) *C. Johnstonii* — Mountains of East Africa, Andes of Colombia.

(2) *C. Jamesonii* — Mountains of East Africa, Northern Andes, Santo Domingo.

(3) *C. obliqua* — Mountains of East Africa and Mexico.

(4) *C. hirsuta* — Mountains of East Africa (Kilimajaro to Ethiopia) and West Africa (Cameroon), Mexico, Jamaica, Santo Domingo, West Australia.

(5) *C. caldeirarum* — Azores. *)

The types of dispersal we have just reviewed, and many more of the same nature which space forbids us to study, yield no sense whatever until we accept as most probable that those are in right who believe that Africa and South America once formed a single continental mass, and that the Angiosperms began their migrations from a point in the south of our maps long before the modern continents took on their current shapes. The issue truly at stake, then, is not with "Landbridges," or, worse still, with "Occasional means of dispersal," but essentially with a type of geography entirely different from the geography of our times. It is certainly not by the ministrations of means of the kind that was established in Southwest Africa a peculiar Cretaceous angiospermous flora (21) represented by pollen which, to the exception of *Myrica*, cannot be credited to living forms.

The dispersal of the Myrsinaceae (Fig. 26) speaks for itself, and it should be difficult indeed to believe that the Primulaceae orginated poles apart from this family. Myrsinaceae and Primulaceae belong to the same *genorheitron*,

---

*) We have experienced great difficulties throughout the preparation of this manual in making decisions between conflicting needs. Dispersal is so comprehensive that it is most easy to swamp its student at the outset with a mass of details. Contrariwise, there is the risk that once the student has progressed far enough he may be critical of not having been told everything at the start. The examples of trans-atlantic dispersal given in the main text lend themselves to meaty observations which could be made only at the price of opening lengthy parentheses, here out of place. We may only briefly observe the following, (i) The tracks of *Minthostachys* and *Bystropogon* would be commonplace, if there only was an additional station of either genus, or an intermediate form, recorded from the line Portorico-Guadeloupe. Were this so, the tracks would consist of two orthodox segments, *Canaries-Portorico (or Guadeloupe)*; *Portorico-Colombia/Venezuela-Bolivia*; amply discussed elsewhere in this manual; (ii) Santo Domingo connects in an altogether orthodox manner the African and American range of *Cardamine Jamesonii*. The score would be perfect, if only this species, or a vicariant, were known from the heights of the Cameroon; (iii) The distribution of *Littorella* is of a comparatively rare pattern (see track (12); p. 550), but matched well enough by that of the Empetraceae; the origin of the latter family is safely determined in reference to that of the Ericaceae Ericoideae. The reader will gradually be informed in detail about these, and similar, distributions.

and so long as the former migrate as they do, the latter cannot originate *) in Central Asia. Both these groups, on the contrary, repeat their ultimate origin in the deep south of the modern map.

V. Fagaceae

This family has proved to be a bone of contention throughout the years because *Fagus* lives in the northern hemisphere, *Nothofagus* in the southern, and both genera are supposedly known as fossils from the northern, and southern, hemisphere alike. The elements in the case, consequently, have favored in turn this or that author, some emphasizing the south, others the north of the map. It has only recently been suggested that the Fagaceae originated (22) somewhere in the vicinity of modern Malaysia.

The generic limits of the Fagaceae are fairly cleancut as to certain groups (e.g., *Fagus*), obscure as to certain others (e.g., *Lithocarpus*), one and the same species having been classified under four to six different genera. Needing a reference for this family, we will use the classification of an author (23) who dismembers *Quercus* into more genera than most taxonomists are willing to accept. This classification is serviceable to us here, nevertheless, for it is organically written.

There exists in the vicinity of the Fagaceae a small family, Balanopsidaceae, of which little material has been available to most botanists. This family has often been overlooked for this very reason, and was once at least classified under the Malvales (24) with the note, "This family is doubtfully given place here, and it may be that it should be placed near the *Fagaceae*, as is done by Baillon." We believe that Baillon was right in so doing, which modern systematists are also inclined to grant.

The Balanopsidaceae and Fagaceae tabulate as follows,

A) Balanopsidaceae

(1) *Balanops* — 3 species in New Caledonia, 2 species in Queensland.
(2) *Trilocularia* — 2 species in Fiji.

---

*) We repeatedly discuss the meaning of the term *originate* (see p. 546) in this manual. We understand as *originate*, "to reach morphological maturity." In other words, a group *originates* there, where we might first recognize it as modern plants. In this sense, the Primulaceae did not originate in Asia, because Primulaceae of modern aspect and characters are seen in the austral regions of the modern world immediately reached by tracks stemming from the angiospermous gates. Primulaceous plants were evidently in being (that is, *had originated*) when these tracks began to operate. See further Casex remota, p. 245.

B) Fagaceae

a) Subf. Fagoideae.

(1) *Fagus* — About 8 species in the North Temperate Zone: Eurasia and Eastern North America.

(2) *Nothofagus* — About 15 species in the southern hemisphere; 3 or more in New Guinea, 12 in New Zealand, Tasmania, Australia, Chile and adjacent Argentina.

b) Subf. Castanoideae

b') Tribe Castaneae

(3) *Castanea* — 12 species in the North Temperate Zone, Eurasia and Eastern North America.

(4) *Castanopsis* — 102 species in Malaysia, the tropical and subtropical Far East; 2 species in northwestern North America (Washington, Oregon, California).

b") Tribe Pasanieae

(5) *Pasania* (*Lithocarpus* of other authors) — 60 species in Malaysia, the tropical and temperate Far East; 1 species in Northwestern North America (Oregon, California).

(6) *Cyclobalanus* (*Lithocarpus* of most authors) — 30 species in Malaysia, the tropical and subtropical Far East.

(7) *Lithocarpus* — 15 species in Malaysia, the tropical and subtropical Far East.

c) Subf. Quercoideae

c') Tribe Cyclobalanopsideae

(8) *Cyclobalanopsis* (*Quercus* of most authors) — 90 species in Malaysia and the Far East.

(9) *Erythrobalanus* (*Quercus* of most authors) — 175 species in North and Central America ranging to northwestern South America; few species in the Far East.

c") Tribe Querceae

(10) *Macrobalanus* (*Quercus* of most authors) — 10 species in Central America.

(11) *Quercus* — 320 species in the North Temperate Zone, Eurasia and North America, ranging as far as Ecuador in South America.

Cast on the map (Fig. 27), — whether in eleven or six genera matters not — this dispersal is transparent. The Balanopsidaceae occupy the center between the southern and the northern massings, which is fitting. The track is conventionally "antarctic" in the Southern Pacific, and conventionally crosses the Northern Pacific between the Far East and a region located between

Southern California and Central America. The outlier that leaves the Far East for the Mediterranean is absolutely standard. The alliance in question visibly streams out of a region in the southwestern or southern Pacific, and departs from this center west and east, north and south in a pattern thoroughly familiar to us. If there is anything worth attention in this distribution, we may find it between Malaysia and Fiji, for it is here that most genera concentrate, and the northern and southern affinities come to a head. The rest is commonplace, and doubtless not worthy of a protracted discussion.

The lone species of *Fagus* endemic to Formosa, *F. Hayatae*, has peculiarly delicate husks reminiscent of *Nothofagus*, and further suggests the Chinese *F. Engleriana* (Kweichow, Hupeh, Szechuan, Yunnan, Shensi). This indicates that *Fagus* might have separated from *Nothofagus* in the range between New Guinea or Fiji, and Formosa, and reached China from Formosa. This indication agrees with the rest of the evidence, as we are soon to learn.

Quercus, or true oaks, is represented in Malaysia and warmer Eastern Asia by a group usually identified as subgenus, or section, *Cyclobalanopsis*. This group consists of about 90 species (25), a third at least of which are localized in Indochina. *Cyclobalanopsis* merges with *Lithocarpus* through at least three species, namely,

(1) *Q. argentata* — Malacca, Sumatra, Bangka, Anambas Islands, Borneo.
(2) *Q. nivea* — Borneo (Sarawak).
(3) *Q. subsericea* — Bangka, Borneo (Sarawak).

Let us observe, then, that *Quercus* and *Lithocarpus* — therefore also *Pasania* — merge together in a Western Malaysian range between Borneo and Sumatra.

Quercus sect. *Cyclobalanopsis* typically ranges in the manner illustrated by the following sections,

A) *Brevistylae*

(1) *Q. Miyagii* — Riu-kiu Islands to Yakushima Island.
(2) *Q. yayeyamensis* — Riu-kiu Islands.
(3) *Q. patkoiensis* — Assam.
(4) *Q. brevistyla* — Malacca.

B) *Flavescentes*

(5) *Q. Championii* — Formosa, Hongkong, Kwantung.
(6) *Q. Poilanei* — Kwangsi, Annam.
(7) *Q. flavescens* — Annam (Island of Tré).

C) *Pachylomae*

(8) *Q. pachyloma* — Formosa, Fukien, Kweichow.
(9) *Q. tomentosicupula* — Formosa.
(10) *Q. Camusae* — Tonkin, Annam.

(11)   *Q. auricoma* — Tonkin.

(12)   *Q. blaoensis* — Annam.

(13)   *Q. baniensis* — Annam.

(14)   *Q. Ramsbottomii* — Burma, Tenasserim.

D) *Semiserratae*

(15)   *Q. Edithae* — Hongkong, Annam.

(16)   *Q. semiserrata* — Burma, Silhet, Assam, Tenasserim, Siam, Annam, Malacca, Bangka, Sumatra.

(17)   *Q. xanthoclada* — Tonkin, Laos, Upper Burma.

(18)   *Q. Fleuryi* — Tonkin.

(19)   *Q. donnaiensis* — Annam.

(20)   *Q. cambodiensis* — Cambodja.

(21)   *Q. langbianensis* — Annam.

(22)   *Q. Elmeri* — Northeastern Borneo (Tawao).

E) *Turbinatae*

(23)   *Q. nubium* — Japan (Kyushu, Shikohu, Honshu), Formosa, China (Fukien, Chekiang, Kiangsi, Hunan, Kweichow, Szechuan).

(24)   *Q. oxyodon* — Assam, China (Hupeh, Yunnan, Szechuan).

(25)   *Q. Fargesii* — China (Szechuan).

(26)   *Q. dilacerata* — Tonkin.

(27)   *Q. Merrillii* — Philippines (Palawan).

(28)   *Q. Hillebrandii* — Burma.

(29)   *Q. Thomsoniana* — Sikkim, Bhutan.

(30)   *Q. Gambleana* — Eastern Bengal.

(31)   *Q. oidocarpa* — Siam, Malacca, Sumatra, Bangka, Borneo (Sarawak).

(32)   *Q. Hendersoniana* — Malacca.

(33)   *Q. turbinata* — Bangka, Java.

(34)   *Q. lineata* — Sumatra, Java.

This dispersal is homologous as to fundamentals of that of *Coriaria* (see p. 71), despite its calling into play a much larger number of species and a broader range. The main channel takes the migration from the open Pacific of our maps directly to a point of penetration located in the vicinity of Formosa. From this point the *genorheitron* streams inland, or still by sea, in the direction of, (i) Japan; (ii) The Himalayas; (iii) Central China; (iv) Western Malaysia. The arm of the track that points most directly southward is oriented toward the Island of Bangka, immediately off the southeastern coast of Sumatra. As it proceeds toward Bangka from the direction of Formosa, this tracks frays out. It reaches eastward the island of Palawan in the Philippines, where occurs the lone true Oak so far reported from this archipelago; westward the petty island of Tré off the Indochinese coast, and Indochina generally. Right where this track ends (triangle Malacca-Borneo-Sumatra, in the main), are endemic three species, which, as we have seen, connect most intimately *Quercus, Pasania, Lithocarpus.* Let us notice that a dispersal of this type, reaching Malaysia from a general direction north to south is standard. It is standard, not because it is straight from "Holarctis", but because tracks that have nothing to do with this fictitious land veer southward to reach their ultimate destination in the manner just shown *).

Quercus is, alas, universally spoken of as "Holarctic," which is erroneous, because, (a) The *genorheitron* of the Fagaceae centers in the Pacific; (b) The Balanopsidaceae are native to Queensland and Fiji; (c) *Nothofagus* is antarctic throughout (New Guinea through Southeastern Australia and New Zealand to Magellania and Central Chile). A genus cannot be "Holarctic" with this background of affinities and dispersal. Moreover, no modern angiosperm ever was, or could be, "Holarctic" in the phylogenetic and phytogeographic meaning of the term, though it might be so in the geographic.

Between New Guinea, Formosa and the continental Far East the track of *Quercus* is identically the same as that of *Coriaria.* There is a critically established record (26) of a true oak (*Quercus* sect. *Cyclobalanopsis*: *Q. Guppyi*) from a small island in the Solomons, which proves, that the track of this genus, like *Coriaria's,* streamed to Formosa along a channel in the modern Pacific off the Philippines **). It is consequently correct to visualize for *Quercus* a point of entrance into the Far East and Malaysia *from the direction of the Pacific,* not from a thoroughly fictitious "Holarctis."

So strong is the current prejudice in favor of *Quercus* being "Holarctic," nevertheless, that we will give here close attention to migrations in the Pacific which can help us in deciding with finality the status of *Quercus* in dispersal. Migrations of the same or homologous pattern are discussed throughout these pages, but we will concentrate on some of them purposefully right here.

Lepechinia is a genus of the Labiatae consisting of about 50 species mostly scattered over the western portion of the Americas between Chile and Mexico.

---

*) We cannot run too far ahead of the score, and force upon the reader conclusions which, being as yet unsupported by massive evidence, he may not be disposed to accept just now. Anticipating somewhat on what is to come (see p. 424, for instance) we may point out that *Quercus* is typically rooted in the "Sunda Coign" of Malaysia. This "Coign" could be reached both from the Western Pacific and the north (*Quercus*), or the south from Northern Australia (see part of *Stylidium,* etc.; p. 433), or finally, from the west (Dipterocarpaceae, *Protium;* see p. 422, 388).

**) Migrations riding the open Pacific of our maps off the Philippines are anything but unusual. See further *Styrax* (p. 328), which, unlike *Quercus* and *Coriaria,* may well ultimately come to Malaysia from the west. Let us repeat that nothing of Malaysian dispersal can be critically understood *until and unless the genorheitral, and immediate, tracks of Malaysian plants are finally and completely oriented,* whether in origin from the Indian Ocean, or the Pacific. This is a "must" without which, as this manual shows, no discussion of Malaysian dispersal can make sense. The case is precisely the same with the *Caribbeans* as it is with *Malaysia,* both these centers being fed ultimately at least from two oceans (*Malaysia:* Indian and Pacific Oceans; *Caribbeans:* Pacific and Atlantic Oceans).

This genus has about 35 species (27) in South America, where its tracks can be followed without difficulty *).

A species of *Lepechinia* allied with the Chilean *L. chamaedryoides* (28) is in the Mascarenes. This record is by no means as extraordinary as it is deemed to be, because *Menodora* and *Turnera*, (see p. 46, 358), both essentially „American" genera, also are in South Africa, the Mascarenes and Seychelles, as we know.

Lepechinia floribunda turns up at the Sierra de la Ventana in Buenos Aires, which is an important phytogeographic region. From this point on, seemingly, *Lepechinia* is seen in Chile (about 3 species), Eastern Brazil to Rio Janeiro (1 species), Bolivia, Peru, Ecuador, Colombia and Venezuela, Central America and Mexico. This genus was in the Americas long before the rise of the Andes, for it takes the conventional path from the Pacific to the Atlantic shore characteristic of many groups of this kind, witness,

(1)   *L. bullata* — Colombia (Cauca, Caldas, Cundinamarca, Santander, Norte de Santander), Venezuela (Caracas) **); and further occurs endemic to ancient rejuvenated alpine cores, for example,

(2)   *L. humilis* — Colombia (Magdalena: Sierra Nevada de Santa Marta). One of the species of *Lepechinia* runs the following track,

(3)   *L. hastata* — Islands of Revilla Gigedo, Baja California, Southern California, Hawaii.

This dispersal shows that, as modern maps read, *Lepechinia* follows a double track, (a) *Overland* from Chile through Central America to Mexico; (b) *Oversea* between the Mascarenes and the Sierra de la Ventana (or Chile), and again between western South America, Revilla Gigedo, Baja California and Hawaii.

The Labiate genus *Teucrium* is dispersed in South America thus (27),

(1)   *T. Grisebachii* — Argentina (Córdoba, San Luis, Catamarca).

(2)   *T. bicolor* — Chile (Coquimbo, Aconcagua, Santiago).

(3)   *T. nudicaule* — Chile (Coquimbo, Antofagasta, Atacama).

(4)   *T. laevigatum* — Brazil (Rio Grande do Sul), Uruguay, Argentina (Entre Rios, Santa Fé).

(5)   *T. tenuipes* — Galápagos (Charles and Chatham Islands).

(6)   *T. vesicarium* — Argentina (Buenos Aires, Entre Rios, Misiones, Chaco, Salta), Uruguay, Paraguay, Brazil (Rio Grande do Sul, Santa Catharina, Rio Janeiro), Ecuador, Colombia (Magdalena, Atlàntico).

---

*) Long months after this had been written, there was published (Epling in Brittonia 6 : 352. 1948) a synopsis of the Tribe Lepechinieae, in which the range and affinities of *Lepechinia* are amply and ably discussed. In the same publication (op. cit. 5 : 491. 1946) was contributed a study of the Labiate genus *Teucrium* by two well known specialists on that family. The new data brought to the record are purposely disregarded in this manual, as an indication that absolutely up-to-date taxonomic records have little or no influence upon interpretations of migration made in the light of rationally assumed generalities. The reader may easily reach his own conclusion, comparing the data upon which we have originally relied with the data newly contributed by the works cited in this note.

**) The use of „Caracas" is objectionable in this sense. Caracas is in reality the main city of a small district officially known as *Distrito Federal* (D. F., in abbreviation). We ought consequently to write as follows, *Venezuela (Distrito Federal)*, also in order to be consistent with entries elsewhere in these pages, in which we give the names of Venezuelan States (Aragua, Mérida, etc.). We maintain, nevertheless, direct reference to „Caracas", feeling that this reference more readily orients readers not familiar with the details of Venezuelan political geography.

This dispersal reveals that *Teucrium*, like *Lepechinia*, also fares along a double track, (a) *Overland* between Argentina (Buenos Aires), Eastern Brazil to Rio Janeiro, and Chile; between Ecuador and Northeastern Colombia; (b) *Oversea* between Northern Chile and Ecuador, entering the Galápagos along this route *).

Tracks of this kind (Fig. 28), *overland* and *oversea* are common (see for instance *Cuscuta, Evolvulus, Ephedra*). The *oversea* channel manifestly fares along the following stations, Galápagos, Revilla Gigedo, Hawaii.

Hawaii can normally be reached in the course of migrations of this kind, and be reached this time *from the New World*, witness the dispersal of the hydrophyllaceous genus *Nama* (29),

(1) *N. dichotomum* — Subsp. *eudichotomum*: Northwestern Argentina, Northern Chile, Southern and Central Peru, Bolivia, Ecuador, Guatemala, Mexico (San Luís Potosí, Coahuila, Chihuahua), United States (Arizona, New Mexico, Colorado).

(2) *N. jamaicense* — Lesser Antilles, Portorico, Jamaica, Haiti, Santo Domingo, Bahamas, Bermuda, Florida, Texas; Guatemala, Mexico (Yucatan, San Luís Potosí, Coahuila, Tamaulipas).

(3) *N. undulatum* — Northwestern Argentina (Córdoba, Tucuman, Salta); Mexico (Oaxaca to Nuevo León, Cohauila, Chihuahua).

(4) *N. humifusum* — Mexico (Baja California, Comondu Island), United States (Southern California).

(5) *N. sandvicense* — Hawaii (Oahu, Maui, Kauai); var. *laysanicum*: Hawaii (Laysan Island; see Fig. 28a, with star).

It will be observed that *N. dichotomum* is once more disconnected, though rather more shortly this time, between Ecuador and Guatemala.

In conclusion, (a) A channel of migration competent to reach the westernmost Hawaiian islands (Laysan) takes its origin in South Africa or the islands adjacent (Mascarenes, Madagascar); (b) This channel, following a crumbled „antarctic" range, enters the New World — in the present case — in Eastern Argentina **); (c) From the point of entrance, this channel follows a double route *inland* and *oversea* of which the latter reaches the Galapagos Islands and Revilla Gigedo, also Hawaii.

---

*) Fearing to drown the reader at the outset amid a mass of details, we refrain as a rule from immediately discussing everything suggested by the course of a track and all the alternatives to which a track may give rise. We single out a main elementary aspect of dispersal at first, and do so pointedly. It may be noticed, however that this double „tracking," east and west in South America has profound significance, for South America indeed consists of phytogeographic domains, each with an individuality of its own. Western South America is not wholly the same as eastern South America to a phytogeographer. Characteristically, the migrations of the Labiatae which adhere to western South America are also those which reveal as a rule disconnections clear across the Pacific. The reason of this is given elsewhere (see p. 228). A vast crumbled range is in play, which once extended the western South American coasts much to the west, biting into what is now the open Pacific. See further the dispersal of *Nama*, next discussed. Much remains to be written on this issue.

**) In other cases (see discussion of *Juania*, p. 455), this channel points to Juan Fernandez Islands in the Pacific. Thus it is shown that the Magellanian gate formely mentioned (see footnote p. 63) is indeed operative east and west of South America, and intimately connected in this instance at least with the African gate.

It necessarily follows that, (a) Either landconnections of which no trace is left on our maps once brought together South America and the Mascarenes, not only, but the western shore of the Americas extended much farther west into the Pacific that it does today, ultimately leading to Hawaii; or, (b) The whole of dispersal must be entrusted to agencies entirely beyond rationalization, birds, winds, etc. flying unfathomable paths.

The former of these alternatives is necessarily to be accepted if we intend to have phytogeography *as a science*. The latter makes no sense, for it is not to be understood *how truly casual agencies of dispersal could constantly operate along the same paths and in the same manner*. Identity of consequences presupposes identity of prime movers in sober thinking.

Having cleared the eastern shores of the Pacific, and connected these shores with part at least of the African gate of angiospermy, let us turn next to the southern shores of this ocean.

The rubiaceous genus *Coprosma* (30) forms large part of the shrubby vegetation of New Zealand. It totals about 90 species, ranging as the following selection shows,

(1)   *C. Hookeri* — Juan Fernandez (Masatierra).

(2)   *C. pyrifolia* — Juan Fernandez (Masatierra, Masafuera).

(3)   *C. benefica* — Pitcairn Island.

(4)   *C. rapensis* — Rapa Island.

(5)   *C. oceanica* — Tuamotus Islands.

(6)   *C. taitensis* — Tahiti Island.

(7)   *C. strigulosa* — Savaii.

(8)   *C. persicifolia* — Fiji.

(9)   *C. novaehebridae* — New Hebrides.

(10)   *C. ciliata* — Campbell, Auckland, Antipodes Islands, New Zealand (South Island).

(11)   *C. acerosa* — Chatham Island, New Zealand.

(12)   *C. pumila* — Stewart Island, New Zealand, Australia (Victoria), Tasmania.

(13)   *C. Tadgellii* — Australia (Victoria).

(14)   *C. papuensis* — New Guinea (Mt. Scratchley).

(15)   *C. Wollastonii* — New Guinea (Mt. Carstensz).

(16)   *C. celebica* — Central Celebes.

(17)   *C. Hookeri* — Borneo (Mt. Kinabalu).

(18)   *C. sundana* — Eastern Java.

(19)   *C. montana* — Hawaii.

This dispersal (Fig. 29) shows, (a) An antarctic range in the Southern Pacific which fed *Coprosma* to all lands between Juan Fernandez and Eastern Java, either overland or by easy access *via* the sea, meaning over peninsular or insular "Landbridges"; (b) This range could possibly reach the Hawaiian archipelago (or its geologic equivalent) from the west (i. e., along the line Eastern Malaysia-Hawaii); (c) *Santalum* (see p. 143) migrates like *Coprosma*;

(d) The trend of this type of dispersal is all westward from the Americas (Juan Fernandez to New Zealand, Australia, New Guinea, Malaysia, ultimately veering again eastward to reach Hawaii) *).

The antarctic range in question is replaced today by an immensity of oceanic waters. It is clear, however, that it had real existence in the dim geological past, not only, but its outlines were such as to create a dividing line of dispersal somewhere in the immediate vicinity of the Marquesas, as proved by the following piperaceous species,

(1) *Peperomia marchionensis* — Marquesas; "Evidently related to *P. circinata* Link of Brazil" (31).

(2) *Peperomia rapensis* — Rapa; "Evidently allied to *P. latifolia* Miquel of Hawaii" (31).

(3) *Piper latifolium* — Marquesas, Society, Austral and Cook Islands, Fiji, Timor. **).

It will readily be seen that a center in the neighbourhood of the Marquesas fed either species (*P. latifolium*) or *genorheitra* to, (a) Brazil, (b) Hawaii; (c) Timor.

We may indeed go further. A good authority on the Piperaceae (32;) p. 8 informs us as follows, "The large subgenus *Sphaerocarpidium* (of *Peperomia*) is, in the main, composed of Central and South American species...... It appears that many of the species found in other parts of Polynesia and in the Galápagos and Juan Fernandez islands likewise belong to this subgenus...... The available evidence indicates, therefore, that the Polynesian species, including those of Hawaii, are more closely related with the species of the predominantingly American subgenus *Sphaerocarpidium* than with those of Malaysia, as represented by the Philippine species."

This text reveals that the component of *Peperomia* subg. *Sphaerocarpidium* in the Pacific gravitates on the whole toward the New World. The migration which *streams away* from the Americas as regards *Coprosma*, on the contrary *comes* to the New World in the case of *Peperomia*, both *Coprosma* and *Peperomia* being strongly represented in the Pacific meantime.

---

*) It may be noticed that *Nertera*, a rubiaceous genus close to *Coprosma*, occurs in the Polynesian and Malaysian range of the latter, not only, but also in Western America.

**) Months after this had been written, an article came to our notice (SKOTTSBERG, C. — „Peperomia berteroana Miq. and *P. tristanensis* Christoph., an interesting case of disjunction"; Meddel. Göteb. Bot. Trädg. 16: 250. 1946), illustrating the close affinity between a species of Juan Fernandez and Tristan da Cunha. Affinities of this kind are matched, of course, in other genera (see *Empetrum*, p. 161; see also discussion of the angiospermous gates, p. 56), and are a mere aspect of the standard dispersal between Africa and South America which is dealt with at length throughout these pages. The Piperaceae are alien, so for as known, to Tasmania, Victoria, South and West Australia, but occur in Eastern Polynesia, New Zealand, Chatham, Kermadec, Norfolk and Lord Howe Islands which is anything but surprising (see, e. g., *Plantago*, p. 285). Professor Skottsberg stresses the fact that part of *Peperomia* in Hawaii (Subg. *Hawaiiana)* has „Asiatic" rather than „American" affinities, which changes nothing in our conclusions (see discussion of *Phrygilanthus*, *Plantago*, p. 146). Characteristically enough, *Peperomia Urvilleana* (Kermadec, Lord Howe, Norfolk Islands, New Zealand) has affinities both toward *P. Ventenatii* of Java (see discussion of *Geniostoma*, p. 103) and *P. Berteroana* of Juan Fernandez (see discussion of *Santalum*, p. 143). In conclusion, nothing is novel or truly striking in the fact that Tristan da Cunha and Juan Fernandez number in their flora similar forms of *Peperomia*, and the existence of forms so distributed merely confirms our previous conclusions.

Juan Fernandez occupies a key-position in this migrational whirlpool. As the palms are to show us (see *Juania*; p. 455), Juan Fernandez is immediately connected with a massive center of origin of angiospermy at the Mascarenes. Juan Fernandez is also one of the starting points, when not the lone starting point, of the dispersal of *Peperomia* toward the heart of South America, witness,

(1) *P. Berteroana* — Juan Fernandez (Masafuera).

(2) *P. Skottsbergii* — Juan Fernandez (Masafuera).

(3) *P. margaritifera* — Juan Fernandez (Masatierra).

(4) *P. fernandeziana* — Juan Fernandez (Masatierra, Masafuera), Chile (Valdivia, Fray Jorge) *).

Fray Jorge is a well known Chilean locality wherein thrive under Lat. 30° 40' South (33) ferns and phanerogams of clean cut "antarctic" affinity, such as normally occurs only 500 and 600 kilometers to the south. The four species of *Peperomia* in question, therefore, lead from Juan Fernandez, and the Pacific, straight toward Northern Chile and the rest of South America.

We may conclude, then, that there is not the slightest doubt as to "antarctic" connections having once brought together the Mascarenes, Eastern Argentina, Chile, Juan Fernandez, the islands of Southern Polynesia, New Zealand, Tasmania, Antipodes, Macquarie, Kerguelen Islands. It is also certain that these connections sent the genorheitra *of the angiosperms northward to Malaysia, Hawaii, etc., in other words, to the northern hemisphere throughout.*

Having cleared the outlines of two sides of the Pacific, east and south, we are ready now to turn westward. The loganiaceous genus *Geniostoma* and its allies will make light for us in these quarters. We stress *Geniostoma* here, but abundant examples of groups that migrate like *Geniostoma* are scattered throughout the pages of this book. It is impossible not to be repetitious when dealing with dispersal, for all the essential evidence could be written in less than ten pages. *The balance of this evidence, thousands of pages if one cared to write them, would add nothing to the weight of these ten.*

Geniostoma numbers about a dozen species in Madagascar and the Mas-

---

*) In an article cited in the preceding footnote, Prof. Skottsberg states that only *P. fernandeziana* belongs to Subg. *Sphaerocarpidium*, and that *P. Berteroana* may be more intimately related to „Distant Old World species than to tropical American ones," although Prof. YUNCKER regards it, too, as a member of that subgenus.

Considering that *P. Berteroana* has an immediate ally in *P. tristanensis*, and that the Juan Fernandez palm *Juania* (see p. 455) has affinities in the direction of Africa, Prof. SKOTTSBERG may be entirely right. However, the discussion we give elsewhere In this manual (see p. 457) of the systematic position of *Juania* makes it abundantly clear that phytogeography is not vitally affected by these observations, which merely confirm the inference suggested to us by *Juania*. Professor YUNCKER, on his part, is beyond doubt in the right in asserting that *Peperomia* subg. *Sphaerocarpidium* gravitates toward the New World, for *P. fernandeziana* would alone prove this conclusively, even if it were to be shown that forms doubtfully belonging to this subgenus (e.g., *P. Berteroana*) have alliances also in other directions. Professor SKOTTSBERG further states that *Peperomia* is scarce in Africa, and that the African species have drupes pointing to affinities in the direction of Malaysia. This is of course less than mysterious, for a track Africa-Malaysia is standard throughout. We regret we cannot give the matter more than a footnote. *Croton*, too, is comparatively scarce in Africa, while its African species mostly show affinities toward the Far East.

carenes. This outlier would seem to be isolated throughout (see also *Vaccinium*, p. 169). Its affinities with the rest of the genus are speculative, and nothing is in the literature which justifies discussing them at this juncture.

Geniostoma teems in the Pacific, witness the following selection of species,

(1) *G. astylum* — Tahiti.

(2) *G. rapaense* — Rapa.

(3) *G. clavatum* — Raiatea.

(4) *G. Fleischmannii* — Samoa, Savaii.

(5) *G. macrophyllum* — Fiji.

(6) *G. foetens* — New Caledonia.

(7) *G. petiolosum* — Norfolk and Lord Howe Island.

(8) *G. micranthum* — Mariannes and Micronesia.

(9) *G. stenurum* — Carolines.

(10) *G. fagraeoides* — Bonin (Ogasawara) Islands.

(11) *G. glabrum* — Bonin (Ogasawara) Islands.

(12) *G. ligustrifolium* — New Zealand.

Leaving the Pacific, *Geniostoma* drives straight to Australia and Malaysia thus,

(1) *G. australianum* — Queensland (Atherton Tableland).

(2) *G. acutifolium* — New Guinea.

(3) *G. moluccanum* — Amboina.

(4) *G. avene* — Poeloe Gebeh.

(5) *G. celebicum* — Celebes.

(6) *G. oblongifolium* — Java.

(7) *G. arboreum* — Java, Sumatra.

(8) *G. Ramosii* — Philippines (Siargao Island).

(9) *G. mindanaense* — Philippines (Mindanao, Luzon: Sorsogon).

(10) *G. Cumingianum* — Philippines (Negros, Luzon: Sorsogon, Camarines Sur, Tayabas, Laguna, Zambales, Bontoc, Benguet; Mindoro).

(11) *G. brevipes* — Philippines (Mindanao; Luzon: Sorsogon; Leyte).

(12) *G. batanenses* — Philippines (Batan Islands; Luzon: Ilocos Norte).

(13) *G. kasyotoense* — Kashoto (Samasana Island *).

Few plants are more instructive than *Geniostoma*, for the direction and origin of the dispersal reaching Malaysia are fairly clear this time, (a) *Geniostoma australianum* is close to *G. petiolosum* and *G. rupestre* (Tahiti, Rarotonga, Tonga, Samoa, Fiji, New Hebrides, New Caledonia); and *G. arboreum* (Java, Sumatra) is akin to *G. ligustrifolium* (New Zealand). This identifies a line of

---

*) The nomenclature of this, and other petty islands off the southeastern coast of Formosa, gives rise to difficulties on account of alternative spellings. These islands are described in Japanese work (e.g., S. SASAKI, Dept. of Forestry, Govt. Research Inst. Taihoku, Rept. 21. 1937) as the „Islandsseries of Kotosyo" and generally made to include Kotosyo (Kotosho, Koto-Sho, etc.) also known as *Botel Tobago*; Syo-kotosyo (Shokotosho) or *Little Botel Tobago*; Kwasyoto (Kashoto, Kasyoto, etc.) or *Samasana*. Authoritative basic spellings (see S. GERR, Gazetteer Japan. Place Names, 1942) are (a) *Kashoto* for Samasana Island; (b) *Kotosho* for Botel Tobago proper. All these islands belong to the same main phytogeographic node, which is fortunate, as it often proves impossible to discriminate unless actual specimens carefully identified as to locality are available.

dispersal from Southeastern Polynesia (Rapa-Tahiti) running to a center, Fiji-Samoa, out of which the *genorheitron* is fed to Micronesia (see birds for homologous migrations; p. 70), Lord Howe Island, and through the latter to New Zealand, Northern Queensland and directly to Western Malaysia across Australia; (b) The Philippines are massively invested from the Pacific, and a connection is immediately established between these islands, the Batan Islands and Botel Tobago. This connections further invests the Bonin (Ogasawara) Islands, and may reach Borneo (Mt. Kinabalu: *Geniostoma* sp. in herbarium of the Arnold Arboretum of Harvard University).

Plotted on the map (Fig. 30) *) these connections further suggest, (a) New Zealand has a flora consisting of two elements, "cold" or "antarctic" and "warm" or "Polynesian." The former (see *Coprosma*, etc., and Fig. 29) reached New Zealand over an antarctic shore immediately south of these islands. The latter which *Geniostoma* (see also *Vaccinium*, p. 169) pointedly exemplifies, came to New Zealand on the contrary from a secondary center of evolution in the north (see also *Ilex*, p. 311) investing Tahiti, Tonga, Samoa, New Caledonia and Fiji, and penetrating New Zealand (also Queensland and New South Wales) from the immediate region of Norfolk and Lord Howe Islands, which is well known (34) as an important node of re-distribution, b) The Philippines were in contact with lands to the north and west through an involved geography stressing at least two ties, Luzon-Botel Tobago-Bonin (Ogasawara) Islands; Mindoro and Palawan-Northern Borneo-Hainan-Southeastern China. This geography gives every indication of having undergone sweeping changes in the course of geological time, and critical studies of phytogeography and ecology are necessary for its ultimate elucidation.

Geniostoma is unknown in Hawaii, but is there replaced by a close ally, *Labordea*, which numbers perhaps as many as twenty species. These two genera constitute with a third, *Logania*, the following tribe

(A) Loganioideae Loganieae

(1) *Geniostoma* — Madagascar and Mascarenes, massive in the latter; Polynesia, New Zealand, Australia, Malaysia to the Bonin (Ogasawara) Islands.

---

*) In this map (Fig. 30, *b*?), we outline two alternative tracks leading directly from the Indian Ocean to Western Polynesia. We suggest in the caption that these tracks might be responsible for feeding the Polynesian and Malaysian massings of *Geniostoma* from the west. This we do less because we believe that these massings were actually fed entirely from the Indian Ocean than because we intend to provide an alternative were it to be proved that one, or more, of the species of *Geniostoma* in this sector are immediately allied with the "stranded" outlier of this genus in Africa. This is a not to be ruled out aprioristically, but it strikes us, at a glance, as less than a strong probability.

**) We regret we can give all too little space to the phytogeography of the Philippines. Much has been written on its subject which merely has confused issues by a fundamental failure of orientation in the tracks to and from this archipelago, and, generally, because of lack of familiarity with the rudiments of dispersal. This archipelago is one of the most interesting domains of the world, and a book ought to be written upon its biogeography.

(2) *Labordea* — Hawaii.

(3) *Logania* — Australia throughout, New Zealand.

Another tribe of this family tabulates as follows,

(B) Loganioideae Antonieae

(1) *Usteria* — Tropical West Africa.

(2) *Bonyunia* — Brazil (Minas Geräes), Venezuela (Pacaraima Mts.)

(3) *Antonia* — Brazil, British Guiana.

(4) *Norrisia* — Malacca, Sumatra.

We may now conclude, reaching the end of our summary for the Pacific, as follows, (a) The center of origin of the Loganiaceae might be set south of Africa. The *genorheitron* left this center to go to the Mascarenes and Madagascar at one end; Southeastern Polynesia at the other; (b) From Southeastern Polynesia, the *genorheitron* fared westward finally to reach the region of Samoa and Fiji. It radiated from this region to New Zealand and Eastern Australia *via* Norfolk and Lord Howe Islands; to Malaysia generally, reaching northward toward the Bonin Islands; (c) The origin of *Labordea* is to be sought in a secondary center approximately located within the triangle Samoa-Philippines-Hawaii; (d) The *genorheitron* of the Antonieae immediately departed from the ancestral home south of Africa to invade West Africa, northern South America, at one hand, Malacca and Sumatra at the other *).

All these migrations are standard, and no longer require comment.

With this on record, we may at last return to *Quercus* and the Fagaceae, remarking; (a) *Nothofagus* is native to Magellania and Chile, New Zealand, Southeastern Australia and New Guinea This genus migrated consequently northward from a point in the "antarctic" shore ringing the Pacific to the south (see *Coprosma*). It is possible that it reached New Guinea through Eastern Australia; (b) *Quercus* originated very nearly in the same secondary center that gave birth to *Labordea*. Unlike this genus, however, it fared westward into Malaysia and continental Asia; eastward to the New World. The connections between these regions have been illustrated in preceding pages (see *Lysimachia, Perrottetia, Distylium, Vallesia,* p. 54) and it has also been shown that the region of Hawaii could be invaded from the east, i.e. from the New World

---

*) The reader is doubtless well aware that we do not moot in this summary the possibility that part of *Geniostoma* could reach Western Polynesia directly from the African Gate, i.e., the modern Indian Ocean. We do not take this possibility as strong because of reasons already detailed in the next-to-last footnote. However, the reader will do well comparing the dispersal of *Geniostoma* with that of *Vaccinium* (see p. 169), *Olea* (see p. 354) and *Protium* (see p. 390). The reader will duly see that, whatever be the alternative he may prefer in the course of the tracks, he will *always* face standard, thoroughly well tested contingencies in dispersal. We believe the reader may further agree that we interpret phytogeography in a manner which is both rigid *as to main premises*, flexible *as to details*. This, we believe, speaks for the ultimate soundness of our general treatment.

(see *Lepechinia, Nama*), which assures in principle perfect transit from Hawaii, or the lands which in a dim past stood for this archipelago, to the Americas and Asia.

In conclusion, it is *obvious* that *Quercus* need not be "Holarctic", and even less are "Holarctic" the Fagaceae.

To bring Formosa in touch with Spain is easy, as follows,

A) *Quercus* Sect. *Mesobalanus*

a) Subsect. *Malacolepides*

(1) *Q. dentata* — Formosa, China, Japan, Corea, Manchuria.

b) Subsect. *Ponticae*

(2) *Q. pontica* — Caucasus, Armenia.

c) Subsect. *Macrantherae*

(3) *Q. macranthera* — Persia, Armenia, Caucasus.
(4) *Q. Farnetto* — Asia Minor, Balkans, Transsylvania to Northern Italy.
(5) *Q. Toza* — France, Spain, Morocco *).

It does not seem necessary to pad the record in order to show that the disconnection of this group in the Himalayan range is probably due to local factors, but, if this disconnection must be bridged, we are surfeited with riches. Numerous oaks thrive in the Himalayas, witness for instance these six species of *Quercus* sect. *Lepidobalanus,*

(1) *Q. mongolica* — China, Manchuria, Corea, Eastern Siberia, Mongolia.
(2) *Q. aliena* — Manchuria, Mongolia, Corea, Japan, China.
(3) *Q. Kingiana* — Siam, Burma.
(4) *Q. leucothrichophora* — Himalayas throughout (Assam to Kashmir).
(5) *Q. sessilis* — All Europe to Sweden, Norway and Scotland, northward to about 60° Lat., southward to Crimea and the Caucasus; Mediterranean from Greece to Portugal through Italy, Corsica and Sardinia.
(6) *Q. infectoria* — Iran (Persia), Mesopotamia, Kurdistan, Asia Minor, Cyprus, Palestina, Morocco.

---

*) Indeed, abundance of plants connect Formosa with the Mediterranean or its immediate approaches, witness, for instance the Formosan *Euphorbia calonesiaca* which is distinguishable mostly by range from *E. orientalis* (Armenia, Western Caucasus); the euphorbiaceous genus *Mercurialis*; etc.

It would not seem that jays carried acorns to all these places, though ptero-dactyls of a rather early generation might have done so.

A monographer of the American Oaks (35) is of the opinion that the oldest forms of *Quercus* in the New World belong to *Quercus* sect. *Protobalanus*, which is distributed thus,

(1) *Q. chrysolepis* — Mountains of California into Oregon and Washington.

(2) *Q. tomentella* — Baja California (Guadalupe Island), California (Islands of San Clemente, Santa Rosa, Santa Cruz, Santa Catalina).

(3) *Q. Palmeri* — Baja California, California (San Diego Co.).

(4) *Q. Wilcoxii* — Arizona, Northwestern Mexico.

(5) *Q. vacciniifolia* — California to Oregon and Nevada.

The specific limits are so hazy that taxonomists are at hopeless odds as to what is a "good species" in this group. As for ourselves, we are agreed that forms so distributed may rank among the most ancient oaks in the New World, for they range at the abutment of the transpacific bridge in the approximate latitude of Hawaii. This is the reason why petty islands such as these loom so large in taxonomy and migration alike.

The oaks can be leisurely followed from the abutment of this bridge to their ultimate destination, as follows,

(1) *Q. peninsularis* — Baja California.

(2) *Q. tepicana* — Mexico (Nayarit).

(3) *Q. Eduardii* — Mexico (Nayarit, Durango).

(4) *Q. conspersa* — Mexico (Chiapas) to Guatemala.

(5) *Q. bumelioides* — Guatemala to Panama.

(6) *Q. tolimensis* — Colombia (Tolima).

In conclusion, nothing whatever is in the record of the fagaceous dispersal that may call for serious controversy. Plotted on the map (Fig. 27) in full freedom from preconceptions, academic issues, etc., this dispersal tells us that a *genorheitron* common to Fagaceae and Balanopsidaceae — we regret we can say no more here of their phylogeny and morphology — found in a land roughly contained within Formosa, the Eastern Himalayas, the Central Philippines and Java ideal grounds of secondary evolution. This *genorheitron* also took advantage of paths well worn to reach Eurasia throughout, and the Americas, both in the north (*Quercus, Fagus,* etc.) and the south *(Nothofagus)*. Obviously, coming to light from geologic shores in the Southern Pacific, this genorheitron dispersed *from the Pacific,* not from any fabulous "Holarctis," which is to us a truism rather than a genuine discovery.

It seems desirable to handle phytogeography on the basis of cleancut concepts and factual records. So handled, dispersal becomes easy, and monotonous in addition, because the same underlying causes (i.e., ancient geography) always bring forward the same results (i.e., tracks common to great many related and unrelated forms). Properly trained, a new generation of botanists will accept as self-evident conclusions which an older generations still is inclined to view as novel and "revolutionary."

VI. Magnoliaceae

The Magnoliaceae and their lesser segregates, Schisandraceae, Winteraceae, etc., closely repeat the migrations of the Fagaceae, and could be dispensed with if much had not been written to exaggerate their significance. This group is currently used as a paragon of "primitiveness" of the Angiosperms, and has honored place in literature as an authentic "Holarctic" phylum. We may deal with it less for all these reasons than because certain of the aspects of its dispersal in Malaysia are interesting.

Malaysia is indeed one of the most important phytogeographic zones of the world, and five continents can easily be reached (Fig. 31) from its boundaries. Strange to say, this region so intimately connected with every corner of the earth has so far been studied almost exclusively as a "local problem." It is obvious that the results thus secured have not lived up to anticipations.

Taxonomic and systematic botany treat the Magnoliaceae in different ways, but an informative conspectus may be arranged by grouping their genera around three key-forms, *Drimys*, *Magnolia* and *Illicium*, as follows,

A) *Drimys* affinity

(1) *Drimys* — 1 species in Celebes, Philippines and Borneo, 29 species in New Guinea, 6 species in Australia and Tasmania, 4 species in the Americas (Fuegia to Eastern Brazil, Venezuela, Guianas (Mt. Roraima), Central America and Mexico, mostly along the Andes).

(2) *Bubbia* — 19 species in New Guinea, 8 species in New Caledonia, 2 species in Queensland, 1 species in Lord Howe Island.

(3) *Belliolum* — 4 species in the Solomon Islands, 4 species in New Caledonia.

(4) *Pseudowintera* — 2 species in New Zealand and Stewart Island.

(5) *Exospermum* — 2 species in New Caledonia.

(6) *Zygogynum* — 6 species in New Caledonia.

B) *Magnolia* affinity

(7) *Magnolia* — About 50 species in Malaysia, Indochina, China, India and Japan; about 20 species in the United States, Mexico, Central America and the West Indies.

(8) *Talauma* — About 45 species in Malaysia, Indochina, China, India; about 12 species in the West Indies, Mexico and Central America, Colombia, Venezuela and Eastern Brazil.

(9) *Manglietia* — About 40 species in Malaysia, Indochina, China, India and China.

(10) *Michelia* — About 50 species in Malaysia, Indochina, China, India and Japan.

(11) *Elmerrillia* — 6 species in New Guinea, Celebes, Philippines.

(12) *Kmeria* — 3 species in Indochina and China.

(13) *Alcinandra* — 2 species in Indochina and India.

(14) *Pachylarnax* — 2 species in Malacca and Indochina.

(15) *Aromadendron* — 2 species in Java.

(16) *Liriodendron* — 1 species in China, 1 species in Eastern North America.

(17) *Paramichelia* — 1 species in Siam, China, India.

(18) *Svenhedinia* — 1 species in Cuba.

C) *Illicium* affinity

(19) *Illicium* — About 30 species in Malaysia, Indochina, India, Burma and China; 2 species in Mexico and the Eastern United States.

(20) *Schisandra* — About 24 species in Western Malaysia, Indochina, India and China; 1 species in the Eastern United States.

(21) *Kadsura* — About 25 species in Borneo, Indochina, India, China. It is clear by nothing deeper than a glance at the record that this affinity, (a) Has an "antarctic" (*Drimys*) group to match *Nothofagus* and the Balanopsidaceae. This component is larger and more varied in the Magnoliaceae than it is in the Fagaceae. *Nothofagus*, doubtless, was *the* modern genus in the dimmest night of the ages; (b) Reaches the New World in the manner of *Quercus, Lithocarpus* etc.; (c) Occurs in Malaysia with various genera, as do the Fagaceae. On these three counts, the dispersal of the Magnoliaceae is homologous of that of the Fagaceae. It may be observed that, as a matter of fact, all groups native to both sides of the Pacific, and not detected at the African gate, can be assumed to disperse in the same general manner *). This assumption requires verification case by case, but is valid as a working hypothesis. Transpacific migrations in the latitude of Hawaii are absolutely standard.

---

*) We suggest that, purely as a test of the methods and principles advanced in this manual, the reader rejects assumptions of the kind, and visualizes the origin of the Fagaceae and Magnoliaceae from a center of the Indian Ocean (e.g., the African gate of angiospermy). The reader will readily find that he can rationalize in principle the distribution of these two families as follows, (a) The *Nothofagus-Drimys* affinity departs from the African gate by the path of *Lepechinia* and feeds first South America northward. It next proceeds westward through the Southern Pacific to New Zealand, Australia and the Philippines along the lines of *Coprosma*; (b) The balance of the *genorheitra* involved leaves the African gate to hit Malaysia from the west; the Himalayas from the south, ultimately reaching Europe, Japan and crossing the Northern Pacific further to enter the Americas. Standard tracks to fit these contigencies are illustrated throughout the pages of this manual.

It could be said that, this being the case, we cannot be sure whether Magnoliaceae and Fagaceae did, or did not, actually originate in the Western Pacific rather than in the Indian Ocean. We will observe, (a) Considering that great many angiospermous groups originated by centers in either one or both of these oceans, it matters little indeed to dispersal, as such, whether one or two families more go one way or the other. That which primarily counts is that dispersal be understood *in an orderly manner, by standard tracks, and in the light of considerations that always leave us full scope to rationally choose among alternatives and hypotheses;* (b) The probability that Fagaceae and Magnoliaceae originated in the Indian Ocean is most remote, considering that neither one of these families is recorded from Africa, Madagascar and the Mascarenes.

Drimys piperita, the lone species within Malaysia proper migrates from Celebes to the Philippines sending meantime an outlier to Northeastern Borneo (Mt. Kinabalu). This track, too, is standard *), and can easily be brought back to New Guinea and Australia through the allies of *D. piperita*, as follows,

(1) *D. insipida* — Southeastern Queensland, Eastern New South Wales (Alpine references: Mt. Tambourine, Mistake Range, Mt. Spurgeon).

(2) *D. membranea* — Northwestern Queensland (Alpine references: Mt. Bartle Frere, Thornton Peak, Mt. Bellenden Ker).

(3) *D. hatamensis* — New Guinea (Alpine references: Mt. Tafa, Wharton Range, Lake Habbema, Arfak Mts.).

(4) *D. macrantha* — New Guinea (Alpine references: Wharton Range).

(5) *D. piperita* — Celebes (Alpine reference: Mt. Lompobatang), Philippines throughout (Alpine references: Mt. Apo, Mt. Urdaneta, Cuernos de Dumaguete, Canlaon Volcano, Mt. Halcón, Mt. Banahao, Mt. Mariveles, Mt. Data, Mt. Pulog, Mt. Pauai, Mt. Caua), Borneo (Alpine reference: Mt. Kinabalu).

It is manifest that the channel of migration followed by this genus is studded by mountains of every description, volcanic or not (e.g., Canlaon, Mt. Halcón), high and low (Mt. Kinabalu, Mt. Tambourine), and that some of these heights (e.g., Mt. Bartle Frere, Mt. Spurgeon) are more important from the standpoint of dispersal than from that of pure geography. Some of these landmarks are well known to all naturalists alike, witness Mt. Data (36) out of which came in a single haul a most remarkable crop of small mammals **).

A track such as we have just illustrated is by no means long, and the futility of studying dispersal as a local Malaysian problem is readily demonstrated with reference to certain plants of Celebes (37), as follows,

(I) *Styphelia (Leucopogon) suaveolens* — Southwestern Celebes (Mt. Bonthain), Philippines (Mindanao: Davao, Mt. Apo), Borneo (Mt. Kinabalu), Timor (Mt. Fetin).

This species is intimately allied with *S. pungens* of Eastern Java, and *S. wetarensis* of the Island of Wetar. The genus itself is large, for it occurs in Southeastern Polynesia (Marquesas, Rapa), Hawaii and Western Australia, with forms of the same species, *S. Tameiameiae*, endemic in the Marquesas, the Society Islands and Hawaii. The track of *S. suaveolens* is interesting be-

---

*) As we reveal in later pages (see p. 425), the range of *Drimys* is rooted in the "Papua Coign" of Malaysia. Worthy of careful notice is the fact that this "Coign" is shown by *Drimys* (and swarms of homologous migrants) to reach all the way southward to New Zealand at least.

**) It is a matter of regret that we cannot conduct comparative studies of dispersal in plant- and animal-life within the cover of this manual. Only recently came to our hands a study by VANDEL, A., on the distribution of the Isopods Crustaceans *(Oniscoidea)* (in Bull. Biol. France Belg. 79: 221. 1945), which deserves indeed extensive review. In Vandel's opinion, the *Oniscoidea* originated in "Gondwana" at the end of the Palaeozoic, or at the outset of the Mesozoic. We credit the inception of active angiospermous migrations to an era slightly later than the Mid-Mesozoic (i.e., Mid- to Late Jurassic). The type of dispersal shown by Vandel might serve *as well* — sometimes *perfectly as well* — for the Angiosperms. It is characteristic of the confusion now reigning in the field of biogeography that the *factual data* used by Vandel are discarded as of no significance by zoologists chasing the wisp-o'-the-will of "Extraordinary capacities for transoceanic dispersal," and such pabulum.

cause it runs a straight line from Timor to Celebes and the Philippines. Timor is a connecting link both with Java to the west and Celebes in the north, and re-distributes tracks that may come in from Australia or the African Gate.

(II) *Geranium ardjunense* — Southwestern Celebes (Mt. Bonthain), Timor, Java, Sumatra.

In this case, the track is thrown to the west, and Java with Sumatra are its beneficiaries. The Philippines, so far as known, are not in the picture.

A reliable authority informs us that *G. ardjunense* is very close to *G. nepalense* (37), and may in reality be a form of the latter. Another authority, maintains these two species distinct, on the contrary, under different sections having the following range (38),

A) Sect. Striatae

(1) *G. nepalense* — Himalayas (Sikkim), China (Kweichow, Yunnan, Szechuan), Japan (Honshu), Southern India (Nilghiri Mts., Pulney Hills), Ceylon, Nepal, Afghanistan.

(2) *G. Wallichianum* — Western Himalayas, Afghanistan, Corea.

(3) *G. rubifolium* — Himalayas.

(4) *G. Krameri* — Central China, Corea, Northern Japan.

(5) *G. Wilfordii* — China, Manchuria, Eastern Siberia, Southern Japan.

(6) *G. tripartitum* — Corea.

(7) *G. gracile* — Caucasus, Armenia.

(8) *G. striatum* — Caucasus, Asia Minor, Balkans, Southern and Central Italy.

(9) *G. nodosum* — Balkans to the Pyrenees through Northern Italy and Central France (Cevennes).

(10) *G. resimum* — Mexico (Nayarit).

(11) *G. Hernandezii* — Mexico (Oaxaca, Chiapas).

B) Sect. Australiensia

(1) *G. ardjunense* — Timor, Celebes, Sumatra, Java.

(2) *G. Traversii* — Chatham Island.

(3) *G. microphyllum* — New Zealand, Tasmania, Southeastern Australia; "Introduced" in Chile.

In our opinion, this dispersal leaves little to imagination. It takes its start somewhere between New Zealand and Tasmania, and invades Timor from Australia, next following to Java and Sumatra. Between these islands and Southern China the track is disconnected, perhaps on account of missing records, or, quite as probably, because tracks are wont to be disconnected in these quarters *).

A student of dispersal soon learns that plant-migrations disregard time and distance alike as factors of no consequence, and is necessarily bound to assume that Spain is next door to Celebes; Mexico but a step remote from Queensland. This student knows one map, *the world's*, and in function of this one map interprets all others. Investigators not informed of these essentials may hotly argue, whether *Geranium ardjunense* has affinities only in the north or the south of the map of the Far East, and may believe that their arguments call into play fundamentals. A phytogeographer will not share this delusion; he sees *Geranium ardjunense* as a single link only within a long, continuous chain of affinities and migrations stretching south to north between Chatham Island and the Himalayas. Naturally, a phytogeographer freely visualizes affinities for this species *either* south or north, or *both* south and north, not only, but is ready to understand that debaters seemingly at hopeless odds about classification discuss nothing in reality which is even mediocrely important. Numerous are the groups that formal classification cannot handle, which is demonstrated in the pages of this manual. Taxonomy labors under definite limitations, and by stressing this or that character in preference of another, classification can be made to speak the voice of any and all botanists. Ultimately, truth cannot be the plaything of opinion so tenuous and so personal; truth lies beyond the scope of formulae, names and minor arguments; it can ultimately be seen only in the light of certain fixed generalities of approach and method, around which formulae, names and minor arguments must of necessity revolve. Patently, nature "classifies" *Geranium ardjunense* in many ways, which all a phytogeographer easily understands, and this is enough for him to know.

(III) *Buchnera urticifolia* — Var. *typica:* Queensland, Thursday Island, New Guinea, Amboina, Boeroe, Celebes, Timor, Soemba, Java, Sumatra; Var. *glaberrima:* Celebes, Lombok, Philippines (Culion, Luzon).

Buchnera is a large scrophulariaceous genus, but we need not check every one of its type-localities to know it comes to Malaysia from the west, i.e. from the direction of the Indian Ocean. As we finger through these localities, we see the names of South Africa, Ethiopia, India, Indochina cropping up together with those of Brazil, Florida, Ontario. We know, then, that *Buchnera* is on a par with *Evolvulus*. The track is here the usual V- or trident-shaped

---

*) Regions visited by major geologic disturbances are frequently centers of disconnection. Classic is the instance of *Parrotia persica*, an hamamelidaceous plant isolated in a comparatively moist, warm region immediately south of the Caspian Sea. *Parrotia* thus survives along a line, Southern China-Mediterranean, which is tenanted by widely disconnected forms of *Platanus, Liquidambar, Cercis,* etc., none of which — so far as is known to us — now occurs in the region where *Parrotia* still thrives. This line can be shown (see *Prunus;* p. 332) to have been deeply modified by the Himalayan uplift and other geological disturbances. In other cases, a region greatly disturbed by geologic revolutions (e.g., Malaysia) may reveal disconnections in the range of a species, which disconnections (see *Potentilla anserina;* p. 147) are filled by vicariants. In conclusion, disconnections may be *absolute (Platanus, Liquidambar,* etc.) or *relative (Potentilla anserina)*. Are absolute the disconnections in which no vicariants exist to fill the gap in the track; relative the others.

affair that invests the New World and the Far East departing from the African angiospermous gate.

The two varieties of *B. urticifolia* would seem to have sound geographic basis. One of them, *typica*, comes in like the myrsinaceous genus *Maesa*, and takes Sumatra, Java, the Lesser Soenda Islands from the south, spreading to Thursday Island, a speck of land just north of Queensland, which for some reason or other is a landmark in migration. The other variety, *glaberrima*, neatly hinges on the track of *typica* at Lombok and from here proceeds northward to Culion and Luzon. This is all very simple, and Eastern Canada can be reached from Lombok in no time at all *).

The range, and intergeneric affinities, of *Magnolia* are illustrated by the following species,

(1) *M. liliiflora* — China, Japan, Southern Kuriles.

(2) *M. pulneyensis* — Southern India (Pulney Hills).

(3) *M. parviflora* — China, Quelpaert Island, Corea, Japan.

(4) *M. amoena* — China; allied with *M. stellata* (Japan).

(5) *M. Biondii* — China (Shensi); allied with *M. salicifolia* (Japan).

(6) *M. paenetalauma* — China; allied with *M. Championii* (China).

(7) *M. nitida* — China; allied with *M. Pealiana* (Assam).

(8) *M. pachyphylla* — Philippines (Palawan); allied with *M. Championii* and *M. Coco* (China)

(9) *M. pulgarensis* — Philippines (Palawan); allied with *M. Championii* and *M. Coco* (China: Kwantung, Kwangsi).

(10) *M. persuaveolens* — Borneo (Mt. Kinabalu); allied with *M. pachyphylla* (Philippines: Palawan) **).

(11) *M. Macklottii* — Sumatra, Western Java; allied with *M. Gustavii* and *M. Pealiana*, both of Assam.

(12) *M. annamensis* — Indochina (Annam); allied with *M. Griffithii* (Burma, Assam) and *M. Maingayi* (Malacca).

(13) *M. Clemensiorum* — Indochina (Annam); allied with *M. Coco* (China).

This tabulation cannot be better than informative, because a comprehensive monograph of this large genus is as yet not available. Even within its meager outline, nevertheless, this tabulation indicates that *Magnolia* ranges as far north as the Southern Kuriles, and extends southward and westward to Southern India and Malaysia. Immediate affinities connect eastern Central China (Shensi) with Japan to match (see p. 487) standard boundaries revealed by *Juniperus*. Malacca stands in relations with Burma and Assam; Palawan of the Philippines is immediately involved with Annam, Borneo (Mt. Kinabalu) and Southeastern China; Assam further is correlated with Sumatra and Western

---

*) The range Culion-Luzon is significant. See the discussion of the Philippine tracks of *Stackhousia* (p. 341) and *Drosera* (p. 435).

**) This alliance is relevant, for it highlights an immediate connection between Northeastern Borneo and beyond, and Palawan. This connection may well date from the first age of angiospermous dispersal, later on spoken of at length in the pages of this manual.

Java, indeed, a capital tie (see *Primula*, p. 83; *Schoenoxiphium-Cobresia* affinity p. 237, etc.). The sum total of the distribution is keyed to a line *Formosa—Assam*, which is normal for a group that invades the Far East and Malaysia from the direction of the Pacific, and probably uses the range Formosa—Luzon as a base to fan out inland throughout.

As we have learned, the triangle Malacca—Southeastern Sumatra—Western Borneo is of considerable significance as a center of evolution of the Fagaceae, because within this triangle occur the species that bring together as one *Quercus, Lithocarpus* and *Pasania* *). North of this triangle opens another center, which is quite as important as regards the magnoliaceous alliance. The generic lines among *Magnolia, Talauma* and *Manglietia* become so thoroughly blurred in Indochina (Tonkin to Cambodja) that authors of equal authority (39, 40) are at hopeless odd in their classifications. The same species changes genus back and forth, nor there is the slightest assurance that the latest in taxonomy is to endure a long while.

We reach, then, with these two centers a zone roughly contained within Tonkin, Malacca, Sumatra and Borneo, wherein Fagaceae and Magnoliaceae are not entirely out of the genorheitral stage. Their generic limits are at this point fluid still, so fluid indeed that taxonomists are hard put to pin them down. This is not surprising, because other regions of the modern map (for example, Southern California, Central Chile, Madagascar) are in the same case, and for the same reasons. As a matter of fact, it can be shown that the zone in which the *genorheitra* of these families are ill defined reaches farther north than the centers we have just outlined. The Southern Chinese *Magnolia Coco* and *M. Championii* are intermediate between *Magnolia* and *Talauma* because they exhibit the technical characters of the fruit of the former, but the habit and vegetative characters of the latter. Further: *Magnolia* and *Manglietia* are currently separated in taxonomy by their ovulation (2 ovules per carpel in *Magnolia*, 4 or more in *Manglietia*), but *Magnolia Paenetalauma* (China: Hainan) has the lowermost carpels of the fruiting heads bearing occasionally more than two ovules.

If we refer these "taxonomic headaches" to a map, we have before us (see Fig. 32) a range including Hainan—Palawan—Annam—Malacca—Sumatra—Java—Borneo, which is patently of the utmost significance for the secondary evolution within Malaysia of groups so large and important as the Fagaceae and one Magnoliaceae. Right by this range opens a second one (see Fig. 32) not less important, because it rates as the hinge of one of the main doors of access into Malaysia and the Far East of many *genorheitra* coming in from the direction of the open Pacific. Ranges of the kind find their counterpart, as we know, in Southern California and Northern Mexico at the opposite side of the same ocean.

---

*) Characteristically, this very same triangle is also relevant for the dispersal of mangroves (see p. 402) and their peculiar plays around Java. In addition, this triangle is crossed, sometimes in disconnection (see p. 438) by forms that stream northward from a point meridional to the Lesser Sunda Islands, hitting certain petty islands in the vicinity of Cochinchina and Siam. Accordingly, this triangle is important both as regards primary and secondary dispersal.

A detailed study of Malaysian dispersal is, unfortunately, here out of question. Other groups can be found, however, to match the behaviour of those mentioned, and one of them is indeed worth notice. *Tilia* has a strong center of evolution in Southeastern China and lands adjacent, rich in archaic forms. The southermost species of *Tilia* occurs in Indochina (Tonkin: Region of Chapa, here *T. mesembrinos*); *Tilia's* lone direct ally, *Schoutenia* (41) is distributed thus,

(1) *S. hypoleuca* — Indochina (Cochinchina, Cambodja, Annam).
(2) *S. peregrina* — Siam.
(3) *S. ovata* — Java, Borneo.
(4) *S. Kunstleri* — Malacca, Java.
(5) *S. glomerata* — Malacca.
(6) *S. accrescens* — Malacca.

We regret we cannot explore the dispersal of the Tiliaceae at length, but it is obvious (which the reader may verify) that the Tiliaceare are one of the families which issued from the African gate, meaning, in the last resort, a continent that filled at one time the whole of the modern Indian Ocean. The reader will hear a great deal more of this lost Gondwana as our work progresses, and may notice meantime (see p. 318) that it is by modern Madagascar that come to a head the *genorheitra* of Flacourtiaceae, Passifloraceae, Caricaceae, etc., further observing that Flacourtiaceae and Tiliaceae are demonstrably closely allied. The dispersal taking place to Malaysia, then, reveals that a marked center of evolution exists for *Schoutenia* in Malacca — which is standard also of other groups —, and that *Schoutenia*, the immediate "tropical" tilioid genus occurs *south of Annam*, while the "temperate" *Tilia* begins its run immediately *north of Annam*. The region we have stressed as a sensitive center of evolution and re-distribution of the quercoid and magnolioid alliance is accordingly also involved as regards the immediate tilioid *genorheitron*.

It could be shown at the price of enlarging this manual much beyond its allotted limits that very much the same ranges are critical in the distribution of, e.g., the Lauraceae. Indochina is a factual bridge between the "tropical" south and the "temperate" north. Obviously, it is to be excluded that the climate of these regions was changed by wild "Polar Shifts" in the Wegenerian manner in the whole of the eras that began with the Late Jurassic. Had these changes occurred, the facts of nature could not be what they are *).

---

*) So precise and repetitious is the evidence that the climate of Malaysia and adjacent Far East remained practically constant ever after the Late Jurassic, at least, that we cannot understand the work of authors (e.g. JARMOLENKO, A. V., in, Materials on the History of the Flora and Vegetation of the USSR 1: 375. 1941) who stress "The great importance of the displacement of the solar zones and the corresponding changes in the solar climate for palaeogeography." It is probable that some of the original northern lands gained in extension northward while losing southward in the course of geological time, or were "floated" northward somewhat, but it seems to us sure that the poles never "wandered" in the manner of Koeppen & Wegener. There is no evidence whatsoever to favor this "wandering" at least *in the whole range between New Zealand and Japan.* Authors who believe the contrary should take care to square their conclusions with the facts revealed by regions other than those with which they might be familiar in the first, and perhaps last, place.

To summarize, the observations just made lead us to conclude as follows, (a) The South China Sea did not exist at all when the first angiospermous *genorheitra* came into Malaysia, nor did it exist ages after then. A land stood where this sea now stretches, and this land was the seat, seemingly, of a mighty hydrography (cf. "Molengraff River"; Fig. 32). This land served a center of secondary evolution and redistribution of the *genorheitra* arriving to Malaysia, because it had a varied ecology and different climates, both on account of latitude and topography; (b) The supposedly "Himalayan" flora of the Sumatran Gajolands (42) and certain regions of Luzon of the Philippines is in actuality a mixture of "cold" forms from two directions, (i) The west, that is to say, the Indian Ocean; (ii) The east, that is to say, the Pacific Ocean; travelling standard tracks throughout. Reference to Fig. 64 proves that conventional phytogeographers confusely identify as "Himalayan" elements of possibly very different origins, and thus stultify the basic run of the major channels of distribution inside and outside Malaysia alike; (c) The "Australian" flora of Mt. Kinabalu, Mt. Gunong Tapis, etc., came on the contrary either from Northern Australia or New Guinea, which is to say always from the direction of the Pacific; (d) Malaysia and the Caribbeans have homologous position in dispersal, for both are invaded by channels of migrations stemming in the main from two oceans; (e) Malaysia and the Caribbeans are secondary, not primary centers of evolution. This proposition can easily be proved true as regards Malaysia in reference to *Tilia* and *Schoutenia*. The primary center of the evolution of the Tiliaceae lies in one and the same range, where also came to light the Flacourtiaceae, Sterculiaceae and Elaeocarpaceae, which rate as the Tiliaceae's nearest allies. We know where the Flacourtiaceae originated (see p. 318), and the Elaeocarpaceae are in addition (19) *conventionally* "antarctic." Nothing is here to justify the statements that the origin of this broad *genorheitron* is either "Malaysian" or "Holarctic." However, there can be no doubt as to *Schoutenia* having matured as a modern genus — which is to say, *to have ultimately originated* — in the lands fronting the modern South China Sea. *Tilia*, on its part, must have reached Indonesia when in a comparatively advanced state of evolution. At that time, *Tilia* was formed well enough to dispatch additional strong specific nuclei to the modern Balkans and the Caucasus, and the New World. At that time, too, *Tilia* merely "crossed" Malaysia to establish a strong center of speciation, possibly sectional affinities, in modern Southeastern China. There is accordingly absolute parallelism between the trends of evolution within Malaysia of Fagaceae, Magnoliaceae and Tiliaceae, three families which, very much mistakenly, are assigned to different levels of "primitiveness" by most systematists.

Finally to dispose of the notion that the Tiliaceae can be "Holarctic", we may conveniently refer to a monotypic genus of this family, *Hainania*, localized in the Chinese Island of Hainan. This genus has affinities with *Pityranthe, Christiania* and *Tahitia*. These genera belong to a Subf. Brownlowioideae (41) which is distributed along the following lines,

A) Tribe Berryeae

(1) *Berrya* — 3 species in India, Ceylon, Andaman Islands, Christmas Island, Indochina, Burma, Java.

(2) *Carpodiptera* — 3 species in East Africa (Zanzibar and the Comoros), 4 in the West Indies (Cuba, Haiti, Saint-Vincent, Grenada) and British Honduras.

B) Tribe Christianieae

(3) *Asterophorum* — 1 species in Ecuador.

(4) *Tahitia* — 1 species in Tahiti.

(5) *Christiania* — 1 species in Madagascar; 1 species in the Congo, Upper Sudan, Cameroon, Senegambia, Brazil (Amazonas), British Guiana (Mt. Roraima).

C) Tribe Brownlowieae

(6) *Brownlowia* — 17 species in Northern India to New Guinea; Malacca, Borneo, Philippines, Moluccas.

(7) *Diplodiscus* — 2 species in the Philippines.

D) Tribe Pentaceae

(8) *Pityranthes* — 1 species in Ceylon.

(9) *Pentace* — 1 species in the Philippines.

This is a perfect pattern of distribution from the African Gate matched throughout the pages of this manual in scores of other groups. The stations classic of this pattern follow in approved uninterrupted sequence, Madagascar, Comoros, Zanzibar, Congo, Upper Sudan, Cameroon, Senegambia, Brazil (Amazonas), Guiana (Mt. Roraima), Lesser Antilles, Haiti, Cuba, British Honduras at one end; Ceylon, India, Andaman Islands, Christmas Island, Burma, Indochina, Malacca, Java, Borneo, Philippines Moluccas, New Guinea, at the other. The lone jarring note is contributed by two small genera in one and the same tribe, *Asterophorum* and *Tahitia*. These two genera are in such a position as to suggest that Eastern Polynesia and Ecuador were reached immediately from the Eastern Pacific, that is to say, by an extension of the Magellanian gate over a now crumbled range projecting from the New World far into the modern ocean. An outlier of this nature in a group so distributed as the Brownlowioideae is normal not only, but immediate ties between the Far East and western South America further are known in the Lardizabalaceae and Chloranthaceae, reviewed elsewhere in this manual. *In brief, nothing is ever new in dispersal once the prime movers are understood.* On the strength of observations contained in a coming chapter (see p. 189), we may here suggest the hypothesis that the Tiliaceae are primarily near-mangroves, remarking that the distribution of the Brownlowioideae agrees in the main with the migrations of this kind of plants. We may also anticipate here the inference that through the Flacourtiaceae (see p. 318) the Brownlowioideae are allied by *genorheitron* with the Passifloraceae. Considering that the Passifloraceae might have invaded the Americas from the Eastern Pacific, the behaviour of *Asterophorum* and *Tahitia* is all the more understandable. This behavior is further proof that we are not entirely mistaken in postulating a considerable accretion of what is now "Andean" America toward the Eastern Pacific. Were we to be told that geologists and geophysicists do not agree *at present* with us on the score of this accretion, we would not alter our opinion at all but wait, in the anticipation that masses of plants cannot lie. *To us, plant-life is a geological layer of the earth.*

When speaking of the fluid generic limits of the Magnoliaceae endemic to certain lands fronting the South China Sea we mentioned *Talauma*, remarking that the Southern Chinese *Magnolia Coco* and *M. Championii* are intermediate between that genus and *Magnolia*. Peculiarly, it would seem that authentic *Talauma* is not known from China at all. We advance this statement with some doubt, considering the uncertainty of the classification of the Magnoliaceae, but it seems well supported from all the evidence that has been available to us. It is not imprudent to affirm, at any rate, that if *Talauma* occurs in China, it is there scarce.

The fact, we repeat, is peculiar, because climatic reasons cannot be its root. *Talauma Hodgsonii*, for example, is distributed to Indochina (Laos), the Eastern Himalayas and Northern Bengal; and the Eastern Himalayas, broadly speaking, harbor a type of flora which has a great deal in common with the Chinese province of Yunnan. There is at least a tenuous indication that the factual exclusion of *Talauma* from China stems from a type of migration at variance with that of other magnoliaceous genera. The Indochinese province of Laos has a peculiar flora, and the occurrence of the northernmost species of *Talauma* within its limits may be pregnant with hidden meanings. We regret we can say no more at present.

Whatever the case be in regard of *T. Hodgsonii*, *Talauma* is not abundant in Indochina, either, where we find,

(1) *T. nhathrangensis* — Annam.
(2) *T. Duperreana* — Cambodja.
(3) *T. Candollei* — Cambodja, Siam, Sumatra, Java.

South of Indochina, however, *Talauma* comes into its own, as follows,

(1) *T. gracilior* — Malacca; allied with *T. Candollei* (see above).
(2) *T. peninsularis* — Malacca.
(3) *T. betongensis* — Malacca, Southern Siam.
(4) *T. villosa* — Malacca; allied with *T. intonsa* (Borneo: Sarawak).
(5) *T. singaporensis* — Malacca, Bangka, Simaloer Island.

(6) *T. rubra* — Sumatra (Gajolands).

(7) *T. soemboensis* — Lesser Sunda Islands (Soemba); allied with *T. oblanceolata* (Malacca).

(8) *T. Elmeri* — Borneo (Tawao).

(9) *T. gigantifolia* — Borneo (Tawao).

These localities are exceedingly interesting, singly and as a whole, and we regret that too little is known of this genus and its distribution as yet. Merely as an introduction, we remark, (a) Tawao is a significant center in dispersal, which invites attention; (b) Some connection, as yet unexplored (see, however, footnote p. 139), exists between the Gajolands and Laos, not only, but Luzon; (c) Malacca is a center of secondary speciation of *Talauma* as it is of *Schoutenia;* (d) While the dispersal of *Magnolia* is to the north, that of *Talauma* is manifestly to the south. There is a zone of taxonomic twilight between these two genera in Southern China, but they clearly are "ecotypes" separable as to climatic preferences (*Magnolia* is "cold," *Talauma* "tropical").

The question, then, arises, how the archetypes of *Talauma* reached Malaysia. These archetypes, unlike those of *Tilia* and *Schoutenia*, are manifestly from the Pacific, but it is not likely that they came in through Formosa and Luzon on the last lap of their migrations. The door of their entrance is to be sought elsewhere.

On the purely Pacific side of Malaysia, *Talauma* is by no means unknown, witness the following,

(1) *T. oreadum* — New Guinea (Arfak Mts.).

(2) *T. gitingensis* — Philippines throughout from the Babuyan Islands to Mindanao and Palawan; Northeastern Borneo.

(3) *T. reticulata* — Philippines (Dinagat Island).

If we are not greatly mistaken, the archetypes of *Talauma* invaded Malaysia taking their start from the vicinity of modern New Guinea. They next entered the Southern Philippines (Dinagat Island), and from this point on passed to Northeastern Borneo, next fanning out to Soemba, at one hand, Malacca at the other. This is one channel of penetration *). A second might have been established to the north, running from the Babuyan Islands or Hainan to Palawan, next to Laos and the Himalayas. In brief, regardless of details, *Talauma* came into Malaysia piercing the front of the Philippines. It is possible that further reports of the genus from the islands fronting the Banda Sea reveal still another approach to Malacca, running this time directly from New Guinea to Soemba, and from this point on to Malacca. Whatever the case, and its details, it is fairly sure that *Talauma* invaded Malaysia using a sector south of the one overrun by the archetypes of *Magnolia* In other words, at the time when the magnoliaceous *genorheitron* invaded Malaysia from the Pacific, the modern genera already existed in embryo, as differentiating

---

*) The reader may compare this channel with the tracks run by *Buchnera* (see p. 113), *Stackhousia* (see p. 341), and *Drosera* (see p. 436). Obvious inferences will be suggested for his attention by the facts on record.

"ecotypes" moving within certain geographic, ecologic and climatic belts to their own convenience.

Before closing, let us consider the remaining three magnoliaceous genera, *Manglietia*, *Elmerrillia* and *Michelia*.

Manglietia, as we know, is connected with *Magnolia* by a species endemic to Hainan (*Magnolia paenetalauma*). The center of maximum speciation of *Manglietia* is, so far as known, located in the Chinese province of Kwangsi. This genus is not in the Philippines, nor in Borneo, but appears to reach Western Malaysia like *Schisandra* (see Fig. 34).

Michelia has one species at least in New Guinea (*M. Forbesii*). A single species in two forms (*M. compressa* and *M. philippinensis*) widely ranges from Mindanao through Cebu, Luzon, Formosa and the Riu-Kius to Central Japan. *Michelia* is further in China, and from China ranges far and wide to Cambodja and Ceylon (*M. mediocris*: Hainan, Kwangsi, Tonkin, Cambodja; *M. Doltsopa*: Yunnan, Eastern Himalayas; *M. nilagirica*: Southern India, Ceylon), finally reaching Western Malaysia (*M. montana*: Malacca, Sumatra, Java; *M. Scortechinii*: Malacca, Sumatra; *M. sumatrae*: Sumatra).

Elmerillia, which is a close ally of *Michelia*, is distributed in this manner,

(1)   *E. papuana* — Northeastern New Guinea, Seroei Islands.
(2)   *E. ovalis* — Celebes, Philippines (Mindanao, Leyte).
(3)   *E. pubescens* — Philippines (Mindanao).
(4)   *E. mollis* — Southeastern Borneo.
(5)   *E.? sericea* — New Guinea.
(6)   *E.? Vrieseana* — Celebes.

We may now conclude on the strength of the facts before us. These facts are far as yet from complete, and whatever we are to say must be qualified on this account. However, it is probable that the distributions of *Magnolia*, *Talauma*, *Manglietia*, *Michelia* and *Elmerrillia* known to us are broadly representative, so that the margin of error in our conclusions cannot be such as to invalidate them throughout.

We see the following as sure or most probable, (a) The magnoliaceous *genorheitron* has its origin in the Pacific, and the modern genera began to take recognizable form (that is, originated in a first time) in a center off modern New Guinea (see 1, Fig. 33); (b) From this center, they closed on Malaysia in two groups already in process of advanced evolution. One of these groups formed by *Talauma* and *Elmerrillia* ultimately invaded Malaysia on a front between two points, Luzon and Soemba; the other, constituted by *Magnolia* and *Manglietia* came to the Far East and Malaysia north of Luzon, possibly along a front opening between Formosa and the Riu-Kiu Islands, or, alternatively, through Luzon and Formosa; (c) *Michelia* seems to have retained a foothold in New Guinea (*M. Forbesii*), which is significant. It penetrated the Far East along the front Mindanao-Formosa (or the Riu-Kiu Islands); (d) *Magnolia*, *Manglietia* and *Michelia* reached in time Western Malaysia mostly from the north. It is not at all likely that they ever crossed over to this sector of Malaysia in the direct manner of *Talauma;* (f) *Talauma*, on the contrary, invaded Western Malaysia by a nearly straight line, departing in the main from a point intermediate between New Guinea and the Philippines; (g) *Elmerrillia*, a close ally of *Michelia* came into Eastern Malaysia, along lines comparable with *Talauma's*. This is significant, for it tends to suggest, if nothing more, that *Michelia* and *Elmerrillia* parted company precisely in the region which served *Talauma* as a base to invade Eastern Malaysia; (h) When these migrations were effected there was as yet no "Wallace's Line," or similar boundary to separate Malaysia in halves (*Papualand* and *Sundaland*), nor any South China Sea, which is proof that these migrations are anterior to the disruption of Malaysia (see further p. 123); (i) It is probable that some mighty orogeny running north to south in, or by, what is now the South China Sea eased the southern migrations into Western Malaysia of "cold" genera like *Magnolia*. Ties between the Gajolands and Laos remain to be fully explored. (j) Once in Malaysia and the Far East evolution ran its course, but at a comparatively slow clip, in itself an indication that Malaysia and the Far East are not centers of *genorheitral* origins. The segregation was mainly effected along "ecotypical" lines, with slight morphologic changes on the whole. Evolution and segregation could be accurately diagnosed and interpreted, if we possessed data of ecology and actual range which are not available to us now, and hardly transpire in the literature. *Michelia* may have special ecological requirements depending less upon climate than other edaphic factors. Climate on the contrary would seem to have been the dominant factor of segregation in the remaining groups. As stated, this factor was operative before the affinity of *Magnolia* had set foot in the Far East and Malaysia, because the dispersal was in its remote origin mainly molded by climatic considerations, which influenced the modern genera to enter lands north or south at the first.

One of our conclusions has been to the effect that when the archetypes of the Fagaceae, Magnoliaceae and Tiliaceae first reached what is now Malaysia, Malaysia was as yet not rent asunder in halves by some "Wallace's Line," nor was the South China Sea in existence. Dealing with the fresh-water-fishes of the Philippines — which are known mainly from the islands of Palawan and Mindanao — an ichtyologist (43) writes as follows, "It is self evident that Palawan and Mindoro were populated by Cyprinidae from Borneo, via Balabac. It is likewise that the migrations must have occurred when Palawan had a greater area than at present and probably at a time when the southern end of the China Sea was land, and a great river flowed northward, draining Borneo and Sumatra. At that time with large streams in Palawan, it would have been comparatively easy for fresh-water minnows to extend their range northward. The relatively large number of endemic species in Palawan shows that this condition ceased long ago . . . . The Cyprinidae of Mindanao entered from Borneo over a Sulu land bridge. Cyprinidae occur on Tawitawi and Basilan, but Jolo, the only other Sulu island with permanent fresh-water creeks, apparently lacks Cyprinidae, as numerous collectors have failed to obtain them there; this is probably due to the greater amount of volcanic activity on this mall land mass in the Pleistocene or Recent time . . . . The endemic species at Zamboanga, the nearest point of Mindanao to Borneo, furnish proof that connections ceased a long time ago, and the numerous changes in elevation that most of the Sulu islands have undergone have destroyed the Cyprinidae if any ever occurred there . . . . We may be reasonably certain that if there ever was any connection between Formosa and Luzon it was a very ancient one, and was broken a long time ago, so that the only land connection by which fresh-water fishes could arrive in the Philippines was that with Borneo."

Palawan, then, was flanking in the east a large valley run through by a mighty system of rivers, possibly having its headwaters between Sumatra and Borneo, or part in this range, part in Southern China. The traces of this hydrography would seem indeed to survive etched in the bottom of the modern South China Sea as the so called "Molengraff River" (44; see Fig. 32). It is currently believed that this hydrography together with the lands adjacent to it was drowned only in the Tertiary, but dispersal indicates it might have been otherwise. Malaysia, as dispersal intimates, must have been in active disruption at an early date, as consequence most likely of the breaking up of Gondwana, and we should be inclined to think that the valley we have sketched in former paragraphs was drowned, or somehow altered by onrushing sea-waters, not later than the beginning of the Tertiary, when indeed not earlier still. The trends in the discussion by the ichtyologist just quoted also imply a belief on his part that connections between Borneo and the Philippines crumbled "Long ago" (meaning in reality the collapse of an earlier type of geography). "Long ago" is indeed a long time, as geologic eras are necessarily figured out *).

The presence of fresh-water fishes is used as a test whether an island is "continental" or "oceanic," it being assumed that an "oceanic" island cannot have true fresh-water fishes. Were this criterium correct, Jolo must be "oceanic," Luzon "oceanic", Mindoro possibly half "oceanic" and half "continental" on account of its possessing only one species of minnows. Palawan, contrariwise, is to be hailed as fully "continental," so also Mindanao. This is all useless, the transparent byproduct of preconceived definitions as to "what" must be "which" in nature.

The remaining Asiatic genera of the affinity centering around *Magnolia, Kmeria, Alcinandra, Pachylarnax, Aromadendron, Paramichelia* hardly require discussion. Had we reliable data regarding their ecology, ranges in detail, morphology and taxonomy we might try to interpret them for the pur-

---

*) We cannot unduly extend the compass of these observations. We observe, nevertheless, that in the Sulu Archipelago two islands, Tawitawi and Basilan, have Cyprinidae, whilst a third island, geographically intermediate between them, Jolo, has none of these fishes. Curious, then, is the fact that Tawitawi is a strategic point in migrations effected by plants at the time when Malaysia was in active crumbling already. See the dispersal of the dipterocarpoid *Vatica*, p. 428. We are certain that most fruitful comparisons can be instituted among all branches of biogeography as soon as biogeographers are weaned out of the faith in the "occasional" and the "miraculous" which, alas, is connatural to many of their current "interpretations".

poses of phytogeography. So much is clearly out of the question just now.

Liriodendron is in two species, *L. chinense* (East Central China: Kiangsi, Hupeh); *L. Tulipifera* (Florida, Southern Alabama, Mississipi, Louisiana, Tennessee, North Carolina, Missouri, Northeastern Arkansas, northward to Canada (Southern Ontario), Southern Michigan, Vermont, Massachusetts). Little need be stated of this genus. Its dispersal is classic, and in our opinion was effected straight between the southern approaches of Japan and the Southern Appalachians and Ozarks of the U.S.A.; Southern Ontario is the normal term northward of migrations of this sort. It may well be that fossil records of *Liriodendron* north of the line California-Arkansas stem out of forms originally different from *L. Tulipifera*, which has all the earmarks of a relic. It will be observed that *Platanus* (see p. 139) still retains a strong component in the Southwestern United States and adjacent Mexico, while sending a species, *P. occidentalis* eastward and northward very much distributed in the manner of *L. Tulipifera*. Characteristically, some authors recognize a distinct variety of *Platanus occidentalis* (*P. occidentalis* var. *glabrata; P. glabrata* of other authors) distributed to Western Texas and adjacent Mexico (Coahuila and Nuevo León). In brief, the original dispersals of the Platanaceae and *Liriodendron* closely match, at least within the American range, which suggests an homologous general trend of transpacific migration. We believe that fossil records of *Liriodendron* (and *Platanus*) north and south of a belt running between California and Arkansas stem from secondary migrations out of this belt. As an alternative, we might consider the possibility that Alaska and Southern California were reached simultaneously over a track "fanning out" from the direction of the modern Pacific. The main point at stake remains, in our deliberate judgement, that Alaska never was *the* "Landbridge" between the Far East and the New World, which its present geographic position tends to suggest. Obviously, to admit the probability of a fanning out of the tracks of genera migrating like *Platanus* and *Liriodendron* to Alaska and Southern California simultaneously, we must accept as selfevident the existence of landconnections of the most intimate nature between the Far East and the New World almost all along the front Alaska—South California.

We think the dispersal of the affinity centering around *Drimys* need not be given painstaking discussion. The reader will, merely from glancing the record over, conclude that the affinity in question is so outspokenly "antarctic" as to leave no doubt as to its dispersal. *Drimys* is a classic "Australian" element, which moving in the pattern of migration of elements of the sort reaches Celebes and the Philippines, not only, but Eastern Borneo as well. In the New World, *Drimys* stresses the western half of South America, which is conventional, dispatching — to match in this *Araucaria* — an outlier to Eastern Brazil. It is altogether possible that this outlier fed the "mesas" of the Guianas (Mt. Roraima), because, as we know, Eastern Brazil and the Guianas were as a rule more intimately related among themselves than either was in origin with western South America.

It is clear that *Drimys* and its allies are out of the Western Polynesian and Magellanian gates, and that the center wherein they parted company with the affinity of *Magnolia* is situated primarily slightly off, and possibly south of, modern New Guinea. This is also the center which acted to separate the affinity of *Quercus* from that *Nothofagus*, indeed one of the great centers of angiospermous origins in the Western Pacific. As the reader is to see (p. 438), parallel if less important centers of evolution exist in the Pacific north of this one. This is standard, for the same happens in the Indian Ocean, the Atlantic and the Eastern Pacific that takes place in the Western Pacific. Doors of entrances, and secondary centers of origins and distribution are staggered north to south throughout these bodies of water.

On the western side of the Pacific, *Illicium* is first seen in the Philippines. It is unknown, or at least unreported, from New Guinea. The Philippines species are the following,

(1) *I. philippinense* — Luzon (Zambales), Mindoro (Mt. Halcón, Mt. Tapulao).

(2) *I. montanum* — Luzon (Bontoc) *).

More species are in Formosa, namely, *I. arborescens, I. daibuense, I. leucanthum, I. randaiense*. One of them, *I. japonicum* extends from Formosa to the Riu-Kiu Islands, Southern and Central Japan, Southeastern China (Kwantung). In China not less than eight endemic species are reported, one at least, *I. Griffithii*, reaching also the Himalayas. *Illicium Griffithii* is said to be very close to *I. cambodianum* (Burma, Tonkin, Annam, Malacca), and to stand as the core of a galaxy of forms that ultimately ranges as far as Sumatra and Borneo.

A dispersal of this type is read without difficulty, because it belongs with *Magnolia's* in fundamentals. The track comes in from the direction of the Pacific, and enters the Far East somewhere between the Riu-Kius and Luzon, or Formosa and Luzon. Having secured a foothold right in these centers, the migration spreads farther to Japan, the Himalayas and Western Malaysia. Luzon feeds Mindoro; Borneo might be reached either from Mindoro or Western Malaysia (Sumatra, in the main), in wholly standard patterns of secondary migration. It will be observed that Palawan is left untouched by *Illicium*, but Mindoro is invaded which confirms previous indications in our hands that Palawan had its most intimate ties in an earlier epoch of the Malaysian and Far Eastern dispersal with Northeast Borneo, Hainan and Southern China than with Luzon. Formosa, Luzon, Mindoro stood together; Northeast Borneo, Palawan, Hainan, Southern China and Indochina seemingly

---

*) The correlation between heights and forms of very ancient dispersal is so intimate that, rejuvenated or not, the heights necessarily appear to a phytogeographer like *pieces of land geologically old lifted up together with most of the plant-world they contained from the very start*. The "mesas" of the Guianas are a classic instance in point. One might say that there is a law, to the effect that everything now level will sooner or later either be lifted or sunk. There seems to be some intimate correlation between the processes of mountain-making and continent-making. Our thoughts are bold and perhaps crude. The reader may take them for what they are worth. We might perhaps refine them in the pages of a coming work.

formed a system of their own. Such a "zoning out" as this is reminiscent of the basic differences between western and eastern South America; western and eastern Africa; western and eastern Australia; and contributes final evidence to the effect that when the angiospermous *genorheitra* began to reach lands still on the modern maps these lands resembled in no way those of the present. On the strength of the inferences supplied by the dispersal of *Illicium* and *Magnolia* we may surmise that we face two different types of local distribution, as follows, (a) *Magnolia* Type; Southeastern China—Hainan—Palawan— possibly Northeast Borneo; (b) *Illicium* Type: Formosa—Luzon (Bontoc)— Mindoro (Mt. Halcón, primarily) — Borneo (Mt. Kinabalu, primarily) *).

Magnolia and *Talauma* range southward from Mexico as follows,

A) Magnolia

(1) *M. Schiedeana* — Mexico (Jalisco, Sinaloa, Nayarit, Veracruz).

(2) *M. dealbata* — Mexico (Veracruz, Oaxaca).

(3) *M. guatemalensis* — Guatemala.

(4) *M. cubensis* — Cuba (Santa Clara).

(5) *M. domingensis* — Santo Domingo.

(6) *M. Ekmanii* — Santo Domingo.

(7) *M. emarginata* — Santo Domingo.

(8) *M. pallescens* — Santo Domingo.

(9) *M. poasana* — Costarica.

(10) *M. Allenii* — Panama.

(11) *M. sororum* — Panama.

---

*) The data we used in the main text to compile the distribution of *Illicium* were secured from sources of indifferent value (indices, floras, uncritical herbarium material, etc.). These data have been entirely superseded by a recent monograph (A. C. SMITH in Sargentia 7: 1. 1947), which credits to *Illicium* not less than fortytwo species. Whether all these species are "good" in the conventional sense remains to be seen. We admit to be skeptical of work done in the herbarium without immediate reference to the field. At any rate, Smith's monograph has no power to alter our basic conclusions. According to this taxonomist, *Illicium* is indeed unknown from Palawan, and known in the Philippines only in Luzon and Mindoro. It teems on the contrary in Southern China (a perfectly normal occurrence) and ranges to Japan, Corea, the Himalayas, ultimately reaching southward to Borneo, Malacca and Sumatra (Mt. Atjeh), which all is standard in the light of the distribution of the *Magnolia* affinity, and plant-life in general.

Baffled by the disorder in the current classification of *Schisandra* and *Kadsura*, but satisfied that nothing truly startling was contained in their dispersal, we omitted these two genera from reckoning altogether in the main text after the original mention of their general distribution. In the work cited, Smith supplies two maps to show the distribution of *Schisandra* and *Kadsura*, which we reproduce in their main outlines, Fig. 34. It will easily be seen that *Schisandra's* is a classic case of migration which, landing in the Far East from the direction of the Pacific on the front Eastern China (Anhwei)-Formosa, next proceeds to work southward and westward very much in the *Magnolia-Manglietia* pattern. *Kadsura*, on the contrary, disperses in the *Talauma* pattern, and the front of its penetration into Malaysia is pointedly stressed to lie at Mindanao of the Philippines. It is here that come together the "northern" Sect. *Eukadsura* and the "southern" Sect. *Sarcocarpon*. Mindanao, moreover, faces just across the South China Sea the range of Sect. *Cosbaea* (Southern China, Eastern Burma, Siam, Indochina, in the main). We have no comment to offer, for the evidence speaks for itself. It is peculiar that Smith (op. cit., 160 fig. 30) in a scheme of theoretical phylogeny based on androecial characters derives *Eukadsura* and *Sarcocarpon* from a single "ancestor", setting out *Cosbea* on its own, and possibly as more "primitive." Dispersal bears out the surmise that *Eukadsura* and *Sarcocarpon* are close, while *Cosbaea* developed very much in a center of its own, and arose from a speculative part of the *genorheitron*. So far we agree with Smith, but not beyond, for "primitiveness" is more easily assumed than proved when one does not know which one might be the initial form.

B) Talauma

(1)   *T. mexicana* — Mexico (Michoacàn, Oaxaca, Morelos, Veracruz).
(2)   *T. gloriensis* — Costarica.
(3)   *T. sambuensis* — Panama.
(4)   *T. dodecapetala* — Lesser Antilles.
(5)   *T. amazonica* — Brazil (Parà).
(6)   *T. fragrantissima* — Brazil (Rio Janeiro: Serra dos Orgãos).
(7)   *T. ovata* — Eastern Brazil.

A close ally of *Talauma*, the monotypic *Svenhedinia minor*, is endemic to Cuba (Oriente), and the track of these two genera points to an original invasion from the Pacific effected across Southern California or Western Mexico, extending to Cuba, and next veering southward through the Lesser Antilles to reach the Amazonas and the "Serras" of Eastern Brazil. The Lesser Antilles are an occasional bridge leading to South America from two directions, Mexico and West Africa, not only but a potential highway between Guiana and the Eastern United States *).

The very same staggering, north and south, apparent in the Far Eastern and Malaysian end of *Magnolia* and *Talauma* **) is repeated in the New World. In the Americas, *Magnolia* ranges south to Panama quite as it does reach Malacca, Sumatra, Borneo and Java in Malaysia. However, these southern extensions of range are mostly secured by taking advantage of alpine or subalpine connections, if not in the immediate present, in the geological past (see the strong subalpine to alpine speciation of *Magnolia* in Santo Domingo and Haiti). The home of *Magnolia* remains in the warm-temperate to temperate north on both sides of the Pacific, not less than nine American species (*M. acuminata, M. cordata, M. grandiflora, M. Ashei, M. virginiana, M. tripetala, M. macrophylla, M. Fraseri, M. pyramidata*) being endemic to a range roughly agreeing with that of *Liriodendron Tulipifera*. Two of these species, *M. acuminata* and *M. cordata* are so distinct as to be treated as a separate genus by various taxonomists, which tends to suggest that the *genorheitron* reached the

---

*) As in the case of *Illicium* (see preceding footnote), the data used in the main text for the American distribution of *Magnolia* and *Talauma* are largely superseded by a recent monographic work (Howard, in Bull. Torrey Cl. 75: 335. 1948) on the West Indian Magnoliaceae. The author of this paper views *Svenhedinia*, for instance, as a mere synonym of *Talauma*. We have personal knowledge on the basis of material still in herbarium that *Talauma*, or a form close to it, also occurs in Venezuela (costal cordillera of Aragua, and on Mt. Roraima). Although in themselves interesting, the new data mentioned in this note are not considered because of reasons stated in the preceding footnote. Methods and approaches to phytogeography, or any other science, effective only in the light of the very latest discovered or reported are not commendable and useful. *Sound methods and approaches are bound to work even under a slight disadvantage.* A manual like the present should not age within a couple of years.

**) According to HOWARD (cited in the preceding footnote), *Talauma* is represented in the West Indies by three species, two (*T. minor* and *T. truncata*) localized in Eastern Cuba, one (*T. dodecapetala*) endemic to the Guadeloupe, St. Vincent and Dominica. The range of the last is *exceedingly* interesting and, if we are not mistaken, altogether rare as regards groups coming in from the Pacific. This range supports the opinion, expressed elsewhere in these pages (see p. 392; Fig. 71), that there once was a continuous front, *Texas (or Mexico)-Cuba-Guadeloupe*. We regret that so poor and confused is the taxonomic record of *Talauma* that we cannot explore the dispersal of this genus in the New World throughout. The matter will be thoroughly studied in the pages of a coming manual on animal-dispersal. See further p. 331 fn. 1.

New World in a state of incipient generic or sectional segregation. Moreover, certain of the American Magnolias are openly "southern" (e.g., *M. pyramidata*, possibly a relic-form), while others (characteristically, *M. acuminata*) range freely northward, ultimately stopping, as does *Liriodendron*, in Southern Ontario. A Mexican form, *M. dealbata* (9) is so close to the North American *M. macrophylla* (northern limit of range in Northern Carolina) as to be only doubtfully separable from it. There is no imprudence, accordingly, in assuming that *Magnolia* reached the New World across the Pacific approximately in modern California, next followed along the line California—Georgia, and ultimately spread southward and northward from this line. To this genus apply the same remarks that are pertinent in regard of *Liriodendron* and *Platanus*.

Illicium is represented in the New World *) by two species *I. parviflorum* (Florida to Georgia, Santo Domingo), *I. floridanum* (Florida to Louisiana and Alabama). Jointly taken with *Magnolia's*, the distribution of *Illicium* reveals that the United States had most intimate connections with Mexico, Cuba and Santo Domingo, which may be said to have been integral part of the modern U.S.A. in the earliest stages of angiospermous migrations. *Illicium floridanum* is known (45) to associate in the Knox Hill region of Florida (the lone locality on record from this state) with *Stewartia* (*Stuartia*) *Malacodendron* and *Magnolia Ashei*. *Stewartia* is a theaceous genus also known from China (Kiangsi, Hupeh, Szechuan); *Magnolia Ashei* is close to *M. macrophylla*, and it seems probable that all these plants were associated together when they first reached the New World, if not as the modern species, at least as well differentiated forms already approaching them. Peculiar also is the fact that *Liriodendron* occurs in Kiangsi and Hupeh, two of the Chinese provinces wherein *Stewartia* is endemic. In brief, there is no justification whatever for assuming that the Magnoliaceae, and *Stewartia* as well, fared to the New World by crossing Alaska. On the contrary, the evidence is precise to the effect that the connections binding the Far East and the New World had their inception, as the modern map of the Far East reads, at a point closely agreeing with 30° Lat. North (line Hupeh—Amamioshima; see also p. 438), and reached the approximate boundary of Mexico and the United States *at the very same latitude*. These are coordinates *in the modern maps*, which may not perfectly agree with the coordinates of palaeogeographic maps as regards the same lands. However, the evidence speaks for itself, and we may not neglect it to fabulate on the contrary about "Extraordinary capacities for transoceanic dispersal," "Floatsam jetsam Methods of Conveyance," and academic recipes of the same ilk. It

---

*) The taxonomy of *Illicium* in the New World has been entirely recast, together with that in the Old, by A. C. SMITH (Sargentia 7: 1. 1947), who recognizes five species in the Southeastern United States (Florida, Georgia, Louisiana), Mexico (Veracruz), Cuba and Haiti. Smith's monograph reached our hands after the main text was written. We take no account of its data for reasons given in preceding footnotes. It will be seen that using data old or new, the dispersal of *Illicium* in the Americas can easily be correlated with that of *Magnolia*. The same holds good in the Far East and Malaysia, as to the general trends of the dispersal, when not all its local aspects.

is manifest that, had not Southern California and Northern Mexico undergone extensive drying in the course of their geological history, we would there find very much the same flora which reigns today in Florida, Cuba, Santo Domingo and the coastal Southeastern United States as far north as the Carolinas. Naturally, we may look for fossils of Cretaceous and Tertiary age to match the plants which no longer live there, but still occur eastward. *Proof is here that living floras elucidate fossil floras (at least, beginning with the Cretaceous) not the contrary.* Further considering that the modern flora of the South-western United States and adjacent Mexico is rich in very old xerophilous or subxerophilous forms (e.g., *Thamnosma;* see p. 91), it proves necessary to conclude that radical changes in floristic composition took place in those regions, mesophytes being duly replaced by tendential xerophytes, which latter had ranges in the geologic past not too far from the regions they later came to occupy. We are doubtless very far in these quarters from the specious simplicity of causes and effects, or the nebulous recourse to mysteries in order to elucidate conundrums apparent, alas, in hopeful contributions on the flora of the Southwestern United States *).

Schisandra is represented in the United States by a single species, *S. glabra* endemic to Louisiana, Florida, Mississipi, Alabama, Tennessee, Georgia, South Carolina, Arkansas **), manifestly a relic, which adds little to what we already know of the magnoliaceous dispersal in the New World.

---

*) IVAN M. JOHNSTON's paper (see Jour. Arnold Arb. 21: 356. 1940) is an excellent sample of this type of literature.

**) A. C. SMITH, in the work cited in previous footnotes, indicates that the distribution given above may not be satisfactory. For our immediate purposes, it seems sufficient. Interesting is the fact that this taxonomist puts *S. glabra* together with *S. chinensis*, *S. repanda* and *S. bicolor* in a special Section *Maximowicxia*, giving the following distributions,

(1) *S. chinensis* — Sakhalin, Japan (Hokkaido, Honshu), U.S.S.R. (Primorskaia: Ussuri, Amur), Corea, China (Kirin, Chahar, Jehol, Hopei, Shansi).

(2) *S. repanda* — Japan (Honshu, Shikoku, Kyushu), Quelpaert Island.

(3) *S. bicolor* — China (Northwestern Chekiang).

This range is ampler than either *Stewartia*'s or *Liriodendron*'s. The starting point of the transpacific track to the New World is contained within a triangle, Sakhalin — Regions of the Amur — Chekiang. Chekiang is of course close to Hupeh and Kiangsi, but the latitudinal range involved between extremes approaches 20°. However, as the map reveals (see Fig. 34), the center of the dispersal of *Schisandra* in Asia is very near to the Tropic of Cancer, that is, to the latitude of 30° N. which, based upon *Liriodendron* and *Stewartia*, we had suggested as the possible starting point of the transpacific channel of migration to the New World — *as modern maps read.* It is clear that litoral Central China, Japan and parts of Corea are relics out of the original lands connected most immediately with the transpacific track on the western side of the modern Pacific.

Concerning the dispersal of *Kadsura* (and *Schisandra*), Smith (op. cit. 161) writes, "It seems reason-able to assume that *Kadsura* and the closely allied *Schisandra* had their center of origin within the area delimited above *(See Fig. 34).* One may note that the focal point of the three sections of *Kadsura* includes Indo-China and Siam, the only countries in which all the sections are known to occur. Actually not many species are known from these two countries, which is perhaps a reflection upon the amount of collecting done in them." We observe, (a) The dispersal of these two genera is not comparable, which the map, Fig. 34, reveals; (b) The origin of these two genera is not to be sought in Asia or Malaysia, but in a point now in the Pacific, off and south of New Guinea; (c) Siam, in particular, and Indo-China might now have been collected extensively, but the track of *Schisandra* which hardly interests these regions shows that they were avoided in harmony with the broad trends of the migration. There are indications that Siam and Central Indochina stood at some time or other within a "Monsoon-climate" belt, which would be fatal to a mesophytic group of the *Schisandra* and *Kadsura* type. Smith's broad generalizations ought to be cautioned against as a matter of principles, for they focus the issue under a wrong light to begin with. This issue remains to be studied in detail but on the strength of the generalities here stated.

In conclusion, the dispersal of the Magnoliaceae is far from ununderstandable once its prime movers are grasped This dispersal is in most respects comparable to that of the Fagaceae, the essentials in play being actually the same. Everything in this dispersal is standard, for everything agrees at some point or other with the dispersal of great many other plants. Valuable — though to be matched in other groups — are the hints which the Magnoliaceae supply as to the nature and extent of transpacific migrations between the Far East and the New World. We should caution the readers, nevertheless, against assuming that the *genorheitra* that are common to China, for instance and the United States, fared first to China to evolve therein, and next only crossed the Pacific. *It is fairly sure on the contrary that both China and the New World were duly reached simultaneously by a net of channels fanning out from a point located approximately by New Guinea in our maps (Neocaledonian Center; see Fig. 67).*

This chapter is perhaps most conveniently concluded with renewed reference to a map (Fig. 31) which show how Malaysia is connected at one stroke with the rest of the world. It is obvious that no problem in Malaysian phytogeography can be studied as a local issue. *That which is of Malaysia is of the world throughout.*

Neither Fagaceae nor Magnoliaceae give indication of ever having used the African gate, that is, of having once colonized an ancient continent that occupied the modern Indian Ocean. It is certain, however, that they had *genorheitral* ties in those quarters, so that we may well hear that somewhere in Africa fossils came to light suggesting something of the morphology of Fagaceae and Magnoliaceae. *There is not a very great deal of difference, for example between certain cupules of the Fagaceae and certain pyxidia of the Lecythidaceae.*

BIBLIOGRAPHY

Chapter III

(1) BRYAN, W. H. — The relationship of the Australian continent to the Pacific Ocean-Now and in the Past; Jour. Proceed. Soc. New South Wales 78: 42. 1944.

(2) SKOTTSBERG, C. — The flora of the Hawaiian Islands and the history of the Pacific Basin; Proceed. Sixth Pacific Sc. Congr. 4: 685. 1939.

(3) MAYR, E. — The origin and the history of the bird fauna of Polynesia; Proceed. Sixth Pacific Sc. Congr. 4: 197. 1939.

(4) WOOLNOUGH, W. G. — The continental origin of Fiji; Proceed. Linn. Soc. New South Wales 28; 484. 1903.

(5) LOUBIÈRE, A. — Sur la structure d'un bois silicifié de la Nouvelle-Calédonie; Bull. Soc. Bot. France 82: 620. 1936.

(6) BANDULSKA, H. — see (25), Chapt. I.

(7) GOOD, R. D'O. — The Geography of the genus Coriaria; New Phytol. 29: 170. 1930 (see further, Oliver, W. R. B. — The genus Coriaria in New Zealand; Records Dominion Mus. 1: 21. 1942).

(8) KUMAZAWA, M. — Ranzania Japonica (Berberidac.) Its morphology, biology and systematic affinities; Jap. Jour. Bot. 9: 56. 1937.

(9) STANDLEY, P. C. — Trees and Shrubs of Mexico; Contr. U. S. National Herb. 23 (I). 1920.

(10) DUSÉN, P. — The vegetation of Western Patagonia; Repts. Princeton Univ. Expeds. Patagonia 1896—1899, Bot. 8 (1): 6. 1903.

(11) SCHNEIDER, C. — Die Gattung Berberis (Euberberis); Bull. Herb. Boiss, 11, 5: 33. 1905.

(12) VAN STEENIS, C. G. G. J. — On the origin of the Malaysian Mountain Flora, Pt. 1, Bull. Jard. Bot. Buitenzorg, III, 13: 135. 1934.

(13) SMITH, A. C. — A taxonomic review of Trochodendron and Tetracentron; Jour. Arn. old Arb. 26: 123. 1945. Bailey, I. W. & Nast, C. G. — Morphology and relationships of *Trochodendron* and *Tetracentron* 1. Stem, root, and leaf; ibid. 143; Nast, C. G. & Bailey, I. W. — Morphology and Relationships of *Trochodendron* and *Tetracentron* 11. Inflorescence, flower, and fruit; ibid. 267.

(14) CROIZAT, L. — See (19), Chapt. II.

(15) UELBRICH, E. — Ueber die systematische Gliederung und geographi-

132

sche Verbreitung der Gattung *Anemone*; Engler Bot. Jahrb. 37: 172. 1905.

(16) SMITH, W. W. & FLETCHER, H. R. — The Genus *Primula*; Section Farinosae; Trans. Roy. Soc. Edinburgh 61 (1): 1. 1943.

(17) VAN STEENIS, C. G. G. J. — see (20), Chapt. I.

(18) BERRY, E. W. — See (9), Chapt. I.

(19) SKOTTSBERG, C. — See (12), Chapt. I.

(20) MEZ, C. — *Myrsinaceae;* E. & P. Pflanzenr. 9 (iv. 236). 1902.

(21) KIRCHHEIMER, F. — On pollen from tne upper Cretaceous Dysodil of Banke, Namaqualand (South Africa); Trans. Roy. Soc. South Afr. 21: 41. 1932.

(22) WULFF, E. M. — Istoriceskaia Geografia Rastenii 525. 1944 (*in Russian*).

(23) SCHWARZ, O. — Entwurf zu einem natürlichen System der Cupuliferen und der Gattung Quercus L.; Notizbl. Bot. Gart. Mus. Berlin-Dahlem 13. 1. 1936.

(24) BESSEY, C. E. — The phylogenetic Taxonomy of Flowering Plants; Ann. Missouri Bot. Gard. 2: 109. 1915.

(25) CAMUS, A. — Genre *Quercus*, sous-genre *Cyclobalanopsis* [Encycl. Écon. Sylvic. VI: Les Chênes] 1: 158. 1938.

(26) MARKGRAF, F. — Die Eichen Neu-Guineas; Engler Bot. Jahrb. 59: 61. 1924.

(27) EPLING, C. — Synopsis of the South American *Labiatae;* Fedde Repertorium, Beiheft. 85 (1): 1. 1935.

(28) EPLING, C. — The distribution of the American *Labiatae;* Proceed. Sixth Pacific Sc. Congr. 4: 571. 1939.

(29) BRAND, A. — *Hydrophyllaceae;* E. & P. Pflanzenr. 59 (iv. 251). 1913.

(30) OLIVER, W. R. B. — The genus *Coprosma;* Bull. Bishop Mus. 132. 1935.

(31) BROWN, F. B. H. — Flora of Southeastern Polynesia, III, Dicotyledons; Bull. Bishop Mus. 130. 1935.

(32) YUNCKER, T. G. — Revision of the Hawaiian species of Peperomia; Bull. Bishop Mus. 112. 1933.

(33) PHILIPPI, F. — A visit to the northernmost forest of Chile; Journ. Bot. 22: 201. 1884.

(34) OLIVER, W. R. B. — The vegetation and flora of Lord Howe Island; Trans. [Proceed.] New Zealand Inst. 49: 94. 1917.

(35) TRELEASE, W. — The American Oaks; Mem. National Acad. Sc. [Washington] 20: 1. 1924.

(36) DICKERSON, R. E. — Mammals of the Philippines; Bureau of Science, Manila P. I., Monogr. No. 21. (Distribution of Life in the Philippines): 273. 1928.

(37) LAM, H. J. — Contribution to our knowledge of the flora of Celebes (coll. C. Monod de Froideville) and of some other Malaysian Islands; Blumea 5: 554. 1945.

(38) KNUTH, R. — *Geraniaceae;* E. & P. Pflanzenr. 53 (iv. 129). 1912.

(39) GAGNEPAIN, F. — Magnoliacées; Suppl. Flore Génér. Indo-Chine 1: 29. 1938.

(40) DANDY, J. E. — New or noteworthy Chinese *Magnolieae;* Notes Bot. Gard. Edinburgh 16: 121. 1928 — New *Magnolieae* from China and Indo-China; Jour. Bot. 68: 204. 1930.

(41) BURRET, M. — Beiträge zur Kenntnis der Tiliaceen; Notizbl. Bot. Gart. Mus. Berlin-Dahlem 9. 592. 1926.

(42) VAN STEENIS, C. G. G. J. — Exploraties in de Gajo-Landen: Algemeene Resultaten der Losir-Expeditie 1937; Tijdschr. Nederl. Aardrijkskund. Genootsch. 55: 728. 1938.

(43) HERRE, A. W. C. T. — True fresh-water Fishes of the Philippines; Bureau of Science Manila, P. 1, Monogr. No. 21 (Distribution of life in the Philippines): 242. 1928.

(44) DICKERSON, R. E. — A drowned Pleistocene stream and other Asian evidence bearing upon the lowering of the sea-level during the Ice Age; Bicentennial Conference Univ. Pennsylvania Shiftings of Sea Floors and Coast Lines: 13. 1941.

(45) SMALL, J. K. — Manual of the Southeastern Flora. 1933. Refer to p. 876.

CHAPTER IV

INTERCONTINENTAL DISPERSAL II.

We continue in this chapter the analysis of patterns of dispersal initiated in the preceding pages. The method we follow is always the same, and the results we secure do not vary, either, though the plants we investigate are often unrelated from the standpoint of systematic botany and enormously distant by mileage. We take this as evidence that our method answers practical needs, and as such again submit it to the reader.

I. Garryaceae and Leitneriaceae

The Garryaceae consist of a lone monotype, *Garrya*, distributed in this manner,

(1) *G. elliptica* — United States (California to Oregon).

(2) *G. Fremontii* — United States (California to Oregon).

(3) *G. Wrightii* — United States (New Mexico, Texas), Mexico (Sonora, Chihuahua).

(4) *G. flavescens* — United States (Arizona, Nevada, New Mexico, Utah).

(5) *G. buxifolia* — United States (California).

(6) *G. ovata* — United States (New Mexico, Texas), Mexico (Chihuahua, Guanajuato, San Luís Potosí, Puebla).

(7) *G. Veatchii* — United States (Southern California), Mexico (Baja California: Cedros Island).

(8) *G. salicifolia* — Mexico (Baja California).

(9) *G. glaberrima* — Mexico (Jalisco).

(10) *G. longifolia* — Mexico (Morelos).

(11) *G. laurifolia* — Mexico (Jalisco, Vera Cruz, Chihuahua, Guanajuato, Chiapas), Guatemala.

(12) *G. Faydenii* — Eastern Cuba, Jamaica. *)

---

*) As the reader well knows, we do not intend to run ahead of the evidence ever, and to force upon the reader conclusions which he may justifiably deem to be unsupported as yet. In our opinion, however, the range of *G. Faydenii* is significant. A range *Eastern Cuba-Jamaica* is a fraction of the track run by *Talauma (Mexico to Lesser Antilles;* see p. 127 fn.; Fig. 71), and is not unprecedented among the Cactaceae (see p. 368). This range, too, matches in essential that of *Washingtonia* and its immediate allies (see p. 443) from the Western United States to Cuba. Had *Garrya* widely penetrated the Carib-

Read at face value, this pattern of dispersal yields no clue as to the origin of the track. Mexico can be reached from the Mediterranean (see *Thamnosma* p. 91), quite as much as from the Far East, or, better to say, a center in the Pacific feeding migration simultaneously to the Far East and the New World (see *Magnolia* and *Liriodendron*, p. 126). Records from Cedros Island, Baja California, and a manifest stress on the western sector of the United States and Mexico throughout, may be construed, nevertheless, as a presumption that the main channel of migration of the garryaceous *genorheitron* comes into the New World from the direction of the modern Pacific.

The affinities of the Garryaceae are obscure, and we are not prepared at this hour to indorse fully the current belief that Garryaceae and Cornaceae are related. Accepting this belief, however, for the sake of exploring it from the standpoint of dispersal, we notice that only recently genus *Corokia* was formally transferred from the Cornaceae to the Saxifragaceae. In this new disposition, *Corokia* is credited to the Saxifragaceae Argophylleae, which consist of the following genera,

(1) *Corokia* (*Lautea*) — 2 species in Southeastern Polynesia (island of Rapa), 3 species in New Zealand.

(2) *Argophyllum* — 7 species in New Caledonia, 3 species in Eastern Australia.

(3) *Carpodetus* — 6 species in New Zealand and New Guinea.

(4) *Colmeiroa* — 1 species in Lord Howe Island.

(5) *Berenice* — 1 species in the Mascarenes. *)

This group is manifestly "antarctic," to use conventional language, and contains a striking, though by no means unusual, tie between the Mascarenes and the Pacific. It is normal that forms occur in the Pacific, which bring together different families (e.g., Ericaceae and Epacridaceae, Umbelliferae and Araliaceae, Labiatae and Verbenaceae, etc.). Accordingly, the Argophylleae, with or without *Corokia*, give rise to no exception from the standpoint of phytogeography. Moreover, the Pacific easily ranks as a major center of *genorheitral* origins (see also, 2), and the Saxifragaceae cover a multitude of forms having oftentimes uncertain systematic and taxonomic position.

If the Cornaceae are indeed related with the Garryaceae, and it is in the deep south of the Pacific that Cornaceae and Saxifragaceae meet, we are bound to conclude that the *genorheitron* of these three families repeats its origins — as does that of the Fagaceae and Magnoliaceae — from a base-line situated in the Southern Pacific. This being the case, it follows as a very strong probability

---

beans, we should find it most likely in a region between Eastern Colombia and Western Venezuela, and from this point on southward to Peru or Bolivia (see dispersal of *Schoenocaulon*, p. 375). In brief — and here purely as our own opinion — we feel that the Garryaceae invaded the New World from the direction of the Pacific indeed, and heavily lost of the Caribbean range, when the lands formerly occupying this region came to crumble. *Garrya Faydenii* is a typical relic, characteristically matching from the phytogeographic standpoint the odd relic-genera of the Cactaceae (see p. 369) in Eastern Mexico.
*) It may not be amiss to remark that direct ties can obtain between the Mascarenes and Lord Howe Island *via* Malacca, as revealed by *Korthalsella Opuntia* (see p. 142, in particular). See further trends in the dispersal of Balanophoraceae, p. 40.

that the prototypes of the Garryaceae reached the United States, Mexico and the Caribbeans from the west (i.e., the Pacific Ocean) in origin. This conclusion agrees with the distribution itself, manifestly most heavy in the direction of the western sector of the New World. It may further be observed that, this being the case, the most probable line of access to the Caribbeans from the Pacific was in a region now occupied by California, New Mexico, Sonora, Chihuahua, Texas. As an alternative, it might be suggested that the Caribbeans could be reached through a southern channel running from Jalisco to San Luis Potosí and Vera Cruz, or, finally, directly through Chiapas and Guatemala. The first of the channels here outlined has most probabilities in its favor, and is doubtless vital as regard the migrations of plants such as Magnoliaceae, Platanaceae, etc.

Supposing now that the affinities of the Garryaceae do not lie in the direction of the Saxifragaceae and Cornaceae, we are left to moot two alternatives, namely, (a) The garryaceous *genorheitron* is from the Pacific; or (b) It stems on the contrary from the Atlantic, and, ultimately, the Indian Ocean.

As these alternatives are both standard, we may freely accept the one which a precise study in the phylogeny of the Garryaceae shall advise. As stated, our mind is open on the subject of the ultimate affinities cf this peculiar family. Purely as phytogeographers, we might agree without further that their migrations indeed stem from a center in the Pacific.

Seemingly quite isolated among living plant is *Leitneria floridana*, monotypic representative of the Leitneriaceae distributed in the Southeastern United States (Florida, Texas, Georgia, Arkansas, Southeastern Missouri). A dispersal of this type is classic of forms having their New World center of range in the Southern Appalachians and Ozarks. Southeastern Missouri is normally one of the ends of the migration (the other is Southern Ontario in Canada) for groups of this nature.

There can be but little doubt, then, that *Leitneria* belonged to the Appalachian and Ozarkian flora from the start, which means in the present case from the Late Jurassic onward. The Appalachian-Ozarkian center could be reached (see *Styrax*, p. 328) in two ways, from the Pacific or the Atlantic. In our ignorance of the *genorheitral* ties of *Leitneria* we have no statement to make, whether this genus reached the center in question by one channel or the other. We suspect, nevertheless, that *Leitneria* came the same way as did the Garryaceae, that is, from the west, and would be inclined to look for their remote ancestors in the vicinity of the hamamelidoid-saxifragoid *genorheitron*.

It is worthy of notice as a general principles that tracks and *genorheitron* are members of one and the same equation. When both the members are known, the equation can be solved purely as a matter of routine in the manner exemplified throughout the pages of this manual. When only one is identified, the other may be worked out as the nature of the issue dictates. As a rule, the range can be learned without difficulty by having access to a good herbarium, or facilities such as an adequate library is bound to provide. A study of *geno-*

rheitral and systematic affinities may offer, on the contrary, serious difficulties whenever the investigation cannot be conducted at first hand. Too much reliance is usually placed in widely consulted texts upon so called "Principles of classification" which are belied by the facts of dispersal, and lead as a matter of course to anything but a genuinely natural sequence of affinities.

Granting that one of the members of the equation is unknown or doubtful, as the case is now, the other may be freely assumed by making reference to a number of alternatives. These alternatives might wait for their solution, but, once correctly stated, will some day be disposed of in a rational manner as the sum of evidence is ultimately to advise. This is a near-mathematical method of handling dispersal, and the only one worth serious thought. Adherence to preconceived theories, putative authoritative statements, surmises as to "Floating Continents," "Pendulations," "Species senescence," etc., never need enter our preoccupations, because the problems require facts for a solution. In the case under immediate consideration, the problem will be on its way to a satisfactory solution the moment we are safe in our mind as to the affinities of Garryaceae and Leitneriaceae alike.

II. Cercis

The dispersal of *Cercis* is typical of migrations commonly accepted as proof that the Angiosperms originated in "Holarctis."

This genus ranges in the following manner,

(1)   *C. Griffithii* — Afghanistan.

(2)   *C. Siliquastrum* — Caucasus, Asia Minor, Balkans, Dalmatia, Italy, Southern France.

(3)   *C. racemosa* — China (Hupeh, Szechuan).

(4)   *C. chinensis* — China (Szechuan and Kwantung to Jehol).

(5)   *C. occidentalis* — United States (California, Arizona, Nevada, Utah).

(6)   *C. canadensis* — United States (Texas to Oklahoma, Kansas and Connecticut), Mexico (San Luís Potosí northward).

On the strength of the migrations we have already studied, and those we will further explore in the pages of this manual, we venture as a preliminary the platitude that *Cercis* entered the New World either from the Pacific or the Atlantic. We regret that we have not firsthand knowledge of the sum total of the intergeneric affinities, and we are deprived of the lights we could secure by consulting a standard monograph of this genus throughout the whole range. Had we knowledge of these affinities we might forthwith decide the issue, how *Cercis* entered the New World, purely on the ground of phylogeny.

Taking our clue strictly from the phytogeographical aspects of the dispersal in play (that is, the tracks, as the tracks read on the map), we might be inclined to assume that *Cercis* came into the New World in the manner of the Magnoliaceae, that is, from the Pacific. The Asiatic end of the transpacific channel is correctly located (Hupeh to Jehol), and *Cercis* tenants in the United States and

Mexico old lands notoriously open to migrations from the direction of the Pacific. Unlike *Magnolia*, however, *Cercis* is not an arrant mesophyte, and this accounts for it range throughout regions where *Magnolia* is nowhere to be seen alive in the present.

So much tentatively settled, the question arises how *Cercis* reached the Eurasian ranges it inhabits. For the purpose let us study the group to which it belongs. This group (Caesalpinioideae Bauhinieae of current classification) is so constituted,

(1) *Cercis* — As detailed.

(2) *Bauhinia* — Pantropic.

(3) *Acioa* (*Griffonia*) — Angola, Fernando Po, Cameroon, Nigeria, Liberia, Ivory Coast.

Bauhinia cannot be explored in the present state of our knowledge. This is a very large genus, of which no monograph throughout the whole of the range was ever written. *Acioa* supplies us with a definite center of secondary evolution in West Africa.

Had we a single station of *Acioa* in East Africa — or could we know of the existence there of a group of *Bauhinia* with characters pointing toward *Acioa*, by no means an impossibility — we would be set in our quest, because we would forthwith have a net of standard tracks departing from the African gate of angiospermy.

However, as coming discussions will reveal, there is an interplay in dispersal between Afghanistan and West Africa (see p. 277), not only, but the Mediterranean as well, which absolves us from the charge of rashness, if we state that *Cercis* originated from a center in the modern Indian Ocean, of which we will learn a great deal in the pages to follow. We know this center already, as a matter of fact.

The dispersal may be drawn accordingly, from the Indian Ocean to Afghanistan and the Far East at one hand, West Africa and the Mediterranean at the other. We suggest the possibility that groups of *Bauhinia* intermediate to *Acioa* may be found in East Africa. *) We might further add that groups of the same nature could occur, (a) In Brazil and the Guianas, which are normally reached by a further extension westward of the track that deposited *Acioa* in its present range; (b) In the Himalayas, Southern China or Malaysia.

Our mind remains wholly open on the score. It is plain, at any rate, on the face of the migrations already on record and on the strength of the evidence scattered throughout the pages of this manual, that we have no reason what-

---

*) Long after this had been written, we came to learn that one species seemingly of *Acioa* is known outside of West Africa. This species, *A. Goetzeana* (see ENGLER in Bot. Jahrb. 30: 316. 1902) is characterised by free stamens and differs in this respect from the rest of the genus. *Acioa Goetzeana* is endemic to Nyasaland (Livingstone Mts.).

In the dispersal of *Acioa* (see HUTCHINSON & DALZIEL, Flora of West Tropical Africa, 1927-1936) frequently recur such locality-names as Oban, Calabar, Eket, the meaning of which will be clearer with reference to our notes under the Connaraceae, etc. (see p. 189). It is clear that the dispersal of the affinity of *Cercis* is keyed to the main channel East Africa-West Africa and that the northern dispersal was probably achieved along the line West Africa-Afghanistan-Burma which is spoken of in various chapters (see p. 61, for instance) of this manual.

ever to assume that *Cercis* is from "Holarctis." Were this assumed, we would face the none too easy taks to show how *Acioa* and *Bauhinia* achieved their present distribution from that thoroughly fictitious land. This task would lead us to devise congruous tracks, and it should eventually be left to us to show how these tracks work by comparison with the tracks, for example, of *Magnolia*, *Quercus*, etc.

III. Platanaceae

The monogeneric Platanaceae range in this pattern of dispersal,

(1) *P. orientalis* — Asia Minor, Southern Balkans.

(2) *P. Kerrii* — Indochina (Laos).

(3) *P. racemosa* — Mexico (Baja California), United States (California).

(4) *P. Lindeniana* — Central Mexico.

(5) *P. mexicana* — Central and Northeastern Mexico.

(6) *P. glabrata* — United States (Texas), Mexico (Coahuila, Nuevo León).

(7) *P. Wrightii* — Mexico (Sonora), United States (Arizona, New Mexico).

(8) *P. occidentalis* — United States (Florida and Texas to Nebraska and New Hampshire), Canada (Ontario).

Platanus Kerrii is a comparatively recent addition to the record *) which implements it according to anticipations.

The same question obtains as regards *Platanus* which we have tackled in dealing with *Cercis*. *Platanus* could be in the New World from the west (Pacific Ocean) and the east (Atlantic Ocean) alike, and, though the issue is in itself secondary, it needs be approached as a preliminary.

We will observe that the stations in the American distribution of *Platanus* are altogether conventional for a group that came into the New World from the west. California and Mexico are very heavily invested, and the speciation taking place in these regions is such as to suggest a very long evolution, not only, but evolution in disturbed ranges favoring the multiplication of species **) It is remarkable as a matter of fact, that a form so patently archaic, and

---

*) The author of this species (GAGNEPAIN, in Bull. Soc. Bot. France 86: 301, 1939) comments that the leaves are lanceolate and unlobed, but concludes a discussion of the characters of the inflorescence as follows, "C'est un *Platanus* sans aucune doute." We have not seen specimens of this noteworthy plant, but we are not inclined to question Gagnepain's conclusions, because, (a) *P. Kerrii* supplies the lone record in the Far East which is absolutely standard for a genus so distributed; (b) The flora of Laos is one of the most interesting of Eastern Asia. We have commented elsewhere (see p. 120) on the ties between Laos, the Gajolands and the Northern Philippines. An unusual form would certainly not be out of place in Laos.

**) It is likely that the speciation of *Platanus* in the southern U.S.A. and Mexico was in reality brought about by the very same order of factors which influenced that of *Menodora* (see p. 358; Fig. 68), certain groups of *Croton* etc. Instead of violently disturbed ranges we had in these quarters regions that by gradual emersion cut off sizeable bodies of waters inland in a first time. Later on, these waters disappeared by desiccation. Marked alterations in climate, precipitation, altitude attended these changes which called for speciation. *Platanus, Menodora, Croton, Frankenia*, marine Monocotyledons (see footnote p. 410) may all successfully be used in fathoming past outlines of land and sea. We feel that correlations between phytogeography and geology *of astounding precision in details* can be easily effected by critical investigations such as we here sketch out. We have little doubt but that a chronology of speciation in the genera and groups mentioned can be provided by making reference to datable geologic sequences. This chronology would be invaluable as a check upon trends in phylogeny and the ultimate roots of speciation in time and space.

so rigidly defined as to inflorescence and flower, could speciate in relative abundance, even though the species may not all be equally well marked. *Platanus occidentalis* takes us to the usual Ozarkian-Appalachian cradle reached by the Magnoliaceae we have just studied, and uncounted plants as well which we are to study only in very small part here. On its way to this cradle, *P. occidentalis* left behind a very close relative, or indeed variety, *P. glabrata* endemic to Western Texas and Mexico (Coahuila and Nuevo León). Hardly anything could be so clearly indicative of the trends of the dispersal, which the sum total of the evidence shows to have taken place in origin from west to east, that is, from the Pacific Ocean toward the Atlantic, not the other way around.

We will then conclude, without for this taking the matter for sure, only most probable, that *Platanus*, like *Cercis*, *Quercus*, *Magnolia*, etc., is in North America and Mexico from the west.

Were the contrary true by the farthest stretch of imagination, still nothing much would change. The Caribbeans, Mexico and North America were open to invasion by the Angiosperms both from the Atlantic and the Pacific, and groups are well known (see Styracaceae, for instance, p. 329) which used both approaches to enter the New World. We would merely lead the tracks into North America from the direction of the Mediterranean instead of from the Pacific. The track from the Far East to the New World being indeed longer than those from the Atlantic, nobody could protest who believes that mileage is of account in dispersal. Moreover, migrations from the Far East into the United States are taken as a matter of course, so great being the evidence in their favor that no one may seriously venture to misconstrue the facts and their implications. It is beyond us that, this being the case, and the evidence such as *Thamnosma*, *Cneorum*, etc. (see p. 91) pointedly illustrate, voices may be heard to deny that North America was also entered from the Atlantic's direction. In brief, we believe that *Platanus* reached the United States precisely as *Magnolia* did — merely to mention an example — and further repeat that nothing much would follow, if it were shown that *Platanus*, on the contrary, reached the same States in the manner of *Thamnosma*. We cannot be bitterly petty when big oceans and uncounted centuries are at stake.

Beyond comparison more important than the issue, how *Platanus* invaded the New World, is the question, where does *Platanus* tie in *genorheitral* affinities. This question can be comfortably answered. The *genorheitron* of the Platanaceae is beyond doubt closely allied with, when indeed not identically the same as, that of the Hamamelidaceae. The starting points of the dispersal of the Hamamelidaceae are easily identified with reference to two genera of the this family, *Ostrearia* and *Dicoryphe*. These genera are allied (3); the former is endemic to Australia (Queensland), the latter to Madagascar. It is according-ly manifest that the Hamamelidaceae issue from one gate of angiospermy certainly, the African, and possibly also from part of the Western Polynesian (*Neocaledonian Center;* see Fig. 67). In other words, the center of gravity of the Hamamelidaceae lies in the Indian Ocean *primarily*, which could further be demonstrated with regard to their entire dispersal. We may add that we are of the opinion that the Hamamelidaceae are related by *genorheitron* with the South African Myrothamnaceae, and fairly sure that Hamamelidaceae and Buxaceae are close by consanguineity. As we review the Myrothamnaceae and Buxaceae elsewhere, we may here drop discussion without more.

Once the center of gravity of the Platanaceae is known to lie in the Indian Ocean, their dispersal becomes transparent, because Asia Minor and Laos can very easily be reached from this Ocean in a thoroughly conventional manner, as the reader already knows, and is ever better to see as our work develops.

We might be shown abundant fossils of *Platanus* from "Holarctis" as evidence that we are mistaken. We will answer, (a) *From the Cretaceous onward at least, dispersal of live plants is to elucidate fossil records, not the other way around.* We cannot stultify the testimony of massive arrays of live plants, which we can study any way we fancy, to run after the evidence of a handful of more or less indifferent petrifacts. This we judge to be basic; (b) "Holarctis" can even at this hour be reached from Canada or the Balkans, and there is no reason why it could not have been reached by homogously located stations in the Early Cretaceous, when, indeed, *Platanus* was already thoroughly modern as a genus, and the Angiosperms had massively permeated all northern lands still on the map from starting points manifestly in the south of modern cartography.

IV. Loranthaceae: **Korthalsella** and **Phrygilanthus**

Korthalsella and *Phrygilanthus* exemplify migrations which are thoroughly mysterious unless viewed correctly at the start. A careful study of these genera is urged upon authors who still believe in "Holarctic" origins, and are satisfied, whether as ornithologists, mammalogists or botanists, that dispersal is the work of "Casual agencies," and landbridges are no longer *à la mode*. Doubtless, their theories will be all the more successful if they can account for the dispersal of these genera, without pronouncing the whole a gigantic "Taxonomic faux-pas" or something to be gotten hold of in a mood of cleverness and prophecy.

Korthalsella is distributed (4) *) in this manner,

---

*) In a more recent work (4), the same author has added various species of *Korthalsella* to the Polynesian record, as follows,

  (a) *K. rapensis* — Island of Rapa.
  (b) *K. Fueuana* — Marquesas Islands.
  (c) *K. Mumfordii* — Marquesas Islands.
  (d) *K. Margaretae* — Austral Island, Henderson Island.
  (e) *K. Lepinii* — Tahiti.

The subsectional affinity of these species is left unstated and for this reason we do not include them in the tabulation given in the main text. Nothing is in their range which may alter our conclusions.

A) Sect. **Eukorthalsella**

(1) *K. salicornioides* — New Zealand.
(2) *K. Honeana* — Fiji.
(3) *K. striata* — Caledonia.
(4) *K. Remyana* — Hawaii.
(5) *K. Dacrydii* — Java, Malacca.
(6) *K. madagascarica* — Madagascar.

B) Sect. **Heterixia**

(7) *K. geminata* — Borneo.
(8) *K. amentacea* — New Caledonia.
(9) *K. Lindsayi* — New Zealand.

C) Sect. **Bifaria**

c') Subsect. I

(10) *K. aoraiensis* — Tahiti.
(11) *K. cylindrica* — Hawaii.
(12) *K. disticha* — Norfolk Island.
(13) *K. dichotoma* — New Caledonia.
(14) *K. Opuntia* — *Type-form:* Ethiopia, Ceylon, India, Southern China, Formosa, Japan, Corea; Siam, Indochina, Sumatra (Gajolands, etc.); Malacca (Mt. Ophir, etc.), Java, Philippines (*Luzon:* Benguet on Mt. Pauai, Mt. Pulog; *Zambales* on Mt. Tapulao; *Negros* on Canlaon Volcano), Australia (Queensland, New South Wales), Lord Howe Island — Var. *fasciculata:* Afghanistan, Western Himalayas, Southwestern China — Var. *Bojeri:* Mascarenes (Mauritius) — Var. *Gaudichaudii:* Mascarenes, Madagascar, Comoros, South Africa.

c") Subsect. II.

(15) *K. Commersonii* — Madagascar.
(16) *K. australis* — Australia (Queensland).
(17) *K. rubescens* — Eastern Polynesia (Henderson Island, Tahiti).
(18) *K. platycaula* — Eastern Polynesia (Austral Islands, Marquesas, Tahiti), Fiji, Hawaii.
(19) *K. complanata* — Henderson Island, Hawaii.
(20) *K. latissima* — Hawaii.
This tabulation is from a work hardly ten years old, and out of the classi-

fication of a botanist who prefers as a rule smaller generic units. It is therefore clear that the generic limits here given are not unduly broad; that each separate affinity is closely knit, and that the distribution, if not absolutely complete, is doubtless thoroughly representative.

This dispersal can be globally understood from a starting point by the African gate of angiospermy (Fig. 35). The map is clear enough not to require comment. It might be pointed out that we seem this time to neglect completely the possibility that the Polynesian massing could be fed directly from the Western Polynesian and Magellanian gates in addition of the African. This neglect is intentional, because we believe that no alternatives are necessary. Not to rush here ahead of the subject, we discuss once again *K. Opuntia* in later pages (see p. 356) of this manual, showing that this species reached the Coral Sea by a track streaming from the Indian into the Pacific Ocean. We believe that what holds good for *K. Opuntia* and its immediate alliance is also valid as regards Sect. *Eukorthalsella*, Sect. *Heterixia* and Sect. *Bifaria* Subsect. II, all of which contain relevant records in Madagascar, Java, Malacca, Borneo. These records give a definite bridgehead from the Indian toward the Pacific Ocean.

Alternative tracks (a, a', a'') are suggested in Fig. 35 for the migrations to Hawaii, which the reader knows already.

Finally, we may observe that, as we have stated when discussing the Myrsinaceae (see p. 85), the fact that certain species of *Korthalsella* might originate by the Western Polynesian, even Magellanian, gate would not subvert the fundamentals upon which our interpretation rests. As all the gates are connected within a southern range, it means comparatively little (see also discussion of *Astelia*, p. 58) whether a group uses one or more gates.

It will be noticed that the range of *K. Opuntia*, South Africa or *Madagascar-Ethiopia* is in disconnection, which points to Ethiopia having been reached from a crumbled range in the Indian Ocean rather than overland. *)

Worthy of notice is also the range of var. *fasciculata* (Afghanistan, Western Himalayas, Southwestern China) which points once again to a direct invasion of modern Asia from a crumbled region in the Indian Ocean, and is reminiscent of the dispersal of *Cercis* in certain essentials.

The alternatives and comparisons of detail which this dispersal suggests are left for the present to the imagination of the reader. He will remark, no doubt, that the track of *K. Opuntia* proceeds from the Mascarenes and Madagascar to India, Southern China and Formosa which is comparable to the course of *Lysimachia* (see p. 53), and, as we shall see, of *Lobelia* (p. 208).

Juan Fernandez and Chile are the seat of the Myzodendraceae, a small family in the affinity of the Santalaceae. *Santalum* (5) itself ranges as follows,

---

*) *Korthalsella* is a group that, not unlike the Sapindaceae Cossignieae (see p. 49), runs both "cold" and "warm" tracks.

A) Sect. **Polynesiaca**

(1) *S. fernandezianum* — Juan Fernandez.
(2) *S. marchionense* — Marquesas.
(3) *S. insulare* — Tahiti.
(4) *S. multiflorum* — Society Islands.

B) Sect. **Hawaiiensia**

(5) *S. hendersonense* — Henderson Island.
(6) *S. ellipticum* — Hawaii.
(7) *S. paniculatum* — Hawaii.
(8) *S. Pilgeri* — Hawaii.

C) Sect. **Eusantalum**

(9) *S. Yasii* — Fiji.
(10) *S. austrocaledonicum* — New Caledonia.
(11) *S. lanceolatum* — Southern and Northern Australia.
(12) *S. papuanum* — New Guinea.
(13) *S. album* — Timor, India?
(14) *S. boninense* — Bonin (Ogasawara) Islands (Of this section?).
(15) *S. Freycinetianum* — Hawaii.

Comments are unnecessary, because this is the dispersal of *Coprosma* repeated over again. Suggestions that there never were "landbridges" in the Pacific; that all this is the result of "Extraordinary capacities for transoceanic dispersal," that Fiji is "oceanic," etc. are *pure nonsense*. The truth is that the *genorheitron* of the Loranthaceae and Santalaceae originated precisely in crumbled geologic ranges, *where all other angiosperms first originated*. *Coprosma, Santalum, Korthalsella, Geniostoma, Primula, Berberis, Anemone*, etc., all go back to these ranges in the end, which are the ultimate center of angiospermy. Taking advantage of the same geological landconnections, these and other uncounted plants all fared to lands which still exist on our maps, and so did well before the inception of the Cretaceous. We know, as a matter of fact, that modern angiospermous plants were actively replacing an older type of flora in North America at the very end of the Jurassic. Obviously, animals followed in the wake of plants, so that the migrations of the ones must perforce be comparable as to essentials with those of the others. Zoologists who fail to see the point, and insist that botanists who believe in "Gondwana" and "landbridges" are lacking in seriousness had perhaps better get hold of the vegetable record in order to square it with their own. Perfect examples of animal distribution to match that of plants will meet them everywhere.

Several authors have discussed the origin of *Santalum album*, debating whether this economic plant is actually endemic to India or was there introduced at an early date. Strictly to judge from the range (Timor and, doubtfully, India), a phytogeographer would be inclined to conclude that the Sandalwood is not truly native of the Gangetic peninsula. The reason for this must be that a plant endemic both to Timor and India (with or without Ceylon) should occur somewhere in the Lesser Soenda Islands, Celebes, Java and Northern Australia, which are stepping stones along the normal track, Timor-India. If not endemic to all these points, the Sandalwood should at least be known from some of them, where ambiental conditions are favorable to its survival. It is a fact, moreover that, as the record shows, *Santalum*, to use conventional language, is essentially an "Oceanic" genus. Plants of this origin easily reach Timor and the islands in its vicinity (witness the track of *Piper latifolium*, as follows, Marquesas, Austral, Cook, Society Islands, Fiji, Timor), but it is unusual that they range as far west as India. In conclusion, the question whether the Sandalwood is endemic to India cannot be answered on the strength of academic arguments. It has to be settled by a critical consideration of ecology, palaeobotanical studies and the like, which all may in the end offset the evidence from pure phytogeography. This evidence, as stated, is to the effect that *Santalum album* is probably not truly native to India.

Phrygilanthus, an ally of *Korthalsella*, is typically distributed (6) thus,

A) Sect. **Muellerina**

(1) *P. Raoulii* — New Zealand.
(2) *P. eucalyptoides* — Eastern Australia (Victoria to Queensland).
(3) *P. celastroides* — Eastern Australia (Victoria to Queensland).

B) Sect. **Hookerella**

(4) *P. tenuiflorus* — New Zealand.

C) Sect. **Furcilla**

(5) *P. myrtifolius* — Eastern Australia (New South Wales, Queensland).
(6) *P. Bidwillii* — Eastern Australia (Queensland).
(7) *P. novoguineensis* — New Guinea.
(8) *P. obtusifolius* — Philippines (Luzon: Bataan; Mindanao).

D) Sect. **Martiella**

(9) *P. Palmeri* — Mexico (Jalisco, Puebla, Chihuahua).

E) Sect. **Euphrygilanthus**

e′) Subsect. **Dipodophyllum**

(10)   *P. Diguetii* — Baja California.

e″) Subsect. **Cymosophrygilanthus**

(11) *P. Berteroi* — Juan Fernandez.
(12) *P. mapirensis* — Bolivia.
(13) *P. heterophyllus* — Chile, Peru.

The face of this dispersal is such that it may consistently give rise to two interpretations, (i) Leaving the base-line of distribution in the South Pacific, *Phrygilanthus* followed tracks *a′* (see Fig. 36). So doing it went to Juan Fernandez and Western South America at one hand; New Zealand, Australia, New Guinea, the Philippines and lastly Lower California and Mexico across the Pacific at the other; (ii) Leaving this same base-line, and over tracks *a′* (*right side of map*) and *a*, *Phrygilanthus* reached the whole of Western America at one stroke; meantime streaming in a westerly direction, the genus went to New Zealand, Australia, New Guinea and the Philippines, here to stop without ever having crossed the Northern Pacific.

There are botanists who not only believe that taxonomy is a necessary adjunct of phytogeography — which no one will seriously dispute — but go further, and insist that, in some way or other, phytogeography must forever play second fiddle to classification. This, as we know, is far from correct. *Phrygilanthus* proves, quite as much as does *Juania* (see p. 455), that there are cases in which taxonomy must play second fiddle to phytogeography.

In the classification we use, *Phrygilanthus Diguetii* is treated as a member of Sect. *Euphrygilanthus* subsect. *Dipodophyllum*. In this position, this species is believed to be typical of *Phrygilanthus* in general, for it is credited to Sect. *Euphrygilanthus* in the first place.

Other botanists, on the contrary, treat *P. Diguetii*, as the type of a distinct genus, *Dipodophyllum*, and other still as a species of the genus *Phoradendron*. Among the last is an author (7) who believes that *Phrygilanthus* numbers no more than two species in Mexico, namely,

(1)   *P. sonorae* — Baja California, Sonora.
(2)   *P. Palmeri* — Jalisco, Puebla, Chihuahua.

The face of the taxonomic record is such that we must reach the conclusion that the Loranthaceae in this group endemic to Western Mexico cannot be formally classified at all. They can, of course, be assigned to this or that group, or genus, but there is no author who can persuade all others to the effect that his taxonomy has greatest weight. As regards these plants in this range, classification breaks down. Let us remark that it is not only the Loranthaceae that behave in this manner in this range. The Lobeliaceae do the same, witness their genus *Palmerella* (see p. 208), which defies the best efforts of those trying to effect a final taxonomic disposition.

As noticed, there are botanists who believe that the phytogeographers ought to stand by impotent while these controversies go on in the field of taxonomy (see, for example, p. 536), endlessly waiting for the "perfect classification" to be written. In reality, the shoe is on the other foot, and in such cases as this taxonomists ought to listen to what phytogeographers have reason to say.

Phytogeographers know, (a) That both shores of the Pacific (see *Perrottetia, Lepechinia*, p. 54, 98) were open to angiospermous invasion at a very early date, and that sizeable lands once occupied part of this ocean, which no longer are to be seen on modern maps; (b) That long before the genera and species of the modern world came into existence, their *genorheitra* were in active migration; (c) That, consequently, it is not surprising to find regions of the modern world — regions, as it were, of strategic significance for migration — in which the ancient *genorheitra* deposited, we may say, forms that fit as yet no modern genus and survive to plague classification.

Knowing this, and apprised that Southern and Baja California stand at the abutment of an ancient bridge connecting the Far East with the New World (see *Distylium, Perrottetia, Lepechinia, Lysimachia*, etc., etc.), the phytogeographer is quite ready to believe that the abutment of this bridge still retains forms of flora which are neither entirely *Phrygilanthus* nor *Phoradendron*, and for this very reason defy the efforts of taxonomists who stubbornly wish to classify them with finality under one or the other modern genus. Likewise, the phytogeographer accepts as a matter of plain evidence the fact that certain tracts of the modern earth, witness the sandstones of Southeastern Venezuela, the Benguet plateau of the Philippines, Mount Bavi in Indochina, etc., etc., still represent fragments of the earth as it once was. These fragments retain forms of plantlife no longer to be seen elsewhere. In brief, here issues are met with that taxonomists can never hope to settle with finality but phytogeographers, on the contrary, can easily understand.

V. **Potentilla** and homologous migrants

Potentilla is a large rosaceous group represented in Malaysia (8) by at least two different groups, *Anserinae* and *Sundaicae*.

The type of the former is a species distributed (9) in this manner,

(1)   *P. anserina* — Chile, Tasmania, New Zealand; Southern China, Tibet, Himalayas throughout, Manchuria, Japan, Siberia throughout, Caucasus, Iran (Persia), Syria (Lebanon), Europe generally northward from the Pyrenees, Southern Italy and the Southern Balkans; Azores; North America throughout southward to California, Arizona, New Mexico, Iowa, New Jersey; Greenland.

The distribution of *P. anserina* is interrupted by a wide geographic gap between New Zealand and the Western Himalayas. Such a gap notwithstanding, this distribution is shown by a glance at the map to match in every fundamental the tracks of *Euphrasia*, *Carex*, etc. In reality, the type of dispersal of all these plants is strictly homologous, because the gap alluded to is occupied by various species, all closely related among themselves and with their parent-form, *P. anserina*. This gap, then, is more apparent than factual, for *P. anserina* still fills it with subordinate taxonomic entities within its close affinity. These vicariants are,

(2) *P. papuana* — New Guinea.

(3) *P. Foersteriana* — New Guinea.

(4) *P. parvula* — New Guinea (Mt. Carstenz), Central Celebes, Borneo (Mt. Kinabalu).

(5) *P. adinophylla* — New Guinea.

(6) *P. habbemana* — New Guinea.

(7) *P. simulans* — New Guinea.

(8) *P. microphylla* — New Guinea, Himalayas.

(9) *P. leuconota* — New Guinea, Central Celebes, British North Borneo.

To this same affinity further belong,

(10) *P. tatsienluensis* — China, Szechuan.

(11) *P. peduncularis* — China (Yunnan), Eastern Himalayas.

Critically viewed, then, the track of this group can be interpreted without difficulty. The migration takes its inception by modern Chile, right there, where another rosaceous plant, *Fragaria chiloensis*, begins its course northward eventually to end, unchanged or but little changed in the taxonomic sense, at Hawaii. From Chile onward, *P. anserina* streams to Tasmania and New Zealand, and next migrates in a manner homologous of *Euphrasia*. Malaysia is a secondary center of evolution and speciation for the group of *P. anserina* the like it is for the Fagaceae, Magnoliaceae, etc. Obviously, it is not surprising that *P. anserina* is replaced between New Guinea and the Himalayas by vicariants. Taken only at its face the taxonomic record implies that *P. anserina* is unknown in Malaysia, but read as to its deeper substances this very same record reveals that this species is not so much unknown as modified in the range New Guinea-Himalayas. We should not be surprised at all, as a matter of fact, if a taxonomist were some day to inform us that the petty vicariants of *P. anserina* endemic to Malaysia are at bottom outright varieties or subspecies of *P. anserina*, thus even furnishing a formally perfect record of dispersal for this species between Tasmania and India.

In the immediate vicinity of *P. anserina*, and in the same group *Anserinae* according to certain authorities (8), is another complex of about a dozen species, the *Rupestres* of other taxonomists (9), distributed thus,

(1) *P. Mooniana* (*P. polyphylla*) — Java, Eastern Himalayas, Ceylon; Var. *kinabaluensis*: Borneo (Mt. Kinabalu).

(2) *P. fulgens* — Himalayas.

(3) *P. tianshanica* — Central Asia.

(4) *P. geoides* — Caucasus, Crimea.

(5) P. poterifolia — Iran (Persia).

(6) P. calycina — Asia Minor, Thrace, Mt. Athos.

(7) P. rupestris — Siberia, Caucasus, Asia Minor, Crimea, Europe (except Northern Scandinavia, Northern Germany, Danemark, Scotland, Ireland, Holland, Western France, Southern Italy, Greece); North America (Washington, Oregon, California, Montana, Wyoming, Nevada).

(8) P. arguta — Canada (Northwestern Territories to New Brunswick), Colorado, Maryland.

(9) P. glandulosa — British Columbia, Alberta, Washington, Oregon, California, Montana, Wyoming, Nevada, Idaho, Utah, Colorado, New Mexico, South Dakota.

(10) P. rhomboidea — Washington, Oregon, Montana, Nevada.

(11) P. cuneifolia — California.

We may not have at hand a complete record of this group, but it appears, as the data now available read, that (a) North America is reached this time from a track in the high north which splits to enter Eastern and Western North America alike, but does not touch Japan at all. This track, consequently, originates somewhere in Central Asia; *) (b) Europe is entered by a conventional channel running through Iran (Persia) and the Caucasus from the direction of the Himalayas. The dispersal is not coastal along the Atlantic; (c) The *immediate origin* of *this particular group* is to be sought somewhere in the high mountains of Malaysia.

A Malaysian geographic origin for this particular affinity is probable, because, (a) *Potentilla Mooniana* and its group are closely allied with *P. anserina* and its vicariants. Malaysia is the region where *P. anserina* disappears and new species out of its kinship replace it. Malaysia, therefore, is an important *secondary* (but not more than *secondary*) *center* of evolution for *Potentilla*; (b) *Sibbaldia*, a segregate from *Potentilla*, can hardly be extricated from the latter in New Guinea. However, *Sibbaldia* becomes a "good" genus in the Far East, which suggests that its characters harden, as it were, the farther away it moves from New Guinea. Homologous cases of "taxonomic ripening" are in the Magnoliaceae (see *Magnolia, Talauma, Manglietia;* p. 115), and in the Euphorbiaceae. *Glochidion* is a "good" genus in the Far East, but an indifferent one in Polynesia. Controversial populations of *Phyllanthus* suggesting evolution toward *Glochidion* are known moreover from Madagascar; (c) *Potentilla Mooniana* is represented in Malaysia by two well marked forms. The var.

---

*) We insist throughout the pages of this manual that Alaska is not *the* landbridge between the Old and New World. In certain of our maps (see Fig. 25, 31, 34, 36, for instance) we show connections between the Far East and the Americas using a channel much to the south of modern Alaska. In other maps (e.g. Fig. 37, 40, etc.) we drive the tracks through Alaska. We warn the reader that this discrepancy does not necessarily mean that we believe that Alaska is the landbridge involved even in maps such as Figs. 37, 40 etc. Being bound to use maps indicative *of trends of dispersal* on a large scale we cannot attempt to portray most faithfully every detail of distribution, particularly so when to ascertain the ultimate point of landing in the New World of transpacific tracks we should indulge in lengthy discussion of *local* American distribution. The reader is constantly to refer the maps to the text, which we have amply state in the very introduction of this manual (see p. 6), and to have a definite understanding of the limitations under which we are forced to labor as regards secondary issues *of all kinds.*

kinabaluensis is endemic to a range known for its "australian" components (Mt. Kinabalu); the typical form, on the contrary, is native to the "himalayan" alpine flora of Java, and is restricted to a range a few square yards broad on the slopes of Mt. Papandajan. The population endemic to this spot is identically the same as in Ceylon *).

The two group we have review thus far give rise to little difficulty of interpretation, and their affinities and dispersal are reminiscent of those of certain Geraniaceae (see p. 113). Altogether different is the case with the *Sundaicae*, a large group that migrates in this manner,

(1) *P. sundaica (P. Wallichiana)* — Java, Northern Sumatra, Central and Western Himalayas (Kashmir to Bhutan), India, Ceylon, China, Western Manchuria, Corea, Japan; var. *ternata:* Eastern Himalayas.

(2) *P. monanthes* — Himalayas (Kashmir to Sikkim).

(3) *P. Regeliana* — Turkestan.

(4) *P. desertorum* — Turkestan, Western Himalayas, Altai Mts., Northern Mongolia.

(5) *P. centigrana* — Manchuria, Northern Corea, Japan.

(6) *P. asperrima* — Transbaikalia, Northeastern Siberia.

(7) *P. cryptotaenia* — Corea, Eastern Siberia (Amur, Ussuri), Japan (Honshu, Hokkaido).

(8) *P. Bungei* — Northern Persia, Northern Syria.

(9) *P. Kotschyana* — Armenia, Syria, Asia Minor.

(10) *P. intermedia* — Central and Northern Russia, Southern Scandinavia, Eastern Siberia (Irkutsk).

(11) *P. supina* — Northern hemisphere throughout to the exception of the arctic regions; var. *aegyptiaca:* Steppes and deserts of Northern Africa, Asia, Southern Russia; var. *egibbosa:* Europa, Asia, absent from America; var. *paradoxa:* Eastern Siberia (Amur), North America.

(12) *P. norvegica* — Northern hemisphere throughout to the exception of the arctic regions; absent or "introduced" in England, Holland, Belgium, France, Southern and Southeastern Europe, doubtful for the Caucasus; var. *hirsuta:* Labrador to Alaska and New Mexico; West Indies (Santo Domingo).

---

*) There cannot be serious question as to the *P. Mooniana* affinity having "Malaysian" origin in the *strictly geographic* sense of the term. However, origins of the kind are not the last word in dispersal. Trusting the classification of the author who classifies this affinity and the *P. anserina* group together, we automatically visualize for both a *genorheitron* located in the Southern Pacific (Chile to Tasmania), which is normal. It might be true, on the other hand, that the *P. Mooniana* affinity, and the *P. anserina* group are not genuinely consanguineous — which is at least suggested by the differences in taxonomic opinion we have commented upon — and that the former stems, on the contrary, from a center in the Indian Ocean, as do the *Sundaicae* next reviewed. In this case, it is to be assumed that the *genorheitron* of the *P. Mooniana* affinity is immediately from the Indian Ocean, not mediately from the Pacific, with Ceylon, Java and the Himalayas reached together with Iran and Central Asia in the very first place.

The issue is very finely spun, for it primarily rests upon affinities in phylogeny of which we cannot be here the judges. It will be noticed, however, that whatever be the case, the *genorheitra*, whether out of the African gate (Indian Ocean), or the Western Polynesian and Magellanian one (Pacific Ocean), all become confluent within the deep south, and the question boils down to what gate was used by this or that group. A question of this nature is not of principle, purely practical. We may solve it any way we like, depending whether we follow this or that taxonomist. See also the case of *Geranium ardjunense*, p. 112.

(13) *P. biennis* — British Columbia to Arizona and California.

(14) *P. Newberryi* — Washington, Oregon, California.

(15) *P. millegrana* — Washington, Colorado, Nevada, New Mexico, Mississipi, Illinois.

(16) *P. pentandra* — Manitoba, Alberta, North Dakota, Minnesota, Nebraska, Kansas, Missouri, Iowa.

(17) *P. rivalis* — Saskatchewan to Oregon and Mexico.

(18) *P. michoacana* — Mexico (Michoacán).

(19) *P. Richardii* — Mexico (Veracruz, Mexico).

(20) *P. heterosepala* — Mexico, Guatemala, Colombia (Magdalena).

(21) *P. Dombeyi* — Ecuador, Peru.

This affinity does not compare with the group of *P. anserina* either as to origin or habitat-requirements. The track emerges seemingly from nowhere, streaming to the Western Himalayas and Central Asia in the first place, where it branches to Europe, Asia and North America, ultimately to reach the Caribbeans and Peru. The forms in this group are not alpines, but dwellers of steppes and open grounds generally, which only in the tropics do ascend to the heights.

Potentilla sundaica is absent from Borneo and New Guinea, and confined on the contrary to regions open to invasion from the west, that is, the direction of the Indian Ocean.

This dispersal readily suggests that *Potentilla* follows in its wanderings precisely the two channels of migration (see p. 80) suggested by *Anemone, Berberis* and *Primula*. In other words: A segment of this genus (the *Anserinae* group) disperses from a base-line in the Southern Pacific, which it reaches from the usual center of angiospermy located between the Kerguelen Islands and Fuegia. The *genorheitron* of this affinity parts company from that of *Fragaria* in the vicinity of modern Chile, and proceeds along an "antarctic" geologic shore to New Zealand and Tasmania, next migrating northward from these regions. The dispersal, consequently, closely follows the lines of that of *Euphrasia*, as previously pointed out.

Meantime, another segment of *Potentilla*, the *Sundaicae*, leaves the same angiospermous center faring along a crumbled range in the Indian Ocean to Northern East Africa and Asia. The two segments eventually come into renewed contact in the northern hemisphere generally.

This interpretation is in agreement with all the types of dispersal so far seen, and may stand on its own merits, for it strikes us as the single one that accounts for the conventionally "antarctic" *P. anserina* and the seemingly "holarctic" group of *P. sundaica* in a rational and consistent manner. So strong are the current prejudices in favor of "holarctic" origins for the Rosaceae, however, that we will elaborate the matter at greater length than necessity requires in our opinion.

The Rosaceae are by no means "holarctic," for, not to insist upon the dispersal of *Fragaria* and *P. anserina*, their tribe Sanguisorbeae is so composed,

(1) *Alchemilla* — Chile (Atacama) northward to Mexico, Northern Africa, Europe.

(2) *Agrimonia* — Northern hemisphere generally.

(3) *Spenceria* — Western China.

(4) *Leucosidea* — South Africa.

(5) *Aremonia* — Mediterranean Europe.

(6) *Hagenia* — Ethiopia.

(7) *Sanguisorba* — Northern hemisphere.

(8) *Poterium* — Mediterranean Europe.

(9) *Margyricarpus* — South American Andes.

(10) *Tetraglochin* — Chile.

(11) *Polylepis* — South American Andes.

(12) *Acaena* — Southern hemisphere throughout northward to the high mountains of the tropics.

(13) *Cliffortia* — South Africa.

(14) *Bencomia* — Madera, Canary Islands.

The classification we follow (10) is admittedly old, and less than complete, but what it reveals speaks for itself whether the genera are ten or a hundred, and the species fifty or a thousand. *Acaena anserinifolia* (*A. Sanguisorbae*) for example, is endemic to New Guinea, and is the main form of a complex with the following distribution (11),

(1) *A. anserinifolia* — New Zealand; Chatham, Stewart, Auckland Islands; Tasmania, Southeastern Australia.

(2) *A. insularis* — Amsterdam Island.

(3) *A. sarmentosa* — Tristan da Cunha, Nightingale and Inaccessible Islands; probably the Cape.

The distribution of these three species is such, that no one may assert they could originate in "Holarctis". *Acaena*, as a matter of the current records of phytogeography, is uncontroversially accepted as a shining example of "Antarctic" migration .The tracks of *Acaena* highlight in addition centers in the southern hemisphere of the utmost significance as regards the dispersal of *Potentilla*. The complex of *Acaena* under discussion stretches primarily between two outposts, Amsterdam Island and Tristan da Cunha. Out of these outposts the migrations of *Acaena* stream to reach New Zealand and New Guinea. This lap of the track, consequently, most likely *) duplicates the first and most important sector of migration of *Potentilla anserina*, not only, but gives some account, why *Fragaria chiloensis* turns up in Magellania. No stream of migration seemingly issues from Amsterdam Island this time but we know that this center has important connections in the north of the Indian Ocean (see *Philica*, p. 58).

---

*) We are careful to qualify our assumption as *most likely* only. We cannot rule out altogether the possibility that *Acaena* did reach New Zealand in the first place from the direction of the modern Indian Ocean rather than that of the modern Pacific Ocean. The face of the dispersal of *Potentilla anserina* is such, on the contrary, as to demand no qualification. Whatever be here the case in *its details*, it is not easy to demonstrate "Holarctic" origins quite as much for *Potentilla* as for *Acaena*.

The three species of *Acaena* here discussed, then, contain essential elements of distribution showing that our assumptions as regard the tracks of *Potentilla* are not beyond sober reason. Facts bear being correlated whenever the evidence requires, and the method whereby correlation is effected can be judged only by the results it achieves in the end. As the reader is to see, and has indeed already seen, frequent are the migrations that issue from the Indian Ocean pointing to Central Asia. Patently, the Rosaceae are not "Holarctic" because nothing is more certaintly "Antarctic" than *Acaena* and *Cliffortia* in the first place.

Potentilla has a group, *Persicae*, consisting of about 19 species, 11 of which are endemic to Iran (Persia) ,with 4 more or less narrowly localized in the proximity of Mt. Elburs. The 8 specie that range outside Iran (Persia) are distributed as follows: 1 common to Persia, Armenia and Asia Minor; 2 in the mountains of Central Asia; 2 in Asia Minor; 1 in the Caucasus; 1 in the Moroccan Atlas; 1 in the mountains of Southwestern Spain. The massive speciation taking place in Iran presupposes a lively *genorheitron* coming to this region early enough to allow time for evolution to work itself out in full. It is therefore manifest that Iran was reached by the ancestors of *Potentilla* precisely at the same epoch when New Guinea was invaded by the *genorheitron* of the affinity of *P. anserina*. We have, then, two secondary centers of evolution in the northern hemisphere, one in Malaysia, the other in the Near East, which is reminiscent of *Cobresia* (see p. 237), not only, but tallies in a sense with the conclusion by other investigators (see p. 60) that the mountains of Western Eurasia harbor seemingly primitive forms of the Compositae. These two centers are coaeval, both are secondary not primary, and were reached by independent streams of migration working in the Pacific, the Indian Ocean or Africa. All the evidence from dispersal and phylogeny conspires to the same ends, and we will understand nothing of it at all if we fail to account for dispersal and phylogeny at the same time, and by the same premises.

Another group of *Potentilla*, *Tormentillae*, ranges as follows,

(1) *P. indica* — India, China, Formosa.

(2) *P. Hemsleyana* — China (Hupeh).

(3) *P. simulatrix* — China (Shansi, Jehol), Mongolia.

(4) *P. flagellaris* — China, Manchuria, Corea, Mongolia, Siberia throughout.

(5) *P. canadensis* — Minnesota to North Carolina and Arkansas.

(6) *P. reptans* — Ethiopia, Egypt, Algeria, Afghanistan, Western Himalayas (Kashmir), Iran (Persia), Turkestan, Western Siberia, Armenia, Asia Minor, Syria (Lebanon).

(7) *P. Tormentilla* — Caucasus, Western Siberia, Altai Mts., Northern and Central Europe to the Alps and Pyrenees, Russia to the Urals.

(8) *P. procumbens* — Azores, Newfoundland, Cape Breton, Madera, Gibraltar, Pyrenees, Northwestern and Central France, England, Danemark, Southern Scandinavia, Belgium, Holland, Western Germany to Central Russia.

This classification is by an author who seemingly takes a very broad view of specific limits, but this is not of the essence of the matter to interest us. The point is that this dispersal reveals eddies of migrations well worth considering in detail.

Part of this affinity streams eastward (*P. indica, P. simulatrix, P. flagellaris*), part westward (*P. Tormentilla*), the parting of the way being set at a line Himalayas-Altais which is precisely what it should be. *Potentilla reptans* streams from Ethiopia to Algeria at one end, to the Western Himalayas, Siberia and the Near East at the other. We may admit, as an alternative, that the migration is straight from Ethiopia to the Western Himalayas, Siberia, the Near East and Northern Africa in geographic sequence, but this alternative has not the weight of probabilities in its favors. It has not them, because from a region located between Algeria and Madera depart two tracks, (a) To Western Europe generally. This is a track of outspoken "Atlantic" pattern, exemplified by *P. procumbens*, and matched, as we are to see, by migrations in *Erica;* (b) To the Eastern United States (*P. canadensis, P. procumbens* in part). In conclusion, all the migrations we have so far described in *Potentilla* can easily be understood along the lines suggested in the map (Fig. 37) that illustrates the distribution of the Groups *Anserinae* and *Sundaicae* *). It will also be seen that this map accounts for the dispersal of the Sanguisorbeae. This map, therefore, is standard for the Rosaceae, the transpacific crossing of some of their members (see *Osteomeles* p. 439) normally falling within the same pattern. We may not expatiate on the far-reaching and most interesting migrations of *Rubus*, merely mention (12) that the Mongolian *R. purpureus* is allied with the Southern Australian *R. macropodus*, and that the Chinese and Himalayan *R. lucens* and *R. Hookeri* have affinities in Mexico and Peru. This is precisely what we may expect not in the Rosaceae only, but any other large family. The whole, of course, makes no sense to those who insist that dispersal is from "Holarctis," or effected by "Extraordinary capacities for transoceanic dispersal.'

As the Ericaceae will better show, dispersal on both sides of the Atlantic is commonplace. At least 25 species in different families (13) are common to Southeastern Newfoundland or Cape Breton and Europe, all of which are characteristic of acid peats and siliceous soils. *Potentilla procumbens* is one of these species, and its presence in Newfoundland and Cape Breton automatic-

---

*) It is an unwritten rule among authors who use the English language, in particular, to do everything possible in order to standardize their figures, diagrams etc. The reader has undoubtedly noticed that in this manual we depart from this rule. This we do, not because we believe this rule to be unsound, rather the contrary. We prefer, however, to retain full freedom in this regard here. Accordingly, we sometimes figure all the angiospermous gates with their connections, etc., while on other occasions indicating merely one gate, or none at all. As we have amply discussed the purposes of our iconography (see p. 6), pointing out that it is used mainly to convey certain general ideas, we see no reason for constantly interfering with the thoughts of the readers themselves. It will soon be clear to them that, standardized or not, our figures convey definite, indeed very repetitious ideas. These ideas are doubtless standardized themselves around certain elementary concepts, which need not constantly be stressed graphically in order to be clear. Referring to Fig. 37 in particular, the reader will see that the purposeful omission in this case of symbols for the Magellanian and Western Polynesian gates (fully outlined on the contrary in Fig. 25, for example) changes nothing of the substances under discussion.

ally accounts also for the presence of *P. canadensis* in the United States. A classic example of circumatlantic distribution is in *Drosera*,

(1) *D. intermedia* — Europe: Northern Spain and Portugal, France, England, Ireland (locally abundant in the west, rare in the east), Belgium, Holland, Germany northward to Danemark and Sweden, eastward to the Carpathians, Russia, Asia Minor. *America:* Cuba and Santo Domingo, Newfoundland to Texas in all the states on the Atlantic front, inland to the Great Lakes region and Tennessee.

Patterns of distribution such as this are oftentimes used by students of northern floras as material to support tenuous theories of "transatlantic" crossings in high latitudes during the "Pleistocene" or the "Late Tertiary". These students can easily credit the "Ice Age" with marvellous effects upon dispersal because they overlook as a rule everything which might have gone before this golden age of confuse and confusing phytogeography. As we well know, there is strong indication to the effect that the shores once connecting South America and Africa were used by numerous unrelated groups to travel between the Caribbeans and the Mediterranean. As the map further shows (Fig. 38), the Atlantic shores of the Old and New World can be reached in a number of different ways, and from every point of the compass. The tracks outlined in this map are standard. They agree with main routes of migration we have already seen, and others still we are to study. Distributions in the high north, consequently, have a far larger scope than many authors suppose (see further discussion, p. 458). The initiates may follow only with mild amusement the theories of botanists who try to rationalize these distributions in the light of what they know of the floras of Boston, tidal Virginia, Gopher Prairie, etc. The map reveals that, as a matter of straight fact, both shores of the Atlantic, east and west, can be simultaneously reached by the same species, without this species ever actually crossing the Atlantic, but departing on the contrary from a standard phytogeographic center (*Altai Node*) in the Altai Mountains of Central Asia and their vicinity.

Next to settle the issue, by what tracks *D. intermedia* reached both shores of the Atlantic, we ought to know with precision whether this species is allied with the "boreal" *D. rotundifolia* or the "southern" *D. capillaris*. Considering that we are not informed of the matter here *), we may suggest two tenable alternatives, as follows, (a) The alliance of *D. intermedia* is with *D. rotundifolia*. In this case, the former might have travelled over the channels of migration generally outlined on the map of Fig. 38 (broken lines departing from

---

*) Nothing should be easier than to secure the data needed. For the purpose, (a) We ought to secure a reliable monographic work of *Drosera*; (b) Trace back to their gates the *genorheitra* of the various major units constituting the genus (subgenera, sections, etc.); (c) Pinpoint the tracks in their main outlines, next in detail. This will give us all the essentials required to handle the problem in the light of phytogeography. Next, we could secure ecologic data regarding the species of immediate interest to us; information of general nature about morphology, geologic features, etc., and cast the whole within the frame of a single broad discussion. Teachers of botany and phytogeography, not to mention ecology, cytology, geology, etc., can devise any number of such problems as the one just outlined. It is to be hoped that phytogeography will at last come into its own because of a much needed devotion to facts rather than to vague generalities and confuse theories.

the Altai Mountains); (b) The alliance of *D. intermedia*, on the contrary, lies in the direction of *D. capillaris*. In this event, the former might have followed the track indicated in Fig. 38 by a broken dotted line. As a third subsidiary hypothesis, *D. intermedia* could take the channel of migration which appears in Fig. 38 as a continuous line crossing the high north.

In submitting these hypotheses we do not intend to exclude *as an utter impossibility* that *D. intermedia* might have fared directly across the Atlantic between Newfoundland and Ireland, or the other way around. However, to follow this path *D. intermedia* must first have reached either one of these outposts, or one of them at the very least. *This is to say that even transatlantic migrations in the high north cannot be viewed too narrowly, because migrations of the sort are but part, as a rule, of much wider distributional circuits. A basic tenet of sound phytogeography is that the student of migrations must strive at first to map out the main streams of dispersal, only in a second time working out their details.* Too much, alas, has been written by botanists with parochial preoccupations, who managed to clutter the records of dispersal with half-baked theories of what plants ought to do, but in reality never did. We are firmly convinced that no one may hope to write on phytogeography successfully who is unable, or unwilling, to visualize at the start the whole of the earth as his own field.

It is convenient to follow this discussion with a brief review of the dispersal of the genera in the affinity of *Primula*. In this affinity (14) are six genera, namely, *Dionysia, Douglasia, Androsace, Cortusa, Stimpsonia* and *Ardisiandra*.

Dionysia consists of about twenty species which but sparingly range outside of Iran (Persia) to Afghanistan, Mesopotamia, Armenia and the Caspian shores. This dispersal ranks, of course, as a good match of that of the group *Persicae* of *Potentilla*. Iran is an important center of distribution in its own right. Not to insist upon such aggregates as *Dionysia* and the *Potentilla Persicae*, we find in Iran plants of definite tropical affinity. Some, like *Parrotia persica* of the Hamamelidaceae survive in a narrow strip of land immediately south of the Caspian Sea, and owe their origin to a migrational stream reaching this point either from the Far East (see *Coriaria;* p. 71) or Africa (see *Sapindaceae Kolreuterieae;* p. 61). Other, like the sterculiaceous genus *Glossostemon* occur, on the contrary to the south of the Iranian highlands (Western Iran, Mesopotamia). Peculiarly, this genus is said to have some affinity toward the African tropical genus *Sparrmannia* of the Tiliaceae. African ties also stand revealed by *Euphorbia larica* *), endemic to Southern Persia. Broadly

---

*) We have strong reason to believe that the "African" connections of *Euphorbia larica*, endemic to Southern Iran, would eventually lead us all over the channels of migration followed by *Thamnosma, Oligomeris* (see p. 91) etc., etc. Though the American species in the affinity of the Mexican *E. antisyphilitica* have appendiculate glands to the cyathium, still they seem to us to be related in the end with *E. lignosa* of Southwestern Africa and *E. larica*. A channel leading through central Africa between the Atlantic and Indian Oceans is further highlighted by the affinities between *E. sudanica* (West Africa) and the large group of the Indian *E. neriifolia*, which group may be traced, seemingly, as far eastward as New Guinea (*E. complanata*). Patent "Crossings" of the Atlantic are highlighted further by *Jatropha* (see *J. gossypiifolia*), *Croton* (see *C. lobatus*), *Dalechampia* (see *D. scandens* and its multi-

speaking, it may be stated that Iran and its geographic approaches preserve to this day relics of warm floras having affinities in the Far East and Africa, while the Iranian highlands proper served as a center of major speciation for groups such as *Dionysia* and *Potentilla*. It is a matter of regret for us that no more can be said of this interesting subject, but we may at least remark that legends such as that of an ancient "Garden of Eden" in Mesopotamia are not without foundation. A "Garden" of the kind could readily come into being at that point of the map, for right here crossed their courses most important tracks of migration from Africa and the tropical Far East.

Douglasia falls into two groups, as follows,

A) Subg. **Eudouglasia**

(1) *D. montana* — United States (Oregon, Montana, Wyoming).
(2) *D. arctica* — Canada (Northwestern Territories).
(3) *D. nivalis* — United States (Washington), Canada (British Columbia, Alberta).
(4) *D. dentata* — United States (Washington).
(5) *D. laevigata* — United States (Oregon, Washington).

B) Subg. **Gregoria**

(6) *D. Vitaliana* — Alps, Pyrenees, Southern Spain, Central Italy.

The dispersal would not alter, if Subg. *Gregoria* were to be treated as a full-fledged genus. Whether in one genus or two genera, the affinity we face is a single one, and its distribution is bound to be interpreted with reference to the same factors. Two main hypotheses offer themselves, as follows, (a) One of the massings, whether American or European, originated from the other by immediate transatlantic migration; (b) The American and European massings came to the parting of their ways in the genorheitral stage in a center by Central Asia, next reaching their ultimate ends by transparent routes overland.

The former alternative is most unlikely. The American massing is manifestly rooted in the west of the United States, in a region, that is, within immediate, most usual reach of tracks coming in from Asia. The European species is wanting a clear "atlantic" dispersal, though it reaches the Pyrenees and South-

tudinous forms), *Amanoa* (as genus), *Ricinus*, etc., etc. A more diluted line of affinities running south to north in the Atlantic is revealed by certain similarities among Moroccan and Canarian succulent Euphorbias (e. g., *E. resinifera* and *E. canariensis*) and forms of unsettled taxonomic status we have seen from Angola. A special development in tree-like succulent Euphorbias (group *Abyssinicae*) pinpoints on its part all the "Great Rift" from South Africa to Nubia. The euphorbiaceous worlds of South Africa and Madagascar could be compared only in an extensive critical treatment, which we regret to be out of our present possibilities. As the reader perceives, the mass of usable material before us is so great that only *very small part of it* can be introduced in this manual.

ern Spain. The latter is doubtless a more tenable hypothesis, which again calls into play certain possibilities, namely, (a) *Douglasia* has part in the flora of Asia, though it still remains to be formally discovered; (b) There exists in Central Asia, the Himalayas, Afghanistan, Southwestern China, or some other region in the vicinity of those just mentioned, a group, now treated as a genus other than *Douglasia*, which has in actuality characters derived from it. In other words, the disconnection existing between the American and European massings of *Douglasia* is filled by a vicariant; (d) Central Asia is not suited ecologically to the survival of *Douglasia*.

The first and last of these possibilities, the last especially, are unlikely. The second is such that we may not critically approach here, but remains to be thoroughly explored, in the anticipation we may learn something of interest from the standpoint of systematic botany and dispersal alike. "Parallel Evolution" and "Polyphiletism" are often spoken of in biological work, but so loosely and so vaguely as normally to lack serious meaning. *Douglasia*, perhaps, might be a case of "Parallel Evolution," if it could be established that Subg. *Eudouglasia* and Subg. *Gregoria* developed in distinct geographic centers. Even so, these two subgenera would come back to a single identifiable *genorheitron*, renewed proof that, to be meaningful, "Parallel Evolution" and "Polyphiletism" must rest upon something tangible, not vague statements and theories.

Androsace is a large genus of very difficult classification. The following species may be taken as representative of its range.

(1) *A. saxifragifolia* *) — Western Himalayas, Indochina (Tonkin), Philippines (Luzon), Formosa, China, Riu-kiu, Japan, Corea, Manchuria.

(2) *A. Chamaesjasme* — Var. *typica:* Central Asia (Alatau Mts.), high mountains of Europe, Tibet, China, Eastern Siberia to Colorado; Var. *carinata:* Colorado; Var. *capitata:* Kuriles; Var. *paramushirensis:* Kuriles (Island of Paramushiro); Var. *ciliata:* Novaya Zemlia; Var. *coronata:* Tibet.

(3) *A. villosa* — Var. *typica:* Altais Mts. to Persia, Transbaikalia and Siberia, Caucasus, Crimea, the mountains of Europe (Carpathians, Balkans, Alps, Apennines, Pyrenees); Var. *robusta:* Afghanistan, Western Himalayas, Tibet; Var. *bisulca:* China (Szechuan); Var. *incana:* Central Asia (Dahuria).

(4) *A. caespitosa* — Iran (Persia), Eastern Siberia.

(5) *A. septentrionalis* — Circumboreal to the exclusion of Greenland, mostly in the high north; in America in North Ellesmere and North Baffin Lands.

The distribution of this genus is transparent. *Androsace* fully connects this time the ranges of *Dionysia* and *Douglasia*, and the parting of the ways to which the *genorheitron* of all these genera belong may be approximately set in the region of the Altai Mountains in Central Asia.

Cortusa consists of but two species, which extend from Afghanistan and

---

*) This species has been re-named *Androsace umbellata.* For comments on the score of this, and similar changes in nomenclature see CORNER in Gard. Bull. Straits Settl. 10: 19, 29, 30, 71, 73 etc. 1939.

Turkestan to the Eastern Himalayas, the Altais, China, Japan and the high mountains of Europe. *Stimpsonia* is monotypic in Southeastern China (Fukien) and the Riu-Kiu Archipelago. *Ardisiandra* migrates on its part (15) in the following manner,

(1)   *A. orientalis* — Southern Rhodesia, Tanganyka, Eastern Belgian Congo, Uganda, Kenya.

(2)   *A. Stolzii* — Tanganyka.

(3)   *A. Engleri* — Tanganyka, Uganda.

(4)   *A. primuloides* — Tanganyka.

(5)   *A. sibthorpioides* — Cameroon.

With this before us, we may now draw the conclusion which the facts warrant. The like Fig. 39 reveals, the dispersal of all these genera takes its most logical departure from the African gate of angiospermy, and from this gate onward proceeds along standard lines throughout. The Altai Mountains (see also Fig. 38) rank foremost as a center of redistribution in the Northern Hemisphere, west and east. The dispersal of *Primula* (see Fig. 40) falls in with that of the Primulaceous genera shown in Fig. 39. Further considering that Myrsinaceae (see p. 85) and Primulaceae are inextricably mixed in phylogeny, it proves impossible to dismiss the Magellanian outlier of *Primula* as hardly worthy of reckoning *). This outlier is in reality fundamental. It will further be observed that the dispersal of the affinity of *Primula* parallels the distribution of *Lysimachia*. The tracks of *Primula*, and its affinity, most likely depart like those of *Lysimachia* (see p. 53, Fig. 17) from a point in the vicinity of the modern Mascarenes, but the migrations of the latter keep to a southern course by comparison with those of the former. *Primula* and its allies point first toward the heart of Central Asia (see also *Sapindaceae Koelreuterieae*, p. 61), and redistribute themselves east and west from an Altaian node, leaving in Iran the large component of *Dionysia* which took the tableland of that country as its center of speciation. A tiny outlier of *Primula* fared to Magellania, and it should be downright unwarranted to believe that this outlier travelled southward along the Andes; when the primuloid *genorheitron* first stirred in migration the Andes were as yet unknown on the map. *Lysimachia*, on the contrary, kept a southermost course as it befits a genus of warm preferences. It left the Mascarenes to hit India, Southern China, the Bonin Islands (Ogasawara), Hawaii, and reached in the end the New World. The Myrsinaceae ranged on their own, as we have seen (see p. 90), but they, too, departed

---

*) Like *Carex*, *Primula* is a stupenduous genus, worth the attention of keen phytogeographers. We regret we cannot deal with its dispersal in a fitting manner in this manual. Even a casual glance at the map of its distribution reveals, (a) The range is beyond doubt influenced by definite ecological factors (see absence of *Primula* in the Russian steppes); (b) These factors may have become operative, however, much later than the original migrations (see absence of *Primula* in the Tibetan highlands and the Gobi, but its distribution peripheral to these ranges, which suggests immediate effects from the Himalayan uplift in the Tertiary; see further relic-distribution in Arabia, etc.); (c) The „Andean bridge", so called, leading „cold" flora northward had chequered and obscure geological history. *Primula* could not use it, while *Saxifraga* did. The Magellanian gate, consequently, operated only in part, and its influence was lessened by geologic factors, seemingly, operative more or less in what is now Central Chile.

from standard gates of angiospermy, following next standard tracks. In brief: the whole, Primulaceae and Myrsinaceae, is orderly, we should say, conventional in migration, and the basic tracks and centers we submit for the attention of the reader take care of both families as one, thus respecting the integrity of the *genorheitron* common to both. These tracks rationally account for the presence of *Primula* in Magellania, Java, Sumatra; for the large speciation of *Dionysia* in Iran, etc.

It stands to reason that, not unlike the Rosaceae and Primulaceae, *Potentilla* used standard channels of migration, too. One of these channels runs in the Pacific (*P. anserina*), the other in the Indian Ocean (*P. sundaica* and its affinity).

VI. Ericaceae

The classification of the Ericaceae is exceedingly controversial, both as regards generic and specific limits, and no recent monograph is available for the whole of this family. Lack of more suitable bibliography forces us to rely upon an old work (16) of the shortcomings of which we are keenly aware. It will be seen, however, that modern phytogeography can be written also with the help of nothing better than ancient works.

(a) Ericaceae Ericoideae

This subfamily is well defined on the whole, and its dispersal (Fig. 41) closely knit. There is no agreement among botanists as regards its generic limits. An author (17) believes that there are twenty genera represented in the flora of South Africa, but another (18) cuts down this figure to eleven. The classification of the latter reads as follows,

(1) *Salaxis* (*Coccospermum, Lagenocarpus, Lepterica*) — South Africa.
(2) *Codonostigma* (*Syndesmanthus*) — South Africa.
(3) *Scyphogene* (*Coilostigma*) — South Africa.
(4) *Sympieza* (*Aniserica*) — South Africa.
(5) *Simochilus* (*Anomalanthus, Syndesmanthus*) — South Africa.
(6) *Eremia* (*Acrostemon, Grisebachia, Hexastemon, Thoracospermum, Thamnus*) — South Africa.
(7) *Macnabia* (*Erica* in part) — South Africa.
(8) *Ericinella* — South and Tropical Africa.
(9) *Blaeria* — South and Tropical Africa.
(10) *Philippia* — South and Tropical Africa.
(11) *Erica* — South and Tropical Africa, Europe, North America.

To this list of genera restricted in the main to Africa should be added two extra-African genera, namely.

(12) *Calluna* — Mediterranean and Europe generally to Western Siberia, the Caucasus, Asia Minor, North America.

(13) *Bruckenthalia* — Balkans (Albania, Macedonia, Jugoslavia, Carpathian Mountains).

Before beginning to discuss the Ericoideae, we ought to place on record an observation of interest for the whole of the Ericaceae. Available descriptions of genera such as *Coccospermum* and *Lagenocarpus* convey a positive suggestion that forms of the kind are reasonably close to the Empetraceae. This suggestion is all the more reasonable, in that it has been established by studies in embryology (19) that Ericaceae and Empetraceae are related by genorheitral ties. We may not discuss here at length the dispersal of the Empetraceae, at the most remark that this dispersal is "bipolar" in the South Atlantic and Magellania, and all the hyperboreal region, as regards *Empetrum;* conventionally "atlantic" (that is, effected to lands adjacent the Atlantic both in Europe and North America) as concerns *Corema* and *Ceratiola*. The modern Atlantic, consequently, is the keystone of the empetraceous migrations, which (see p. 92) suggest those of *Littorella*. Considering that the Empetraceae are keyed up to the southern ranges of the South Atlantic, and that the Ericaceae Ericoideae are massive in South Africa, we may readily conclude that embryology and phytogeography tally to drive home the conclusion that Ericaceae and Empetraceae belong the to same *genorheitron*. A further study of these two families may be left to some of the readers of these pages, in the assurance that their toils will be amply rewarded.

As the map shows (Fig. 42; 20), the distribution of the species of *Philippia* in the immediate affinity of *P. Mannii* does not stream from South Africa northward to agree with the modern continental outline. Taking its start from the Mascarenes, this track next invades Madagascar, and following through the Comoros finally enters Continental Africa in Usambara *). From this region onward, the track streams westward across the Congo, eventually to stop at the Cameroon and the islands of the Gulf of Guinea. It is manifest by the map that Nyasaland and the Eastern Cape are simultaneously invaded from the direction of Madagascar.

Though unreported from Madagascar or the Mascarenes, *Blaeria* migrates somewhat like *Philippia*. *Blaeria* neatly breaks into two units, however, a smaller one of about nine species (Subg. *Eublaeria*) endemic to the Cape, and a much larger one of twentyone (Subg. *Blaeriastrum*) which extends as far south as a line run approximately from Nyasaland to Angola.

Homologous dispersal occurs in the Tiliaceae (genus *Sparrmannia*), and were space no consideration with us numerous other homologous migrants could be dealt with in these pages. The iridaceous genus *Aristea*, for example **) contains one of the most telling examples of invasion of Africa from the east. This genus falls in three massings, (a) *Cape:* 25 species, mostly localized relic-

---

*) Usambara and Somaliland are most interesting regions of Africa for the phytogeographer. They still require study.

**) This example is from Weimarck's meaty study of dispersal in South Africa (in Lunds Univ. Arsskr., N. F., Avd. 2, 36 (1). 1940).

endemics; (b) *Drakensberg (Eastern Transvaal) and African Mountains:* 18 species; (c) *Madagascar:* 6 species.

One of the species of Madagascar, *A. cladocarpa* has isolated position, but all others are most nearly related with forms in the Drakensberg Mountains and the heights, generally, of East Africa. The Madagascan *A. nitida* is germane of *A. Goetzei* (Tanganyka: Uluguru), and this latter stands in turn related with *A. alata* (Rhodesia, Nyasaland: Mt. Mlanje, Tanganyka, Kenya, Belgian Congo, Ethiopia, Cameroon, Northern Nigeria).

If we study this dispersal from a point of vantage south and east of Madagascar we readily give ourselves reason of all its peculiarities, not less than of the peculiarities of the distribution of the ericaceous and tiliaceous groups formerly mentioned. To begin with, the Cape has in this type of dispersal very much the same situation which West Australia holds in another sector of the modern world. The Cape is a blind alley, fed in the main from the east by a door opening between modern Natal and Madagascar. It is a domain which has remained in an "oceanic" position, to borrow the term employed by the author from whom we have taken the example of *Aristea*, and out of geologic vicissitudes of its own has caused a rich speciation to take place in the ranks of the ancient *genorheitra* that fell its lot, the archetypes of *Erica* being one of them.

In stressing the fact that the Cape was fed by a door opening between Natal and Madagascar in the main, we intend to state our belief that this door was doubtless not the only one active. The Cape was also fed from other sources, probably in the west, some of them perhaps very important but as yet most obscure (see fossil pollen from Southwest Africa; p. 93). We feel confident that students of the details of the flora of South Africa may eventually add a great deal to the few words here stated as regards the doors that poured angiospermous life in what is now the Cape region *)

A track leading through Tanganyka to the Cameroon is one of the most impórtant in dispersal. This channel of migration is ultimately reponsible for having poured into Northeastern South America, and the West Indies, a powerful stream of angiospermous *genorheitra*. This channel was in addition followed by certain ferns (see p. 500) of a primitive type. The stretch Tanganyka-Cameroons rates in phytogeography as a quick, direct link between tropical South America and the Far East which stems from the heart of an ancient continent once filling the whole of the Indian Ocean.

Aristea alata highlights this basic track from beginning to end. The stations follow along its course without break, Rhodesia; Mount Mlanje in Nyasaland, a classic landmark of African dispersal; Tanganyka and Kenya; Belgian Congo with an outlier in Ethiopia; finally, Cameroon, and Northern Nigeria. It will

---

*) A critical comparison of the succulent euphorbiaceous groups of Madagascar and the Cape would be most interesting. We are satisfied that the Cape (and South Africa) *Euphorbia* sect. *Treisia* has affinity with the Madagascan groups of *Euphorbia* centering around *E. splendens* and *E. lophogona*. This affinity is not immediately obvious, but clear nevertheless. The two groups in question are supposedly to be strictly local. *Drosera regia* and similar elements, on the contrary, may not stem from a track coming to the Cape from the east. See further, p. 58.

be observed that this highway first proceeds in a northerly direction, and only later veers westward. It may be a coincidence but, if at all so, a remarkable one that, (a) The dispersal of the cycadaceous genus *Encephalartos* (see Fig. 93) is shaped approximately in the same manner as the track of *A. alata*, and an additional cycadaceous relic, *Stangeria*, is narrowly localized in Natal; (b) Natal is directly connected with a lost range in the southern waters of the Atlantic and Indian Ocean by certain Liverworts (see p. 509); (d) Tracks directly running from Rhodesia and adjacent regions to Angola seem to belong in some majority to plants of moist ranges (e.g., *Drosera, Triumfetta;* see for this p. 419), one of their genera, *Genlisea*, having possibly originated (see p. 351 for distribution) right here. Things geographical shape up in these quarters as if the dispersal of *Aristea alata* (and that of *Encephalartos* as well) pinpoint an ancient shore open to the west, and next the south, upon a region that probably stood in alternative rhythms of emersion and submersion, a shallow sea or large lakes, in sum. This sea, or lakes, had islands or peninsulae in the west (heights of Angola), and in the south (the Cape).

So much, then, for the direction of the tracks coming to Africa in the first place. *These tracks— let us repeat — do not follow modern continental outlines at all. They invest Africa from a center located in the modern Indian Ocean which agrees with part of what we have identified as the African gate of angiospermy. So much also for the geography of the lands that later became modern Africa.* That these lands were first reached by angiospermous life in ages long past — the Late Jurassic as most probable — can further be shown with reference to the taxonomic behavior of certain ericaceous genera. *Philippia,* for example, is so difficult from the taxonomic angle that authors of equal experience radically disagree as to the number of its species. One believes, for instance (20), that these are 32 in Madagascar alone, while another (22) is of the opinion they are only half as many. It is further known that in *Philippia* and *Blaeria* occur very distinct and stable species usually endemic to a single alpine locality, perhaps only a few crags, while the forms of the lowlands are genetically unstable and for this reason most difficult to classify. That groups behaving in this manner may be ancient no botanist will venture to deny.

Erica is massive in the Cape and regions immediately adjacent. Here occur some 500 of its species, distributed in five subgenera totalling to about fortyone sections; outside of the Cape and its immediate vicinity *Erica* has no more than 25 to 30 species.

Three of the species alien to the Cape are distributed as follows,

(1) *E. arborea* — Eastern Afrika (Tanganyka: *Kilimajaro;* Uganda: *Ruwenzori;* Ethiopia), Tunisia, Algeria, Morocco, Canary Islands, Madera, Portugal, Spain, Italy to the Alps, Southern France, Corsica, Tyrol, Dalmatia, Crete, Samos, Mytilene, Greece, Macedonia, Asia Minor. *)

---

*) Specialized investigations cannot have place in the page of this manual. We suggest that the dispersal of *Erica arborea* be compared with that (see p. 90) of *Myrsine africana*. Both these dispersals meet in the

(2) *E. cinerea* — Northern Spain and Portugal, Western France and North-western Italy, Ireland to the Faer-Oes, Southwestern Norway to Bergen; North America (Nantucket Island northward probably to Newfoundland in scattered coastal colonies.)

(3) *E. terminalis* — Morocco, Southern Spain, Southern Italy, Sardinia, Corsica.

These three species are allied from the standpoint of phylogeny. Their dispersals give a track, which barring an understandable disconnection in the Sahara, leads all the way from Eastern Africa to, respectively, Newfoundland and Norway. Some taxonomists are inclined to think that *E. cinerea* is adventitious in the New World, but we believe, on the contrary, it is a legitimate relic-form, not at all a casual migrant (see further dispersal of *Drosera intermedia* and *Calluna*, p. 155, 166).

The distribution of the three species just discussed suggests important observations. These observations require, however, certain preliminary remarks, as follows: (a) *Erica arborea* is allied with *E. Whyteana*, endemic to Mt. Mlanje in Nyasaland. Another of the Heaths native to this mount, *E. Johnstoniana*, is related to *E. Solandra* which occurs in Natal (Van Reenens Pass) and the Southern Cape (George Division). *Erica trichoclada* of Natal exhibits affinity in the direction of *E. leptoclada*, found only in the Cape (Piquetberg, Paarl and Caledon Divisions), at a distance of about 800 miles from the point where *E. trichoclada* is known to occur; (b) The track of *E. arborea* begins to run at the Kilimajaro and Ruwenzori (Tanganyka and Uganda, respectively), leading from this center on to Ethiopia first. It is probable that the affinities underscored above are not the only ones among Heaths of Northern and Southern South Africa, and that, indeed, these affinities are far more pervasive than we know at the moment. *Erica arborea* might perhaps occur in some isolated stations to the immediate south of the heights mentioned, but we have reason to believe that these heights mark in reality the inception of the regular track northward.

We observe, (a) A chain of affinities can be traced all the way from Tanganyka and Uganda *(E. arborea)* through Mt. Mlanje in Nyasaland *(E. Whyteana, E. Johnstoniana)* as far as Natal *(E. Solandra, E. trichoclada)* and the Cape *(E. leptoclada)*. This chain of affinities makes it possible to tie together the Cape, Norway, Newfoundland at one stroke; (b) The sector Nyasaland (Mt. Mlanje in particular) — Tanganyka (Mt. Kilimajaro especially), is a region of transition, a very sensitive hinge of distribution indeed; (c) This sector has intimate ties with Madagascar; (d) Tanganyka is the starting points of two tracks, (i) Leading through the Congo to Cameroon, often beyond to the West Indies, the Guianas and Eastern South America generally; (ii) Faring to Ethiopia first and next through a speculative path to the Mediterranean

---

Macaronesian domain (Azores, Canary Islands, Madera), but the channels followed by *Erica* and *Myrsine* are so different that much is to be learned from their study. Clearly, we face forms with different ecological requirements in origin.

throughout; (e) Ethiopia, though in itself the first station of track (ii), can be invested by elements using track (i), which is proved by *Aristea's* course.

In conclusions, Mt. Mlanje is a cardinal landmark in dispersal. From its slopes we can drive tracks northward or southward to reach hyperboreal lands or the relic-ranges of the Cape, not only, but freely to travel to the Indian Ocean through Madagascar and the Mascarenes. Bergen, Quebec, Capetown, Tananarive are all within reach from this one mountain.

The stretch between Tanganyka and Ethiopia (we might almost say the Kilimajaro and the Ruwenzori) is not less vital to dispersal. It is here that the decision is finally made whether a plant is to be native to Manáos or Athens, because it is here that two of the main tracks in all the phytogeographic world take their ultimate start.

We comment that the track of such a form as *Erica arborea* is speculative between Ethiopia and the Mediterranean. This is true, as such, but reference to *Olea* (see p. 354) cuts short the speculative element. *Olea* reveals (see, in particular, *O. Laperrinei*) that the mountains of the Hoggar and Tibesti in Central Sahara are in intimate connection with the Macaronesian domain (Madera, for instance); Cyrenaica and points beyond such as Sicily, the Southern Balkans, possibly Crete and Cyprus; Ethiopia itself.

Plant-life stemming from a center situated in the waters of the modern Indian Ocean (i.e., originating at the eastern end of the African gate of angiospermy) can, consequently, reach the New World anywhere between the Eastern United States and the northern boundary of Uruguay by two main tracks, as follows, (i) *Indian Ocean—Tanganyka (Ethiopia)—Congo—Cameroon—America;* or and (ii) *Indian Ocean—Ethiopia—Central Sahara—Macaronesia (Mediterranean)—America.* Plant-life from the same center further can reach the whole of Central Asia, the Far East, the High North throughout, Malaysia, all lands fronting the Western Pacific generally, Polynesia and, finally, the New World again by a number of standard tracks outlined throughout the pages of this manual. These tracks may run the first lap of their course over oceanic disconnections, or, sometimes, partly overland witness *Primula* sect. *Floribundae*, which leaves Ethiopia to reach Arabia (Sinai, heights of Southeastern Arabia), Beluchistan, Afghanistan and regions adjacent. *)

The tracks just outlined, or mentioned, call into play nothing that may be said to be speculative or theoretical. They are drawn on the strength of the

---

*) We regret we may not fully discuss numerous interesting issues of detail concerning the course of the primary channels of migration briefly outlined in the main text. The reader ought to be warned, at any rate, that the flora of the centers reached by these tracks may contain elements that originally fared over them, together with elements of a much later derivation. With immediate reference to the flora of the Hoggar in the Central Sahara (see MAIRE, R., in „Études sur la Flore et la Végétation du Sahara Central" 2 [Mission du Hoggar, No. 3: Mém. Soc. Hist. Nat. Afrique Nord]: 48 *et seq.* 1933), we have some reason to affirm that *Myrtus Nivellii, Olea Laperrinii Cupressus Dupreziana,* etc., belong to the vegetation that first reached the Hoggar. Elements like *Calotropis procera, Ifloga spicata,* etc., may, on the contrary, be part of vegetation which reached the Hoggar when this region began to undergo active drying-up. It is accordingly easy to argue that the flora of the Central Sahara has „Definite affinities with that of Arabia, Iran, Western India, etc.," which nobody would attempt to deny. However, the ultimate origin of this flora, and the course of the original primary tracks, are much sooner determined

factual record of living plants, and critical comparisons of their affinities. They must be accepted, consequently, very much as they read, and if discussed must be strictly discussed on a factual basis. *It stands to reason that a point close by the heart of the modern Indian Ocean could not occupy the capital position it holds in respect of dispersal throughout the world, if a continental mass had not stood there in the geological past. A massive body of evidence argues for the existence of this land, drawn from sciences wholly unrelated with botany, and botany further contributes its own mite of confirmation. If we agree to identify this land as Gondwana, the affirmation is in order, without restrictions, qualifications or apologies, that Gondwana is not a theory, but a tangible fact. Gondwana has crumbled out of our maps, but it lives through, and by, the dispersal it has called into being on these very same maps.*

Calluna is distributed (23) as the map (Fig. 43) shows. In our firm opinion, its petty New World outliers reached America much sooner over a track leading from Macaronesia to the Caribbeans, thencefrom to Mexico and North-eastern America as far as the Great Lakes, than by crossing the Atlantic in high latitudes in the „Glacial Age." The tracks of *Calluna* are standard, for they tally with the general course of the migrations of *Erica*. *Calluna* is *Erica's* vicariant in the north. Seen by a morphologist and a taxonomist the genera are here patently two, but approached by a phylogenist and a phytogeographer the affinity in play manifestly stems from one and the same *genorheitron*. *Calluna* perhaps "originated" in the north *in the purely geographic sense of the term, certainly not in other senses.* We advisely specify *perhaps*, because *Calluna* might very well have originated — *even geographically* — in ranges to the south of the Mediterranean, where unfavorable ecological conditions later on blotted it out. Let it be repeated that the term, originate, cannot be loosely used ever.

The dispersal of *Erica* subg. *Pentaptera* (Fig. 43) is interesting because it contributes additional inferences to the effect that *Erica* indeed crossed the modern Sahara toward the Mediterranean, and highlights notorious connections (24) between the flora of Cyrenaica and that of the Eastern Mediterranean generally. The monotypic genus *Bruckenthalia* (Fig. 43) is confined on its part to Northern Asia Minor and the Balkans, to centers, that is, most ancient both from the standpoint of dispersal and palaeogeography. Migrations of the Ericoideae from centers primarily located in "Holarctis," West Africa (25) or the like are clearly wholly out of question.

---

in reference to elements like *Myrtus, Olea, Cupressus,* possibly *Erica* (which last, most likely, was also part of the original Hoggar flora). Guilty of serious blunders are authors who, dimissing these elements uncritically as "relics," compile pure statistical tabulations, which ultimately lead them to conclude that the flora of a certain region is massively "Indian," "Arabian," etc. Blunders of this kind are very common as regards the flora of the so called "Oceanic" islands.

(c) Ericaceae Gaultherioideae

These Ericaceae (Fig. 44, 44 A) are dispersed, always according to the same authority, in this fashion,

(1) *Wittsteinia* — Australia (Victoria).

(2) *Pernettya* — Fuegia to Mexico, Galápagos Islands, New Zealand, Tasmania.

(3) *Gaultheria* — New World northward from Brazil and Chile; New Zealand, Tasmania, Australia, New Guinea, Malaysia to Malacca, India, Burma, Formosa, Japan.

(4) *Diplycosia* — New Guinea, Philippines, Malaysia generally to Malacca, Burma, India, China.

(5) *Chiogenes* — Japan to Atlantic North America, British Columbia and Idaho.

Nothing deeper than a glance at the distribution — whether this be based on the records made by genera taxonomically "perfect," or not, does not matter — reveals that this subfamily has the Pacific, not the Indian Ocean, as its immediate field of action. The Gaultherioideae are unreported from Africa, which should rule out any connection between them and the African gate (Indian Ocean; Gondwana). However, the *genorheitron* of the Ericoideae, Andromedoideae and Vaccinioideae turns up at the African gate, not only, but the Vaccinioideae are quite active in the Pacific, as we soon as to see. If we further consider that a *genorheitron* thoroughly well rooted in the Indian Ocean and the African gate (see Passifloraceae, Caricaceae and Achariaceae, p. 316), may dispatch one of its component immediately to the Pacific with no trace of it appearing in Africa *(Passiflora)*, we cannot be charged with rashness because we believe that the immediate *genorheitron* of the Gaultherioideae behaved like that of the *Passiflora* affinity. Accordingly, we draw the migrations of the Gaultherioideae from a point in the Indian Ocean (African gate) south of Africa to the vicinity of Magellania (Magellanian gate), where the *genorheitron* begins to unload *(Pernettya,* in part) *). The migration next flows to the Western Pacific *(Pernettya,* again in part, in Tasmania and New Zealand). From the Western Pacific, the track veers northward in the main, crosses Malaysia, sends an outlier to India, speeds further through the Far East to Japan, and ends its run hitting the New World. It is most likely that the whole of the Americas is reached from the north southward in the manner of *Magnolia* and *Talauma* (see p. 126) rather than by some cryptic outlier faring through the Atlantic from the Indian Ocean.

This pattern of migration, then, is tame and conventional. It has much suggesting *Euphrasia's* (see p. 19, Fig. 6), and a most obdurate will-to-believe is necessary to claim that the Gaultherioideae are "Holarctic".

One of the most interesting aspect of the migrations of the Gaultherioideae is the occurrence of a species of *Pernettya* in the Juan Fernandez Archipelago

---

*) See also Fig. 44 A, using a base-line of dispersal in the Pacific as alternative.

and the Galápagos Islands alike. A track of this nature (Juan Fernandez-Galápagos) is nothing unusual, of course, because dispersals of this extent and nature are commonplace between the vicinity of Juan Fernandez and Hawaii, not to speak of the Galápagos and the Revilla Gigedos (see *Lepechinia, Fragaria, Cuscuta, Nama,* alternatives in the range of *Phrygilanthus,* etc.).

In our opinion, the presence of a species of *Pernettya* endemic to the Galápagos (*P. Howellii:* Indefatigable Island, on summit of Mt. Crocker) disposes forthwith and most finally of any claim that these islands are "oceanic," and were colonized by "Casual means of dispersal." These facile assumptions are belied by the circumstance that *Pernettya* travels a well worn track to reach the Galápagos, and belongs beyond doubt to the ancient flora of those islands. The well authenticated occurrence of a species of this nature on an island tells us that this island may have undergone many geologic changes in the course of its age-long history without for this ever becoming "oceanic." Obviously, a long and involved history of submersions and emersions with attending climatic and edaphic changes tend to destroy all delicate endemics, those most particularly rooted in set climaxes, and to favor the survival and spread of that part of the local flora which is essentially "weedy." Magnificent rain-forests with but a thin fringe of "weedy" or "secondary" growth at their margins can be blotted out by slight climatic changes to the very exception of this "weedy" growth which in due course of time occupies all the range previously held by the rain-forest. A small island seldom offers the varied ecology which is a prerequisite of the existence of a rich and multifarious flora. When setting foot upon an island of this description, the casual observer is greeted at every turn by plants common to most tropical shores or their immediate hinterland, and all too readily argues that the very few ancient endemisms he finds are insignificant, perhaps the byproduct of the casual visit of a bird, lizard or such denizen of forgotten worlds. *The truth is quite to the contrary,* because a single one well authenticated endemism of the nature of *Pernettya Howellii* sets to naught such arguments. No discussion is possible on the score in our ripe judgement, because rigorous phytogeography and precise ecology dispose of it with finality. In reality, the degradation attending the ancient autochthonous flora of such islands as the Galápagos, Revilla Gigedo, Saint-Helena, Mauritius etc., etc., is homologous in essence of the destruction which in geologic eras past befell floras that the Angiosperms were called upon to replace. *Plants live by associations and associations tend to establish climaxes, which is to say that all plants gradually acquire habits threatening in the end their own survival.*

In conclusion, the Galápagos are definitely not "oceanic," because a handful of ancient endemisms is enough to prove otherwise. The Galápagos lie within a definite major track of dispersal, to which *Pernettya* bears uncontrovertible testimony.

(d) Ericaceae Vaccinioideae

The classification of the Vaccinioideae is quite as obscure, if indeed not more so, than that of the ericoid subfamilies we have reviewed thus far (see as proof, 29 and 30). We should be utterly lost if we were to attempt to straighten up this classification as the condition of using it successfully in phytogeographical work. Fortunately for us, the distribution of this subfamily as such, offers little difficulty.

Vaccinium, a very large and most controversial genus, may conveniently be elected here as standard (Fig. 45) of the dispersal of the Vaccinioideae. *Vaccinium* is represented in Africa by about half a dozen species, of which three to five (26) are endemic to Madagascar. One species, *V. africanum*, is native to Continental Africa (Nyasaland: Mt. Mlanje); a second *(V. Exul)* occurs in the same general vicinity (Transvaal, near Barberton) *)

It readily transpires that *Vaccinium* occupies in the Dark Continent the very same initial position held by *Philippia*, *Blaeria* and *Agauria*. We mean by this statement that *Vaccinium* entered Africa over and across the very same connections which allowed entrance to the three genera named above. It is clear, accordingly, that *Vaccinium* used the African gate of angiospermy, too.

Vaccinium, however, remained dormant in Africa. It speciated but very sparingly in this continent and no track to match *Erica*, *Philippia*, *Blaeria* and *Agauria* issued from the small initial lodgment which *Vaccinium* effected in the Dark Continent. According to present indications, the *Vaccinium* populations endemic to Ceylon are unrelated with those of Africa which suggests that no track ran for this genus between Madagascar and Ceylon. **)

Magellania and Fuegia offer, as such, ideal conditions to *Vaccinium*, if we understand as "ideal" the conditions which this genus now prefers in the Northern Hemisphere. Strange to say, *Vaccinium* is unknown to the flora of these regions so that we are bound to conclude that this genus never reached the southermost tip of the Americas at all.

Considering that *Vaccinium* is endemic to Africa, but here inactive; that the African *Vaccinium* is not related (according to present indications, be this clear) with the Ceylonese ***); that *Vaccinium* is unknown in Magellania and

---

*) The region of Barberton is most interesting from the standpoint of phytogeography. We regret we are not at all informed of the ecology of the species of *Vaccinium* endemic here and at Mt. Mlanje. It should be interesting to learn, if at all possible, why *Vaccinium* stopped between Barberton and Mt. Mlanje, when, in theory, it easily could run the whole of the tracks of *Erica* from this center.

**) Purely as a matter of personal impression, we are inclined to assume as a working hypothesis that a track from the islands off the eastern coast of South Africa to Ceylon begins as a rule much sooner at the Mascarenes than in Madagascar. In other words, it might be that a form common to Ceylon (and further on India, etc.), Madagascar, and the Mascarenes, reached Madagascar by a secondary migration only, leaving the main track to run to Ceylon directly from the Mascarenes.

***) Let us suppose it could be proved in the future that the species of *Vaccinium* endemic to Ceylon are, wholly or partly, allied with the African contingent. This would merely add a standard track (Madagascar—India) to the net of the dispersal of *Vaccinium*, but in no way invalidate our conclusions, even less our approaches to the issues of phytogeography.

Fuegia, and reaches its meridional range in South America (31) with *V. thibaudioides* native to Bolivia (Cochabamba); that, lastly, *Vaccinium* is thoroughly well represented in the Northern Hemisphere and even part of the Southern; considering all this, we are bound to conclude that this genus runs its main tracks to and in the Pacific, the Western Pacific of our maps to be precise.

As a matter of fact, *Vaccinium* tenants the Pacific, and we meet it there, where those who still think of it as "holarctic" would not anticipate its presence. Five species are endemic to Hawaii, while others variously range in Southeastern Polynesia (*V. cereum* in Tahiti, Society Islands, Rarotonga, Tubuai; *V. rapae* in Rapa; *V. raiateense* in Raiatea; *V. Macgillivrayi* in the New Hebrides). All these species, belong to a single affinity, Sect. *Macropelma*. Still another species, this time of unknown sectional affinities, is native to the island of Savaii.

It may be sheer coincidence that this species, *V. Whitmeei*, turns up in this particular island, but, if so, this coincidence is worth notice. Savai lies by Samoa, which latter even a firm believer in the miraculous power of "Casual transoceanic means of distribution" (see p. 70) visualizes as a primary center of avifauna throughout Polynesia. It is further well known that Samoa, Fiji, New Caledonia rank as fundamental outposts from the standpoint of phytogeography. These islands doubtless bear to a world that was significant relationships, and to this they owe their faunal and floristic peculiarities.

There can be no doubt, that the Southern Pacific of our maps — not any fabulous "Holarctis" or "Caribea" — is a fundamental center of evolution for the Ericaceae. *Wittsteinia*, a monotypic genus intermediate between the Gaultherioideae and the Vaccinioideae, is endemic to Southern Australia. *Pernettya* and *Gaultheria* become confluent (27, 28) in the vicinity of certain species of New Zealand that are markedly unstable as regards carpic characters. *Pernettya nana* has peculiarities of the anthers suggestive of the American populations of the genus. *Prionotes*, a genus of two species, one in Fuegia, the other in Tasmania, is believed on good authority (32) to be the connecting link between the Ericaceae and the Epacridaceae. Nothing of this surprising, for the Pacific is the ocean where vagaries of the kind are usual (sec p. 62, 208).

If we now consider the cleancut ties that bind the Pacific with the Indian Ocean (see *Lepechinia*, *Juania*, *Protium*, etc.), and the fact that these ties sometimes work without Continental Africa being called into play at all, we readily understand why the Ericaceae now stress the Indian Ocean, then again the Pacific; sometimes turn up massively in Africa, or there are not seen at all.

If *Vaccinium* is not in Magellania but otherwise turns up in the Pacific, it is in the Western Pacific that we must seek the inception of its migrations northward. Indeed, here we find the main track that leads northward in approved fashion. *Vaccinium*, still scanty in New Caledonia, swells in a mighty *crescendo* of speciation in New Guinea, Malaysia and the Far East. Firmly established in these quarters, and in this behaving very much like *Euphrasia* (see p. 19), *Vaccinium* takes all the north by storm. Re-routing at the *Altai Node* is plainly seen in the dispersal of certain of its species, as,

(1) *V. Vitis-Idea* — Himalayas and Altais, *westward* to Siberia, Scandinavia, Iceland and the whole of the Europe generally to the Pyrenees, Alps and Northern Balkans; *eastward* to Siberia inclusive of Novaya Zemlia, Kamchatka, Sakhalin, Japan, North America to Greenland, Minnesota and Massachusetts.

(2) *V. Myrtillus* — Mongolia to Transbaikalia, *westward* to Europe in the latitude 71°, ranging southward to the mountains of Italy, Spain and Portugal; to the Balkans exclusive of Greece, and the Caucasus; *eastward* to North America from Alaska to Colorado and Utah.

As usual, the tracks of *Vaccinium* in the New World remain (see 31, for instance) for us here to an extent speculative. Clear are the origins of the American components of such species as *V. Vitis-Idea* and *V. Myrtillus*, which plainly entered the New World from the direction of the Northern Pacific. It is altogether likely in addition that certain populations of the genus in the Caribbeans and South America used a connection much to the south. Involved is the pageant of secondary inter-american distribution, and much remains to be done to unravel all its tangled skens. Other problems remain for us to tackle as regards the distribution of *Vaccinium* across the Atlantic to Europe (see p. 449), or from Europe across the same Ocean to the New World. Let it be repeated that these are indeed the issues which demand precise taxonomy as a preliminary, for it is only by extant affinities that we may identify secondary tracks and their course. Primary tracks can be plotted out on the skimpiest material. For example, learning that a group of plants turns up in Ceylon, South Africa and Eastern Brazil we forthwith understand that this group has used a gate of angiospermy different from that of another group reported from Juan Fernandez, New Zealand and the Philippines. We need no more than three records on the map to gauge with fair accuracy in the matter. Far more exacting is the task, for example, of investigating whether a species endemic to Ireland came to its present abode from the east or the west. To achieve this we need to have clear before us the affinities of this plant, east and west, not only, but to understand how it might have varied in the course of its phylogeny.

(e) Ericaceae Andromedoideae

The Ericaceae Andromedoideae are distributed (Fig. 46) in this manner,

(1) *Agauria* — Madagascar, Mascarenes, Nysaland, Tanganyka, Uganda, Kenya, Eastern Belgian Congo, Angola, Cameroon.

(2) *Orphanidesia* — Northeastern Asia Minor, Iran (Persia).

(3) *Enkianthus* — Himalayas, Indochina, China, Japan.

(4) *Cassiope* — Himalayas, Burma, Siberia to Japan and Alaska; North America to Greenland, Nevada, California; northern Europe to Scandinavia.

(5) *Andromeda* — Northern Asia to Central and Northern Europe, North America to Idaho, Indiana, New Jersey.

(6) *Chamaedaphne* — Northern Eurasia, North America to British Columbia, Illinois and Georgia.

(7) *Epigea* — Japan, North America to Wisconsin, Ohio, Georgia, Tennessee.

(8) *Lyonia (Xolysma)* Western Himalayas to Malacca, Indochina, Burma; massive in the West Indies, ranging to Mexico and Atlantic North America.

(9) *Leucothoe* — Himalayas, China, Japan; the Americas to Eastern Brazil and Bolivia.

(10) *Oxydendron* — Atlantic North America.

The classification of this group is controversial, which is the norm with the Ericaceae. Certain of our readers will doubtless criticize the conspectus we find necessary to accept, and we will not attempt to meet their strictures as between taxonomists. Speaking purely as phytogeographers, we feel confident that the discussion next to follow is to show that Latin generic names might as well be replaced by figures or other symbols, without this altering our conclusions in the least. We thoroughly understand that taxonomists specialized in the classification of any given group are shocked, when they are asked to face records that may not meet everything they know. Indeed, we understand this because we have some familiarity with certain groups ourselves, and are aware at first glance whether the classification we might be shown is antiquated, perhaps even contemptible, or up to the latest standards of treatment.

However, phytogeography is not taxonomy. A problem in phytogeography essentially consists of the statement of a set of localities together with the request that these localities be interpreted as regards groups A, B, C, etc. This is far from saying that correct taxonomy may not be very useful — indeed at times essential — which we have repeatedly stated. This is merely to say that each case is to be handled upon its merits, and that the merits of the case under immediate consideration are not such as to demand that we know everything of the formal classification of the Ericaceae Andromedoideae (or Andromedeae), if we intend to interpret the mainsprings of their dispersal.

Agauria, to begin with, yields (21) a perfect African track running all the way from the Mascarenes to the Cameroon, not only, but to Angola (*Agauria pyrifolia*: Mascarenes, Madagascar, Northern Rhodesia, Nyasaland, Tanganyka, Uganda, Cameroon, Fernando Po, Angola (Huilla, Benguela)). Patterns of dispersal to meet *Agauria's* are already known to us from the Ericoideae, which reveals that patterns of the kind are standard, and uniformly begin from a center east of the Mascarenes in the Indian Ocean. If anything is noteworthy in *Agauria's* tracks, this is the fact that *A. pyrifolia* reaches Angola. This outlier might have been established, (a) By a direct connection from the Cameroon and the islands in the Gulf of Guinea (Fernando Po); or, (b) Immediately by a secondary track crossing the whole of Africa from

Rhodesia or Nyasaland, or coming in from the South Atlantic. Both alternatives are possible, and we should like to discuss them at length, which space forbids to do here. We may at the most state our belief that direct connections between the Islands in the Gulf of Guinea, Saint-Helena, Tristan da Cunha, Eastern Argentina (Sierra de la Ventana) and, finally, Southwest Africa and Angola were anything but impossible. On the strenght of the dispersal of *Phylica* (see p. 58), we may further state that these connections also immediately interested Madagascar and the Mascarenes, not to mention South Africa.

Safe in the knowledge that *Agauria* migrates in unison with the Ericoideae, and further occurs in ranges not unknown in the dispersal of the Vaccinioideae — therefore is rooted by the African gate of angiospermy — it is left for us to work the balance of the migrations from this gate onward. Our next step, then, is to follow them in standard fashion to the front Asia Minor-Himalayas, with Iran *(Orphanidesia)* in full focus. Precedents for this step are scattered throughout the pages of this manual, and need not be forever repeated.

Well rooted between Asia Minor and the Himalayas, the dispersal invest Eurasia and the Far East, together with Malaysia in wholly standard fashion, reaching at one hand Japan, at the other Northern Europe. From Japan, the track speeds eastward across the Pacific to invade the north of the New World, next veering southward in the direction of Mexico and the Caribbeans.

Everything is so conventional in this type of distribution that comments may be dispensed with, lest the reader rightfully takes us to task for abusing his patience, and constantly harping upon the very same. If there is anything worth passing notice, this is the fact that *Lyonia* has a center of massive speciation in the Caribbeans. As regard this center, we may suggest two alternatives, (a) This center was fed from the North American massing; and/or, (b) On the contrary, this center was brought into being by a track out of West Africa, immediately reaching the Caribbeans across what is now Atlantic Ocean.

To settle as between these two alternatives, we ought to know a great deal more of the taxonomic status of the Caribbean species, and, in particular, ascertain whether peradventure these species are allied in some manner or other with *Agauria*. Our mind is wholly open on the score, but we know that *Agauria salicifolia*, for example, was once treated as *Leucothoe*, not only, but that *Leucothoe racemosa* also was credited to *Lyonia*. In brief, it is transparent that the generic limits among *Agauria*, *Leucothoe* and *Lyonia* are not airtight. A genus like *Lyonia*, massive in the West Indies, and a genus like *Leucothoe* that ranges to Bolivia and Eastern Brazil (the latter particularly), may both be involved in the genorheitral ties of *Agauria*, perhaps more deeply involved that current taxonomy is ready to accept.

Our mind is open on the score, we repeat, but, as phytogeographers, we are bound to underscore the possibility that the Andromedoideae consist of archetypes fed the New World both across the Pacific — which clearly is the case with *Epigea*, for instance — and the Atlantic — which is an alternative worth reckoning with, as regards *Lyonia* and *Leucothoe* at the very least.

The reader may understand that if we designate *Epigea* as A, *Agauria* as B, *Lyonia* as C and *Leucothoe* as D, and refer A, B, C, D to such centers as Japan and the Cameroon, we have a neat problem in phytogeography on our hands which is to be solved with the statement that, indeed, it is possible that A, B, C, D are in the New World in part from Japan, and perhaps in part from the Cameroon.

We might go further. We are not aware that specialists in the classification of the Andromedoideae (or Andromedeae) ever seriously mooted the possibility that the American components of this subfamily (or tribe) stem in part from West Africa. We have heard that some debate the question, whether this subfamily (or tribe) may not have originated in "Brasilia," or "Caribea," or such other academic center of "origins," which we believe to be beside the point. The sum total of the dispersal of this subfamily (or tribe) rules out "Brasilia," "Caribea," etc., in no uncertain terms, as it does, of course, the wholly fabulous "Holarctis." It seems to us possible, then, that we have as yet no adequate systematic and taxonomic knowledge of the Andromedoideae (or Andromedeae) to meet minor contingencies in their dispersal, and for this reason we do not worry as yet about a "perfect" classification of this group.

(f) Ericaceae Arbutoideae

Though important from the standpoint of taxonomy and horticulture, the Arbutoideae are not especially interesting as regards distribution. The classification of this group is, as usual, controversial in the extreme. Since, however, we cannot worry about "perfect" classification at present — witness our previous notes — we will accept as indicative of the distribution of the Arbutoideae *) the following.

(1) *Arbutus* — Northern Africa, the Mediterranean, Atlantic Europe to Ireland.

(2) *Arctostaphylos* — Northern Eurasia and North America to Mexico.

To handle this type of distribution with success, we ought to know exactly the genorheitral ties of the Arbutoideae in the direction of the Andromedoideae and Gaultherioideae. Fully informed as to this, we could drive the tracks to their ultimate ends and decide whether, (a) The dispersal was effected to

---

*) Genus *Arctous* of this Subfamily (or Tribe) ranges with three species in Europe (from Scandinavia to the Alps); Asia (Altai Mts.; China: Kansu, Szechuan; Corea, Japan, Kamchatka, Sakhalin, Aleutian Islands); North America (Alaska to Greenland, New Hampshire and British Columbia). This dispersal is keyed to the Altaian node, and from this node on spreads to Europe and North America. Comparable dispersal is to be found in the Andromedoideae which conveys at least a tenuous suggestion that the same genorheitron might have been immediately common to both groups, Arbutoideae and Andromedoideae. It will be noticed that the distribution of *Arctous* has no final bearing upon the problem, how *Arbutus* reached North America, whether from the Pacific or the Atlantic.

the New World (and Eurasia generally) from a secondary center of migration in the heart of Central Asia (see Andromedoideae), the dispersal itself being rooted whole in the African gate of angiospermy; (b) This dispersal stemmed, on the contrary, by tracks running in the Atlantic throughout (see for reference the Ericoideae), the African gate always being in play; (c) This dispersal, eventually, was made to the Old and New World alike from a point originally located in the lands adjacent to the Western Pacific, the Western Polynesian gate being called into question most immediately.

Whatever the case be, the migrations always would go back to a center in the Indian Ocean; so much not because we are partial to these waters, but because it is at least within reason to suppose that *Arbutus* and *Arctostaphylos* could not originate at the North Pole, when the whole of the Ericaceae never had anything to do with so frigid a cradle.

In brief, the Arbutoideae hold in regard of another ericaceous group, whether Andromedoideae or Gaultherioideae, etc., the same position which *Calluna* has in respect of *Erica*. *Calluna* if of the north, *Erica* of the south, but both *Calluna* and *Erica* jointly originated — in the precise sense of the term — there, where they first stirred out of the genorheitral matrix.

We may add the remark that, when we offer alternatives such as we have done in the present case, we do not intend to have these alternatives stand as exclusive. Indeed, the contrary is our intention. We intend these alternatives to stand as leads in a free investigation which we cannot perfect in these pages. Whether one or the other is true in the absolute, or both are true in part, is none of our concern. In reality, we would not offer alternatives if we knew beforehand how the matter stands.

(g) Ericaceae Rhododendroideae

The classification of *Rhododendron (Azalea)*, keystone of the Rhododendroideae, is incredibly confuse. Much as we would like to discuss this genus at length, holding it as most interesting in regard of certain details of dispersal, we cannot do so, because we would necessarily get involved into frightful complications of pure taxonomy and nomenclature. We will refer to *Rhododendron*, to the exclusion of other genera of its subfamily, mostly to discuss the origin of its track.

We take no interest here in other genera of the Rhododendroideae because we do not believe *Tripetaleia* (2 species in Japan), *Elliottia* (1 species in Eastern North America), *Cladothamnus* (1 species in Northwestern America), *Ledum* (3 or more species in Boreal Eurasia and America), etc., hold in store anything worthwhile for us at the moment. We state this, because we cannot we concerned with secondary aspects of dispersal, and for this reason may bot give time to a consideration of the distribution of *Ledum*, for example, nhich is worthwhile only as regards certain problems in the High North.

The most interesting problem in general phytogeography which the Rhododendroideae offer for our attention is, as intimated, in the inception of the track of *Rhododendron*. The southermost species of the genus is *R. Lochae*, endemic to Queensland.

Believers in angiospermous "holarctic" origins will assume as a matter of course that Queensland is the last station of *Rhododendron* from "Holarctis" going southward. We, on the contrary, hold fast to the belief that Queensland is the first station of this genus going northward.

This we do, because we are by now aware that the Southern Pacific is the center wherein basic affinities in the Ericaceae come to a head (see p. 170), not only, but because we know that this family is also active in the Indian Ocean, and we can connect the two oceans by standard channels. In other words, we can make sense in the dispersal of the Ericaceae — and Epacridaceae as well, if need be — taking our start from the south, not the north of the modern map.

It is an axiom to us — an axiom which the reader will at least accept as a very strong probability, when not a certainty, after closing the covers of this manual — that we cannot saunter all over the map if we wish to understand dispersal as something orderly and precise. We believe it is necessary to understand dispersal as orderly and precise, because nothing is ever disorderly and imprecise in nature.

To us, it is an impossibility that *Rhododendron* is from "Holarctis," when the whole of the Ericaceae is from a point of the map diametrically opposed to this academic figment. Even more so, to us it seems impossible that one genus is from "Holarctis," when the whole of the Angiosperms orderly behaves and moves around a center which had absolutely nothing to do with this mythic land.

One of the genera of the Vaccinioideae, *Agapetes* (33) turns up with two out of its many species (namely, *A. Meiniana* and *A. queenslandica*) in Fiji and Queensland, respectively. This being so, there is absolutely no reason, why a species of *Rhododendron* should not also be endemic to Queensland, and be endemic there out of tracks that are standard for another ericaceous form such as *Agapetes*; out of an ocean which contains the connecting-link between the Gaultherioideae and the Vaccinioideae, the Ericaceae and the Epacridaceae.

True, *Rhododendron* is not seen anywhere else, to present knowledge, but Queensland, and is unknown in Magellania and the Mascarenes. In a kingdom, however, were migrations like *Lepechinia's*, *Passiflora's*, the Saxifragaceae Argophylleae's, etc., etc., are conventional, we may not reverse the entire course of the angiospermous migrations to make *Rhododendron* "holarctic". Cause and effects stand to each other in definite relationships in science. In science, *Rhododendron* cannot be proved to be "Holarctic" by a statement of opinion. It must be shown that the normal run of phytogeographic causes and effects, and their interrelations, are such as to demand that this genus be "Holarctic." This done as a preliminary, opinions may be heard with patience and respect.

Rooting *Rhododendron* in Queensland as a matter of course, we lead its migration northward in any one of great many standard manners. For example, Queensland is not too far from New Guinea. *Dimorphanthera*, an ally of *Agapetes*, is massive indeed in New Guinea, but contains also certain species that freely invade Malaysia, as follows,

(1) *D. moluccana* — Boeroe Island.

(2) *D. pulchra* — Amboina, Ceram.

(3) *D. mindanaensis* — Philippines (Mindanao).

(4) *D. apoana* — Philippines (Mindanao, Negros, Panay).

As does *Dimorphanthera*, so does *Rhododendron* of which a large number of species are already known to exist in New Guinea. *).

From New Guinea onward, throughout Malaysia and the Far East, *Rhododendron* gains momentum until it sets foot on the Asiatic mainland. Here its speciation soon achieves stupenduous dimensions, and from Tropical Asia *Rhododendron* can be followed along wholly standard paths of dispersal throughout the whole of Eurasia and much of North America. *Rhododendron aureum*, for instance, endemic to Japan, Manchuria and the Siberian and Mongolian heights, is nearest *R. caucasicum* of the Caucasus. The *Ponticum* Series, according to one of the authorities on the taxonomy of the genus **) includes in the Subseries *Ponticum* three North American species *(R. californicum, R. catawbiense, R. maximum)* and an European one (*R. ponticum*: Asia Minor, Caucasus, Armenia, Southeastern Spain, Portugal). In the same *Ponticum* Series but in the different Subseries *Caucasium* are met species such as these,

(1) *R. Smirnowii* — Caucasus.

(2) *R. adenopodum* — China (Szechuan, Hupeh).

(3) *R. yakusimanum* — Japan (Island of Yakushima; see p. 438).

The *Lapponicum* series described as a "Large natural group" reveals the following,

(4) *R. lapponicum* — Lapland.

(5) *R. parvifolium* — Altai Mts. to Kamchatka and Sakhalin.

(6) *R. setosum* — Himalayas (Sikkim), Tibet.

(7) *R. chryseum* — China (Yunnan).

The *Vaccinioides* Series which, according to the author we follow, is "Very distinct and natural," contains these species in several,

(8) *R. emarginatum* — China (Yunnan).

(9) *R. Kawakamii* — Formosa.

(10) *R. Vidalii* — Philippines (Luzon: Bontoc, Benguet), Batan.

---

*) According to the „Rhododendron Book", App. I, 1935, not less than 104 species are known from New Guinea, 4 in Celebes, 26 in the Philippines, 29 in Borneo, 5 in Sumatra (2 common with Japan), 10 in Java, 14 in Malacca.

**) HUTCHINSON, J., „The species of Rhododendron", 1930.

In brief, these patterns of distribution are standard for track driven from the south northward all the way from Queensland and New Guinea to Lapland, the Caucasus, Portugal, California and the Southeastern United States. While there is no doubt, that many of the ancestors of *Rhododendron* reached the New World from across the Pacific, it is open to question whether the Subseries *Ponticum* did not, on the contrary, migrate from Portugal to the Appalachian center of the Southeastern United States and California.

A peculiar aspect of the dispersal of *Rhododendron* is in the avoidance of African ranges, and in the fact that the southern limit of the distribution of the genus in North America seems closely to agree with the political boundary between the United States and Mexico. The first of these aspects is far from obscure, in view of the circumstance that the track is not from the Indian Ocean, *but from the Pacific*, and streams westward clearing tropical and subtropical lands to the south. The normal route, consequently, is from the Himalayas directly to the Caucausus and Asia Minor, next entering the Balkans, finally and from this point Portugal. The second aspect, involving North America, is less clear. It might stem from palaeogeographic conditions such as to discourage migrations into Mexico and the West Indies. These conditions cannot be suspected, and even less known by approximation, until and unless we are thoroughly well informed of the intergeneric ties of *Rhododendron* in the New World and Europe. If a study of these ties should lead us to conclude that subseries *Ponticum*, for instance, reached the Appalachian node of the U.S.A. from Europe across the Atlantic, we might perhaps understand that the ancestors of this Subseries were greeted, when arriving in the New World, by conditions unlike those found by *Magnolia* and *Platanus*, which came to America from the direction of the Pacific, and occur now in Mexico at least.

VII. Certain peculiarities of Polynesian tracks

As a sequel to our study of the Ericaceae, and before concluding as regards their dispersal in general, we may give cursory attention to certain aspects of migration that interest the tracks of this family, and other Angiosperms as well, in the Pacific. These aspects deserve consideration. Seemingly whimsical migrations take place throughout the Southern Pacific, and we have just seen a case like *Rhododendron*, which turns up in Queensland as if from nowhere. Moreover, *Vaccinium* is not in Magellania but in Southern Polynesia; *Pernettya* is native to Magellania, Juan Fernandez, New Zealand and the Galapágos Islands, not to Southern Polynesia. *Coprosma* (see p. 101) is endemic to Juan Fernandez and South Eastern Polynesia but unknown in Magellania. Examples of this same nature could be given without end.

The current explanation is that these peculiarities of distribution are due to "Casual agencies of dispersal." The Polynesian islands are supposed to be "oceanic," and to have been colonized by plant-life in the course of migrations effected without regard of "Landbridges". It is easy under the circumstances to ask "Casual agencies of dispersal" to perform any way we like. Naturally, these "Agencies" will not account for the fact that the flora of New Zealand, for example, contains two elements, a "cold" and a "warm" one, nor clear up the reasons why the latter, though admittedly most ancient, is after aeons of geologic time still unfit for the climate it must put up with.

Dispersal can be critically studied only by making reference to dispersal. It is a rule of scientific phytogeography that the oddities of the distribution of one group; the seemingly fantastic disconnections marring the dispersal of a genus, etc., can always be understood by taking a sufficient number of examples of parallel cases and comparing them in the light of common sense and known palaeogeographic and ecologic factors. Assumptions, theories, etc., are not needed, purely facts and maps.

The rule just outlined has been applied in this manual (see p. 39) in an unobtrusive manner to show that disconnections in the tracks of certain species of *Carex* were easily filled in by existing records of other species. We take this procedure to be correct in its general aspects, because it is axiomatic that two or more plants can pass where one can. Be it clear that this does not mean that forms having unlike ecologic preferences can all fare the same tracks. The contrary is indeed true. However, the fact remains that a connection between two points of the map is never the monopoly of a single plant. The method we outline is a bottom statistical, and strives to reconstruct — so far as possible — ancient geography by a study of the tracks that first overran it, and still persist on modern maps. The reader knows by now that this is not at all impossible.

The examples we can presently study are admittedly few, perhaps too few. We introduce them, nevertheless, in the anticipation that the trail blazed in this manual will be followed in due course of time by students of dispersal, using an evidence beyond comparison more massive than the one we are privileged to use at this point.

Ascarina is a genus of the Chloranthaceae, a family which suggests to us a measure of kinship with the Gnetaceae. This genus is distributed in the following manner,

(1) *A. lanceolata* — Kermadec Islands, Rarotonga Island, Samoa, Fiji, New Hebrides.

(2) *A. lucida* — Stewart Island, New Zealand.

(3) *A. raiateensis* — Raiatea.

(4) *A. subfalcata* — Raiatea.

(5) *A. chloranthoides* — Society Islands, Tahiti.

(6) *A. polystachya* — Society Islands, Tahiti.

(7) *A. alticola* — New Caledonia.

(8) *A. rubricaulis* — New Caledonia.

(9) *A. Solmsiana* — New Caledonia.

The record is, as always, not complete, perhaps imperfect. It is probable, on the other hand, that it is representative of the trends in the dispersal of this genus. It is clear that this dispersal has two main massings, in Southeastern Polynesia (Raiatea, Tahiti, Society Islands generally), and New Caledonia, respectively.

The largest genus of the Chloranthaceae is *Hedyosmum*. *Hedyosmum* is both in the New World (continental South America and the West Indies, ranging as far north as Mexico) and in a few centers of the Far East. Its distribution is accordingly strikingly reminiscent of that of the Lardizabalaceae. The base-line of the chloranthaceous migrations is beyond doubt to be sought in the Southern Pacific, and it is an even guess that the Far Eastern and American massings are connected through the Northern (via China-Mexico) or the Southern Pacific. As a matter of personal opinion, and with reference this time to the dispersal of the Coriariaceae, we are perhaps more inclined to believe that *Hedyosmum* is connected in the Far East and New World by a now crumbled range in the Southern Pacific. We admit, at the same time, that both channels, Northern and Southern Pacific, might have been used by the Chloranthaceae.

Whatever be the case in its details, it stands within reason that *Ascarina* migrated from a point in the Southern Pacific of our maps. We may assume that this point lies near Tahiti, next to draw the tracks from this center on to Samoa, Fiji, the New Hebrides and New Caledonia, at one hand; the Kermadec Islands, Stewart Island and New Zealand at the other. We may contrariwise believe that Stewart Island was reached first, then New Zealand and the Kermadecs, Tahiti and New Caledonia. Whatever we do, we are bound to connect Stewart Island, the Kermadecs, Tahiti, Samoa, Fiji, New Hebrides and New Caledonia by a track also touching New Zealand. The outline of this track will come very close to matching the geologic shore indicated in these quarters (see p. 69) by geophysicists, and the line limiting a fairly rich orchid flora. It may be that Tahiti was not solidly connected within a single landmass with Stewart Island, though good evidence is at hand (see p. 103) that an "antarctic" shore ran all the way between New Zealand and the New World immediately interesting such modern islands as Rapa, Sala y Gomez Island and the Juan Fernandez. In brief, whatever the details be, the tracks of *Ascarina* are certainly not casual. They agree with data in the hand of geophysicists and plant-geographers which, if a "coincidence" at all, is worth noticing.

We are not exactly informed as to the nature of *Ascarina*, whether it is a "cold" or "warm" element of the New Zealand flora. To judge from the Chloranthaceae in general, and the sum total of the distribution of *Ascarina* in particular, we would say that *Ascarina* is a "warm" element in that flora.

We regret we cannot give here to pure Neozelandian issues of distribution all the attention that these issues doubtless deserve. New Zealand is one of the most interesting domains in the world, and a book could easily be written on the subject of its flora, its composition and ties. We suggest, however, that the coexistence of two floras in New Zealand, "cold" and "warm", bears a measure of elucidation if only we understand that, (a) New Zealand was reached *immediately from the south* by certain elements of which *Nothofagus* and *Coprosma* may well be part. This is a "cold" flora; (b) Contrariwise, New Zealand was reached *immediately from the northeast, north, possibly even the northwest* by certain elements of which *Ascarina* is probably one. This was a "warm" flora, not because this flora had never touched in origin the antarctic shore known to the archetypes of *Nothofagus* and *Coprosma,* but because this flora had ripened up, as it were, in such centers as the vicinity of Tahiti, Fiji, New Caledonia etc. In brief, New Zealand stood between the jaws of two ranges, a "cold" one in the south; a "warm" one in the north, and was fed by one or the other, depending upon the nature of its connections in different geologic eras, and its own climate through the ages. As regards the tracks of *Ascarina, Ascarina* could be a "cold" element in New Zealand had its track immediately reached New Zealand from Stewart Island, next to branch off to the Kermadecs and Tahiti at one hand; New Caledonia at the other. By exactly an opposite route (i.e., a track running from Tahiti or New Caledonia to Stewart Island through the body of New Zealand), *Ascarina* could be a "warm" element. The matter resolves itself in a play of alternatives and possibilities in the end, all of them to be studied in the light of concrete factors of dispersal and ecology. The Kermadecs, at any rate, seem to mark the limits in our days of a truly "tropical" flora southward.

Melicytus is a violaceous genus in five species, as follows,

(1) *M. lanceolatus* — Stewart Island, New Zealand.

(2) *M. macrophyllus* — New Zealand.

(3) *M. micranthus* — New Zealand.

(4) *M. ramiflorus* —Kermadec Islands, Tonga (Eua), Samoa, Fiji, Norfolk Island, New Zealand.

(5) *M. fasciger* — Fiji.

This type of dispersal, judged on its face, offers to us a choice among alternatives such as these, (a) The track begins to run by Stewart Island, south of New Zealand. It next immediately invades New Zealand, and here it branches; parts goes to Norfolk Island and ultimately Fiji, while the balance streams to the Kermadecs, Eua and Samoa; (b) The track begins at the Kermadec Islands and fares from this point on to Samoa, Fiji, Norfolk Island and New Zealand as far south as Stewart Island; (c) The track has its start by a center Norfolk Island—Fiji—Eua (Tonga)—Samoa, and from this center invades New Zealand and the Kermadecs.

As we have no means to settle among these alternatives, let us further study the dispersal of a second genus of the same Subtribe to which *Melicytus* belongs, namely *Hymenanthera.*

Hymenanthera is distributed in this manner,

(1) *H. latifolia* — Lord Howe and Norfolk Islands.

(2) *H. novaezelandia* — New Zealand (North Island: Three Kings Island, Great Barrier Island, Cuvier Island).

(3) *H. chathamica* — Chatham Island, New Zealand (North Island).

(4) *H. crassifolia* — New Zealand (both islands).

(5) *H. obovata* — New Zealand (both islands).

(6) *H. dentata* — New Zealand (both islands), Stewart Island.

(7) *H. angustifolia (H. dentata)* — Victoria, South Australia, Tasmania.

It is obvious by the face of the dispersal that we deal here with two "eco-types" of generic rank associating within a single subtribe. *Melicytus* holds a northern and warmer range by comparison with *Hymenanthera*, which latter reaches as far south as Tasmania and Stewart Island. The separation between these two genera, however, is not sharp enough as to prevent species of the one (e.g., *M. lanceolatus*) from sharing exactly the same range known from species (e.g., *H. dentata*) of the other. Rather more significant is here the fact, we believe, that *Melicytus* ranges to Tonga and not to Australia, while *Hymenanthera* ranges to Australia and Tasmania, but not to Tonga.

Were we consequently to rationalize the dispersal on its face, we would say that the *genorheitron* common to both these genera was rooted in origin in a center Lord Howe Island—Norfolk Island—Fiji—Samoa—Tonga which is notoriously old lands in these quarters.

Leaving this center, this *genorheitron* fared westward to Southern Australia and Tasmania; eastward to New Zealand by various minor tracks. One of these tracks can easily be reconstructed along the lines suggested by the peculiar ranges of *H. novaezelandiae* and *H. chathamica;* this track drives from a point intermediate to Lord Howe and Norfolk Island straight to the northern coasts of North Island of New Zealand. A second track, and a more roundabout one, may be Norfolk Island—Kermadec Islands—Chatham Island—Stewart Island, touching New Zealand by one or more approaches.

In conclusion, New Zealand stands as a land-bastion surrounded by a half-crumbled insular arc, Norfolk Island—Kermadec Islands—Chatham Island—Bounty Island—Antipodes Island—Campbell and Auckland Islands—Macquarie Island. Whether this arc is actually single, or made out of lesser arcs and their branches (e.g., Lord Howe Island—Three Kings Island—Chatham Island, matching the arc Norfolk Island—Kermadec Islands—Chatham Island etc., described above) we do not know. The fact seems to us clear, however, that dispersal speaks for New Zealand being the *largest remnant* of some land once massive in these quarters. The Tasman Sea which divides New Zealand from Australia is doubtless a very old feature of the map, one which the *genorheitron* of *Hymenanthera* and *Melicytus* could bridge only by bypassing from a point of vantage *in the north* (line Lord Howe—Norfolk Island).

It is clear, on the other hand, that this same Tasman Sea could be bridged from a point of vantage *in the south* by a host of forms of genuine ,"antarctic" description, which reach both New Zealand and Tasmania without ever starting from the point wherefrom *Melicytus* and *Hymenanthera* began their tracks. In brief, so far as dispersal proves, the Tasman Sea, though wide open south and north by the present map, is in reality a mediterranean very closely hemmed in by lands all around. These lands are not all on the map today, but had they not been on the maps of the past, *dispersal could never be today what it is.* The Tasman Sea, then, is not without homologous features so far as dispersal reveals. These features are the Mozambique Channel and the Banda Sea, not only, but the Caribbean Sea as well.

We cannot hope to fathom in the pages of this manual all the problems of Polynesian and Western Pacific dispersal because these problems are legions *in detail.* We believe, however, we will begin to simplify matters considerably as a preliminary to studying these details if we understand the position held by the Tasman Sea in dispersal (see further p. 263).

In order better to grasp the origin and distribution of the group of which *Melicytus* and *Hymenanthera* are part it is convenient to review here the Violaceae Rinoreae (34) which tabulate as follows,

A) Hymenantherinae

(1) *Hymenanthera* — 7 species in New Zealand, Tasmania, Southern Australia, Chatham Island, Lord Howe and Norfolk Islands.
(2) *Melicytus* — 5 species in New Zealand, Stewart Island, Kermadec Island, Norfolk Island, Tonga (Eua), Samoa, Fiji.

B) Isodendriinae

(3) *Isodendrion* — 5 species in Hawaii.
(4) *Amphirrox* — 5 species in Brazil (Bahia, São Paulo, Amazonas) Guianas.
(5) *Paypayrola* — 7 species in Brazil (Rio de Janeiro to Pará), Guianas, Eastern Peru.

C) Rinoreinae

(6) *Rinorea* — About 260 species throughout the tropics to the exception of Australia.
(7) *Allexis* — 3 species in Western Africa (Gabun to Southern Nigeria).
(8) *Gloeospermum* — 7 species in Brazil (Amazonas), Colombia, Eastern Peru.

The species in *Rinorea* break down approximately as follows, 22 in Madagascar and the Comores; 4 in Southeastern Africa; 19 in Eastern Africa; 97 in Western Africa; 70 in the Tropical Far East and Malaysia; 5 in Polynesia; 10 in Central America and the Caribbeans; 42 in South America.

It is transparent that the critical group in this Tribe are the Isodendriinae which associate as one Hawaii and Tropical South America. This is an unusual tie. The balance of the dispersal as represented by the Rinoreinae is trite. This Subtribe migrates from the African gate westward in conventional manner, and the secondary center of origins it has in West Africa is standard too, as we are soon to learn from the Connaraceae, Dichapetalaceae, etc.

The Hymenantherinae cannot certainly be styled a "cold" group (see p. 247[fn.]) and what we have learned of their dispersal indicates that they never travelled an "antarctic" path as did, for instance, the Restionaceae and Centrolepidaceae (see p. 359). Running ahead of our subject somewhat, we may state that the Rinoreinae are homologous in migration of the Oleaceae (see *Olea* and *Notelaea;* p. 353) in certain respects, and that their *genorheitron*, strongly rooted in the Indian Ocean (Rinoreinae) furnished the *genorheitron* also to the Pacific (Hymenantherinae) by a track which it is usually hard to plumb, but *Korthalsella Opuntia* (see p. 142) makes it possible for us to determine in the main. This track fares south of Malacca going toward the very same center out of which *Melicytus* and *Hymenanthera* sallied forth to reach Tasmania and New Zealand, respectively.

The very same track that went from the Indian Ocean to the Western Pacific laden with the *genorheitron* of the Hymenantherinae carried with itself, also the archetype of *Isodendrion*. We do not know this genus at first hand, but, to judge from dispersal, we should think it might be intermediate between the Rinoreinae and Hymenantherinae.

In sum, the Isodendriinae might exaggerate a disconnection West Africa—Burma not unusual in dispersal. They might carry this disconnection all the way between South America and Hawaii. Let us observe that so much need not strike us as fantastic. The same species of *Cuscuta* (see p. 220) has forms in the West Indies and Burma, and we know enough by now not to see a miracle in the fact that a track leads further from Burma or Southern China to Hawaii.

Returning now to the Pacific; We may summarize the dispersal of the Rinoreae, and illustrate (Fig. 47) at the same time the fundamentals of Pacific dispersal by a few strokes of the pen. We will not be overrash if, adding to this map the tracks of *Vaccinium* in Polynesia and of the Restionaceae, in part, we will suggest that Eastern Polynesia could also be reached immediately by a base-line of distribution set in the Southern Pacific to agree with a geologic antarctic shore.

It is obvious by the map that New Zealand could easily be reached by "warm" (e.g., *Hymenantherinae*) and "cold" (e.g., *Nothofagus*) forms, not only, but it is further clear (see p. 285) why certain groups of *Plantago* could hold New Zealand within a two-pronged track without for this invading it at all.

VIII. Connaraceae and Dichapetalaceae

A check-list of the angiospermous families most interesting from the standpoint of distribution could be prepared only with difficulty, because everyone of these families is worthwhile for some reason or other. Certain among them illustrate peculiar disconnections, others cast sharp light on *genorheitral* affinities, others still run cleancut tracks throughout. In the end, none of them is unrevealing or useless, so that our greatest difficulty in preparing this manual has been selecting few examples only for brief discussion.

Were such a check-list to be drawn up regardless of odds, we should not fail to include the Connaraceae among the first five most interesting families. This group is seldom heard of, and best known only to botanists specialized in tropical flora, yet has capital position in systematic botany, for its morphology points toward the Leguminosae and Rosaceae alike. It follows that a single *genorheitron* belongs to these three families, and the origin of none can be determined without taking into account that of the remaining two. Unlike the Leguminosae and Rosaceae, however, the Connaraceae are a comparatively small group, and for this reason can be analyzed with ease. There is good ground to believe that other affinities are involved in this *genorheitron*, and it is not to be excluded that certain families of unsettled position, witness Polygalaceae, may repay scrutiny by comparison with the Connaraceae.

The monographer of the Connaraceae (35; p. 21) believes that this family is of "American" origin, although he grants at the same breath that its most "primitive" modern genera are to be looked for in Africa. He further identifies (35; p. 14) three "Centers of evolution", (a) West Africa, which is richest as regards numbers of species; (b) Southwestern Malaysia and the Philippines; (c) Southern Brazil and the Amazonas.

To us, the presence of three centers so located does not foretell "American" origins in the least. To us, this type of dispersal bespeaks migrations from the African angiospermous gate, for only this gate — indeed, the cardinal gate of angiospermy — is best fit to feed these three centers simultaneously.

We may not undertake to discuss the Leguminosae which are one of the largest angiospermous families, and the subject only of a thoroughly outdated monographic work. We may point out, nevertheless, that one of their groups (32), *Sophora* sect. *Edwardsia*, has cleancut circumpolar range in the south (Mascarenes, Lord Howe Island, New Zealand, Chatham, Rapa Islands, Austral Islands (Raivavae), Hawaii, Juan Fernandez, Southern Chile, Gough Island) which is in itself significant. Quite as significant is the distribution of the phyllodine *Acacia* (see p. 52), which have in Western Australia their major center of evolution. Moreover, another leguminous group, Podalyrieae, is massively Australian, and the affinity of *Cercis* (see p. 137) consists of the pantropic genus *Bahuinia* and the African *Acioa* (*Griffonia*). None of these data agrees in the least with an "American" origin of the *genorheitron* common to Connaraceae and Leguminosae.

We have reviewed certain Rosaceae (see p. 147), and their dispersal led us to assume for them an origin which matches that of all other angiosperms. This origin is doubtless not "American". It only remains for us to learn whether the dispersal of the Connaraceae can be made to square up with that of the

Rosaceae and Leguminosae, or is basically "American," therefore at odds with that of the remaining Angiosperms.

It is not impossible that the Dichapetalaceae are in some measure allied with the Connaraceae. They have certainly no affinity with the Euphorbiaceae which many systematists believe to be their next-of-kins. At any rate, the Dichapetalaceae are distributed (36) very much as do the Connaraceae, that is, as follows,

(1) *Dichapetalum* — About 200 species; 3 Sections *(Rhopalocarpus, Brachystephanium, Tapurinia)* restricted to Western Africa; 1 Section *(Eudichapetalum)* wide-ranging through Tropical Africa (massive in Western Africa), the West Indies, Amazonas and Central America, the Far East, reaching China and Fiji.

(2) *Stephanopodium* — 4 species in the Amazonas and Peru.

(3) *Tapura* — 12 species; 5 in South America (Amazonas, Guianas, Eastern Peru) and the West Indies; 2 in Western Peru.

(4) *Gonypetalum* — 3 species in the Amazonas to Peru.

It is transparent that the dispersal of the Dichapetalaceae is wholly conventional from the African gate (Indian Ocean) to the Far East and Western Polynesia, at one hand; the Amazonian basin, the West Indies and Tropical America generally, at the other. If there is anything remarkable in this type of distribution, this is the fact that three out of the four sections of *Dichapetalum* are narrowly localized in Western Africa. There is something here well worth investigating.

The Connaraceae themselves consists of 24 genera. One only, *Connarus*, is pantropic, and the largest in the family with about 120 species. The subgeneric limits are somewhat less than clear, but the American species of subg. *Connarellus* stand out as a distinct complex, well characterized by the anatomy of the fruit-wall.

Connarus is comparatively poor in Africa, where it has probably less than 20 species. These species fall into three sections of subg. *Euconnarus* which is entirely restricted to the Old World.

One of the sections native to Africa, *Afromphalobium*, is monotypic, and represented by *C. Staudtii* (Southern Nigeria: Oban, Eket; Cameroon, Gabun, Eastern Belgian Congo). A second section, *Omphalobium*, tabulates as follows,

(1) *C. africanus* — Islands of the Gulf of Guinea (São Thomé), French Congo, Cameroon, Southern Nigeria (Eket, Lagos), Western Africa generally to Senegambia.

(2) *C. Sapinii* — Belgian Congo.

(3) *C. monocarpus* — Ceylon, Southern India.

The third section native in part to Africa, *Neuroconnarus*, is fairly large (about 35 species). Little less than half of this group is restricted to the Dark Continent, massing, as usual, in Western Africa. Some species reach Eastern Africa, without for this necessarily losing contact with Western Africa (for example, *C. longestipitatus:* Angola, Belgian Congo, Tanganyka, Uganda).

The balance of the group belongs to the Far East, and can be traced from India eastward as far as New Guinea and the Palau Islands. Western Malaysia, the Indochinese Peninsula generally and the Philippines are well stocked, and some species have significant ranges, witness, *)

(4) *C. culionensis* — Philippines (Tawitawi, Palawan, Busuanga, Culion).

(5) *C. palawanensis* — Philippines (Palawan).

(6) *C. balsahensis* — Philippines (Palawan).

(7) *C. mutabilis* — Sumatra, Bangka, Billiton, Java, Borneo.

(8) *C. Championii* — Ceylon.

(9) *Gaudichaudii* — Moluccas, Boeroe, New Guinea, Solomon Islands, Palau Islands.

It is obvious that the dispersal of *Connarus* is effected from Africa both westward (Subg. *Connarellus*) and eastward, and has much to suggest that of the Bombacaceae (see p. 319 **)). *Connarus* invades Malaysia over a broad front, but despite the breadth of this front (Eastern Himalayas to Northern Australia), this genus clings to tracks well known to us from *Prunus*, its ally *Pygeum, Barringtonia,* etc. In short, there is definitely nothing in all this which bespeaks "American" origins, and it is quite clear that *Connarus* did nothing which *Prunus, Pygeum, Cassipourea, Barringtonia,* the Dipterocarpaceae, etc., failed to do.

In addition to *Connarus,* the Connaraceae consists of 23 genera. Six at least of these genera, *Jollydora, Manotes, Hemandradenia, Spiropetalum, Paxia* and *Santaloidella* are entirely restricted to Western Africa, and the first ranks as type of a special subfamily characterized by caulocarpic inflorescences and indehiscent 2-seeded follicles. Its three species range as follows,

(1) *J. glandulosa* — Cameroon.

(2) *J. Pierrei* — Gabun, Cameroon.

(3) *J. Duparquetiana* — Gabun, Cameroon, Southern Nigeria (Old Calabar).

Cnestis consists of 37 species, and is believed by the monographer of the family to be nearest the archetype. The distribution of this genus is significant, and forthwith belies "American" origin. It tabulates as follows,

---

*) We cannot drown the reader at first with details, but suggest he duly compares at the proper time the dispersal of these species with that of the Dipterocarpaceae and Barringtoniaceae (see p. 422, 412). He will readily learn that *Connarus* behaves like the former in the Far East and Malaysia; like the latter, in the Far East, Malaysia and West Africa. *Connarus Championii* most likely belongs to the primary dispersal; *Connarus Gaudichaudii* possibly to the secondary (see, for instance, *Vatica papuana,* p. 428). Most interesting are *C. culionensis, C. palawanensis* and *C. balsahensis;* Palawan could be both a primary and a secondary center of distribution, though in this case it probably is of the latter. *Connarus mutabilis* arose by a region (see p. 402) notorious as concerns mangroves and their vagaries around Java.
**) The Bombacaceae, *Connarus* and *Croton* are good examples of peripheral evolution trending away from Africa.

A) Sect. *Eucnestis*

a) Subsect. *Brevipetalae*

(1) *C. ferruginea* — Angola, Fernando Po, Belgian Congo, Cameroon, Nigeria (Old Calabar, Oban, Eket, Lagos, etc.) to the Senegal.

(2) *C. grandifoliolata* — Angola, Lower Congo, Gabun.

(3) *C. pseudoracemosa* — Gabun.

(4) *C. Mannii* — Cameroon, Nigeria (Old Calabar).

(5) *C. macrantha* — Nigeria (Old Calabar, Oban).

a') Subsect. *Aequipetalae*

(6) *C. lurida* — Madagascar.

(7) *C. Boiviniana* — Madagascar.

(8) *C. polyphylla* — Madagascar.

(9) *C. glabra* — Madagascar, Mascarenes.

(10) *C. natalensis* — Pondoland, Natal, Rhodesia, Gazaland.

(11) *C. racemosa* — Liberia, Sierra Leone.

(12) *C. Palala* — Andaman Islands, Lower Burma, Siam, Indochina (Laos to Cochinchina), Malacca, Sumatra, Borneo (to Tawao).

(13) *C. diffusa* — Philippines (Luban, Burias, Cebu, Luzon: Ilocos Sur, Union, Pangasinan, Bataan, Laguna, Tayabas).

B) Sect. *Ceratocnestis*

b) Subsect. *Longipetalae*

(14) *C. ugandensis* — Eastern Africa (Uganda).

(15) *C. confertiflora* — Zanzibar, Tanganyka.

(16) *C. calocarpa* — Tanganyka.

(17) *C. riparia* — Tanganyka.

(18) *C. Lescrauwaetii* — Angola, Belgian Congo, Gabun.

(19) *C. grandiflora* — Angola, Belgian Congo.

(20) *C. Claessensii* — Belgian Congo.

(21) *C. Sapinii* — Belgian Congo.

(22) *C. Mildbraedii* — Belgian Congo.

(23) *C. urens* — Belgian Congo, Gabun.

(24) *C. gabunensis* — Belgian Congo, Gabun.

(25) *C. Pynaertii* — Belgian Congo, Cameroon.

(26) *C. iomalla* — Cabinda, Gabun, Cameroon.

(27) *C. leucantha* — French Congo, Cameroon.

(28) *C. calantha* — Cameroon.

(29) *C. cinnabarina* — Cameroon, Nigeria.

(30) *C. aurantiaca* — Cameroon, Nigeria (Oban).

(31) *C. grisea* — Gabun, Cameroon, Nigeria (Old Calabar, Oban).

(32) *C. longiflora* — Nigeria (Oban).

(33) *C. corniculata* — French Guinea, Liberia, Senegal.

(34) *C. Dinklagei* — Liberia.

b) Subsect. *Macrosepalae*

(35) *C. congolana* — Cabinda, Belgian Congo, Cameroon.

(36) *C. agelaeoides* — Gabun.

(37) *C. macrophylla* — French Congo, Cameroon.

We are not at all certain that *Cnestis* is a genuine archetype of the Connaraceae, though our knowledge of this family is not such at present as to authorize us to pass final judgement. We are definitely certain on the other hand, that this genus does not have "American" origins. *Cnestis* strikes us as one of the very many groups we met throughout our work (e.g., Bombacaceae, Dipterocarpaceae, Tiliaceae, Barringtoniaceae) which answer the description of "Near-mangroves." *) These groups were favored throughout eras of widespread geological submersions and emersions. *Cnestis* is beyond doubt tied to centers which are typical of the first age of angiospermous dispersal (e.g., Madagascar, Mascarenes, Tanganyka, Angola, Cameroon), but the massive speciation showing up in the Congo, the region of Oban and Eket in Nigeria etc., is partly secondary, and effected in the epoch when Western and Central Africa was in active process of drying following emersion out of shallow inland seas or leakes. *Cnestis Palala* is an arrant near-mangrove, when not a mangrove throughout.

Cnestis, not unlike the Dichapetalaceae at large, and a multitude of other Angiosperms, originated in a center which now lies in the Indian Ocean (African gate of angiospermy; more accurately, the eastern end of this gate, which corresponds to Gondwana in our terminology). In a prior stage, *Cnestis*, to repeat, ran the usual tracks of all true tropical plants rooted in this part of the Late Jurassic world. It sent westward archetypes that reached Tanganyka and Uganda, the Congo, Cameroon and the islands in the Gulf of Guinea, not to mention Natal, and may still occur in Madagascar and the Mascarenes (possibly, *Cnestis glabra*). Also westward, and beyond the modern Atlantic, were dispatched the prototypes of *Connarus* subg. *Connarellus*.

*) We understand as „Near-mangroves" plants of the hot tropical lowlands competent to stand long periods of drought interrupted by torrential rains, and, possibly also, times of active root-submersion. Forms of this kind seem to evolve readily in the direction of adaptation to the tropical moist forest; the conventional „Monsoon Forest" up to a level of about 1000 meters (ca. 3300 feet); the subxerophilous parkland. *Plants of this nature and indifferent psammophytes and heliophytes are exceedingly common in the genorheitral background of the Angiosperms.* As we better explain in a coming chapter (see p. 424), these plants had favorable conditions, therefore migrated and speciated widely in the epochs of wide submersions that attended the breaking up of the old geography typified by Gondwana. This epoch began, it seems, in the Mid-Cretaceous and lasted to the inception of the Tertiary.

Meantime, other archetypes migrated eastward to Ceylon and Southern India (possibly including *C. monocarpus*), and the Far East to the Philippines. From the very start, the Connaraceae must have reached West Africa in strength, for it here is that were rooted the ancestors of genera such as *Jollydora*, *Manotes*, *Hemandradenia*, *Spiropetalum*, *Paxia* and *Santaloidella*. We feel confident that a critical study of the forms and groups here mentioned might make it possible to identify with some accuracy the archetypes that took part in this first stage of the migration. We are inclined to believe that it is in this first stage of angiospermous dispersal that was colonized the riverine delta which is now partly preserved in the sandstone "mesas" of the Pacaraima Mountains of Venezuela. It is known (37) for example that the Guinean ochnaceous monotype *Fleurydora* is directly allied with *Poecilandra*, which latter is represented in the Pacaraima orogeny by such species as *P. sclerophylla* (Mt. Roraima, Mt. Auyan-Tepui) *P. retusa* (Mt. Roraima, Mt. Auyan-Tepui). The migrations of this first angiospermous era coincide with the emersion of the lands later to become the Pacaraima, which took place in the Late Jurassic or very Early Cretaceous.

Established along the classic track leading from the heart of Gondwana to West Africa and beyond, as far at least as the "mesas" of the Guianas, the ancestors of *Gnestis* were at first distributed very much like such groups as *Agauria*, *Blaeria* subg. *Blaeriastrum*, and the cycadaceous *Encephalartos*. In the lands that are now Asia and Malaysia, these ancestors undoubtedly tenanted, as said, Ceylon, parts of Southern India, perhaps centers in Sumatra, the Moluccas and the Solomon Islands.

When the ancient geography began to break up, and the former continental shapes to alter to acquire present outlines, *Cnestis* undertook a new period of travelling and colonizing, which brought about active speciation. The lands that once stood under water in what is now Central and West Africa (see p. 421), to repeat, gradually came to emersion. Meantime Malaysia began active disruption. *Cnestis*, not unlike *Barringtonia* and the Dipterocarpaceae Dipterocarpoideae *) took immediate advantage of the new conditions. Archetypes strewn along the ancient track began to run secondary tracks to colonize lands freshly emerged. We find accordingly whole groups (e.g., Sect. *Ceratocnestis*) massively speciating in the Congo, Gabun, the Cameroon, Nigeria, etc., and such localities cropping up in the dispersal as Old Calabar, Eket, Oban, etc., which are repeated in the distribution of aggregates (e.g., Bombacaceae, Barringtoniaceae) homologous in history and migrations of the Connaraceae. The exact why of the monotonous cropping up of these localities in the dispersals of this type is not definitely known to us. We surmise, however, that they stand for bodies of waters that in time were left to dry wholly inland, and for grounds accordingly that were open to massive re-colonization

---

*) It seems that the African Dipterocarpaceae Monotoideae did not take active part in the migrations and speciation of this second period of angiospermous history. Their ecology rather suggests that they continued to live under conditions agreable, for instance, to the Ericaceae.

on the part of mangroves and "Near-mangroves." Localities like Oban, Eket, Old Calabar Koulikoro, etc., in West African dispersal mean exactly the same as do the Andaman Islands, Tawao, Tawitawi and Palawan (this last in a second time, not in a first that immediately interested the Magnoliaceae and Fagaceae; see p. 123) in Far Eastern distribution. Classic in our opinion is the track run by *Cnestis Palala* (Andaman Islands, Lower Burma, Siam, Indochina (Laos to Cochinchina), Malacca, Sumatra, Borneo as far as Tawao), which points to some zone of ancient faulting and transgression possibly having a geologic climate of the "Monsoon" description. It may be a "coincidence," but if so a peculiar one, that it is precisely in Siam and Indochina (see footnote p. 129) that certain Magnoliaceae are quite scarce.

Genera occur in the Connaraceae, witness *Roureopsis*, in which ranges typical of the period just discussed are underscored both in Africa and the Far East. *Roureopsis* is distributed in this manner,

(1) *R. obliquifoliata* — Belgian and French Congo, Cameron, Nigeria (Oban, Eket), Spanish Guinea, Cabinda, Angola.

(2) *R. Thonneri* — Belgian Congo, Cabinda.

(3) *R. birmanica* — Northern Burma.

(4) *R. breviracemosa* — Northern Burma.

(5) *R. stenopetala* — Southern Burma, Siam, Southern Indochina, Malacca.

(6) *R. asplenifolia* — Malacca, Sumatra.

(7) *R. pubinervis* — Malacca, Lingga, Bangka, Sumatra.

(8) *R. javanica* — Java, Northern Borneo.

In this dispersal appears a classic disconnection, West Africa-Burma, sometimes bridged in the mangroves (see p. 418) by stations in Afghanistan, and reaching on occasion as far as the Philippines. Doubtless, some weighty geological reason lurks in the background of types of dispersal of this nature, which some day we may bring to light *). The dispersal of *Agauria* (see p. 171) gave us reason to surmise that some ancient connections southward from the Cameroon ran as far as certain heights in Angola (Huilla, for example) and points beyond. While we are not so bold as to venture statements, still we cannot conceal our impression that species like *Roureopsis obliquifoliata* and *R. Thonneri* had their immediate origin when the modern coast of West Africa began to take shape. At the other end of the compass, the vicinity of Malacca, Southern Indochina, Siam, Burma must have represented in part a region, as stated, of geologic disturbance, perhaps a line of faulting, which the dispersal of certain species of *Barringtonia* (see p. 418) further suggests to have extended as far south as Australia. The importance of Burma in this

---

*) Things shape up very much as if there had been a zone of lowlands by a shore running the following course, West Africa—Sahara—Arabia—Mesopotamia—Afghanistan—India—Burma—Central and Southern Indochina—Malacca—Sumatra (and Bangka)—Java—Northern Australia. Whether the shore in question was actually of a sea, or merely a depression with occasional bodies of water we may not venture to say. We believe that it should be possible by comparative studies of plant-life to drive home conclusions having a sound basis in fact.

scheme of things is underscored by a monotypic genus, *Schellenbergia* (*S. sterculiifolia:* Lower Burma). Characteristically, *Castanola* which has eleven species in the Far East and Malaysia has one in Western Africa, so distributed,

(1) *C. paradoxa* — Belgian Congo, Gabun, Nigeria (Oban, Eket, Gold Coast, Sierra Leone).

Proof that the Connaraceae, which actively migrated in the second epoch of angiospermous dispersal, had already migrated in the first can readily be had in two species of *Cnestidium,*

(1) *C. rufescens* — Colombia (Bolívar), Panamá, Nicaragua, Honduras Guatemala, Mexico (Tabasco), British Honduras, Cuba.

(2) *C. guianense* — French and Dutch Guianas, British Guiana (Mt. Roraima).

This track is a classic of dispersal. It hits the Guianas and carries an archetype to Mt. Roraima, farther ranging along the modern northern shore of South America to Colombia (Bolivar). It also hits Cuba, and glides ahead to enter Mexico through British Honduras, which rates as a standard approach to Mexico from the West Indies. Central America is invested either through Mexico or through Colombia.

The fundamentals in the distribution of *Cnestidium* are repeated in that of *Rourea glabra* which ranges in this manner,

(1) *R. glabra* — Cuba, Jamaica; Venezuela (Bolívar, Guárico, Zulia: Sierra de Perijá), Colombia (Magdalena, Bolívar); Panamá, Costarica, Nicaragua, San Salvador, Honduras, Guatemala, Mexico (Tabasco, Guerrero, Colima, Jalisco, Nayarit, Veracruz, Tamaulipas); Brazil (Pará, Bahia, Pernambuco, Rio de Janeiro) *).

It will be observed that the Pacaraima Heights (Mt. Roraima, for example) are out of this dispersal, though it is not at all to be excluded that *R. glabra* may some day be collected also there. If we miss these heights, we have in focus the Venezuelan Sierra de Perijá, which may be in part coaeval with the Pacaraima, and certainly ranks as an old orogeny,probably Pre-Andean as to remote origins. Jamaica, too, is in focus, which is normal, for Jamaica is a very old Caribbean center. Santo Domingo ought to be in picture but is not, which is remarkable. Cuba is there, however, which is to be expected. The Lesser Antilles are out, because the track is to the south of them, and Cuba and Jamaica were most likely reached not directly from the east, but mediately from the south (line Colombia — Jamaica — Cuba; see p. 345). Central America and Mexico are conventionally invested. Let us further notice that the track does not reach southward from Colombia, though it reaches southward through Northern Brazil to Rio de Janeiro. This is conventional, for the two halves of South America, western and eastern, were by

---

*) The distribution of this species should be compared with that of certain forms of *Protium* (see p. 388), noticing that while *Rourea* reaches the Caribbeans and Brazil *from the east*, *Protium* hits them *from the west*. It will be seen that, whether from the east or the west, *Rourea* and *Protium* could not invade the same lands had not these lands existed in so tangible a form as to stand open to plant-life from all the quarters of the compass.

then perhaps connected in the north. The track invested this landconnection full blast, taking in Brazil which most immediately fronted its point of original issuance (West Africa), but stopped in the regions of Colombia immediately facing the Caribbeans, next again to veer northward to Central America. It is most likely that Mexico was invaded in part from Cuba, and in part again from Guatemala. It seems possible that a study of the forms in *Rourea glabra* may further refine our knowledge of secondary ranges, therefore perfect out understanding of the course of the track just outlined.

The ranges highlighted in the course of the migrations of *Rourea* are also fundamental in the dispersal of the Cactaceae in the affinity of *Cereus* and certain groups of *Croton*, represented by such forms as *C. platanifolius* (Madagascar), *C. gossypiifolius* (Trinidad, Coastal Venezuela, with a vicariant in Colombia, *C. Funckianus*), *C. panamensis* (Panamá and Central America), *C. Draco* (Mexico), *C. celtidifolius* (Eastern Brazil). Peculiarly, these forms of *Croton* are alien to the West Indies, which confirms that the West Indies (the Lesser Antilles particularly), were left out of the course of the main track which closely adhered to the coasts of South America fronting the Caribbeans, and further ranged eastward and northward to Central America and Mexico. We regret we cannot here undertake a discussion of the affinity of *Cereus*, merely remark that true *Cereus* is not endemic to the Lesser Antilles and, possibly, the West Indies in general, being here replaced by vicariants (*Pilocereus* etc.) having allies in coastal South America and Brazil. Clearly, a secondary center of speciation and genus-making was evolved for the Cactaceae along the line Grenada-Pernambuco, and this center in a later time (see also p. 370) fed northward genera in the alliance of *Cereus*.

Most interesting would be a discussion of forms of *Croton* endemic to Cuba and Santo Domingo, whose gross morphology is immediately reminiscent of the Sterculiaceae Lasiopetaleae of Western Australia, and, possibly, Madagascar. This discussion, however, cannot have place here. Suffice it to say that the tracks of *Rourea glabra* are anything but whimsical, but answer on the contrary basic realities of phytogeography and palaeogeography.

We are, then, at the antipodes of anything savoring of "Casual means of dispersal," and the like. We face something orderly, precise, *so orderly and so precise that all plants with comparable ecology and similar background speak the same language.* The seemingly fantastic disconnections in *Castanola* and *Roureopsis* obey a precise plan against a background of palaeogeography and phylogeny calling for the breaking up of a continent, Gondwana, and the oncoming of a new geography. Names such as Eket, Oban, Tawitawi clearly speak a message of their own.

The Connaraceae, then, migrated along wholly conventionally tracks in a first period of their history which is the Late Jurassic and Early Cretaceous. They again migrated, still along conventional lines, in a second period of their history between the Mid-Cretaceous and the first eras of the Tertiary.

The Connaraceae are certainly not "American" in origin. The reader knows why.

IX. Erythroxylaceae and Linaceae.

Erythroxylaceae and Linaceae are a perennial bone of contention for systematists. Some authors limit the former to two genera, *Erythroxylon* and *Aneulophus;* others, on the contrary, place under it *Ixonanthes*, *Phyllocosmus*, *Ochthocosmus* and sundry more or less important genera in addition, which systematists of a different mind bring under the Linaceae.

The Linaceae themselves are extraordinarily variable in habit, *Hugonia*, for example, suggesting anything but the common flax, *Linum*. On every count, then, it proves difficult to discriminate between Erythroxylaceae and Linaceae. It should be feasible to confine the latter to a narrow group of genera centering around *Linum* disposing of the balance of the affinity under an enlarged family Erythroxylaceae. As we shall see, the dispersal of *Linum* and its immediate affinity would favor in appearance this disposition.

In our opinion, clear affinities exist between *Linum* and the insectivorous and near-insectivorous *Byblis*, *Roridula* and *Drosophyllum* and these affinities are such that we may not ignore them much as we may wish. *Byblis*, on its part, is doubtless allied with the pittosporaceous *Cheiranthera*, and we are not at all sure that *Cheiranthera* and its group are actually remote from the small family Tremandraceae, which certain authors place in classification close by the Polygalaceae.

Drosophyllum is integral part of the Droseraceae. *Parnassia*, a most interesting genus, is constantly being tossed back and forth between the Droseraceae and the Saxifragaceae. The Saxifragaceae in turn are notoriously related with the Pittosporaceae. Linaceae and Erythroxylaceae are not to be certainly separated in phylogeny from Geraniaceae and Oxalidaceae. The last have some measure of kinship with the Sterculiaceae, and the Sterculiaceae themselves are notorious members of a single affinity comprising Euphorbiaceae, Malvaceae, Bombacaceae, Tiliaceae. Beyond doubt, Euphorbiaceae and Tiliaceae led us toward the Flacourtiaceae, which last reach by easy steps the Passifloraceae and Caricaceae.

It goes without saying that the bonds of consanguineity among the groups mentioned above are not equally strong throughout. Some of these bonds need not be taken up in formal systematic and taxonomic work, but none of them can be wholly ignored in this work, either. It is accordingly clear to us that classification can never be made entirely "natural" because names and forms of nomenclature can never cope with the intricacies of genuine natural bonds. *Evolution and formal classification are essentially antithetic, which is so palpable a truism we must marvel why it goes so often unheeded.*

In this persuasion, we are certainly not inclined to take issue with this or that systematist on the ground that by using his right to free opinion he understands the Erythroxylaceae and Linaceae in a "broad" rather than a "narrow" sense. We will certainly not argue whether these two families are to be cut up one way or the other, because so long as no manifest error is involved we must be tolerant of opinion in a field wherein opinion rules supreme. We believe no more in chromosome counts than we do in old fashioned visual evidence as a final means of discriminating in creation, because we are sure that a contemplation of life in evolution and motion — a genuine biological viewpoint, that is — is essentially a form of personal philosophy. Even less will we think of contributing a "Phytogeographic scheme of classification" of the Erythroxylaceae and Linaceae, for, by overstressing phytogeography, we may make classification even more "artificial" than it now is. Moreover, we are not particularly observant of formalities, considering that our work leads us by its very essence to break down formal limits born out of time and space.

According to a standard treatment (38, 39), the Erythroxylaceae consist of but two genera, *Aneulophus* and *Erythroxylon* (or *Erythroxylum*). The latter is industrially important as the source of the drug *coca* and its therapeutic derivatives.

Aneulophus is quickly disposed of, for it consists of a lone species, *A. africana*, endemic to Western Africa (Gabun). This part of the world, then, receives its dues at the very first, as it does in the dispersal of the Connaraceae. With Gabun on the map before us we may safely anticipate migrations to the New World and the Far East in the ranks of *Erythroxylon* which is a large genus of about 200 species in 19 sections.

These 19 sections, and some of their species having relevant range, tabulate as follows;

A) Sect. *Pogonophorum*

1 species in Brazil (Rio Janeiro).

B) Sect. *Macrocalyx*

8 species in Tropical America from Mexico (Veracruz) to Bolivia (La Paz), Eastern Brazil and the Guianas.

C) Sect. *Rhabdophyllum*

42 species in Tropical America, ranging from the West Indies and Central America to Paraguay, Northern Argentina, Eastern Brazil and the Amazonas generally.

A perfect tie between the Guianas, significant regions of Venezuela (Sierra de Perijà, etc.), and the West Indies is highlighted by these two species,

(1) *E. rufum* — British Guianas, Venezuela (Zulia: Sierra de Perijà; Falcón: Cerro Santa Ana; Caracas; Amazonas), Cuba, Hispaniola.

(2) *E. squamatum* — French and British Guianas, Trinidad, Grenada, Martinique, Dominica, Guadeloupe, Montserrat.

The ties involved in this dispersal are standard as we will better learn in coming pages devoted to a study of the Cactaceae, Cyrillaceae, Protieae, etc. The Sierra de Perijá and Cerro Santa Ana stand close by the abutment on South American soil of a fundamental track, *Bolivia-Cuba*, which track oftentimes frays eastward to reach the Guianas. *Erythroxylum squamatum* has typical dispersal from the Guianas northward. It is seldom, however, that this dispersal reaches that far north with a single species common to the Guianas and Montserrat.

D) Sect. *Leptogramme*

5 species from Eastern Brazil to Peru (Loreto) and Colombia (Cauca).

E) Sect. *Heterogyne*

7 species in the West Indies, Central America and Southern Mexico.

F) Sect. *Archerythroxylon*

57 species in Tropical America ranging among the Bahamas; to Western Mexico (Nayarit), Northwestern Argentina (Tucumán) and Uruguay.

Interesting is the dispersal of the following species,

(3) *E. coelophlebium* — Eastern Brazil (São Paulo, Rio Janeiro, Minas Geräes); Var. *brevifolium:* Rio Janeiro; Var. *petiolatum:* São Paulo, Rio Janeiro, Minas Geräes; Var. *Grisebachii:* Dutch Guiana.

(4) *E. novogranatense* — Colombia (Cauca, Tolima, Cundinamarca, Magdalena (Sierra Nevada de Santa Marta)); Var. *macrophyllum:* Venezuela (Amazonas, Bolívar); Var. *tobagense:* Tobago; Var. *microphyllum:* Southeastern Brazil.

(5) *E. roraimae* — British Guiana (Mt. Roraima).

These ranges are matched most closely in the dispersal of the Connaraceae, as we may readily learn.

G) Sect. *Mastigophorum*

1 species in Brazil (Bahia).

H) Sect. *Microphyllum*

12 species in Tropical America from the West Indies and Central America to Northern Argentina.

I) Sect. *Gonocladus*

6 species in Madagascar and the Comoros.

J) Sect. *Sethia*

1 species in Ceylon and Southern India ranging to Eastern Bengal.

K) Sect. *Lagynocarpus*

15 species in Madagascar, the Comoros and Tropical Africa; with the following interesting ranges,

(6) *E. pyriformis* — Madagascar, Nossibé.

(7) *E. Dekindtii* — Angola (Huilla).

(8) *E. emarginatum* — Nyasaland (Mt. Malosa, 1200—1800 meters alt.), British Rhodesia, Tanganyka (Usambara and at the coast on coral-sands), Uganda (Mt. Ruwenzori), Nigeria, Ivory Coast; Var. *angustifolium:* Nyasaland; Var. *caffrum:* Natal, Angola.

This gives perfect penetration of continental Africa from the east, that is, from the Indian Ocean of the modern map. *Erythroxylon emarginatum* exactly repeats the performance of *Philippia* and *Cassipourea*. It occurs both as an alpine and as a plant of the sea-shore. Like *Philippia (P. mafiensis)* it occasionally thrives at the coast in coral-sand, and is accordingly an arrant calciphile, the meaning of which is soon to appear.

L) Sect. *Coelocarpus*

14 species in Asia, the Far East generally, Northern Australia eastward to New Caledonia; 2 species in Africa.

Significant are these species,

(9) *E. pictum* — Eastern Cape (Albany and Komgha Districts), Natal.

(10) *E. delagoense* — Mozambique, Zanzibar.

(11) *E. lanceolatum* — Ceylon, Southern India.

(12) *E. latifolium* — Sumatra, Lingga, Bangka, Billiton, Northern Borneo (Sarawak).

(13) *E. cambodjanum* — Cambodja.

(14) *E. cuneatum* — Southern Burma, Malacca, Sumatra, Java; Var. *bancanum:* Bangka, Luzon, Ternate, Ceram, Halmahera.

(15) *E. ecarinatum* — Celebes, Ceram, Amboina, Boeroe, New Guinea.

(16) *E. ellipticum* — North Australia (Groote Eylandt).

(17) *E. iwahigense* — Philippines (Palawan).

(18) *E. platyphyllum* — Philippines (Mindanao).

(19) *E. novocaledonicum* — New Caledonia.

This group, so far as we can gather, is of xerophytes of the sea-shore or the open grassland, but adapts itself as an alpine. The habitats include, (a) Coral-sands; (b) Probably alpine grasslands; (c) Secondary forest on limestone *(E. cuneatum);* (d) Thickets back of the mangrove *(E. platyphyllum);* (e) The "Monsoon-Belt" generally *(E. cambodianum).*

M) Sect. *Eurysepalum*

1 species in Madagascar.

N) Sect. *Venelia*

5 species in Madagascar, the Comoros and Tanganyka, one of them at least *(E. platycladum:* Madagascar, Comoros, Tanganyka) a litoraneous calciphile.

O) Sect. *Pachylobus*

9 species in Madagascar, the Comoros, Mascarenes, Seychelles, Tanganyka, Uganda, Kenya, Sudan.

P) Sect. *Schistophyllum*

1 species in Madagascar.

Q) Sect. *Oxystigma*

1 species in the Eastern Himalayas, Burma, Southern China (Yunnan).

The phytogeographical aspect of this dispersal is not the element that can interest us most at this juncture. Patently, the tracks follow the customary main lines of all groups which leave the African gate to reach the Americas and the Far East. The element which commands our attention is sooner ecological. In other words, we are dealing with calciphiles, and the dispersal — though molded in the first place by standard tracks common to all manners of plants — is in a second time materially modified by the ambiental requirements of the plants taking part in it.

It is dangerous to discuss subjects of this nature without having first-hand knowledge of all factors in play. Hints in the literature may be precious, but on account of their being contributed by authors who not always understand the significance of certain factors, these hints cannot be trusted to the hilt. Indeed, it seems that we deal with arrant calciphiles, capable of secondary adaptations in the sense outlined above, and this we will believe. Trusting the evidence, then, as far as the evidence may lead us, we will briefly comment upon certain aspects of the tracks of the Erythroxylaceae that seems to agree with, or be at variance from those of the Connaraceae.

For example, we may repeat here the record of distribution of three species, as follows,

(1) *E. roraimae* — British Guiana (Mt. Roraima).

(2) *E. coelophlebium* — Var. *Grisebachii:* Dutch Guiana; Var. *macrophyllum:* Venezuela (Amazonas, Bolívar); *typic form:* Eastern Brazil (São Paulo, Rio de Janeiro, Minas Geräes).

(3) *E. novogranadense* — Var. *macrophyllum:* Venezuela (Amazonas, Bolívar); Var. *tobagense:* Tobago; Var. *microphyllum:* Southeastern Brazil; *typic form:* Colombia (Cauca, Tolima, Cundinamarca, Magdalena (Sierra Nevada de Santa Marta).

If we compare the dispersal of these three species with that of *Rourea glabra* (see p. 192) we readily are aware we face comparable migrations. The northern coast of South America is invested in standard manner. The Guianas, parts of Venezuela, the coast of Colombia on the Caribbeans are covered, including what is probably a rejuvenated orogeny (Sierra Nevada de Santa Marta). Eastern Brazil is invaded. The track is shorter than *Rourea glabra*'s, but homologous in fundamentals. It is manifest furthermore, that the three species of *Erythroxylon* just reviewed break down into varieties rich in phytogeographic meaning. It seems possible that we witness a genuine incipient speciation.

Another species migrates in this manner,

(1) *E. suave* — Var. *typicum:* St.-Thomas, Portorico, Santo Domingo, Cuba, Bahamas; Var. *aneurum:* Bahamas; Var. *jamaicense:* Jamaica; Var. *compactum:* Mexico (Puebla).

This is a peculiarly interesting example of dispersal. It is perspicuous from the standpoint of pure phytogeography, for it runs in standard fashion from Mexico to a classic region of break between the Greater and Lesser Antilles. From the viewpoint of ecology, on the other hand, much would be learned if the entire range could be carefully explored in order to ascertain the nature of the habitats throughout, from Mexico to the Bahamas. The Bahamas are doubtless an old angiospermous center, strategically located between the Southeastern United States, Florida in particular, and some of the Antilles. It is likely that the Bahamas are in addition an important center of secondary evolution. As we are soon to learn, *Linum* exhibits a massive speciation in these islands, part of which at least cannot well be primary. Calciphiles like *Ery-*

throxylon and *Linum* would probably take advantage of geological sequences exposing limestone of ancient and recent origin. In sum, a precise knowledge of the habitats of these two genera in the Bahamas, and a critical investigation of their interspecific and intersectional relations is highly desirable.

An authoritative treatment of the Linaceae (40) reveals the following,

A) Subf. Linoideae

a) Tribe Hugonieae

(1) *Hebepetalum* — 1 or 2 species in the Amazonas, Guianas and Eastern Peru.

(2) *Hugonia* — 35 species in Madagascar, Mascarenes, Cameroon, and the Far East generally.

(3) *Durandea* — About 15 species in Eastern Malaysia, New Guinea, New Caledonia.

(4) *Philbournea* — 2 species; *P. magnifolia:* Northern Borneo (Sarawak); *P. palawanica* (Philippines: Palawan).

(5) *Indoroucheria* — 3 species in Indochina, Malacca, Java, Sumatra, Borneo.

(6) *Roucheria* — 5 species in Brazil, Guianas, Colombia southward to Eastern Bolivia.

a') Tribe Anisadenieae

(7) *Anisadenia* — 2 species from the Himalayas to Central China.

The Hugonieae migrate in approved fashion within a triangle Bolivia—Madagascar—New Caledonia in a manner typical of *genorheitra* out of the African gate. We would like to have precise information of *Philbournea* which is endemic to an interesting range. This range may be very ancient, that is, established in the first epoch of angiospermous dispersal, or, contrariwise, be not older than the period of active Malaysian disruption.

The Anisadenieae are a group which ought to be studied with great care, in order to determine whether it is actually intermediate between Hugonieae and Lineae. The range, *as such*, is uninteresting and hardly likely to be successfully discussed in the ignorance of the morphology and phylogeny of the forms involved. We offer brief comments upon it at the end of our review of the Linaceae.

The third Tribe of the Linoideae is the affinity of *Linum* proper which is distributed in the following manner,

a'') Eulineae *)

(8) *Reinwardtia* — 2 species in India, China, Java; "Naturalized" in the West Indies (Martinica and Guadeloupe).

(9) *Tirpitzia* — 1 species in China and Tonkin.

(10) *Hesperolinum* — 10 species in the Western United States (California).

(11) *Radiola* — 1 species in the mountains of Tropical Africa, extending to Madera, the Canary Islands, the Mediterranean generally to Southern Russia, England and Norway.

(12) *Linum* **) — About 200 species in five sections, as follows,

i) Sect. *Eulinum* — 1 species in Tasmania and Southern Australia *(L. marginale),* 1 species in the Altais *(L. pallescens),* 2 species in North America *(L. Lewisii:* Alaska to California and Texas; *L. pratense:* Saskatchewan to Arizona and Texas); the balance scattered throughout Eurasia (Europe to Mesopotamia and China).

ii) Sect. *Linastrum* — 1 species *(L. gallicum)* in the Mediterranean, ranging from Madera and the Canary Islands to Northern Africa, Italy, France, Syria and Crimea; a form *(L. gallicum* var. *abyssinicum)* in Ethiopia, Tanganyka (Kilimajaro) and the Cameroon; 4 species in the Cape, one *L. africanum* a shrub to 1 meter tall, another, *L. quadrifolium,* part-woody; other species in the Mediterranean to Greece and Croatia *(L. maritimum);* India *(L. mysoren-se);* South America from Chile to Peru and Eastern Brazil *(L. Macraei, L. scoparium, L. andicolum, L. littorale, L. organense).*

iii) Sect. *Cathartolinum* — 1 species *(L. catharticum)* in Northern Africa, the Canary Islands, northward to Scandinavia (68° Lat.), the Mediterranean to Persia and the Caucasus; the balance, about 50 species, in the Bahamas *(here: L. corallicola, L. bahamense, L. Bracei, L. lignosum, L. Curtissii),* Central America (Guatemala) and North America throughout from Mexico to Utah, Wyoming, Saskatchewan and the Eastern United States.

iv) Sect. *Syllinum* — The Mediterranean to Southern Russia, Persia and the Caucasus.

v) Sect. *Cliococca* — 1 species, *L. selaginoides,* in Chile, Uruguay, Brazil and Peru.

The dispersal of the Eulineae is arresting and, at first sight, widely at variance from the remaining Tribes of the Linoideae.

To understand it, let us begin with Sect. *Cathartolinum.* This group has a base-line of distribution between Iran (Persia) and the Canaries. This base-

---

*) We abstain as a rule from commenting upon nomenclature. We notice by exception that *Eulineae* is illegitimate under the International Rules. We maintain capitalization of epithets against a recently accepted Recommendation believing this Recommendation to be unwarranted.

**) We follow the same authority in the classification of *Linum* whom we take for our guide as regards that of the Linaceae. See, however, for instance notes on *Linum* by Ciferri (in Atti Ist. Bot. Univ. Pavia, Lab. Critt., Ser. 5, 2: 142. 1944). This author finds that the ethiopian forms of *Linum* are intermediate between those of the Mediterranean and India, in a manner which the reader of this manual will understand without further. There is homology, as a matter of fact, between the affinities in *Linum* and *Olea* within the triangle Ethiopia—India—Mediterranean.

line feeds migrations northward in Europe as far as Scandinavia; holds the Mediterranean and Northern Africa; dispatches a component of definite calciphilous archetypes (later to become some 50 modern species) to the Bahamas; invades the Eastern United States, not only, but Central America, Mexico and the Western United States.

All this is standard, and we may refer it to a great many precedents (see *Thamnosma, Oligomeris, Minthostachys* and *Bystropogon, Andrachne, Croton, Cneorum,* etc., etc.). It rates as a classic instance of transatlantic migration.

Section Linastrum ties Sect. *Cathartolinum* with the south of the map. *Linum gallicum* ranges from Madera to Crimea and Syria; one of its forms is in Ethiopia, Tanganyka, and the Cameroon. Other of its allies are in South Africa, India, Eastern Brazil and America generally to Peru and Chile. A special Section, *Cliococca,* stresses once more Brazil, Uruguay, Peru and Chile.

Section *Eulinum,* lastly, is in Tasmania and Southern Australia (1 species), Eurasia and Central Asia (Altais Mts.; here *L. pallescens*), North America (Alaska to Saskatchewan, California and Texas).

The dispersal of *Linum,* then, is to be reconstructed as follows, (a) *Cathartolinum* migrates very much like the ericoid genus *Calluna* (see p. 166); (b) *Linastrum* connects the range of *Cathartolinum* with the southern hemisphere both in the Old and New World. Noteworthy is the circumstance that the tie *Western Africa—Eastern America* is worked by this group — possibly, when not certainly — south of the usual range *Cameroon—Northeastern Brazil.* Strong are the indications that parts of *Linastrum* might have evolved in Southwestern Africa, reaching the New World from this point. Migrations of a similar pattern (see *Genlisea,* p. 351) will come under review in coming pages; (c) *Eulinum* has a pattern of migration strongly reminiscent of that of certain *Carex* (e.g., *Carex pyrenaica*). The track takes its inception in Tasmania and Southern Australia, and next reaches in disconnection the Altai Node, ultimately to migrate from this Node eastward and westward in a conventional manner. Well worthy of attention is the fact, though endemic to Australia and Continental Africa, *Linum* does not seem to occur in Madagascar and the Mascarenes. The significance of this is to appear fully in coming pages (see p. 349). We may state here as a preliminary that *Linum* has mostly antarctic, not Gondwanic, origins.

Radiola, Hesperolinum, Tirpitzia can be grafted without the slightest difficulty upon the net of tracks we have just outlined. Most instructive is the distribution of *Reinwardtia.* Two of its species are in the Far East, but also occur in Martinica and Guadeloupe. Faced by this dispersal, the monographer we follow, quite arbitrarily rules out the second massing as "Adventitious." In reality, there is every probability that this massing is quite as authentic as the one in the Far East, because *Linum* is both in the Far East and the Bahamas, and there is certainly no reason why *Reinwardtia* should not also be endemic to the Far East and the French West Indies. Rather, the dispersal of *Reinwardtia* is significant because it conveys at least a hint (confirmed also in the dispersal of *Prunus;* see p. 326) that Martinica and the Guadeloupe stand clear in the abutment of the track that comes to the Caribbeans from the Mediterranean. In other words, connections between the Old and New World, calling for a "transatlantic crossing" are effected between these regions of the modern map, (a) *French West Indies and the Mediterranean (range Southern Portugal—Canary Islands as the door to or from the Mediterranean);* (b) *Guianas (and Northern Brazil) and the Ivory Coast (and the Cameroon).* The former is used by groups that are not strictly tropical, and may lead further to Socotra and Western India (see *Thamnosma*) on one hand, Mexico and the Western United States on the other; the latter is taken advantage of by tropical groups most particularly. It leads to the northern coast of South America as far as Central America and Mexico (see *Rourea glabra,* etc.), to Eastern Brazil generally, the whole of the modern Amazons as far west as the foothills of the Peruvian and Bolivian highlands. At the other hand, this connection may reach the whole of the Far East, Malaysia, New Guinea, New Caledonia, Fiji and Samoa through Central and East Africa. Still a third connection runs south of these two, which has very intimate relations with Angola and South Eastern Brazil. It is probably through the last connection that *Linastrum* and *Cliococca* fared to the New World from South Africa.

As we have seen, *Linum* migrates in Africa very much in the manner of *Erica,* avoiding, that is, Madagascar and the Mascarenes. Moreover, unlike the balance of the Linaceae, *Linum* is a genus of potentially "cold" to "temperate" nature. Quite different is the case with *Hugonia,* for example.

We may safely affirm on the strength of these factors that *Linum* and *Hugonia* (and, together with *Hugonia,* probably the balance of the Linaceae outside of *Linum*) originated in different centers. These centers we will definitely identify in coming pages (see for the discussion, p. 349). As a preliminary, and for the record, we will merely state here that *Linum* is from the *Afroantarctic,* the balance of the Linaceae mostly on the contrary the *Gondwanic Triangle.* These two centers stand oftentimes in very intimate connection (for instance, in the case of the Ericaceae, see p. 160), which is reason why we do not believe it is possible to use different origins from one or the other to effect taxonomic and systematic discrimination of absolute value.

There is reason to affirm, nevertheless, that *Linum* is a well marked offshoot of the traditional Linoideae. This offshoot doubtlessly originated in, and dispersed from, a center of origin different from that of the Hugonieae. On this ground, as well as on ground of morphology, it should be possible to exclude the Hugonieae from the Linoideae. However, the Anisadenieae remain as a most critical group. The range of the two species in *Anisadenia* (Himalayas to Central China) may include forms that precisely originated in the center of *Linum* (the *Afroantarctic Triangle,* to be definite) and reached India and China in the very first upsurge of angiospermous migration (see for a parallel case certain groups in affinity of the Compositae Crepidinae, p. 59). It is then possible that the Anisadenieae (which we regret not to know at all from authentic live or dried material) stand as the connecting link between the Hugonieae and Lineae, and that all thought of confining the Linaceae to *Linum* had better be abandoned.

As we have noticed, certain forms of *Linum* are woody and perennial, while other are not so. Were we to believe certain theorists we ought to assume that the former are "primitive," the latter "derivative." An assumption of the kind would be ridiculous — we believe — because it is beyond reason that anything can be assumed on the "primitiveness" and "derivativeness" of forms that had existence *prior to the inception of the Cretaceous*, on the ground that certain of the tissues of these forms are "woody" rather than "non-woody".

The Cyperaceae are massively herbaceous, yet it is difficult to conceive of any other groups quite as old as this family in the record of angiospermy. The survival of certain "woody" Graminaceae and Cyperaceae in ranges like Western Australia, West Africa, the Guianas indicates that, granted steady climatic conditions in ranges free from geological disturbances throughout long ages, even arrant herbs tend to become "woody." Above all, the meaning of "woody" ought to be critically elaborated before using this term as an argument to debate "primitiveness." A botanist familiar with the Compositae in the Northern Hemisphere and the Tropics knows very well that the seemingly "weedy" forms of cold to cold-temperate climates differ from the "shrubs" and small "trees" of the same family abundant in the Tropics only by degrees. "Woody" and "herbaceous," consequently, are most often aspects of secondary adaptation.

It is true that much has been written to prove by the use of arguments based as a rule upon preconceived definitions that "woody" plants are primitive. Authors who put their faith in these notions are usually mislead by narrow preoccupation with certain phases of wood-anatomy, but less than well informed about other, perhaps more important aspects of botany.

It is understandable that, climatic conditions being what they are, the *genorheitron* of *Euphorbia*, for example, delivered to the Canary Islands (and here retained to this day) xerophytic and subxerophytic types, while it evolved in colder Eurasia perennials from the rootstock, and annuals in the deserts of Central Asia and Africa. The conditions affecting this adaptation are potentially so varied and variable, and so thoroughly interlocking with hereditary tendencies and fixed characters, that academic discussions as to herbs being "derivative" and woody plants "primitive" are necessarily wide of the mark. In our opinion, the tuberous underground "roots," and the "buttresses" of "wood-parenchyma" which characterize certain forms (e.g., *Manihot;* 41) as against certain others (e.g., *Tarrietia;* 42) differ from the physiological standpoint only in degree, and are closer from the anatomical than current opinions are ready to grant. In considerations of these and analogous factors, we should not be inclined to indulge in wide generalization as to "woody structures," setting up definitions and technical manipulations hopefully to replace deeper and better rounded thought *).

Next to the Lineae is another the Linoideae's tribe, as follows,

a''') Tribe Nectaropetaleae

(13) *Peglera* — 2 species in South Africa (Cape: Kentani District; Mozambique).

(14) *Nectaropetalum* — 3 species in Uganda, the Congo, Zululand.

This Tribe, the last in the Linoideae, is distributed in a manner that does not conflict with the trends in the dispersal of *Linum*.

Its evolution is clearly bound with a standard center, later to be pointedly identified in these pages (see p. 349), which includes large parts of South and Central Africa once the seat of broad bodies of salt or fresh water. The Cape is clearly entered from the east, not the south, which confirms the indications in the dispersal of the Ericaceae and other groups previously reviewed in this manual.

The Linaceae Linoideae are followed in conventional classification by the Linaceae Ctenolophonoideae, which latter tabulate in this manner,

---

*) It is not in the scope of this manual to discuss subjects that are unessential, and it is not excluded that a book of ours may next be written to deal with aspects of morphology and classification which must now be left out of reckoning. It might be mentioned, nevertheless, that E. W. SINNOTT and I. W. BAILEY are joint, or separate, authors of a number of contributions on the phylogeny of the Angiosperms (in Ann. Bot. 28: 547. 1914; in Amer. Jour. Bot. 1: 303, 441. 1914; in op. cit. 2: 1. 1915; in op. cit. 3: 24. 1916). In one of these papers (Ann. Bot. 28: 597. 1914) the statement is made that „Evidence from phytogeography also supports the contention that the most ancient Angiosperms were woody. There is a great preponderance of herbs in temperate regions and of woody plants in the tropics. The latter climate probably approaches more nearly to that under which Angiosperms first appeared." In another contribution by the same authors (Amer. Jour. Bot. 2: 20. 1915) the following is written, „Among woody plants, the multilacunar (more ancient) nodal type predominates in temperate regions and the unilacunar (more recent) in the tropics. The palmate lobed leaf among such plants is almost entirely confined to temperate regions. These facts in company with others, indicate that the Angiosperms first appeared under a climate more temperate than tropical; a climate in the Mesozoic probably found only in the uplands."

It is evident that these authors are not quite sure as to the climate which favored the Angiosperms most in the beginning. Speaking of „herbs" (Ann. Bot. 28: 599. 1914), the two authors claim that the „Great majority" were probably developed in the landmasses of the northern temperate zone, first of all in the mountains. The same form of vegetation arose, however, also in the mountainous regions of the tropics and the southern hemisphere, always according to these authors. This is not all, by any means, for they further surmise that „Herbs" were evolved under tropical conditions in response to the alternation of wet and dry seasons or „For other causes". Further still, other „Herbs" originated in their opinion in deserts or dry regions. To cap the discussion, the two authors record their belief that a „Comparatively small body of herbs" had independent origin in certain „Antarctic lands," which are spoken of by these authors as synonymous of the „Temperate regions of the Southern Hemisphere." It is obvious that the authors in question are no better satisfied as to the „origin of herbs" than they are about the climatic conditions under which the Angiosperms first „evolved." It remains to be seen what they understand as „Herbs," and what is their concept of the supposedly primitive leaf. Forms in the phyllodine Acacias show, for instance, that a „phyllode" may be, (a) a purely cortical expansion of the stem; or, (b) a structure originating from an „abscission layer;" the one being anything but the homologous of the other in a precise anatomical and morphological sense. In short, the statement that „Phytogeography" is in favor of these vague generalizations and none too successful assumptions cannot carry weight. It so happens that, according to the authors in question, the Erythroxylaceae are „unilacunar," the Linaceae, contrariwise „trilacunar," which all does not seem to rhyme to much in genuine scientific phylogeny one way or the other.

B) Subf. Ctenolophonoideae

(15) *Ctenolophon* — 3 species in Malaysia (Malacca, Sumatra, Borneo, Philippines (Mindanao, Leyte, Samar)); 1 species *(C. Englerianum)* in West Africa (Angola).

This disconnection is repeated, as we know, in the dispersal of genus *Castanola* of the Connaraceae. A similar break occurs also in the tracks of various families (Sterculiaceae, Barringtoniaceae, etc.), not only, but can be detected, though not on the map this time, by the affinities for example that bind *Cochlospermum angolense* to *C. Gillivraei* (Northern Australia, New Guinea). In brief, this disconnection is tame; it stems from palaeogeographic factors that forced it upon unrelated groups as one.

Moved by breaks of the kind, a taxonomist interested in the flora of the Philippines (43; p. 81) hastens to bring to our attention the following records of dispersal,

(1) *Combretodendron* — 1 species in the Philippines; balance of the genus in Africa.

(2) *Guadua* — 1 species in the Philippines; balance of the genus in Tropical America.

(3) *Uncinia rupestris* — Magellania, New Zealand, Tasmania, Philippines.

(4) *Angraecum* — 1 species in the Philippines; balance of the genus in Madagascar and Africa.

(5) *Spatiphyllum* — 1 species in the Philippines and Moluccas; balance of the genus in Tropical America.

(6) *Erythrophloemum* — 1 species in Southern China; 1 species in Indochina; 1 species in Northern Australia; 1 species in the Philippines.

(7) *Phrygilanthus* — 1 species in the Philippines; balance of the genus in New Guinea and Australia.

(8) *Omphalea* — 1 species in Australia, 4 species in the Philippines (3 common to Borneo), the balance in Tropical America.

(9) *Villaresia* — 1 species in Polynesia, 2 species in Java, 1 species in the Philippines; balance of the genus in Tropical America.

The taxonomist cited next follows in this manner, "No explanation is offered regarding the anomalous distribution of the representatives of these genera except that the generic range was apparently attained at an early date and that the outlying species have persisted as relic types. In some cases these widely distributed genera may be of dyphiletic or polyphiletic origin."

We observe, (a) Phytogeography makes no sense until and unless these patterns of distribution are explained; (b) These patterns are not anomalous; (c) It is true that the generic range was attained "At an early date," but phytogeography is blind until and unless this early date is known at least with some approximation; (d) It is certainly erroneous that the genera listed are "Diphyletic" or "Polyphiletic." These terms, quite as "Relic types," etc., are meaningless unless they are critically assumed and pertinently understood.

Next to the Ctenolophonoideae stands by rote the last Subfamilies of the Linaceae, as follows,

C) Subf. **Ixonanthoideae**

(16) *Ochthocosmus* — 11 species in Angola, Congo, Sudan, Cameroon, Lower Guinea, the Amazonas, Guianas and the Pacaraima Mountains (Mt. Roraima, Mt. Auyan-Tepui).

(17) *Ixonanthes* — 12 species in the Tropical Far East and Malaysia (Eastern Himalayas, Indochina, Southern China, Malacca, Sumatra, Borneo, Philippines) to New Guinea.

D) Subf. **Humirioideae** (see further p. 387)

(18) *Humiria* — 4 species in Tropical South America.

(19) *Saccoglottis* — 17 species in Tropical South America (Eastern Brazil, Amazonas, Guianas, Venezuela); 1 species in Western Africa (Gabun to Sierra Leone).

These two Subfamilies require no elucidation. They tell us that the tracks involved are constantly the same, away from Gondwana to the New World through Central Africa, and to the Far East. There can be no doubt as to the age of these tracks, because they are active between the Pacaraima Mountains and Western Africa, which forthwith stamps them as of the first angiospermous age of migration. This means the Late Jurassic.

Manifestly, there is no need to stress the fact that nothing whatever in the patterns of dispersal just reviewed demands the ministrations of "Extraordinary means of distribution," or recourse to such things as "Holarctis," "Polyphiletism," "Polyploidism," "Species Senescence." We feel under no necessity of discussing what we understand as "Landbridges," and why we like "Landbridges" or, perhaps, we do not like them. It seems we can go by, without being cornered, and asked whether we believe in "Climatic changes by Shifts of the Poles," or relie precisely on the contrary. *The stuff we handle is fluid, be this granted, but orderly, simple, precise, repetitious to the point of utter boredom.* Given three stations in a track, or hearing of such localities as Oban, Eket, Mt. Roraima, Mérida, Tawao, we can forthwith know at least by approximation where we stand. We can lay a finger on the map and show where the decision was made in the Late Jurassic to dispatch a certain archetype to Bolivia, the Congo or Japan. We can take in hand huge masses of flora, break them apart, study a species quite as much as a *genorheitron* involving ten families. If all these be our own dreams those who are awake will soon tell us why.

X. Lobelia

The taxonomy and systematy of this genus, and the Lobeliaceae generally, is in a state of confusion, but we may hope to show once again that phytogeography stands above synonymies, shifts of nomenclature, and the like.

A large contingent of shrubby or arborescent Lobelias is endemic to the mountains of Eastern Africa, and would seem indeed to be narrowly restricted to these mountains. The forms in question fall under Sect. *Rhyncopetalum*. Various systematists believe that this group is allied with Sect. *Tylomium* and Sect. *Tupa*, the former endemic to the West Indies, the latter to Chile. It is also suggested (44) that sect. *Rhyncopetalum* includes two at least of the species of Southeastern Brazil, *L. exaltata* and *L. thapsoidea*. The lobeliaceous genus *Isotoma* is massively Australian but allied with *Brighamia* of Hawaii. *Trimeris* restricted to the island of Saint-Helena shows affinity with the large American genus *Siphocampylus*. A controversial form, *Palmerella*, is endemic to the neighbourhood of Baja California, and it is not excluded that Sect. *Tylomium* has a representative, *L. boninensis*, in the Bonin (Ogasawara) Islands. The Pacific, lastly, is alive with group of controversial taxonomic and systematic status, *Brighamia*, *Clermontia*, *Rollandia*, *Delissea*, *Cyanea*, *Trematocarpus*, *Trematolobelia* in Hawaii; *Apetahia*, *Cyrtandroidea*, *Sclerotheca* in the Society and Marquesas Islands, and Rapa.

This sound very involved, but is simple, on the contrary. If the limits of Sect. *Rhyncopetalum* are in themselves obscure, the dispersal of species that surely belong to this group is transparent, on the contrary, witness,

(1) *L. rhyncopetalum* — Ethiopia.

(2) *L. Leschenaultiana* — Ceylon, Southern India.

(3) *L. nicotianifolia* — Ceylon, Southern India.

(4) *L. trichandra* — Southern India.

(5) *L. Doniana* — Eastern India.

(6) *L. pyramidalis* — Eastern India.

(7) *L. philippinensis* — Philippines (Luzon: Bontoc, Benguet, Ifugao, Lepanto, Zambales), Celebes.

(8) *L. yuccoides* — Hawaii.

(9) *L. hypoleuca* — Hawaii.

These species run the same track as does *Lysimachia mauritiana* (see Fig. 17; p. 53), and a group which migrates in this manner is quite apt to take further advantage of a transpacific bridge in the higher north. It is accordingly understandable that the Lobeliaceae are represented by Sect. *Tylomium* in the Bonin (Ogasawara) Islands and the Caribbeans. It is moreover normal that the flora of Baja California, already enriched by forms which are no more *Phrygilanthus* than *Phoradendron* (see p. 146), a special section of *Aesculus*, etc., should further include an unclassifiable lobeliaceous genus (or subgenus, or section) *Palmerella*.

It is also clear that a group that leaves Africa (whether the Mascarenes, or

Tanganyka, or Ethiopia) *) to travel as far afloat as Hawaii *eastward*, may easily reach the New World westward from the African gate, or, mediately, eastward across the Northern Pacific from Asia. It is then understandable that forms with affinity toward Sect. *Rhyncopetalum* are in Eastern Brazil, and a section, *Tupa*, is in Chile which is in its turn allied to said section. As a matter of fact, Eastern Brazil and Chile can also be reached directly from Africa.

The Pacific is replete with controversial genera of the Compositae, and it is but normal that it should suffer from a pletora of controversial genera of the Lobeliaceae, for both these families are part of one and the same very broad *genorheitron*. Leaving the African gate of angiospermy to speed westward toward Chile and Australia, the lobeliaceous *genorheitron* duly unburdened itself of primitive forms (ill-digested genera, we might as well call them) all along the Polynesian front, so doing by tracks already revealed to us by *Geniostoma, Vaccinium, Pernettya, Ascarina, Melicytus*, etc.

It is clear, lastly, that a direct track may lead from the antarctic cradle across the Southern Atlantic to the New World. This track follows of necessity a route that proceeds in the direction of Saint-Helena prior to its hitting the Americas, and we need not be surprised on this account that a genus endemic to Saint-Helena, *Trimeris*, is allied to a much larger "American" group, *Siphocampylus*. The former is merely a straggler out of the *genorheitral* stream that begot the latter, and Saint-Helena is not made "oceanic" by forms of this nature, far from it.

The conclusion therefore follows that *Lobelia*, and its *genorheitron*, migrated in the Pacific and Indian Oceans in the manner of *Lysimachia, Potentilla*, etc., and in the Atlantic in that of *Turnera*, Rapateaceae, etc. In short, the whole of these migrations (Fig. 48) is standard throughout and very simple.

We may add to this conclusion another. It is certainly false that phytogeography is dependent on taxonomy, for taxonomy may depend, on the contrary, upon phytogeography. It is definitely erroneous to believe that plays of nomenclature are of the essence of botany. Only a brand of lore that has lost touch with the realities of thought can mistake formulae for higher substances that formulae cannot represent.

---

*) The record in our hands is still too crude to make it possible for us to be concerned even with important details. We remark, nevertheless, that these arboreous Lobelias, typic of Central African vegetation in the highlands, seem first to turn up in Tanganyka, Uganda and Ethiopia. They appear consequently right in a region of break (see p. 164), which is also the one that witnesses the beginning of the tracks of *Erica arborea*. This very same region is of interest in the migrations of *Primula, Berberis, Populus* etc., which all turn up in this corner of Africa as if from nowhere. Believers in the „Holarctic" origins of the Angiosperms never gave these centers a thought. It is clear, on the other hand, that these centers are fundamental, whether we run the routes of the migrations north to south or the other way around. We believe that some mighty orogeny once ran between the Mascarenes and Ceylon, and that this orogeny immediately interested also Tanganyka, Uganda and Ethiopia. In brief, we feel that arborescent Lobelias could already be seen on the heights of Gondwana in the Late Jurassic, together with *Erica, Berberis, Primula, Acer*, the Sapindaceae Koelreuterieae, etc., etc.

XI. Alnus, Acer and the Salicaceae

Many are the phytogeographers who believe that "Antarctica" is the corner-stone of angiospermous dispersal. None of them, so far as we know, has ever maintained that *Alnus* could derive from the depths of the southern hemis-phere. This genus is now so thoroughly committed to the northern half of the earth that the idea of pronouncing it "antarctic" seems to most absurd, and fit only to invite unanswerable criticism on the part of devotees of "Holarctis," and "Extraordinary means of transoceanic dispersal."

These devotees are mistaken, of course, for *Alnus* to begin with is by no means unknown to South America, witness,

(1) *A. jorullensis* (and closely related forms) — Argentina (Tucumán), Bolivia (Larecaja, Santa Cruz), Peru (Puno, Libertad, Cuzco, Lima, Junín, Lima, Huánuco, Amazonas, Anchacs, Cajamarca), Ecuador, Colombia (Na-riño, Cundinamarca, Santander, Norte de Santander), Venezuela (Mérida, Táchira), Panamá, Costarica, Guatemala, Mexico (Chiapas, Michoacán, Veracruz, Jalisco, Sinaloa).

(2) *A. arguta* — Guatemala, Mexico (Chiapas, Oaxaca, Veracruz, Chi-huahua, Tamaulipas).

(3) *A. glabrata* — Mexico (Oaxaca, Hidalgo, Guanajuato, Durango).

(4) *A. Pringlei* — Mexico (Michoacán, Jalisco).

(5) *A. firmifolia* — Mexico (Morelos, Mexico).

(6) *A. oblongifolia* — Mexico (Nayarit, Durango, Sonora, Chihuahua).

The distribution of this group of *Alnus* matches in all essentials (see p. 358) that of *Menodora*, and is standard for a great number of unrelated genera. We cannot postulate that *Alnus* migrated from "Holarctis" southward by an "Andean Bridge," because there never was an "Andean Bridge" answering the outline of the Andes in our maps, not only, but *Alnus* had reached the New World when the Andes were as yet to begin active rise. In brief, *Alnus* came to lands later to become the Americas when these lands utterly differed from the Americas we know today. To argue against this is futile.

Menodora, as we know, keeps a precarious toehold in South Africa. So does *Alnus*. Its South African record is omitted outright from the maps showing dispersal (45) or run out as adventitious *), because this record does not agree with current preconceptions. As we know, the same fate was meted out to embarassing outliers of *Rubus*, *Reinwardtia*, etc., in short to everything of which academic lore could not approve. This was done at times (see p. 283) with the accompaniment of statements of astounding boldness. Of *Alnus glutinosa*, the supposed "European" Alder, writes a good authority on the flora of South Africa the following, **) "The common Alder *(Alnus glutinosa)* is found throughout the colony, apparently wild, but whether truly so or not I

---

*) This, for instance, is the opinion of WINKLER (in E. & P. Pflanzenr. 19 (iv. 61): 115. 1904).
**) See HARVEY, as quoted by Skan in Dyer's Fl. Cap. 5 (2): 574. 1925.

cannot say." The common Alder then is scattered throughout the Cape at least "Apparently wild."

It remains now for us to form some idea of the manner in which *Alnus* could attain its present dispersal by migrations from the general direction of the Indian Ocean. Though in itself interesting, the *Alnus* outlier in the Cape is too weak a peg upon which to hang the whole of the modern distribution.

The fact that a species of *Alnus* endemic to such North American States as Delaware, Maryland and Oklahoma *(A. maritima)* is related with a species *(A. nitida)* endemic to the Kashmir and Punjab is of one cloth with the circumstance that precise affinities connect the Ozarkian *Andrachne phyllanthoides* and the Caucasian *A. colchica*, not only, but Himalayan species as well in the *A. colchica* affinity. The Texan *Euphorbia Peplidion* and *E. texana* are beyond doubt precisely allied with archetypes from Asia Minor. *Calluna* has outliers throughout North America as far inland as the Great Lakes. Not to extend this enumeration beyond the necessary, we may conclude that there is not the slightest doubt as to a cool to cold mesophytic flora having ranged on both sides of the Atlantic from ages most remote. *Alnus maritima* and *A. nitida* might very well have belonged to this flora. Moreover *A. incana* and *A. sinuata* are common to Eurasia and the Northern New World which clinches the case for migrations in *Alnus* of transoceanic scope and nature.

It may be inquired whether the connections in *Alnus* between the New and Old World were all secured through the Atlantic. This question could better be answered by a specialist in the classification of *Alnus* than by ourselves. We are not familiar with the intergeneric relationships under *Alnus* and have not investigated the complex ecologic factors that may make it possible for us to assure that American species "A" has ties in Japan, while American species "B" can directly be traced to Asia Minor or Europe on account of certain peculiarities of morphology. As phytogeographers, we know that standard channels of migrations crossed the Atlantic and the Pacific alike in eras past, when these oceans were not what they are now.

Leaving aside the question, whether *Alnus* reached North America *all* by transatlantic, or *all* by transpacific migrations, or *part* by transatlantic and transpacific migrations (which last we think is the most probable), we will concentrate upon the distribution of that part of the genus which ranges most particularly in Mexico, Central and South America.

One aspect of this migration immediately invites our attention. *Alnus* gives the West Indies a wide berth, and does not reach in Venezuela beyond the western states of Mérida and Táchira. In other words, only the western half of South America is invested together with Guatemala and Mexico.

This means, almost of a certainty, that *Alnus* did not reach the New World by a channel faring through the body of Central Africa to West Africa, and ultimately, the Americas. The fact that *Alnus* is not a "tropical" plant does not mean much. We know that *Erica* is in East Africa, *Vaccinium* in the Transvaal and Polynesia, *Xolysma* in the West Indies, *Rhododendron* in Queensland, and since *Alnus* currently associates in the north with all these genera it might have put up at least an appearance on the outskirts of this channel. *Populus* and *Salix* do so, as we shall see. Moreover an outlier of *Alnus* is reported from the Cape.

Ruling out a transatlantic crossing by this channel, we are left to consider other tracks. One of these tracks, of course, runs in the Northern Atlantic, and we have mentioned it already. This track accounts for the affinities of *A. maritima* and *A. nitida*, as we know, but hardly explains the peculiarities of the distribution of *Alnus* in Western Venezuela. This distribution is "Andean," as conventional language has it.

Distributions of this nature, highlighting such localities as Tucumán, Larecaja, Puno, Sonora, Chihuahua point to an invasion of the archetypes of *A. jorullensis* and its allies from the direction of the Pacific. We may not be sure, but things look as though *Alnus* primarily came in by a track hitting either Bolivia or Western Mexico in the first place. In conclusion, the whole of the migrations of *Alnus* may be summarized as follows, (a) The genus departs from a point in the Indian Ocean to reach first a front roughly agreeing with a line, Asia Minor-Western China. The Cape outlier is specifically connected with Europe, which repeats certain aspects of the migrations of *Erica* and *Orobanche;* (b) Migrating from the line Asia Minor—Western China, and strongly established in a range in Southwestern China, *Alnus* in thoroughly standard fashion reaches the whole of the north by tracks that may run in part across the modern Atlantic and the modern Pacific; (c) The affinity centering around *A. jorullensis* came to Western America by either one of two routes, (i) From the Far East in modern maps by a transpacific migration to Mexico first and foremost, next ranging southward (compare with *Quercus*, which ranges at least as far south as Colombia, and is not scarce in Mexico and Central America); or, (ii) From a point in the south to Bolivia in the first place, next migrating northward (compare with *Passiflora*, the Palms in the affinity of *Juania, Lepechinia*, Celastraceae, etc.) Of these alternatives the latter strikes us as perhaps the more probable. South Africa rates as a center of dispersal and evolution of a host of unrelated, at times, peculiar plants (see *Myrothamnus*, Buxaceae, etc.), not only but in the dim Jurassic and very Early Cretaceous past had fast relations with the western end of South America, all the way up to Mexico and the United States (see *Thamnosma, Oligomeris, Menodora*, etc., etc.). We see no reason whatever for believing that *Alnus* is adventitious in South Africa, rather, are inclined to read the record for what it says. *Alnus* was in these quarters together with *Salix*, for example, with which it still associates to this hour, in a very dim geologic past.

It could be objected that *Alnus* is a "very ancient form," and that it might have migrated by some channels of its own, not at all agreeing with standard angiospermous routes. We will answer, (a) *Salix* and *Populus* accepted by common consent as coaeval with *Alnus*, not only, but the Gnetaceae which are as yet not angiospermous, all migrate by standard angiospermous channels.

Similar channels as it will further be shown, are followed in addition by Conifers, Ferns, Liverworts, etc. There is consequently no justification whatever for *Alnus* having migrated entirely on its own. As a matter of fact *Alnus* follows standard routes throughout Western America and in the whole of the Northern Hemisphere; (b) It could be denied that the Cape outlier proves that *Alnus* ever was in the deep south of our maps. If so, and if this outlier is truly "Adventitious" (which we do not believe), the dispersal remains on the whole unchanged. *Alnus* reaches as far south as Tucumán in Argentina, which, as modern latitudes read, is not incompatible with the geographic position *Alnus* has in the Cape. Moreover, the sum total of a distribution never was basically changed by the shift of one record in the tracks, be this record in itself relevant.

To us, the problem of the point of entrance of *Alnus* in Western America is not to be solved by denials or affirmations *a priori*. We outline here certain alternatives, and incline to favor one as most probable on the very face of the dispersal. A solution of the problem, however, how *Alnus* reached Western America, is to be sought by careful critical studies in taxonomy, systematic botany and ecology. Meantime, we believe the map (Fig. 49) here attached is not without support in fact.

It stands to reason that since *Alnus* and *Betula* are part of one and the same family Betulaceae, we have no reason whatever to accept for the Betulaceae an "Holarctic" origin, no more that we can believe that *Rhododendron* is "Holarctic," when the rest of the Ericaceae is clearly "Antarctic" in the conventional sense of the term. We cannot review here the Betulaceae, but readers can do this for themselves, and, we think, reach the conclusion that the Betulaceae, too, migrate by thoroughly standard channels to every station they now occupy on the modern nap.

Acer is one of the greatest mainstays of the theory of "Holarctic" angiospermous origins. There is no book, popular or otherwise, which, asserting that the Angiosperm are ultimately native to lands in the north, fails to produce figures of fossil *Acer*. It is argued as a rule that since *Acer* already was in the boreal hemisphere in the Early Cretaceous, it must be in this hemisphere that it "originated." Texts of the kind affirm as a matter of course that the origin of the Angiosperms are somewhat "mysterious."

The truth to us is other. A petrifact of the Early Cretaceous proves absolutely nothing which runs contrary to the evidence furnished by living plants. This ranks with us as argument number one. *Acer* was in the north by the Early Cretaceous in a manner, and by channels, which the tracks of living *Acer* illustrate. Petrifacts may be very valuable in certain respects, but certainly not in all.

General argument number two with us is that there is no angiospermous family which stands unrelated by *genorheitron* (or Order, as formal taxonomy has it, more or less accurately). Origins cannot be assumed but must be proved

To prove origins, *genorheitra* must be studied in the very first place, not single genera or families.

The Aceraceae consist of *Acer* and *Dipteronia*. These two genera, the latter most particularly, are related most intimately with the Sapindaceae. The Sapindaceae are thoroughly well known in the deep south of the map (see p. 49), and rank as one of the largest angiospermous families. The Aceraceae, consequently are but the boreal rump of the Sapindaceae, and successfully to maintain that the Aceraceae were born in "Holarctis", we are bound to undertake the preliminary impossible task of "holarcticizing" also the Sapindaceae. We cannot review the entire distribution of the Sapindaceae here because of obvious reasons of space, but suggest that some of the readers undertake the task, in the anticipation that our statements will not be found wholly out of the mark.

It has been reported (46) that a fossil wood, *Aceroxylon*, peculiarly suggestive of the living *Acer campestre*, was found in a bed together with dinosaurian bones in Madagascar. We have not studied this remarkable finding in person, so may not vouch for it. We know, however, that pollen from a world of plants no longer existing is reported from Southwest Africa (see p. 93), so that we have no reason to question the Madagascan petrifact. We know very well, perhaps too well, the petrifacts in the north. It is time that we give those in the south some attention.

It may be a coincidence, but if so a remarkable coincidence, that one of the very few groups, perhaps the lone group, which is definitely not tropical, or "warm" at least, in the Sapindaceae is the affinity centering around *Koelreuteria*. This affinity, Sapindaceae Koelreuterieae *) is distributed in this manner,

A) Koelreuteria

(1) *K. paniculata* — China (Szechuan, Shensi, Jehol), Corea, Japan.
(2) *K. minor* — China (Kwantung).
(3) *K. bipinnata* — China (Yunnan, Kweichow, Hupeh).
(4) *K. integrifolia* — China (Kwantung, Hunan).
(5) *K. Henryi* — Formosa.
(6) *K. formosana* — Formosa.

B) Stocksia

(1) *S. brahuica* — Beluchistan, Afghanistan, Iran (Persia).

---

*) We follow here the latest monographic work by RADLKOFER in Engl. & Prantl, Pflanzenr. iv. 165 (Bd. 2): 1328. 1934.

C) Erythrophysa

(1) *E. aesculina* — Madagascar *).

(2) *E. undulata* — Northwestern Cape (Kamiesberg), Southwestern Africa (Namaland).

This dispersal leaves nothing to imagination, and, as the reader well knows, is thoroughly standard. A track of this nature could take *Alnus* from the Cape anywhere north, and, doubtless, *Aceroxylon* to the very regions where *Acer campestre* occurs today (47), namely, Iran (Persia), Asia Minor, Northern Africa, the Mediterranean generally, and Europe to Russia and Scandinavia. The nearest ally of *A. campestre* is *A. Mono*, endemic to Northern Japan. Peculiarly, our authority believes that the group of *Acer* nearest the original sapindaceous *genorheitron* is precisely the one typified by *A. campestre* If all this is coincidence, it is worthy of notice. *Dipteronia* is also well and properly located in China (*D. Dyeriana:* Yunnan; *D. sinensis:* Szechuan, Hupeh, Shensi). *Acer* has a species in Malaysia (*A. niveum:* Sumatra, Java, Celebes, Philippines), the distribution of which is doubtless not at all incompatible with the run of a track stemming from a point in the modern Indian Ocean.

We believe that *Acer* originated where the Indian Ocean stands now and, like *Tilia* (see p. 116), fared to Malaysia, Asia generally, and the rest of the northern hemisphere following two main tracks; one pointing toward Malaysia, the other toward Iran (Persia).

Once in the north, *Acer* followed standard channels throughout. Our authority tells us the following, for instance, (a) Series *Spicata* has 1 species in the Himalayas *(A. caudatum)*; 2 in the Far East *(A. urukundense)*; 1 in the Southeastern United States (*A. spicatum)*; (b) Section *Lithocarpa* has 3 species in the Himalayas *(A. villosum, A. Thompsonii, A. Schoenermarkiae)*; 1 in Pacific North America *(A. macrophyllum)*; (c) Series *Palmata* has 1 species in Central China *(A. ceriferum)*; at least 11 species in Manchuria, Corea and Japan; 1 in Pacific North America *(A. circinnatum)*; (d) Section *Rubra* has 1 species in Japan *(A. pycnanthum)*; 2 species in Atlantic North America *(A. rubrum, A. Drummondii)*; (e) Wholly American in Mexico and the United States, on the Pacific and Atlantic alike, is Sect. *Saccharina* with at least 8 species.

This pattern of dispersal conveys strong hints to the effect that the migrations to the New World were mostly, when not exclusively, effected across the Pacific in a manner reminiscent of *Magnolia*. China and the Mediterranean rank as major centers of secondary speciation. Europe can easily be connected with the Far East through Series *Tatarica*, for instance (Mediterranean and Central Europe: *A. tataricum;* Central Asia: *A. Semenowii;* Central China: *sp. nov.;* Manchuria and Corea: *A. Ginnala;* Japan: *A. subintegrum*). In brief, all these migrations are conventional throughout.

---

*) This station is noteworthy, for it is seldom indeed that Madagascar turns up in the record of dispersal of a "cold" group. However, the Sapindaceae are massively "warm," so that the exception established by the Koelreuterieae rather than contradict may *perhaps* confirm the rule that genuinely "cold" forms are not in Madagascar. See further *Korthalsella*, etc. p. 141, 239 fn.

Acer, in conclusion, never originated in "Holarctis," though it reached "Holarctis" in the Early Cretaceous. This is not against reason and evidence, for by the end of the Jurassic modern Angiosperms had reached the immediate vicinity of Washington D. C. in the United States. This is attested by the fossils of the Potomac Beds (see p. 30), and wholly agrees with the evidence from modern dispersal. Our authority (47; p. 370) believes that "The roots of the genus *Acer* vanish in the depths of the pre-Eocene period." We agree, if these depths are carried back *at least as far as the Early Jurassic*. We are very far from agreeing with our authority, on the other hand, when she stresses the Tertiary as a very important epoch in the evolution of *Acer* except perhaps as to species-making.

Salix calls to our mind the cold north and "Holarctis" most directly. Evidently, we are not well informed about the dispersal of *Salix* in Africa if we allow these fancies to get hold of our imagination, for *Salix* is in the Dark Continent as follows,

(1) *S. madagascariensis* — Madagascar.
(2) *S. capensis* — Rhodesia, Transvaal, Natal, Cape.
(3) *S. Woodii* — Transvaal (Drakensberg Mts.), Natal.
(4) *S. crateradenia* — Bechuanaland.
(5) *S. hirsuta* — Cape, Namaqualand.
(6) *S. Safsaf* — Tanganyka, Ethiopia, Sudan, Egypt; Var. *cyathipoda:* Rhodesia, Tanganyka, Ethiopia.
(7) *S. Hutchinsii* — Uganda.
(8) *S. Murielii* — Sudan.
(9) *S. Schweinfurthii* — Sudan.
(10) *S. adamauensis* — Cameroon.
(11) *S. camerunensis* — Cameroon.
(12) *S. Ledermannii* — Cameroon.
(13) *S. nigerica* — Northern Nigeria.
(14) *S. Chevalieri* — French Sudan.

This is a perfect example of invasion of Africa from the east, that is, in an absolutely standard manner. The track that leads to the Drakensberg Mts. and Natal penetrates the Cape, which is conventional; the route that begins in Tanganyka and Ethiopia, and extends southward to Rhodesia, ultimately leads to West Africa, which is altogether true to pattern. In brief, *Salix* is one of the genera which after an early invasion of the Dark Continent in the Late Jurassic (see the stations Madagascar, Drakensberg Mts., Natal, Tanganyka, Uganda, Ethiopia, Cameroon) found propitious times further to migrate in a second epoch of angiospermous dispersal. To this second epoch may belong stations such as the Sudan, Egypt, Bechuanaland and Northern Nigeria. This list of stations is tentative, but may be useful as a lead in coming studies.

Various authors (48, 49) credit part at least of African *Salix* to Sect. *Humboldtianae*, or its immediate nomenclatural equivalents, which includes *S.*

Bonplandiana and *S. Humboldtiana* (Cuba and Mexico to Uruguay and Chile). Also within the close affinity of Sect. *Humboldtianae* is *S. tetrasperma*, widespread in the Tropical Far East from India, Burma and Southern China to the Philippines, Borneo, Java. *Salix Safsaf* is connected through *S. acmophylla* (Iran (Persia), Syria, Armenia) with *S. babylonica* distributed to nearly all of Europe and Siberia.

Salix is certainly not "Holarctic." *Salix* is a conventional migrant from Gondwana to continents still on the maps of the modern worlds. It runs standard tracks in Africa; crosses over from Africa to the New World in a wholly ordinary manner; has commonplace distribution from Rhodesia, Tanganyka, Ethiopia to India, Burma, Southern China, the Philippines, Borneo, Java at one hand, Iran (Persia), Asia Minor, Eurasia generally at the other; occurs in Madagascar.

We may here safely forego studying *Salix* in the north and high north, in the assurance beforehand that the tracks and centers of speciation in that range are bound to be wholly standard. This is necessarily the case, given the commonplace distribution which belongs to *Salix* in the south.

Salix is a classic instance of seemingly "Holarctic" and "cold" genus which, critically studied, is immediately seen to be neither "Holarctic" nor "cold." As a matter of fact, the tracks of *S. tetrasperma* seem to espouse a range of geologic lowlands in India, Burma, etc. lying in a "Monsoon" climatic belt.

Before dismissing *Salix* altogether we may take notice of a conumdrum of conventional plant-geography which no author seems thus far to have solved. *Salix reticulata* is reported to be circumpolar to the exception of Greenland, which is commonly rated as right peculiar in a genus of implicit "Holarctic" pedigree. Reference to Fig. 37, 38, however, will reveal that Greenland is at the receiving end of at least two tracks, one running from the Eurasian shore of the Arctic Ocean; the other from the direction of Alaska and Baffin Land. Both, or either one of these tracks may reach Greenland, or none at all. Obviously, plants that fare these two tracks may be in every corner of the High North but Greenland, if it so happens that their course stops short of Greenland.

Populus, the lone other important genus of the Salicaceae (*Chosenia*, a petty ally of *Salix* is in the Far East), turns up at first in Tanganyka and Uganda, the classic region of break which also marks the inceptions of the run northward of *Erica arborea*, *Primula*, *Berberis*, etc. The species that tenants this range is *P. ilicifolia*, a close ally of *P. euphratica* (Lybian Desert, Algeria, Morocco, Turkestan, Central Asia, China: Sinkiang, Kansu, Jehol). Both these species are consectional with *P. pruinosa* (Turkestan, Sinkiang) under Sect. *Turanga*.

We admit not seeing a very great difference between *Salix* and *Populus*. The former for reasons of its own, but not all mysterious in principle, managed to reach the Drakensberg Mts., which is to say to vault clear across the range Tanganyka-Transvaal (or Nyasaland). Having vaulted over this range, *Salix* was fated, we should almost say, to reach the Cape. *Populus*, on the contrary, failed to manage this range, and turned up just beyond its northern end in Tanganyka and Uganda. In brief, *Salix* and *Populus* are both from the African gate of angiospermy and both could have been seen in the ranges of Gondwana in the Late Jurassic, ready to start active migrations toward lands still with us. It seems worth remarking that *P. euphratica* is exceedingly polymorphous, and forms occur in which the foliage in quite willow-like.

The range achieved in time by *Populus* northward, is well illustrated by Sect. *Aegeiros*. This group contains such species as these,

(1) *P. nigra* — Iran (Persia), Afghanistan, Turkestan, Europe generally, Northern Africa, China (Sinkiang, Kansu, Shensi, Hupeh, Shantung).

(2) *P. Wislizenii* — Northwestern Mexico, Western Texas, Colorado.

(3) *P. Macdougalii* — Southern California, Nevada, Arizona.

(4) *P. Fremontii* — Baja California, Western California.

(5) *P. deltoides (P. balsamifera)* — North and South Dakota to Oklahoma and Texas; along the Atlantic coast from Florida to Quebec.

Readers accustomed to think of *Populus* as a "cold" genus straight from "Holarctis" may well inquire what is *Populus* doing in British East Africa and Baja California. Our answer will be that *Populus* is doing there precisely the same as that which *Platanus* does in Laos and Texas; *Acer* in the Philippines and California; *Magnolia* in Sumatra and North Carolina; *Salix* in Chile and Borneo; *Alnus* in the Cape and Bolivia; and so on without end. Dispersal is very repetitious, indeed boring.

XII. Cuscuta

The dispersal of *Cuscuta* interests phytogeography because it is both very narrow and very broad. In Malaysia, this genus is restricted to Java, and seemingly alien to the flora of Borneo and the Philippines. It is elsewhere pandemic to the exception of the high north.

Cuscuta is further interesting for another reason. This parasite is often distributed with the seeds of its hosts, and it might be thought that man-made records interfere with the natural range to the extent that the latter can no longer be read with accuracy. This, as we shall see, is far from being the case.

Cuscuta has four species in Malaysia (50), as follows,

(1) *C. timorensis* — Java, Timor, Wetar Island; East Africa.

(2) *C. reflexa* — Java; Afghanistan, Ceylon, India, China.

(3) *C. campestris* — Java; America generally, South Africa, China, Japan, Australia, Fiji, New Caledonia, Tahiti.

(4) *C. australis* — Sumatra, Java; New Guinea, Malacca, Australia Central Asia to India, Japan.

These species are distributed in a manner that leaves nothing to imagination. Their dispersal calls for three standard channels of migration, all rooted in the modern Indian Ocean (eastern end of the African gate, *Gondwana*), as follows, (a) *A track primarily pointing toward Java, and the islands of Wetar and Timor.* This is essentially a "dry" route, followed by plants of the "Monsoon ecology." Beyond Timor its course normally run in either one (or both) of two directions, (i) The Philippines; (ii) Northern Australia (Thursday Island is a frequent station in this type of dispersal), New Guinea, New Caledonia, Fiji; (b) *A track streaming in the direction of the line Afghanistan— Ceylon.* This channel passes on toward the Altai Node in the case of "cold" flora; fares on the contrary to Burma, sections of Indochina south of Laos, Western Malacca and beyond (Java, Northern Australia, to cross Track (a)) in the case of "Monsoon" plants; (c) *A track moving into Africa and the New World* along lines which we will identify in the coming pages. This track may call into play vicariants (e.g., *C. reflexa* as described, *C. cassythoides* in South Africa).

This dispersal is accordingly standard, not only, but such as to suggest at a glance that *Cuscuta* does not belong to moist tropical formations and associations. *Cuscuta* might never have reached Borneo or the Philippines, or conversely have died out there on account of secondary unsuitable ecology. The issue of ecology is one we may not hope to discuss in these pages beyond the roughest generalities. It is well known (51), for instance, that Eastern Malacca — a region which suggests the wet tropics at a glance because of its position on the map — has a "Monsoon-climate", and for this very reason harbors plants *(e.g., Sterculia foetida, Kleinhovia hospita, Tournefortia argentea, Pisonia excelsa)* foreign to the moister rest of the peninsula.

Cuscuta campestris belongs to Subsect. *Arvenses* which is distributed (50) in this manner,

(1)   *C. Tatei* — Central Australia.

(2)   *C. Stuckertii* — Argentina (Córdoba).

(3)   *C. gymnocarpa* — Galápagos Islands.

(4)   *C. campestris* — Java; Argentina northward to the West Indies and the United States (most abundant in the west); Australia, Polynesia (New Caledonia, Fiji, Tahiti); Africa to Natal; Europe; China, Japan.

(5)   *C. glabrior* — Northern Mexico, Southwestern United States to Texas.

(6)   *C. pentagona* — United States (most abundant in the east).

(7)   *C. Harperi* — United States (Alabama).

(8)   *C. plattensis* — United States (Wyoming).

Cuscuta australis is one of the species of Subsect. *Platycarpae* migrating on the whole as follows,

(1)   *C. bifurcata* — South Africa (Eastern Cape to the Transvaal).

(2)   *C. cordofana* — Africa (Madagascar, Tanganyka, Uganda, Belgian Congo, Upper Sudan, Liberia, Sierra Leone, Nigeria).

(3)   *C. Schlecteri* — West Africa (Cameroon).

(4)   *C. australis* — Australia (New South Wales, Queensland), New Guinea, Java, India, Eastern China, Manchuria, Corea, Japan, Turkestan, Europe.

(5)   *C. victoriana* — Australia (Victoria, South Australia).

(6)   *C. cristata* — Argentina, Uruguay.

(7)   *C. obtusiflora* — Paraguay, Uruguay, Central Brazil, Peru, Ecuador, Colombia; Var. *glandulosa:* West Indies, Northern Mexico, Southern United States; Var. *latiloba:* Burma.

(8)   *C. polygonorum* — Eastern North America to Ontario, Minnesota and Nebraska.

The monographer of *Cuscuta* whom we follow believes that Subsect. *Platycarpae* is the most primitive group of the genus. He further thinks that this subsection gave origin to Subsect. *Arvenses*, which released in turn three other subsections, *Californicae, Subinclusae, Denticulatae.*

If these opinions are at all correct, we are bound to find them confirmed by dispersal, for distribution and *genorheitron* are facets of the same prism. Let us observe, meantime, that the American *C. obtusiflora* has one variety, *latiloba* endemic to Burma. This would be ruled out by some as a "Taxonomic faux-pas," and might perhaps be credited to a faulty record, or an error of interpretation. We may not take too much for granted, however, for it is conceivable that the competent botanist author of this record was well aware of its extraordinary character, and he is entitled at least to the benefit of the doubt. It well may be, after all, that a purely American species has a purely Asiatic variety. As we know (see p. 358), a purely American species of *Menodora* also has a purely South African variety; dispersal is replete with seemingly oddities of the same nature.

The three subsections stemming out of Subsect. *Arvenses* tabulate as follows,

A) Californicae

(1)   *C. insquamata* — Southern Bolivia.

(2)   *C. californica* — Baja California to Washington; Vars. *papillosa* and *apiculata:* Southern and Southwestern California.

(3)   *C. occidentalis* — California, Oregon, Washington, inland to Western Colorado.

(4)   *C. brachycalyx* — California.

(5)   *C. Jepsonii* — Northern California.

(6)   *C. sandwichiana* — Hawaii.

B) Subinclusae

(1)   *C. micrantha* — Chile (Coquimbo, Santiago); var. *latiflora:* Chile (Concón, Atacama).

(2)   *C. salina* — Mexico to British Columbia, Utah and Arizona.

(3)   *C. subinclusa* — Mexico to Oregon.

(4)   *C. Suskdorfii* — Oregon, Washington.

C) Denticulatae

(1) *C. microstyla* — Chile (Antuco).
(2) *C. Veatchii* — Baja California.
(3) *C. denticulata* — California, Nevada, Utah.

So far advanced in our compilations from the record, let us take stock of the data immediately before us in order that we may proceed in close agreement with the facts. Three Subsections, then, *Californicae, Subinclusae, Denticulatae,* are purely Western American. The track begins to run in Central Chile and Southern Bolivia which is standard. Wide disconnections occur between Chile and California (e.g., *Denticulatae*), which is also normal. The drift of the migrations is so markedly to the west, that Hawaii is reached by at least one species. In conclusion, we face a type of dispersal having all the essential characters of *Lepechinia* and *Nama,* which is to say, altogether conventional migrations.

Subsection *Arvenses,* which our authority believes to be ancestral for the three Subsections just mentioned, also has a very strong Western American component *(C. glabrior, C. pentagona, C. Harperi, C. plattensis,* part of *C. campestris).* The track begins to run in Argentine (Córdoba), and the generally western drift of the dispersal is underscored by the presence of a species endemic in the Galápagos. In addition, Subsect. *Arvenses* ranges to Polynesia and Central Australia *(C. campestris, C. Tatei),* and further runs a standard channel of migration in the Indian Ocean (Natal, the Far East, Europe).

This conspectus of dispersal suggests the following, (a) We are in the right, rejecting every suggestion that the Galápagos, the Revilla Gigedo, Hawaii are "oceanic" outposts colonized by "Casual agencies of dispersal." These outposts belong in this case to distributions of essentially Western American (conventionally, "Andean" and "Pacific American") pattern. It is therefore consistent with the evidence that, as we have repeatedly stated, the New World had coasts once running much farther west than those in our times. The existence of an essentially Western American pattern of migration is vouched for by the facts in the record, and the face of the map. The position held in dispersal by Juan Fernandez, Galápagos, Revilla Gigedo, the minor islands off the coast of Baja California and Southern California, Hawaii dovetails throughout with this western pattern. Frequent disconnections between Chile and California become rational when viewed in connection with this pattern. In brief, the whole of the dispersal in the South and North Pacific fall in harmoniously with the same factors, and we may thus affirm without fear, that there once were lands, large lands, west of the New World in our maps, quite as certainly as there once was a continent in the modern Indian Ocean; (b) It is altogether possible that Subsect. *Arvenses* is primitive in regard of Subsects. *Californicae, Subinclusae* and *Denticulatae,* because the tracks of the first are more comprehensive than those of the last three, which essentially rate as a secondary American development. We known (see *Juania,* etc.)

that definite connections exist between the African gate of angiospermy (Natal is a critical locality in the range of *C. campestris*) and a region of South America fairly closely corresponding to the range, Juan Fernandez—Central Chile. We know that the African gate sends genorheitral streams also to the Far East and Eurasia generally. In this understanding we may correctly visualize the immediate affinity of *C. campestris* as the one which is strategically located in relation to the sum total of the distribution now before us. *Cuscuta Tatei*, probably, reached Central Australia from a center in the immediate vicinity of Java. As an alternative, it might have come to the same region directly from the west (cf. *Pelargonium* Sect. *Peristera*) or, perhaps, the Pacific. The migrations of this particular species suggest merely details.

Always in the opinion of the authority whom we follow, Subsect. *Platycarpae* is the most primitive group in *Cuscuta*. The dispersal of this groups. brings before us the following, (a) A perfect track connecting the modern Indian Ocean with West Africa (*C. cordofana*). This channel of migration is standard; (b) A secondary center of speciation in the south of South Africa connected in dispersal with the Transvaal (*C. bifurcata*). This too, is standard; (c) A secondary center of speciation in West Africa (*C. Schlechteri*). This also is wholly conventional; (d) A center of speciation in Australia (*C. victoriana*). This, though less frequent, is not at all unknown; (e) A center of speciation located with a variety (*C. obtusiflora* var. *glandulosa*) in the West Indies, Northern Mexico and the Southern United States, extending with the type-form of the same species (*C. obtusiflora*) prevailingly to the west of modern South America (Colombia, Ecuador, Peru), and from here running seemingly eastward through Paraguay and Uruguay to Central Brazil. This is absolutely standard, the West Indian massing being properly grafted on an extension of the track running between the Indian Ocean and West Africa; (f) A center of speciation in Argentina and Uruguay (*C. cristata*), which is a logical consequence of such migration as the one described just above; (g) A center of speciation in the Eastern United States (*C. polygonorum*) ranging as far inland as Nebraska and Minnesota and as far north as Ontario. This is a thoroughly consistent extension of the massing in the West Indies and the Appalachian-Ozarkian center vouched for by *C. obtusiflora* var. *glandulosa;* (h) A seemingly fantastic disconnection involved by a variety of *C. obtusiflora*, *latiloba*, native to Burma. This disconnection, on the contrary, is not at all fantastic. A range, West Africa—Burma with an absolute intervening break, or scattered connections surviving in Afghanistan, India, is standard. The track of *C. obtusiflora* ties in manifestly with the center of speciation in West Africa which is the antechamber of the Caribbean massing. Accordingly, *C. obtusiflora* var. *latiloba* is a strayer which left a region roughly corresponding to West Africa in our maps to run a *standard* track to Burma, while *C. obtusiflora* var. *glandulosa* was running on its part another *standard* track to the West Indies. The case matches the linaceous genus *Reinwardtia's* in every respect, not only, but can be compared to the distribution of the *Cercis-*

Acioa complex (see p. 138). This complex, as we know, calls for a consideration of West Africa, China and the Southeastern United States as one. (i) An Eurasian range (*C. australis*) reaching to colder Asia (Manchuria) but essentially part of a migration involving the temperate and tropical Far East and Malaysia to Corea and Japan at one hand, Europe at the other. Interesting is the line in this migration running from Java to Queensland and New Guinea, possibly reaching New South Wales from the northwest. This migration is of standard type in every one of its main aspects.

In brief, *Cuscuta* Subsect. *Platycarpae* easily ranks as a classic example of dispersal, worth of being introduced to beginners in the study of phytogeography in a very first time. This dispersal is thoroughly well balanced, even to the extent of a seemingly fantastic disconnection. The species here brought in play are not many, and all significant. Were we try to compound an example of our own, using symbols instead of specific names to exemplify standard distribution we could do no better.

It is obvious that the migrations of *Cuscuta* are uniformly keyed to Gondwana (eastern end of the African gate of angiospermy) and contain the essentials of the dispersal of the Passifloraceae-Caricaceae affinity (see p. 316). *Passiflora* indeed seems to match in distribution the status of such Subsections as *Californicae*, *Subinclusae*, *Denticulatae*.

In conclusion, the *genorheitron* of *Cuscuta* stirs at first by southwestern Gondwana, and from this point invades the whole of Eurasia, Malaysia and part of Australia in standard fashion. It further invades Africa by a standard channel of migration ultimately leading to West Africa, and beyond to the Caribbeans and the New World. Meantime, a genorheitral stream glides unobtrusively by a conventional "antarctic" track through the modern South Atlantic toward southern South America which it reaches in the vicinity of Central Chile. Once in Central Chile, this genorheitral stream takes advantage of lands now crumbled in the Eastern Pacific to penetrate the Galápagos and Hawaii, not to mention the whole of Western America in our maps.

Needless to say, everything in the dispersal of *Cuscuta* is standard (Fig. 50), or easily to be understood (range of *C. obtusiflora* var. *latiloba*) by reference to standard elements of dispersal.

Cuscuta, perhaps more than any other genus, lends itself to easy generalizations about "Dispersal by casual agencies." It seeds are small, light and inclosed within berries which might be supposed to be eaten by birds. *Cuscuta* contains species that are parasitic on plants of economic significance, and for this reason may be said to scatter widely on account of man-made factors. In sum, if there ever could be a record of dispersal purposely made to be meaningless on account of haphazard interventions of birds, animals generally and man, this record ought to belong to *Cuscuta*.

The dispersal of *Cuscuta*, on the contrary, is one of the most beautiful and best knit. Dispersal suggests that the authority from whom we derived certain statements regarding the primitiveness of some of the groups of the genus is well informed. His taxonomy and phytogeography work in unison. Cases of the kind are not rare. Indeed, we may go further, and affirm that classification, though limited in manners we have explored, is true to the fundamental realities of biology as a matter of course. The human eye may be liable to error, which is understandable, but grasps withal the fundaments of nature most surely. Our difficulties are much less with what the eye sees than with what the mind tries to make out (or not to make out at all) of visual perceptions.

XIII. Cruciferae and Capparidaceae

Cruciferae and Capparidaceae are both large families which would bear being characterized as vicariants. The former are essentially "cold" to "cool", the latter "warm" to "tropical". Their potential of migration is exactly the same but the ranges that belong to either are unlike, in the main, on account of different requirements.

These two families are notoriously related. Recently, an author stated (52) that the cruciferous tribe Stanleyeae is "primitive" for Cruciferae and Capparidaceae alike, because it contains forms intermediate between the two.

This opinion, whether correct or erroneous, is challenging. The Stanleyeae are restricted to the Northern Hemisphere, and it could accordingly be argued that the origin of the Cruciferae and Capparidaceae is to be sought in "Holarctis," no longer in the deep south. As a matter of fact, it is on opinions of this very same cloth that rest supposedly authoritative statements to the effect that the Cyperaceae and the Compositae Crepidinae "originated" in Central Asia.

The Stanleyeae consist of four small genera (53, 54) distributed in the following manner,

(1) *Stanleya* — 6 species in the Western United States, ranging eastward to the Dakotas, Kansas, Western Texas.

(2) *Warea* — 4 species in the Eastern United States; 3 in Florida, 1 further ranging to Alabama and South Carolina.

(3) *Chlorocrambe* — 1 species in Oregon and Utah.

(4) *Macropodium* — 2 species in Central and Eastern Siberia, from the Altais eastward to Mongolia and the region of the Ussuri.

At a glance, we recognize here three massings, (a) Western American — *Stanleya* and *Chlorocrambe;* (b) Eastern American — *Warea;* (c) Asiatic — *Macropodium.* Nothing can be safely advanced on the face of this dispersal because too many are the hypotheses it suggests. It is not clear, for example, whether the American massings are from the Pacific or the Atlantic, wholly or part, or whether, perhaps, the Asiatic center was not fed this time from the New World. Above all, these four genera appear to stand isolated in the Northern Hemisphere. Obviously, the first step required on our part is that we investigate the migrations of the *genorheitron* to which these four genera may belong.

The Cruciferae Stanleyeae are commonly believed to be most directly allied with the Capparidaceae Cleomoideae, which are distributed (55) in this manner,

(1) *Cleome* — Pantropical with about 200 species.

(2) *Justago* — 1 species in Queensland and Northern Australia.

(3) *Gynandropsis* — About 20 species in warmer America; 1 species ranging to the Old World tropics.

(4) *Tetratelia* — 4 species in Mozambique, Bechuanaland, Transvaal.

(5) *Physostemon* — 7 species in the Guianas, Brazil, Mexico, Colombia, Paraguay, Argentina (Misiones).

(6) *Dactylaena* — 6 species in the West Indies, South America generally to Brazil and Argentina.

(7) *Haptocarpum* — 1 species in Eastern Brazil (Bahia).

(8) *Wislizenia* — 2 species in Mexico, 10 species in the Western United States.

(9) *Isomeris* — 1 species in Mexico (Sonora) and California.

(10) *Oxystylis* — 1 species in California.

(11) *Cristatella* — 2 species in the United States (Texas, Arkansas, Iowa, Dakotas, Colorado).

(12) *Cleomella* — About 20 species in the Western United States, eastward to the Rocky Mountains.

Considering that the Cruciferae are "cold" vicariants of the "warm" Capparidaceae, not only, but the Stanleyeae are nearest the Cleomoideae, we will act within reason if we compare the dispersal of the two groups together as a potential unit at first. Effecting this comparison, we notice, (a) *Cleome* could be fruitfully analyzed if we had detailed records of specific dispersal at hand. As matters stand, we can hope to do little with it; (b) However, *Justago* and *Gynandropsis* yield an Australian and South African range, respectively; (c) Better still, *Gynandropsis* repeats the behavior of the Turneraceae, *Menodora*, etc., because most of its species are in the New World, only a small rump remaining outside of it.

The last two observations bring us forthwith to consider the probability that the dispersal of the Cleomoideae was effected in origin from a center near South Africa. The reader no longer need be told in detail why this is so.

The balance of the Cleomoideae is quartered in the New World. A dominant western massing can immediately be made out in North America *(Oxystylis, Wislizenia, Isomeris, Cleomella)*, but one genus at least *(Cristatella)* might be rooted in the east. It will be noticed that this exactly matches the position of *Stanleya, Chlorocrambe* and *Warea* in the same region. The whole of the South American dispersal is confuse, however, as the records in our hands read, because we cannot be sure it stems from the Caribbeans rather than a point of entrance into South America from the direction of the Pacific.

Everything taken under consideration, we may conclude nevertheless as follows, (a) The Cleomoideae are from the African gate; (b) The trends in their dispersal appear to match those in the distribution of *Cuscuta;* (c) It is likely that the American massing is strongly western; (d) The dispersal of the Stanleyeae is consistent — at least as to the American end — with a genorheitral origin common between the Stanleyeae and the Cleomoideae; (e) It is transparent that the Stanleyeae occupy in regard of the Cleomoideae precisely the very same position which the Cruciferae hold by comparison with the Capparidaceae. In both cases the former (Cruciferae, Stanleyeae) are northern vicariants of the latter (Capparidaceae, Cleomoideae), and come into immediate contiguity in regions climatically tolerable to both; (f) The Stanleyeae cannot be "Holarctic" on account of the consanguineity that connects them with the Cleomoideae. *)

Our review of the Cleomoideae has yielded at least a suggestion that (not to mention the wide-ranging *Cleome*) the distribution of this subfamily to the New World called into play a channel ultimately leading to the end of South America from the west (that is, from the direction of the Pacific). We have further learned, purely on a factual basis, that the Stanleyeae tend to mass in the west of the United States. In sum, we seemingly face an invasion of the New World, both by Cleomoideae and Stanleyeae, effected from the Pacific rather than from the Atlantic.

If this invasion actually took place in the manner indicated, it is not beyond reason to anticipate that other groups of the Cruciferae repeat it, that is, run tracks entering the New World from the direction of the Pacific Ocean as also do certain groups under *Cuscuta,* for example. To verify, or disprove, this surmise let us go, then, to the records of cruciferous dispersal. We do not need theories, only facts.

The Cruciferae are a very large family. At least twentyone of their genera are small, not only, but scattered throughout comparatively petty tribal aggregates. These groups are negatively related in the sense that they fit well nowhere. It is consequently within reason to view them as odd-and-ends of the cruciferous ultimate *genorheitron,* which must rate today as relics, either because of the intervening loss of a former wider range, or because they never could achieve wide speciation in the geological past.

These odd genera are distributed in this manner;

A) Pringleeae

(1) *Pringlea* — 1 species in the Crozet and Kerguelen Islands.

B) Romanschulzieae

(2) *Romanschulzia* — 4 species ranging between Guatemala and Mexico.

---

*) We regret indeed being bound by precedent and convention to give so much time to the issue "Holarctis" v.s. "Antarctis", which we see as trifling at bottom.

C) Streptantheae

(3) *Streptanthus* — 3 species in Arkansas and Texas.

(4) *Streptanthella* — 1 species in Western Mexico and the Western United States.

(5) *Disaccanthus* — 6 species in Northern Mexico and the Western United States.

(6) *Cartiera* — 6 species in the Western United States.

(7) *Agianthus* — 1 species in Southern California.

(8) *Mitophyllum* — 1 species in California and Nevada.

(9) *Microsemia* — 1 species in California.

(10) *Euklisia* — 10 species massive in California and Oregon.

(11) *Icianthus* — 1 species in Oklahoma and Texas.

(12) *Pleiocardia* — 6 species in California.

(13) *Caulanthus* — 8 species in the Western United States.

(14) *Stanfordia* — 1 species in California.

D) Cremolobeae

(15) *Cremolobus* — 10 species from Chile to Ecuador.

(16) *Loxoptera* — 1 species in Peru.

(17) *Urbanodoxa* — 1 species in Peru.

(18) *Menonvillea* — 10 species in Patagonia and Chile.

(19) *Hexaptera* — 12 species in Patagonia, Argentina, Chile.

(20) *Decaptera* — 1 species in Chile.

E) Chamireae

(21) *Chamira* — 1 species in the Southwestern Cape.

We have learned in an early stage of our work that a definite phytogeographic tie runs between the Mascarenes and Central Chile. Referred to the modern map, this tie consists of a belt of oceanic waters about 15° Lat. wide. By bringing *Pringlea* into the record, this belt is widened by the entire difference in latitude between the Kerguelens and the Mascarenes. In other words, the phytogeographic tie making contact among the Mascarenes, Kerguelens, Crozet Island and Patagonia is no longer 15°, but 25° Lat. wide. This is sizeable belt, which suggests that the tie in question had as its background lands much larger than the petty islands just mentioned. We may go further, and remark that these lands had a definite physiography which can still be felt in dispersal. Though the Kerguelens are much nearer Africa than Magellania, still they seem to make easier contacts with the latter than with the former. The suggestion contained to this effect in the distribution of the petty genera listed above is repeated, as we are to see, by certain Cyperaceae.

The dispersal of these genera clings to the western half of the New World. There are seemingly exceptions (e.g., *Streptanthus* in Arkansas and Texas; *Icianthus* in Oklahoma and Texas) to agree with certain aspects of the migrations of the Cleomoideae (e.g. *Cristatella* in Colorado, Dakotas, Iowa, Arkansas, Texas), but it is altogether likely that these exceptions are in reality standard cases of migrations from the Pacific cryptically disturbed in a second time by the rise of the Rocky Mountains.

Dispersal like that of the Romanschulzieae, Streptantheae and Cremolobeae is usually disposed of with the statement that the genera involved travelled taking advantage of an "Andean bridge." Believers in "Holarctis" claim, as a matter of course, that the travelling took place from north going south. Their opponents argue the contrary. Faced by concrete cases which run against theory, both parties in the discussion vaguely refer, as a rule, to factors unknown which, someday perhaps, will be understood in the light of "Floating Continents." The continents usually float in phytogeographic work on thin fare of this kind.

We understand the facts otherwise. It is certain that the Andes rose most actively in the Tertiary, and are still actively rising under our own eyes. It is crystalclear that the Angiosperms had already achieved wide dispersal, and a thoroughly modern aspect, when the Andes began to stir. Under the circumstances, it is impossible to conceive of an "Andean Bridge" *in the abstract*. It is impossible to conceive by the light of reason of factors that could impel hosts of petty cruciferous genera and subgenera to "originate" upon heights of recent geologic vintage; there is no rational account, why an "Andean Bridge" of academic proportions allowed certain genera to pass whole (e.g., *Saxifraga*), while shattering irretrievably the Cruciferae; it is not to be understood why huge disconnections consistently cripple this "Andean Bridge" between Chile and Mexico; the prime movers are not to be grasped, why the same shrubs are common to the subxerophilous and xerophilous domains of Argentina, Chile and the U.S.A., but are seen nowhere inbetween, despite the occurrence of certain narrow ranges, at least, suitable to their survival inbetween the geographic extremes; there is no justification for the fact that the "Andean Bridge" should not have coaxed *Thamnosma* to reach Bolivia: it is beyond sober reason that the Galápagos can be dubbed "oceanic", when their heights contain manifest ancient relics, and these islands are normally reached by standard tracks faring the Eastern Pacific; it is unscientific through and through to reject as of no account a community of flora, light as this may be, between the Revilla Gigedo Islands and Hawaii. In sum, to have the "Andean Bridge" perform by academic fiats, factual evidence is to be swept out massively; the absurd must be called into play to account for the "mysterious."

The self-evident truth is to us that in the Late Jurassic, possibly much earlier, the American shores ran much farther west than they do now. The New World had by then entirely different outlines than it came to acquire, and to retain to our days, in the process of general reshuffling of lands and seas which took place most actively in the Cretaceous. The western longitudinal half of the New World, which is now a thin "Andean" or "Cordilleran" strip, was in epochs bygone a considerable bulge of lands into the Eastern Pacific, not only, but had no Andes as yet to show such at least as we know.

We might be taken to task for running our own scheme of palaeogeography. We will answer that palaeogeographers and geologists stop today at the shore, while we can freely sail the blue in the wake of channels of migrations that have all the hardness of a shore. Plantlife, we may repeat, is a geological layer of the earth.

The lands which in the dim geological past reached far and deep into the Eastern Pacific were in origin colonized by an angiospermous flora of which Palms, Graminaceae, Leguminosae, *Peperomia, Ephedra, Lepechinia, Cuscuta, Ilex, Menodora, Nama* all were part. The Cruciferae also teemed there, as did the Compositae. Seen from this distance, the flora in question appears to have a thin one, but a careful consideration of the evidence shows this to be most likely erroneous. Many Chilean forms *(Aextoxicon, Lardizabala, Adenopeltis,*)* etc. undoubtedly had part in this flora together with *Primula, Berberis,* etc. Even stronger, though more subtle and cryptic, is the evidence from the plant-world of Western Colombia (e.g., Chocó, Gorgona Island) and coastal Ecuador (Guayas, Manabí, Esmeraldas), to the effect that in these regions thrive plants of arresting affinities, sometimes immediately pointing in the direction of the Far East (e.g., *Phyllanthus lacerilobus*). Farther still: a strictly "Amazonian" genus like the euphorbiaceous *Mabea* can in the last resort be credited to no affinity, except the purely "Oceanic" *Homalanthus.* Patently, *Mabea,* though "Amazonian," has nothing whatever to do as to geographic origins with the "Amazonian" euphorbiaceous genus *Amanoa,* which is endemic to Brazil and West Africa both. A mighty evidence, direct and inferential, all comes to the same focus. Masses of flora came to what we understand today as America directly from lands now crumbled, but once mighty, in the direction of the west.

When these lands began to crumble in the Cretaceous much of their ancient flora must have died out without return. We have no idea as to what could be lost, but fossil pollen from Southwest Africa (see p. 93), which cannot be credited to plants now extant with but insignificant exceptions, proves that whole floras might easily come to extinction, of which the cryptic Chilean *Aextoxicon* is most likely a fragment. The plants that most easily survived were weeds such as *Lepechinia,* Cruciferae, Compositae; xerophiles such as *Ephedra;* ubiquists like *Peperomia;* forms in sum capable of quick ecological orientation.

---

*) In the classification of ENGLER & PRANTL, the Euphorbiaceae Adenopeltinae consist of three genera, (a) *Adenopeltis;* 1 species in Chile; (b) *Colliguaja:* 4 species in Chile, 1 in Paraguay, Uruguay and Southern Brazil; (c) *Dalembertia:* 4 species in Mexico. It will be observed, (i) The range is Western American; (ii) The range is disconnected between Chile and Mexico; (iii) The tie Chile-Southern Brazil manifest in *Colliguaja* is also stressed by *Araucaria.*

As the crumbling of these lands went on, the survivors were rolled eastward This is to say that the "Andean Bridge" was perhaps in some case taken by storm from the south, but in great many cases invaded directly from the west. The plants tossed, as it were, on the American shore as an immediate result of the crumbling going on in the west were ultimately caught in the Andean uplift, which was the last act — probably the effect — of the immense revolution that had gone on before. In the course of this revolution were "floated" eastward what we may call the bones of a multitude of groups of the Cruciferae, the very same petty genera we have discussed above. They might have speciated to some extent in their new surroundings (e.g., *Hexaptera, Menonvillea, Euklisia*), but it seems sure that they reached the ranges where the Andes were to grow already as the modern genera that they are still.

As Africa was colonized by angiospermy mostly from the east as an aftermath of sweeping geologic changes having their main (though by no means their sole) theater in the Indian Ocean, so was the New World overrun by massive streams of the same plants coming in from the west, as an aftermath of sweeping geologic changes having their main theater in the Eastern Pacific. Polynesia, Australasia, Malaysia, the Eastern Far East were on their part massively colonised because of homologous happenings in the Western Pacific (see also p. 242).

If we intend to stultify the evidence held before our eyes by the factual record of dispersal, we can do no better than to run the course of migration in such a manner as to conform with modern outlines of lands and seas. It is certainly impossible to square up dispersal with modern maps, because we deal with unlike quantities the moment we try to do this. Present day geography, and dispersal effected on the strength of factors operative in the geography of the past, cannot agree ever if we approach dispersal in the only way dispersal becomes rational. Dispersal moved, modern maps are static.

To return now to *Macropodium;* this genus is described (56) as a "Relic of the old Tertiary flora" in Central Asia, but this estimate should be altered to read *of the old Cretaceous or Jurassic flora.* A form having characters intermediate between Cruciferae and Capparidaceae antedates by far the early Cretaceous, when the Angiosperms had already a wealth of modern genera. The estimate in question might be correct, perhaps, if it is meant to the effect that *Macropodium* survived the Pleistocene cold that killed the Mammooth in Central Asia and Siberia.

The affinities that bind *Macropodium* and certain American genera can be explained in at least two ways. One, and the most obvious by the modern map, is that *Macropodium* stems from part of the *genorheitron* of the Stanleyeae which reached Asia from the New World's direction. To support this opinion, the Hydrophyllaceae easily lend one of their genera. The family is massively American, but one of its members, *Romanzoffia,* reveals the following,

(1) *R. unalaschkensis* — Vancouver Island, Alaska, Aleutians, Eastern Siberia. Dispersal of this kind documents a penetration of Siberia from the New

World, as the modern map reads, which — be it said parenthetically — does not mean as yet the truth, as we have just learned.

Whatever the case be, the distribution of *R. unalaschkensis* reads well, and might be satisfactory, in part, as regards this one species. However, the Polemoniaceae, which are allied with the Hydrophyllaceae and entirely American, have by exception three of their species (57) distributed in this manner,

(1) *P. coeruleum* — Southern and Central France ("naturalized" in England, Spain, Portugal, Central Italy), Germany, Scandinavia, Northern Balkans, Hungary, Transylvania, Russia, Caucasus, Urals, Altais, Himalayas, Siberia, Manchuria, Sakhalin, Japan; Alaska and Canada (Northwestern Territories).

(2) *P. lanatum* — Subarctic and Arctic Eurasia to 72° Lat.; Northern Asia to the Altais (subsp. *pulchellum*); Greenland, Canada (Northwestern Territories).

(3) *Phlox sibirica* — Central Asia (Lake Balkash) to Northern Asia and Alaska.

The distribution of these three species might be drawn from the New World to the Old across an "Alaskan Bridge" which looks so alluring on our maps. It might also be contended that England, Spain, Portugal, ruled out mistakenly (which is quite possible) as an adventitious sector of the dispersal of *Polemonium coeruleum* by the authority we follow are in reality genuine "Atlantic distribution." It might accordingly be open to conclusion that *P. coeruleum* reached Europe from the selfsame shore-line in the Atlantic which furthered the migrations of forms like *Oligomeris, Thamnosma,* etc.

These hypotheses are not absurd, far from it — the latter in particular, because the former is doubtless weak — but it seems to us that the best possible explanation is the one which involves the longest disconnection in the modern polemoniaceous tracks. We choose this explanation because it also rationalizes at one stroke the distribution of the Hydrophyllaceae, notorious allies of the Polemoniaceae.

The Polemoniaceae and Hydrophyllaceae are both massively American. The former number but three extra-American species, which we have just studied. The latter have a genus, *Codon,* in the Cape and Southwest Africa, not only, but their type-genus, *Hydrolea,* is distributed to Madagascar, Tanganyka, Congo, Cameroon, Angola, Ceylon, Burma, Siam, Java, Philippines and Queensland. It is then clear that the Hydrophyllaceae, not unlike the Turneraceae and genera such as *Gynandropsis, Menodora, Rhipsalis,* etc., are one of the comparatively numerous "Pseudo-American" groups skimpily rooted in the African gate, but strikingly numerous in the New World, hence the usual error of tagging them out as "American."

In this knowledge, we believe that the three polemoniaceous species in Eurasia, also *Codon* and *Hydrolea,* owe their origin alike to a branch of the track, which issuing from a center in modern Indian Ocean directly went to Asia rather than to America. Obviously, we hold that *Macropodium,* too, fared

this channel, nor any route from the New World leading westward to Central Asia. We believe this, because it is the simplest explanation we can give in the light of standard tracks, not only, but the explanation which accounts equally well for everything in the Stanleyeae, Polemoniaceae and Hydrophyllaceae. The three species of the Polemoniaceae stranded in Asia bear to the sum total of the polemoniaceous dispersal precisely the same relation (see p. 174) as does the genus of the Arbutoideae *Arctous* to its relatives *Arbutus* and *Arctostaphylos*. The relative position of these three ericaceous genera is easily explained by making reference to the ericaceous tracks in general. We see no reason why we should make an appeal to thin "Landbridges" between Alaska and Eastern Siberia in order to rationalize the distribution of an occasional species, or genus, when trusted principles of interpretation make it easy for us to understand at one stroke the distribution of great many of them. In conclusion, the reader may interpret the Asiatic polemoniaceous outlier and the ultimate origin of *Macropodium* in reference to his own lights, if what we state is not satisfactory to him. Nothing fundamental is involved in the fact, whether *Macropodium* and the polemoniaceous Asiatic species fared one track or the other. Their ultimate origins and genorheitral affinities do not alter by a free use we may make of certain minor alternatives in preference to certain other.

BIBLIOGRAPHY

Chapter IV.

(1) ENGLER, A. — *Saxifragaceae:* E. & P. Nat. Pflanzenf. 18a: 214. 1930.

(2) CROIZAT, L. — see (19) Chapt. II.

(3) WHITE, C. T. — Contributions to the Queensland Flora, No. 5; Proceed. Roy. Soc. Queensland 47: 61. 1936.

(4) DANSER, B. H. — A revision of the genus Korthalsella; Bull. Jard. Bot. Buitenzorg, III, 14: 115. 1937; Supplement to the revision of the genus Korthalsella (Lor.); ibid. 16: 329. 1940.

(5) PILGER, R. — *Santalaceae;* E. & P. Nat. Pflanzenf. 16b: 52. 1935.

(6) ENGLER, A. & KRAUSE, K. — *Loranthaceae;* E. & P. Nat. Pflanzenf. 16b: 168. 1935.

(7) STANDLEY, P. C. — Trees and Shrubs of Mexico; Contr. U. S. National Herb. 23 (2). 1922.

(8) VAN STEENIS, C. G. G. J. — see (12) Chapt. III.

(9) WOLF, Th. — Monographie der Gattung Potentilla; Bibl. Bot. 16 (Heft 71). 1908.

(10) FOCKE, W. O. — *Rosaceae;* E. & P. Nat. Pflanzenf. 3 (3): 1. 1888.

(11) BITTER, G. — Die Gattung Acaena; Bibl. Bot. 17 (Heft. 74i—74iv). 1910—1911.

(12) FOCKE, W. O. — Über die natürliche Gliederung und die geographische Verbreitung der Gattung Rubus; Engler Bot. Jahrb. 1: 87. 1880.

(13) FERNALD, M. L. — The geographic affinities of the vascular floras of New England, the Maritime Provinces and Newfoundland; Amer. Jour. Bot. 5: 229. 1918.

(14) PAX, F. & KNUTH, R. — see (10), Chapt. II.

(15) WEIMARCK, H. — Die Verbreitung einiger Afrikanisch-Montanen Pflanzengruppen. III. Die Gattung Ardisiandra; Svensk Bot. Tidskr. 30: 36. 1936.

(16) DRUDE, O. — *Ericaceae;* E. & P. Nat. Pflanzenf. 4 (1): 15. 1891.

(17) BOLUS, H. & GUTHRIE, F., BROWN, N. E. — *Ericaceae;* Dyer's Fl. Cap. 4 (1): 2. 1905—1906.

(18) THONNER, F. — The Flowering Plants of Africa. 1915.

(19) SAMUELSSON, G. — Studien über die Entwicklungsgeschichte einiger Bicornes-Typen; Svensk Bot. Tidskr. 7: 97. 1913; see further: HAGERUP,

234

O. — Studies on the Empetraceae; Dansk. Videnskab. Selsk., Biol. Meddel.
20 (5): 1. 1946.

(20) ALM, C. G. & FRIES, Th. C. E. — Monographie der Gattung Blaeria;
Act. Hort. Berg. 8: 221. 1925 — Monographie der Gattungen Philippia
Klotzsch, Mitrastylus nov. gen. und Ericinella Klotzsch; Svensk. Vetenskap-
sakad. Handl., III, 4 (4): 1. 1927.

(21) SLEUMER, H. — Die Gattung Agauria (D. C.) Hook. f.; Engler Bot.
Jahrb. 69: 374. 1938.

(22) PERRIER DE LA BATHIE, H. — Les Philippia de Madagascar; Arch.
Bot. [Caen] 1 (2): 1. 1927.

(23) BEIJERINCK, W. — Calluna. A monograph of the Scotch Heather. 1940.

(24) TURRILL, W. B. — The Plant-life of the Balkan Peninsula. A phytogeographical study. 1929 (see here p. 414).

(25) CHEVALIER, A. — L'origine géographique et les migrations des Bruy-
ères; Bull. Soc. Bot. France 70: 855. 1923.

(26) PERRIER DE LA BATHIE, H. — Catalogue des Plantes de Madagascar
(edit. Académie Màlgache): Ericaceae et Vacciniaceae. 1934.

(27) CHEESEMAN, T. F. — see (21) Chapt. II.

(28) SKOTTSBERG, C. — A botanical survey of the Falkland Islands; Svensk.
Vetenskapsakad. Handl. 50 (3): 45. 1913.

(29) MACBRIDE, J. F. — Vaccinium and Relatives in the Andes of Peru;
Univ. Wyoming Publs. 11: 37. 1944.

(30) SMITH, A. C. — The American species of Thibaudieae; Contr. U. S.
National Herb. 28: 311. 1932.

(31) SLEUMER, H. — Die Arten der Gattung Vaccinium in Zentral- und
Süd-Amerika; Notizbl. Bot. Gart. Mus. Berlin—Dahlem 13: 111. 1936.

(32) SKOTTSBERG, C. — see (12) Chapt. I.

(33) SLEUMER, H. — Revision der Ericaceen von Neu-Guinea; Engler Bot.
Jahrb. 70: 95, 106. 1939.

(34) MELCHIOR, E. — Violaceae; E. & P. Nat. Pflanzenf. 21: 329. 1925;
see further (21) Chapt. II.

(35) SCHELLENBERG, G. — Connaraceae; E. & P. Pflanzenr. 103 (iv. 127).
1938.

(36) ENGLER, A. & KRAUSE, K. — Dichapetalaceae; E. & P. Nat. Pflanzenf.
19c: 1. 1931.

(37) DWYER, J. D. — A discussion of the ochnaceous genus Fleurydora
A. Chev. and the allied genera of the Luxemburgiaceae; Bull. Torrey Cl.
71: 175. 1944.

(38) SCHULZ, O. E. — Erythroxylaceae; E. & P. Pflanzenr. 29 (iv. 134).
1907.

(39) SCHULZ, O. E. — Erythroxylaceae; E. & P. Nat. Pflanzenf. 19a:
130. 1931.

(40) WINKLER, H. — Linaceae; E. & P. Nat. Pflanzenf. 19a: 82. 1931.

(41) CIFERRI, R. — Saggio di classificazione delle razze di manioca (Mani-

hot Esculenta Crantz); Relazioni e Monografie Agrario-Coloniali No. 44. 1938.

(42) FRANCIS, W. D. — The development of buttresses in Queensland Trees; Proceed. Roy. Soc. Queensland 36: 21. 1925 — The development of the corrugated stems of some Eastern Australian trees; ibid. 38: 62. 1927.

(43) MERRILL, E. D. — An Enumeration of Philippine Flowering Plants 4: 80. 1926.

(44) BENTHAM, G. & HOOKER, J. D. — Genera Plantarum 2: 551. 1876.

(45) SCHMUCKER Th. (editor) — Silva Orbis: Monographs of the International Forestry Center, No. 4: The Tree Species of the northern temperate zone and their distribution. 1942.

(46) LOUBIÈRE, A. — Anatomie comparée d'un bois de Dicotylédone crétacique de Madagascar; Bull. Mus. Hist. Nat. Paris, II, 11; 484. 1939 — De la valeur diagnostique des caractères structuraux dans l'étude comparative des bois vivants et fossiles des Dicotylédones; ibid. 13: 489. 1941.

(47) POJARKOVA, A. I. — (English resumé titled) Botanico-Geographical Survey of the Maples in USSR in connection with the History of the whole Genus Acer; Act. Inst. Bot. Acad. Sc. URSS, Series I, Fasc. 1: 367. 1933.

(48) ANDERSON, N. J. — Monographia Salicum; Svensk. Vetenskapsakad. Handl. 6 (1). 1863.

(49) PAX, F. — Salicaceae; E. & P. Nat. Pflanzenf. 3 (1): 29. 1894.

(50) YUNCKER, T. G. — The genus Cuscuta; Mem. Torrey Bot. Cl. 18: 113. 1932; see further: VAN OOSTROM, S. J. — The Convolvulaceae of Malaysia; Blumea 3: 62. 1938.

(51) CORNER, E. J. H. — Notes on the Systematy and Distribution of Malayan Phanerogams. III; Garden's Bull. Straits Settl. 10: 250. 1939.

(52) JANCHEN, E. — Das System der Cruciferen; Öst. Bot. Zeitschr. 91: 1. 1942.

(53) SCHULZ, O. E. — Cruciferae; E. & P. Nat. Pflanzenf. 17b: 227. 1936.

(54) ROLLINS, R. C. — The Cruciferous genus Stanleya; Lloydia 2: 109. 1939.

(55) PAX, F., & HOFFMANN, K. — Capparidaceae; E. & P. Nat. Pflanzenf. 17b: 146. 1936.

(56) BUSCH, N. A. — Macropodium; Flora URSS 8: 27. 1939.

(57) BRAND, A. — Polemoniaceae; E. & P. Pflanzenr. 27 (iv. 250). 1907.

CHAPTER V

INTERCONTINENTAL DISPERSAL III

So monotonous is the behavior of plants in dispersal that its students can hardly avoid in the end a feeling of boredom. The same features constantly recur upon the map; the same tracks repeat themselves without end. Soon a point is reached, where the budding phytogeographer unconsciously begins to inquire, why there is seemingly no end to the same fare. Soon, too, the advanced student of phytogeography learns to identify a track at a glance, and can almost foretell whether a certain plant is to be looked for, or not, at a definite spot of the modern maps *).

Monotony and repetitiousness must be welcome, however, because reiteration is proof in itself that we have the prime movers of migration well in hand. Were we bound to contrive each time, and as regards each new case, a fresh theory or hypothesis of dispersal; where we forced to give up the distribution of a certain group as beyond rationalization; were we, in brief, constantly being tossed right and left in our quest, we would end with having no explanation of dispersal at all. Correct methods and proper solutions cannot be forever novel, for the elementary reason that these methods and solutions embody essentials to agree with the behavior of things in nature.

We speak of *correct methods* and *proper solutions,* but we are certainly far from claiming that what we offer in these pages is wholly correct and forever proper. It is amply sufficient for us to begin in the right direction, making it clear that what has so far been done in the field of phytogeography generally leaves something to be desired.

I. Cyperaceae

The Sedges match in general interest all other families of seedplants. They are a stupenduous group in their own right, and none other is so completely

---

*) A concrete example of overlooked records duly detected because of general considerations suggested by the course of the tracks is given elsewhere in these pages (see p. 537). It is plain that a phytogeographer who knows that a certain plant is in Madagascar and Ceylon is apt, in the light of the method advanced in this manual, to take for granted — unless the contrary be proven — that this same plant is in Malaysia, Sumatra and Java, particularly. Likewise, a phytogeographer who knows that a plant is in Samoa and Japan will not go after it necessarily in Afghanistan, though he will not be upset if he finds it there, too.

ubiquitous. They cover the earth, and *Carex*, one of their major genera, is endemic practically everywhere. Every track, or combination of tracks, is sooner or later discovered in the cyperaceous dispersal. It is well within the limits of the possible to give a full course in phytogeography with no better material at hand than a map of the world, and the standard monograph of *Carex* (1) to which reference is constantly made in the coming pages.

The generic limits in the Cyperaceae are not always cleancut, and it is comparatively easy to connect Sedges with Lilies through endless intermediates. These interrelations, and the migrations of the remote *genorheitra* in the background of the cyperaceous affinity, are fascinating subjects deserving ample review. Unfortunately, we may only concentrate here upon a few essentials of dispersal that immediately concern the Cyperaceae themselves.

To begin our review, we may deal first with two genera, *Schoenoxiphium* (1, 2) and *Cobresia*. The former is distributed along the following lines,

(1) *S. lanceum* — Cape.

(2) *S. Thunbergii* — Cape.

(3) *S. Ecklonii* — Cape.

(4) *S. Buchananii* — Natal.

(5) *S. rufum* — Cape, Pondoland.

(6) *S. Kunthianum* — Cape, Natal, Transvaal.

(7) *S. Lehmannii* — Cape, Tanganyka?

(8) *S. sparteum* — Cape, Transvaal, Tanganyka (Usambara), British East Africa (Mt. Ruwenzori), Ethiopia.

(9) *S. kobresiodeum* — Northern Sumatra (Gajolands) *).

Let us observe, as a preliminary, that the track run by *Schoenoxiphium* is standard, in the sense that it reaches from East Africa to Malaysia. Dispersals of this type are, as we well know, common. Let us further notice that the Sumatran Gajolands, where is endemic *S. kobresioideum*, are inhabited by a peculiar flora with marked affinities in the direction of the Himalayas. This flora, as a matter of fact, could be said to contain a marked "Holarctic" element. *Schoenoxiphium kobresioideum*, on its part, has characters intermediate between *Schoenoxiphium* and *Cobresia*, which the specific name stresses.

Unlike *Schoenoxiphium*, *Cobresia* is entirely confined to Asia, within the following pattern of distribution,

A) Sect. *Elyna*

(1) *C. schoenoides* — Afghanistan, Asia Minor, Caucasus, Central Asia (Pamir, Altai Mts.), Western Tibet, Eastern Himalayas.

---

*) As we cannot run too much ahead of our subject, we do not stress in the discussion in the main text certain aspects of the dispersal of *Schoenoxiphium*, which the reader will duly see as quite important in time. He should observe that the dispersal is topheavy in the direction of the Cape, Natal and Transvaal; that Madagascar is left out; that the range of *S. sparteum* is disconnected between Transvaal and Usambara. All this is *quite* typical of the dispersal of a "cold" form immediately out of the *Afroantarctic Triangle* (see p. 349), and riding a track in the modern Indian Ocean. This track enters Africa from the south and the east throughout, precisely in the manner shown by Fig. 51.

(2)   *C. Bellardii* — Central Asia (Altai Mts., Alatau), Caucasus, Europe to Italy, the Pyrenees and Norway; Alaska, Greenland, the Rocky Mountains to Oregon and Colorado.

(3)   *C. nitens* — Western Himalayas (Kashmir to Kumaon).

(4)   *C. robusta* — Tibet, China (Kansu).

B) Sect. *Hemicarex*

(5)   *C. pygmaea* — Eastern Himalayas, Tibet.

(6)   *C. trinervia* — Himalayas throughout, Tibet, China (Yunnan).

(7)   *C. graminifolia* — China (Shensi).

C) Sect. *Eucobresia*

(8)   *C. Royleana* — Afghanistan, Asia Minor, Turkestan, Pamir, Himalayas, Tibet, China (Szechuan).

(9)   *C. uncinioides* — Central Asia (Tianshan), Tibet, Central Himalayas, China (Yunnan).

(10)   *C. caricina* — Central Asia (Alatau Mts.), Caucasus, Europe northward from the Alps to the high north; northwestern Himalayas, Alaska, Greenland, Canada.

(11)   *C. Clarkeana* — Eastern Himalayas.

D) Sect. *Pseudocobresia*

(12)   *C. macrantha* — Western Tibet.

E) Affinities unsettled.

(13)   *C. species* — Western Java.

The distribution of *Cobresia* falls into two main groups, one generally restricted to the Himalayas between Western India and China (Sects. *Pseudocobresia* and *Hemicarex*) *); the other wide ranging to Europe and the New World (Sects. *Elyna* and *Eucobresia*). Both these groups occur in the region of the Altais.

---

*) We have no title to discuss the species of *Cobresia* having unsettled status endemic to Western Java. It seems to us, however, that the presence of this species in that region reinforces everything we state concerning *Schoenoxiphium*. In other words, it is probable indeed that a part of the original *genorheitron* of the *Schoenoxiphium-Cobresia* affinity was delivered straight to Western Malaysia, without ever going to the Himalayas in the first place.

The distributions of *Schoenoxiphium* and *Cobresia* accordingly highlight the following centers, (a) East Africa *); (b) The Himalayas and Altais; (c) Sumatra (Gajolands). These centers are the keystones in the dispersal, for, as we know (see p. 171), from the Altais all of the northern hemisphere can be reached by easy steps. The Altais, then, are a cardinal node of secondary dispersal throughout the boreal lands.

The form connecting most immediately *Schoenoxiphium* and *Cobresia* is endemic to the Sumatran Gajolands, which is of course right interesting. *Schoenoxiphium* is very close to *Carex*, and *Carex* merges in turn with *Uncinia*. It stands to reason, consequently, that we cannot retrace the origin of the *genorheitron* of this group by studying only one of the genera involved. We are bound to investigate them all, and to conclude as regards the origin of the *genorheitron* in such a manner that dispersal and phylogeny both receive their dues.

Carex and *Uncinia* will come under scrutiny in the coming pages, in which it will be shown that these genera fully deserve inclusion in the list elsewhere alluded to of "Antarctic" plants common to the Australian and Magellanian worlds. It will be shown that the dispersal of *Carex* and *Uncinia* yields to reason only if approached from a point of vantage in the deep south of modern maps.

It being so that the distribution of *Carex* and *Uncinia* yields to reason only when approached from a point of vantage in the deep south of modern maps, we may not hope to read the tracks of *Schoenoxiphium* and *Cobresia* from the north going southward.

We will, on the contrary, read these tracks the way we have always done when dealing with all manners of plants, that is, going from the south northward.

Doing this, as already noticed, we have before us tracks of *Schoenoxiphium* and *Cobresia* that are standard throughout in their main outline, tracks, that is, quite as commonplace as those of *Evolvulus* (see p. 24), for example.

The question remains before us to settle, how were reached the Sumatran Gajolands. This question can be settled with reference to anyone of three hypotheses, and their possible combinations, as follows, (a) The track moving west to east generally (as the map now reads) first reached Malaysia (Sumatra), next swerving from Sumatra northward to the Himalayas; (b) The track,

---

*) As of a rule, we do not intend to swamp the reader with a swarm of consideration and details which, though oftentimes highly interesting, may distract his attention from the substances immediately brought to focus. This rule is not free of drawbacks, but we have adopted it after some thought. We will invite the reader's attention here, nevertheless, to the fact that *Schoenoxiphium* does not turn up either in Madagascar or the Mascarenes, and, though "warm" by actual dispersal, has close ties with the definitely "cool" to "cold" *Cobresia*. We observe in this regard that not few are the groups (e.g., *Erica, Linum*) which, not unlike the *Schoenoxiphium-Cobresia* affinity, (a) Are unknown in Madagascar and the Mascarenes; (b) Though "warm" or "tropical" by reason of part at least of their dispersal, easily turn "cool" to "cold" when migrating to the north. *There is, then, a definite connection between a group being potentially "cold" or "cool" and its not turning up in Madagascar and the Mascarenes, despite its dwelling elsewhere in Africa or Tropical Asia.* This connection is not without exceptions, of course, but well deserving of attention. See also p. 215fn.

on the contrary, always streaming in the same direction, first reached the Himalayas, then moved on to Sumatra; (c) The Himalayas and Sumatra were reached simultaneously from the southwest.

These three hypotheses, and their possible combinations, involve each nothing essential. Referred to the general orientation of the track (Africa-Asia (Malaysia especially)), they tell us something highly interesting, however. They tells us that the putative "Holarctic" or "Himalayan" element in the flora of Gajolands may never have seen either "Holarctis" or the Himalayas. This element, on the contrary, might have come straight from the direction of the Indian Ocean, and invested Sumatra, the Himalayas, and everything "Holarctic" of the modern earth almost simultaneously, when indeed not simultaneously, in a fan-shaped pattern of distribution — opening up west to east.

In the course of our work, we made frequent reference to *genorheitra* reaching Central Asia directly from a range now crumbled, but once existing, in the Indian Ocean of modern maps. This manner of putting the evidence, supposing, that is, that a *genorheitron* reached Asia out the blue waters of a modern ocean is bound, we know, to arouse misgivings in most anyone who begins to study distribution, and is still accustomed to look to modern maps, or their like, as the standards of migration. Bewildered by conflicting "theories" of "floating" continents; needled by a plethora of learned papers, some of which decry, other on the contrary extol "Landbridges," the beginner in phytogeography is as a rule inclined to theorize before acting. Normally, he is unwilling or unprepared to take a map, the records of dispersal and, purely and simply, draw first on the map the tracks which the taxonomic records pinpoint. This beginner inquires in the first place, what do we think of Wegener's cerebrations, etc., etc., deeming this to be essential, *which it certainly is not*.

We may imagine what we like about the tracks of *Schoenoxiphium* and *Cobresia* without for this ever getting away from certain essentials, as follows, (a) The migrations of these two genera are standard when compared with the migrations of uncounted other plants; (b) These two genera belong to one and the same close *genorheitron* together with *Carex* and *Uncinia*.

If, this being firm before us, we analyze the dispersal of *Schoenoxiphium* and *Cobresia*, we are bound to reach certain conclusions. These conclusions may not necessarily seem to us all equally probable, but that all are worthy of serious thought we may not doubt. These are the conclusions, (a) The *genorheitron* of *Schoenoxiphium* and *Cobresia* migrated to the Himalayas and Sumatra directly from Eastern Africa, following a path which is of lands on modern maps. This path goes — as modern maps read — from the Cape to Ethiopia, next vaulting over a disconnection to reach Afghanistan; (b) This *genorheitron*, on the contrary, freely migrated from a gate of angiospermy located slightly off modern Africa, in the Indian Ocean of our maps, and dispatched from this gate a genorheitral branch which matured as

Schoenoxiphium both in Eastern Africa and the Gajolands; as *Cobresia* in the Himalayas generally and Altais, next streaming onward to boreal lands in the New and Old World.

We will not force upon the reader the acceptation of one or the other of these alternatives, for we think they are immaterial in the end. Personally, we incline toward the latter, nevertheless, because of the following reasons, (a) To reach the Gajolands and Africa, *Schoenoxiphium* ran a track over part at least of the modern Indian Ocean; (b) There is good evidence to the effect that when travelling between Africa and the Gajolands, the *genorheitron* of the *Schoenoxiphium*— *Cobresia* affinity was still plastic, witness the characters of *S. kobresiodeum;* (c) The distributions of *Schoenoxiphium* and *Cobresia* are consistent with the hypothesis that these genera are, in a definite sense, ecotypes out of the same *genorheitron,* one of fairly "warm" tendencies *(Schoenoxiphium)*, the other of "cold" *(Cobresia)*. Let us not forget that similar breaks in genorheitral masses, forming "cold" and "warm" affinities, are definitely known to have taken place in the Compositae (see p. 59), part of these Compositae being now endemic to the Himalayas and the very same regions where *Cobresia* occurs; (d) Very numerous are the groups — and many of them indeed are discussed in the pages of this manual — which reach Central Asia (and the Americas as well) from a point somewhere in the Indian Ocean, be this point very far, or very near, the modern African coasts does not matter.

In conclusion, we regard the *Schoenoxiphium-Cobresia* case as a mere accident in dispersal. This group is interesting, doubtless, because it has one species, *Schoenoxiphium kobresioideum,* strategically located, both as such (i.e., in disconnection; the disconnection calling into play what a beginner in phytogeography is apt to take for a stupendous mileage), and in relation to the flora in which it partakes (Gajolands; these highlands contain both "Himalayan" elements and "Australian" plants in mixture). This group, however, tells us nothing whatever that is unique. If we only consider that dispersal manifestly took place in origin over maps of the world that was, not the maps of the world that is (indeed, a truism), we have no reason to prefer as the starting point of a channel of migration a piece of land rather than a piece of ocean of the modern maps. *That starting point is advised, which is consistent with the whole of the dispersal, and brings dispersal to book in a rational manner, both as to itself and as to every other plant.* If we intend to know what plants did in dispersal we had better take their records in hand first, and next endorse the consequences which seem to us to be within sober reason. There is no other approach to knowledge.

It seems to us clear that the *genorheitron* of the *Schoenoxiphium-Cobresia* alliance massively invaded Asia, and so it did at more than one point. The evidence is good that it reached the Gajolands of Sumatra from the Indian Ocean's direction. Quite as good is the evidence that also through the *modern* Indian Ocean it invaded Asia, first reaching the Western Himalayas and Altais.

Thus we believe, according to sound evidence in our lights, that the base line of the distribution is squarely set across the Indian Ocean, and rooted in the African gate of angiospermy. The map we offer (Fig. 51) answers in our judgement the essentials in play, and we leave it for the reader to modify it in such details as he may choose.

Seeing that the Indian Ocean of our maps behaves as solid land as regards distribution, we are quite willing to believe — as do many geophysicists and palaeogeographers — that this Ocean was once occupied by lands, *Gondwana*, that is, of much current terminology. How these lands disappeared, whether by "floating," or "sinking," etc., etc., does not concern us *in the least*. We take at face value what plant-life manifests, and that ends the matter for us here.

It is apparent that as the modern Indian Ocean was once of lands, we can speak of tracks as regards the *Schoenoxiphium-Cobresia* affinity only with the understanding that these tracks are not linear channels of migration thinly strewn along certain unsubstantial "Landbridges." It is patent that Asia was massively invaded from the south, quite as much as Africa was widely entered from the east, and that *Gondwana* — always to use current terminology — was not a land of shadows and makebelieve. *It was something indeed quite as tangible as lands are today.* The *genorheitra* that streamed out of it had to cope with phytogeographic and ecologic factors quite as weighty as those in the world today, which explains why these *genorheitra* took different paths to reach modern Africa, America, Asia, Australia and Malaysia *).

One more observation, a very important one, finds place here.

By following dispersal *from the south northward*, that is to say, by going from "Antarctic" to "Holarctic" lands, we have been able to retrace certain definite channels of migration, not only, but to learn that these channels are patronized by all manners of plants. By so doing, we have rationalized, perhaps, what was before a wild array of records and a rather indifferent show of theories.

Let us now suppose that we are hopelessly in the wrong, and that dispersal is effected *from the north southward* on the contrary. If this be so, we believe that the tracks we have first mapped are not at all erased from the face of the earth. These tracks are beyond doubt convenient, if nothing better.

If the tracks in question are convenient, standard, and apt to rationalize what formerly stood as irrational — and these tracks were first identified *from the south northward* — we may hold that the dispersal proceeding over these tracks was *in fact* also effected *from the south northward*. We reach the same conclusion, this time on other grounds, also elsewhere in the pages of this manual.

If, however, there are phytogeographers unwilling to agree with us, and willing on the contrary to use the tracks we have mapped *from the north southward*, in adherence to some theory of "Holarctis," we will stand patiently by when these phytogeographers shall turn our maps and tracks upside down. This will be indeed their privilege, and bystanders will eventually judge.

---

*) See also discussion, p. 230.

We will not enter into controversy ever as to such issues as this. Factual tracks, however orientated, always are to count for more than trackless learned theories.

Schoenoxiphium, as we have learned, reached the Gajolands of Western Sumatra *from the west* (direction of the modern Indian Ocean). Other Cyperaceae landed this time in the very same spot *from the east* (direction of the modern Pacific), witness the following distributions,

A) *Oreolobus* (Fig. 52)

(1)   *O. obtusangulus* — Magellania to Ecuador along the Andes.
(2)   *O. strictus* — New Zealand, Stewart Island.
(3)   *O. pumilio* — New Zealand, Australia, New Guinea.
(4)   *O. pectinatus* — New Zealand, Auckland and Campbell Islands.
(5)   *O. Clemensiae* — New Guinea.
(6)   *O. ambiguus* — New Guinea, Borneo (Mt. Kinabalu).
(7)   *O. Kuekenthalii* — Sumatra (Gajolands), Malacca (Pahang: Gunong Tahan; Perak: Gunong Kerbau).
(8)   *O. furcatus* — Hawaii.

B) *Schoenus* (in part)

(1)   *S. antarcticus* — Fuegia.
(2)   *S. pauciflorus* — New Zealand.
(3)   *S. philippinensis* — Var. *typicus:* Luzon (Benguet, Tayabas, Laguna, Negros, Mindanao); Var. *kinabaluensis:* New Guinea (Wharton Range, Arfak Mts.), Southwestern Celebes (Mt. Bonthain), Borneo (Mt. Kinabalu); Var. *pachystylus:* Northern Sumatra (Gajolands).

These patterns of distributions are important, because, (a) They definitely prove that Malaysia can be invaded by different *genorheitra*, or, indeed, *genorheitra* of the same affinity, both from the east and the west, the Indian and Pacific Ocean; (b) They establish that when these *genorheitra* reached Malaysia, the Gajolands already existed very much in their present geographic position, if by then still connected with lands that bore to modern Sumatra no likeness. In other words, the Gajolands seem to be very ancient, quite as ancient, when not older, as, for example, the sandstone "mesas" of the Guianas in America; (c) They prove that it is certainly and absolutely futile to deal with phytogeography while making appeal to conventional adjectives of "origin." While no such adjective exists that can be applied to *Schoenoxiphium*, *Schoenus* and *Oreolobus* are beyond doubt to be identified as "Australian," "Antarctic" or "Old Oceanic" in the academic terminology of dispersal. *Cobresia*, on its part, is certainly "Himalayan" and "Holarctic" in said terminology. It follows that academic phytogeography recognizes in Western Sumatra and Western Java the presence of at least five "elements," namely, "Antarctic," "Holarctic," "Himalayan," "Australian," and — to cap the deal most fittingly — "Of unknown origins." Manifestly, science cannot thrive upon plays of words of this hollow sort. It should be feasible to theorize by a track running all the way, for example, between New Zealand and Spain, that New Zealand was "invaded" by "Holarctic" elements, quite as much as to argue that Spain was "colonized" by "Antarctic" ones, and to conduct the theorizing to the bitterest, though most unconstructive, end. *This is a path which we will never enter.* To us only *precise* words or adjectives have scientific meaning, and to speak competently of origins we must know — *and know in the very first place* — where to begin, and where to end.

It is obvious that a group like the Cyperaceae is not wanting in interest. Fingering through its records of distribution, we find something else worthy of attention, witness the dispersal of *Carex* Sect. *Elongatae,*

(1) *C. stellulata* — New Zealand, Australia; Europe, the high north excepted; North America (Saskatchewan to California, Newfoundland, Maryland); var. *omiana:* Japan.

(2) *C. remota* — Europe to 61° Lat., Caucasus, Iran (Persia), Syria; subsp. *alta* var. *brizopyrum:* Java, Sumatra (West Coast: Mt. Kerintji), Himalayas, China (Szechuan); subsp. *Rochebrunii:* Himalayas (Sikkim), China (Yunnan), Japan.

(3) *C. elongata* — Caucasus, Northern and Central Europe.

(4) *C. laeviculmis* — Northern Japan, Kamchatka, Alaska, British Columbia to California.

(5) *C. Deweyana* — Central Japan, North America (British Columbia to Utah, California and Nova Scotia).

(6) *C. arcta* — North America (British Columbia to New Brunswick, Oregon and New York).

(7) *C. illota* — Western North America (British Columbia to California).

(8) *C. Bolanderi* — Western North America (British Columbia to California).

(9) *C. seorsa* — Eastern North America (Massachusetts to Delaware).

(10) *C. bromoides* — Eastern North America (Nova Scotia to Florida).

(11) *C. Durandii* — Costarica.

(12) *C. Bonplandii* — Columbia, Ecuador, Peru, Bolivia.

The dispersal of *Carex* Sect. *Elongatae* brings before us in the eloquent language of the record, the following, (a) *Carex stellulata* is disconnected between New Zealand, Australia and Europe. This many mean that the two massings of this species are disconnected absolutely (see for *absolute* and *relative* disconnections footnote p. 113), or, conversely, that the break in the range is relative, there being microspecies in the immediate affinity of *C. stellulata* to fill the gap. We do not know which one of these alternative is correct, but, based on the generalities of the case, we may at least speculate within reason (see also discussion of *Potentilla anserina,* p. 147) on the nature of truly primary and secondary species in the light of phytogeography; (b) *Carex stellulata* also follows a circumpacific track to the west leading to Japan and, ultimately,

North America generally. This dispersal is homologous, on the whole, of *Euphrasia* (see p. 29), and it is accordingly absurd to claim that *Euphrasia* is phylogenetically less "primitive" than *Carex*, or the other way around, when it can be shown that both these genera fared together through the ancient maps of the dim past; (c) *Carex remota* could be said to "originate" in Malaysia, the Far or Near East or the Mediterranean at the whim of a conventional phytogeographer. In reality, this species is part of the sectional *genorheitron* released within Malaysia (see *Potentilla anserina* once more, p. 147), and its true ultimate origin remains set there, where the whole of the Cyperaceae had theirs. In order to understand the interrelations between *C. remota* and all other forms in its affinity — which is essential to the phytogeographer and phylogenist alike — we need to know, doubtless, where the whole *genorheitron* originate in the first place; where the *genorheitron* of all the sectional affinity took its true origin. *Carex remota*, as such, is meaningless, unless we intend to study it from the standpoint of narrow taxonomy. Narrow taxonomy, whether by cytology or other means, is not the best that botany can contribute toward the progress of the biological sciences.

Most interesting is the circumstance that, though well, or fairly well, represented in Europe and the Americas, and firmly rooted by Australia and New Zealand, *Carex* sect. *Elongatae* is but sparingly known from Malaysia, which would seem to be the obvious "Landbridge" to connect the extremes, Australia, Europe and America, in the whole track.

As we discuss elsewhere (see p. 356) channels of migration between the Coral Sea and the East Atlantic, showing that these channels bypass the Soenda while skirting it southward, we need not stop to explain in detail why Java and Sumatra are logical stations along such a track as that leading *C. remota* westward. This species significantly left traces of its passage by the Soenda, and it is altogether possible that it reached the Himalayas and Southern China from Java and Sumatra. We know, as a matter of fact, that homologous, though differently oriented, migrations were run by *Schoenoxiphium*, *Primula*, etc.

Not less interesting is the fact that the track of *C. stellulata* is disconnected all the way between New Zealand, Australia and Japan. We will have occasion of discussing migrations of this sort elsewhere in this manual (see p. 438), but may meantime state here that migrations of the kind bypass Malaysia in the east, and keep off Malaysia by travelling in what is now the open Pacific. So much is, of course, the *primary* corollary of the fact, that the shores of the Western Pacific once ran much to the east of their present limits. *Absolute homology thus obtain between the two shores of the Pacific, west and east, in dispersal. Both were far closer to the center of this ocean in the past than they are at present.*

As this is the case, we may consistently anticipate that wide disconnections, involved by the crumbling of these shores, greet us both in the East and West Pacific. This anticipation is fully borne out by the facts. Disconnections between Chile and Mexico are of normal occurrence in the East Pacific; breaks in dispersal between regions by the Coral Sea and Japan, or its immediate vicinity, are quite as common.

Carex sect. *Acutae* subsect. *Cryptocarpae* is distributed (Fig. 52) as follows,

(1)   *C. Darwinii* — Fuegia, Magellania, Southern Chile, Argentina; var. *urolepis:* Patagonia, Chatham Island.

(2)   *C. ternaria* — New Zealand, Auckland Island.

(3)   *C. subdola* — New Zealand.

(4)   *C. Stokesii* — Southeastern Polynesia (Rapa).

(5)   *C. Feani* — Southeastern Polynesia (Marquesas).

(6)   *C. Rechingeri* — Samoa.

(7)   *C. sandwichensis* — Hawaii.

(8)   *C. Middendorfii* — Japan, Sakhalin, Eastern Siberia (Amur).

(9)   *C. tuminensis* — Northern Corea, Sakhalin.

(10)   *C. subspathacea* — Sakhalin; circumpolar, southward to Norway and Hudson Bay.

(11)   *C. Lyngbyei* — Southwestern Manchuria and Central Japan to the Bering Strait; circumpolar along the seacoasts, southward to Norway and Oregon.

(12)   *C. salina* — Arctic and subarctic Eurasia, southward to Scotland, Danemark, California and New England along the seacoasts.

(13)   *C. maritima* — Europe and North America along the seacoasts in Scandinavia, Finland, Labrador, Newfoundland, Nova Scotia, Quebec and New England.

(14)   *C. crinita* — Atlantic North America (Newfoundland to Florida, inland to Minnesota and Texas).

(15)   *C. Schottii* — Vancouver Island to California.

(16)   *C. magnifica* — Alaska to California.

(17)   *C. laciniata* — Oregon to Utah and California.

It will be observed that his subsection is "bipolar," that is, consists of two massings, one in the deep south (*Carex Darwinii* to *C. Rechingeri* in the tabulation), the other in the north including Hawaii.

It is probable that a group somewhere exists in the Cyperaceae which is immediately allied with this subsection, and occupies a position intermediate between the two massings highlighted above. In other words, as, broadly speaking, the Capparidaceae, lie between the boreal and austral zone prevailingly occupied by the Cruciferae, so probably do certain Sedges, allied with Subsect. *Cryptocarpae*, range in the warmer Pacific and adjacent tropical domains.

Whether this probability is verified, or not, we do not know at present, though we incline to the opinion that it might be shown true. That, which interests us at this juncture is other. At this juncture, we are interested in establishing the fact that *Carex* Subsect. *Cryptocarpae* crossed the Pacific south to north in a manner which is not apparent by the modern map. The dispersal of this group is quite as free, "oceanic" we might say, as that of *Korthalsella* (see p. 141). There is a difference between the two, however. *Korthalsella* is a "warm" element that hugs the tropics, *Carex* Subsect. *Cryptocarpae* a "cold" one that lightly vaults the waters of an immense ocean to find, north and south, the ranges it prefers. Obviously, as to *potential of migration*, *Korthalsella* and *Carex* Subsect. *Cryptocarpae* are on a par, and we can understand the dispersal of both, or none at all.

To account for the "bipolarism" of *Carex* Subsect. *Cryptocarpae* we are driven to accept as consistent with the evidence, that the *genorheitron* fared south to north over a crumbled shore in the Pacific, to be exact, the Western Pacific of our maps. We accept this surmise to be consistent with the evidence because, (a) We know from geophysics that an ancient Pacific shore ran much to the east of the present shores of the Old World abutting on this ocean; (b) We have positive evidence from plant-dispersal in general that this shore was indeed much used in migration.

With this before us as an introduction, we may consistently argue further. Both "cold" and "warm" tracks *) run in Africa and the western half of the Indian Ocean. In these quarters, an empiric criterium of great value to decide whether a group is "cold" or "warm" is to be found in the fact whether this group turns up, or does not turn up, in Madagascar and the Mascarenes. A group endemic to these islands is usually a "warm" one. This group, as to ultimate origins, is rooted almost necessarily within the *Gondwanic Triangle* (see p. 349) or its immediate vicinity.

The very same order of facts obtains in the Pacific that rules in the Indian Ocean. "Cold" are here the groups which originate by Tasmania and the south of New Zealand, or Magellania; "warm" on the contrary those which can be traced to Eastern Polynesia, New Caledonia, Eastern Brazil or Bolivia. Though not without exception, these principles are usually correct, at least most useful.

Carex subsect. *Cryptocarpae* is doubtless a "cold" element. It might yet be found in ranges typical of this element in Malaysia (heights in the Northern Philippines, Northeastern Borneo, New Guinea, perhaps Celebes), but might not be found there at all, because a track compatible with "cold" elements ran off the Philippines, in the Western Pacific, all the way at least between New Guinea and Japan.

We repeat that this track is not speculative. It necessarily has place, because of the fact that the shores of the Western Pacific once stretched much farther east than they do at present. We refer to this *as a fact*, because geophysics

---

*) Although we frequently refer to this subject at various points, we may additionally make here a concrete statement. When speaking of "cold" and "warm" tracks and groups, we intend tracks and groups taking care, and made of, plant-life which can ultimately adapt itself to colder climates, or is contrariwise restricted to warm ones exclusively. To elucidate: The Ericaceae are widespread in the tropics, but they are potentially, when not always actually, a "cold" group (see *Erica, Rhododendron, Vaccinium*). Not so the Ochnaceae and Cochlospermaceae, for instance, which are exclusively a "warm" aggregate. *Linum*, as we have seen, is a "cold" genus; *Hugonia* strictly a "warm" one, etc. As shown elsewhere (see p. 352, etc.) the ultimate origin of "cold" and "warm" elements is not the same by the map. To exemplify; out of a single family, Ericaceae, *Erica*, a "cold" or potentially "cold" element originated in the *Afroantarctic Triangle*; *Philippia* a "warm" one in the *Gondwanic Triangle*.

and dispersal are absolutely agreed on the point, nor could dispersal ever make sense if this were not *a fact*. Moreover, without this shore, Fagaceae, Balanopsidaceae and Magnoliaceae never could originate there, where they did. As two at least of these families are potentially "cold" (Fagaceae: see *Quercus;* Magnoliaceae: see *Magnolia*), it seems well established that their *genorheitron* reached the region of ultimate invasion of the Far East and Malaysia (approximately, the vicinity of New Guinea) from a center farther south, to match the austral center out of which *Erica* and *Linum* emerged to begin their treks northward.

Our viewpoint may be attacked, on the ground that is does not agree with the "wandering poles" of the Wegenerian cosmography. Attacks of the kind do not strike us as worthy of attention, for reference to a wrong theory is no reason. Only then may we listen, when the facts we bring to the record shall be explained, wholly or even be in part, by "theories" other than ours. The Wegenerian cosmogony conflicts with the facts, purely and simply (see also p. 512), and as such has no status in science.

It may be pointed out further that fossil records prove this time that Europe and Greenland once had "warm" floras. We will answer, (a) *Erica* and *Linum*, to mention obvious examples, both occur in the tropics, yet both are demonstrably "cold" groups; *Cochlospermum* (or *Amoureuxia*) belongs to a definite "warm" family, yet lived in the Tertiary about Lat. 41° South in Argentine. It is accordingly clear by the facts that we can have very little idea of what is genuinely "warm" and "cold" when we theorize on the strength of preconceived notions. A dweller of Natal will implicitly assume that *Vaccinium* is a "warm" group; an inhabitant of Finland, that this same genus is "cold"; (b) We are inclined to think that the nature of an aggregate, whether "cold" or "warm," can be ultimately gauged only *by its ability to withstand excesses of temperature*, not by the circumstances that this aggregate occurs inside, or/and outside, the geographical tropics. Accordingly, the presence of Lauraceae in the Tertiary of Europe is no evidence that Europe was by then in the tropics; rather, that Europe had a milder climate by then than it has now. Lauraceae occur to day in the temperate to cold United States. Moreover — which strikes us as fundamental — Europe need not have wandered all over the map in the Tertiary — following certain Wegenerian "poles" — to become much colder in the Pleistocene than it was in the Eocene. *The Tertiary was an active period of orogeny, not only, but of material growth of all lands northward.* The latitude of Tibet was doubtless less cold than it is now, when there were neither Himalayas, nor Tibetan Highlands on the spot; (c) A limited measure of "floating" northward of certain lands, like Greenland, is indeed a possibility, but this possibility does not justify theoretical excesses; (d) We discuss elsewhere (see p. 363) how "cold" flora could cross the tropics, and do so by factual examples. These examples do not at all agree with the theories of Wegener or similar authors.

A map to illustrate the dispersal of *Oreolobus, Carex* sect. *Elongatae,*

Carex subsect. *Cryptocarpae* and *Schoenoxiphium* (Fig. 52) reveals nothing with which the reader of this manual is by now unfamiliar. None of these genera ever originated in "Holarctis," and their tracks take no account whatever of distances, modern oceans, etc. Something indeed paradoxical — in a way — greets us at this juncture. Plants like *Carex*, *Oreolobus*, etc. absolutely demand land *to grow*, but these very same plants absolutely disregard lands *to migrate*. How could this ever be, is something that none of the theorists of dispersal ever seems to have seriously asked, even less answered. The answer is, of course, that modern plants migrated upon maps in the past that do not at all answer the maps of the present.

Uncinia consists of close to two dozens species distributed in this manner (Fig. 53),

A) Subg. *Euuncinia*

a') Sect. *Platyandrae*

(1) *U. brevicaulis* — Amsterdam and Saint Paul Islands, Tristan da Cunha, Falkland Islands, Western Patagonia, Chile, (Valvidia, Llanquihue), Juan Fernandez.

(2) *U. Douglasii* — Juan Fernandez.

(3) *U. costata* — Juan Fernandez.

(4) *U. hamata* — Eastern Brazil (Minas Geräes), Argentina (Tucumán), Colombia, Ecuador, Venezuela, Jamaica, Costarica, Guatemala, Mexico (Oaxaca, Veracruz).

(5) *U. phleoides* — Juan Fernandez, Argentina (Sierra Achala de Córdoba, Sierra de la Ventana), Chile to Colombia along the Andes.

(6) *U. multifaria* — Chile (Valdivia, Antuco).

(7) *U. erinacea* — Chile (Guaitecas Islands, Chiloé, Coronel), Argentina (Patagonia).

a'') Sect. *Stenandra*

(8) *U. macrophylla* — Western Patagonia, Southern Chile.

(9) *U. Lechleriana* — Fuegia, Magellania.

(10) *U. macrolepis* — Magellania, New Zealand.

(11) *U. Negeri* — Chile (Valdivia).

(12) *U. fuscovaginata* — New Zealand.

(13) *U. purpurata* — New Zealand.

(14) *U. caespitosa* — New Zealand, Chatham and Stewart Islands.

(15) *U. pedicellata* — New Zealand.

(16) *U. uncinata* — New Zealand, Hawaii.

(17) *U. leptostachys* — New Zealand.

(18) *U. riparia* — Australia (Victoria, New South Wales), Tasmania,

Antipodes Island, Lord Howe Island, New Guinea; Var. *Hookeri:* Campbell and Auckland Islands, New Guinea, Borneo (Mt. Kinabalu); Var. *affinis:* New Zealand, Chatham Island.

(19) *U. rubra* — New Zealand.

(20) *U. rupestris* — Tasmania, New Zealand, Stewart Island, New Guinea, Philippines (Luzon: Benguet, Laguna), Mindanao (Davao).

(21) *U. compacta* — Kerguelen and Amsterdam Islands, Tasmania, New Zealand, Macquarie Island, Australia (Victoria), New Guinea.

(22) *U. tenuis* — Juan Fernandez, Fuegia, Magellania, Chile to Costarica.

(23) *U. tenella* — New Zealand, Tasmania, Australia (Victoria).

(24) *U. filiformis* — New Zealand.

B) Subg. *Pseudocarex*

(25) *U. Kingii* — Fuegia, Magellania.

The dispersal of *Uncinia* is doubtless not the mightiest of the far-travelled cyperaceous hosts, nor is it the most whimsical in a group that plays fast and loose with modern geography. This dispersal contains, on the other hand, elements to invite forthwith the attention of the keen phytogeographer. Though endemic to islands in the Indian Ocean (Amsterdam and Saint-Paul) *east of Africa, Uncinia* does not enter the Dark Continent at all. The trends of its distribution are all *to the west from Africa,* quite as much as the trends of the dispersal of *Coprosma* are all *to the west from South America.* Nothing of this could ever be as it is, had dispersal been effected by casual means of dissemination. Should it be claimed that this is because winds and marine currents drive westward in high southern latitudes, the prompt retort would be that *Schoenoxiphium* hit Africa all right, and that *Coprosma*' ally, *Nertera,* and the Cyperaceae just went, whole or part, the opposite way of *Coprosma.*

The group of *Carex* most closelly allied with *Uncinia* is *Carex* Sect. *Unciniae-formes* (Fig. 53), which ranges in this manner,

A) Subsect. *Aciculares*

(1) *C. trichodes* — Fuegia, Chile.

(2) *C. caduca* — Fuegia.

(3) *C. acicularis* — New Zealand, Tasmania, Australia (Victoria).

(4) *C. vallis-pulchrae* — Falkland Islands, Chile, Argentina (Mendoza).

B) Subsect. *Capitellatae*

(5) *C. rara* — Central Celebes, Philippines, Luzon (Benguet), Northern Borneo, Western Java, Sumatra (Gajolands), India (Assam, Nilghiri Mts.),

Ceylon; Var. *biwensis:* China (Hupeh), Japan; Subsp. *capillacea:* Australia (New South Wales), India (Sikkim to Bhutan), China, Japan, Sakhalin, Manchuria; var. *nana:* Northern Japan (Hokkaido), Sakhalin.

(6) *C. fulta* — Japan.
(7) *C. uda* — Japan, Eastern Siberia (Ussuri, Amur).
(8) *C. litorhyncha* — China (Yunnan).
(9) *C. Oenoei* — Corea, Northern Japan, Eastern Siberia (Ussuri).
(10) *C. capitellata* — Northern Asia Minor.

C) Subsect. *Callistachys*

(11) *C. pyrenaica* — New Zealand, Japan, Siberia at the Bering Strait to Alaska, British Columbia to Washington, Oregon, Idaho, Utah, Colorado; Northwestern Iran (Persia), Caucasus, Asia Minor; Europe: Transsylvania to the mountains of Northern Spain; Var. *cephalotes:* Australia (Victoria: Mt. Kosciusko), New Zealand; Var. *altior:* Japan; Var. *articulata:* Europe (Transsylvania).

(12) *C. nigricans* — Alaska to Colorado and California.

D) Subsect. *Macrostylae*

(13) *C. pulicaris* — Northern and Central Europe southward to Central Russia, Transsylvania, Northern Jugoslavia, the Alps and Pyrenees.

(14) *C. macrostyla* — Northern Spain.
(15) *C. peregrina* — Madera.

E) Subsect. *Pauciflorae*

(16) *C. microglochin* — Northwestern Himalayas, Central Asia (Altai Mts., Dauria), China (Szechuan), Kuriles, Western Canada to Colorado, Greenland; Northern Europe to Iceland, southward to Lithuania, the Carpathians and Alps; Var. *oligantha:* Falkland Islands, Fuegia, Patagonia.

(17) *C. Lyonii* — Western Canada.

(18) *C. pauciflora* — Sakhalin to Siberia, Kamchatka, Alaska, Newfoundland, Michigan, Washington; Persia; Northern Europe southward to England, Central France, the Alps and Sudeten, Transsylvania, Poland.

(19) *C. parva* — Himalayas to Eastern Tibet and China (Yunnan), Central Asia from the Turkestan to the Alatau Mts.

Uncinia and *Carex* sect. *Unciniaeformes* are a single affinity from the standpoint of morphology and phylogeny alike. They, obviously, belong to the same *genorheitron.* They, too, mass in the Pacific, and migrate from this ocean onward to the far corners of the earth. Unlike *Carex* sect. *Unciniaeformes,* however, *Uncinia* retains a toehold in the Southern Atlantic (Tristan da Cunha) and Indian Ocean (Amsterdam and Saint Paul Islands).

As we know, these three islands belong to African gate of angiospermy, and plant-life endemic to them can be shown to reach Africa without seemingly difficulty, witness *Phylica*. Africa can also be reached from the Kerguelen Islands (see *Anisothecium Hookeri*, p. 509), south of Amsterdam and Saint Paul Islands. A certain "something", on the other hand, is active as regards *Uncinia* (and *Carex* sect. *Unciniaeformes* by unavoidable implication) that bars these two groups from reaching the Dark Continent. They stream mightily westward to the Pacific, and do so with such a wealth of forms that it seems impossible to admit that the track running to the Pacific from Amsterdam, Saint Paul and Tristan da Cunha is a thin trickle out of a large *genorheitron*. It is the whole *genorheitron*, on the contrary, which avoids Africa and shifts its course westward in a manner to defy modern geography *).

The very same "something", alluded to above, which acted as a bar against *Uncinia* and *Carex* sect. *Unciniaeformes*, relented, we may well say, as regards *Schoenoxiphium* and its immediate allies; *Carex* subg. *Indocarex* and *Cobresia*. This "something", then, *was selective* behaving differently in the case of this or that group.

Considering that this "something" has nothing to do with modern sea-currents, winds and the like, we may not be beyond sober reason if we understand it as features of lands and seas that are no longer on our maps. Sober reason dictates our decision to believe the most probable. The cyperaceous *genorheitron* is manifestly so ancient, and its migrations so diversified and so overwhelmingly large that it should be absurd to suppose it began to disperse in a late age of angiospermy.

In our opinion, it is most probably, sure we would venture to say, that an ancient "antarctic" shore connected every land in the deep south of our maps; not only, but that the lands, now vanished, but once extant between the approaches of the Kerguelen Islands and Magellania, are *the ultimate hub of angiospermous dispersal*. It is from this center that all the angiospermous gates can be shown to have been fed the *genorheitra* of the seedplants **).

If our opinion is at all correct — the like we believe it is, not only on the ground of the immediate discussion here conducted, but on the strength of facts and conclusions reached throughout the pages of this manual — we ought to find some support for it in the modern dispersal throughout the southermost insular outposts of our maps. We believe, as a matter of fact, that this support can be found.

South Georgia, an island located between Fuegia and Graham Land, is hardly more than a high mountain range, 95 miles long and 20 miles wide,

---

*) It seems useless to discuss such a case as this in the light of what sea-currents, drift, etc., could and might do. The "Ineffectiveness of alien seed dispersal in South Africa", for example, is well established in the pages of MUIR's work (Union of South Africa, Bot. Survey Mem. 16: 7. 1937).

**) This statement is to be understood with reference to the maps of our own world only. It is not intended to stand in the absolute.

still fairly heavily glaciated. Its flora (3) consists of but 18 vascular plants, of which 17 are further endemic to Fuegia, not unexpectedly, but 13 range on the contrary to Kerguelen and Macquarie Islands and the lesser insular outposts south of New Zealand. The Kerguelen Islands harbor a flora of about 30 vascular plants, of which 6 are endemic, representing well marked genera in two cases, *Pringlea* and *Lyallia*. With one exception (*Cotula plumosa* of the Compositae), the 24 non-endemic plants of the Kerguelen Islands are found also in Fuegia. Twelve of them, moreover, occur also in Macquarie Island.

The minor islands by New Zealand (Stewart, Auckland, Campbell, Antipodes, The Snares, etc.) have a flora of about 194 species of which fully 53 are described as endemic. According to an oft quoted sources (3) this flora consists of three elements, (a) "Ancient endemic" represented by the unique araliaceous genus *Stilbocarpa* *), *Pleurophyllum*, *Celmisia* Sect. *Ionopsis;* (b) "Fuegian" composed of *Colobanthus, Abrotanella, Phyllacne*, etc.; (c) "Recent" which supposedly emigrated to the islands under discussion from New Zealand **).

It will readily be seen that the flora of the outposts in the deep south just mentioned, South Georgia, Kerguelen Islands, etc., is a strange mixture of unique endemics and wide-ranging species. We interpret the origin of this flora in the following manner: These islands are fragments of lands once far larger and far richer in diversified plant-life. This flora was depauperated by a long geologic history of emersions, submersions and the like, including peculiar climatic conditions, until only two parts of it survived to an extent, (a) "Weedy" plants, ubiquists, that is, capable of ready adaptation to changing conditions, at home even in transitory seres; (b) Ancient endemisms such as *Pringlea, Lyallia, Stilbocarpa*, etc., of which but few managed to survive after the destruction of the original climaxes.

We believe that, when not correct in every detail, this interpretation of the dualism in the flora under discussion has much in its favor. There is no doubt, for example, that the plant-life of the Kerguelen Islands lost in time part of its components, because fossils of vanished *Araucaria* and dicotyledonous forms (4) are known from these remote outposts. It should be ludicrous to insist

---

*) We have seen material neither of *Stilbocarpa* nor of *Myodocarpus*, the New Caledonian endemic which, according to BAUMANN (in Ber. Schweiz. Bot. Gesellsch. 56: 13. 1946) is exactly intermediate between the Umbelliferae and Araliaceae. We regret that the vastness of these two families, and the lack of modern monographic work to include all their subdivisions, made it impossible for us to review them in the pages of this manual. We feel that these two groups may profitably form the subject of an important work in phylogeny, morphology and dispersal, and hope to see in time a comparative investigation made of *Stilbocarpa* and *Myodocarpus*.

**) It does not seem necessary to discuss here the ideas of the author referred to. Terms such as "Ancient endemic," "Fuegian" and "Recent" mean very little, when indeed anything at all. All these elements are ancient in about the same measure, so that distinctions effected between "Ancient endemic" and "Recent" elements in the premises are nugatory. The adjective "Fuegian" has at best descriptive significance. The assumption that the "Recent" element migrated from New Zealand might be true to the extent of some occasional "warm" element (see, for instance, *Melicytus* and *Hymenanthera*, p. 181) originally delivered to New Zealand, and the southern islands from the north, but is unmitigatedly false as regards forms of genuine "antarctic" distribution, like Restionaceae for example. In brief, adjectives of this kind confuse rather than elucidate matters. Unfortunately, they are the staple upon which all too much phytogeographical work thrives nowadays.

that these dead plants reached the Kerguelen Islands in ages past by "float-sam-jetsam methods," or the like, when the *whole of dispersal,* past and present, flatly stands against this notion.

It is moreover manifest that the same vicissitudes — and their effects — here postulated to account for the flora of the southern islands in play readily account for the "weedy" flora and the few striking endemics of the Galápagos Islands (see also p. 168). Like the Kerguelens, the Galápagos lost, by becoming eventually insular, masses of their autochtonous flora to the gain of "weedy" vegetation. As the Kerguelens were part, together with Fuegia, Macquarie Island, etc., of an ancient antarctic continent, the Kerguelens and these islands retained bonds of common "weedy" flora; in the same manner, the Galápagos, being once part of the Americas, kept in addition to a few of their own endemics a large contingent of "weedy" forms common to themselves and the adjacent New World. In sum, we believe that the very same factors conspired to modify the flora *of all islands,* cutting down in every case the highly "endemic" element by reason of depriving this element, wholly or in part, of suitable habitats, thereby increasing as a corollary the "weedy" forms. We may indeed go further, and conclude that these very same factors, were responsible for the destruction of the ancient floras replaced in time by the Angiosperms. Almost the whole of the earth was at one time or other — beginning with the Late Jurassic — profoundly disturbed by geological events influencing its temperature, immediate ecology, etc. etc. Thus, very nearly the whole of the earth witnessed during the Late Jurassic, Cretaceous and Tertiary changes that one or more times, destroyed the old climaxes, always favoring the dispersal of forms of ready adaptability and "weedy" behavior. If we study the angiospermous *genorheitra* we cannot fail to detect in them precisely those tendencies that connotate forms of the kind, and to conclude that the majority of these *genorheitra* were mangroves or near-mangroves at first. Proof of this (see p. 476), is in the fact that the Cycadaceae survived, (a) In geologically most ancient, usually narrow ranges (e.g., eastern Mexico, sections of Cuba, Natal, Queensland); (b) Contrariwise, in wide sectors, but then only when the forms in play (e.g. *Cycas,* a classic instance) had the nature of mangroves or near-mangroves *).

In conclusion, the history of vegetation throughout is molded by the same factors. Climaxes are in every cases established, which call for comparatively slow, well ordered, methods of natural reposition. These climaxes are in time destructively attacked by various changes stemming from geological revolutions. Seres then obtain, in which dominate forms that do not rely upon slow methods of natural reposition, and are highly plastic as to ecological requirements. When the geological revolutions spend themselves, the flora tends to become static, and to establish once more set climaxes. This lasts until a new revolution is on, and so on without end.

---

*) See note p. 189, for a definition of these terms.

The reader may ask of himself, what would happen to the modern Angiosperms if profound geological changes were to begin right now, such as may be fit to recast the face of the earth entirely in the course of the next hundred millions years. He may conclude, we assume, that all the Angiosperms now living wedded to peculiar habitats would be in danger of dying out, while the "weeds" among them would carry the day, not only, but find ample opportunity for migrating and evolving further. *There is no reason to believe that what is common sense today could be nonsense in the geological past.*

Much has been made in the literature of phytogeography of the "Antarctic" floras and their "problems." We think that much of this is artificial, and answers at bottom purely academic preoccupations. As we show throughout the pages of this manual, it is very difficult to draw the line between what is "antarctic" and what is "non-antarctic" in the history of angiospermous origins, for all the angiospermous *genorheitra* belong in the last resort to the southern hemisphere, and include in certain cases (e.g., Ericaceae, Linaceae) both "antarctic" and "non-antarctic" groups.

That there was a continuous "antarctic" shore running in the dim geological past to latitudes much higher than the present is so thoroughly well established that denials of the fact are not to be honored with a retort. If this geological shore could be freely used by hosts of angiosperms now thinly represented in, or absent altogether from, the deep south of our maps, this very same shore could all the more easily be used by forms still endemic today to this same deep south. In short, the distinction between "antarctic" and "non-antarctic" plant-worlds wears altogether thin in the end, so that the "problems" of the "antarctic" floras — quite like those of their "holarctic" counterparts (see, furthermore, p. 357) — are but part and parcel of the tale, *how all plants migrated everywhere.* The theory that a nebulous "Glacial Epoch" killed off very nearly all the ancient vegetation of the antarctic islands is shown to be false by the comparatively large plant-world still endemic to South Georgia, indeed a sizeable flora, if we consider that it thrives in what is now hardly better and a glaciated mountain straight out of the ocean. "Glaciations" of unspecified age that crash through the landscape and wipe out everything in sight are the stock-in-trade of imaginative academic phytogeography, not the sober mainstay of factual, competent investigations of ecology and phytogeography.

Finally, the strictly "antarctic" floras of the present tell nothing that supposedly "non-antarctic" plants, fossil and living alike, fail to reveal. The history of the flora in the islands of the deep south agrees in every essential with that of the islands to the immediate north. At a glance, there should be no reason why certain small insular domains, like the Mascarenes and New Caledonia, should loom so large in the tale of dispersal. A beginning at least of explanation readily forthcomes, however, when we open our eyes to the realization that these islands owe their preeminency to their standing right to the north, or in the vicinity, of centers wherefrom the angiospermous *genorheitra* issued to begin their dispersal on the maps of the present. The position of the Mascarenes and New Caledonia in dispersal, then, is a corollary of that of the Kerguelens, and the other way around. If there had never been an Antarctica much larger than the present one*); if there had never been a continental mass in the Indian Ocean; if the prototypes of the Angiosperms had not originated by this Antarctica and continental mass; the Mascarenes and New Caledonia could not have achieved the peculiar position which is theirs in phytogeography, nor could the Kerguelen Islands exhibit *both* their fossils and their striking endemisms. Our knowledge of dispersal would be beyond comparison poorer, on the whole, if the few square miles of land represented by insular outposts such as the Mascarenes, New Caledonia, Juan Fernandez, Tristan da Cunha, Chatham, Rapa, etc., had followed under-sea the ancient continental masses of which they were a part.

It is hard to believe that so relevant an array of facts and circumstances is lightly cast aside by authors who, in the very end, cannot give a single valid reason because they dislike "Land-bridges," and like, on the contrary, "Holarctis." One of these authors, when referring to the Kerguelen Islands (5), comments that Darwin suggested they had been originally stocked by seed brought in by icebergs, but Hooker believed, on the contrary, they were remnants of formerly much larger landmasses. This author eventually agrees that, perhaps, Hooker might be in the right. Taking this declaration for the candid acknowledgement it evidently is, we have the right to anticipate that the author in question, admitting that the "Kerguelen problem" cannot be solved by icebergs, ducks, fishes etc., etc. carrying seeds all over the map, is at last ready to see the light and to approach this problem the way Hooker did. This anticipation comes to nothing, nevertheless. The author in question ends believing that the "Kerguelen problem" is an "exception," and with this belief folds up the argument.

To wind up this chapter, let us return to the point from which we started, the Gajolands of Northern Sumatra. Although not wholly free from volcanism of later date, the Gajolands contain a core of very ancient lands, and for this very reason rank, in a sense, as the counterpart of the relic-islands strewn in the deep south of the modern maps. All these dots on the map — whether hemmed in by more recent land-accretions, therefore conventionally "continental," or left free by themselves within an immensity of waters, therefore "oceanic" — keep to this day the record of ages long ago bygone. It is here that the clock of time stood still. We say this in the understanding that, had the Gajolands been thoroughly disturbed by later geological events, they still need not have lost their ancient flora altogether. Java was so disturbed, yet also in Java occur significant relics, such as *Alisma natans*.

It has been contended (6) that the flora of the Malaysian volcanoes is young, because these volcanoes appear to be geologically recent. This contention

---

*) Be it remembered that the distinction between "Antarctica" and "Gondwana" may wear quite thin. See also p. 361.

seems unanswerable. It has meet, however, the pointed and well taken observation (7) that this is not necessarily true, because supposed young mountains many be in reality old heights rejuvenated. It is indeed probable (8) that the alpine flora of the Hawaiian volcanoes antedates the coming of age of these very same fiery heights.

II. Cochlospermaceae and Oxalidaceae

Cochlospermaceae and *Biophytum*, a genus of the Oxalidaceae will give us ample means to test several aspects of dispersal in the Americas, Africa and the Far East jointly. We will migrate from one continent to the other at ease, as always, and stop to consider local problems of distribution of a general interest which will come under more stringent discussion in coming pages.

The Cochlospermaceae (9, 10) range as follows,

A) *Cochlospermum*

a) Sect. *Eucochlospermum*

(1)   *C. angolense* — Angola.
(2)   *C. intermedium* — Cameroon.
(3)   *C. niloticum* — Equatorial Africa (Kordofan).
(4)   *C. tinctorium* — Sudan.
(5)   *C. Planchonii* — Cameroon, Togo, Nigeria.
(6)   *C. regium* — Brazil, Paraguay.
(7)   *C. Codinae* — Brazil (Pará).
(8)   *C. Gossypium* — India, Siam, Cambodja, Lesser Soenda Islands (Bali).
(9)   *C. Gregorii* — Northern Australia.
(10) *C. heteroneurum* — Northwestern Australia.
(11) *C. Gillivraei* — Northern Australia and islands in the Gulf of Carpentaria, Southern New Guinea.

b) Sect. *Diporandra*

(12) *C. tetraporum* — Bolivia.
(13) *C. orinocense* — Amazonian Brazil and Venezuela.
(14) *C. Wentii* — Guiana.
(15) *C. Parkeri* — Guiana.
(16) *C. pavieaefolium* — Guiana.

B) *Amoureuxia*

(1) *A. unipora* — Bolivia (Santa Cruz).
(2) *A. colombiana* — Colombia (Tolima).
(3) *A. Schiedeana* — Mexico (Veracruz).
(4) *A. Gonzalesii* — Mexico (Sinaloa).
(5) *A. malvaefolia* — Mexico (Chihuahua).
(6) *A. palmatifolia* — Mexico (Sonora), United States (Arizona).
(7) *A. Schiedeana* — Mexico (Veracruz).
(8) *A. Wrightii* — Mexico (Veracruz, Tamaulipas, Coahuila, Nuevo León), United States (Arizona, Southwestern Texas).

C) *Sphaerosepalum*

(1) *S. alternifolium* — Madagascar.
(2) *S. coriaceum* — Southeastern Madagascar.
(3) *S. Louvelii* — Northeastern Madagascar.
(4) *S. madagascariense* — Northwestern Madagascar.

The main outline of this dispersal (Fig. 54) is of the simplest. The migration is keyed to the usual center in the modern Indian Ocean, and streams out to the New World and Far East most conventionally, so that we need not discuss it at length. The reader has already seen abundance of homologous distribution. Quite interesting, on the contrary, are certain phases of the phylogeny and dispersal of the various genera making up the family.

Whether *Sphaerosepalum* still retains the characters if an archetype we do not know. All we are sure of is that a region near Madagascar on modern map is the genorheitral hub of the Flacourtiaceae (see p. 318), which are the Cochlospermaceae's immediate allies. It is reasonable to suppose because of this that *Sphaerosepalum* retains archaic characters. Doubtless, this genus is geographically nearest the center of origin of the *genorheitron*. It might have advanced far on evolutionary lines of its own, and no longer be factually close to the ancestors, however, which we may not forget when critically appraising primitiveness in function of geographic situation.

Neither *Cochlospermum* nor *Amoureuxia* occur in Madagascar. The center of gravity of the former lies in Western Africa, and the dispersal is effected from this center eastward along a line, Sudan—India—Siam—Cambodja—Lesser Soenda. This line, typically followed by "Moonson" plants, is parallel of another, Cameroon—Afghanistan—India—Siam—Southern Indochina—Lesser Soenda Islands, which we already know. In brief, a wide belt of geologic "Moonson" climate and ecology leads from Western Africa straight to Northern Australia and the Philippines. This line, as we have also suggested elsewhere, gives indication of shallow grounds and shores along its course, and highlights one of the major phytogeographic ranges of the past, whether Late

Cretaceous or very early Tertiary we may not say. It seems probable, nevertheless, that this zone is anterior to the Himalayan uplift, and may have dried out precisely on its account. It is difficult to visualize anything but marine conditions in part at least of its course, if we consider that mangroves are stranded in Afghanistan (see p. 418). Let us notice that it is along this zone that occur spectacular disconnections immediately involving Western Africa and the Far East, not only, but on occasion the whole range between the West Indies and Burma (see *Cuscuta*, p. 220). Disconnections of this nature have been noticed by authors earlier than ourselves (see p. 38), but it does not seem that their true nature and significance has ever been fully understood.

It is clear, at any rate, that Western Africa is one of the major centers of secondary origins for a world of vegetation (e.g., Connaraceae, Leguminosae, Cochlospermaceae) of "Moonson" and "Near-mangrove" status. It is from West Africa that the Amazonian valley received a very large part of its original angiospermous world. Because of this, the region of Loreto in Peru, and the Amazonian Hylaea generally, have phytogeographic bonds with Siam, Indochina, "Moonson" Malaysia and Northern Australia. In a sweeping generalization, it is possible to conceive of a single continuous ecologic and climatic belt running either in the Late Cretaceous or very Early Tertiary all the way from the modern Amazonian Hylea to Northern Australia. The vacuum left in the records of vegetation by the modern Sahara and the Arabian wastes detracts much from our knowledge, but a great deal can be learned and understood, if we give ourselves a correct account of the significance of Western Africa in dispersal, and its interrelations with the Far East and the New World.

West of Western Africa, *Cochlospermum* is found in Pará (Brazil), which is precisely where it ought to be with regard both to its ecological requirements and the course of the track. Naturally, we find it further in all Amazonian Brazil and Venezuela, in the Guianas, Bolivia and Paraguay, which all could easily be reached by the usual channel connecting Western Africa with a base-line of ultimate American approach strewn between the Caribbeans and Pernambuco. This base-line, as noticed already, is a very important secondary center of origin of the Cactaceae in the immediate affinity of *Cereus*.

As regards its dispersal in the New World, then, *Cochlospermum* may properly be defined as a classic example of "Amazonian element." It answers this connotation both as to track and ecology, if only we understand that *Cochlospermum* is seemingly much less a dweller of the riverine "Gallery-Forest" than of the dry hinterland away from the river, and the banks which the river periodically inundates.

The dispersal of *Amoureuxia* is noteworthy on two counts, (a) This genus is unknown to date in the West Indies, but teems in Mexico and part of the adjacent United States (Arizona, Southwestern Texas); (b) It is not seen in Venezuela, coastal Colombia or Central America.

These factors are seemingly at fundamental odds. We interpret them, however, in the light of trusted generalities, as follows, (a) The track is actually disconnected between Western Colombia and Southeastern Mexico (Veracruz). Disconnections of this, and similar nature, are, as we know, wholly standard; (b) Mexico was not entered by a channel running through the modern West Indies, but by a track running west of Central America in what is now the open Pacific. This track, too, is standard. Homologous dispersal is, for example, in *Schoenocaulon* (see p. 375).

Our surmises might be altered by new reports, but, believing the extant record of the dispersal of *Amoureuxia* to be representative, we are under the impression that, even so, they will not be radically upset. In our opinion, and as logical corollary of the course of the probable migration, we believe that *Amoureuxia* originated out of the generalized *genorheitron* of the Cochlospermaceae somewhere in the vicinity of modern Bolivia. We might recall that this vicinity is precisely the point invaded by the Piperaceae (see p. 103) streaming in from the Pacific. It is consistent with the fact to think that, as *Peperomia* could come into the New World at this point, so could *Amoureuxia* leave the New World by this point to enter a now crumbled range in the Eastern Pacific ultimately to reach Mexico. In brief, Bolivia is a capital hinge of dispersal both ways.

Let us observe that we can follow the whole of the cochlospermaceous migrations very nearly overland, and reach Bolivia from the Indian Ocean by channels of dispersal so repetitious that doubts as to the existence of these channels are impossible. Contrariwise, we cannot follow overland the migrations of plants that came to Bolivia, or similar ranges, from the Pacific. Landgeography which helps us in the former case, hinders us in the latter. However, the sum total of the dispersals moving from the Pacific to Bolivia, or reaching Bolivia from the direction of the Atlantic, tells us in no uncertain terms that tracks now far at sea in the Pacific are definitely as factual as tracks now wedded to lands such as the whole of Brazil and Africa. *Land and sea are consequently synonymous insofar as dispersal is concerned, which is to say that — as the modern map reads — we need not trust the former more than the latter.*

In the last analysis, we visualize for the Cochlospermaceae the following centers of secondary origin and distribution, (a) *Western Africa* — This center is immediately connected with the Far East and Australia by a belt such as we have previously described. This connection accounts for the fact that *Cochlospermum angolense* (Angola) is close to *C. Gillivraei* (Northern Australia and New Guinea). Fernando Po in the Gulf of Guinea is thus next door to Thursday Island in the Strait of Torres. We may add that Angola and Northern Australia can further be connected by a channel, Cameroon—Tanganyka—Java, this channel and the one previously described being alternative much sooner than exclusive, because plants world follow the one or the other mostly depending upon ecological requirements. *Cochlospermum* came into its own in this center, and leaving this center further migrated to the Amazonian valley; (b) *Bolivia (and Western Colombia)*. Here originated *Amoureuxia*. This center stands into contact with Mexico mostly by tracks now running in the open Pacific of our maps. The disconnection Colombia—Mexico in the dispersal of *Amoureuxia* is homologous of the break Chile—California in other groups; (c) *Madagascar—Sphaerosepalum* is rooted in this center that, insofar as the Cochlospermaceae, would seem to have been quiescent.

It is manifest that the cochlospermaceous *genorheitron* widely migrated among these three centers, and possibly the Far East in the first stage of angiospermous dispersal, but speciated most intensively in the Late Cretaceous and, perhaps, the Early Tertiary.

There remain for brief consideration two secondary aspects of the cochlospermaceous dispersal.

A petrifact, *Cochlospermum praevitifolium*, was discovered in a seemingly Tertiary flora in the region of Río Pichileufú (Argentina) at about 41° Lat. South *) farther south, that is, than the southernmost limit of the living range. This is mildly interesting, of course, but reveals nothing apt to change our understanding of the living distribution of the Cochlospermaceae (and Eleocarpaceae as well). The petrifact in question merely shows that prior to the Mid-Tertiary *Cochlospermum* (or more likely still, *Amoureuxia*) had ranged south to the general vicinity of the Neuquén, Argentina, there acquiring an extension of distribution, which climatic changes most likely attending the further uplift of the Andes latter blotted out.

In our opinion, petrifacts of this nature and age merely confirm the principle that the dispersal of all fossils later than the Late Jurassic, at least, is illustrated by the living dispersal of the Angiosperms, not the other way around. As regards the Angiosperms, at least, palaeobotany is a mere appendage of phytogeography within the limits stated.

It has been debated (11), whether *Cochlospermum Gossypium* is truly native to Bali, and the conclusions has been reached in a first time that it is so. In a second time **), this conclusions was controverted. We are short of pertinent data on the score, but is seems to us that a plant which, like *C. Gossypium*, runs the typical "Monsoon" track India—Siam—Cambodja, and is demonstrably allied with a group that came into flower in Western Africa, may be truly native to Bali. Whatever be the case in detail, if this species is not native to Bali it might well occur there, which closes the issues insofar as we are concerned.

Biophytum is an oxalidaceous genus of about 50 species none of which is endemic to the "antarctic" floras of Australia, Chile, New Zealand, Polynesia

---

*) It seems to us evident that BERRY, the author of this "species" (Geol. Soc. Amer., Spec. Paps. No. 12, 1938) compounded it with leaves of *Cochlospermum* or *Amoureuxia*, and the fruit of a plant that impresses us as *Sloanea* (Eleocarpaceae). *Sloanea* could very well be in this range. We regret we cannot deal here with the distribution of the Eleocarpaceae.

**) According to a verbal communication by Dr. C. G. G. J. VAN STEENIS (December, 1946), *C. Gossypium* is introduced to Bali.

and the Cape. So much does not mean, of course, that the Oxalidaceae or their close allies Geraniaceae are late comers on the angiospermous stage, witness a few species of

A) *Oxalis* sect. *Acetosellae* (12)

(1)   *O. magellanica* (including *O. lactea*) — Strait of Magellan, Western Patagonia, Chile (Rio Manso), Southern Bolivia; New Zealand, Tasmania, Australia (Victoria) *).

(2)   *O. acetosella* — Northwestern Himalayas to Bhutan, Central Asia (Altai Mts.); Eastern China, Corea, Sakhalin, Manchuria, Eastern Siberia, Japan, all of Europe, to the exception of the high north, to the Mediterranean; Saskatchewan to Nova Scotia and North Carolina.

(3)   *O. oregana* — Western United States (Washington to California).

(4)   *O. Griffithii* — Central and Eastern Himalayas, China (Yunnan, Shensi), Formosa, Japan.

This dispersal has evidently much in common with that of *Carex*, with Magellania, Tasmania and the Altai Mountains all in focus. Distributions of this nature were not laid down in the Tertiary, a paltry yesterday in the history of plant-life.

The Geraniaceae (13) are quite as old, as revealed by the following examples,

A) Tribe Dirachmeae

(1)   *Dirachma* — 1 species in Socotra.

B) Tribe Vivianieae

(2)   *Viviania* — About 30 species in Southern Brazil, Uruguay, Argentina, Chile;

and representative species of *Pelargonium* sect. *Peristera*,

(1)   *P. iocastum* — Southwestern Cape.

(2)   *P. procumbens* — Cape (Riversdale, Knysna, Port Elizabeth), Natal.

(3)   *P. althaeoides* — South Africa (Karroo, Namaqualand).

(4)   *P. mossambicense* — Mozambique.

(5)   *P. inodorum* — New Zealand, Tasmania, New South Wales, West Australia.

---

*) The complex centering around *O. magellanica* should be carefully investigated. The dispersal is manifestly "antarctic", because Magellania and Tasmania are both immediately invested. However, the American range reaches all the way to Southern Bolivia, which is oftentimes the point of inception of tracks seemingly divorced from any connection with Magellania and Tasmania, therefore not "antarctic." It is by a critical study of forms such as this species that we may hope to make light upon issues of detail which cannot be fully discussed here.

(6) *P. grossularioides* — Cape (Karroo, Riversdale, Port Elizabeth); "Introduced" in Tristan da Cunha and Southern India (Nilghiri Mts.); var. *madagascariense:* Madagascar.

Dirachmeae and Vivianeae each consist of a lone genus and it is obvious that they both go back to the same genorheitral node in origin. A tie between Chile and Socotra is nothing extraordinary. *Thamnosma* showed us that California and Socotra are immediately bound in dispersal, and ranges or disconnections calling into play California and Chile are anything but unusual. In brief California, Chile and Socotra are within easy distributional reach one of the other. Further considering that Chile is also in contact with the Mascarenes, we may draw on the modern map a pertinent quadrangle California—Chile—Mascarenes—Socotra, which leaves nothing to imagination from the standpoint of dispersal. It should be useless to discuss whether these localities were connected by "Landbridges" a few miles wide in the Miocene, because the map speaks eloquently enough to dismiss arguments of the sort as futile.

Pelargonium sect. *Peristera* binds the Cape immediately with classic "Antarctic" centers such as New Zealand, Tasmania, New South Wales and West Australia. The tie is so obviously in the deep south as the modern map reads that it is most likely it was forged in high southern latitudes throughout. However, the tracks of *Pelargonium grossularioides* suggest at least the alternative of another channel of migration starting from a point near modern Madagascar to the Cape (the Karroo is an interesting locality), the South Indian Nilghiris, and northern West Australia. The dispersal would then have taken place from West Australia to New South Wales, Tasmania and New Zealand. The authority whom we follow dismisses the records of *P. grossularioides* in Tristan da Cunha and Southern India (Nilghiris) ad due to introduction. We think otherwise. A range, Madagascar (here a variety, *madagascariense*), Cape, Tristan da Cunha, Southern India is perfectly consistent (see for essentials, *Phylica*, for instance). In brief, whatever be the details, and the possible alternatives of the purely Australian dispersal, it is clear that *Pelargonium* sect. *Peristera* is a classic instance of direct tie between the Cape and New Zealand. Ties of the same order turn up in the distribution of the Proteaceae, which we regret not to have the means of studying in detail in these pages (see, however, notes p. 359, 361).

Biophytum is distributed as follows,

A) Sect. *Prolifera*

(1) *B. molle* — Madagascar.
(2) *B. Forsythii* — Madagascar.
(3) *B. Perrieri* — Madagascar.
(4) *B. Commersonii* — Madagascar.
(5) *B. robustum* — Madagascar.

(6) *B. macropodum* — Madagascar.
(7) *B. myriophyllum* — Madagascar
(8) *B. aeschynomenifolium* — Madagascar.
(9) *B. castum* — Brazil (Ceará).
(10) *B. proliferum* — Ceylon.
(11) *B. nudum* — Ceylon, Southern India (Pulney Hills).
(12) *B. intermedium* — Ceylon, Southern India (Pulney Hills).
(13) *B. polyphyllum* — Southern India (Nilghiri Mts.).
(14) *B. Thorelianum* — Siam, Indochina (Laos, Cambodja, Cochinchina), Southern China.
(15) *B. Esquirolii* — Southern China (Kweichow).

B) Sect. *Grandifoliolata*

(16) *B. somnians* — Brazil (Rio Negro).
(17) *B. amazonicum* — Peru (Loreto: Iquitos).
(18) *B. Foxii* — Peru.

C) Sect. *Orbicularia*

(19) *B. macrorrhizum* — Africa (Rhodesia).
(20) *B. crassipes* — Africa (Rhodesia, Tanganyka).
(21) *B. Kessneri* — Congo.
(22) *B. Ringoeti* — Congo.
(23) *B. Zenkeri* — Congo, Cameroon, Gabun.
(24) *B. sessile* — Tropical and subtropical Africa throughout, Madagascar, India from the south to the Eastern Himalayas, Siam, Indochina (Cochinchina), Java, Philippines (Luzon: Bontoc, Benguet, Lepanto; Panay, Mindanao), Queensland, New Guinea.

D) Sect. *Sensitiva*

(25) *B. mimosella* — Madagascar.
(26) *B. Hildebrandtii* — Madagascar.
(27) *B. albizzioides* — Madagascar, Comores.
(28) *B. abyssinicum* — Angola, Nyasaland, Tanganyka, Ethiopia, Eritrea.
(29) *B. Reinwardtii* — Ceylon, India, Southern China.
(30) *B. madurense* — India, Malaysia (Island of Madoera).
(31) *B. adianthoides* — Tenasserim, Indochina (Laos), Malacca.
(32) *B. albiflorum* — New Guinea.
(33) *B. sensitivum* — Mascarenes, Nyasaland, Tanganyka, Eastern Congo,

Sierra Leone, Southern India to the Himalayas, Tenasserim, Andaman Islands, Burma, Northern Siam, Indochina (Cambodja), Southern China, Formosa, Philippines (Luzon to Mindanao, Sulu Archipelago), Java, Timor, Ternate, Amboina.

E) Sect. *Dendroidea*

(34) *B. Talbotii* — Cameroon, Liberia, Southern Nigeria.

(35) *B. ferrugineum* — Bolivia (La Paz).

(36) *B. bolivianum* — Bolivia (La Paz).

(37) *B. mapirense* — Bolivia (La Paz).

(38) *B. globuliflorum* — Bolivia (La Paz), Eastern Peru.

(39) *B. peruvianum* — Bolivia (La Paz), Peru (Huánuco, Loreto), Brazil (Acre).

(40) *B. Boussingaultii* — Ecuador, Colombia (Cundinamarca, Antioquia, Madgalena).

(41) *B. lindsaefolium* — Colombia.

(42) *B. gracile* — Colombia.

(43) *B. antioquiense* — Colombia (Antioquia).

(44) *B. Tessmannii* — Peru (Loreto: Iquitos).

(45) *B. dormiens* — Colombia (Caquetá: Cerro de Aracoara).

(46) *B. calophyllum* — Brazil (Amazonas: by the Vaupés), Venezuela (Amazonas).

(47) *B. casiquiarense* — Venezuela (Amazonas: Río Guainía).

(48) *B. Passargei* — Venezuela (Bolívar: Caura).

(49) *B. dendroides* — Brazil to Mexico through Peru, Colombia, Guatemala.

The distribution of *Biophytum* is valuable because it closely resembles that of *Cochlospermum* and its allies, while at the same time offering characters of its own. Section *Prolifera* is well knit; strongly rooted in Madagascar, the track points straight to Southern India through Ceylon, then follows the classic "Monsoon" path to Siam, Indochina ultimately reaching Southern China. Only one species is American in this group *(B. castum)*, and this species is well and properly located in Northeastern Brazil (Ceará). While Sect. *Grandifoliolata* is purely "Amazonian" (which is conventional in a genus of this dispersal and ecology), Sect. *Orbicularia,* on the contrary, is wholly African and Far Eastern. A secondary center of evolution in this group is situated between the Congo and the Cameroon, but this center is far from possessing the importance it has in *Cochlospermum. Biophytum* rather wanders in Africa in the manner of *Triumfetta* (see p. 419). The differences in behavior between *Biophytum* and *Cochlospermum* are not difficult to grasp, if we only consider that *Biophytum* is wedded to an ancient center of origins by Madagascar (see Sect. *Prolifera*), while *Cochlospermum* actually drew its immediate being in a secondary capital center of evolution and dispersal located in

Western Africa. The starting point of ultimate migrations (the stress is on *ultimate*), consequently, is not exactly the same for these two genera, despite their ecological affinities.

The dispersal of *B. sensitivum* is classic of a "Near-mangrove" (see note p. 189 for definition) which the course taken by its tracks reveals at a glance (notice the Andamans, Burma, Northern Siam, Cambodja, Timor, in the record). None of the species under Sect. *Sensitiva* is endemic to the New World, which is consistent with the trends of the migration all pointing eastward as far as New Guinea. It will be noticed that Madagascar, the Mascarenes and Comoros, and a sector at least of East Africa (Nyasaland, Tanganyka, Ethiopia, Eritrea) are integral part of a distribution which is largely Far Eastern and, even more, Malaysian. Other African localities (Eastern Congo, Sierra Leone, Angola) turn up in the distribution of this Section, but the possibility is strong that these were largely reached in the course of secondary distribution.

Section *Dendroidea* inverts the score of Sect. *Prolifera*. It is all American to the exception of one species *(B. Talbotii)* endemic to West Africa. The track points to Amazonian Brazil in the first place, and always in the watershed of the Amazonas follows to Eastern Peru (Loreto), Venezuela (Bolívar, Amazonas), Colombia (Caquetá), which is standard. However, this very same track invests other centers *(Peru:* Huánuco; *Bolivia:* La Paz; *Colombia:* Cundinamarca, Antioquia, Madgalena; *Ecuador; Guatemala; Mexico)*, which are not necessarily "Amazonian." It is clear that the archetypes of Sect. *Dendroidea* had reached the New World long before the Andean uplift, because we find them today still in such classic pre-Andean "cerros" as the Cerro de Aracoara (see Fig. 55), and in one of the oldest bastions of northwestern South America (Colombia: Antioquia). Moreover, a strong secondary center of speciation is in evidence in the region of La Paz, which we here regret to dismiss without knowing exactly the whole range the species endemic to this region, whether all "alpine," "subalpine," or of the lowlands. At any rate, it seems well established that Sect. *Dendroidea* tenants not only the lowlands of the Amazonas but regions higher up, and in one case at least Ecuador (here *B. Boussingaultii*, which significantly further extends to Cundinamarca, Antioquia and Magdalena). It is accordingly clear that this Section "crossed" over the modern Andes almost to the Pacific. This "crossing" was effected, naturally long before the Andes began their spectacular rise, so that it rates as illusory. Plainly, *Biophytum* was thoroughly well rooted in these quarters long before the Late Cretaceous and the Tertiary, that is, when the Andes were as yet not.

To summarize the dispersal of *B. dendroides* we use the formula "Brazil to Mexico through Peru, Colombia and Guatemala," which has the advantage of being descriptive as the modern map reads. However, this formula is not correct as to underlying substances. Peru and Colombia might be "Amazonian" or not, which we do not exactly know, and this species might have reached them directly from Brazil in the course of its normal spread in the

Amazonian Basin. Guatemala and Mexico are a different quantity. Seeing that Ecuador is in focus, we might surmise, and do so within reason, that Guatemala and Mexico were reached by *Biophytum* exactly in the same way as by *Amoureuxia* (see p. 258), that is to say, by a track running directly in the modern Pacific between Colombia (or Peru) and Guatemala. In brief, it is manifest that the American invasion of *Biophytum* Sect. *Dendroidea* is not purely "Amazonian" in character. It involves something else, for which conventional phytogeography has yet to invent a term. *This something else is a wholly Pre-Andean penetration of centers beyond the Amazonian Basin, most likely reaching to now crumbled lands in the Eastern Pacific.*

In sum, *Biophytum*, unlike the Cochlospermaceae, runs tracks that do not essentially trend westward. This genus breaks up into well knit phytogeographic units, as follows, (a) *Section Prolifera* — Basically Gondwanic, i.e., centering as the map now reads in the Indian Ocean. Only one species occurs in Northeastern Brazil. Madagascar is strongly underscored, and so are the Nilghiris of Southern India, and Ceylon. This is altogether conventional. Interesting is the range in Kweichow, a Chinese province rich in endemisms. *Biophytum Thorelianum* runs the classic "Monsoon" track, Siam, Indochina; (b) *Section Grandifoliolata* — Strictly "Amazonian," but stressing the Peruvian Amazonas most; (c) *Section Orbicularia* — Once again Gondwanic, but underscoring modern Continental Africa which Sect. *Prolifera* does not touch; (d) *Section Sensitiva* — Strongly Gondwanic, but intermediate as to range between Sect. *Prolifera* and Sect. *Orbicularia;* (e) *Section Dendroidea* — Seemingly "Amazonian," but in reality demonstrably rooted in the New World long before the Andean uplift, therefore antedating the actual origin of the modern Amazonian Basin. In direct opposition to Sect. *Prolifera*, this group contains one African species (*B. Talbotii:* Cameroon, Liberia, Southern Nigeria). It will be noticed by comparing *B. castum* with *B. Talbotii* that we secure the total range, *Ceará—Liberia—Southern Nigeria—Cameroon*, which is illustrative of the geography involved in ancient transatlantic connections between the New and Old World. This connection is not the usual one West Indies—Northern Africa formerly (see p. 92) discussed.

It is likely that further additions of range will be made to the score which we have used to tabulate our conclusions. It seems plain, nevertheless, that even with the present score before us, we can secure reasonable interrelations between systematic botany and phytogeography.

It brooks no discussion that the hub of the dispersal of *Biophytum* is the African gate of angiospermy (actually, its eastern end, the now crumbled continent of Gondwana).

III. Gnetaceae and Monimiaceae

Through the study of migrations analyzed in the preceding pages, we have visualized the former existence of landmasses now crumbled, but once filling the whole of the modern Indian Ocean (Gondwana), and further connecting in the deep south of our maps such extremes as the Kerguelens, Tristan da Cunha, Fuegia and points beyond to the west in the Pacific.

The patterns of dispersal we are about to discuss will further confirm what we have already learned, and contribute fresh evidence that all channels of distribution move in unison, constantly repeating their origin from land-masses now no longer in existence, and taking advantage of connections of which the modern maps contains hardly a trace. As we have gradually learned, we cannot trust modern continental outlines in our work. Africa is not approached from the south as much as from the east in the main, and the basic channels of dispersal that overrun the Dark Continent are oriented in certain vital cases almost parallel with the longitudes. The "Andean Bridge" is not a straightforward bridge at all. The "Alaskan Bridge", the "Central American Bridge" mean nothing material to dispersal. Living distributions speak with deadly accuracy against all these "Bridges" once we learn how to understand them.

The Gnetaceae are controversial Gymnosperms, and this is written by one of their students (14; p. 182), "The relationship of the Gnetales to the other Gymnosperms and to the Angiosperms are so obscure that there is little use in discussing them at large . . . . We have good reason to believe that the three existing genera are the remnants of an ancient race once more numerously represented than it is now; even this is open to question . . . . Their relations to the rest of the Gymnosperms, as to the Angiosperms, are quite obscure; those obtaining between the still living members of the group are but little clearer. While we need not yet despair of finding evidence sufficient to illum-inate some, perhaps, all, of these obscurities, it cannot be denied that, at the moment, no clear light has been thrown upon any of them."

We do not wholly share the opinions of this writer, and believe that, on the contrary, some at least of the affinities of the Gnetaceae are possibly to be found in the direction of the Monimiaceae (see *Kibara*) and the Chloranthaceae (see *Hedyosmum* and *Ascarina*) *). Agnosticism, vague generalities and pious hopes are not genuine scientific caution. We have come to fear words, and the question, whether the Gnetaceae have, or do not have, an "angiospermous" style is seriously being debated, without our realizing that it is *we who are responsible for the very set of characters whereby an "angiospermous" style is supposedly characterized. Thus, we are stopped on our way by academic*

---

*) In a most interesting work recently come to our hands ("A New System of the Cormophyta," in Blumea 6: 282. 1947—1948), Prof. H. J. LAM voices opinions on the general subject of angiospermy similar to ours, which is gratifying. He further refers to HALLIER's opinion, that the Gnetales are related to the Loranthaceae and Santalaceae, and *Ephedra* to the "Hamamelidaceeengattungen" *Casuarina* and *Myrothamnus*. We are not in the position to discuss the matter at length here, but may state, or restate, in this footnote certain opinions of ours, as follows, (a) Monimiaceae and Chloranthaceae are doubtless near the Gnetaceae; (b) As we have already suggested (in Bull. Torrey Bot. Cl. 74: 60. 1947), the Hamamelidaceae are a group in which the "sexualization" to maleness of certain "perianth-parts" leads away from the true "Amentiferae" (*Juglans, Quercus*, etc.) toward families with a supposedly typical "Euanthium." Obviously, the Hamamelidaceae may be assimilated to most any form, of the Angiosperms at the price of stressing one or the other of their characters *and tendencies;* (c) Hallier

definitions of our own making. As we have pointed out elsewhere (20), and we better expect to see in a coming book, the question, whether the Gnetaceae have, or not, an "angiospermous" style has very little value.

The suggestion that the Gnetaceae are allied with the Monimiaceae and Chloranthaceae should be followed to the extent at least to learn whether the dispersal of these three families is compatible with a common phylogeny. This we will do in coming pages, opening meantime with the distribution of the Gnetaceae.

The Gnetaceae consist of three genera, *Ephedra*, *Welwitschia*, and *Gnetum*. The classification of *Ephedra* is still unsettled as regards certain regions (South America, Europe in part), up-to-date in other (North America, 15; Soviet Russia, 16); *Welwitschia* is relic genus of one species; *Gnetum* was only recently monographed (17) at length

These three genera range as follows,

A) *Ephedra*

a) Sect. *Alatae*

a') Subsect. *Tropidolepides*

(1) *E. lomatolepis* — Central Asia (Lake Balkash).

(2) *E. Przewalskii* — Central Asia.

(3) *E. strobilacea* — Iran (Persia), Turkestan.

(4) *E. alata* — Sinai, Palestine, Egypt, Central Sahara (Hoggar), Tunisia, Morocco.

a'') Subsect. *Habrolepides*

(5) *E. trifurca* — Southwestern Texas and New Mexico to California and Mexico (Baja California, Sonora, Coahuila, Chihuahua).

(6) *E. Torreyana* — Arizona, Nevada, Texas, Colorado.

(7) *E. multiflora* — Chile (Atacama).

---

himself probably had a dim understanding of the fact, for, as shrewdly noticed by Prof. Lam (op. cit. 289, footnote), he indifferently spoke of *Casuarina* and *Myrothamnus* as „Hamamelidaceengattungen" or „Amentaceengattungen"; (d) The Buxaceae are beyond doubt closely allied with the Hamamelidaceae. We also believe that the Myrothamnaceae belong to the affinity of the Hamamelidaceae, though our opinion is based upon the literature, not the study at first hand of adequate material; (e) *Casuarina* is a much isolated group, both as to „leaf" and inflorescences. Its female cone would be come truly „coniferous" by a slight change in the orientation of the scales forming the „ovary"; (f) Loranthaceae and Santalaceae belong to one of the widest most and involved angiospermous genorheitron. We think they are related with the Celastraceae and certain Saxifragaceae in the affinity of *Brexia* in the first place, but would not now reject as possible some very diluted affinities of theirs with the Gnetales.

b) Sect. *Asarca*

a') Subsect. *Asarca*

(8) *E. californica* — California, Arizona, Mexico (Baja California).

(9) *E. aspera* — Texas, New Mexico, Arizona, California, Mexico (Baja California, Zacatecas, Coahuila, Chihuahua, Tamaulipas, San Luís Potosí).

c) Sect. *Pseudobaccatae*

c') Subsect. *Scandentes*

(10) *E. Alte* — Somaliland, Nubia to the Red Sea, Egypt, Sinai, Syria, westward to Cyrenaica.

(11) *E. foliata* — Arabia, Afghanistan, Western India, Iran (Persia), Turkestan.

(12) *E. Fedtschenkoi* — Central Asia (Pamir, Tianshan), Tibet.

(13) *E. ciliata* — Central Asia (Pamir), Persia.

(14) *E. procera* — Western Himalayas, Iran (Persia), Caucasus, Asia Minor, Greece and the Southern Balkans.

(15) *E. campylopoda* — Thrace to Dalmatia, Greece, Crete.

(16) *E. altissima* — Tunisia, Algeria, Morocco, Central Sahara (Hoggar), Canary Islands.

(17) *E. fragilis* — Sicily, Mediterranean coasts to the exception of continental Italy and France, Tunisia, Algeria, Southern Spain, Canaries.

c'') Subsect. *Pachycladae*

(18) *E. sarcocarpa* — Beluchistan, Afghanistan.

(19) *E. pachyclada* — Iran (Persia), Beluchistan.

(20) *E. intermedia* — Central Asia (Turkestan, Pamir, Alatau), Western Siberia, Iran (Persia), Tibet, China (Sinkiang, Kansu), Mongolia.

c''') Subsect. *Leptocladae*

(21) *E. monosperma* — Central Asia to the Altai Mts., Siberia throughout to the Ussuri, Mongolia, Tibet, China (Sinkiang, Kansu).

(22) *E. Gerardiana* — Himalayas (Kashmir to Sikkim), Afghanistan, Tibet, China (Yunnan, Szechuan).

(23) *E. equisetina* — Western Siberia, Central Asia, Mongolia, China (Sinkiang, Shensi, Kansu, Hopei, Jehol, Shantung).

(24) *E. distachya* — Central Asia to arctic Western Siberia and the Caucasus; Europe along the Mediterranean coasts from Thrace and Bulgaria to Dalmatia, Italy, Corsica and France; western coasts of France to Brittany.

(25) *E. major (E. nebrodensis)* — Himalayas, Afghanistan, Central Asia to the Tianshan Mts., westward to the Canaries across the Mediterranean, Tunisia, Algeria, Morocco, Central Sahara (Hoggar).

(26) *E. helvetica* — Southeastern Switzerland; Italy (Northwestern Piedmont).

c'''') Subsect. *Antisyphiliticae*

(27) *E. ochreata* — Argentina (Patagonia, Mendoza).

(28) *E. Tweediana* — Argentina (Patagonia, Buenos Aires), Uruguay.

(29) *E. gracilis* — Chile (Aconcagua, Atacama).

(30) *E. triandra* — Brazil (Rio Grande do Sul), Uruguay, Argentina (Córdoba, Catamarca), Bolivia (Tarija).

(31) *E. americana* — Patagonia to Colombia through Chile, Bolivia, Peru and Ecuador.

(32) *E. antisyphilitica* — Texas, Oklahoma, Mexico (San Luís Potosí, Nuevo León).

(33) *E. nevadensis* — Arizona, Nevada, California, Utah, Oregon.

Welwitschia is quickly disposed of, for it consists of a lone species,

(1) *W. mirabilis* — Angola, South-West Africa (approximately between 14° and 23° Lat. S.).

Gnetum is a fairly large genus, a good monograph of which is fortunately extant. It is dispersed in this manner,

A) Sect. *Gnemonomorphi*

a') Subsect. *Micrognemones*

(1) *G. africanum* — Angola, Congo, Gabun, Cameroon.
(2) *G. Buchholzianum* — Cameroon.

a'') Subsect. *Araneognemones*

(3) *G. Schwackeanum* — Brazil (Amazonas, Pará).
(4) *G. venosum* — Brazil (Amazonas, Pará).
(5) *G. Leyboldii* — Brazil (Amazonas, Pará).
(6) *G. nodiflorum* — Guianas, Brazil (Amazonas, Pará).
(7) *G. urens* — Dutch and French Guianas, Brazil (Pará), Venezuela (Amazonas).
(8) *G. paniculatum* — British and French Guianas, Brazil (Amazonas: by the Vaupés), Colombia (Río Guainia).

a''') Subsect. *Eugnemones*

(9) *G. Gnemon* — *Var. Griffithii:* Malacca, Tenasserim, Burma, Assam; Var. *Brunonianum:* Malacca, Tenasserim, Burma, Assam, Anambas Islands, Northern Borneo; Var. *tenerum:* Malacca, Borneo; Var. *gracile:* Celebes; Var. *silvestre:* Celebes, Ceram, Amboina, New Guinea, Bismarck and Solomon Islands, Fiji; Var. *domesticum:* Soemba, Celebes, Amboina, Morotai, Philippines (Luzon: Central Luzon to Sorsogon; Palawan, Polillo, Panay, Leyte, Sibuyan, Mindoro, Bucas Grandes, Mindanao, Siargao), New Guinea, Bismarck and Solomon Islands.

(10) *G. costatum* — New Guinea.

B) Sect. *Cylindrostachys*

b') Subsect. *Stipitati*

(11) *G. Ula* — Southern India (Malabar, Nilghiri Mts.).

(12) *G. contractum* — Southern India (Travancore).

(13) *G. oblongum* — India (Bengal), Burma, Pegu, Tenasserim.

(14) *G. montanum* — India (Sikkim, Assam), Burma, Tenasserim, Northern Siam, Tonkin, Southern China (Yunnan).

(15) *G. arboreum* — Philippines (Luzon: Rizal, Tayabas).

(16) *G. tenuifolium* — Malacca, Sumatra.

(17) *G. latifolium* — Var. *macropodum:* Andaman and Nicobar Islands; Var. *funiculare:* Andaman Islands, Malacca, Simaloer Island, Sumatra, Java; Var. *minus:* Philippines (Luzon: Benguet); Var. *Blumei:* Malacca, Borneo, Java, Celebes, Philippines (Luzon: Benguet to Sorsogon; Palawan, Mindoro, Polillo, Panay, Leyte, Mindanao), Ceram, Amboina, Batjan, Kei Islands. New Guinea; Var. *laxifrutescens:* Philippines (Luzon, Sibuyan), Kei Islands, Bismarck Archipelago.

b'') Subsect. *Sessiles*

(18) *G. gnemonoides* — Malacca, Borneo, Celebes, Philippines (Basilan), Halmahera, New Guinea, Bismarck Archipelago.

(19) *G. cuspidatum* — Malacca, Siam, Sumatra, Java, Borneo, Celebes, Talaud Islands.

(20) *G. macrostachyum* — Malacca, Siam, Tenasserim, Sumatra, Java, Borneo, New Guinea.

(21) *G. microcarpum* — Forma *silvestre:* Malacca, Siam; Anambas, Riau (Riouw), Lingga, Bangka Islands, Sumatra; Forma *campestre:* Malacca.

(22) *G. oxycarpum* — Mentawai Islands.

(23) *G: Ridleyi* — Malacca.

(24) *G. Loerzinghii* — Sumatra.

(25) *G. neglectum* — Borneo.

(26) *G. Klossii* — Northern Borneo.

(27) *G. diminutum* — Borneo.

(28) *G. leptostachyum* — Var. *robustum:* Eastern Borneo; Var. *tenue:* Borneo.

The dispersal of *Ephedra* falls into sectional and subsectional groups, strongly knit as such from the standpoint of modern geography. It consists of two massings, one in the Old, the other in the New World.

It is apparent at first glance that Subsect. *Scandentes* is dispersed in a manner homologous of *Populus* (see p. 217). *Ephedra Alte* is first seen to the south in Somaliland, while *Populus ilicifolia* turns up in British East Africa. The difference is hardly worth mentioning *), for, as we know, the great break in East African dispersal is made south of Tanganyka; north of Tanganyka, we are right on the threshold of the Mediterranean not to speak of Norway and Newfoundland. *Ephedra* is patently a xerophyte and it is easy to imagine a landscape with ecology to fit both this genus and *Populus*. It is likely that many of the readers of these pages know such a landscape at first hand, in the Western United States or Soviet Russia, for instance. A landscape of this ecology, doubtless existed somewhere in the modern Indian Ocean, in the Late Jurassic, not too far, either, from modern British East Africa and Somaliland. Here could be seen *Ephedra, Populus, Salix* happily consorting with a world of other plants which time and patience would make it possible for us to identify to a sizeable score.

Characteristic of the New World dispersal of *Ephedra* is the range, for example, of Subsect. *Habrolepides*. Northern Mexico and the adjacent United States are firmly held *(E. trifurca, E. Torreyana)*, and a species *(E. multiflora)* further turns up in Chile (Atacama) across and beyond a wide disconnection typical of Western American dispersal. A disconnection of the same nature appears in Subsect. *Antisyphiliticae* together with a classic outlier to Southern Brazil. This outlier is to be compared with homologous dispersal in the Conifers *(Araucaria)* and the Euphorbiaceae *(Colliguaya)*. Peculiar instructive, however, is the ephedroid track in this regard, because (see *E. triandra*) it gives us a series of stations leading almost without break from Bolivia (Tarija) to Brazil (Rio Grande du Sul) through Argentina (Catamarca, Córdoba) and Uruguay. This outlier was probably established by a penetration into the New World from the Pacific in the case of *Araucaria* (see p. 481), and most

---

*) Though correct within its immediate purpose, this statement needs qualification in other regards. The range of *E. Alte* is, on the contrary, exceedingly interesting in the sense that it contributes final evidence to Somaliland having received flora *immediately from a crumbled Gondwanic center in the Indian Ocean*. Thus, the flora of Somaliland may contain, and does indeed contain according to the indications in our hands, at least two distinct elements. One of these elements reaches Somaliland from continental Africa, therefore binds Somaliland most directly to lands such as Tanganyka, Uganda, etc.; the other, on the contrary, invades Somaliland from the direction of crumbled Gondwanic outposts in the east, and has no immediate affinities in the Dark Continent. We regret we cannot give more time to this question, but hope to see other workers busy on the biogeography of these quarters in the light of the generalities here supplied.

likely also in that of *Colliguaya*. The Bolivian region interested (Tarija) is fully within range of one of the classic doors (see *Peperomia*, p. 103; *Ilex* Subg. *Yrbonia*, p. 309) opening into South America from the direction of the west (Eastern Pacific). The reader will notice that this penetration crosses the modern Andes outright, does not follow them, which indicates that our interpretation of the dispersal of certain cruciferous groups (see p. 228) and homologous forms is consistent with the facts here on record. It might further be added that the line Copiapó (Chile) — Larioja (Argentine) cuts short the southern spread of *Croton* in these quarters, which all confirms that the theory of an academic "Andean Bridge" hardly stands searching investigation.

To return to *Ephedra:* The ranges of *Thamnosma* and *Oligomeris* (see p. 91) give us ample precedents to connect the Mexican and Texan range of *Ephedra* immediately with the East African by a line running almost parallel with the longitudes, not only, but further to bring the entire distribution of *Ephedra* into immediate contact with the modern South African range of *Welwitschia* (see *Thamnosma* subg. *Palaeothamnosma, Oligomeris* sect. *Holopetalum*). Once again we face (see p. 263) the quadrangle California—Chile—Mascarenes—Socotra mentioned in former pages.

Using the precedents set by *Thamnosma* and *Oligomeris*, not only, but other plants as well in the Euphorbiaceae, Nyctaginaceae etc., we easily could connect, then, the American and Afro-Asiatic range of *Ephedra* by a direct line California-Somaliland. This, however, we choose not to do for a number of reasons, as follows, (a) The New World range of *Ephedra* is typically Western American, which is to say that there is every reason to believe it was effected by migrations in the East Pacific, not the Atlantic; (b) The outlier established by *E. triandra* in Eastern Brazil is characteristic of migrants from the Pacific; (c) *Ephedra* is not in West Africa. One of its nearest approaches to the Atlantic *) is in Morocco and the Central Sahara, but the species involved in this approach *(E. alata)* is subsectionally allied with a group, Subsect. *Tropidolepides*, neatly ranging eastward to Lake Balkash and Iran (Persia); (d) Were it ever so that *E. alata* is the connecting-link between Subsect. *Tropidolepides* and Subsect. *Habrolepides*, we still would face, and be bound to account for, the disconnection Mexico—Chile yawning in the latter, that is, be involved in a problem calling into play a classic Western American disconnection much sooner than a putative transatlantic dispersal; (e) Though not necessarily, a range Baja California—Sonora—Chihuahua—Coahuila—Tamaulipas—San Luís Potosí is prevailingly established from the west, not the east; (f) The "bipolarism" of *Carex incurva* resolves itself (see p. 463) in a three-cornered play among Afghanistan, a center in the southern Indian Ocean of our maps, and Chile, which has all the elements involved in the distribution of *Ephedra;*

---

*) The nearest absolute approach is in reality at the Canary Islands, and occurs in Subsect. *Leptocladae* later discussed. We use Subsect. *Tropidolepides* here because there is almost no difference in mileage between Morocco and the Canary Islands, and the latter Subsection is more convenient. It will be seen, at any rate, that the course of Subsect. *Leptocladae* is all to the east of the Canary Islands, precisely as the trend of Subsect. *Tropidolepides* is all to the east of Morocco.

(g) It is patent by the position of *Welwitschia* on the map that the Ephedraceae have cryptic ties in the direction of South Africa, which we cannot disregard while working out the dispersal of other ephedraceous groups.

In our deliberate opinion, the two massings in the living range of *Ephedra* are connected (Fig. 56) by a base-line in the deep south of modern maps. In sum, these two massings are keyed up, in a wholly standard manner, to a center in the southwest of the modern Indian Ocean which feeds as one Somaliland and Patagonia. From Somaliland and Patagonia the tracks are standard, so we need not work them here in detail any longer. It will be observed that the classic disconnection Chile—Mexico met with in *Ephedra* is precisely explained having recourse to the same generalities which make rational the distribution of *Lepechinia, Nama, Menodora, Evolvulus, Juania, Cossignia, Llagunoa*, etc., not only, but materially assist in the task of disposing of troublesome aspects of the migrations of Piperaceae, Cruciferae, Euphorbiaceae, Cactaceae, Cochlospermaceae, Oxalidaceae, etc. We could without difficulty expatiate on this same subject, bringing to the fore an evidence beyond comparison more bulky than the one space allows us to introduce in these pages. *A book thoroughly subversive of current notions on the whole of American distribution could most easily be written in addition to, and as a sequence of this manual.* Plants do speak with unerring assurance; once understood, their language goes through the walls of the dourest academy as if thin sheets of paper.

We are free of *Ephedra* and *Welwitschia* at one stroke. Both these genera are bracketed by the same line in the deep south well known to us from the dispersal of such groups as, for example, *Phylica* (see p. 58). Once so bracketed, these genera become tame and quite standard from the standpoint of dispersal.

In the east, the destinies of *Ephedra* are evidently tied in with the existence of xerophytic to subxerophytic domains, as they do in the west. However, an interesting track is to be picked up in the distribution of Subsect. *Leptocladae*, which involves something more than deserts, and elucidates connections between the Far East (China: Yunnan, Szechuan, Shantung, Jehol; Siberia to the Ussuri) and Macaronesia (Canary Islands) of capital importance in the dispersal of great many plants. We know this track already, as a matter of fact, but *Ephedra* Subsect. *Leptocladae* is too cleancut a specimen to be missed here.

This track cuts trough a front Afghanistan—Himalayas (see run of *E. major)* further reaching to the Altais *(E. monosperma)*. Out of this triangle it fans out in two main directions. Eastward it reaches to the Ussuri and China all the way between Yunnan and Jehol. Westward, it travels by easy stages to Western Siberia in the high north, Central Asia generally, and the Caucasus. It next follows to the Balkans in Thrace and Bulgaria; to Dalmatia, Italy and the Mediterranean islands (Corsica); to France, here faring northward as far as Brittany. A species *E. helvetica* turns up in Southeastern Switzerland and

Northwestern Piedmont in Italy. Meantime, the track also takes in Mediterranean Africa from Tunisia to Morocco, reaching the Central Sahara (Hoggar) to the south, ultimately to stop at the Canary Islands.

This channel of dispersal is standard, and further leads from the Canary Islands and their general vicinity to the West Indies (the French West Indies most directly; see Linaceae and *Prunus*, p. 200, 326). The American abutment of this channel is set between the Appalachian—Ozarkian center in the Southeastern United States and Guadeloupe.

Ephedra helvetica has a significance in dispersal much greater than an occasional record of locality. Right in Western Piedmont is found the "alpine" *Euphorbia Valliniana* endemic to a few peaks. Peculiarly, this Spurge is most intimately allied with *E. Potaninii* of Eastern Mongolia, and both these species fall in with the track of *Ephedra* Subsect. *Leptocladae*, which may be allowed to speak its own message without comments of our own. It is strange that, seemingly, no *Ephedra* of this group is in the Iberian Peninsula. The track appears to fork in Western Piedmont, or its vicinity, to glide at one hand to Western France (Brittany), at the other to fade into nothingness unless it be to hit Tunisia and Algeria straight. Peculiarly again, the French vicinity of Western Piedmont is known to harbor certain elements, which, like *Euphorbia serrata*, have the following range, Southwestern France, Switzerland (Ticino), Italy (Liguria, Sardinia, Naples), Northern Africa, the Canary Islands and the whole of the Iberian Peninsula. This Spurge is ruderal in character, therefore its range might be to an extent modified by introduction. This notwithstanding, this range is noteworthy. It might be added, better to outline the problem at hand before closing these notes, that a South African Fern, *Pellaea hastata* occurs isolated on a few alpine crags in Catalunia (Eastern Spain). In brief, there is room to suspect that right between Northeastern Spain and Northwestern Italy abut two tracks, one from Mongolia and the Himalayas *(Ephedra helvetica);* the other from South Africa *(Pellaea hastata, Erica ssp.)*, which latter gives every indication of having touched the Hoggar. Also from the Hoggar (see *Erica*, p. 160; *Olea*, p. 354) the Canary Islands, Cyrenaica, Malta, Sicily, Crete, the Balkans, Cyprus are very easily reached. Patently, we would miss a very great deal of necessary evidence, were we to strive to "work out" *Ephedra helvetica* and *Euphorbia Valliniana* as "Alpine problems" of Tertiary, perhaps Pliocene vintage.

Had we exact knowledge of the ecology of *Gnetum*, we would indeed like to deal with it at great length. This genus falls into two massings, a western and an eastern one, which are disconnected by a wide gap (Fig. 56), as the tabulation of the dispersal reveals. To bridge this disconnections is one of the most important problem in the dispersal of *Gnetum*.

The western massing consists once again of two sectors. One, Subsect. *Micrognemones*, is in two species restricted to the triangle Angola—Cameroon—Congo; the other, Subsect. *Araneognemones*, has eight species, of typic "Amazonian" distribution (Guianas, Brazil (Pará, Amazonas), Venezuela (Amazonas)). The whole of this massing is perfectly consistent within itself, but geographically isolated throughout.

The eastern massing, peculiarly, is shorn of ranges in East Africa, Madagascar, the Mascarenes, all the standard stations to which we are accustomed to look in the very first place. *Gnetum* might, perhaps, be found also there in the future, but the evidence is so far negative, and this is arresting.

We pick up the westermost outpost of the eastern sector of *Gnetum* in the Deccan (Travancore, Malabar, Nilghiri Mts.), but the genus does not seem to be in Ceylon, which is once more peculiar, but tallies well with the fact that *Gnetum* is not in the Mascarenes and Madagascar, either. From the Deccan onward, we have no difficulty in following the gnetoid tracks always in an eastern main trend. We read in the record of localities, Bengal, Sikkim, Assam, Burma, Southern China (Yunnan), Tenasserim, Andamans and Nicobars, Siam, Philippines, Malacca, Sumatra, Borneo, Java, Soemba, Celebes, Ceram, Amboina, New Guinea, Halmahera, Bismarck Archipelago, Solomon Islands, Fiji, etc.

The localities in this record are not all equally perspicuous, but the outlines of a standard ancient track can easily be culled out such as, Burma, Andamans and Nicobars, Siam, Indochina, Malacca, the Lesser Soenda, Philippines. This track is rooted by the "Sunda Coign" (see p. 424), and as we are to learn, is a classic starting point of further migrations eastward in the mangrove style. This track, too, is very easily connected with the west of Africa (see, e.g., p. 138), so that we are fully justified in bringing together the two ends of *Gnetum's* dispersal in West Africa and West India by a line immediately connecting these two regions.

The migrations of *Gnetum*, it will be noticed, stream constantly eastward from India, in the manner typical of certain mangroves and near-mangroves. The eastern trend of these migrations might be charged by some to the fact that westward from India only ranges exist unsuitable to a "tropical" plant like *Gnetum*. This cannot be the truth, however, because a mangrove, *Cassipourea*, does indeed migrate southward and westward from India (see p. 408); moreover *Gnetum* is not in Ceylon which cannot be dismissed as insignificant.

The fact is that westward from a line Sikkim—Travancore in India extends a zone of mighty geological convulsions which destroyed ancient flora wholesale; left both mangroves and palms stranded in Afghanistan; immediately isolated south of the Caspian Sea a venerable hamamelidoid relic *(Parrottia persica)*. Also westward from India runs a powerful stream of "cold" forms which points straight in the direction of the Altai Node faring through a gap, *Afghanistan—Western Himalayas*. In short, it is right in this vicinity that a mighty channel of dispersal connecting the Amazonian Hylea and Fiji through West Africa is slashed across at approximately right angle by a second, and nearly as mighty, channel running from the Indian Ocean, Tanganyka, Ethiopia and Eastern Arabia in the direction of the Altai Mountains. *This is the crucial crossroads of a world gone out of our geography aeons ago.* It is here that something took place which tore Gondwana asunder; twisted the highlands of Iran, Afghanistan and Beluchistan with cosmic force; stamped upon the modern map of Africa a "Great Rift." When all this came to pass it was too late already to stop *Gnetum*, which by then had safely travelled all the way between West Africa and India. The same crossroads had also been safely negotiated by then by Connaraceae, Sterculiaceae, Combretaceae, Cuscutaceae, etc.

Gnetum is native to the Philippines in the highlands of Northern Luzon (Benguet, etc.). These highlands, open alike to plants from the west and east, rank as one of the geographic cornerstones of Northeastern Gondwana, and are coaeval of other land-blocks of the same dim age, the Gajolands of Sumatra, possibly Laos, the provinces of South China generally, and a multitude of "alpine" centers (e.g., Mt. Bavi in Indochina, Mt. Kinabalu in Borneo, Mt. Ophir in Malacca, etc.) which in the Far East stand as a match of such others as the Drakensberg, Mt. Mlanje, Mt. Kenya, Mt. Ruwenzori, Mt. Kilimajaro in Africa, and — in the New World — of the Pre-Andean heights underscored (see Fig. 55) elsewhere. A book could easily be written — and ought indeed be written — on the flora of these heights, and their connections all the way from Venezuela to Australia. Many of these heights were seemingly rejuvenated by later uplifts, mostly of Late Cretaceous and Tertiary vintage, attending the general process of orogeny to which we owe the Alps, Himalayas, Andes, Rocky Mountains, etc. which creates the illusion that these heights are young, while the contrary is amply proved by their plant-life *).

Gnetum found in the breaking up of Malaysia precisely the same opportunities of wide migrations eastward which are underscored in a coming chapter (see p. 424) devoted to the epochs of angiospermous dispersal. *Gnetum* took advantage of these opportunities very much as did *Cycas*, Dipterocarpaceae, Barringtoniaceae, Tiliaceae, Combretaceae, and, in general, the hosts of plants for which an era of active submersion and emersion, life in seres rather than in climaxes — we should say — was favorable rather than otherwise. *Gnetum*, then, precisely as *Cycas*, managed happily to survive the remaking of the earth's surface which caught up sooner or later with the plants of its own kind, the hemi-angiospermous clan. While the fossil beds of India, for example, yield the dead score of its near-congeners, the multitudinous array of forms that tried to become Angiosperms but fell short of it, *Gnetum* survived and survived lustily, though it, too, died out *between West Africa and India.* The fact it died out here shows, we believe, that had *Gnetum* not been a near-mangrove it might have followed to extinction in India or India's immediate vicinity all other hemi-angiosperms. *Ephedra*, too, survived for which the worst of xerophilous habitats does not hold death. So did *Welwitschia* manage not to become extinct, a unique form which never had an orthodox growing point.

In what part of Gondwana, then, did *Gnetum* originate? Our answer is

---

*) We regret that we cannot discuss the subject in detail right in this manual; space forbids.

that it most probably originated there, where the threads of *Ephedra* and *Welwitschia* come to light, in a center south of Africa, as the map (Fig. 56) reveals. We qualify our statement as most probable only, because we have no means of knowing for sure where all these plants first acquired their modern outline, that is to say, originated in the precise sense of the term. The Gnetaceae are, in this regard, one with the Conifers and the Ferns. We can reconstruct their tracks, and prove that these tracks match those of the Angiosperms throughout, which is a truism, granted that nature never drew two sets of map, one to please *Gnetum*, the other to satisfy *Rosa*. We cannot, however, be certain as to the focus of origin. In the case of *Gnetum* very sound evidence (e.g., absence of *Gnetum* in the modern Indian Ocean south of a line Southern Deccan—Timor, for instance) indicates that *Gnetum* first reached West Africa from a track leading straight from an antarctic center to Angola and the Cameroon *). This center is very close by a mysterious fossil flora (see p. 93) of Southwest Africa; close by the living range of *Welwitschia;* certainly intermediate between the extremes of the dispersal of *Ephedra* in the south. However, *Gnetum* might have worked its way to this center from India and West Africa southward. This is not probable in the light of the dispersal of *Ephedra*, which is beyond doubt of a definite "angiospermous" character, but rates as a remote possibility, nevertheless, which we might not rule out, in consideration of the fact that we have absolutely no theory to offer which we care to defend at all costs. *All we are interested is in facts, and the methods whereby the facts can be made sensible and amenable to a handful of set, workable principles.* The tracks of *Gnetum*, when not those of *Ephedra*, elucidate the reasons why this genus survived the onset of the ages and their revolutions. These track perfectly tally with those of uncounted Angiosperms in Malaysia, India, West Africa, Brazil, the Guianas and Venezuela to the very extent that they bear witness to the same geological events. Genuine origins, however, are another matter, which demands a definite knowledge of *genorheitra* in addition to an insight of migration.   We mentioned the Chloranthaceae and Monimiaceae, *Kibara* in particular, as allies of the Gnetaceae, and it remains for us to account be it so briefly for their migrations.

The Chloranthaceae are "Antarctic" in the conventional sense of the term. The base-line of their dispersal is squarely set to cross the Pacific, and both the Far East and the New World (the Western New World most particularly) were simultaneously reached by southern connections of which our atlases no longer give indication. That these connections crossed the path of *Ephedra* at least somewhere in the Southeastern Pacific is selfevident by the map.

Kibara is a monimiaceous genus that ranges between the Nicobar Islands andous New Guinea. It reckons not less than 15 species in New Guinea, and about 8 in the Philippines. It is represented in Celebes by 3 to 4 species at least and occurs further widespread throughout Malaysia. It ranges, consequently,

---

*) So much, of course, *insofar as modern maps show*, which is about all we may hope for in matters of this sort.

though somewhat narrower in spread (Nicobars to New Guinea instead of Southern India to Fiji), closely matches that of *Gnetum*. Were we to judge by the modern map, we ought to say that *Kibara* came into being in a last stage at the Andaman arc, a zone of faulting of considerable significance in the further evolution of angiospermous plant of the "Near-mangroves" and mangrove type.

The affinity of *Kibara* is Tribe Mollinedieae (18) distributed approximately as follows (see Fig. 5),

(1) *Macropeplus* — Eastern Brazil.

(2) *Mollinedia* — Tropical South America to Trinidad and Mexico, massive in Eastern Brazil.

(3) *Macrotorus* — Eastern Brazil.

(4) *Ephippiandra* — Madagascar.

(5) *Matthaea* — Malacca, Borneo, Sumatra, Java, New Guinea.

(6) *Steganthera* — Celebes, New Guinea.

(7) *Anthobembix* — New Guinea.

(8) *Tetrasynandra* — Queensland, New South Wales.

(9) *Wilkiea* — Queensland, New South Wales.

(10) *Lauterbachia* — New Guinea.

The genera of this group are not large. Seven out of eleven, including *Kibara*, are localized with in a range reaching from Malacca to New South Wales. Two are in Eastern Brazil, one is in addition widespread throughout the West Indies and the whole of Tropical South America including Mexico. One is in Madagascar, none in West Africa.

This distribution is manifestly out of balance because it lacks a Western African connection between Eastern Brazil and the Far East *). This connection is furnished by the African massing of *Gnetum*, as we know, but is not wholly supplied by the presence of *Ephippiandra* in Madagascar in the case of the Mollinedieae. Whatever we may do, it seems clear that we cannot fully rationalize the distribution of the affinity of *Kibara* without supplying a Western Africa component which has disappeared somehow. We are not certain that this component may not in reality be supplied by some other Tribe of the Monimiaceae which, perhaps with sound morphologic reason, is not recognized as part of the Mollinedieae by current systematic and taxonomic treatments. Whatever be the case, a thoroughgoing exploration of the Monimiaceae in the affinity of *Kibara* remains to be conducted before a decision is reached. We observe meantime that, not to speak of the missing Western African group in the affinity of *Kibara* and the presence of *Ephippiandra* in Madagascar, the two dispersals, Gnetaceae's and Mollinedieae's agree in, (a) Stressing Malaysia; (b) Highlighting Eastern Brazil. If these similarities do not

---

*) In the coming discussion of the Celastraceae (see p. 306, comments under the Trypterygioideae) we ventilate parenthetically the possibility that the Mollinedieae might have entered the New World from the Eastern Pacific. This would do away with the necessity of balancing the modern distribution by an intervening West African massing. As we have no records of locality and competent data to support this possibility, we place it before the reader merely for guidance in coming studies.

justify the statement that Gnetaceae and Mollinedieae are homologous migrants throughout and allies, they at least do not rule out the possibility that these two groups have a measure of consanguineity by origin, as we believe they do.

We may not hope to deal here at length with the distribution of the Monimiaceae, but can at least summarize the general situation of their tribal components, as follows (see Fig. 5),

A) Trimenieae

(1) *Xymalos* — Natal, Transvaal, Tanganyka, Cameroon.
(2) *Piptocalyx* — New South Wales.
(3) *Trimenia* — Fiji.

B) Monimieae

(1) *Tambourissa* — Mascarenes, Madagascar, Comoros, (not in Java).
(2) *Monimia* — Mascarenes, Madagascar.
(3) *Hennecartia* — Eastern Brazil, Paraguay.
(4) *Palmeria* — Queensland, New Guinea.

C) Hortonieae

(1) *Peumus* — Chile.
(2) *Hedycarya* — Tonga, Samoa, Fiji, New Caledonia, New Zealand, Australia (Victoria, New South Wales).
(3) *Amborella* — New Caledonia.
(4) *Levieria* — New Guinea, Queensland.
(5) *Hortonia* — Ceylon.

D) Laurelieae

(1) *Laurelia* — 1 species in Chile (42° to 34° Lat. S.), 1 species in New Zealand.
(2) *Atherosperma* — Tasmania, Australia (Victoria, New South Wales, Queensland).
(3) *Doryphora* — New South Wales.
(4) *Daphnandra* — New South Wales, Queensland.
(5) *Nemuaron* — New Caledonia.

E) Siparuneae

(1) *Siparuna* — Brazil to Mexico throughout South America and the West Indies.

(2) *Glossocalyx* — Cameroon, Gabun.

It is obvious by the face of the record that this family is a group of staggering antiquity. Certain of the main threads in the dispersal can easily be made out, as follows, (a) A cleancut "antarctic" tie (Laurelieae, *Peumus* and *Hedycaria*); (b) A definite Western African—American massing (Siparuneae); (c) A manifest insistence upon Western Polynesia (Laurelieae, Monimieae, Hortonieae, Trimenieae, Mollinedieae); (d) A strong center of evolution in Africa, involving as one Mascarenes, Comoros, Madagascar, Natal, Transvaal, Tanganyka and Cameroon. All these aspects of migration are conventional, but the composition of certain Tribes is no longer standard, witness the Hortonieae, which would be well balanced as "antarctic" *(Peumus, Hedycarya, Amborella, Levieria)* but for *Hortonia* (Ceylon).

Much remains to be done about the taxonomy and systematy of the Monimiaceae which may somewhat change the picture now in our hands. This picture, however, can be safely drawn in its main outlines right now. While not altogether fragmentary, the Monimiaceae are not as strongly and harmoniously knit in dispersal as, for instance, *Biophytum* (see p. 263). They have a long history of wanderings in the Southern Pacific beyond the shadow of a doubt. This history also has a special chapter in the Indian Ocean, witness *Tambourissa, Monimia, Xymalos,* and another chapter still in Western Africa and the "Amazonian" sector, as revealed by the Siparuneae.

In our opinion, the distributional and systematic status of the Monimiaceae is certainly such as to commend to our attention the possibility that this family has roots way back in the night of the ages, and represents the offshot of a *genorheitron* which well might have been related with such "hemi-angiospermous" forms as the Gnetaceae.

With Gnetaceae and Monimiaceae we reach a level where the problem of origins, in the narrowest sense of the term, begins necessarily to lose some of its cogency and sharpness. The migrations before us, however, speak eloquent language. All of these migrations are connected by standard base-lines in the deep south of our maps, which we can actually see in the case of the Monimiaceae, and infer with considerable assurance in that of *Ephedra.* In dealing with the Monimiaceae we are reminded of the living and fossil range of certain Podocarpaceae, including *Acmopyle* *); of lands, that is, antedating the modern continents by aeons of geological time. We see the Monimiaceae simultaneously using the three standard gates of angiospermy, African, Magellanian and Western Polynesian, and thus no longer may we hope to pinpoint their cradle, whether in the heart of the Indian Ocean or, rather, in the

---

*) See FLORIN, R., in Svensk Bot. Tidskr. 34: 117. 1940.

Pacific. This family drew its dispersal somewhere from the south of our maps; from this center it streamed out at one stroke by all three angiospermous gate. This done, it clung to centers of further genus and species-making still clear on our maps (Mascarenes—Natal; Western Africa—Brazil; Queensland—New Caledonia—Fiji; etc.), which, as seen by us, are land-blocks out of continents that were. In this light, we assume, the phylogeny and dispersal of the Monimiaceae can be brought to reason, without for this yielding all the details of its remote past. It should no longer be impossible to rationalize the ties which the Monimiaceae maintain with their northern offshot, Caly-canthaceae, which latter seem to be lost somewhere in "Holarctis" with but two genera, *Calycanthus* and *Meratia*, distributed from the Far East to the United States (*China:* Szechuan, Hupeh, Fukien, Kiangsi, Kiangsu to Shensi and Jehol; *United States:* California, Florida to Alabama and Pennsylvania). It is readily to be seen that the lands colonized in the north by the Calycan-thaceae are not a bit younger, in the fullest meaning of the term, than the lands still held in the grip of the Monimiaceae in the south. Perfect is the parallelism between the connection Kiangsi-California *) and the ties Came-roon—Eastern Brazil; New Zealand—Chile, apparent in the dispersal, respec-tively, of Calycanthaceae and Monimieae.

With this before us, it is time to listen to Berry's querulous comments (19; p. 36) to the effect that, "If *Laurelia* came to Antarctica, or it crossed Ant-arctica, why did not some of the other genera also, such as *Mollenedia* and *Siparuna*? Why did not *Laurelia* reach Australia inasmuch as *Hedycarya* is common to Australia . . . .? As it will be seen we arrive at no conclusion, and it seems to me that we cannot hope to solve problems of this sort until we have sufficient facts, no matter how clever we are in speculation or prophesy."

Patently, we arrive at no conclusion if we have no ideas, and forever wait for facts to turn up, which will "prove" the things we wish according to the academic preconceptions we nurture. We form these preconceptions, and next insist that nothing is ever "proof" which fails to come up to their "standards." This is unfair to the facts. The issues which are placed before us by dispersal are not handled to a solution by "cleverness" or "prophesy," but by the applic-ation to the facts in the record of certain general principles and ideas. These issues are not to be allowed to pile up dust because we indulge in academic fancies meantime. Had mathematicians waited for facts to explain themselves we still would reckon today on the fingers of two hands. Strange to say, the author of the comment quoted above is so blissfully unaware of the basic realities in play, that he describes as "Tortuous" (19; p. 41) the mind of a co-worker who rationalizes distribution appealing to "Land-bridges," as if only minds out of gear can believe in such connections. He also pens the fol-lowing (19; p. 41) "The lauraceous genus *Phoebe*, as delimited by systematists, is found in South America and the East Indies and has an outlying species in

---

*) See further notes on transpacific connections, p. 70.

the Canary Islands. As remarked elsewhere, this apparent wide range is probably explained by a taxonomic fauxpas and requires no explanation."

This statement is indeed most peculiar; if the record no longer squares with preconceptions, preconceptions are to be treasured and the record is to fly off the window as a taxonomic *fauxpas*. If *Phoebe* (or similar lauraceous genus) is in the Canaries, and its presence there rates as a taxonomic *fauxpas*, we no longer know what to think of patterns of dispersal so baffling at first sight as those of *Notelaea, Myrsine africana, Cuscuta, Protium*, etc. It is not all to the good that the writings of the author quoted are circulated as "authoritative."

In conclusion, we reach with the Monimiaceae the threshold of angiospermy in a group of the most remote antiquity. This group is such as to be possibly related by ultimate genorheitral ties with the Gnetaceae as revealed both by dispersal and morphology. The issue of "angiospermy" and "gymnospermy" has been entangled, as we have already noticed, in theoretical definitions wide of the mark. Seed-coats produced into a *tubillus* occur in the Urticaceae (20). In the Geraniaceae, for instance (21), the germinating pollen may not be conducted to the ovules by a conventional style, but by degraded tissue of conventional "axile" nature. "Stigmatic" are in a sense the scales in the cone of *Araucaria* upon which the pollen may freely germinate, and indeed tarry very long before entering the micropyle. Chalazogamy is rife in many families (Casuarinaceae, Juglandaceae, Betulaceae, etc.). In conclusion, the "gymnospermy" of the Gnetaceae is a matter of limits, a function — we should say — of our own definitions which, they badly need revision and restatement *).

IV. Plantago and the Labiatae Prasioideae

Plantago is a large pandemic genus which, like *Carex*, could all by itself furnish material for a primer in dispersal. Although widespread in different habitats, this genus can with some assurance be traced back to the seashore for its ultimate origins. *Plantago*, instead of being "derivative", is one of the groups that were sharply differentiated in the earliest epochs of angiospermy, and had a long history of migrations and speciation behind them by the Mid-Cretaceous.

One of the subdivisions of *Plantago*, Sect. *Palaeopsyllium*, is commonly accepted as archetypal, and is distributed (22) along the following lines (Fig. 57),

---

*) After this was written, an article appeared by Prof. P. MARTENS (in Bull Classe Scs. (Acad. Roy. Belgique), 5 sér., 33: 919. 1948) dealing with the fundamentals of angiospermy. It is evident that we are not the only ones aware of the facts and considerations expressed in the main text. „Unorthodoxy" is ultimately seen as necessary by those who at first received botany the like it was handed to them, but later thought about it on their own. See also footnote p. 268, dealing with a paper by Prof. LAM.

(1) P. robusta — Saint-Helena.

(2) P. tanalensis — Madagascar.

(3) P. laxiflora — South Africa (Cape, Pondoland).

(4) P. longissima — South Africa (Cape to Transvaal).

(5) P. remota — South Africa (Cape, Transvaal).

(6) P. Fischeri — Eastern Africa (Tanganyka; Kilimajaro).

(7) P. palmata — Western Africa (Fernando Po, Cameroon), Eastern Africa (Tanganyka to Ethiopia).

(8) P. Dielsiana — South America (Uruguay, Southeastern Brazil northward to Rio Janeiro).

(9) P. fernandezia — Juan Fernandez.

(10) P. eriopoda — Northern Mexico, Central United States, Canada.

(11) P. Tweedyi — North America (Utah, Wyoming).

(12) P. cordata — Eastern North America.

(13) P. sparsiflora — Eastern North America.

(14—21) P. princeps, P. glabrifolia, P. hawaiiensis, P. Hillebrandii, P. melanochrous, P. pachyphylla, P. Krajnai, P. Grayana — Hawaii.

(22) P. rapensis — Eastern Polynesia (Rapa).

(23) P. rupicola — Eastern Polynesia (Rapa).

(24) P. aucklandica — Auckland Island.

(25) P. Hedleyi — Lord Howe Island.

(26) P. Cornutii — Europe (Western Black Sea, Adriatic, Southern France).

A marked dispersal in the Eastern Pacific, and the Pacific in general, is immediately obvious by the face of the map. On that side of the waters, the track begins in Juan Fernandez, dispatching an outlier to Uruguay and South-eastern Brazil *). After a disconnection all the way to Mexico, the track is once again picked up leading to a standard center of the Western United States (Utah and Wyoming. These two states and Colorado are center of "cold" speciation. See discussion of Carex incurva, p. 463). Veering next to the open Pacific, the track conveys these Plantains to Hawaii, where they turn up in a massive concentration. All of this is standard, and demands no comment.

Always in the Pacific but this time to the west, Sect. Palaeopsyllium first shows up at the island of Rapa. Its next stations bracket New Zealand most neatly between a pair of prongs pointing to Auckland and Lord Howe Island, respectively. Vagaries of the kind (see p. 181) are standard contingencies of Western Pacific dispersal. New Zealand is left untouched so far as the record now reads.

Beyond New Zealand Plantago Sect. Palaeopsyllium fades out, but it is possible that, as we shall see, it does not wholly disappear, though it is no longer to be followed on our maps.

---

*) This outlier reaches unusually far north (i.e., Rio de Janeiro). See further in the main text for possible alternatives.

The scene next shifts to the Indian Ocean, as the map reads. Section *Palaeo-psyllium* is in Madagascar, South Africa, Tanganyka, Ethiopia, Fernando Po, Cameroon, which is all standard. The record in Saint-Helena might have been established directly from Fernando Po, or as an alternative from the direction of South Africa. Both are possible hypotheses. A third hypothesis is that this record actually leads to Brazil by a secondary track, South Africa—Saint Helena—Uruguay or Eastern Brazil. Matters of the kind cannot be settled unless by a stringent study of immediate intersectional affinities which we cannot undertake here. Everything of dispersal can be elucidated at a price.

There remains to account for the presence of one species in Europe, *P. Cornutii*, which we believe to have been established by direction connection between the southwestern sector of the modern Indian Ocean and the Eastern Mediterranean. Although the reader is by now amply informed on the score of tracks faring in such a manner as to fit the case under immediate study (see *Erica*, Andromedoideae, *Potentilla*, Primulaceae, Sapindaceae Koelreuterieae, etc., etc.) we will bring to the record here the tracks of a species, and part of a genus, as follows,

A) *Orobanche* (23)

(1)   *O. Mutelii* — Vars. *typica* and *spissa:* Northern Ethiopia (Eritrea), Egypt, Lybia (Cyrenaica), Tunisia, Algeria, Morocco, Canary Islands; Cyprus and the Aegean Islands (Rhodes, Samos, Kalymnos, Karpathos); Mesopotamia, Southern Persia, Asia Minor, Caucasus; Spain, Portugal, Southern France, Italy and adjacent islands, Dalmatia, Jugoslavia, Greece, Crimea, Southern Russia; Var. *sinaica:* Ethiopia, Egypt, Portugal, Syria, Southeastern Iran; Var. *angustiflora:* Malta, Pantelleria, Sicily, Southern Italy; Var. *interjecta:* Syria, Caucasus; Var. *interrupta:* Cape, Ethiopia, Somaliland.

B) *Pimpinella* (24)

(1)   *P. peregrina* — Ethiopia, Asia Minor, Southern Europe.
(2)   *P. tenuissima* — Northern Ethiopia (Eritrea); a close ally of *P. erio-carpa* (Syria, Mesopotamia).
(3)   *P. caffra* — Eastern Cape, Natal, Southern Ethiopia.

Dispersal of this type speaks for itself at first glance. Ethiopia is the region in *Orobanche* that marks the inception of the northern distribution, which is orthodox. Altogether consistent is the track of var. *interrupta;* this track issues from a point in the Southwestern Indian Ocean of our maps *), and

---

*) This point is identified in a later part of this manual as the *Afroantarctic Triangle* (see p. 357), *Orobanche* is a „cold" genus, and this species behaves within absolute norm while entering the Cape. Somaliland and Ethiopia, but meantime avoiding Madagascar and the Mascarenes. Compare the migrations of this species in the Indian Ocean with those of *Schoenoxiphium* and *Ephedra*. It should not be extraordinary if a species distributed like *O. Mutelii* has an outlier in Java or Malacca. *Linaria alpina*, endemic both to Malacca and the Pyrenees, is a case in point.

faring northward, as does the whole of the species, sends a sharp branch westward to hit Somaliland and Ethiopia. Quite as telling is the track of *Pimpinella caffra*, which rates as a mere variant of that of *Orobanche Mutelii* var. *interrupta*. The affinities of *P. tenuissima* toward *P. eriocarpa* are of one cloth as those exhibited by *Linum* (see footnote, p. 201) through *Linum* wanders farther east than Mesopotamia to India.

This review may close our summary of the dispersal of *Plantago* sect. *Palaeopsyllium*. This dispersal use all the standard angiospermous gates, so that we cannot say whether these plants acquired modern generic outlines in the Western Pacific rather than in the Indian Ocean, though the latter is more probable. Of a certainty, the archetypes widely fared by southern channels in the manner of the Restionaceae and Centrolepidaceae.

These migrations play fast and loose with the Pacific which is supposed to be one of the deepest oceans on earth. Considering that truly deep waters measure deeper than 3000 meters, it seems useful to submit a map (Fig. 58) illustrative of the readings of this depth. It will be noticed that one of the greatest measured depths (about 7635 meters) faces straight the coast of Northern Chile *in a region which countless patterns of dispersal treat as solid land*. Inasmuch as abyssal depths of between 5000 and 7000 meters are recorded also in Malaysia (Sulu and Banda Sea, respectively), *in regions which countless patterns of dispersal treat as solid land*, we are free to conclude with some reason, that *these depths were gouged out of the ocean's bottom after, not before, the Angiosperms had begun to migrate*. There is some inconsistency in the thought of authors who drive tracks within Malaysia to span the Sulu and Banda Seas without heed of depth, because these seas are narrow on our maps, but seem to become panicky at the idea of running a track straight across the Pacific, because this ocean is wide in our atlases. Depth is depth, and if we are stopped by a paltry 3000 meters in one direction we must be all the more stopped by some 7000 meters in another.

Depth, however, cannot stop us at all, because the evidence from dispersal is formal to the effect that neither depth nor width of ocean have the power of chocking off a track. In the sight of a phytogeographer, modern lands merely connotate points in the map where it is possible for plants now living to grow. The modern atlas cannot be trusted, when plants speaking through dispersal affirm that the atlas is in the wrong. Moreover, the facts in dispersal must be forever stronger than any current geophysical theory. If these facts speak one language, and these theories another, the former must be believed, imposing as may seem at the moment the latter. Phytogeography is a necessary branch of geophysics, as a matter of fact, because phytogeography has in its power to keep the record straight by a constant application of facts when other magnets are in danger of losing their head.

Taking our bow to convention, and believing for the moment, as convention requires, that depths of about 3000 meters are as yet not serious, we return to bathymetric readings on the map. We notice that the abyssal depth fronting

Chile is taken within the jaws of two lines of much lesser depth, and seemingly faced by a reading of merely 3300 meters. *In brief, if on a larger scale, the Chilean Deep is homologous for us of the „troughs" in the Banda and Sulu Seas, no more, no less.* The Galápagos and Desaventuradas Islands which stand as the jaws just mentioned are certainly not "oceanic" by position, which the flora of the former confirms most clearly (see p. 306). Between Clipperton and Easter Island there run two chains of depths not exceeding 3000 meters. The immediate connections suggested by dispersal to have existed between Easter Island and Eastern Polynesia (Ducie, Henderson, Pitcairn Islands, the Tuamotus) are anything but impossible as the modern bathymetric record reads. Most suggestive are the readings around Hawaii. The ocean is partly over 3000 meters deep between the islands of the coast of Southern California and a center north of Hawaii, but the ocean's floor comes up in at least two points to 3400 and 3700 meters. More interesting still is the fact that the line we are in course of discussing does not interest Hawaii, but leaves this archipelago to the south, immediately tying in at the Crespo (or Roca de Plata) Reef (Fig. 58, *b*) with a line of lesser depths coming down from Commander's Island (Fig. 58, *b'*). The system so formed hinges at Krusenstern Reef (Fig. 58, b") with shallows going to the Mariannes (Fig. 58, b''').

It is obvious that connections made between Southeastern China and California need not cover Hawaii, if the suggestive lines of lesser depth just outlined mean anything at all. These connections — even as the modern map reads — leave Hawaii in an "insular" position, and possibly entered by plantlife only from the immediate direction of the Crespo (or Roca de Plata) Reef.

As we are convinced that modern depths mean nothing at all in regard of dispersal, we may drop the subject without further quite satisfied that it has been shown here that the Pacific is not the abyssal ocean it is believed to be. *The future will take care of itself, when phytogeography, strictly run by facts and cogent principles, will become an integral part of the study of geophysics.*

To return to *Plantago:* Sect. *Palaeopsyllium* intergrades, according to our authority, to *Plantago* sect. *Polyneuron* through two alpine Javanese species, *P. Hasskarliana (P. rubens)* and *P. incisa. Plantago* sect. *Polyneuron*, on its part, consists of about 20 species distributed over a range illustrated by the following selection of binomials,

(1) *P. major* — Subsp. *Eumajor:* Western Himalayas, Tibet, Western Siberia to Europe throughout; Subsp. *pleiosperma:* Angola, Ethiopia, New South Wales, Argentina, Uruguay, Southern Brazil; China (Szechuan, Honan, Kansu, Hopei), Corea (Quelpaert Island); Caucasus, Syria; Morocco, Azores, Europe throughout; var. *paludosa:* Chile and through South America to the West Indies, Mexico and the United States; Africa (Ethiopia, Cabo Verde, Madera, Morocco, Egypt), Europe throughout; China (Shensi), Philippines, Southeastern Polynesia (Rapa), Hawaii; var. *scopulorum:* Cape, Egypt, Asia Minor, Dalmatia, Western United States to Canada (Manitoba).

(2) *P. asiatica* — Var. *recta:* Java; Var. *angusta:* Central Himalayas;

Var. *densiuscula:* China, Formosa, Riu-kiu, Japan; Var. *laxa:* Central Himalayas, China; Var. *brevior:* China, Eastern Siberia.

(3) *P. Aitchinsonii* — Afghanistan.

(4) *P. himalaica* — Western Himalayas.

(5) *P. Hasskarlii* — Java.

(6) *P. incisa* — Java.

(7) *P. Sawadaii* — Formosa.

(8) *P. Ruegelii* — Ontario to North Dakota and Texas.

This dispersal is seemingly wild in the case of *P. major,* and it is not to be excluded that certain of its record are due to introduction. The face of the dispersal, nevertheless, stresses the following, (a) Java, the Riu-kius and the Central Himalayas stand together in the range of *P. asiatica.* Java is surely a center of secondary speciation; (b) *Plantago major* var. *scopulorum* appears to obey no law being distributed from the Cape to Dalmatia, at one hand, Manitoba, at the other, but rigorously follows standard channels, on the contrary. Reference to the tracks of *Orobanche* and *Pimpinella* will forthwith make contact between South Africa and Dalmatia, while reference to the tracks of *Menodora* will without difficulty bring together the Cape and centers in North America not too far from Manitoba. Tracks of the kind are characteristic in the first place of "cold" forms of which *P. major* is doubtless one. Variety *paludosa* of the same subspecies runs in a belt from Mexico, the United States and West Indies to Africa north of Ethiopia and China. This track frays out to Europe, the whole of South America to Chile. The stations in Rapa and Hawaii are in such a position that we may not credit them to a definite stream of dispersal. They may tie in with Chile quite as much as with Australia, always granted that neither is due to introduction; (c) Subspecies *pleiosperma* is in two massings, southern (Angola, New South Wales, Argentina, Uruguay, Southern Brazil) and northern (China, Corea, Caucasus, Asia Minor, Europe, Macaronesia), with Ethiopia normally in center between the two.

In brief, while the less widespread entities in this group still are consistently knit together (e.g., *P. asiatica, P. Ruegelii*), and suggestive specific segregates exist (e.g. *P. Aitchinsonii* in Afghanistan), the dispersal of such a species as *P. major* no longer yields tracks in the details but lines among *centers of massing,* purely and simply. These lines and centers could all be resolved into tracks of conventional patterns at the cost of conducting specialized investigations in taxonomy and ecology, but tracks of the kind cannot as such be critically read out of the face of the distribution. Suffice it for our immediate purposes to remark that *Plantago* sect. *Polyneuron* has a center of secondary origins in Java, which center is connected with Japan and the Riu-kius, at one hand, the Himalayas at the other. A tenuous indications exists, accordingly that, while the intergrading between this Section and Sect. *Palaeopsyllium* is factually completed in Java *(P. Hasskarliana, P. incisa),* this intergrading might stem on the contrary from a center occupying the whole of the range Rapa Island—Norfolk Island—Riu-kius. With this possibility before us — tenuous as this possibility may be — we are no longer sure that the whole of the species of Sect. *Palaeopsyllium* in Hawaii harks back in origin to a tie Juan Fernandez—Hawaii. Part of these species might have reached Hawaii in origin immediately from the direction of the axis Java—Riu-kius. In sum, the problem involved by the Hawaiian specific massing is much sooner an issue of Peripacific dispersal than a consideration, whether *Plantago* in Hawaii is from one direction only. Appearances say that this genus is in Hawaii from the direction of Juan Fernandez, but there may be more here than hits the casual eye. We would not argue the problem of the Hawaiian *Plantago* before having at hand all the facts of taxonomy and ecology that may elucidate its fundamentals, and, above all, would never contend that these plants represent in Hawaii and "American" rather than a "Polynesian," "Far Eastern," or "Old Pacific" element. To begin with, Juan Fernandez is an archipelago rating "American" *as the modern map reads*, no more. Were it to be proved that *Plantago* reached Hawaii from Juan Fernandez most immediately, we could not say that *Plantago* is an "American" component of the Hawaiian flora. We might merely state that *Plantago* fared over a track leading from the southeastern Pacific of our days to Hawaii across a crumbled range in this part of that ocean. That would be all.

Two authors came to debate (25) the origin of the Labiatae Prasioideae in Hawaii, affirming or denying that these Labiatae were, respectively, an "Old World" or an "American" component of the flora of this archipelago. The debate in question proved futile, as could be anticipated.

The Labiatae Prasioideae range (26) as follows,

(1) *Johowia* — Juan Fernandez.
(2) *Phyllostegia* — Tahiti, Hawaii.
(3) *Stenogyne* — Hawaii.
(4) *Haplostachys* — Hawaii.
(5) *Gomphostemma* — Java, Malacca, Burma, Eastern Himalayas, China.
(6) *Bostrychanthera* — China.
(7) *Prasium* — Near East, Mediterranean, Canary Islands.

Johowia was not credited to the Labiatae Prasioideae in former taxonomic work, and it is probable that the two authors in question were not adequately informed of its position at the time of their debate. As the record now reads, it is perspicuous that the Prasioideae could reach Hawaii both from the Far East *(Gomphostemma* and *Bostrychanthera)* or a center Tahiti-Juan Fernandez *(Phyllostegia—Johowia)*. In reality, the addition of *Johowia* to the record adds little to it, because *Phyllostegia* is indicative enough by itself. Authors who argue whether the Prasioideae are an "American," "Old Pacific," "Far Eastern," etc., element of the Hawaiian flora argue nothing better than words. Were it to be certainly and absolutely established that the Hawaiian Prasioideae are most intimately related with *Johowia* this would not mean that these Prasioideae are an "American" element in the plant-world of Hawaii, because Juan Fernandez is not America to a phytogeographer, but a domain very much of its own, intermediate between Polynesia and the New World. Farther still, it is doubtful whether anyone well informed on the score is to hold that *Prasium* is an "Hawaiian" element in the flora of the Canary Islands; or to insist that *Stenogyne* is a "Macaronesian" component in the plant-world of Hawaii. However, there should be no reason whatever by the standards now current in phytogeography to condemn even obvious nonsense. In a world of phytogeography which believes in "Occasional means of dissemination"; has no means whatever to orient tracks, and no understanding of *genorheitra;* candidly confesses — as it is done among us — to have no idea whatever, how the Sarraceniaceae acquired their distribution; in this world, in sum, it should be altogether feasible to identify anything and everything any way one may like.

To investigate the *genorheitron* of the Labiatae is a serious task. This family has a component in Australia, Prostantheroideae, which is not quite typic of the rest of the affinity. A New Zealand monotype, *Teucridium*, suggests characters intermediate between Labiatae and Verbenaceae. The Verbenaceae, on their part, are represented also in Australia by a peculiar Subfamily Chloantoideae, that merges in the end with the large affinity of *Clerodendron* through the New Guinean monotype *Archboldia*.

Nesogenes an aberrant genus traditionally long retained under the Verbenaceae, is distributed as follows,

(1) *N. glandulosa* — Madagascar.
(2) *N. prostrata* — Agalega Island.
(3) *N. Dupontii* — Aldabra and Assumption Islands.
(4) *N. decumbens* — Mascarenes (Rodriguez Island).
(5) *N. africana* — Tanganyka.
(6) *N. euphrasioides* — Eastern Polynesia (Tuamotus and Ducie Island).

This genus is in two massings, one of five species centering between Madagascar and Tanganyka and ranging to petty islands in the Western Indian Ocean; the other in Eastern Polynesia. This seems staggering, but possibly rates as a modification of the tracks run by *Lepechinia*, *Juania*, Saxifragaceae Argophylleae, Sapindaceae Cossignieae etc., to the extent that Ducie Island and the Tuamotus are called into play rather than Chile or other Polynesian outposts *).

Nesogenes has some affinities toward the Acanthaceae and the "aberrant" South African petty family Stilbinaceae (or Stilbaceae). Characteristically, Socotra harbors a massive world of Acanthaceae, and it is also in Socotra that first turn up to the south the Globulariaceae, which from Socotra run a course taking them to the Mediterranean, Europe and Macaronesia. In this affinity many systematists recognize another small family, Phrymaceae which ranges with two species, or perhaps forms of the same species, to Eastern Asia and the

---

*) This interpretation we take as probable because of the facts stated. However, the reader is warned further to refer to such migrations as those of *Triumfetta* (see p. 419), which show that Polynesia can beyond doubt be reached by plants of the shore (like *Nesogenes* and *Triumfetta*) by tracks crossing directly all the Indian, and large sectors of the Pacific Ocean.

Eastern United States to Canada. In brief, a practically uninterrupted chain of intermediates can be aligned to associate Labiatae with Scrophulariaceae, Verbenaceae, Avicenniaceae, Acanthaceae, etc., etc. Cardinal in the dispersal of these groups are the lines, Madagascar—Socotra; Chile—Mascarenes; Rodriguez—Ducie Island and Tuamotus; New Zealand—Australia—New Guinea; Juan Fernandez—Hawaii; Hawaii—China—Canary Islands; etc., etc. Doubtless, the southern waters of the modern world are firmly held in the grip of these plants, and to argue whether a certain one of their group is in Hawaii "American" rather than "Asiatic", for example, is to argue words that make very little sense as regards the whole. A point is reached where there are before us no longer families, but hazily defined groups of families; no longer tracks but massive arrays of centers of genus- and species-making; no longer a geography of the present but multitudinous geographies of the past gradually changing to become the geography we know. In the end, we face no longer the history of the Labiatae but the history of the Angiosperms, a colossal mass of life in time and space. If we come to this with anything but a very broad mind, and the trusted guidance of principles and methods soaring much above the casual, the "mysterious," the petty, we are fated to fail even before beginning. *Let us thoroughly understand this.*

V. Carpinus

This genus, a member of the Betulaceae and for this reason an ally of *Alnus* (see p. 210), consists of between thirty and forty species in two sections, *Distegocarpus* and *Eucarpinus*.

Section *Distegocarpus* consists of four species (27, 28, 29), as follows,

(1)   *C. Fangiana* — China (Szechuan, Kweichow, Kwangsi).

(2)   *C. mollis* — China (Szechuan).

(3)   *C. cordata* — China (Szechuan, Hupeh, Anhwei, Honan, Shensi, Shansi, Hopei, Kirin), Corea (including Quelpaert Island), Northern Manchuria, Eastern Siberia (Vladivostock region), Japan (Honshu, Hokkaido).

(4)   *C. japonica* — Japan (Honshu).

This dispersal is cleancut along one of the major tracks in the Far East. This track begins in Southwestern China (Szechuan or Yunnan), and streams in the general direction of Japan across continental China. Migrations along this very same channel, or its variants, take place in many unrelated groups such as Berberidaceae, Magnoliaceae and Conifers.

Section *Eucarpinus* covers the balance of the genus and ranges in the manner illustrated by this selection of species,

(5)   *C. faginea* — Western Himalayas (Punjab, Kumaon).

(6)   *C. macrocarpa* — Northeastern Iran (Persia).

(7)   *C. oxycarpa* — Southern Caucasus.

(8)   *C. Betulus* — Near East, Caucasus, Crimea, Southeastern and Central Europe.

(9) *C. caroliniana* — Guatemala and Northern Mexico to Texas, Florida, Kansas, Nebraska, Minnesota, Quebec.

(10) *C. viminea* — Himalayas (Punjab to Assam), Northern Burma, Indochina (Tonkin to Annam), China (Yunnan, Kwangsi, Kwantung, Anhwei).

(11) *C. Tschonoskii* — China (Szechuan, Anhwei, Chekiang), Corea (Quelpaert Island), Japan (Honshu, Hokkaido).

(12) *C. Tsiangiana* — China (Kweichow, Hupeh, Honan).

(13) *C. putoensis* — China (Chekiang).

(14) *C. pubescens* — China (Yunnan), Indochina (Tonkin).

(15) *C. Kawakamii* — Formosa.

(16) *C. kweichowensis* — China (Kweichow).

(17) *C. Tanakeana* — Japan (Shikoku).

China easily comes first in point of species as no less than twentythree are on record within its boundaries. The hinge of the dispersal is located in the Western Himalayas, for the migrations proceed from this center both east and west.

It will be observed that unlike *Alnus* which freely ranges northward, and *Betula* which stresses the north most strongly, *Carpinus* closely adheres to a latitudinal belt between 30° and 45° Lat. North. Also at variance with *Alnus*, *Carpinus* in the New World stops its southern course in Guatemala.

Alnus, Carpinus, Betula are currently advertised as "proof" of the "Holarctic" origin of the Angiosperms together with *Cercis, Platanus, Fagus, Magnolia, Quercus, Fraxinus, Salix, Populus*, etc. As we have already reviewed some of these genera, we may begin to draw the line as to what is "Holarctic."

Certain of the groups in question are beyond doubt not at all "Holarctic." *Salix* is not, nor is *Populus*, which we have ascertained debating dispersal on purely factual bases. *Cercis* and *Platanus* also never were born in a fabulous north. The former ties in most certainly with African centers, and runs a track slightly to the north of one of the major channels of migration between Western Africa and the Far East. *Platanus* is connected by *genorheitron* with the Hamamelidaceae which are not "Holarctic" in the least. *Magnolia, Quercus* and *Fagus* demonstrably originate in a center located in the waters of the Western Pacific, and are not "Holarctic." *Fraxinus* we have briefly review in other pages. *Acer* is too much sapindaceous to be "Holarctic" in the least, and there may be of it fossil records in Madagascar.

As we have gradually progressed in our work, and acquired mastery over the fundamental channels of dispersal throughout the modern world, we have come to realize *as an essential* that dispersal could never be what it is, had it not been for the former existence of certain continental landmasses of which our maps *seemingly* contain no trace. *Seemingly* is used here on purpose, because sizeable sectors of these landmasses have become part of the modern continents in a manner which seems to us fairly obvious. For example, when we speak of Madagascar and the Mascarenes as "African", we speak by the modern map, of course. In our opinion, however, these islands are not "African" but fragments out of the continent which once filled the modern Indian Ocean. East Africa also strikes us as largely made out of pieces from this "crumbled" landmass. The New World is certainly a crazy quilt of lands that came together to give the Americas we know. As phytogeographers, we are not going to be embroiled in theories, how continents "crumbled", and how they "floated", but, always as phytogeographers, we have right to assert that that which dispersal tells us is correct. Aside from such ancient lands which appear to have been bodily incorporated in the making up of the modern continents, it is probable that part of the very stuff of the primitive continents was in some manner or other transferred mostly northward to compound the huge land accretions there.

In the light of this evidence, controverting unconstructive affirmations as to the "Holarctic" origin of the angiospermous hosts, it seems a waste of time still to argue that the Betulaceae, last bulwark of the "Holarctic" school, are not "Holarctic," even if we can not prove here that they are otherwise. This is not because we are sponsors of the "Antarctic" theory, but because phytogeography will never make sense until the last of "Holarctis" has been sent to its limbs. We believe, as a matter of fact, that great many are the Angiosperms which originated in Gondwana which is not yet "Antarctis." At the very bottom of our thought we visualize much of the phylogeny and dispersal of the Angiosperms as byproduct of an immense geological revolution which, begun in the Late Jurassic ended but with minor posterior changes in the Early Tertiary. The crumbling of the "American" shore all along the Eastern Pacific rolled in toward the New World all manners of Angiosperms to meet Angiosperms from the opposite direction.

Considering that in the triangle Fiji—New Caledonia—Queensland turn up peculiar "ranunculoid" relics, and families such as the Balanopsidaceae, we would not reject as certainly mistaken the suggestion that Fagaceae and Magnoliaceae actually originated as the modern families — in the person of certain early archetypes *) — rather more outside this triangle than close by the north and east of it. We would not dispute the "theory" that the archetypes of the archetypes — that is, "something" *later to become fagaceous and magnoliaceous but as yet not such* — first saw the light as a direct result of the Permo-Carboniferous glaciations somewhere between the points occupied in the modern maps by India and Timor or India and Japan. We would not at this stage of our thinking take serious concern with these suggestions and "theories," rather admit that nobody as yet knows anything of the truth in the matter, others or ourselves alike. We would reject most firmly, however, anything and everything which would turn into a sorry mess the fundamentals upon which rests a consistent reasonable interpretation of modern angiospermous mi-

---

*) We hold that these archetypes were genuine Hemi-angiosperms, very much on a „Bennettitean" level of evolution. See further next page.

gration. It is in precise thought in fundamentals that we are interested, and we believe that this thinking is to be defended to the utmost against encroachment by "Holarcticism" which is a doctrine meaningless and corrupt as to fundamentals. Rejection of factual evidence on the ground that this evidence is a "Taxonomic fauxpas;" appeals to perpetual waiting on the ground that we will never know anything until we have more "facts;" academic definitions concocted beforehand; "proofs" based upon misinterpreted fossil evidence, which we can better interpret by going from life to petrifacts, seem to us most undesirable, and pointedly to be rejected.

Carpinus ties nowhere in the south, for it is last seen in Guatemala. How did it reach Guatemala, then, and how did *Alnus* reach the whole of Western South America? This we cannot answer definitely. We can nevertheless suggest hypotheses for further study, noticing that the theory of an "Andean Landbridge" which would serve plant-life to migrate right along the north to south axis of modern lands does not stand under scrutiny based upon factual evidence. *Carpinus* and *Alnus*, to the extent at least of certain species, must have filtered into the New World from the Pacific. Did they "originate" in the deep south of the modern Pacific? This we do not know, either. Would we welcome the presence of well authenticated fossils of these two genera in Madagascar and Patagonia? Yes, beyond doubt. Could we connect these fossils in dispersal with the modern ranges? We think we could.

We believe that the *genorheitron* of the Hamamelidaceae is the most promising phylogenetic node to connect the modern families with an "Euanthium" (read, a conventional flower) with the "amentiferous" clan. The hamamelidaceous genus genus *Sinowilsonia* is still part "amentiferous". As we have pointed out elsewhere *), the "flower" of the hamamelidaceous *Rhodoleia Championii* is in reality exactly intermediate between flower and inflorescence, a congested female "ament" with appendages sexualized into maleness for stamens. While we are far from entering an open statement to the effect that the "Amentiferae" are next-of-kins of the Hamamelidaceae — a statement which would bear considerable qualification as regards *Quercus*, when not *Fagus*, for instance — still we believe that Juglandaceae, Betulaceae and Faga-.ceae are nearer the living Hamamelidaceae than any other group. The ultimate *genorheitron* lies on the level of the "Hemi-angiosperms," if this term be allowable in our own meaning.

While we may not bring here a full review of the hamamelidaceous distribution, we may state, (a) Out of the 25 genera currently recognized in the family, only *Trichocladus*, *Dicoryphe* and *Ostrearia* are outside what we may describe as the "*Carpinus* belt" that is to say, a zone which essentially includes temperate North America and Europe, the warm-temperate and tropical Far East and Malaysia. *Carpinus* is definitely more on the temperate than on the warm side of the fence, but its dispersal is not at basic odds with that of

---

*) See Bull. Torrey Bot. Cl. 74: 71—73, fig. 9, 10. 1947.

the Hamamelidaceae in fundamentals; (b) *Trichocladus*, numbers about half a dozen species in Eastern Africa within the strip Cape—Natal—Nyasaland—Ethiopia. Here is the base-line of dispersal, with records to the west made by extension of the basic range. *Dicoryphe* has about a dozen species massive in Eastern Madagascar and the Comoros. *Ostrearia*, which has affinities with *Dicoryphe* is seemingly monotypic in Northeastern Australia (Queensland); (c) No hamamelidaceous plant is living in South America within present knowledge. Fossils have been recorded which do not bear being discussed. It is not impossible that a group having living genera in the Cape and Australia further ranged in the geological past to Patagonia, but fossils cannot be accepted as evidence, until and unless they are identified by precise methods. It is also not impossible that some hamamelidaceous form may turns up in Colombia or the West Indies, but it is plain that the Hamamelidaceae never ran a major track to the south of the New World or the Caribbeans. (d) As *Platanus* turns up in Laos (see p. 139) with a peculiar form, so does an hamamelidaceous monotype, *Mytilaria*, live in Laos. Slightly north of Laos (region of Chapa, Tonkin) is seen the first species of *Tilia* in the Far East; (e) Seemingly very few hamamelidaceous plants are in the Philippines, one of which is the monotypic *Embolanthera* in Palawan; (f) Subtropical China is the region where the Hamamelidaceae are most abundant; (g) This family runs a thin line from the Mediterranean to the Far East and North America (to Central America) with genus *Liquidambar;* (h) One of its genera, *Parrottia*, is a relic narrowly localized south of the Caspian; (i) Various of its genera, e.g. *Hamamelis, Liquidambar, Distylium* run cleancut tracks between the Far East and the Southeastern United States; (j) It is most probable that Buxaceae and Myrothamnaceae are immediately allies of the Hamamelidaceae. The latter is a petty family restricted to South Africa. The Buxaceae are next reviewed in this manual.

On the strength of these data, we notice, (a) The origin of the family is essentially Gondwanic. We mean, it is localized in and around the modern Indian Ocean; (b) The main track runs through Laos to Southern China. This track feeds the temperate and northern subtropical New World, at one hand, reaches to the Mediterranean at the other. A region which is generally crippled by disconnection is tenanted in the case of the Hamamelidaceae by a relic, *Parrottia*, allied with a genus still in the Himalayas *(Parrotiopsis);* (c) The dispersal of *Liquidambar* is homologous of that of *Platanus, Cercis, Carpinus, Acer, Fraxinus*, etc., which rank as paragons of "Holarcticism."

In substance, the Hamamelidaceae would be a classic "Holarctic" family, were it not so that they have retained instead of lost about 10% of their modern components. This rump, *Trichocladus, Dicoryphe, Ostrearia*, is conclusive to the effect that the Hamamelidaceae are not all "Holarctic." Their original center of origin and dispersal is located in a cleancut manner in the modern Indian Ocean. Let us notice that the shore of the ancient continent once occupying this ocean ran as far as Western Polynesian of our days, which accounts for the isolated position of *Ostrearia* in Queensland *).

As we have documented in several occasion the ties that bind this ocean with the Pacific we need no longer tarry on the score.

In our opinion, *Alnus, Carpinus, Juglans* and *Betula* most likely originated as did the Hamamelidaceae. They are quite as "Holarctic" as *Liquidambar, Populus, Salix, Cercis, Rhododendron,* etc.

A map of the dispersal of *Carpinus* speaks for itself (Fig. 59), and this dispersal may further be compared with that of such genera as *Cercis* and the Ericaceae Rhododendroideae.

---

*) Notice the homology with *Rhododendron*, p. 175.

BIBLIOGRAPHY

Chapter V

(1) KUEKENTHAL, G. — *Cyperaceae-Caricoideae;* E. & P. Pflanzenr. 38 (iv. 20). 1909 — Neue Cyperaceen aus dem Malayschen und Papuanischen Gebiet; Bull. Jard. Bot. Buitenzorg, III, 16: 300. 1940.

(2) LEVYNS, M. R. — A comparative study of the inflorescence in four species of Schoenoxiphium and its significance in relation to Carex and its allies; Jour. South Afr. Bot. 11: 79. 1945.

(3) CHEESEMAN, T. F. — The vascular Flora of Macquarie Island; Australasian-Antarctic Expedition 1911—1914, Sc. Repts. Ser. C., Zool. & Bot., 7 (3). 1919.

(4) SEWARD, A. C., & CONWAY, V. — A phytogeographical problem: Fossil Plants in the Kerguelen Archipelago; Ann. Bot. 48: 715. 1934.

(5) RIDLEY, H. N. — The dispersal of plants throughout the world. 1930.

(6) BECCARI, O. — Malesia. Raccolta di Osservazioni Botaniche intorno alle piante dell'arcipelago indo-malese e papuano 1 (3): 220. 1878.

(7) STAPF, O. — On the flora of Mt. Kinabalu in North Borneo; Trans. Linn. Soc. (London), II, Bot., 4: 114. 1894.

(8) SKOTTSBERG, C. — Remarks on the flora of the high Hawaiian volcanoes; Act. Hort. Gothob. 6: 47. 1930.

(9) PILGER, R. — *Cochlospermaceae;* E. & P. Nat. Pflanzenf. 21: 316. 1920.

(10) SPRAGUE, T. A. — A revision of Amoureuxia; Kew Bull. 1922: 97. 1922.

(11) VAN STEENIS, C. G. G. J. — The Cochlospermaceae of the Netherlands Indies; Bull. Jard. Bot. Buitenzorg, III, 13: 519. 1936 — Addenda, ibid. iv. 1936.

(12) KNUTH, R. — *Oxalidaceae;* E. & P. Pflanzenr. 95 (iv. 130). 1930; E. & P. Nat. Pflanzenf. 19a: 11. 1931.

(13) KNUTH, R. — *Geraniaceae;* E. & P. Pflanzenr. 53 (iv. 129). 1912; E. & P. Nat. Pflanzenr. 19a: 43. 1931.

(14) PEARSON, H. H. W. — Gnetales; Cambridge Botanical Handbooks. 1929 (see p. 182).

(15) CUTLER, H. C. — Monograph of the North American species of the genus Ephedra; Ann. Missouri Bot. Gard. 26: 373. 1939.

(16) BOBROV, E. G. — Ephedra; Flora URSS 1: 195. 1934.

(17) MARKGRAF, F. — Monographie der Gattung Gnetum; Bull. Jard. Bot. Buitenzorg, III, 10: 407. 1930.

(18) PERKINS, J. & GILG, E. — *Monimiaceae;* E. & P. Pflanzenr. 4 (iv. 101). 1901 — PERKINS, J. — *Monimiaceae;* E. & P. Pflanzenr. 49 (iv. 101. Nachtr.). 1911.

(19) BERRY, E. W. — see (9) Chapt. I.

(20) CROIZAT, L. — see (19) Chapt. I.

(21) GUILLAUMIN, A. — Recherches sur la constitution de l'ovaire des Géraniacées à fruit rostré; Ann. Sc. Nat. IX, 19: 33. 1914.

(22) PILGER, R. — *Plantaginaceae;* E. & P. Pflanzenr. 102 (iv. 269). 1937.

(23) BECK-MANNAGETTA, G. — *Orobanchaceae;* E. & P. Pflanzenr. 96 (iv. 261). 1930.

(24) NORMAN, G. — The Pimpinellas of Tropical Africa; Jour. Linn. Soc. (London), Bot., 47: 583. 1927.

(25) SKOTTSBERG, C. — Juan Fernandez and Hawaii. A phytogeographical discussion; Bishop Mus. Bull. 16: 1. 1925.

(26) EPLING, C. — see (28) Chapt. III.

(27) WINKLER, H. — *Betulaceae;* E. & P. Pflanzenr. 19 (iv. 61). 1904.

(28) HU, H. H. — A review of the genus Carpinus in China; Sunyatsenia 1: 103. 1933.

(29) REHDER, A. — Manual of Cultivated Trees and Shrubs, ed. 2, 1940.

CHAPTER VI

INTERCONTINENTAL DISPERSAL IV

Several families, or groups of less comprehensive status, Celastraceae, Aquifoliaceae, Bombacaceae, Passifloraceae, Cyrillaceae, Rosaceae Prunoideae, Oleaceae, Caricaceae, Buxaceae, Salvadoraceae and Stackhousiaceae are reviewed in this chapter. The discussion is, as usual, strictly confined to essentials.

Some of these groups *(Prunus, Olea, Carica, Ceiba, Durio)* are interesting from the economic standpoint. As it will be seen, they disperse exactly as do plants without economic significance, which is proof that utilitarian investigations of origin and migration cannot dispense with sound phytogeographic generalities. This principle has not always received full attention, with the result that the dispersal of some of the most useful plants still demands the attention of well trained investigators.

All the families and groups covered in this chapter are interesting for some reason or other. The Cyrillaceae, for instance, contain telling examples of purely American dispersal, not only, but have *genorheitral* ties with several petty families in three other continents. The Celastraceae are a family that is so large and so broadly dispersed as to discourage at the very first the thought that their migrations can be brought to book along simple, standard lines. It will be shown that these appearances are belied by facts, and that the size and complexity of a group are no bar against the fruitful application of few simple generalities.

The dispersal of the Aquifoliaceae stresses the Pacific almost to the exclusion of all other centers, which is to challenge our belief that the Angiosperms repeat their origin in the main from a now crumbled center south of Africa. The Bombacaceae suggests more questions that we may now answer, but yield final evidence that phytogeography can make the best even of unsatisfactory taxonomic records.

The Buxaceae further show that migrations bear being rationalized even when we cannot definitely say whether these migrations call into play, or not, transpacific crossings in the north. We give the possibility of these crossings a mere footnote, which is less bold than it may seem.

It will be apparent throughout this chapter, as it has been in the preceeding ones, that phytogeography approaches all groups alike, large as well as small, in the light of few conservative generalities, and that these generalities constantly return reasonable answers. We make no claim for these generalities beyond the fact that they are workable, and feel confident that readers who may not subscribe to them, wholly or partly, will in due course contribute their own solutions, *always keeping in mind that dispersal cannot be handled piecemeal, and that the solution proposed for a family or group must be workable for all other families and groups alike.*

I. Celastraceae, Aquifoliaceae, Passifloraceae and allies.

Celastraceae, Aquifoliaceae and Passifloraceae are among the most important of the angiospermous families. All three are significant from the standpoint of floral morphology and phylogeny, but the systematic position of none of them is believed to be clear. Their taxonomy is often controversial, and their dispersal seemingly whimsical and unfathomable. These, then, are groups which challenge reason and imagination alike, and have honored place in investigation.

A skillful taxonomists who has of the flora of Madagascar intimate first hand-knowledge, writes (1) that two genera endemic to this island, *Brexia* and *Brexiella* are intermediate between the Celastraceae and Saxifragaceae Escallioniodeae, and should be placed under the former. This opinion is emphatically rejected by the foremost authority on the Celastraceae (2; p. 105), who avers that this family and the Saxifragaceae are absolutely unrelated. Commenting on this rejection (3), the author first cited reiterates his belief, challenges as less than successful the classification proposed by his opponent, and goes on record describing three new genera, *Brexiopsis*, *Hartogiopsis* and *Euonymopsis* placing under them some species which, he maintains, his opponent is unable to classify.

The picture before us is anything but clear, and further to confuse it, a botanist who has of the floras of the Pacific first-hand knowledge (4; p. 157) comments on *Celastrus* as follows, "Over 150 species have been described, centering definitely with related genera in Asia . . . . According to Berry the genus *Celastrus* Linné is the largest fossil genus of the family. Though its present center of distribution lies in the uplands of southeastern Asia and the East Indies, its history shows that the ancient stock was cosmopolitan and very abundant in the Tertiary of America and Europa. It is highly probable that it originated in America at the close of the Upper Cretaceous or somewhat earlier."

As colored by various opinions, the score now before us tallies accordingly as follows, (a) The Celastraceae merge with the Saxifragaceae in Madagascar; (b) The two families are, on the contrary, absolutely unrelated; (c) *Celastrus*, the type-genus of the Celastraceae, originated most likely in America in the Cretaceous, but the family was once cosmopolitan, though its present center of distribution is in the uplands of Southeastern Asia and the East Indies.

None of these opinions can be accepted as true, even as probable, if it conflicts with the record from dispersal. Migration and evolution proceed jointly, and *Celastrus* cannot have originated in America if tracks and *genorheitron* argue to the contrary.

The Celastraceae tabulate (2) in this manner,

A) Subf. Campylostemonoideae

(1) *Campylostemon* — 10 species in Africa (Congo, Tanganyka, Cameroon, Islands of the Gulf of Guinea).

B) Subf. Celastroideae

b′) Tribe Euonymeae

(2) *Torralbasia* — 2 species in Cuba, Hispaniola.

(3) *Monimopetalum* — 1 species in China (Anhwei).

(4) *Euonymus* — About 170 species; Subg. *Scyteuonymus:* Sect. *Orientales:* 35 species in India, China, Eastern Asia; Sect. *Malaicae:* 30 species, in Australia (1 species), Malaysia, Southern China; Sect. *Echinatae:* 16 species in Mexico (1 species), India, Malaysia, the Far East; Sect. *Cornutae:* 3 species in India, China; Sect. *Glomeratae* 8 species, in Madagascar (1 species), India and Malaysia; Sect. *Multiovulatae:* 5 species, in Madagascar (1 species), India, Central China — Subg. *Lepteuonymus:* Sect. *Tuberculatae:* About 5 species in Mexico and the United States; Sect. *Lophocarpae:* 20 species in Europe and the Near East (2 species), North America (2 species), the Himalayas, the Far East; Sect. *Globosae:* About 4 species in Japan and the Western United States; Sect. *Pterocarpae:* 15 species, in Europe (1 species), the Near East (2 species), the Himalayas (1 species), Eastern Asia; Subg. *Naneuonymus:* Sect. *Nanoides:* 4 species from Eastern Europe to Central China.

(5) *Hedraianthera* — 1 species in Australia (New South Wales, Queensland).

(6) *Glyptopetalum* — 24 species in India, Indochina, Philippines.

(7) *Microtropis* (incl. *Otherodendron*) — About 70 species in Ceylon, India, Burma, Siam, Indochina, China, Formosa, Japan, Malaysia; 2 in Mexico, 2 in Guatemala, 1 in Nicaragua, 2 in Costarica.

b″) Tribe Celastreae

(8) *Hypsophila* — 3 species in Australia (Queensland).

(9) *Denhamia* — 4 species in Australia (Northern Australia, Queensland).

(10) *Celastrus* — Subg. *Paniculatae:* About 4 species in Fiji, New Caledonia, Malaysia, the Far East to India, North America; Subg. *Axillares:* About 20 species in India, China, Corea, Manchuria, Japan, Formosa; Subg. *Sempervirentes:* About 25 species, in Madagascar (1 species, *C. madagascariensis*), Tropical South America (1 species, *C. racemosa*), Mexico (1 species, *C. Pringlei*), Australia (1 species, *C. australis*), New Guinea, the tropical Far East to China and India.

(11) *Maytenus* — About 200 species; Subg. *Pseudocelastrus:* 2 species in Australia *(M. disperma, M. bilocularis)*; Subg. *Eumaytenus:* Sect. *Trichomatophylla:* 2 species in Southern and Tropical Africa, India; Sect. *Stenophylla:* 1 species in Australia; Sect. *Coriifolia:* 1 species in Brazil (Bahia); Sect. *Laxiflora:* 1 species in New Caledonia, 10 species in Indochina, Central Africa; Sect. *Theoides:* 14 species in Tropical South America (Peru, Ecuador, Brazil); Sect. *Scytophylla:* 3 species in South Africa; Sect. *Umbelliformes:* 7 species in South and Tropical Africa; Sect. *Tricerma:* 7 species in Argentina, Uruguay, Paraguay, Bolivia, Galápagos Islands, Mexico to Baja California; Sect. *Magnifolia:* 1 species in Brazil (Amazonas); Sect. *Fasciculata:* 7 species in Madera and the Canary Islands, India, Ceylon, Bismarck Archipelago (here, *M. Rapakir*); Sect. *Densiflora:* 15 species in South and Tropical Africa; Sect. *Oxyphyllae:* 17 species in Tropical and Subtropical America (Uruguay, Paraguay, Bolivia, Brazil); Sect. *Pachyphylla:* 52 species in Tropical America, massive in Brazil and the West Indies; Sect. *Leptophylla:* 38 species in Tropical America (Brazil, Ecuador, Jamaica, Trinidad); Sect. *Microphylla:* 26 species in Tropical America (Patagonia, Chile, Argentina, Uruguay, Brazil, Peru, West Indies to the Bahamas).

(12) *Moya* — 4 species in Argentina, Bolivia.

(13) *Gymnosporia* — 80 species from Southern Africa to Spain, scattered through Madagascar, Mascarenes, Central and Eastern Africa, Arabia, India, China, Philippines, eastward to the Carolines, Fiji, Samoa.

(14) *Hexaspora* — 1 species in Queensland.

(15) *Psammomoya* — 2 species in Western Australia.

(16) *Putterlickia* — 2 species in South Africa.

(17) *Catha* — 1 species in Southeastern Africa, Ethiopia, Arabia.

(18) *Menepetalum* — 6 species in New Caledonia.

(19) *Salaciopsis* — 2 species in New Caledonia.

(20) *Pterocelastrus* — 6 species in South Africa (Kalahari, Natal).

(21) *Polycardia* — 9 species in Madagascar.

(22) *Kurrimia* — 8 species in India, Malaysia, New Guinea.

(23) *Pachistima* — 5 species in Mexico, the United States, Canada.

b''') Tribe Lophopetaleae

(24) *Lophopetalum* — 4 species in India, Cochinchina, Borneo.

(25) *Solenospermum* — 20 species in Malaysia and New Guinea.

(26) *Peripterygia* — 1 species in New Caledonia.

(27) *Kokoona* — 5 species in India, Burma, Malacca, Borneo, Philippines.

C) Subf. Tripterygioideae

(28) *Tripterygium* — 3 species in China, Corea, Japan, Formosa.

(29) *Ptelidium* — 1 species in Madagascar.

(30) *Zinowiewia* — 1 (or 6) species in Central America and Mexico.

(31) *Plenckia* — Subg. *Austroplenckia:* 3 species in Brazil, Paraguay; Subg. *Viposia:* 1 species in Northern Argentina and Bolivia.

(32) *Wimmeria* — 12 species in Central America, Mexico.

D) Subf. Cassinoideae

d') Tribe Cassineae

(33) *Cheiloclinium* — 2 species in Brazil (Minas Geraës, Amazonas).

(34) *Elaeodendron* — About 30 species; Sect. *Rubentia:* Africa, Madagascar, Mascarenes, eastward to India, Borneo, Philippines, New Caledonia; Sect. *Andropetaleia:* South Africa, Eastern Africa, Western Africa, New Caledonia, Norfolk and Lord Howe Islands, Australia; Venezuela (Nueva Esparta), West Indies (Haiti).

(35) *Cassine* — 7 species in South Africa, Madagascar.

(36) *Mystroxylon* — 20 species in Southern and Tropical Africa, Madagascar.

(37) *Herya* — 1 species in the Mascarenes (Réunion).

(38) *Maurocenia* — 1 species in the Cape.

(39) *Hartogia* — 3 species in South Africa, Madagascar, Somaliland.

(40) *Pleurostylia* — 6 species in South Africa, Mascarenes, Angola, Cameroon, Nyasaland, Ceylon, India, Cochinchina, New Caledonia, Queensland.

(41) *Lauridia* — 1 species in South Africa.

(42) *Gyminda* — 3 species in Central America, Mexico, West Indies.

(43) *Tetrasiphon* — 1 species in Jamaica.

(44) *Rhacoma* — 14 species; in Madagascar (1 species), Colombia, West Indies to the Bahamas and Florida.

(45) *Myginda* — 9 species in Colombia, Venezuela, Central America, Cuba.

(46) *Fraunhofera* — 1 species in Brazil.

(47) *Goniodiscus* — 1 species in Brazil (Amazonas).

(48) *Acanthothamnus* — 1 species in Mexico.

(49) *Mortonia* — 7 species in Mexico and the United States to Texas and Utah.

(50) *Forsellesia* — 3 species in the United States from California and Texas to Colorado, Washington.

(51) *Schaefferia* — 16 species in Uruguay, Argentina, Brazil, Bolivia, Peru, Colombia, Mexico, West Indies to New Mexico and Florida.

(52) *Ortosphenia* — 1 species in Mexico.

d") Tribe Perrottetieae

(53) *Perrottetia* — 15 species in Queensland, New Guinea, Moluccas, Celebes, Philippines, Northern Borneo, Formosa, Central China, Hawaii, Mexico, Colombia, Venezuela.

E) Subf. Goupioideae

(54) *Goupia* — 3 species in Amazonian Brazil and Guianas.

F) Subf. Siphonodonoideae

(55) *Siphonodon* — 6 species in Australia (New South Wales, Queensland), Celebes, Borneo, Philippines, Java, Indochina.

G) Affinity unsettled

(56) *Canotia* — 1 species in Arizona, California, Northwestern Mexico.

So pervasive and all encompassing is the living distribution of this family that to speak of tracks in the conventional sense is out of the question. We face in reality centers of massing and secondary origins scattered all over the face of the earth except the colder regions in the north and south. The best we can do is to bring together these centers by standard channels, and to interpret them as simply as possible.

To begin with, *Microtropis* and *Perrottetia* run a transpacific channel from the Far East to the New World, which in the case of the latter is perfect on account of every essential station in the traject being highlighted by one or more species. The track of *Perrottetia* starts in Queensland and moves along in a thoroughly standard sequence, New Guinea—Moluccas—Celebes—Philippines and Northern Borneo—Formosa—Central China—Hawaii—Mexico—Colombia—Venezuela. No doubt, this genus originated by the Western Polynesian gate, that is to say, in a now crumbled range of the Western Pacific closely connected by palaeogeographic ties with Gondwana. If there is anything to catch the eye in *Perrottetia*'s tracks, this is the strong possibility that this track suffers a standard break between Mexico and Colombia. Though standard, this break is always worth noticing. *Microtropis* departs from Ceylon, and follows the "Monsoon" channel, Burma—Siam—Indochina at first. We find it further in Central America, precisely in a range which *Perrottetia* does not touch at all, or may only lightly enter. Whether this is a coincidence, or, as we suspect, a result of these two genera having fared to the New World by slightly different ecological channels remains to be seen. At any rate, we know without further that both the Pacific and the Indian Ocean are involved in the celastraceous net of tracks.

Campylostemon is the mainstay of a monotypic Subfamily which stresses a Western African center reached by a standard track having its inception in Tanganyka. Given the importance of Western Africa in the phylogeny and distribution of plants like the Celastraceae it is not surprising that a Subfamily should be in the Congo, Cameroon and the islands in the Gulf of Guinea. *Microtropis* and *Campylostemon* both come a head in the Indian Ocean. The African gate then is in full play.

Finding Western Africa strongly underscored by a Subfamily, we need not be surprised if another Subfamily *Goupioideae* quite as forcefully highlights the range of the New World ("Amazonian" Brazil) which is the classic counterpart of Western Africa in dispersal. It is certain that the Celastraceae were both in Western Africa and the southeastern New World in the first flush of angiospermous migrations.

The Tripterygioideae speak for themselves. *Tripterygium* is in the Far East. *Ptelidium* in Madagascar, *Zinowiewia*, *Plenckia* and *Wimmeria* in the Americas, Minus an Australian outlier, this is the dispersal of the Monimiaceae Mollinedieae (see p. 280). However, there is a suggestion at least in the American dispersal of the celastroid genera that, unlike the "Amazonian" Mollinedieae, these genera invaded America from the south of the East Pacific. Madagascar, then would be central to China and Bolivia. Could it be possible that the Mollinedieae also followed this route, and became "Amazonian" by the devious device of entering the New World from the south of the Eastern Pacific? This is worth learning, but to learn it a detailed conspectus of the ecology and repartition of the American mollinedioid species is a prerequisite, which is not now in our hands.

Euonymus runs two main tracks, (a) African gate (Madagascar — See Sect. *Glomeratae* and *Multiovulatae*) to India and the Far East generally, including Malaysia and Northern Australia; (b) The Himalayas to the Far East, at one hand, Europe, Mexico and the United States at the other. The migrations to the New World may be both transpacific and transatlantic, which better data than those now in our hands would make it possible to ascertain. At any rate, the whole of this massive genus migrates by these two main channels, also well known in the dispersal of *Ilex*, next to be studied.

In *Maytenus*, a perfect Western American distribution is exemplified by Sect. *Tricerma*, which is disconnected in standard fashion between the Galápagos and Mexico. *Patently, the Galápagos are not "oceanic" but a normal station over a much travelled route. Maytenus, Cuscuta, Pernettya* mean more as to the origin of the flora of these islands than a host of indifferent pantropic forms. Possibly by the same route such regions as Bolivia, Paraguay, Uruguay, Peru, Ecuador and Brazil (a most interesting record as regards the possibility previously underscored that the Mollinediae came to Brazil by the Eastern Pacific) were reached by the species of Sect. *Theoides* and Sect. *Oxyphillae*. Section *Microphylla* contributes more questions than we may now conveniently answer. All the records in its distribution (Patagonia, Chile, Argentina, Uruguay, Peru, Brazil) may be credited to the channel revealed by Sect. *Tricerma*, but it is less sure than the West Indies and the Bahamas were reached by this channel. They might have been reached by an extension of the Macaronesian outposts of Sect. *Fasciculata*. However, it should be strange that Sect. *Microphylla* travelled to the West Indies and Bahama one way; to Patagonia and Chile the other. We need taxonomic and distributional data to work out this puzzle, a fascinating one, for it would set us aright as to the ultimate origin of certain forms in the West Indies. Let us notice meantime that the West Indies can easily be reached from the Pacific (see *Magnolia, Talauma, Washingtonia,* etc.).

A genus running tracks like *Maytenus* may easily wander into the Pacific from the Magellanian gate to reach New Caledonia. However, it would seem that *Maytenus* is in the Western Pacific also from the Indian Ocean (Sect. *Laxiflorae*). Naturally, *Maytenus* reaches India from East and Tropical Africa at ease (Sect. *Trichomatophylla*).

Gymnosporia runs a seemingly fantastic track from South Africa, Madagascar and the Mascarenes to Spain, Arabia, India, China, Philippines ultimate reaching the Carolines, Fiji and Samoa. This is a classic "dry" track, however, noteworthy in this case only on account of its unusual reach but typical, for instance, of the Salvadoraceae next reviewed. Parts of this track are commonly seen in the dispersal of "Monsoon" plants faring from West Africa to Indochina, the Philippines and Northern Australia. Better than a track, this could be called a xerophilous or subxerophilous front of geological age. *Catha, Pterocelastrus* run part of this track in South and Eastern Africa to Arabia. *Moya,* on the contrary, acts within the channels of *Maytenus* Sect. *Tricerma*. It is curious that so little part of the Celastreae centers in West Africa. However, this may be due to the fact that the Campylostemonoideae are the vicariants of the Celastreae in West Africa. Obviously, the Campylostemonoideae tie in with some other branch of the celastroid *genorheitron,* and this branch may well be represented by the modern Celastreae most immediately. The Campylostemonoideae are, at any rate, crucial in any searching study of the phylogeny and distribution of the Celastraceae.

The Lophopetaleae are typically "Malaysian," and invite no comment. The Cassinoideae cover the Perrottetieae, typified by the sole *Perrottetia* which we have already reviewed, and the Cassineae. The Cassineae are a strong group, thoroughly well rooted in the Indian Ocean, therefore at home also in Africa *(Elaeodendron* in part; *Cassine, Mystroxylon, Herya, Mauro-cenia, Hartogia, Lauridia)*. Naturally they fare *(Pleurostylia)* a track to Malaysia and the Far East which reaches to Queensland and New Caledonia, while at the same time spreading to the Cameroon and Angola from the usual center situated between the Mascarenes and South Africa. *Elaeodendron* goes *Pleurostylia* one better. It dispatches a section, *Rubentia,* from the Mascarenes and Madagascar to Africa generally, India, Borneo, Philippines and New Caledonia. Another of its section nearly exactly repeats certain vagaries of *Reinwardtia* and *Cuscuta* (see p. 202, 218). It leaves Africa massively to stream eastward by an unusual channel which cuts south of Malaysia throughout, and goes straight to New Caledonia, Norfolk and Lord Howe Island (see, however, *Notelaea,* p. 355); meantime, and most likely from West Africa this group sends a thin outlier to the Venezuelan island of Margarita and Santo Domingo (Haiti). *Rhacoma* repeats the play, for it retains one species in Madagascar, while sending about thirteen to the West Indies as far as Bahama and Florida and Colombia. The American component of the Cassineae, as usual, is difficult to interpret, whether part (and then what part) from the East Pacific *(Schaefferia, Mortonia, Forsellesia,* perhaps) or the Atlantic (most likely by the face of the record, *Myginda).* We are brought to wonder whether a cryptic form of "parallel development" might not have taken place, in the sense that the archetypes of the genera just mentioned part went to the Southeastern Pacific, part remained in Africa, later converging — by then matured as contribal throughout in the modern sense — on the New World from the direction of the two modern oceans, Pacific and Atlantic.

Canotia which has unsettled affinities is suggestive of typical Western American migration. *Siphonodon* *) type of a strongly marked subfamily Siphonodonoideae proves, together with *Perrottetia,* that a capital center of secondary development of the Celastraceae is located in Eastern Australia.

In conclusion, the seemingly fantastic record of localities made by the Celastraceae in their wanderings can be brought to book as follows, (a) Marked centers of secondary evolution are set in Western Africa, Eastern Brazil and Eastern Australia, not to mention the usual region Madagascar—Mascarenes; (b) A track taking the New World by storm from the Eastern Pacific is well defined; (c) The usual track from the African gate to the Far East and Malaysia is in full vigor; (d) A standard channel of migration approximately between Lat. 30° and 45° Lat. North rings the world. This is the supposedly "Holarctic" belt in which move such genera as *Carpinus, Platanus, Cercis,* etc. (e) This belt is paralleled by another, xerophilous this time, stretching between Spain and Samoa. This belt merges with a second, well known to us from numerous plants, hardly shown by the Celastraceae, which goes from West Africa to Malaysia and Northern Australia. *In brief, these belts are climatic and edaphic ranges of one and the same continental shore and immediate hinterland. These*

---

*) See CROIZAT in Lilloa 13: 31. 1947.

belts were vitally disturbed at least at three points (a) In the Sahara and the Mediterranean; (b) In Iran and Afghanistan; (c) In Tibet. This is to say that the ranges previously established were modified and shattered by the revolutions attending the uprise of the Alps, Caucasus and Himalayas and the formation of the "Great Rift" of East Africa. These revolutions are part of the general reshuffling and lands and seas which dismembered Gondwana, and gave our continents their modern outlines.

Like the Celastraceae, the Aquifoliaceae (5) are thoroughly wide-ranging. Like that family, too, the Aquifoliaceae were modern in the dimmest night of the ages, because their dispersal is patently hoary with aeons of wanderings throughout lands and seas that bore to ours no likeness at all. Conventional systematics look upon the Aquifoliaceae as anything but "primitive," but conventional systematics are in this respect at odds with phytogeography, which is bound to take for granted that *Ilex* existed as a modern genus quite as early as did *Quercus* or *Magnolia*. It is exceedingly difficult, in our opinion, to place the Aquifoliaceae as to remote affinity. The query occurred to our mind, whether the Cneoraceae might not be their allies, but our thoughts never went farther. Patently, the Aquifoliaceae are one of the basic angiospermous *genorheitra*.

The Aquifoliaceae consists of but three genera, *Phelline* (10 species in New Guinea), *Nemopanthus* (1 species in Eastern North America) and the large *Ilex*, which alone contains close to 500 species scattered throughout the world. Some of these species are valuable as ornamental or industrial plants (*Ilex paraguariensis*, the source of "mate").

Ilex breaks down into five subgenera, *Byronia, Rybonia, Yrbonia, Prinus* and *Euilex*, some of which are treated as distinct genera by authors other than the one we here follow.

The dispersal of the four subgenera outside of *Ilex* tabulates as follows,

A) *Byronia*

About 20 species ranging from Tahiti to Northern Australia, China, India, Hawaii; 1 species (*I. anomala*) common to Tahiti and Hawaii.

B) *Rybonia*

1 species (*I. oppositifolia*) in Borneo (Mt. Kinabalu).

C) *Yrbonia*

1 species (*I. teratopis*) in the Andes of South America (Bolivia and Peru).

D) *Prinus*

(1)   Sect. *Euprinus* — 6 species in Japan, Canada and the Eastern United States.

(2)   Sect. *Prinoides* — 10 species in the Himalayas, China, Philippines, Formosa, Japan, Mexico and the United States (*I. dubia* represented by closely related subspecies or varieties in China, Japan, Mexico and the United States).

It is easily perceived that these four subgenera are keyed up to the Pacific. One species (*I. anomala*), common to Tahiti and Hawaii, reminds us forthwith of *Korthalsella* and *Coprosma*, already reviewed in these pages. A second *I. dubia* is so distributed as to leave us in no doubt that it crossed over from the Far East to the United States through the modern Pacific in a wholly standard manner. Mount Kinabalu is a phytogeographic landmark of the first importance as regards dispersal in the Pacific most particularly. *Yrbonia*, while not manifestly "Old Oceanic" or "Pacific" — to borrow the terminology of conventional phytogeography — is so located as to indicate that it might have reached the New World over a path followed by the Piperaceae (see p. 130), which leads from the Pacific to the Amazon Valley in the end. In brief, this dispersal is originally in the Pacific, and from the Pacific, in the very first place.

This settled, insofar as regards *Byronia*, *Rybonia*, *Yrbonia* and *Prinus*, we need next turn to *Euilex*, the last remaining and largest subgenus of *Ilex*.

Let us break down the major ranges of *Euilex* in the first place, correlating them next with the distribution of the other subgenera. The author we follow believes that Subg. *Byronia* merges with Subg. *Euilex* through the latter's Sect. *Indico-Malaica*. Referred to a map, this means that a single phylogenetic complex is distributed within a range having as extremes, Ceylon—Northern Australia—Tahiti—Hawaii—China *). It is not possible to settle the issue, at a glance, whether this range is *all* from the Western Polynesian gate, or *part* (Ceylon to Northern Australia) from the African. The issue could be settled, of course, at the price of a searching study of the intersectional affinities throughout *Ilex*, concentrating in particular upon the ecology and morphology of the species endemic to Ceylon and Northern Australia. This is impossible for us here. Merely as a matter of opinion, we incline to believe that the range occupied by Subg. *Byronia* and Sect. *Indico-Malaicae* is all (or *very nearly all*) from the Western Polynesian gate. Ceylon, of course, represents an extension westward of the scope of this gate which is not very frequent (see comments under *Santalum*, p. 145), but having regard to the whole of the dispersal we are not inclined to take an exception this time.

The Western Polynesian gate is not the only one active in the Pacific as regards *Ilex* and its allies. The Magellanian gate is also active on its own, carrying the ancestors of Subg. *Yrbonia* to the Andes of Bolivia and Peru. It might be thought that this subgenus reached its present stations along the

---

*) Section *Indio-Malaicae* contains about 18 species distributed in New Guinea, Philippines, Borneo, Sumatra, Malacca, Ceylon. See the main text for the dispersal of Subg. *Byronia*.

mighty channel that connects the modern Indian Ocean (the heart of Gondwana, in reality) with the approaches to the Caribbeans through Central Africa, but this supposition strikes us as remote in the present case. We know (see p. 130), as a matter of fact, that a direct line of migrations point straight from the direction of Juan Fernandez toward modern Bolivia and Peru. This line is not usual, for, not to mention the Piperaceae (see further, p. 103), it was manifestly used by the tribe Cossignieae of the Sapindaceae, as follows,

(1) *Cossignia* — Sect. *Eucossignia:* 2 species in the Mascarenes; Sect. *Melicopsidium:* 1 species in New Caledonia.

(2) *Llagunoa* — 1 species distributed from Peru through Ecuador to Central Colombia.

As we have seen already (*Lepechinia,* p. 98; *Juania,* p. 455; Saxifragaceae Argophylleae, p. 135), ties in dispersal between the Mascarenes and the Southeastern Pacific are anything but rare. There is some chance as regards the Cossigneae, that the dispersal is from the Southeastern Pacific immediately — and mediately from the African gate or its vicinity — to Peru, Ecuador and Colombia. It does appear to be so, for *Cossignia* is both in the Mascarenes and New Caledonia, and *Llagunoa* follows the Andes from Peru to Central Colombia. Subgenus *Yrbonia* of *Ilex,* then, acts within the norm when hitting Bolivia and Peru from the Southeastern Pacific.

A group of Subg. *Euilex,* the Sect. *Lemurenses* next definitely takes us to the Indian Ocean of modern maps, and the African gate of angiospermy. This Section consists of about 12 species of which 11 are distributed to Ceylon, India and China, the remaining *(I. mitis)* ranging, on the contrary, throughout Tropical Africa, South Africa and Madagascar. There is no doubt, then, as to where we stand. This range is standard for the African gate, and standard throughout.

As usual with distributions of this scope, the knottiest problem is in determining how the New World was reached, whether directly from the African gate through the modern Atlantic, south or north, or from the Western Polynesian gate through the modern Pacific. The problem is knotty, because we are almost always lacking in the means to solve it in these pages. That this problem can be solved, nevertheless, goes without saying. All is needed are reliable data regarding affinities and ecology, also printing space and time to give *to details.* It may sound odd that we refer to issues of the kind *as details,* but this is indeed the way we feel about them. Truly essential are the issues which, once left in the dark, turn an orderly science into a disorderly academy of over-timid and over-rash notions alike.

Ilex Subg. *Euilex* Sect. *Aquifolioides* consists of two Subsections, so composed and distributed,

a) Subsect. *Oxyodontae*

(1) *I. dipyrena* (including var. *paucispinosa*) — India (Nilghiris, Manipur), China (Yunnan and central provinces), Formosa.

(2) *I. Pernyi* — India (Manipur), China (Yunnan).

(3) *I. ciliospinosa* — Central China.

(4) *I. cornuta* — China, Quelpaert Island.

(5) *I. integra* — Corea, Japan.

(6) *I. spinigera* — Northern Iran (Persia).

(7) *I. Aquifolium* — The Caucasus and the coasts of Asia Minor as far south as the Gulf of Alexandretta; the whole of Eastern Europe south of the Danube, including the Balkans to the exception of Crete; the whole of Western Europe west of the Rhine; Belgium, the Netherlands, coastal northeastern Germany, Danemark, Norway to the coast as far as ca. Lat. 63°; the British Isles, Northern Scotland excepted; the Iberian Pensula throughout including the Baleares; Corsica; Italy, Sardinia and Sicily; coastal Algeria and Tunisia.

(8) *I. Perado* — Southern Spain (Andalusia: var. *iberica*), Portugal, Azores, Canary Islands, Madera.

b) Subsect. *Insignes*

(9) *I. sikkimensis* — India (Sikkim).

(10) *I. insignis* — India (Himalayas).

(11) *I. latifolia* — Southeastern China, Japan.

(12) *I. borneensis* — Borneo.

This distribution is most interesting. It was primarily effected by a marked sectional *genorheitron* hitting in a prior place (as the map now reads) the Himalayas between Yunnan and the Kashmir across Southern India (Nilghiris), Japan and the Azores. Subsection *Insignes* is a "warm" offshoot of the *genorheitron* in question which roamed as far eastward as Borneo. Borneo, then, stood within reach of an ancestral stream moving onward from the Indian Ocean. So much is worth noticing, because Borneo also was penetrated by the *Ilex genorheitron* from the direction of the modern Pacific, as we have already seen.

It may be useful further to remark that Subsect. *Oxyodontae* follows a channel of migration leading from Southwestern China to Quelpaert Island, Corea and Japan, which channel is standard for the Conifers (see p. 486). Doubtless, then, we are on very ancient grounds.

We believe that Sect. *Aquifolioides* is primarily from the Indian Ocean, and the African gate of angiospermy, because of reasons too obvious to require detailed discussion. We face here essentially a line Azores—Formosa—Japan, and we know that this line, or its easy variants, are precisely the classic range of a host of plants such as *Platanus, Liquidambar, Cercis, Fraxinus, Carpinus,* etc., etc. In brief, this line agrees with the northern limits of an ancient continental mass and was demonstrably reached *from the south.* Let us notice that the Azores can be reached in many ways (see p. 22), and that these islands stand as the westernmost and northernmost end, respectively, of a track followed (see p. 90) by *Myrsine africana.* There is no evidence in the whole dispersal of Sect. *Aquifolioides* that the migrations followed westward, eventually to end in what is now North America.

We say that there is no evidence in this Section for migrations of the kind, taking for granted that the classification we use is wholly correct. However, and without the slightest intention of disputing it, we are also bound to take notice that *I. opaca* of Northeastern America, which at least superficially resembles certain forms of the *Aquifolium* group, belongs to *Ilex* subg. *Euilex* Sect. *Cassinoides* *). This Section is distributed in the manner exemplified by the following selection out of its species,

(1) *I. canariensis* — Canary Islands.
(2) *I. Cassine* — Bahama Islands, Cuba, Eastern United States, Mexico.
(3) *I. opaca* — Eastern United States.
(4) *I. Gale* — Colombia.
(5) *I. Uleana* — Brazil and Peru (Amazonas).
(6) *I. yunnanensis* — Southwestern China.
(7) *I. Sugerokii* — Japan.

In our opinion, Sect. *Cassinoides* departs in a westerly stream of migration (Canary Islands to the Bahamas and Colombia) taking its start precisely there (Azores, Canary Islands, Madera), where Sect. *Aquifolioides* stops its trek westward. In other words, the Macaronesian domain (Azores, Canary Islands, Madera primarily) is the hinge of an entire zone of dispersal in *Ilex* stretching between Mexico and Japan over and across the modern Atlantic, Europe and Asia. We might be in error in assuming that *I. opaca* bears at least diluted kinship in the direction of *I. Aquifolium* and its allies, but we are certainly not in error in relying upon a geologic shore fit to carry *Ilex* at ease between Europe, Northern Africa and the New World. The reader will find elsewhere (see p. 92) examples of factual distribution to document the existence of this shore.

In brief, we strongly incline to believe — rather, to take for sure — that part of the New World *Ilex* stemmed from the African gate, which means from an ancient continent answering the waters of the modern Indian Ocean. This continent, as we already know, fed the New World a wealth of *genorheitra* along a channel opening through modern Central Africa. This same continent, by a path less obvious but by no means mysterious (Himalayas—Caucasus—Near East—Europe—Macaronesian domain (Azores, Madera, Canary Islands)), further fed the New World, sending to it this time such plants as *Ilex*. Notice also that the ranges of Sect. *Aquifolioides* and Sect. *Cassinoides* overlap in the whole stretch, Azores—Japan.

---

*) The systematic position of *I. opaca* and its immediate allies does not seem to be free from doubt. REHDER places this species (Man. Cult. Trees & Shrubs, ed. 2, 549, 1940) under Sect. *Lioprinus* next to Sect. *Aquifolioides*. Always according to Rehder, these two Sections key out as follows, Sect. *Lioprinus:* Flowers or inflorescences solitary, in the axils of the leaves on young branchlets, or at the base of the young branchlets; Sect. *Aquifolioides:* Flowers or inflorescences fascicled, rarely solitary, in the axils of leaves on last year's branchlets. The difference to separate these two Sections is seemingly not vital.

We have before us, then, reasonable assurances to the effect that part at least of *Ilex* in the New World came from the east, and evidence as well that *Ilex*, like *Thamnosma, Oligomeris, Cneorum*, etc., etc. behaves conventionally in dispersal. We are also led to infer that the modern approaches of the West Indies are reached from the direction of the Old World by a double channel of migration. This channel fares through the body of modern Central Africa in one case; approximately through the modern Mediterranean in another case. In both cases, however, this channel ultimately leads straight to the geographic heart of the modern Indian Ocean, which is to say to Gondwana. Gondwana is a term much abused, but it means to us precisely a region, now crumbled, stretching between East Africa and Western Australia, and having easy access to Arabia, India and Western Malaysia of our maps.

As part of *Ilex* came from the Old World to the New *across the modern Atlantic,* so did another part of this very same genus reach the New World *across the modern Pacific. Ilex* Sect. *Aquifolioides* and Sect. *Cassinoides* document the former crossing; *Ilex* Sect. *Prinoides* the latter. *Yrbonia,* moreover, conveys a very strong suggestion that a further crossing to South America was effected from the Southeastern Pacific along a line of penetration, this time, which well may be the same one as that followed by the Piperaceae, and the Sapindaceae Cossignieae.

We might with this drop *Ilex* and the Aquifoliaceae from immediate reckoning, for we have by now mapped out the principal trends in their dispersal and far more exacting data than we now possess are required to carry on a critical investigation of the intersubgeneric and intersectional affinities of *Ilex*. It seems useful, nevertheless, to pay further attention to still another Section of this vast genus.

This Section is the *Microdontae* distributed in various Subsections, the ranges of which are illustrated by the following selection of species,

A) Subsect. *Eumicrodontae*

(1) *I. Krugiana* — West Indies (Hispaniola), Bahamas.
(2) *I. cerasifolia* — Brazil.
(3) *I. amygdalina* — Peru.
(4) *I. brevicuspis* — Argentina (Misiones, Mendoza).

B) Subsect. *Repandae*

(5) *I. tolucana* — Mexico (Baja, California, Central Mexico).
(6) *I. nitida* — Mexico, West Indies.
(7) *I. Urbaniana* — Portorico.
(8) *I. Guayusa* — Columbia, Peru.

(9) *I. paraguariensis* — Brazil (Matto Grosso, Minas Geraës), Paraguay, Uruguay, Northern Argentina.

(10) *I. Mertensii* — Bonin (Ogasawara) Islands; Var. *Volkensiana:* Caroline Islands.

(11) *I. cinerea* — Riu-kiu Islands, Hongkong.

(12) *I. Buergerii* — Japan.

(13) *I. Curranii* — Philippines.

(14) *I. formosana* — Formosa.

(15) *I. tonkiniana* — Indochina.

(16) *I. odorata* — China (Yunnan), India, Java, Celebes.

C) Subsect. *Vomitoriae*

(17) *I. vomitoria* — United States (Texas to Virginia).

D) Subsect. *Sideroxyloides*

(18) *I. Wilsonii* — China (Kiangsi, Hupeh).

(19) *I. sideroxyloides* — Jamaica, Portorico, Lesser Antilles.

(20) *I. divaricata* — Brazil (Amazonas).

We take the distribution of Subsect. *Repandae* to be basic for this group. We believe that this distribution was effected to the New World from *the Pacific* in consideration of the fact that the migration highlights the following points; Hongkong, Riu-kiu Islands, Bonin (Ogasawara) Islands, Caroline Islands, Baja California, Central Mexico. The significance of these localities need no longer be stressed, nor does it seems necessary to insist that crossings of the Pacific along this path are commonplace (see *Distylium, Perrottetia, Washingtonia, Mahonia, Lysimachia, Quercus, Talauma, Magnolia,* etc., etc.). Obviously, the dispersal of Subsects. *Eumicrodontae, Vomitoriae* and *Sideroxyloides* becomes transparent in the light of that of Sect. *Repandae*.

In conclusion — and this is worth indeed being strongly underscored — *Ilex* was invading the New World full blast in remote ages, and so doing from three directions simultaneously, (a) *The modern Atlantic* — The genorheitral segment faring this route directly issued from the African gate, meaning in plain words the ancient continent of Gondwana; (b) *The modern Northern Pacific* — Landbridges were used in these quarters which we have discussed generally in a former chapter (see p. 288); (c) *The modern Southern Pacific* — Landbridges also were taken advantage of in these quarters of which modern maps give no idea. The genorheitral streams active in the Pacific used the Western Polynesian gate (with possible additional infiltrations eastward from the African) and the Magellanian. *Ilex* and its allies, accordingly, issued in the genorheitral stage from all the standard gates of angiospermy.

The pattern of migration of *Ilex* Sects. *Aquifolioides* and *Cassinoides* (or *Lioprinus*) thus largely duplicates the dispersal of groups like *Platanus*, *Fraxinus*, *Liquidambar*, *Acer*, etc., the truest warhorses, in sum, of the believers in the "Holarctic" origins of the Angiosperms. *Ilex* reveals that these groups, too, never originated in "Holarctis," and tells us so with all the trustworthiness of a fellow-traveller of these groups along the ancient northern Gondwanic shore.

The pages of a century of botanical literature monotonously complain, that the origin of the Angiosperms is "Mysterious." We do not believe this to be true. These plants did not appear "Of a sudden," nor did they crazily wander all over the map. They acted, on the contrary, as nature dictated, and it is we, not they, who are responsible for building up "Mysteries," where no "Mystery" ever existed. Obviously, if we begin to assume that *Ilex* is "Derivative in phylogeny;" that the Angiosperms have "Holarctic origins;" that there is neither rhyme nor reason in dispersal; that to understand the fictitious we need pile up much which is absurd outright; we inevitably end by abusing our intellectual means to the point where these means stop functioning in a normal manner. We, not plants, wander about beating meaningless tracks.

The Passifloraceae, and their close allies, Achariaceae and Caricaceae, are distributed (6, 7, 8) in this manner (Fig. 60),

A) Passifloraceae

(1) *Mitostemma* — 2 species in Brazil, Guianas.
(2) *Dilkea* — 4 species in Brazil.
(3) *Crossostemma* — 1 species in West Africa (Sierra Leone).
(4) *Schlechterina* — 1 species in East Africa (Mozambique).
(5) *Machadoa* — 1 species in West Africa (Benguela).
(6) *Triphostemma* — 30 species in East Africa (Transvaal to Kenya).
(7) *Deidhamia* — 8 species in Africa (Madagascar, Cameroon, Gabon).
(8) *Adenia* — 80 species in Africa and Madagascar; about 20 species (Sect. *Microblepharis*) from Madagascar to Ceylon, Malaysia, Formosa, China, Northern Australia.
(9) *Passiflora* — 400 species massive in the Americas; none in Africa; some in Australia, New Zealand; about 15 species in Polynesia (Society Islands, Marshall Islands, Fiji, Samoa, New Hebrides, Norfolk and Lord Howe Islands, New Caledonia).
(10) *Tetrapathea* — 1 species in New Zealand.

B) Achariaceae

(1) *Ceratosycios* — 1 species in Africa (Cape to Natal).

(2) *Acharia* — 1 species in Africa (Cape).

(3) *Guthriea* — 1 species in Africa (Cape).

C) Caricaceae

(1) *Cylicomorpha* — 2 species in Africa (Cameroon, Tanganyka).

(2) *Jacaratia* — 8 species in Tropical America from Paraguay and Bolivia to Mexico.

(3) *Carica* — 40 species in Tropical America.

(4) *Mocinna* — 1 species in Mexico.

This dispersal is transparent. It leaves the usual angiospermous cradle south of Africa to dispatch (a) A standard track to the Far East from Madagascar (*Adenia* sect. *Microblepharis*), which invades Africa throughout (balance of *Adenia*) at the same time *); (b) A standard track leading straight from East to West Africa and further spreading to the tropical New World (*Deidhamia, Cylicomorpha, Mitostemma,* etc.). These two main channels of dispersal operate in conjunction with lesser important ones (e.g., Transvaal to Kenya: *Triphostemma;* Natal to the Cape: *Ceratosycios*), and sundry genera are scattered at strategic localities nearing the course, or the abutment, of the main tracks (e.g., *Crossostemma* in Sierra Leone; *Schlecterina* in Mozambique; *Mocinna* in Mexico). The Achariaceae, on their part, are typical of the southern flora of the Cape (see discussion, p. 52), and it is most likely that *Ceratosycios* points to the original region of entrance (Natal) into the Cape region of the achariaceous *genorheitron **).

---

*) We have underscored throughout the preceding pages the fact that the African gate of angiospermy is operative most frequently in the east of modern Southern Africa, and that the geographic center of the modern Indian Ocean was once occupied by a continent (Gondwana) which fed angiospermous *genorheitra* to Central Asia, the Far East generally and the Americas, through a channel of migration running as a rule across Central Africa of our maps. We have also shown by repeated examples that the Mascarenes and the waters of the Southeastern Pacific are intimately related in dispersal. We have further pointed out, at least once (see p. 49), that the tracks issuing to the New World and the Far East from the primary center of angiospermy and the African gate can be compared — for illustration's sake — to the sides of an immense angle, holding South Africa close by its vertex. We venture to suggest that readers who do not believe in the "Antarctic" origin of the Angiosperms work out patterns of dispersal from the African gate, at first following our method of procedure, then try to evolve their own in opposition to ours. While we are very far from claiming we are constantly in the right, we believe that these readers are bound to agree in the end that dispersal makes sense when driven in the main *from the south northward* — as we do — but yields nothing rational when worked *from the north southward*. In the present case, *Adenia* sect. *Microblepharis* follows what we may describe as the right side of the angle holding Africa, as it were, within its spread. The balance of *Adenia* hits Africa straight, on its part. It is beyond us, how migrations of this type could be worked out successfully from a point in the north.

**) Running here ahead of a discussion to follow (see p. 357), we will point out, (a) The Achariaceae originated in the *Afroantarctic Triangle;* (b) So did the *Passiflora-Tetrapathea* aggregate. This is shown to be true by, (a') The fact that the Achariaceae are narrowly localized in the Cape by a point of entrance from Natal; (b') The circumstance the *Passiflora-Tetrapathea* group cryptically fares from the southwestern Indian Ocean, or its vicinity, to the Pacific.

Adenia, on the contrary, originated in the *Gondwanic Triangle.* Other genera and group originated as follows, (a) *Triphostemma* in the eastern sector of the *Kalaharian Center;* (b) *Crossostemma* and *Schlechterina* in the *Nigerian Center.* The Caricaceae followed a classic track straight from the *Gondwanic Triangle* to the New World through Tanganyka and the Cameroon (cf. *Cassipourea,* p. 405).

As it will be seen, it proves now possible to break down large groups by primary and secondary

Could we rely upon a complete monographic treatment of *Passiflora*, we might guess with success as to the point of entrance of the passifloraceous *genorheitron* in the New World. Without the benefit of such a monograph, and purely as the face of the dispersal reads, we are inclined to surmise that the passifloraceous *genorheitron* invaded the New World from the direction of the modern Pacific, not the Atlantic. Certain species of *Passiflora* are endemic to Polynesia, and an aberrant passifloraceous form *(Tetrapathea)* is narrowly endemic to New Zealand. On this basis, the passifloraceous *genorheitron* might have entered the New World at any one of several regions, possibly several of these regions simultaneously. These regions might be found to lie in Peru, Bolivia and Chile in the very first place.

Paropsia is a genus of controversial status, which the authorities we follow (6, 9) credit to the Flacourtiaceae under a separate tribe Paropsieae. However, this very same genus is classified as passifloraceous outright by another author (10), whose intimate knowledge of the flora of Madagascar (wherein *Paropsia* is endemic with about seven species) compels critical attention. It is certain, at any rate, that a form connecting the Passifloraceae with the Flacourtiaceae is native to Madagascar. Let us notice, then, how properly located is this controversial form in relation to the *genorheitral* center of origin of the Passifloraceae, Achariaceae and Caricaceae. Right in this vicinity thus come to a head the tangled skeins that connect these three families together with the Flacourtiaceae. This is all the more worth noticing, in that the Flacourtiaceae can without difficulty be further connected with the Tiliaceae, Sterculiaceae and Euphorbiaceae. In brief, several of the most important families of the Angiosperms take their origin from a common node which lies in the vicinity of modern Madagascar, but is demonstrably operative further afloat in the Pacific *(Passiflora)*. Accordingly, three modern oceans at least, Indian, Atlantic, Pacific are open to inroads of vegetable life from this single one center. Of the families so involved one at least, Achariaceae, is "stranded" in the Cape, and may properly be viewed as a relic, despite its consisting of three genera and as many species.

Seeing matters in this light, we need no longer worry, whether *Paropsia* ought to be classified with the Passifloraceae or the Flacourtiaceae. We need not worry, likewise, about effecting a "perfect" taxonomic disposition of such genera as *Palmerella* (see p. 208), parts of *Phrygilanthus* (see p. 145), or *Juania* (see p. 455). Taxonomy stops in its tracks, when it comes to deal with forms that left the genorheitral matrix without having advanced far enough as to be "safely" classified one way or the other.

It is further clear that the flora of centers such as Madagascar, the Mascarenes, New Caledonia, etc. is bound to teem with peculiar endemisms. It may

---

origins, which open the door to fruitful comparative studies of systematic botany and phytogeography. Such studies, however, are obviously not the province of this manual.

The Ericaceae and the families under immediate discussion are homologus in the sense that both groups originated part in the *Afroantarctic*, part in the *Gondwanic Triangle*. This is also true of the Linaceae.

be a bold generalization to affirm that these endemisms, whether of one or twenty species, are relics but this affirmation must be true at least in part, perhaps, a large part. Lands steeped in antiquity, and geographically close by the primaeval genorheitral cradles, necessarily harbor forms of the kind. Madagascar has *Paropsia* to show, and New Caledonia such odd plants as *Canacomyrica* and *Dendromecon* *). It is from regions such as these that botany may legitimately expect answers to basic issues which the northern lands and all their floras are not competent to meet.

In conclusion, the ultimate antarctic and Gondwanic origin of the Angiosperms stands verified by such a mass of facts and circumstances that we ought to be blind to question it. The three factors of evolution, time, space and form, all come to head in the deep south of the modern map. They can be worked together harmoniously from this deep south northward, *but in no other rational manner*. Facts turning up in the flora of the Congo, and that of Greenland as well, all lead us invariably to the same angiospermous gates over the same tracks. The whole is orderly, precise, and — we like to repeat — altogether monotonous once its prime movers are safely identified. This is precisely what must be, because, despite its multitudinous aspects, plant-life is simple, and as such obeys simple laws.

II. Bombacaceae

The Bombacaceae are a forbidding group from the taxonomic and nomenclatural viewpoint, but a transparent affinity as regards distribution. It should be difficult, in fact, to find another group that so tellingly illustrates significant aspects of dispersal.

To judge from the literature of taxonomy, *Bombax*, *Gossampinus*, *Pachira* and *Salmalia*, not to mention minor segregates in their vicinity, can hardly be extricated one from another, and the phytogeographer is at times tempted to believe that much of this literature is so finely spun as to leave him in grave doubt, where "perfect" classification is to be had.

Taking *Bombax* in a broad sense, its range is fairly well covered by the following selection of species.

(1) *B. Ceiba* — Venezuela (Bolívar, Aragua), Colombia (Magdalena, Atlántico), Panamá, Costarica, Nicaragua.

(2) *B. aquaticum* — Guianas, Brazil (Pará), Peru (San Martín), Central America, Mexico (Chiapas to Veracruz).

(3) *B. Tussacii* — West Indies (Haiti).

(4) *B. ellipticum* — West Indies (Haiti).

---

*) We have not seen material of *Dendromecon*, but would suggest that it be compared with *Crossosoma* (Southwestern United States). A comparison may yield striking results.

(5) *B. emarginatum* — Cuba.

(6) *B. coriaceum* — Amazonian Colombia (Cerro de Aracoara, Serrania de Chiribiquete).

(7) *B. cyathophorum* — Brazil (São Paulo, Minas Geraës), Paraguay.

(8) *B. gracilipes* — Bolivia (La Paz: "Yungas.")

(9) *B. brevicuspe* — Ivory Coast, Gold Coast, Southern Nigeria, French Cameroon.

(10) *B. Andrieui* — Nigeria (Koulikoro).

(11) *B. costatum* — Nigeria (Koulikoro).

(12) *B. Houardii* — Nigeria (Koulikoro).

(13) *B. kimuensae* — Belgian Congo.

(14) *B. reflexum* — Western Africa, Western Uganda.

(15) *B. rhodognaphalum* — Mozambique, Tanganyka, Zanzibar.

(16) *B. malabaricum* — India, Indochina, Siam, China (Yunnan, Kwantung, Fukien, Formosa), Philippines (Luzon: Abra, Bulacan, Nueva Ecija, Batan, Rizal; Mindoro, Camiguin de Misamis, Mindanao), Java, Sumatra, Timor, Celebes, Tanimbar Laoet, Northern Australia, Queensland, Aroe Islands, New Guinea, Bismarck Archipelago.

(17) *B. insigne* — Burma (Mergui, Tenasserim), Indochina (Laos), China (Yunnan).

(18) *B. anceps* — Indochina (Cochinchina, Cambodja).

This dispersal is cleancut. It has Africa as its point of start, and from Africa reaches the Americas and the Far East generally in standard fashion.

Adansonia matches *Bombax*'s dispersal in several respects. This genus has possibly as many as ten species in Madagascar, plus two other, as follows,

(1) *A. digitata* — Tropical Africa throughout from the Senegal to Rhodesia and Northern Transvaal.

(2) *A. Gregorii* — North Australia.

. The classification is so involved, and so doubtful are the limits of many species that these tabulations are essentially informative. The dispersal is transparent on the contrary, for it constantly returns to the African angiospermous gate.

Adansonia is characteristic of the dry coastal region, savannah forest and open country generally. *Bombax* is also found in comparable habitats, but ranges far inland. It occurs furthermore in swampy forest not liable to periodical inundation, subsaline grounds, and the open forest generally. We may not generalize on the strength of our inadequate records, but certain inferences are at least suggested as probable,

(a) These plants are of great antiquity, and migrated to the New World prior to the Andean uplift; (b) Their original migrations were mostly made by the sea-shore and the dispersal was perfected in the first wave of angiospermous migration (Late Jurassic—Early Cretaceous), not only, but the epoch of wide submersions in the Mid-Cretaceous.

Certain species of *Bombax* are patently relics, witness *B. tenebrosum*, which was discovered in a very dry section of the forest west of Szemao, Yunnan *).

The Far Eastern main species of *Bombax*, *B. malabaricum*, invades Malaysia full tilt, both from the northwest (Northeastern India to Southern China, Formosa, Luzon) and the southwest (Java, Northern Australia, Lesser Soenda, Celebes, Mindanao, New Guinea, Bismarck Archipelago). This is typical mangrove distribution (see further p. 400).

Ceiba, the source of the industrial vegetable silk *kapok*, is represented in the first place by *C. pentandra*. The origin of this valuable tree has given rise, as usual, to discussions, and different authors have affirmed, or denied, its being native to various regions of the New and Old World. Some sources believe that the archetype closely agreed with the "spiny Kapok" of Cameroon and Western Africa.

Ceiba pentandra is widespread throughout the tropics in at least two forms (*C. pentandra* proper (*C. pentandra* var. *indica*): India and the Far East; *C. occidentalis* (*C. pentandra* var. *caribaea*): Africa, the Americas), and is beyond doubt native to Western Africa. While it has been widely planted and cultivated in the New World, certain records in the Amazonas, Portorico and Mexico hardly seem to stem from introduction. Moreover, *Ceiba* occurs in the New World with at least a dozen species, witness,

(1)   *C. acuminata* — Mexico (Chiapas to Baja California, Chihuahua, Tamaulipas).

(2)   *C. Samauna* — Brazil (Amazonas), Peru (San Martín), Bolivia (La Paz, "Yungas").

Under the circumstances, arguments as to *C. pentandra* not being endemic to the New World are pointless. Everything in dispersal points to the contrary being true.

It is claimed that *C. pentandra* is only planted in Asia and the Far East, but this claim, too, is not entirely supported by the evidence. Records from Northern Borneo in the immediate vicinity of Mt. Kinabalu and the Philippine island of Negros (Cuernos de Dumaguete) need be verified further, of course, but speak, as they read, against the hypothesis that *C. pentandra* owes its presence in Malaysia to man. It is far more probable that this species migrated throughout as did *Bombax* and, in part, *Adansonia*. Cleancut disconnections between Western Africa and the Philippines are definitely known (see p. 206, for example), and we are soon to see again which is the track responsible for them.

Bombax, *Adansonia*, *Ceiba* and their lesser segregates are the genera of the Bombacaceae which migrate most widely. All others are restricted to the Americas and the Far East, Africa having none so far as known today. This is not against reason, for the evolution of the bombacaceous *genorheitron* took place

---

*) If the data now available are at all to be trusted, *Gyranthera* is another cleancut relic-form of the Bombacaceae, with two species, *G. darienensis* in Panama, and *G. caribensis* in Venezuela (Miranda, Aragua).

away from the original center of dispersal (African gate: Madagascar—Transvaal). This family is accordingly a classic instance of peripheral evolution *).

Discussing *Prunus* (see p. 326), we point out that New Guinea is at the end of tracks from the African gate, and imply that records beyond New Guinea and its immediate dependencies (Bismarck Archipelago, for example) may stem from the Western Polynesian gate. The Bombacaceae are a probable exception to this. They have a monotypic genus *Maxwellia* in New Caledonia, represented by *M. lepidota*. A second monotype, *Papuodendron lepidotum* is narrowly endemic to Southern and Eastern New Guinea. As the Bombacaceae are otherwise unknown in the Pacific, it appears most probable that New Caledonia received the *genorheitron* that sired *Maxwellia* from the African gate, not from Western Polynesia at all, and that a secondary center of evolution arose between New Caledonia and New Guinea out of archetypes so received in origin.

The course of the tracks heading to New Caledonia, and New Guinea from the west is highlighted by several Malaysian genera, witness the following,

A) *Camptostemon*

(1) *C. Schultzii* — Northern Australia, New Guinea.

(2) *C. guineensis* — Eastern New Guinea.

(3) *C. philippinensis* — Borneo, Celebes, Aroe Islands, Philippines (Mindanao, Basilan, Bohol, Masbate, Panay, Negros, Luzon, Palawan).

This genus points to a track heading toward New Guinea, therefore New Caledonia, between North Australia and the Lesser Soenda, and veering northward to the Philippines from Celebes. This track is peculiar of mangroves and plants of the shore (see p. 400) and it is significant that *C. philippinensis* is one of their number.

A lesser segregate of *Durio* migrates as follows,

B) *Boschia*

(1) *B. Mansonii* — Burma.

(2) *B. acutifolia* — Northeastern Borneo (Tawao).

(3) *B. excelsa* — Northeastern Borneo (Tawao).

(4) *B. Griffithii* — Malacca (Perak, Selangor), Northeastern Borneo (Sandakan).

Still another petty genus yields the following,

---

*) We have pointed out (see p. 113) that the tracks originating from the African gate of angiospermy tend to diverge, V-shape, away from modern Africa, ultimately reaching the Caribbeans and the Far East. It will be seen that the evolution of the Bombacaceae bears out this rationalization, and cannot be accounted for by "Holarctic" migrations. This evolution, on the contrary, is accounted for in a standard manner if we lead the tracks away from the southern half of modern Africa to the Caribbeans and the Far East in the main pattern of the letter V.

C) *Coelostegia*

(1) *C. Griffithii* — Malacca, Sumatra, Bangka and Riouw Islands.

(2) *C. borneensis* — Sumatra, Northeastern Borneo (Tawao).

The tracks of these genera can be read in two ways, (a) From Burma in disconnection to Northeastern Borneo; (b) From Malacca, Bangka, Sumatra to Northern and Northeastern Borneo. The second of these tracks is standard, and pinpoints a region which is important in the dispersal of mangroves, *Quercus* and the affinities of *Magnolia* and *Tilia*. The former alternative, however, is by no means to be neglected. When discussing the magnoliaceous *Talauma* (see p. 119), we have pointed out that the immediate ancestors of this genus cut a straight path from the Philippines to Eastern Bengal avoiding China in the process. Peculiarly, *Talauma* contains these three records,

(1) *T. Elmeri* — Northeastern Borneo (Tawao).

(2) *T. gigantifolia* — Northeastern Borneo (Tawao).

(3) *T. Hodgsonii* — Indochina (Laos), Eastern Himalayas, Northern Bengal; which suggest a direct track between the region of Tawao and the Eastern Himalayas through Northern Indochina *).

All this *might* be a coincidence, but it is not likely to *prove* a coincidence, for it automatically accounts for the Philippines being recipient of forms that do not occur elsewhere in Malaysia. A channel of migration was open between the Eastern Himalayas and a region in Northern Borneo which easily could interest the Philippines (vicinity of Palawan, and Luzon, especially).

Considering that the Bombacaceae are entirely from the African gate, it is clear that all these petty genera were released by the *genorheitron* along a line connecting Burma—Malacca—Northern Australia, which, as we known (see p. 418), happens to be significant for the breaking up of Malaysia during the second half of the Cretaceous. The migrations of *Boschia* and *Talauma* are most likely, the latter's especially, anterior to this epoch, but those of *Camptostemon* fit right in with it.

Durio is a fairly large genus a species of which, *D. zibethinus*, yields a foul-smelling but most palatable fruit. *Durio* is massive in Western Malaysia (Malacca, Sumatra, Eastern Borneo), and some of its species migrate in this manner,

(1) *D. ceylanicus* — Ceylon.

(2) *D. malaccensis* — Burma, Malacca.

(3) *D. testudinarum* — Malacca (vars. *macrophyllus, pinangianus*), Southeastern and Northeastern Borneo, Philippines (Palawan).

The track of the last nearly perfectly matches that of *Prunus Junghuhniana* (see p. 327), and the dispersal is once more three-cornered, Burma—Malacca—

---

*) This track can work both ways, i. e., from the region of Northern Borneo and the Philippines to the Himalayas, or the other way around. Obviously, elements common to Borneo and the Himalayas are not necessarily "Himalayan" and "Holarctic." Their origin, whether from the African or Western Polynesian angiospermous gate, is to be determined against the background of the whole dispersal, not with reference to a secondary track only, such as the one outlined above. See further Figs. 31, 64, 65.

Palawan. Patently, we face here a circuit represented by a complex set-up of lands of which the modern map no longer gives a correct idea. The axis, Malacca (or Sumatra), Northern Borneo, Palawan, is one of the mainstay of Malaysia dispersal, and one which offers comparatively minor disconnections as the modern geography reads. Another track to match is in the north, Burma—Southern China—Hainan—Formosa—Luzon, or Burma—Northern Philippines outright. These quarters are crisscrossed, consequently, by tracks running in a manner that suggest the original presence of lands having different ecology throughout.

In the New World, the Bombacaceae are represented by a wealth of lesser genera. Some of them are typically "Amazonian" in distribution as

A) *Matisia*

(1) *M. bracteolosa* — Brazil (Amazonas).
(2) *M. Bolivari* — Eastern Colombia (Caquetá).
(3) *M. stenopetala* — Eastern Peru (Loreto).
(4) *M. longiflora* — Colombia (Boyacá).

B) *Catostemma*

(1) *C. fragrans* — Guianas.
(2) *C. sclerophyllum* — Brazil (Amazonas).

C) *Scleronema*

(1) *S. Spruceana* — Colombia (Vaupés).

These ranges are representative, for *Matisia*, for example, is a fairly large genus, while *Scleronema* is monotypic. The taxonomy of all these groups is still unsettled, and the standard revision of the Bombacaceae now available (11) unfortunately short of detailed records of locality.

Other American genera place before us a different type of distribution, as, for example,

A) *Cavanillesia*

(1) *C. platanifolia* — Panamá (Darien), Colombia, Peru.
(2) *C. hylogeiton* — Brazil (Acre), Bolivia (La Paz, "Yungas").
(3) *C. umbellata* — Brazil, Peru.

B) *Hampea*

(1) *H. thespesioides* — Colombia (Antioquia).
(2) *H. panamensis* — Panamá.
(3) *H. appendiculata* — Costarica.
(4) *H. trilobata* — Guatemala, British Honduras, Mexico (Yucatán).
(5) *H. tomentosa* — Mexico (Oaxaca).

C) *Neobuchia*

(1) *N. Paulinae* — West Indies (Gonave Island).

D) *Quararibea*

(1) *Q. turbinata* — Eastern Brazil (São Paulo at least), Guianas, West Indies (Grenada, St. Kitts, Portorico, Santo Domingo) *).
(2) *Q. Fieldii* — Mexico (Yucatán).
(3) *Q. guianensis* — Brazil (Amazonas), Peru (Loreto), Guianas.
(4) *Q. magnifica* — Venezuela (Aragua).
(5) *Q. Cacao* — Colombia.
(6) *Q. asterolepis* — Panama.
(7) *Q. funebris* — El Salvador, Guatemala, Mexico (Oaxaca, Puebla, Veracruz).

Quararibea has typical dispersal along a line Lesser Antilles—Eastern Brazil, with further migrations westward to the foothills of the Andes. Mexico and Central America also give certain indication of having been reached this time from the Caribbeans. The remaining genera are local segregates, rather, ecotypes of certain ranges, of which Gonave is noteworthy.

The Bombacaceae contain, like the Dipterocarpaceae (see p. 422), elements of the shore, and the hinterland right behind the shore. They indeed number in their ranks an arrant mangrove *(Camptostemon philippinense)*, and are seen in habitats (open forest not liable to inundation; forest occasionally inundated; slightly swampy soils with a substratum of humus and sand; arid coasts) suggesting, (a) The *genorheitron* originated at the marine shore, and the optimum of speciation was seemingly reached in times of submersion and emersion; (b) The surmise that *Catostemma* is an anomalous genus, intermediate between the Myrtaceae and Theaceae, proved erroneous (12), for, indeed, the technical characters of this genus clearly point to the Bombacaceae. This

---

*) A track leading from the Guianas to the Lesser Antilles is standard (see p. 345), but it is seldom that a track of the kind leads with one and the same species all way from Grenada to Santo Domingo. It is possible that *Q. turbinata* consists in reality of a complex of forms. At any rate, the front *Eastern Brazil—Guianas—Grenada—Santo Domingo* is important as a line of dispersal, and secondary evolution, in these quarters.

surmise, nevertheless, strikes us as a valuable phylogenetic lead, for, were the ovary of *Catostemma fragrans* to accentuate its present tendency to becoming inferior, and the androecium tend so fray more deeply, an important, perhaps a decisive, step, would be taken to turn this species into a member of the Myrtaceae; (c) The Bombacaceae show that a track from the African gate of angiospermy can possibly reach all the way to New Caledonia.

In all these respects, and in their stressing American and Asiatic ranges of undoubted palaeogeographical relevancy (Fig. 61), the Bombacaceae are a group of unusual interest. It is unfortunate that we have of them at present an all too sketchy and controversial record, and that it proves for this reason inexpedient to go beyond what has here been stated. It may be observed, nevertheless, that *Bombax* has a center of speciation in Nigeria (region of Koulikoro) which suggests comparison with another center of speciation of the Connaraceae, Barringtoniaceae, etc., located not too far, in the region of Oban and Eket in Southern Nigeria. All in all, we take for most probable that the bombacaceous *genorheitron* fared from the modern Indian Ocean to the New World over a classic line of migration which crossed Central Africa in the direction of the modern West Indies. This is the reason why this *genorheitron* has a center of heavy speciation in Nigeria.

III. *Prunus* and the Prunoideae

Prunus is a large and economically important genus which centers at present in China, where no less than 135 species are endemic. Taxonomists, as usual, are not agreed concerning its limits, for some define it narrowly, to include only the true Plums, and segregate out of it as separate genera Cherries *(Cerasus)*, Almonds *(Amygdalus)*, Apricots *(Armeniaca)* and other groups *(Laurocerasus, Padus)* of less popular status.

Differences of this kind, naturally, do not interest phytogeography because affinities and ranges remain in the end unchanged whether an alliance is handled as a collection of genera or an array of sections.

The Prunoideae consist of the following genera,

(1) *Prunus* (in a broad sense) — Over 200 species in the Americas throughout, Northern Africa, Europe, Asia, the Far East to New Guinea.

(2) *Pygeum* — About 50 species; 1 throughout tropical Africa from the Transkei (Caffraria) to Cameroon and Ethiopia, the balance in India, the warmer Far East to China and Formosa, Northern Australia (Queensland), Malaysia to the Bismarck Archipelago.

(3) *Maddenia* — 4 species in the Eastern Himalayas, Burma, Southeastern Tibet, China (Szechuan, Hupeh, Kansu, Anhwei).

(4) *Dichomanthes* — 1 species in China (Yunnan, Szechuan).

(5) *Prinsepia* — 4 species from the Western Himalayas (Punjab) to China (Honan, Kiangsu, Shensi, Shansi), Formosa and Eastern Siberia (Ussuri).

Though not at all unknown in tropical regions, the Prunoideae are a potential "cold" to "cool" group, witness the distribution of *Prunus, Prinsepia* and *Maddenia* to ranges like Northern Europe, Tibet, Kansu, Shansi and the Ussuri. We are not positive whether *Pygeum* is wholly alien to Madagascar and the Mascarenes, but it seems that, if perhaps not altogether unknown in this range, *Pygeum* is but very scarcely represented there. All in all, we are inclined to believe that this genus entered Africa immediately by the vicinity of Natal from a center of origins with "antarctic" connections (the *Afroantarctic Triangle*, to be precise; see p. 357) rather than from crumbled lands more to the north, once occupying the modern Indian Ocean. An origin of this kind is wholly harmonious with "cold" to "cool" preferences, witness *Erica, Linum*, etc.

If we are not mistaken, then, we can located the center of origin of the Prunoideae in the manner shown by the map (Fig. 62). Tracks can freely be drawn from this center to invade Eurasia in approved fashion, and further to enter Malaysia and the Pacific, as the map reveals. We will observe that the Prunoideae are one of the groups which like Cyperaceae and Compositae (see *Schoenoxiphium-Cobresia*, p. 237; the Crepidinae, p. 59; etc.) immediately dispatch to the Himalayas and colder Asia beyond, certain affinities (*Prunus* in part, *Maddenia, Dichomanthes, Prinsepia*) that seem to spring from the Indian Ocean as if from nowhere. *Potentilla*, we may recall (see p. 147) is another rosaceous group doing the same.

We very much regret we are short of a good monograph of the whole of *Prunus*, and we cannot for this reason explore the affinities of the strong component which this genus has in South America, whether peradventure part of it did not reach Argentine and Bolivia directly from a point in the Southeastern Pacific first approached by a track taking its start not far from Natal. It is not impossible that part of *Prunus* in the New World followed this southern channel, thus matching the behavior of *Lepechinia, Menodora, Cuscuta*, etc. in fundamentals. We will have occasion to mention this possibility further.

Leaving for the immediate present the broader aspects of the dispersal in play, we will take notice of certain ranges in *Prunus* and *Pygeum*, as follows,

(1) *P. Junghuhniana* — Sumatra, Java, Palawan.

This is the lone species of *Prunus* known from the Philippines (13), and a noteworthy one from the standpoint of distribution. The track drives from Sumatra and Java to Palawan, and is homologous of the channel taking certain forms of *Shorea* (see p. 422) directly from Sumatra to Northern Borneo. This channel belongs most likely to the first age of angiospermous dispersal and reveals, as exemplified by *P. Junghuhniana*, that Palawan is an ancient center of angiospermy in the Philippines. It is in this islands as a matter of fact that forms of the Magnoliaceae, Fagaceae and Hamamelidaceae occur unknown elsewhere in the Philippines. There is no doubt, then, that Palawan could be reached by the Angiosperms at the very first.

(2) *P. javanica* — Sumatra, Java, Northern Borneo, Moluccas (Soela and Boeroe Islands).

This track, too, is suggestive of *Shorea* (see p. 427), and it is possible that the Moluccan range was added to the primitive dispersal (Sumatra, Java, Northern Borneo) only after Malaysia began to undergo disruption.

(3) *P. phaeosticta* — Eastern India, Northern Burma to Indochina, Southern and Central China, Hainan, Formosa.

This range is classic of the first angiospermous age of migration.

(4) *P. Preslii* — Philippines (Mindanao, Leyte, Bohol, Luzon: Sorsogon), Formosa.

This, too, is a very ancient track. The Dipterocarpaceae are not in Formosa, though plentiful in the Philippines, because they found the connections between these regions already interrupted when they came secondarily to migrate in the age of Malaysian disruption. *Prunus Preslii* must have run this channel, accordingly, before this disruption took place.

The Philippines are one of the most interesting Malaysian regions, a bastion of angiospermous dispersal effected in both epochs of migration. We cannot give matters of local import more space in this manual than absolutely necessary, but in view of the circumstance that *Prunus* brought us to consider this archipelago we might perhaps consider some of the major tracks coming to it.

The following major tracks ran to the Philippines in the first age of angiospermous distribution (Late Jurassic to Early Cretaceous), (a) *From the direction of the Indian Ocean (i.e., Gondwana)* — (i) Northern India—Burma —Northern Indochina—Southern China—Formosa—Luzon; (ii) Sumatra (or Java)—Northern Borneo—Palawan—Mindoro—Luzon; (b) *From the direction of the Pacific* — (i[1]) New Guinea—Luzon—Mindoro; (ii[1]) Eastern Australia (or New Guinea)—Celebes—Mindanao or (preferably and then in disconnection) Southern Luzon.

In the second age of migrations, the Philippines were invested full blast from Borneo by channels of migration taking in at the same time Luzon, the Central Philippines and Mindanao.

We warn the reader that this outline is not intended to take care of all details, and to cover every contingency, as, for example, we show elsewhere tracks (see *Vatica papuana*, p. 428; etc.) which glance off the Philippines at Tawitawi, without entering the archipelago. This outline, nevertheless, will be found useful as a reference.

Styrax plays around the Philippines in a manner which has excited wonderment (14), because this genus appears to avoid these islands while freely migrating in Southern and Western Malaysia. To us, the matter is of the simplest.

Styrax "avoids" the Philippines merely because it bypasses them. This genus is not known from the body of these islands, but one of its species, *S. philippinensis*, is native to the Camiguin and Batan Islands barely off the northern tip of Luzon. This species is nearly allied with *S. kotoensis* (Botel Tobago) and *S. japonica* (Southern China, Japan, Corea), which is all as it ought to be. This track, then, is a mere variant of the one outlined just above (see (a) (i)), which hits the islets north of Luzon instead of coming to Luzon *).

While *S. philippinensis* bypasses the main body of the Philippines *in the north*, *S. agrestis* so does *in the south*, witness the latter's range, Cochinchina—Borneo—Celebes—Ceram—New Guinea—Solomon Islands—Caroline Islands. *Styrax agrestis* barely fell short of repeating the deed of *Vatica papuana* which just touched the southermost end of the Sulu Archipelago (Tawitawi) while on its way to New Guinea and points beyond.

We see no reason to wonder at migrations (Fig. 63) of this sort. The reader may easily verify for himself — for this manual contains the necessary elements of judgement — that *Styrax* hit Malaysia from the direction of the Indian Ocean.

Before leaving the *Styrax* affinity (Styracaceae) in the hands of the reader, we may, perhaps, take the liberty to contribute here brief notes to guide his investigation, as follows, (a) The Styracaceae are known in Africa only by one small genus, *Afrostyrax*, endemic to the Cameroon; (b) A genus formerly credited to this family, *Diclidanthera*, but now oftentimes treated as the type of a distinct family, Diclidantheraceae, occurs with about four species in Eastern and Amazonian Brazil, and Eastern Colombia; (c) Many species of *Styrax* are in the Americas all the way from Bolivia and Paraguay to the United States; comparatively few, however, seem to tenant the West Indies; (d) Two of the North American species are worthy of attention, namely, (i) *S. californica* (California and Western Mexico); (ii) *S. americana* (Virginia to Florida, westward to Missouri, Arkansas and Louisiana); the former is hardly better than a form of *S. officinalis* (Asia Minor, Balkans, Italy and Southern France); the latter is related with Far Eastern species; (e) *Halesia*, a styracaceous genus, has one species in Eastern China (*H. Macgregori*), and few species in the Southeastern United States (e.g., *H. carolina*); (f) *Styrax* does not seem to be represented in Ceylon, but occurs in India and becomes massive in the temperate Far East, where are also found hosts of minor styracaceous genera (e.g., *Parastyrax, Alniphillum,* etc.).

We observe, and the reader will verify our observations, (a) The Styracaceae match *Gnetum* in that they have a strong West African and Amazonian massing (*Afrostyrax, Diclidanthera*) but are otherwise unknown in Africa and Ceylon while being massive in India and the Far East; (b) The Styracaceae are largely "warm" but some of their genera are properly to be defined as "cool" (*Halesia, Styrax* in part). The origin of this family is to be sought in a center by the south of Africa, which fed at one hand the West African—Brazilian massing by a channel emerging northward from the South Atlantic; at the other the Far East generally in disconnection from the south of the

---

*) See also *Phoenix Hanceana*, p. 450.

Indian Ocean; (c) *Styrax* reached the United States by both transatlantic and transpacific migrations of the most obvious nature. The tracks of this genus appear to have largely bypassed the West Indies in the south and north alike, which accounts for the paucity of West Indian components in *Styrax*.

Naturally, we believe that *Styrax* bypassed the Philippines, too, closing upon Malaysia from the direction of the Himalayas and Southern China at one hand; of Indochina and Southern Malaysia at the other. The reader may further embroider on the matter. With reference to the dispersal of the Oleaceae (see *Notelaea*, etc., p. 355) we do not rule out the possibility altogether that some styracaceous genus might have originated in the Coral Sea, though the Indian, and Atlantic, Oceans remain as the keystones of the distribution of the Styracaceae in our opinion.

To return now to *Prunus*. This genus is well represented in the Old and New World by a Subsect. *Mesocraspedon* which according to our authority (15) is formed in this manner, and has this distribution, in part,

(1) *P. subcoriacea* — Paraguay, Northeastern Argentina.

(2) *P. tucumanensis* — Northwestern Argentina (Tucumán).

(3) *P. Ulei* — Brazil (Santa Catharina).

(4) *P. myrtifolia* — Var. *reflexa:* Brazil (Rio Janeiro: Serra dos Orgãos); Var. *brasiliensis:* Brazil (Rio Janeiro: Serra do Cubatão); Var. *accumulans:* British Guiana; Var. *typica:* Cuba Jamaica, Portorico, Hispaniola, Lesser Antilles (St. Vincent, St. Kitts, St. Eustatius).

(5) *P. oleifolia* — Paraguay, Bolivia.

(6) *P. amplifolia* — Peru (Puno).

(7) *P. integrifolia* — Peru (Huánuco).

(8) *P. rigida* — Peru (Cajamarca); Var. *subintegra:* Bolivia (La Paz).

(9) *P. recurviflora* — Colombia (Antioquia).

(10) *P. serotina* var. *salicifolia (P. Capuli)* — Ecuador, Colombia (Cundinamarca, Norte de Santander), Mexico (Michoacán and Puebla northward); Var. *typica:* Florida northward to Nova Scotia and Ontario, inland to Eastern South Dakota, Nebraska, Kansas, Oklahoma, Texas.

(11) *P. ilicifolia* — California.

(12) *P. Lyonii* — Baja California, islands of California.

(13) *P. occidentalis* — Guadeloupe, Portorico, Haiti, Jamaica, Cuba.

(14) *P. acutissima* — Guadeloupe.

(15) *P. lusitanica* — Azores, Madera, Canary Islands, Portugal.

(16) *P. Laurocerasus* — Central Balkans, Asia Minor, Caucasus, Iran (Persia).

(17) *P. semiarillata* — Western China (Yunnan, Szechuan).

(18) *P. macrophylla* — Western China (Szechuan, Yunnan), Riu-Kiu, Japan (Honshu).

(19) *P. adenopoda* — Java (Banjoemas, Noesa Kambang Islands).

The composition of this Subsection is most curious, for, not to mention transparent records in the Old World, it covers ranges in the New (Peru, Bolivia, Paraguay, Southern Brazil, Baja California) typical of tracks running a course in the East Pacific. Taking the whole, it might look as though part of this groups invaded the Americas from Bolivia going northward in the main; part again, on the contrary, came in straight upon the New World from the Mediterranean, running both north and south from the Caribbeans.

Most peculiar is the dispersal of *P. serotina*. Variety *typica* is of the Ozarkian—Appalachian Node throughout; var. *salicifolia* on the contrary of Mexico, Northern Colombia, Ecuador. One could readily understand this distribution, if only *P. serotina* occurred in the Bahamas or the Greater Antilles, for then the channel in play would be standard throughout (Florida—Bahamas—Greater Antilles—Eastern and Central Colombia southward). The door of entrance of this track into the Caribbeans would be properly set by the French West Indies *(P. occidentalis, P. acutissima)*, further to run to the Greater Antilles and Mexico in approved fashion. Quite as challenging is the dispersal of *P. myrtifolia*. The range Eastern Brazil—Guianas is consistent, but it is normal that the track running to the Antilles from the Guianas takes in the Windward Islands to Grenada and St.-Kitts, without further extending to the Greater Antilles westward from the Virgin Islands. We could easily understand tracks covering *as one* the ranges of *P. serotina* and *P. myrtifolia*, but can only with some qualms digest the distinct tracks of these two species as offered to us by the author whom we follow *).

In the last analysis, witness the dispersal of the Cochlospermaceae and Oxalidaceae, also of *Protium*, it is altogether possible for *genorheitra* closing upon the New World from the east (i.e., across the modern Atlantic) to reach stations typical of migrations reaching the Americas from the west (i.e., the Pacific's direction), and the other way around. Considering that *Prunus* is not in Western Africa; that Subsect. *Mesocraspedon* has species in the Mediterranean, not only but Macaronesia as well *(P. lusitanica);* we might *tentatively* conclude that this group came to the Caribbeans from the Mediterranean first, hitting the French West Indies in a normal manner, then faring to Mexico through the Greater Antille; to the United States through a disconnected ranges formerly involving the Bahamas; to the whole of South America by a standard tie Greater Antilles—Colombia.

As *Prunus* is a member of the Rosaceae we cannot stand the temptation of giving brief notice to another member of this family, *Rubus*. Working upon the genesis of the *Rubus* flora of Europe, a cytologist **) all too passingly comments that he found similarities between "Some South American species," and what he characterizes as "Four primary species." These species are,

(1) *R. Bollei* — Canary Islands.
(2) *R. ulmifolius* — "Introduced" according to Focke (Bibl. Bot., Heft

*) Plants, at any rate, are not the best material to work out Caribbean dispersal. Birds, as we hope to show, are much more satisfactory for the purpose.
**) GUSTAFSSON in Lunds Univ. Årskr., N. F., Avd. 2, 39 (6): 188. 1943.

83^2. 1914) in Uruguay, Southern Brazil, Argentina, Chile, Ceylon, South Africa (chiefly Natal); "Native" in the opinion of the same author in India (Kashmir), Iran (Persia), Caucasus, Crimea, Syria, Asia Minor, the Balkans and Europe generally to Germany, Northern Africa, Macaronesia (Azores, Canaries, Madera).

(3) *R. tomentosus* — Iran (Persia), Caucasus, Asia Minor, Syria, Transsylvania, Southern and Central Europe.

(4) *R. incanescens* — Algeria, Spain (Aragón), Southern France, Italy (Tuscany to Liguria).

We observe, (a) Focke is doubtless in grave error when crediting *R. ulmifolius* to "introduction" in South America, Ceylon and Natal. This species runs, on the contrary, a classic rosaceous (and generally "cold") track which — rooted in a point south of Africa — invests Natal and South America; Ceylon and Western Eurasia with the Mediterranean and Macaronesia. *This track is shaped V-like* (see furthermore p. 322 fn.) which is to be expected; (b) Gustafsson is altogether right in describing the four species above as "primary," for they have a dispersal consistent with the first age of angiospermous migrations; (c) It is possible that *Prunus* ran also the tracks of *R. ulmifolius*, barring the Natal foothold. Here, however, occurs *Pygeum*, a genus which is oftentimes hard to separate from *Prunus* by strong taxonomic characters; (d) A group of allies of *R. Bollei* (namely, *R. Hochstetterorum* — Azores; *R. Bornmuelleri* — Canary Islands; *R. rhamnifolius* — Central and Western Europe from Portugal to Norway; *R. ramosus* — Portugal, England) yields a perfect "Atlantic" pattern of dispersal. Some of these tracks bear being compared with those of *Erica*, and like the four "Primary species" of Gustafsson, *Erica*, too, originates in the deep south, right there, whence come *R. ulmifolius* in the first place.

The generic segregation between *Prunus* and *Pygeum* is not very strong, as said, and species are known almost intermediate between the two. At any rate, *Prunus* was beyond doubt thoroughly modern, and already in the possession of the generality of its modern range in the Early Tertiary. The Himalayan uplift rent in twain, for example, the range established in an earlier time by *Prunus armeniaca*, witness (see also Fig. 40, *Primula*),

(1) *Western massing* — Tianshan Mountains.

(2) *Eastern massing* — Yunnan, Hupeh, Kiangsu, Kansu, Shansi, Shensi, Hopei, Jehol; var. *holosericea*: Eastern Tibet; var. *Ansu;* Honan, Shantung, Corea.

A further marked center of species-making is located within a triangle, Pamir—Western Himalayas—Altais, wherein occur *P. divaricata*, *P. ferghanica*, *P. bucharica*, *P. spinosissima*, etc. This center feeds in standard fashion Central Asia, Afghanistan, Iran and Europe. *Notice how by following standard tracks we not only rationalize the whole of the migration, but we are led in addition to identify secondary centers of evolution of the utmost value in critical taxonomic and genetic work.*

The "European" *Prunus Avium* and *Tilia cordata* have their factual easternmost limit of dispersal south of the Caspian in the Eastern Punjab (Kulu District). This district is quite as sensitive a hinge of dispersal as is the Kurram Valley of Afghanistan. We regret we cannot say more here (see, however, p. 335) beyond remarking that these points can be reached from standard tracks out of the modern Indian Ocean.

A form of the "Mediterranean" *Prunus spinosa* (Iran to Northern Asia Minor, Southwestern Siberia, Caucasus, Europe generally, Northern Africa) is conflictingly reported as native or adventitious (Pennsylvania to New Jersey and Massachusetts) by taxonomists busy with the flora of the United States. We should welcome careful ecological work to learn which are the climaxes, if any, in which this large shrub has part in the New World. We feel confident, nevertheless, that this *Prunus* is almost certainly native to North America. To prove the contrary, we should be inclined to ask plain evidence to the effect that it was taken by man to the New World as a plant of medicinal or industrial value.

IV. Buxaceae

The Buxaceae are commonly held to be allied with the Euphorbiaceae. This is an error. The Euphorbiaceae belong to a *genorheitron* common to Tiliaceae, Sterculiaceae and, more distantly, Flacourtiaceae and Sapindaceae, while the Buxaceae repeat their origin from a *genorheitron* primarily embracing the Hamamelidaceae and Myrothamnaceae.

Regardless of their affinities, the Buxaceae are significant from the standpoint of dispersal on account of the light they cast on the connections between different continents. They consist of few genera with no more than 50 species, distributed in this manner,

A) Buxeae

(1) *Sarcococca* — Himalayas, Burma, Indochina, China (Yunnan, Szechuan, Kweichow, Hupeh); 1 species *(S. saligna)* in Afghanistan, Ceylon, India, China (Kwantung), Formosa, Philippines (Luzon: Mt. Data), Sumatra, Java, Bali.

(2) *Pachysandra* — 3 species in China, 1 in common with Japan *(P. terminalis:* Szechuan, Hupeh, Yunnan); 1 species *(P. procumbens)* in the Eastern United States (Florida to Louisiana, Kentucky and West Virginia).

(3) *Buxus* (including *Triceras*) — 1 species *(B. Macowanii)* in the East Cape and Caffraria; 3 species in Tropical Africa *(B. nyasica:* Nyasaland, Mt. Mlanje; *B. benguelensis:* Angola; *B. Hildebrandtii:* Ethiopia, Somaliland, Socotra), 1 species in Madagascar *(B. madagascarica)*, 1 species from the Northwestern Himalayas to Europe *(B. sempervirens:* Northwestern Himalayas, Persia, Caucasus, Northern Asia Minor, Central and Southern Balkans, Central and Southern France; Northern Spain; scattered in Southern Asia

Minor, Italy, South Central Europe and the Rhine Valley, Portugal, Algeria); 1 species in Syria, 1 in the Balearic Islands, 9 in the Eastern Himalayas, Southern China, Hainan, Riu-Kius, Japan, 4 in Indochina, 3 in the Philippines (Luzon, Palawan, Mindanao), 3 in Malacca and Western Malaysia *); about 20 species in Cuba, Portorico and the Lesser Antilles (massively of Sect. (or Genus) *Triceras*, and most abundant in Eastern Cuba); 2 species in Jamaica, 1 in British Honduras.

B) Stylocereae

(4) *Notoboxus* — 1 species *(N. natalensis)* in Caffraria and Natal; 1 species *(N. acuminata)* in the Belgian Congo (Region of the Ituri River) and Western Africa (Dahomey).

(5) *Styloceras* — 3 species in Bolivia, Peru, Ecuador and Colombia.

C) Simmondsieae

(6) *Simmondsia* — 1 species *(S. chinensis:* not in China!) in Mexico (Sonora, Baja California), United States (Southern California, Southern Arizona).

Certain aspects of this dispersal stand out in cleancut outlines, other are less transparent. We will review the former first as a preliminary of approaching the latter on a strictly factual basis.

Pachysandra patently runs a standard track between Central China (further involving Japan and Western China) and the Ozarkian—Appalachian node of the Southeastern United States. This track is in no need of comment.

Sarcococca is confined within a standard range having its extremes in Afghanistan, Formosa, Sumatra and Bali. The migrations of this genus are of very ancient age, because it occurs in the "Himalayan" centers of the flora of Luzon (e.g., Mt. Data), not only, but Formosa as well. It is altogether likely that *Sarcococca* streamed from the Himalayas to Kwantung and Formosa, ultimately to invade Luzon. Sumatra, Java and Bali may have been reached by a southern arm of the same main track. It should be interesting to study the ecology of the species called into play by these two secondary branches. At least the latter suggests a "Monsoon" type of vegetation.

---

*) The conspectus we give of *Buxus* is a compilation from various sources (indices, floras, herbarium material, etc.) made necessary by the fact that no comprehensive monograph of the genus is extant. Because of the war, we could not promptly consult HATUSIMA's review (in Jour. Dept. Agr., Kyusu Univ., 6: 262. 1942) of the Far Eastern end of *Buxus*, which, however, did not alter our conclusions of an earlier date. Much as we would like to discuss certain details of the migrations apparent in the distribution of species reviewed by HATUSIMA, we cannot do so because of the lack of ecological and particularized physiographic data, and even more because we cannot take too close an interest here in local issues.

Two species of *Sarcococca* are recorded from Southern Mexico (*S. Conzattii*: Oaxaca) and Guatemala *(S. guatemalensis)*. If the generic disposition is correct, which seems probable, *Sarcococca* has place in the list of the genera having cleancut transpacific migration. *Distylium* of the Hamamelidaceae, the characters of which convey at least some suggestion of the Buxaceae, is another of these migrants.

Buxus has a cleancut Gondwanic component (i.e., a group rooted in origin and dispersal in the Indian Ocean and lands immediately adjacent), which occurs in Nyasaland (typic is the station at Mt. Mlanje), Ethiopia, Somaliland and Socotra. If the dispersal stands precisely in the terms here described, we have a very interesting variant of migration, which skips Tanganyka and Uganda to turn up in Somaliland and Socotra, ultimately reaching Nyasaland*). We have seen a similar case in another part of our work (see p. 273), and believe something is here of potentially great significance for the obscure distributional history of the Afghanistan—Socotra—Somaliland triangle. This triangle has definite immediate connection with the Caribbeans and lands adjacent (see *Thamnosma* and the nyctaginaceous *Selinocarpus*), not to mention South Africa. Noteworthy, though not unusual, are the records of *Buxus Macowanii*, pointing to a penetration of the Cape from the northeast; and *B. benguelensis*. The latter is potentially rich of meaning, but the data in our hands do not even justify open surmises. It may be noticed that these two species are in the range of *Notobuxus* and the unique *Myrothamnus flabellifolius* which, like the Buxaceae, we believe to be related by *genorheitron* with the Hamamelidaceae. A tenuous indication might be read in the record just discussed that the *genorheitron* of *Buxus* is deeply committed to the south of South Africa. Precise distributional records are needed to go farther, which are not to be had today.

The Asiatic and Eurasian component of *Buxus* does not invite lengthy comment. It is standard for a genus having a Gondwanic component. The absence of *B. sempervirens* from Afghanistan (where *Sarcococca* occurs) may be significant. It is in this general vicinity (Kurram Valley in Afghanistan; Kulu Valley in Northwestern India; see p. 332) that are oftentimes broken the threads between "European" and "Asiatic" dispersal. It is also through this very same range (see p. 275) that runs a powerful stream of "cold" *genorheitra* pointing toward the Altais and the Far East generally. Here, then, is a region of break, parallel of that between Tanganyka and Nyasaland in the south.

This closes the conventional aspects of the buxaceous distribution. Though conventional, these aspects — as noticed — may still be quite interesting for what they say, and they not say alike.

If the Stylocereae are well formed in two genera, *Notobuxus* and *Styloceras*, and the records we have are representative of the ranges, it seems clear that the Buxaceae ran a track in the deep south of the Atlantic, which track in

---

*) Bird-dispersal yields homologous patterns. More about this will be said in a coming book.

wholly standard fashion could reach Bolivia, Ecuador, Peru and Colombia from the Eastern Pacific. That a track of this kind might well have been run is further strongly suggested by the typically Western American range of Simmondsia.

Buxus Sect. Triceras (or genus Triceras) could be easily interpreted on the strength of a reliable monograph of the Buxaceae. As this monograph is not yet extant, we are brought to moot possibilities, in this vein, (a) If Triceras is germane of the segment of Buxus in the immediate affinity of B. sempervirens, Triceras might have reached the West Indies from the Mediterranean over a now crumbled range involving Macaronesia; (b) If Triceras is allied, on the contrary, with Simmondsia and Styloceras, its origin is to be sought in the Pacific; (d) If Triceras is related with Sarcococca, it also might have invaded the West Indies from the Pacific; (e) Triceras might perhaps be connected with a buxaceous center in the Indian Ocean on account of the discovery of forms as yet unreported in Central and Western Africa; (f) Finally, the archetypes of Triceras might have reached the West Indies directly from an occult center of early origins in the south of the modern Atlantic. A precedent for a channel of migration of this nature could be found in the dispersal of the Empetraceae, Littorella, Myrsine africana, perhaps the Cneoraceae that turn up in the Caribbeans unaccountably in appearance, and doubtless belong to some relic-genorheitron.

As our choice, a choice based on guess and impression, no more, we should think that the last hypothesis is the most probable. The Buxaceae are so deeply committed to the region which is now South Africa, and so peculiarly situated in Angola and the vicinity of the Ituri River in the Congo, that we can easily imagine that the archetypes of Triceras came to the Caribbeans directly from a genorheitral focus situated in the south of the modern Atlantic.

We have brought the distribution of the Buxaceae to these pages as a typical example of interpretation of dispersal effected on the strength of indifferently compiled data. Whenever decisions prove impossible or unadvisable, we offer the reader alternatives for further consideration. Though the face of the record of the Buxaceae does not read at all like the dispersal of Phylica (see p. 58), still we believe that the ultimate base of dispersal is the same in both. To speak the language originally introduced in this manual, the Buxaceae seemingly used both ends of the African gate, and certainly at least the eastern, which stamps them as conventionally Gondwanic and altogether standard in dispersal.

The dispersal of the Buxaceae further lends itself to observations that are of general interest. We have commented, for example, that Sarcococca reaches the "Himalayan" center of flora in the Northern Philippines. Reference to a diagram (Fig. 64) shows without further how unsatisfactory is the term "Himalayan" in regard of Malaysia. In this region, "Himalayan" is anything and everything which in some way or other is common to the Philippines or Sumatra (here the Gajolands above all) and the Himalayas. Such a community could be established, to begin with, by tracks coming in from the Pacific or streaming from the Himalayas as a byproduct of original migration from the Indian Ocean. It is further possible to conceive of any number of contingencies, whereby a Malaysian center can have contact with the Himalayas. Under the circumstances to identify anything in phytogeography as "Himalayan" without a previous rigorous study of the tracks involved is obviously objectionable.

The distribution of *Buxus sempervirens* in Europe is effected by a classic line of penetration which departing from the Caucasus points almost due in the direction of Dalmatia, and from Dalmatia veers toward the Pyrenees. Variants of this line commonly occur, such as the one previously identified (see p. 275) in the migrations of *Ephedra*. Both *Buxus sempervirens* and *Ephedra* can be brought back to a starting point located within the triangle Ethiopia—Ceylon—Caucasus.

It is readily apparent that plants other than those just mentioned, e.g., *Carex pyrenaica*, *Coriaria* go back to a starting point which has nothing to do with the one just described. Their starting point is situated in the Southwestern Pacific. Quite as much, the starting points of such plants as *Cneorum*, *Littorella*, *Erica* are other than those of *Carex* and *Coriaria*. In brief, a world of "European endemics" becomes ultimately "European" out of starting points in dispersal that are absolutely antipodal.

A discussion of European distribution cannot find place in these pages, because it would call for extensive considerations of live and fossil material alike. *Rhododendron ponticum*, for instance, mainly runs a track leading from Syria through the Aegean ultimately to the Iberian Peninsula, the Canary Islands and Azores. However, this species (16) is also known from some isolated station in Thrace and was reported fossil in the Alps. It is accordingly manifest that its range varied throughout the course of time, and the task of reconstructing it in detail would be most laborious and time-consuming.

Ruling out a consideration of particulars, we may contribute here at least a rough outline of the major channels of distribution that come to Europe from all over the compass. It will immediately be seen that there is no more an "European" than there can be a "Malaysian" phytogeography. Everything in dispersal belongs to, and with, everything else.

The major channels of migrations (Fig. 65) coming to Europe are the following (a) *From the Altais westward and mostly northward* — The *genorheitra* fed at first the Altai Node coming in either from the Western Polynesian *(Rhododendron)* or African gate (Ericaceae Andromedoideae) alike. This track reaches the highest north; (b) *From the Himalayas mostly westward*. This track, too, is fed from the African *(Buxus)* or Polynesian *(Carex)* gate. Huge disconnections usually mar the course of this channel of migration between the Western Himalayas and Asia Minor; (c) *From the region Tanganyka—Ethiopia (ultimately also the Cape), mostly northward*. This track runs two main branches, (i) This branch *(Erica)* involves the Central Sahara (Hoggar),

Cyrenaica, Macaronesia in the first place. Leaving Cyrenaica it easily passes on to Sicily and the Aegean; (ii) Are called here into play *(Orobanche, Pimpinella)* the Sinai, Egypt and generally the Eastern Mediterranean. This track takes its inception uniformly by the African gate of angiospermy in the Indian Ocean; (d) *From the West Indies mostly eastward.* Typical of this route is the European end of *Cneorum;* (e) *From South America and the Southern Atlantic generally.* Here fall the Malvaceae Malopeae, *Littorella* and *Empetrum;* (f) *From North America generally.* Certain species of *Vaccinium* may belong to this stream.

The ultimate origins of the plants reaching Europe over "American" channels (i.e., tracks (d) to (f)) are to be sought for case by case. Supposedly classic "Pontic," "Sarmatic" etc., elements in the flora of Europe may belong to tracks (a), (b), (c) alike. Typical "Atlantic" dispersal is furnished by plants faring over tracks (c), (d), (e), (f), etc.

It is transparent that terms such as we have mentioned, "Pontic," "Sarmatic," "Atlantic" are misbegotten. Two concepts are usually confused as one by those who employ them, namely, (a) *Geographic.* This is the concept most immediately involved when "Pontic, ' for instance, is taken to refer to a form which "originates" from the ancient Pontus or modern Black Sea. *Tilia,* of instance, is a "Pontic" element in this sense; (b) *Ecological.* We stress this idea when we understand as "Pontic" certain evergreen prevailingly sclerophillous elements, actually or supposedly dominant in certain formations of the Caucasus. In this sense, *Tilia* can no longer be "Pontic."

These notes are not intended to sketch even broadly, the phytogeography of Europe, an impossible task within the limits of this manual. They are directed to show that it is mostly useless to speak of an "European" flora made up of "Pontic" elements, etc., when we intend to use forceful, precise language. This terminology has been endlessly manipulated and altered by authors who have sought in description means to convey ideas that description is unable to return; because of this, technology has come to swallow up genuine knowledge.

V. Salvadoraceae

The Salvadoraceae are distributed (17) in this manner,

A) *Azima*

(1) *A. angustifolia* — Madagascar.

(2) *A. spinosissima* — Southwest Africa.

(3) *A. tetracantha* — Cape, Angola, Southwest Africa, Madagascar, Comoros, Eastern Africa, Southeastern Arabia.

(4) *A. sarmentosa* — Burma, Indochina, China (Hainan), Java, Madoera, Sumbawa, Philippines (Mindanao: Zamboanga, Davao; Luzon: Rizal, Union, Ilocos Norte).

B) *Dobera*

(1)   *D. loranthifolia* — Eastern Africa (Mozambique, Tanganyka).

(2)   *D. Macalusoi* — Somaliland.

(3)   *D. glabra* — Upper Sudan, Ethiopia, Somaliland, Arabia, Western India (Sindh).

C) *Salvadora*

(1)   *S. angustifolia* — Madagascar, Aldabra Island.

(2)   *S. australis* — Natal, Transvaal, Mozambique.

(3)   *S. persica* — Southwest Africa, Angola, Tanganyka, Upper Sudan, Western Africa to Senegal, Ethiopia, Somaliland, Egypt, Arabia, Palestine, Syria, Iran (Persia), Baluchistan, Western India (Sindh, Punjab), Southeastern India, Ceylon.

(4)   *S. oleoides* — Western India (Sindh, Punjab).

The Salvadoraceae are xerophytes or subxerophytes to judge by the face of their dispersal. Tracks of theirs that read *Burma—Indochina—Java—Madoera—Sumbawa* in the east are a fair match of tracks spelling *Upper Sudan—Somaliland—Arabia—Sindh* in the west. It is a reasonable anticipation that plants of this nature, and faring by tracks of this description, would take prompt advantage of even comparatively late events, such as the desiccation of Africa and India in historical times *) and the recent Tertiary, to acquire increased ranges throughout.

If this is possibly the case with other plants, it seems plain that the Salvadoraceae are not in the same category. The record of their distribution shows that their spread follow a generally eastern trend. *Salvadora persica*, one the "wides" in the Salvadoraceae, occurs in Western Africa to the Senegal, but none other of the dozen species, or so, in its affinity does the same.

In the western sector of Africa this family tenants relatively little; Southwest Africa with three species, one endemic *(Azima spinosissima)*, which shows that the dispersal is not of altogether secondary nature in this range; Angola with two species, both "wides."

The question, why the Salvadoraceae failed to cross over to the New World can be answered, then, with the statement that this family never had in Western Africa a *substantial* center of secondary, even less of primary, distribution and evolution. If we slash Africa with a diagonal line running from Sudan to Southwest Africa, the Salvadoraceae are all of the east of this line, but for occasional outliers in Angola and the general vicinity of the Senegal.

It might be objected that *Thamnosma* and *Oligomeris* actually ran all the way from the Caribbeans regions to South Africa and Socotra. We will answer that there is no doubt as to the range of *Thamnosma* being primary in all these

---

*) Good evidence to this effect is contained in "The Kalahari or Thirstland Redemption" by SCHWARZ, E. H. L. (undated, Maskew Miller publs., Capetown), and an article by RHANDAWA (in Jour. Bombay Nat. Hist. Soc. 45: 558. 1945).

centers, witness the comparatively strong speciation and the isolated position which Socotra generally occupies. As to *Oligomeris* we know that certain of its ranges are primary (e.g., in Southern Africa), but we are forced to believe that *O. linifolia*, the hero of the dispersal mentioned above is a straight weed that was strategically located in origin to take advantage both ways of the events that lead to the separation of Africa and South America. In conclusion, *primary migration* is one thing, *secondary migration* another, and we will understand nothing of either, to begin with, if we uncritically associate *Oligomeris linifolia* with the whole of *Thamnosma* and the Salvadoraceae.

It stands to reason that as the Salvadoraceae never established themselves firmly in Western Africa by *a primary migration*, they could not take advantage there of events favoring a *secondary one*. One of their species, *S. persica*, came indeed in time to the very shore of the Atlantic in Western Africa, but too late to cross over into the New World.

It is accordingly perspicuous that dispersal is not by indiscriminate agencies, but on the contrary by centers originally reached in migration which, as conditions and circumstances dictate, may become active in a second time. Sarracenia contains a classic example (see p. 468) of genuine secondary spread from a point originally reached by primary migration. The difference in time between these two migrations in *Sarracenia* is not less than the whole difference between the Late Jurassic and the Pleistocene. However, in order that the latter could eventuate in the Pleistocene, possibly even later, it was necessary that the former took place in a prior epoch, which evidence suggest to have been in this case the Late Jurassic.

There is, then, conflict between the modern map and the tables of dispersal as regards the Salvadoraceae. The modern map states that this family "is" in Western Africa; the tables of dispersal say the contrary. Right are the latter, of course.

What are, then, the primary centers of the Salvadoraceae, the centers in other words reached by a primary migration?

To identify these centers involves an element of opinion, but, acting within our lights, we will designate them.

It is clear in the first place that the Salvadoraceae were rooted in the south of South Africa *ab initio*. Perfectly consistent, and closely knit, is the range of *Salvadora australis* (Mozambique, Transvaal, Natal) which through *D. loranthifolia* leads northward from Mozambique to Tanganyka. There is no doubt as to Madagascar belonging to the primary migration on account of *Azima angustifolia* and *Salvadora angustifolia*, the latter consistently reaching the Aldabra Islands. Southwest Africa is quite as important as Madagascar, witness *Azima spinosissima*, *A. tetracantha* and *Salvadora persica*. In brief, the range reached in original dispersal by the Salvadoraceae can be confidently sketched as follows, to begin with, Southwest Africa (ranging to Angolo), Cape (most likely entered from Natal), Natal and Transvaal, Mozambique, Madagascar, Tanganyka.

Once in Tanganyka, we are offered a further selection of records all most likely belonging to the primary range, namely, Ethiopia, Somaliland, Southeastern Arabia, Iran (Persia), Ceylon. The balance of the modern range may be left to opinion, but the Sindh is seemingly to be reckoned as a primary center of *Salvadora*, which turns up here with an endemic.

The dispersal of *Azima sarmentosa* is not easily read. Localities like Java and Hainan are most likely of the primary migration, but the evidence in our hands is to meager to vouch for the whole of the track, despite its being well known from other plants. In cases of the kind, it is desirable to undertake studies on the basis of as many homologous migrants as possible.

In conclusion, the Salvadoraceae have certainly not wandered unaccountably all over the map. They have remained very much tied to the original center of their dispersal. This center is essentially Gondwanic, that is, rooted by the African gate and, on the whole, closely wedded to lands fronting the modern Indian Ocean.

VI. Stackhousiaceae

The Stackhousiaceae are a first-class conundrum of systematic botany. We will not venture to discuss their possible affinities, which at least distantly suggest in part the Celastraceae possibly also the Thymeleaceae, but record their known distribution (18), as follows,

A) *Macgregoria*

(1) *M. racemigera* — Central Australia and adjacent regions in Australia.

B) *Stackhousia*

(1) *S. minima* — New Zealand.
(2) *S. pulvinaris* — Tasmania, Victoria, New South Wales.
(3) *S. flava* — Tasmania.
(4) *S. viminea* — All Australia, except the center.
(5) *S. occidentalis* — West Australia.
(6) *S. spathulata* — Tasmania, Australia.
(7) *S. megaloptera* — Central and Western Australia.
(8) *S. intermedia* — Queensland and Northern Australia, New Guinea, Boeroe, Amboina, Celebes, Sumatra, Philippines (Culion, Guimaras, Luzon: Pangasinan, Bontoc, Ilocos Norte, Cagayan), Palau, Carolines.

C) *Tripterococcus*

(1)   *T. Brunonis* — West and Northwestern Australia.

Stackhousia intermedia rides in part a classic track which leads from Queensland and Northern Australia generally to New Guinea, Celebes, the Philippines, at one hand; Sumatra at the other. This track gives us a standard tie between Queensland and the Philippines to begin, and leaves us free to consider next the balance of the dispersal.

The Australia massing contains two hints that, taken at face value, are of some interest. The first is that *S. viminea* occupies all Australia, except the center; the second that the monotypic *Tripterococcus Brunonis* is restricted to West and Northwestern Australia.

In the Stackhousiaceae, then, we have three centers, (a) *Southern and Eastern Australian* — This center takes in Tasmania, New Zealand, and all of Eastern Australia generally; (b) *Western Australian* — Here fall Northwestern and Western Australia; (c) *Extra-Australian* — This covers New Guinea, Malaysia, Philippines, and Micronesia.

There is a close agreement between the range of *S. viminea* and the cycadaceous *Macrozamia* (see Fig. 93), not only, but the speciation of *Frankenia* see footnote, p. 383) gives indication of having been effected by the shore of bodies of water in process of shrinking mostly in the interior, and the south of Australia. Taking our clue from these factors, we would be inclined to suspect that the connection between Western and Eastern Australia was primarily assured through Northern Australia. To put this on the map, we are not inclined to assume for the Stackhousiaceae a direct migration north to south or south to north, rather a point of entrance into Australia so located in the modern Coral Sea as to be intermediate between Tasmania and the Philippines, not only but very nearly facing Northern Australia. It will be observed that, (a) This being the initial set-up, West Australia would be reached from the east or northeast; Tasmania and New Zealand from the northwest; (b) The Stackhousiaceae would normally take place among the "warm" elements in the New Zealand and Tasmanian flora, which seems to be within reason in consideration of the sum total of the stackhousiaceous distribution in warmer lands; (c) The original center of migration would be geographically located in a position intermediate between the extremes.

The track of *S. intermedia*, taken as it stands, is most interesting. It leaves Australia and New Guinea to Celebes and the Moluccas which is standard; from this point on, it runs a seemingly whimsical course to Sumatra, at one hand, the Palau and Caroline Islands at the other, last to hit Guimaras, Culion and Luzon.

It is not difficult to account for a connection between Celebes and Sumatra, but the rest of this track obeys none of the geographical premises of our maps. With the island of Guimaras as the center it describes a nearly perfect arc resting upon the Carolines and Northeastern Luzon through the Palau, Culion and Western Luzon.

We may not settle the matter, what does this arc means in detail, because we are not certain we already have the whole of the range of *S. intermedia* before us. It is clear, nevertheless, that a channel of dispersal could run in the Western Pacific to the Palau and Caroline Islands fit to play in and around the Philippines, and Malaysia generally, in a manner defying the map of today. Let us observe that this channel pointedly ignores Mindanao, which often-times happens in migrations of the first angiospermous age. Migrations of the kind usually invade Luzon in disconnection from Australia or New Guinea.

In conclusion, the tentative inference is in order that the Stackhousiaceae originated from a *genorheitron* which matured in the modern Coral Sea. As to the track of *S. intermedia*, we may safely infer from its behavior and that of other species (e.g. *Styrax agrestis*), that connections existed by Ceram and Celebes which easily led to the Carolines at one hand, and such regions of the Philippines as we have mentioned at the other.

VII. Cyrillaceae and their allies

The Cyrillaceae are a small American family distributed thus (19),

A) *Cliftonia*

(1) *C. monophylla (C. ligustrina)* — Florida, Louisiana, Georgia.

B) *Purdiaea*

(1) *P. microphylla* — Eastern Cuba.
(2) *P. cubensis* — Cuba, Isle of Pines.
(3) *P. stenopetala* — Eastern Cuba.
(4) *P. velutina* — Eastern Cuba.
(5) *P. Shaeferi* — Eastern Cuba.
(6) *P. nutans* — Colombia (Norte de Santander, Santander).
(7) *P. Weberbaueri* — Peru (Amazonas).

C) *Cyrilla*

(1) *C. racemiflora (C. antillana)* — Brazil (Rio Negro), Venezuela, Gui-anas (Mt. Roraima), Trinidad, Lesser Antilles (St. Vincent, Dominica, Gua-deloupe), Portorico, Hispaniola, Jamaica, Cuba, British Honduras, Mexico (Oaxaca), United States (Florida to Texas, Missouri, Southern Virginia).

(2) *C. brevifolia* — Guianas (Mt. Roraima), Venezuela (Southeastern districts: Mt. Duida, Mt. Auyan-Tepui).

(3) *C. Perrottetii* — Lesser Antilles (Guadeloupe).

(4) *C. cubensis* — Eastern Cuba.

(5) *C. nitidissima* — Eastern Cuba.

(6) *C. nipensis* — Eastern Cuba.

(7) *C. parvifolia* — Florida, Louisiana.

(8) *C. arida* — Florida.

D) *Cyrillopsis*

(1) *C. paraensis* — Brazil (Pará: Rio Trombetas).

The dispersal of this family is one of the most interesting in phytogeography.

Cyrilla racemiflora fills by itself the Caribbeans, North and South America to Missouri, Virginia and the Amazonas, respectively. This is most unusual. According to rule, the range of this one species should fall at least in two distinct sectors, each tenanted by a separate species, as follows, (a) Amazonas—Guianas—Trinidad; (b) Greater Antilles (Portorico to Cuba)—Mexico—Southeastern United States. The Lesser Antilles (St. Vincent, Dominica, Guadeloupe) might in reality stand as a third sector of their own, for they may go with some difficulty with range (a); and with difficulty belong to range (b).

If there is a conclusion to be drawn from the dispersal of this one species, and the balance of *Cyrilla*, not only, but *Cliftonia* and *Cyrillopsis*, this is that these three genera reached full maturity along the whole axis *Florida—East Cuba—Brazil*. This axis, or part of it at least, is altogether important for the history of the affinity of *Cereus*, for instance (see p. 370), and it is clear it corresponds with a relevant part — now lost altogether undersea — of the continental connections binding Africa to the New World. The island of Guadeloupe is a classic bridgehead for channels of migrations coming to the New World from the Mediterranean and Macaronesia, as we know; Eastern Cuba is one among the very oldest lands afloat in the Caribbeans; the Pacaraima orogeny was open to angiospermous colonization not later than by the end of the Jurassic or the very Early Cretaceous. In brief, the Cyrillaceae are, like the Cneoraceae, a group which reached maturity out of a very ancient *genorheitron* right between Africa and the New World of yore. We are next to see how the *genorheitron* reached these quarters.

Purdiaea is very strongly rooted in Cuba, Eastern Cuba particularly, which is not surprising in view of the fact that *Cyrilla* has three different species in the same center. The station held by *Purdiaea* in Cuba generally, and the Isles of Pines is also interesting, because this stations lies along the front Lesser Antille—Texas (or adjacent Mexico), which, though unusual, is known at least from the magnoliaceous *Talauma minor* and *Talauma* generally (see p. 119, 331), and further adumbrated by the distribution of some *Prunus* (see p. 331). This channel lies along one of the classic disconnections crossing the modern Caribbean Sea more or less parallel with the longitudes. We discuss these disconnections elsewhere (see p. 368), which probably correspond with lines of ancient faulting, or orogenies long ago vanished out of sight.

Striking is in *Purdiaea*'s dispersal the cleancut track *Cuba—Eastern Colombia—Peru*. We are not now quite clear as to what this track meant in the ancient geography of the Caribbeans and South America generally, whether a marine trough, or a chain of heights, or some other relevant physiographic feature. We are sure, on the other hand, that the ecologic and geographic nature of this channel could be worked out by studying an adequate number of plants faring it.

Plants of the kind are numerous indeed because this channel is a keystone in American phytogeography. Its action, and the centers of secondary speciation to which it gave rise, are responsible for numerous relevant features of the flora of the Colombian—Venezuelan boundary. Its influence is felt as far afloat as Bolivia, Eastern Brazil, Central America, Mexico and the West Indies.

Some of the evidence that bears upon this channel will be reviewed in pages to come. Merely as a preliminary we may give here cursory attention to two unexplained peculiarities of dispersal that interest both Venezuela and Colombia, as follows, (a) *Croton Malambo* is a relic-species of speculative affinity, narrowly ranging by the original type-locality on the northeastern Colombian coastal region. This relic, or a form quite close to it, turned up in disconnection in the Venezuelan state of Barinas. The localities of finding are connected by a diagonal line that slashes clear across the modern Andes; (b) A species of the cunoniaceous genus *Weinmannia* was discovered endemic to the summit of Cerro Santa Ana, a wholly isolated low peak in the Venezuelan state of Falcón. This genus belongs to a group (see p. 385) of wide distribution, well represented in the modern Andes. It further occurs in Venezuela in the western states (e.g., Mérida), the vicinity of Caracas, the Pacaraima orogeny and the sierras of the state of Sucre (Cerro Turumuquire, etc.).

Distributions of the type illustrated by *Croton Malambo*, *Weinmannia* are commonly described as "Andean." This adjective is a misnomer, and the interpretations it fosters are wholly erroneous, because, (a) *Croton Malambo* and *Weinmannia* had reached their present stations geologic eras before the rise of the Andes, which is a late event of American physiography, mostly dating from the second half of the Tertiary; (b) The Cerro Santa Ana, the heights of Sucre and vicinity (e.g., Cerro Turumuquire) cannot be "Andean" by the farthest stretch of imagination.

It is transparent that we face here a pale of geography that has nothing to do with the Andes, though it was later modified by them, and in part bodily overtaken by them. The channel which leads from Eastern Peru to the Greater Antilles makes contact in the general region of the Colombian—Venezuelan modern boundary with an artery leading eastward through coastal Venezuela as far at least as the islands of Margarita and Trinidad, not only, but the Paca-raima System as well. This very same channel feeds outlying tracks westward which eventually reach Guatemala and Southern Mexico (Oaxaca most particularly).

Purdiaea manifestly fared over this main artery between Cuba and Peru. *Cyrilla* on its part crossed over from Cuba to Mexico (Oaxaca) and British Honduras. The Cyrillaceae, then, freely roamed a Caribbean and American world of which our maps contain no longer trace, but dispersal still highlights with massive accuracy. In view of this evidence it is unconceivable that a palaeobotanist should affirm *) that the original colonization of many "North-ern American" plants still persisting in the Antilles and Central America is to be credited to the Late Miocene. Quite as incredible are the statements of certain zoologists (op. cit., 108) that life in the Caribbeans affords no evidence to the effect that the Lesser Antilles were once connected by a "Landbridge" with the Greater Antilles. Strange to say, the author from whose work we cull these opinions — which he accepts as well informed — is well aware (op. cit., 155) *out of the store of his own knowledge* that there existed "Rapid and easy means of communication" between Mexico and the worlds to the east ("Tethyan Provinces") in the Late Jurassic. Peculiar once again, that the very same author is informed that in the Late Jurassic, too, migrants from India could enter the Caribbeans from the direction of the Northern Pacific. *Indeed, the Late Jurassic is the competent epoch in early angiospermous dispersal,* and the connections which Schuchert stresses on his own are precisely those which account for the manner of the angiospermous invasion of the Caribbeans. Characteristic of the utter lack of coordination between the geological and biological sciences is the circumstance that the geologist in point (Schuchert) gives no importance to the fundamental evidence in his hands, but accepts as authoritative the dicta of a palaeobotanist (Berry) which pointedly stultify everything of angiospermous dispersal. *The geologist, as a matter of fact, may readily use this dispersal to reach without difficulty, and with a safe margin of accuracy, conclusions which he culls out of marine records, etc., of a far more obscure and less accessible nature.*

The Cyrillaceae are usually classified near the Aquifoliaceae which is not satisfactory. In our opinion on the contrary, the Cyrillaceae belong to a large and still unexplored *genorheitron*, vital part of which is represented by the living Crypteroniaceae. The Crypteroniaceae consist of the monotypic genus *Crypteronia*, which is distributed as follows,

(1) *C. paniculata* — Himalayas, Burma, Malacca, Sumatra, Java, Soem-bawa, Southeastern Borneo, Philippines (Negros, Luzon: Cagayan, Lepanto, Bontoc, Benguet to Sorsogon).

(2) *C. Griffithii* — Malacca.

---

*) This palaeobotanist is BERRY. See throughout the pages of SCHUCHERT's "Historical Geology of the Antillean-Caribbean Region," 1935, especially pp. 86 *et seq.*

(3) *C. leptostachya* — Northeastern Borneo (Mt. Kinabalu), Philippines (Sorsogon to Laguna).

(4) *C. Cumingii* — Borneo, Philippines (Samar, Leyte, Camiguin de Misamis, Luzon: Sorsogon to Cagayan), New Guinea.

As we are better to see in coming pages, *Crypteronia* is distributed in a conventional manner by two channels (a) Java—Sumatra—Malacca—Burma; (b) New Guinea—Central and Northern Philippines—Northeastern Borneo. Whether these channels were run south to north or north to south we may not state just now, but will return on the matter next again.

In our opinion, there exists a small family in South Africa more or less closely related with the Cyrillaceae and Crypteroniaceae. This family, Heteropyxidaceae, consists of the lone genus *Heteropyxis* which ranges in this manner,

(1) *H. natalensis* — Nyasaland (Mt. Zomba), Natal.

(2) *H. canescens* — Southeastern Transvaal (Barberton).

We believe, then, that Cyrillaceae, Crypteroniaceae and Heteropyxidaceae belong to the same archetypal node, or *genorheitron*, and that the mainsprings of their dispersal (Fig. 66) are to be accounted for by migrations effected to Malaysia and a now crumbled range in the West Indies from South Africa. Migrations of the kind are conventional.

Southeastern Transvaal is the region where *Vaccinium* came to a dead stop on African soil (see p. 169), not only but an important node in the distribution of the cycadaceous genus *Encephalartos* (see p. 479). The occurrence of the Heteropyxidaceae in this very same region, and Nyasaland, is arresting enough to deserve pause.

In and around Barberton of Southeastern Transvaal occur certain species of *Euphorbia*. One of them is *Euphorbia Tirucalli;* the other *E. ingens*. Both these succulent tree-like Spurges follow from Barberton northward along the "Great Rift" or its vicinity. In Natal occurs a peculiar development in tree-like *Euphorbia* having generally thin branches (e.g., *E. Evansii*, *E. grandidens*, *E. triangularis*, possibly *E. ramipressa*), also certain low-growing forms such as *E. passa* and *E. bupleurifolia* that further extend to the Cape.

These Spurges yield the following, (i) *Euphorbia Tirucalli* — Known in Madagascar, Africa, India (?), but part of a group otherwise localized in Madagascar. Forms with diluted affinities toward this group occur in the West Indies, Southeastern Iran, Southwest Africa, but the group itself is beyond question Madagascan in the very first place; (ii) *Euphorbia ingens* — Wholly unknown in Madagascar, not only, but part of an affinity unknown throughout in this island. This affinity is essentially East African, and closely wedded to the "Great Rift." Distant allies of *E. ingens* may occur in West Africa and the Northeastern Himalayas *(E. Royleana);* (iii) *Euphorbia Evansii* and allies — Wholly unknown in Madagascar, and rather narrowly localized in and around Natal; (iv) *Euphorbia passa* — Wholly unknown in Madagascar, but part of an affinity — broadly understood *(E. Caput—Medusae, E. mammil-*

348

laris, E. inconstantia) certain members of which (e.g., *E. Morinii, E. inconstantia*) suggest at least diluted affinities in the direction of *E. lophogona* and its group, which latter are narrowly localized in Madagascar *).

It is plain from this tabulation that Southeastern Transvaal and Natal received Euphorbias (a) From Madagascar *(E. Tirucalli);* (b) From a center floristically wholly unrelated with Madagascar *(E. ingens);* (c) From a center that was not Madagascar but once had connections with that islands (*E. passa* affinity, *sensu lato*).

The Ericaceae Ericoideae yield on their part (see p. 160) the following, (i) *Erica* — Unknown in Madagascar and the Mascarenes; unknown in West Africa; thoroughly well known in the Cape, where a massive speciation took place revealing affinities as far as Natal and Nyasaland. A "break" marked by the appearance of *Erica arborea* occurs in Tanganyka; (ii) *Philippia* — Well known in Madagascar, the Mascarenes and Comoros, not only, but in West Africa and the Islands of the Gulf of Guinea; unknown in the Cape.

The Cycadaceae reveal that *Encephalartos* (see p. 479) is well represented in Transvaal, Natal, Southeastern Cape, East Africa and parts of Central and West Africa north of the Cameroon. Natal is further the seat of the narrowly ranging *Stangeria*. Neither one of these genera occur in Madagascar or the Mascarenes, which are tenanted on their part by *Cycas*.

In the introductory part of this manual we have spoken of the African gate of angiospermy as having two sectors, eastern and western, and referred (see p. 58) to *Phylica* as an illustration. We may now return to this same genus, repeating here its dispersal, as follows,

(1) About 150 species in the immediate Cape region.

(2) *P. mauritiana* — Madagascar, Mascarenes.

(3) *P. emirnensis* — Madagascar; Var. *nyasae:* Nyasaland.

(4) *P. arborea* — Mascarenes (Mauritius), Amsterdam Island, Tristan da Cunha, Inacessible, Gough and Nightingale Islands.

(5) *P. polifolia* — Saint Helena.

Few genera match *Phylica* in interest from the standpoint of phytogeography. The dispersal of *Phylica* yields a cleancut clue to the origin of the archetypal migrations coming to Africa, as follows, (i) Part of the genus defi-

---

*) We speak of Madagascar here, and elsewhere in this manual, without going to the length of specifying whether we intend Eastern, Western or Southern Madagascar which ought to be done in any *precise* work on the flora of this island. We are aware that the best informed opinion (see, e.g., PERRIER DE LA BATHIE in Mém. Acad. Malgache 12: 227 *et seq.* 1932) sharply differentiates in the matter, and we, too, occasionally comment that *Protium*, for example (see p. 388), is localized in Eastern Madagascar and the Mascarenes. The biotic domains of Madagascar and Somaliland are both sharply divided, and so are because of wholly homologous reasons, stemming from ancient geography. We regret we cannot say more here, but point out that East Africa; the portions of Africa immediately adjacent to it; Eastern Madagascar; and Western Madagascar form in reality four parallel "slices" (or blocks), of which three seemingly belong to Gondwana most closely (i.e., the whole of Madagascar, East Africa), while the fourth is a different quantity. East Africa strongly tends to tilt, being higher in the western than eastern longitudinal end. What we state here is *based upon plant-dispersal* solely and completely but, we suspect, will not shock geophysicists and geologists beyond measure. In fact, we are certain well informed geophysicists and well informed phytogeographers are brother-workers, necessarily bound to agree in essentials.

nitely centers in Madagascar and the Mascarenes, and invades Nyasaland from the former; (ii) Another segment colonizes the Cape, which has seemingly no connection with either Madagascar or the Mascarenes; (iii) Still another segment is active in the Atlantic *(P. arborea, P. polifolia)*, which in one case *(P. arborea)* retains a tenuous link with the Mascarenes.

Based upon this evidence; the evidence so far studied throughout this manual; evidence still to come (see p. 509) from the migrations of certain *Hepaticae*, etc.; we believe that two are the centers active in and around South Africa, which centers we have merged together at first under the common designation, *African gate of angiospermy.* These centers are approximately located as follows, (a) In the triangle *Madagascar—Mascarenes—Seychelles.* As this center is the fountainhead of massive streams of migrations reaching the New World, Africa, Asia, Europe, Oceania from the modern Indian Ocean we will designate it as *Gondwanic;* (b) In the triangle *Natal—Kerguelen Islands—Tristan da Cunha.* This center we will understand as *Afroantarctic.*

In pointedly identifying these centers by name (Fig. 67), we are far from implying, even less meaning that these two centers are certainly and absolutely distinct. In reality, these two centers correspond to palaeogeographic continental masses, home of original angiospermy, which — as *Phylica, Euphorbia,* Ericaceae Ericoideae, etc., show — had frequent, massive interchanges. These interchanges, nevertheless, may vary in intensity, not only, but in certain cases do not take place at all. We shall soon see how groups behave, such as Restionaceae and Centrolepidaceae which do not touch the *Gondwanic Triangle.*

The usefulness of a cleancut appreciation of these two centers and their interrelations is readily demonstrated with reference to the Ericaceae. It is patent that Ericaceae Ericoideae and Empetraceae are related by genorheitral ties. The former do not turn up west of the Cape; the latter begin their visible dispersal only westward from this region. The genorheitral tie between the two was forged, then precisely in the range pointedly stressed by the distribution of *Phylica arborea,* that is, within the triangle *Natal—Kerguelen Islands—Tristan da Cunha.*

Philica arborea retains, nevertheless, stations in the Mascarenes. Right in the Mascarenes turns up *Philippia,* which *Erica* does not colonize at all. It is then clear that — as *Phylica arborea* pointedly shows — the archetypes of the Empetraceae and Ericoideae freely mingled over the two triangles, Natal—Kerguelen Islands—Tristan da Cunha; Madagascar—Mascarenes—Seychelles, though *Erica* and the Empetraceae finally stood by the former, *Philippia* the latter. The mingling, then, was not indiscriminate, neither was it such as to blur the outlines of archetypal dispersal. It was just what we would expect from lands, now crumbled in the deep south of the map, having made contacts with lands still extant. As these lands, all of them, had certain geographic connections, and certain types of ecology, these lands did not allow haphazard passage to all plants alike. *In other words, migrations were selective, orderly throughout, whether the lands interested are still on our maps or no longer to be seen on them matters not at all.*

We have included the Seychelles with the Mascarenes and Madagascar to constitute the *Gondwanic Triangle* because of reasons that will more clearly appear in a coming study of the dispersal of the Dipterocarpaceae and *Cassipourea*. Briefly stated, the most important of these reasons is that the Seychelles demonstrably connect Madagascar and the Mascarenes with Ceylon, India and West Africa as well.

It is possible to pass, then, by an uninterrupted chain of outposts from the Kerguelen Islands and Natal to Saint-Helena at one hand, South America and India at the other. In this intelligence, and knowing the behavior of the Ericaceae we cannot be surprised at all in learning that the Ericaceae Andromedoidea have a genus in Africa, *Agauria*, which faithfully migrates in the wake of the ericoid *Philippia*. Even less can we be surprised that this andromedoid genus is the lone one to turn up in Africa, while the balance of the subfamily fares to Asia. This genus, *Agauria*, was distributed in origin — *and throughout lands now wholly lost under the waters of the Indian Ocean* — in a territorial manner that made it necessary for it to follow suit in the wake of *Philippia*. *Agauria*, then, was within reach of the Mascarenes, Madagascar and Comoros when Gondwana began to crumble and the Angiosperms actively to migrate out of it. The rest of its immediate allies was by then somewhere else, within reach of land that took them northward toward the heart of Asia. *Erica* was at the time far remote to the southwest, so that it never came to Madagascar or Asia. In brief, when Gondwana shattered, and the Angiosperms actively moved out of it, saving themselves by whatever land they could reach — which lands is still on our maps — the dispersal of the Ericaceae Ericoideae (*Erica, Philippia* at least) had been already made in such a manner as to influence the whole course of later migrations. By then, too, *Agauria* had been segregated, geographically as well as philogenetically, out of the Andromedoideae. At the same time, *Vaccinium* — as a tiny outlier indeed — had been directed to Natal, while the balance of the *genorheitron* had been already dispatched to Polynesia.

The *Gondwanic* and *Afroantarctic* Triangles responsible for the migrations just sketched contributed a wealth of archetypes, some successful in dispersal, other not, to a third center still which we will designate as *Kalaharian*. This center is highlighted by the peculiar *Myrothamnos*, a genus in two species having affinities toward Saxifragaceae and Hamamelidaceae, which is type of the Myrothamnaceae. *Myrothamnos* is distributed as follows,

A) Sect. *Myosurandra*

(1)   *M. moschatus* — Madagascar.

B) Sect. *Eumyrothamnos*

(2)   *M. flabellifolius* — The whole of South Africa immediately north of the Cape proper to about 6° Lat. South (Transvaal, Natal, Nyasaland, Southwest Africa to Angola).

This center had immediate connections through Angola with West Africa, witness the distribution of the euphorbiaceous genus *Elaeophorbia*,

(1)   *E. acuta* — Transvaal.

(2)   *E. Hiernii* — Angola (Cuanza do Sul).

(3)   *E. drupifera* — Sierra Leone, Gold Coast, Dahomey.

Further still this same center sent streams to Eastern South America and the West Indies, as proved by the dispersal of the lentibulariaceous genus *Genlisea*, sa follows,

(1)   *G. africana* — Angola.

(2)   *G. angolensis* — Angola.

(3)   *G. hispidula* — Natal, Transvaal, Nyasaland.

(4)   *G. subglabra* — Rhodesia, Tanganyka.

(5)   *G. glandulosissima* — Rhodesia.

(6)   *G. Stapfii* — Ivory Coast.

(7)   *G. biloba* — Brazil (São Paulo).

(8)   *G. Luetzelburgii* — Brazil (Bahia, Goyaz).

(9)   *G. minor* — Brazil (Minas Geraës).

(10)   *G. ornata* — Brazil (Minas Geraës).

(11)   *G. pygmaea* — Brazil (Minas Geraës).

(12)   *G. violacea* — Brazil (Minas Geraës).

(13)   *G. aurea* — Brazil (Minas Geraës).

(14)   *G. pusilla* — Eastern Brazil.

(15)   *G. cylindrica* — Eastern Brazil.

(16)   *G. filiformis* — Brazil (Minas Geraës, Bahia), British Guiana (Kaieteur Savanna).

(17)   *G. guianensis* — Guiana.

(18)   *G. roraima* — British Guiana (Mt. Roraima).

(19)   *G. anfractuosa* — British Guiana (Kaieteur Savanna).

(20)   *G. pulchella* — British Guiana (Kaieteur Savanna).

(21)   *G. luteoviridis* — Cuba.

We think that the Rapateaceae, a family with ecologic requirements not incompatible with those of the Lentibulariaceae, originated at first in the same center as did *Genlisea* that is, somewhere between Angola and Mozambique. There is good evidence that this center was mostly with lowlands and water-habitats. Here too may have originated *Notobuxus* (Natal, Congo, Dahomey).

The *Kalaharian* center just sketched made contact still with a fourth center of secondary origins in West Africa, mostly resting in the Cameroon and Nigeria. This center, which we shall designated as *Nigerian*, is responsible for massive genus- and species-making in many groups (see Connaraceae, Leguminosae, etc.) both in primary and secondary dispersal, and stands intimately connected this time, with a New World center investing the mouth of the Amazonas Basin and the sandstones of the Pacaraima System *(Roraiman Center)*.

It will be observed that many migrations through the *Kalaharian* and *Nigerian* centers do not necessarily follow the eastern half of Africa, rather take on straight to the western. These migrations consequently are not homologous of those of genera like *Philippia* or *Cassipourea* which immediately leave the Madagascar—Mascarenes—Seychelles triangle to cut straight across Tanganyka in a westerly path to the Cameroon, feeding meanwhile outliers to East Africa as far south as Nyasaland. It is further obvious that we never can draw fast and hard lines as regards these and similar migrations, though certain main outlines stand out fairly clearly to help us in effecting rational discriminations case by case. For example, it is highly probable that *Myrothamnos* — not unlike *Euphorbia* in the broad affinity of *E. passa* — was simultaneously fed both to Madagascar and the *Kalaharian Center* from the *Afroantarctic Triangle*, further to evolve in that Center rather than in the *Gondwanic Triangle*. Contrariwise, *Philippia* and *Agauria* reached Continental Africa only from the *Gondwanic Triangle*. The interplay of these Triangles and Centers can easily be made out with reference (Fig. 67) to a diagram that shows their respective locations and the outline of certain main tracks, "cold" and "warm" raising from, or going to, them.

Currently treated in the near kinship of the Cyrillaceae, but perhaps not closely related with them even through archetypal bonds, are two other petty families, Corynocarpaceae and Pentaphylacaceae, which are distributed in this manner,

A) Corynocarpaceae (*Corynocarpus* only);

(1) *C. australasica* — Queensland, New Guinea.
(2) *C. dissimilis* — New Caledonia.
(3) *C. similis* — Northern New Hebrides (Torres Island).
(4) *C. laevigata* — Kermadec Islands, New Zealand, Chatham Island.

B) Pentaphylacaceae (*Pentaphylax* only);

(1) *P. malayana* — Malacca.
(2) *P. spicata* — Indochina (Tonkin).
(3) *P. euryoidea* — China (Kwantung, Kiangsi, Hongkong, Hainan).

If these two families are indeed allied with the Cyrillaceae, their dispersal adds little to the picture before us. *Corynocarpus* typically centers in the Coral Sea, one of the hubs of angiospermous origins in the Western Pacific, and most likely reached New Zealand from the north *via* the Kermadec Islands (see p.

181). *Pentaphylax* is a rump dispersed among three points, all of which are of most remote antiquity, and would agree, we believe, with *Crypteronia* in the main trends of its distribution.

Thus, Cyrillaceae and Crypteroniaceae may be connected by Heteropyxidaceae. The ancestral forms of these three families saw the light by the African gate. *Heteropyxis* secured a small landing on Continental Africa which seems to involve the *Afroantarctic* center most directly. *Crypteronia*, on the contrary, migrated — if it did not originate, which we cannot know — by a straight *Gondwanic* path. The Cyrillaceae, by far the largest and most successful end of the genorheitron reached the Caribbeans and Amazonian Brazil by a track running most likely out of the *Kalaharian Center* very much in the manner of *Genlisea*. Let us not forget that the distribution of *Myrsine africana* (see p. 90) contains a cleancut tie Angola-Azores, and that not rare are the groups which migrate taking the modern Atlantic as the axis of their distribution, witness *Littorella*, Empetraceae etc.

VIII. Oleaceae

Two controversial segregates in the immediate vicinity of *Olea*, type-genus of the family we deal with here, are distributed in this manner,

A) *Steganthus*

(1) *S. lancea* — Mascarenes.
(2) *S. urophylla* — Nyasaland.
(3) *S. Welwitschii* — Kenya, Angola, Islands of the Gulf of Guinea.

B) *Leuranthus*

(1) *L. Woodiana* — Natal, Transvaal.

With reference to the statements contained in preceding pages we can rationalize these distributions without difficulty. *Steganthus* might have migrated in either one of two ways, (a) Along the line Mascarenes—Nyasaland—Kenya—Islands in the Gulf of Guinea—Angola; (b) Along the line Mascarenes—Nyasaland—Kenya—Angola—Islands in the Gulf of Guinea. To settle as between these alternatives we ought to know a great deal more of the taxonomy of the genus than we do just now. *Leuranthus*, clearly, is one with *Heteropyxis*.

Olea is well represented in South Africa *(O. africana, O. capensis, O. crassifolia, O. exasperata* etc.) and turns up in West Africa *(O. guineensis)* in addition. *Olea europaea* has unusually transparent and interesting connections with other species, as follows, (a) *Through its var.* maderensis *(Madera)*, with *O. verrucosa* (Natal, Transvaal, Griqualand, Cape Province: Somerset, Graaff Reinet, Paarl, Uitenhage, Albany Divisions, etc.); (b) *Through its var.* nubica *(Nubian hills by the Red Sea)*, with *O. chrysophylla* (Madagascar, Mascarenes, Mozambique, Rhodesia, Tanganyka, Kenya, Ethiopia, Somaliland, Arabia, Upper Sudan). Quite close to *O. europaea* is *O. Laperrinei* (Central Sahara to the southern slopes of the Moroccan Atlas) and *O. cuspidata* (Beluchistan, Afghanistan, Northwestern Himalayas and Punjab, Tibet, China: Yunnan, Szechuan).

It is not difficult to see that *Steganthus, Leuranthus* and *Olea* all originated by the African gate of angiospermy, most likely in the *Gondwanic Triangle* *). *Olea verrucosa* came to a head in the *Kalaharian* center; *Olea chrysophylla* remained *Gondwanic* throughout, running at the best side-lines to Rhodesia. *Olea europaea* is connected with both these species which indicates that all of them, *O. verrucosa, O. chrysophylla* and *O. europaea,* formed a single complex prior to the epoch of active angiospermous migrations to the *Kalaharian* center. The geographic tie among *O. europaea* and its two next-of-kin, *O. verrucosa* and *O. chrysophylla,* may well involve, as a matter of fact archetypal migrations in the modern Atlantic (see var. *maderensis*) and the modern Indian Ocean (see var. *nubica*), a telling reminder that mileage is as nothing in dispersal. In sum, *O. verrucosa, O. chrysophylla, O. europaea* and *O. cuspidata* are broad "ecotypes" of comparatively weak speciation, which readily accounts for the controversies besetting their classification. The African and Mediterranean ends of the dispersal of *Olea* are reminiscent of *Erica,* but being *Afroantarctic* rather than *Gondwanic, Erica* reached neither West Africa nor Tropical Asia.

Olea europaea breaks into two varieties (20) as follows,

(1) Var. *Oleaster* — Mediterranean generally northward to Southern France, coastal Northern Italy, Southern Balkans, Crimea, Caucasus southward to the Moroccan Atlas, Algeria, Tunisia, Lybia, Western Egypt, Palestine, Arabia (Oman).

(2) Var. *maderensis* — Cabo Verde Islands, Canary Islands, Madera.

If we recall that in *Linum,* for instance (see p. 202), forms closely connected with South Africa become "Mediterranean" just north of Tanganyka and Ethiopia **), we cannot fail but to understand that the whole of *O. europaea*

---

*) Not to complicate the text constantly by the addition of references, qualifying clauses, etc., we leave it to the reader to derive the necessary inferences from our statements. For example, *Steganthus* certainly originated in the *Gondwanic Triangle,* which is shown by the face of the dispersal; *Leuranthus* might, conversely, have originated in this *Triangle* or the *Afroantarctic* one, which again the face of the dispersal shows to be in the order of things. As *Menodora,* next to be studied (see p. 358) is in South Africa with a thin outlier, and massively in the New World, we need next infer that *Menodora* is from the *Afroantarctic Triangle.* Thus, the Oleaceae — of which *Steganthus, Leuranthus* and *Menodora* are part, originated both in the *Gondwanic* and *Afroantarctic Triangle,* the ultimate genorheitral origins being to be sought between the two, and, on account of the "antarctic" migrations of *Menodora,* most likely in the latter in a final reckoning.

**) This is an important aspect of dispersal which we cannot consider in detail in the pages of this manual, for it would call for extensive discussion of local floras. The mainsprings in play are, generally speaking, the following; Let us suppose that an archetypal form in *Gondwanic* position (i.e., within the

is the northern end, the boreal variety — we might almost say — of a *commiscuum* involving as one *O. europaea*, *O. verrucosa*, *O. chrysophylla* and *O. cuspidata*.

This approach to the nature of *O. europaea* and its allies in phylogeny and dispersal alike further elucidates another controversial issue. The two varieties of *O. europaea* essentially differ in the inflorescence, which is strictly axillary in var. *Oleaster* and tend on the contrary to form in apical clusters in var. *maderensis*. Strains of *Oleaster* which occasionally revert or mutate to the inflorescence pattern of var. *maderensis* are worthless for cultivation (21). The difference between these two expressions in the inflorescence is easily accounted for by intercalarity (22), but once fixed in phylogeny and accompanied by slight variations in the flower and fruit, these expressions may give rise to forms which easily can be credited to different genera on the basis of strong taxonomic motives. The controversial nature of certain allies of *Olea* may stem, perhaps, from factors of this order.

Always to *Olea* belongs a peculiar group, Section or Subgenus *Gymnelaea*, characterized by a lack of corolla and the hypogynous position of the stamens. This aggregate is distributed in the manner revealed by the following selection of species,

(1) *O. montana* — New Zealand (North and South Island).

(2) *O. apetala* — Norfolk Island, New Zealand (North Island).

(3) *O. paniculata* — Australia (New South Wales, Queensland), Lord Howe Island, New Caledonia.

(4) *O. floribunda* — Bismarck Archipelago.

This group enters New Zealand from the north using a channel that fares through Norfolk and Lord Howe Islands. This channel is commonplace, as such, for it connects New Guinea and its vicinity, also Australia, with New Zealand by a standard connection running through the two small insular outposts just mentioned.

As regards the Oleaceae, however, this very same channel gains importance on account of the dispersal of a genus *Notelaea* which is distributed as the following species indicate,

(1) *N. brachystachya* — New Caledonia.

(2) *N. ligustrina* — Australia (Victoria, New South Wales).

(3) *N. quadristaminea* — Norfolk Island, Lord Howe Island.

(4) *N. excelsa* — Madera, Canary Islands.

Notelaea excelsa is often treated as a monotype, *Picconia excelsa*, which changes nothing of the substances with which we deal just now. The issue at stake is for us to learn by what channel New Caledonia, Australia, New Zea-

---

triangle Madagascar—Mascarenes—Seychelles) dispatches part of itself to South Africa through Tanganyka and Nyasaland to the *Kalaharian Center*, and simultaneously follows directly through Tanganyka and Ethiopia to the Mediterranean. This form in the end will split up into two components, staggered *on the modern map* south to north, and breaking into varieties, or vicariant species, south and north, between Ethiopia and Tanganyka. The *Olea* complex we just reviewed is an excellent example in point.

land, Norfolk and Lord Howe Island stand into contact with Madera and the Canaries.

This channel, of course, is nothing unusual, because it is followed by *Carex*, as we know, which has the same species in New Zealand and the Pyrenees of Spain. However, the mere knowledge of this channel does not elucidate its course fundamentally enough for our purposes. It is no longer adequate for us to learn that the same species is in New Zealand and the Pyrenees. *How* the Pyrenees could be reached from New Zealand is of the essence of our present quest.

This quest can be brought near a solution, when not solved outright, by a consideration of the dispersal of the loranthaceous genus *Korthalsella*. This genus, already reviewed elsewhere (see p. 141) contains a species distributed in this manner, (a) *K. Opuntia* — Type-form: Ethiopia, Ceylon, India, Southern China, Formosa, Japan, Corea; Siam, Indochina, Sumatra (*Gajolands*, etc.); Malacca (*Mt. Ophir*, etc.), Java, Philippines (Luzon: *Benguet, Zambales; Negros*), Australia (New South Wales), Lord Howe Island; — Var. *fasciculata:* Afghanistan, Western Himalayas, Southwestern China; — Var. *Bojeri:* Mascarenes; — Var. *Gaudichaudii:* Mascarenes, Madagascar, Comoros, South Africa.

Korthalsella Opuntia, as such, is a species of *Gondwanic* origin. A certain part of it matured in the *Kalaharian* center, but the balance migrated from the triangle Madagascar—Mascarenes—Seychelles. The dispersal follows from this center in standard fashion, part directly to the Western Himalayas, South-western China and Afghanistan (Var. *fasciculata*), part to Ethiopia, Japan, Australia and Lord Howe Island.

A noteworthy station is met with along the path leading from the Gond-wanic triangle to Australia, namely, Mt. Ophir of Malacca. The connections of this peak with the rest of the world are defined as follows, (a) *Mt. Ophir— Carpathians and Alps: Linaria alpina;* (b) *Mt. Ophir—Mascarenes: Foe-tidia;* (c) *Mt. Ophir—Lord Howe Island: Korthalsella Opuntia.*

We do not exactly know the origin of *Linaria alpina*, which could be ascertained at the price of studying the dispersal of the Scrophulariaceae throughout. This is here impossible, but it seems, considering the abundance of Scrophulariaceae in Africa, that *Linaria* originated by the African gate. *Korthalsella Opuntia* and *Foetidia* doubtless originated by this very same gate. In conclusion, the *Gondwanic Triangle* leads to Spain, and New Zealand alike. As *Korthalsella Opuntia* reached Lord Howe Island from this triangle, so did the archetypes of *Notelaea* fare in the same manner both to the Canaries and Madera and New Zealand.

Obviously, it could be suggested that *Notelaea* reached Macaronesia and New Zealand primarily by a straight antarctic path. This is most unlikely, however, because the Oleaceae themselves are *Gondwanic*, and groups of this nature and origin do not extensively migrate, as a rule, through antarctic lands.

Fraxinus is a large genus, well known throughout the Northern Hemisphere, and often assumed in textbooks as an example of "Holarctic" filiation. That textbooks of the kind are unfortunate in the choice of their examples need not be said. Nothing can ever be "Holarctic" in a family which contains *Olea*, *Notelaea*, *Steganthus*, *Lauranthus*. Moreover, one of the species of *Fraxinus* is distributed in this manner,

(1) *F. Griffithii* — Celebes, Philippines (Luzon: Laguna, Rizal, Bataan, Bontoc, Cagayan, Ilocos Norte), Java, Formosa, Japan, China (Hupeh), Eastern India.

A form which is in Celebes, the Northern Philippines, Formosa, Japan Central China and Eastern India originated by the *Gondwanic* triangle, or its immediate vicinity, and was fed in the first place to the Far East and Malaysia from a point in the modern Indian Ocean. Subordinately, it might originate from still another of the major centers of angiospermous forms situated in the modern Coral Sea, which center we shall designate henceforth (see Fig. 67) as *Neocaledonian*. Be it noticed that the *Neocaledonian Center* is homologous of the *Kalaharian* and *Nigerian* in the sense that many archetypes came to it from the *Gondwanic* and the *Afroantarctic Triangles*. To rate these centers by order of absolute importance is impossible because of reasons which must by now be clear to the reader. We might suggest nevertheless than in order of their relative importance these centers stagger themselves as follows, (I) *Gondwanic;* (II) *Afroantarctic;* (III) *Neocaledonian;* (IV) *Kalaharian;* (V) *Nigerian*. The score may be rounded up by the addition of a last center, (VI) *Magellanian* *), which feeds South and North America by a point located, roughly speaking, between Fuegia and the southern boundary of Bolivia. The *Magellanian* center not unlike the *Kalaharian* is mostly fed from the *Afroantarctic* and *Gondwanic* triangles. We would place the *Magellanian* center next to the *Neocaledonian* in importance. It will be notice that what we have called "Gates of angiospermy" correspond with the triangles, or centers, just mentioned in this measure, (a) *African gate* — Includes both the *Gondwanic* and *Afroantarctic* centers; (b) *Western Polynesian gate* — Agrees in the main with the *Neocaledonian* center; (c) *Magellanian gate* — This is the *Magellanian* center itself.

We have refrained from introducing at the start of our work mention of these centers, or triangles, broadly speaking of gates instead, because we felt that early pointed statements involving details of distribution — weighty as these details might be in themselves — might be misconstrued as unwarranted assumptions entered into on the strength of aprioristic theories of dispersal. The reader may now judge for himself, for he has just enough facts in hand to appraise our own position in regard to factual evidence.

To return to *Fraxinus;* This genus is certainly not "Holarctic," and it should be a pure waste of time and printing to follow its wanderings through

---

*) The map (Fig. 67) shows two other *Centers*, Roraiman and *Ozarkian—Appalachian* often spoken throughout the pages of this manual. These *Centers* require no comment. Though very important, they are essentially local (i.e., American), thus secondary.

the Northern Hemisphere, in order to conclude, for instance, that certain of the Ashes came to the United States from the Far East, which is the case. *Fraxinus* could be worthwhile material for detailed studies in distribution along the boundary between Mexico and the United States, or in the Mediterranean Basin. As these studies are not the province of this manual, we will drop *Fraxinus* without further.

A last genus of the Oleaceae is to receive our attention here, *Menodora*. This group consists (23) of about 17 species. Two of them, *M. africana* and *M. juncea* are restricted to South Africa (Namaqualand, Transvaal, central districts of the Cape), which is to say that these species matured in the *Kalaharian Center*. None of the forms of *Menodora* is on record anywhere else from Africa, which leads to the conclusion that this genus had nothing to do with the *Gondwanic Triangle*. That it might have originated at the very first in the *Afroantarctic Triangle* is a very distinct possibility.

Six species of *Menodora* belong to South America of which *M. robusta* is southernmost (Argentina: Santa Cruz, Chubut) with *M. Hassleriana* (Argentina: Rio Negro, Tucumán; Paraguay) next northward. *Menodora integrifolia* and its variety *trifida* are the forms in this massing which reach farther north (Argentina: Córdoba, San Luís; Uruguay, Paraguay; Bolivia: Santa Cruz, Cochabamba, La Paz).

Nine species lastly occur in North America from the Mexican state of Oaxaca to Baja California, Texas, Southern Colorado, Utah. The range of many of these species is oddly shaped (Fig. 68) in a manner which convey a striking, pointed suggestion that the speciation ran its course at the shore of bodies of water in progress of drying, in a manner homologous of that of *Frankenia* (see note p. 383). The dispersal of species so distributed could be profitably worked out in the field with relation to geological and geophysical features. We are inclined to suspect that the Mexican Highlands once were anything but the tablelands they are today, rather a shallow sea, which divided Eastern from Western Mexico north of Oaxaca. In brief, it may well be that Mexico was "compacted" into its present shape by stresses arising from the Pacific, no doubt, and perhaps from the Caribbeans *). We know no better, and if we voice our suspicion, this is because future will decide whether it is worthy of attention, as we believe, or simply predicated upon impressions that facts are to belie in the end.

The three ranges of *Menodora* in South Africa, South and North America are connected by two species, as follows,

(1)　*M. heterophylla* — Mexico (Tamaulipas, Nuevo León), United States (Texas); var. *australis:* Transvaal, Bechuanaland.

(2)　*M. decemfida* — Central Chile; var. *longifolia:* Mexico (Durango, San Luís Potosí), United States (Texas).

---

*) There is a very striking suggestion in dispersal that modern Mexico, quite as does modern Madagascar, is of two original halves, western and eastern. Could it ever be that Mexico and Madagascar were both compacted into their present shape by homologous geophysical processes?

We observe, (i) Tamaulipas, Nuevo León, San Luís Potosí, not less than Oaxaca, are doubtless one of the oldest landcores in North America. It is in these ranges, as a matter of fact, that is rooted the cycadaceous genus *Dioon* (see p. 477), and a host of most peculiar Cactaceae (see p. 369) (ii) The dispersal of *Menodora* between South Africa and Northern Mexico is a good match of the dispersal of *Thamnosma* (see p. 91); (iii) The track of *M. decemfida* exhibits a classic disconnection Chile—Mexico.

There may be a question, whether *Menodora* reached Chile first from the direction of the Eastern Pacific, or entered Argentina in a prior time from the direction of the Southern Atlantic. This issue is one which can be finally settled only at the price of thoroughoing studies in the flora of Chile, Argentina, Uruguay and Southern Brazil, studies, in sum, which cannot have place here. This issues is a pressing one as regards the Cactaceae (see p. 364) it being possible, for example, that the affinities centering around *Rhipsalis* and *Cereus* invaded the New World from the Atlantic, while other cactoid groups give strong indication of having reached Chile and Bolivia first and foremost. However, it will be observed that, though important, this issue cannot be viewed as primary. At any rate, the tracks of *M. decemfida*, type-form, and *M. integrifolia* read jointly as follows, Central Chile-Northwestern Argentina—Central and Eastern Bolivia—Paraguay—Uruguay—Southern Brazil, which is the track run from the Pacific by such genera as *Araucaria* and *Colliguaya*. It may be, of course, that *Menodora* ran this track in the reverse, that is, from the Atlantic shore first to the Pacific. Be that as it may, it is clear that a channel Central Chile—Southern Brazil is standard, any way it might be run by this or that groups or form.

This channel will some day be studied in detail and critically, but so consi derable is the material which need be digested in order to elucidate all the details of its course that we must regretfully state the problem, refraining all the while from a pointed, immediate discussion. Its difficulties may easily be inferred from the footnote p. 452.

IX. Restionaceae and Centrolepidaceae

Restionaceae and Centrolepidaceae are closely allied. Their members are perennials of herbaceous description, or annuals, most of the latter occurring in the Centrolepidaceae.

Our purpose in reviewing these two families is to investigate the behavior of groups which do not interest at all the *Gondwanic Triangle*, and may for this very reason be described as purely "antarctic."

The Restionaceae are distributed in this manner (24).

(1) *Lyginia* — 1 species in Southwestern Australia.
(2) *Hopkinsia* — 1 species in Southwestern Australia.
(3) *Ecdeiocolea* — 1 species in Southwestern Australia.
(4) *Anarthria* — 6 species in Southwestern Australia.

(5) *Onychosepalum* — 1 species in Southwestern Australia.

(6) *Lepidobolus* — 3 species in South and Southwestern Australia.

(7) *Elegia* — 25 species in South Africa.

(8) *Askidiosperma* — 1 species in South Africa.

(9) *Chondropetalum* — 16 species in South Africa.

(10) *Dielsia* — 1 species in Western Australia.

(11) *Restio* — 90 species in South Africa, 27 in New Zealand and Australia.

(12) *Lepyrodia* — 17 species, mostly local, in Western and Eastern Australia, some in Tasmania, New Zealand, Chatham Island.

(13) *Thamnocortus* — 23 species in the Cape.

(14) *Staberoha* — 5 species in the Southwestern Cape.

(15) *Calopsis* — 10 species in the Cape.

(16) *Leptocarpus* — 12 species in Chile, Australia, India and Cochinchina.

(17) *Lamprocaulos* — 3 species in the Cape.

(18) *Chaetanthus* — 1 species in Southwestern Australia.

(19) *Hypodiscus* — 8 species in South Africa.

(20) *Loxocarya* — 7 species in Western Australia.

(21) *Harperia* — 1 species in Western Australia.

(22) *Anthocortus* — 1 species in the Southwestern Cape.

(23) *Mastersiella* — 10 species in the Southwestern Cape.

(24) *Hypolaena* — 3 species; 1 in Western Australia; 1 in Tasmania, Victoria, South Australia, New South Wales; 1 in Tasmania.

(25) *Calorophus* — 3 species; 1 in Tasmania; 1 in Western Australia; 1 in Tasmania, New Zealand, Victoria, New South Wales, Queensland.

(26) *Cannomois* — 7 species in the Cape.

(27) *Willdenowia* — 12 species in the Cape.

(28) *Ceratocaryum* — 2 species in the Southwestern Cape.

(29) *Phyllocomos* — 1 species in the Southwestern Cape.

The Centrolepidaceae are a smaller aggregate ranging according to the same authority (24) in the following manner,

(1) *Juncella* — 2 species in Tasmania and South Australia.

(2) *Hydatella* — 2 species in Western Australia.

(3) *Brizula* — 5 species in Tasmania and South Australia.

(4) *Aphelia* — 1 species in Tasmania and South Australia.

(5) *Centrolepis* — 25 species; most of them in Southwestern Australia; 3 in Tasmania, 1 in Borneo, 1 in the Philippines, 1 in Tonkin.

(6) *Alepyrum* — 2 species in Auckland and Campbell Islands, Tasmania and New Zealand.

(7) *Gaimardia* — 2 species; 1 in Fuegia, 1 in New Zealand.

We are bound to be struck in the first place by the very large number of genera (particularly in the Restionaceae) compared with the comparatively small total number of species. The average size of a restionaceous genus is 10 species, and olygotypes are anything but scarce in certain regions, such as Western Australia, South Africa, etc. We may discount many of these genera as not uncontroversial — though most of them are currently accepted as "good" — without for this altering the facts. It is plain that the Restionaceae, especially, are far richer in generic types than, proportionally, in species.

It is evident that groups of the kind are of very great antiquity, and that the morphological trends of the *genorheitron* had ample opportunities of working themselves out rather more in time than in space, many being the genera congregated in the same region. Both the Cape and Western Australia are seen to be homologous. Both are storehouses of a peculiar flora of their own, which was left stranded by a first wave of migration to work out its destinities in isolation during ages following. These floras were not administered the fillip of renewed geological changes of great moment, so they lived on carrying to their logical consequences their own genetic powers in answer to a fairly steady or slowly altering ambient. In a sense these floras are relic through and through. So peculiar is their aspect, as a whole, that an experienced eye can readily spot out in herbarium the elements that belong to them.

These floras are rich in different groups and forms on the whole; though tophevay with certain families or groups (e.g., Proteaceae, Ericaceae, Euphorbiaceae, Asclepiadaceae, Aizoaceae in the Cape; "Phyllodine" *Acacia* in Western Australia) they cannot be described as monotonous ever. The fact that these floras caught and retained whole assortments of angiospermous archetypes and *genorheitra*, not only, *but often molded them out in time in the shape of taxonomic groups wholly of their own is sound indication that these floras occur in centers within easy reach of ancestral forms.*

A glance at the record of the Restionaceae and Centrolepidaceae (Fig. 69) tells us that Magellania, Tasmania, New Zealand, Southern Australia, the Cape and certain section of South Africa could freely communicate among themselves without at the time interesting such centers as the *Gondwanic* or *Neocaledonian.* In other words, the channels run by the Restionaceae and Centrolepidaceae are *purely antarctic.* While ranging between the Cape and Southern Australia, these two families neatly bracketed the lands occupying the Indian Ocean (Gondwana) *without entering them,* which is pointed evidence that these lands were not definitely part and parcel of lands more to the south, which latter fully deserve, they, the name of Antarctica. It might be suggested that the separation thus effected between Gondwana and Antarctica is due more to the influence of ecologic barriers than the presence of a cleancut geographic discontinuity; it might be pointed out that the behavior of groups like *Phylica,* for instance, is definite proof that the *Afroantarctic Triangle,* Natal—Kerguelen Islands—Tristan da Cunha, stood in geographic contiguity together with the *Gondwanic Triangle,* Madagascar—Mascarenes—Seychelles. This we will not deny, because it is a fact. However, the facts themselves, as revealed by the distribution of the Restionaceae and Centrolepidaceae, prove that the *Gondwanic Triangle* might not be invested at all; and that pure antarctic migrations are a matter of the record. The behavior of certain groups of the Cyperaceae (see p. 252) warns us that a geographic barrier, or some equivalent feature, active in the immediate vicinity of the Kerguelen Islands could shunt the course of dispersal away from Africa toward Magellania and the Pacific in the first place. Moreover, the dispersal of certain Liverworts (see p. 509, Fig. 97) tells us that Natal could be entered, without for this anything else of Africa being invested in the process. We believe that a rational scheme of philogeny might be worked out for the Restionaceae and Centrolepidaceae, fit to show whether the archetypes are sooner to be found in Africa or Australia or Magellania, which might ultimately help us in elucidating the question, why these families are strictly antarctic in dispersal. Meantime, we have nothing else to do, but to bow to the facts. These two groups are antarctic in the purest meaning of the term, and it is not at all to be ruled out that they actually originate in the *Afroantarctic* center in the first place. Beyond this we might not go just now.

A list has been widely current among phytogeographers (26) of all families of the seedplants which exhibit "Antarctic migrations" involving congeneric species. This list reads, Araliaceae, Araucariaceae, Boraginaceae, Campanulaceae, Cariophyllaceae, Centrolepidaceae, Compositae, Coriariaceae, Cornaceae, Crassulaceae, Cruciferae, Cupressaceae, Cyperaceae, Donatiaceae, Droseraceae, Elaeocarpaceae, Ericaceae, Eucryphiaceae, Fagaceae, Gentianaceae, Geraniaceae, Goodeniaceae, Graminaceae, Halorrhagidaceae, Iridaceae, Juncaceae, Leguminosae, Liliaceae, Loranthaceae, Magnoliaceae, Monimiaceae, Myrtaceae, Oenotheraceae, Oxalidaceae, Plumbaginaceae, Polygonaceae, Primulaceae, Proteaceae, Ranunculaceae, Restionaceae, Rhamnaceae, Rosaceae, Rubiaceae, Scrophulariaceae, Stylidiaceae, Taxaceae, Tetrachondraceae, Thymeleaceae, Umbelliferae. A glance at this list readily shows that the families in the rooster are certainly not "antarctic" in the same measure. For instance, Restionaceae and Centrolepidaceae are genuinely antarctic, while Fagaceae and Magnoliaceae (p. 94, 109) are antarctic only to the extent of two of their genera, *Nothofagus* and *Drimys*, not as regards *Pasania* and *Talauma*. However, as it is manifest that as both Fagaceae and Magnoliaceae contain "cold" to "cool" genera in their ranks — precisely as do the Ericaceae, Linaceae, etc. — we need ultimately conclude that the *genorheitra* arose in the southern tier of the map. Accordingly, the history of Fagaceae and Magnoliaceae is one of ultimate origin in the south, which means the old "antarctic" shore of the geologic Pacific; of migration northward, and settlement by the *Neocaledonian Center*, wherein under warmer skies were finally released genera like *Pasania* and *Talauma*, which later reached ripe maturity in Malaysia (see p. 119). This history, then, parallels that of the Ericaceae, Oleaceae, etc., which, however, saw the light of the day not in the Western Pacific, but the Indian Ocean of modern maps. It stands beyond doubt that the adjective "antarctic" — though less offensive in most cases than its counterpart "holarctic" — can be the source of grave misunderstandings when uncritically assumed. It is true that all major angiospermous groups are "antarctic" to the extent at least that none of them arose in the beginning north of the equator, but it is certainly not true that all these plants are "antarctic" in the same measure. *Olea, Magnolia, Talauma, Gaimardia* are subtly but certainly differently "antarctic."

This settled, we may take notice of the outliers which *Centrolepis* and *Leptocarpus* dispatch northward.

Centrolepis runs a track Australia—Philippines—Borneo—Tonkin which is crippled by wide disconnections but is on the whole suggestive of that of *Drimys. Drimys* does not reach Tonkin, which is not to say that Tonkin is out of reach from the Philippines. *Leptocarpus* does not fare this channel but another, strongly reminiscent of the tracks of *Araucaria* between Chile and Australia in the first place. Leaving Australia *Leptocarpus* cuts straight to Cochinchina, a classic center for all manners of plants wedded to moist habitats (see Stylidiaceae, Podostemonaceae, Lentibulariaceae) and from Cochinchina fares further to India.

Relevant is the fact that both *Centrolepis* and *Leptocarpus* failed to gain a foothold in or around Tonkin and India such as to make it possible for them to continue migrations northward. Had these two genera secured such a foothold, we would doubtless find them all over the temperate, possibly cold, Northern Hemisphere. These two genera, then, exemplify migrations northward from an antarctic shore which barely failed to achieve a wide spread north of the Tropic of the Cancer.

X. Cactaceae

It is commonly held that the Cactaceae are strictly "American" in origin and dispersal, and the opinion is often heard that they are a "derivative" group. This opinion is likely to become less current in the future than it has been up to a recent past, because we have by now certain evidence (see footnote, p. 413) to the effect that *Opuntia* was part and parcel of an Eocene flora of the Western United States. Even so, however, many will persist in believing that the Cactaceae belong to New World lock, stock and barrell.

This belief is false throughout. One at least of the cactaceous genera, *Rhipsalis* is endemic to Tropical Africa, Madagascar, the Mascarenes and Ceylon, not to mention large sectors of Tropical America. It is accordingly pacific that the Cactaceae are one of the very many groups which like *Menodora, Evolvulus*, Turneraceae, Rapateaceae, Bromeliaceae, Hydrophyllaceae, Polemoniaceae, etc., are massive in the New World, but sparingly represented in the Old. To credit groups of the kind with "American" origins steps must be taken as a preliminary which no one trained in precise thinking will encourage. It is required that the non-American components of these groups be dismissed as unworthy of serious thought; it is demanded that these groups themselves be handled *in vacuo*, without regard, that is, of all other groups that migrate homologously; it is necessary to confuse, and to do so badly, all fundamentals of orderly dispersal. At this price, perhaps, the groups in question, Cactaceae included, many be styled "American" in origin.

As to *Rhipsalis*, great many are the authors who assume that this genus was carried to Africa and the islands in the Indian Ocean by birds fond of its sticky, fleshy berries. The least that can be said in this regard is that these authors are none too well informed of the preeminent position held in dispersal by such outposts as Madagascar, the Mascarenes and Ceylon. These authors are ill at ease in addition with rudiments of distribution showing that, as we have indicated, the Cactaceae in general, and *Rhipsalis* in particular, are of one cloth with great many other groups the fruits of which are certainly not sticky, fleshy berries, such as birds may like to consume. These authors make of the Cactaceae a *local* American issue which immediately stultifies the evidence that lies clear and uncontroversial in front of them. Having thus stultified the evidence these same authors conclude that the poles wandered, etc., etc. It should be pointless to debate subjects mishandled from the start, and for this reason we will refrain from giving time and print to the further opinion that the Cactaceae "originated" in the West Indies, and so forth and so on.

We have, then, *as a fact*, the occurrence of *Rhipsalis* in certain sectors of the New World, Africa, Madagascar, the Mascarenes and Ceylon. Also *as a fact*, numerous are the groups which are massive in the New World and poorly represented in the Old. To proceed with due rigor all we need to do is to reconcile the rest of the facts pertaining to the cactaceous distribution with the two facts already set forth. There is no other way in science fit to yield a final elucidation of the migrations of this family.

Rhipsalis is doubtless a Gondwanic genus because it is rooted within the triangle Ceylon—Madagascar—Mascarenes. It well might have reached Brazil and the West Indies from West Africa. *Rhipsalis*, however, is but a minor segment of the Cactaceae, so that the last word is not yet said regarding the distribution and origin of this family, even if we have good reason to believe that we are no longer wholly ignorant of *Rhipsalis'* origin and dispersal.

Where, then, did the Cactaceae originate, and how did they enter the Americas? What is the primary center whereby they gained admission in the New World?

These questions can be reasonably answered, as a preliminary, considering, for instance, the dispersal of two genera in the immediate affinity of *Opuntia*, the very same *Opuntia* which had colonized the Western United States in the Eocene. These genera range (26) in this manner,

A) *Pterocactus*

(1)   *P. Hickenii* — Argentina (Chubut).
(2)   *P. Fischeri* — Argentina (Rio Negro).

(3) *P. pumilus* — Argentina (Chubut).

(4) *P. tuberosus* — Argentina (Mendoza, Córdoba).

B) *Maihuenia*

(1) *M. Poeppigii* — Chile.

(2) *M. Tehuelches* — Southwestern Argentina.

(3) *M. brachydelphis* — Argentina (Mendoza).

(4) *M. Valentinii* — Argentina (Chubut).

(5) *M. patagonica* — Argentina (Patagonia).

With dispersal of this type before us, we are fairly sure as to what we face. Precisely as the Palms *Beccariophoenix*, *Jubaeopsis* and *Jubaea* interplay across the *Afroantarctic Triangle* among Madagascar, Natal and Central Chile, so do *Rhipsalis*, *Pterocactus* and *Maihuenia* likewise interplay among Madagascar, the Mascarenes, Chile and Patagonia. In other words, there is no room to doubt that the Cactaceae are of one cloth with *Menodora* and homologous migrants. Their base-line of distribution squarely rests between Patagonia and the Mascarenes, and if *Rhipsalis* is Gondwanic it is likely, to say the least, that the archetypes of the family itself stem from the *Afroantarctic Triangle*. Unlike the Passifloraceae which are massive in the Old World, and have in the New only *Passiflora*, the Cactaceae are massive in the New World, and have in the Old only *Rhipsalis*. It is all very simple and quite standard. Before rejecting as preposterous the thought that the bulk of the cactaceous *genorheitron* is from the *Afroantarctic Triangle* (Natal—Kerguelen Islands—Tristan da Cunha; see p. 357), let us stop to consider that a form very close to *Passiflora* is endemic to New Zealand (see p. 318), and that allies of the Passifloraceae, the Achariaceae, are narrowly confined to the Cape and its immediate vicinity. In conclusion, the passifloraceous *genorheitron* ran "antarctic" tracks, and it should be impossible to believe that what the "American" and "tropical" *Passiflora* did could not have been done at all by the equally "American" and "tropical" Cactaceae.

Safe in the well documented preliminary that the Cactaceae are homologous by migration of *Menodora*, *Evolvulus*, Rapateaceae, etc., etc., we may proceed factually to study cactoid dispersal in the certainty we will no longer be lost among a welter of "floating continents," "wandering poles," "Occasional means of dispersal," etc., etc. *Opuntia*, for example, yields the following score, typified by a selection of representative species,

A) Subg. *Cylindropuntia*

a) Sect. *Vestitae*

(1) *O. vestita* — Bolivia (La Paz).

(2) *O. hypsophila* — Argentina (Salta).

b) Sect. *Clavarioides*

(3) *O. clavarioides* — Chile.

c) Sect. *Salmianae*

(4) *O. Salmiana* — Northern and Central Argentine, Paraguay, Southern Brazil.
(5) *O. maldonadensis* — Uruguay (Maldonado).

d) Sect. *Subulatae*

(6) *O. subulata* — Chile or Argentina.
(7) *O. exaltata* — Northern Chile, Bolivia, Peru, Ecuador.
(8) *O. pachypus* — Central Chile.
(9) *O. cylindrica* — Peru, Ecuador.

e) Sect. *Miquelianae*

(10) *O. Miquelii* — Chile (Atacama).

f) Sect. *Leptocaulae*

(11) *O. caribaea (O. metuenda)* — Caribbean coasts of Venezuela to Margarita, Santo Domingo.
(12) *O. leptocaulis* — Central Mexico (Puebla) northward (Tamaulipas, Coahuila, Nuevo León) to the United States (Texas, Arizona, New Mexico).

With this on record, we could stop without further, in the assurance we know how *Opuntia* reached the Western United States, and so did in so relatively early an epoch as the Late Cretaceous. The tracks outlined by these six groups are standard throughout. These tracks are standard throughout because, (i) They hit southermost South America and send an outlier to Southeastern Brazil; (ii) They follow in a pseudo-Andean pattern of dispersal northward to Ecuador; (iii) They use a secondary base in Venezuela to invade the West Indies; (iv) They disappear by disconnection between Ecuador and Mexico; (v) Having bridged this disconnection, they again turn up in Southern or Central Mexico and next follow through Eastern Mexico, mostly, to the United States. Tracks of the kind are matched in the migrations of groups which, like *Cuscuta* — merely to mention an example — are wholly unrelated with the Cactaceae, and are made possible by the fact that the coast of Western America once ran much farther west into the Pacific than it does today. Naturally, tracks of the kind are standard and altogether competent to take *Opuntia* to the Western United States from a starting point in the Southeastern Pacific or the South Atlantic reached in origin long before the Late Cretaceous.

Two groups are worthy of notice in *Opuntia* Subg. *Platyopuntia* as follows,

B) Subg. *Platyopuntia*

g) Sect. *Pumilae*

(1) *O. pascoensis* — Peru (Lima).
(2) *O. depauperata* — Venezuela (Caracas, Mérida).
(3) *O. pubescens* — Guatemala, Mexico.
(4) *O. pumila* — Mexico (Oaxaca, Morelos).

h) Sect. *Curassavicae*

(5) *O. curassavica* — Venezuela (Island of Tortuga), Dutch West Indies (Curaçao, Bonaire, Aruba) *).
(6) *O. repens* — Mona Island to Portorico and Ste.-Croix.
(7) *O. Taylori* — Santo Domingo.
(8) *O. borinquensis* — Portorico.
(9) *O. militaris* — Cuba.
(10) *O. Darrahiana* — Bahama (Turks Island).
(11) *O. Drummondii* — United States (Florida).
(12) *O. Tracyi* — United States (Northern Florida, Mississipi, Georgia).
(13) *O. nemoralis* — United States (Texas).

These two Sections illustrate a standard track (Fig. 70) which leaving Central Peru drives on to Venezuela bypassing Ecuador and Colombia, and next follows by two branches, respectively, to Central America and Southern Mexico; Virgin Islands, Cuba, Santo Domingo, Portorico, the Bahamas and the United States. This track is of the commonest, and — whole or in part — occurs in groups (e.g., Cyrillaceae, Liliaceae Melanthioideae) that bear to the Cactaceae no consanguineity whatsoever.

Section *Curassavicae* illustrates in addition a significant break in dispersal taking place between continental Venezuela and the islands immediately facing it. Hardly any of the species if this group belongs to the mainland **); the vast majority is purely "Caribbean" and "North American." This can only mean that Sect. *Curassavicae* draws its origin by a base-line of secondary distribution parallel with, and near by, the Venezuelan coast. The existence

---

*) Distributional data from WAGENAAR—HUMMELINCK, in Rec. Trav. Bot. Néerland, 35 : 34. 1938.

**) None of these species, as a matter of fact, is recorded from the mainland by the authors whose classification is followed here. However, the author of this manual knows a seemingly undescribed form of Sect. *Curassavicae* from the Venezuelan State of Anzoátegui. It might be worthy of notice that the locality of this finding is due south from the island of Tortuga, and that this island stands at the southernmost end of a line of major depths running southward from the Mona Passage. This line of major depths seems to continue southward into the lowlands which, this time on Venezuelan soil, divide the western (cordillera of the coast centering around Caracas) from the eastern (mountains of Sucre and Monagas) heights of Venezuela.

of a center of secondary distribution in this vicinity is further confirmed by the fact that the flora of the islands of Curaçao, Aruba and Bonaire noticeably differs from that of the Venezuelan mainland. Leaving the base-line in question Sect. *Curassavicae* runs an altogether standard track northward along the line Curaçao—Ste. Croix—Bahamas—Florida, which track frays westward to Texas through Cuba.

The break that takes place between the Venezuelan mainland and the group of islands centering around Curaçao is part and parcel of an homologous series of disturbances in these very same quarters. The distribution of *Tetragastris* (see p. 393) reveals on its part a huge disconnection between Central America and the Guianas. Less striking but quite as relevant breaks of the same nature occur between Panamá and Central Venezuela (Carabobo, Yaracuy; see p. 321 fn.). These disconnections always slice through the crumbled channel leading from Eastern Colombia and Western Venezuela to the Greater Antilles in the manner shown by the map attached (Fig. 71). It seems quite clear that the modern Caribbean Sea was created by faultings in part homologous of those that carved out the Indian Ocean. The process of "slicing off" responsible for the disconnection mentioned above would also seem to account for the differences in the flora of the Dutch West Indies as compared with that of coastal Venezuela. These islands and certain petty insular outposts politically belonging to Venezuela (Blanquilla, Los Testigos, Los Frailes, etc.) might stand for a landfront already "sliced off," but as yet not swallowed up, by the Caribbean Sea. The tale of dispersal in the Caribbeans, then, is not so much the tale of "Landbridges" connecting this or that modern island *as the record how a land once filling the extant sea crumbled front by front, thereby creating huge gaps in distribution that can now be filled only oversea.*

Sound evidence to the effect that this manner of interpreting the mainsprings of dispersal in the Caribbeans cannot be wrong in essentials is further forthcoming from the distribution of *Mammillaria*. This large and well marked genus is almost wholly "Mexican" because the huge majority of its species belong to the flora of Mexico. However, the non-Mexican component of *Mammillaria* is significantly dispersed in this manner,

(1) *M. bogotensis* — Colombia (Meta).

(2) *M. mammillaris* — Venezuela in the coastal regions throughout, south to Mérida; Dutch Leeward Islands.

(3) *M. Tamayonis* — Venezuela (Lara).

(4) *M. nivosa* — Mona Island, Virgin Islands to Antigua, Southern Bahamas.

(5) *M. prolifera* — Santo Domingo, Cuba, Gonaive.

(6) *M. Eichlamii* — Guatemala, Honduras.

Mammillaria prolifera is hardly better than a variety of *M. multiceps* endemic to Texas and Northeastern Mexico. Taking this in view, it will be seen that the tracks of *Mammillaria* outside of Mexico are a close match of those of the sections of *Opuntia* just reviewed. The base-line of dispersal is located between Eastern Colombia and Eastern Venezuela, and *M. Tamayonis* stresses most definitely the abutment of the "bridge" leading to the Antilles (see *Purdiaea, Dacryodes*, etc.). Leaving the base-line, the track fares to the West Indies which it invests between Cuba and Antigua, reaching the Bahamas and barely falling short of entering Florida. A branch of the track meantime leads to Central America (Guatemala and Honduras) and through Cuba to Texas and Northeastern Mexico. How close are the dispersals of *Opuntia* and *Mammillaria* can readily be seen from the map (Fig. 70), which map is further standard for the distribution of great many other groups wholly different from the Cactaceae.

Once rooted in Mexico, *Mammillaria* migrates northward and eastward from Oaxaca in the manner clearly revealed by certain of its subdivisions, *Series Candidae*, for instance. Out of approximately fifteen species in this group *) only one, *M. Halbingeri*, is endemic to Southern Oaxaca, the balance running through Tamaulipas, San Luís Potosí, Hidalgo, Queretaro, Guanajuato, Nuevo León, Coahuila, Durango, the Southeastern United States. Parallel with this track runs another exemplified this time by a different *Ancistracanthae*. The inception of this channel also stands in Oaxaca and Hidalgo *(M. zephyranthoides)*, but its further course is to the west, through Guerrero, Sinaloa, Nayarit, Durango, Sonora, Baja California to the Southwestern United States. Lastly, it is necessary to reckon with the fact that the track leading from the boundary now extant between Colombia and Venezuela to the Greater Antilles frays off near Cuba to enter Texas and Northeastern Mexico. Considering that Central and Eastern Mexico are the seat of a swarm of oligotypic or monotypic cactoid genera, some of them unique indeed, such as *Lophophora, Epithelantha, Obregonia, Aztekium, Strombocactus, Ariocarpus, Leuchtenbergia, Pelecyphora, Porfiria, Astrophytum;* and that some of these odd genera range into Texas (but not westward from Texas) as do *Lophophora, Epithelantha, Ariocarpus*, we are bound ultimately to conclude that a now crumbled sector defined by the following points, Texas—Eastern Cuba—Veracruz had fundamental significance in the evolution and dispersal of the Cactaceae. We are far indeed from anything savoring of thin "Landbridges" of Tertiary age when dealing with distributions of this scope and nature. *Things happen as if the modern Caribbean Sea did not exist at all, not only, but its very waters begot peculiar forms now found as relics along its shores.*

Better to understand how precise is dispersal we may compare the ranges of *Opuntia* Sect. *Curassavicae* and *Mammillaria* just reviewed with those, for instance, of certain species of *Pilocereus (Cephalocereus)* as follows,

(1) *P. barbadensis* — Barbados.

(2) *P. nobilis* — Grenada to St. Kitts.

(3) *P. Urbanianus* — Grenada, Martinique, Guadeloupe.

These three species precisely fill the range which *Opuntia* Sect. *Curassa-*

---

*) See, BRAVO, HELIA — Las Cactáceas de Mexico. 1937.

vicae and *Mammillaria* hardly enter. The line, Ste. Croix—Antigua, thus represents a landmark of some significance *). We much regret that the classification of the affinity centering around *Pilocereus* is in a state of utter confusion, which prevents us from critically discussing the ultimate origin of the three species of *Pilocereus* mentioned above, but we are inclined to believe that these forms stem from a powerful secondary center of origin of certain groups in the affinity of *Cereus* located between Eastern Brazil and Grenada This group, like *Rhipsalis*, might have been fed in origin from a line of migration directly hitting Eastern South America from the direction of the modern Atlantic. It is pacific, at any rate, that attempts made at interpreting the "origin" and "dispersal" of the Cactaceae which rely upon "occasional means of transportation," and the like are not worthy of the name of science. *Distribution is precise, repetitious in the first place, never haphazard or irrational.*

As we have seen, Colombia marks the southernmost limit of the distribution of *Mammillaria*. Characteristically, Colombia also marks the northernmost limit of that of *Malacocarpus* and *Frailea*. Colombia, then, stands out as one of the foremost nodes of cactoid dispersal; *it is significant that this node is precisely located in the heart of one of the major channels of dispersal in the New World running from Peru to Cuba.*

Mammillaria is on the whole a well marked genus. It becomes confluent with *Echinocactus*, nevertheless, through a controversial group, *Coryphanta.* *Mammillaria*, *Echinocactus* and *Coryphantha* all essentially belong to the flora of North America. As their counterpart in the flora of South America stand a host of controversial genera including *Malacocarpus* and *Frailea*. These South American genera range in the manner exemplified by the following selection of species,

A) *Malacocarpus*

(1)   M. *Vorwerkianus* — Colombia (Boyacá).
(2)   M. *Sellowii (M. tephracanthus)* — Eastern Argentina, Uruguay, Brazil (Rio Grande do Sul).

B) *Frailea*

(1)   F. *colombiana* — Western Colombia.
(2)   F. *pygmaea* — Argentina (Entre Ríos), Uruguay.
(3)   F. *Grahliana* — Paraguay.

C) *Austrocactus*

(1)   A. *patagonicus* — Argentina (Patagonia).
(2)   A. *Dusenii* — Argentina (Chubut, Río Negro).

---

*) This landmark is overstepped by comparatively very few species of *Opuntia* (e.g., *O. triacantha,* *O. rubescens*); by none of *Mammillaria*. The Lesser Antilles are as a rule poor of cactoid flora. Grenada is nevertheless connected with Venezuela by "*Cereus*" *grenadensis*.

D) *Notocactus*

(1) *N. Ottonis* — Northeastern Argentina, Uruguay, Southern Brazil.

(2) *N. submammulosus* — Paraguay, Northeastern Argentina.

E) *Neoporteria*

(1) *N. subgibbosa* — Central Chile.

(2) *N. napina* — Northern Chile.

F) *Parodia*

(1) *P. catamarcensis* — Argentina (Catamarca).

(2) *P. Maassii* — Northwestern Argentina, Bolivia.

(3) *P. Schebsiana* — Bolivia (Cochambamba).

(4) *P. paraguayensis* — Paraguay (Sierra de Amambay).

(5) *P. brasiliensis* — Brazil (São Paulo).

With this dispersal (Fig. 72) before us we may conclude in a wholly conservative vein as follows; A *genorheitron* laden with the genetic powers of certain genera *(Mammillaria, Echinocactus, Malacocarpus, Parodia)* entered the New World in the south along a standard channel followed by countless plants other than Cactaceae. This *genorheitron* came to the parting of the ways in a center of secondary origins located in modern Colombia. This center of secondary origin is intimately related with a similar center set between the modern coasts of Venezuela and the chain of islands immediately to the north of it. Both these centers lies in the heart of one of the major channels of American dispersal, running the line *Peru—Cuba* in the main. This channel frays westward to reach Central America and Mexico directly from Colombia; Yucatán, Texas and Northeastern Mexico directly from Cuba. This very same channel crosses over to the Bahamas keeping to a course which passes west of Antigua; from the Bahamas this channel can enter Florida and the southeastern United States.

No time need be spent to show that the ranges of *Malacocarpus, Frailea, Austrocactus,* etc., are wholly standard for South America.

If we intend to have added proof that the conclusions just drawn as regards the *Mammillaria—Malacocarpus* affinity are not wide of the mark, we can find this proof in the distribution of *Melocactus (Cactus).* This genus is an ally of *Echinocactus,* therefore a *distant* relative of *Mammillaria* as well. *Melocactus* is distributed in this manner,

(1) *M. Townsendii* — Peru (Lima).

(2) *M. Neryi* — Brazil (Amazonas).

(3) *M. bahiensis* — Brazil (Rio de Janeiro, Bahia, Pernambuco).

(4) *M. obtusipetalus* — Colombia (Cundinamarca).

(5) *M. caesius (amoenus)* — Colombia, Venezuela to near Trinidad.

(6) *M. macracanthus* — Dutch West Indies (Curaçao, Aruba, Bonaire).

(7) *M. Broadwayi* — Tobago.

(8) *M. communis* — Jamaica.

(9) *M. intortus* — Portorico, Southern Bahamas, Virgin Islands, St. Christopher, Antigua, Montserrat, Dominica.

(10) *M. Lemairei* — Santo Domingo.

(11) *M. Harlowii* — Eastern Cuba.

(12) *M. Ruestii* — Honduras.

(13) *M. Maxonii* — Guatemala.

(14) *M. salvador* — Mexico (Oaxaca).

Melocactus does the following, (i) Runs the classic track Peru—Greater Antilles through Colombia; (ii) From Colombia further runs the classic track going to Guatemala, Honduras and Southern Mexico (Oaxaca). Notice that the track stops in Oaxaca, there precisely where begins the main stream of the dispersal of *Mammillaria;* (iii) Breaks into two distinct but most variable species in the range Coastal Venezuela—Dutch West Indies, thus marking the usual secondary center of dispersal located in this region; (iv) Unlike *Mammillaria,* has a center of speciation situated in Brazil (Amazonas, Pernambuco, Bahia, Rio de Janeiro), which center of speciation is most likely responsible for the species endemic to the Lesser Antilles. Typical is the track of *M. intortus* running all the way from Dominica to Portorico and the Southern Bahamas. Notice that Portorico and Guadeloupe stand at bottom merely as an extension of the key-line Ste. Croix—Antigua. In other words, a weighty landmark of dispersal is set by a line that may include Portorico and Guadeloupe, *if broad;* only Ste. Croix and Antigua, *if narrow;* (v) Unlike the origins of *Mammillaria,* which are narrowly contained within Colombia, those of *Melocactus* are to be sought within a belt running from Central Peru to Eastern Brazil. This is to say that the break taking place between the northern and southern end of the cactaceous dispersal is to be sought ultimately within a quadrangle Colombia—Guianas—Peru—Eastern Brazil. This triangle also marks the origin of genera like *Tetragastris* (see p. 393) segregated out of the Protieae, and is accordingly important for all manner of plants, related or unrelated that they be with the Cactaceae.

A well marked group in this family is constituted by a northern genus, *Echinocereus,* and three southern ones, *Echinopsis, Lobivia* and *Trichocereus.* These four genera are distributed in the manner exemplified by the following selection of species,

A) *Echinocereus*

(1) *E. cinerascens* — Mexico (Hidalgo, Mexico D. F.).

(2) *E. pectinatus* — Mexico (Guanajuato to San Luís Potosí and Chihuahua).

(3) *E. Blanckii* — Mexico (Tamaulipas), U.S.A. (Texas).

(4) *E. huitcholensis* — Mexico (Jalisco, Nayarit).

(5) *E. acifer* — Mexico (Durango, Coahuila).

(6) *E. scopulorum* — Mexico (Sinaloa, Sonora).

(7) *E. polyacanthus* — Mexico (Durango, Chihuahua), U.S.A. (Arizona, New Mexico).

(8) *E. pensilis* — Mexico (Baja California).

(9) *E. mojavensis* — U.S.A. (California).

(10) *E. Engelmannii* — Mexico (Sonora, Baja California), U.S.A. (California, Arizona, Nevada, Utah).

(11) *E. triglochidiatus* — U.S.A. (New Mexico, Colorado, Texas).

(12) *E. Baileyi* — U.S.A. (Oklahoma).

B) *Echinopsis*

(1) *E. cordobensis* — Argentina (Córdoba).

(2) *E. Bridgesii* — Bolivia (La Paz).

(3) *E. Eyresii* — Argentina (Entre Ríos), Uruguay.

(4) *E. Meyeri* — Paraguay.

C) *Lobivia*

(1) *L. breviflora* — Argentina (Salta).

(2) *L. caespitosa* — Bolivia (La Paz).

(3) *L. mistiensis* — Peru (Arequipa).

D) *Trichocereus*

(1) *T. coquimbanus* — Northern Chile (Coquimbo).

(2) *T. deserticolus* — Northern Chile (Antofagasta).

(3) *T. lamprochlorus* — Argentina (Córdoba, Mendoza, Jujuy).

(4) *T. Bridgesii* — Bolivia (La Paz).

(5) *T. fascicularis* — Northern Chile to Southern Peru.

(6) *T. cuscoensis* — Peru (Cuzco).

(7) *T. Pachanoi* — Northern Peru, Southern Ecuador.

It is readily seen that these four genera run a track unlike the *Malacocarpus* —*Mammillaria* affinity. This track is essentially "Andean," to use a conventional term. It reaches Uruguay and Paraguay, true, but does not touch Colombia and leaves Venezuela alone. In other words, this track closely adheres to the western half of the Americas, and does not follow at all the classic channel, *Peru—West Indies.*

Naturally, this track becomes disconnected in the range Ecuador—Mexico, which is standard of the groups following similar channels of dispersal.

With this we may close this part of our review of the Cactaceae, affirming in a sober and conservative vein the following, (a) This family is one with many others, all of which are massive in the New World, thinly represented in the Old; (b) The thought of handling it successfully as an "American" problem cannot occur to a student informed of the rudiments of dispersal; (c) The base-line of the distribution is set between the Mascarenes and Patagonia in a wholly standard manner; (d) It is possible that certain genera and groups (*Rhipsalis*, affinity of *Cereus*) entered South America directly from the Atlantic, but it is even more likely that the bulk of the *genorheitron*, on the contrary, invested the tail-end of South America along the Atlantic and the Eastern Pacific; (e) A quadrangle Colombia—Guianas—Peru—Eastern Brazil is critical as a main secondary center of origin and redistribution of the Cactaceae; (f) Certain groups in this family run the classic track *Peru—Greater Antilles*, others the none less classic channel, *Bolivia—Ecuador—Mexico*.

To end the subject, let us take notice of the often heard statement that the "leafy" genera *Pereskia* and *Pereskiopis* are "primitive" in the Cactaceae. This statement, if at all true, ought to be supported by the facts in dispersal.

Here are the facts, retold by the actual distribution of these two genera,

A) *Pereskia*

(1) P. *Weberiana* — Bolivia (Cochabamba).
(2) P. *Sacharosa* — Northwestern Argentina to Paraguay.
(3) P. *Moorei* — Brazil (Matto Grosso).
(4) P. *bahiensis* — Brazil (Bahia).
(5) P. *Zehntneri* — Brazil (Bahia).
(6) P. *Humboldtii* — Northern Peru.
(7) P. *Bleo* — Northern Colombia.
(8) P. *colombiana* — Northern Colombia (Magdalena).
(9) P. *Guamacho* — Venezuela (coastal, and Orinoco Basin).
(10) P. *aculeata* — Northern coasts of South America, West Indies.
(11) P. *grandifolia* — West Indies, Brazil.
(12) P. *cubensis* — Cuba.
(13) P. *portulacifolia* — Santo Domingo.
(14) P. *nicoyana* — Costarica.
(15) P. *autumnalis* — Guatemala.
(16) P. *Conzattii* — Mexico (Oaxaca).
(17) P. *tampicana* — Mexico (Veracruz).

B) *Pereskiopis*

(1) P. *Kellermanii* — Guatemala.
(2) P. *Pititache* — Mexico (Oaxaca, Jalisco).

(3) *P. Chapistle* — Mexico (Oaxaca, Morelos).

(4) *P. rotundifolia* — Mexico (Oaxaca, Morelos).

(5) *P. velutina* — Mexico (Querétaro).

(6) *P. aquosa* — Mexico (Jalisco, Nayarit).

(7) *P. Blakeana* — Mexico (Sinaloa).

(8) *P. Gatesii* — Mexico (Baja California).

It is transparent that these genera run the usual tracks of all other Cactaceae. *Pereskia* is rooted in the range Bolivia—Southern Brazil like the affinity of *Malacocarpus;* it has a center of speciation in Eastern Brazil like *Melocactus;* runs the track Peru and/or Brazil—Greater Antilles once again, like the affinity of *Mammillaria, sensu lato. Pereskiopis* is conventionally strong in Southern Mexico, and from Southern Mexico follows on to Western Mexico as far as Baja California.

The conclusion is inescapable that these two genera may retain archetypal characters — perhaps — but that they are certainly not the most "primitive" cactoid genera, if by "primitive" we understand the forms ancestral to all Cactaceae. These forms are something else, necessarily intermediate among *Leuchtenbergia, Rhipsalis* and *Pereskia.* Attempts made at deriving these forms from the Aizoaceae, Phytolaccaceae, etc., are hardly worthy of serious discussion.

XI. *Schoenocaulon* and its affinity

Schoenocaulon (Sabadilla, Veratrum) is a revealing genus from the phytogeographer's viewpoint. Not less interesting is the broad affinity of which it is part, Liliaceae Melanthioideae. We will review this genus and its alliance briefly in order to derive conclusions soon to transpire.

Schoenocaulon — of which a good monograph is available (27) — is distributed as follows,

(1) *S. dubium* — U.S.A. (Florida).

(2) *S. texanum* — U.S.A. (Texas, New Mexico), Mexico (Chihuahua, Nuevo León, Tamaulipas, Hidalgo, Puebla).

(3) *S. Drummondii* — U.S.A. (Texas), Mexico (San Luís Potosí).

(4) *S. macrocarpum* — Mexico (Nuevo León).

(5) *S. megarhiza* — Mexico (Chihuahua, Sinaloa).

(6) *S. regulare* — Mexico (Durango, Jalisco).

(7) *S. obtusum* — Mexico (Hidalgo).

(8) *S. Coulteri* — Mexico (Hidalgo).

(9) *S. tenue* — Mexico (San Luís Potosí, Morelos).

(10) *S. Mortonii* — Mexico (Michoacán).

(11) *S. Pringlei* — Mexico (Nayarit, Mexico, D. F., Hidalgo, Puebla).

(12) *S. jaliscense* — Mexico (Jalisco).

(13) *S. caricifolium* — Mexico (Coahuila, Hidalgo, Oaxaca).

(14) *S. comatum* — Mexico (San Luís Potosí, Puebla, Oaxaca).

(15) *S. Conzattii* — Mexico (Puebla, Mexico D. F., Hidalgo, Oaxaca).

(16) *S. calcicola* — Mexico (Oaxaca).

(17) *S. tenuifolium* — Mexico (Oaxaca).

(18) *S. Ghiesbrechtii* — Mexico (Chiapas).

(19) *S. yucatanense* — Mexico (Yucatán).

(20) *S. officinale* — Mexico (Colima, Guerrero, Michoacán, Mexico D. F., Morelos, Veracruz, Oaxaca, Chiapas), Guatemala, Honduras, Salvador, Costarica, Venezuela (Western Andean States, Caracas, Amazonas), Peru (Cuzco).

The affinity of the Liliaceae Melanthioideae of which *Schoenocaulon* is part (28) is tribe Veratreae. This tribe is composed and distributed in this manner,

A) Veratreae

(1) *Amianthium (Chroosperma)* — 1 species in the Eastern U.S.A. (Florida to Arkansas and New York).

(2) *Tracyanthus* — 2 species: 1 in Texas; 1 in the Southeastern U.S.A. (Florida to Mississipi and North Carolina).

(3) *Schoenocaulon (Sabadilla)* — As previously detailed.

(4) *Stenanthella* — 2 species: 1 in Sakhalin; 1 in the Western U.S.A. (Oregon, Idaho) and Canada (British Columbia, Alberta).

(5) *Stenanthium* — 2 to 4 species in the Eastern U.S.A. (Florida to Pennsylvania, Arkansas, Ohio) and Mexico.

(6) *Toxiscordion* — 8 species in North America, mostly in the U.S.A.

(7) *Oceanoros* — 1 species in the Eastern U.S.A. (Louisiana to Virginia).

(8) *Zygadenus* — Sect. *Euzygadenus:* 1 species in the Eastern U.S.A. (Florida to Louisiana and Virginia) — Sect. *Anticlea:* 1 species in Siberia, 7 species in North America and Mexico (*Z. elegans* ranging from the Behring Strait to Mexico).

(9) *Melanthium* — 3 species in the Eastern U.S.A. (Florida to Rhode Island and Minnesota).

(10) *Veratrum* — About 46 species; Subg. *Pseudomelanthium:* 3 species in the Eastern U.S.A. (Florida to Virginia); Subg. *Euveratrum* Sect. *Alboveratrum:* About 6 species in Northern Eurasia from the Pyrenees to Japan; 1 in the Eastern U.S.A. to Eastern Canada (Florida to Minnesota and New Brunswick); Sect. *Fuscoveratrum:* 2 species (1 in the Alps, Carpathians, Northern Balkans, Central Asia to Northern China and the Ussuri; 1 in Formosa); Subg. *Pseudoanticlea:* 9 species in the Far East (Burma, China, Yunnan to Manchuria, Eastern Siberia, Corea, Japan, Sakhalin).

With the pertinent facts of the record before us, we observe, (a) *As to the Veratreae* — This tribe is one of the very many groups which reach as a preliminary a point in Central Asia *(Altai Node)*, and from this point onward migrate both east and west to the New World and Europe. It is certain that the

Veratreae entered North America by transpacific tracks (cf. *Zygadenus* and *Veratrum*). There is no indication, on the contrary that even part of this group came to the Northern New World by transatlantic migrations (cf. *Veratrum* Sect. *Fuscoveratrum* ranging between Formosa and the Alps only); (b) *As to Schoenocaulon* — As part and parcel of the Veratreae, the archetypes of this genus reached the New World by transpacific channels. Mexico was heavily invested from the start at a point which the sum total of the evidence suggests to have stood within the triangle Chiapas—Jalisco—Veracruz. It is worthy of notice that while *Schoenocaulon*'s allies, *Zygadenus* and *Veratrum*, are both well known in Arizona, *Schoenocaulon* itself is not seen westward from New Mexico. All this is consistent with a track running eastward and westward in the main from a center such as we have individuated above.

So much for the generalities of the dispersal of these two groups. To consider now in detail certain aspects of the distribution of *Schoenocaulon;* (a) *Schoenocaulon officinale* is rooted in Southern Mexico and Central America. It next turns up in Venezuela over a disconnected range of standard nature, right by the abutment of the classic channel *Peru—Greater Antilles* on continental South American soil. This species follows this channel, very much in the manner of *Purdiaea* and *Dacryodes*, but in one direction only, going, that is, southward to Peru; (b) It is clear that the sum total of the tracks of *Schoenocaulon* holds the Caribbeans in a two-pronged track (triangle Central Mexico—Florida—Venezuela) without, however, interesting any of the West Indian insular outposts. In this, *Schoenocaulon* basically behaves like *Gaylussacia* and the Sarraceniaceae. In other words, *Schoenocaulon* is a classic instance of a genus which hits the New World from the direction of the Pacific but leaves the West Indies untouched. As we know, other genera (e.g., the *Washingtonia* affinity, most likely *Sporolobus*) behave differently. It stands to reason that *Schoenocaulon*, or forms closely allied to it, might be looked for in Cuba at least, for Cuba lies straight by a track, Texas—Florida and Venezuela—Santo Domingo. Naturally, were *Schoenocaulon* ever to be found in Cuba, Jamaica or Santo Domingo (a distinct possibility, as such) the interpretation we give of its migrations would not change in the least. Possible findings of the sort would merely add their weight to the evidence such as we have understood it.

Although it seems no longer necessary to give time and print to considerations fit to prove that the Veratreae are not "Holarctic" in origin, we will briefly consider here certain of their allies under the Melanthioideae. A tribe of this subfamily, for instance, ranges in this manner,

B) Uvularieae

(11) *Kreysigia* — 2 species in Eastern Australia.

(12) *Schelhammera* — 2 species in Eastern Australia.

(13) *Gloriosa* — 4/5 species: *G. superba* in Africa (East and Central Africa

to Nigeria, Togo, Guinea, Senegambia) and Asia (Ceylon, India, Siam, Cochinchina, Malaysia); *G. virescens* in Tropical Africa; *G. speciosa* in East Africa (Ethiopia, Eritrea, Somaliland); *G. minor* in East Africa (Somaliland).

(14) *Littonia* — 7 species in Natal, Pondoland, Transvaal, Angola, Tanganyka, Congo, Somaliland.

(15) *Sandersonia* — 1 species in Natal, Pondoland, Griqualand.

(16) *Uvularia* — 2/3 species in the Eastern U.S.A. and Canada (Florida to Oklahoma, North Dakota, Quebec and New Hampshire).

This tribe is perfectly balanced to the extent of the first five genera listed. *Kreysigia* and *Schelhammera* are in Eastern Australia; *Gloriosa* is conventionally Gondwanic; *Littonia* and *Sandersonia* stress the *Kalaharian Center* of secondary evolutions first and foremost. In sum, the center of origin of this aggregate is doubtless located in the modern Indian Ocean in a wholly standard manner. *Uvularia* is the seemingly jarring note in an otherwise altogether harmonious picture.

However, the author whose classification we follow tells us that the Uvularieae are quite close the Tofieldieae by anatomy and embryology. The Tofieldieae are on their part distributed as follows,

C) Tofieldieae

(17) *Tofieldia* — About 18 species in the north temperate to cold zone (Europe, Asia, (Siberia, Tibet, China, Japan, the Himalayas), North America).

(18) *Pleea* — 1 species in the Southeastern U.S.A. (Florida to North Carolina).

(19) *Narthecium* — 4 species in North America, Northern Europe, Japan.

(20) *Nietneria* — 1 species in British Guiana (Mt. Roraima).

The Tofieldieae neatly reverse the score of the Uvularieae. *Tofieldia, Pleea, Narthecium* are boreal and harmonious in their distribution. *Nietneria* is seemingly misplaced in this group.

However, if we consider Tofieldieae and Uvularieae jointly — as a broad affinity, that is, which is the proper way of viewing the facts on record — we readily understand that *Uvularia* is no more "mysterious" than *Nietneria*. The distribution of these two tribes is perspicuous. The *genorheitron* is Gondwanic, which is to say originated in what is now the Indian Ocean. Leaving this center of origin, the affinity in question went to, (i) South Africa; (ii) West Africa; (iii) Central Asia; in an altogether trite pattern of migration. It is most likely that the track faring to Central Asia ran close by Somaliland on its way to the Himalayas and Central Asia, because Somaliland is an important center of speciation of *Gloriosa* and *Littonia*, proof by itself that Somaliland was reached by *genorheitral* streams in highly plastic conditions.

Once in Western Africa (Nigeria—Guinea), the *genorheitron* crossed without difficulty and in an altogether standard manner to the *Roraiman (Nietneria)* and the *Ozarkian—Appalachian Centers (Uvularia)*.

In conclusion, the distribution of *Uvularia* and *Nietneria*, respectively, which seem "mysterious" when considered apart, become more than transparent when seen jointly. Two "mysteries" brought together become fairly clear, which is proof again that nothing in phytogeography can be viewed from the "specialized" point of view. A student of *Gloriosa* and *Littonia* working in South Africa may break his head trying to figure out why these two genera are allies of *Uvularia* right at the same time, when a North American botanist interested in *Narthecium* is ready to dismiss the range of *Nietneria* as a deep "mystery." Would these students only get together, and compare notes they could not fail to understand that the "mysteries" baffling them are not mysteries at all.

Nine out of ten of the "mysteries" and "doubts" besetting botany are exactly of one cloth with the "mysteries" involved in the dispersal of Uvularia *and* Nietneria. *Everybody in this science seems to be digging his own hole, and to go ever deeper into it to the ultimate stultification and sterilization of everything and anything he may happen to find.*

A peculiarly interesting monotypic group of the Melanthioideae is the following,

D) Hewardieae

(21) *Hewardia* — 1 species in Tasmania.

Our authority comments as follows, "Die sehr isoliert stehende Gattung besitzt durchaus den Habitus einer Iridacee und kann vielleicht als eine Urform der Iridaceentypus mit noch oberstaendigem Ovar angesehen werden." Clearly, the Liliaceae are one more of the groups to become confluent with another family (Iridaceae in this case) in the deep south of the map *). The Liliaceae are not "Holarctic" by a long throw.

A group of genera of the Melanthioideae which seemingly originated mostly, when not entirely, in the *Afroantarctic Triangle* is the following,

E) Anguillarieae

(22) *Reya* — 1 species in Southern Australia and Tasmania.

(23) *Androcymbium* — 33 species in South Africa. Two as follows, (i) *A. striatum:* Transvaal, Angola, East Africa to Ethiopia and Western Somaliland; (b) *A. punctatum:* Macaronesia, North Africa (Morocco to Egypt) — var. *palestinum:* Palestine, Sinai.

(24) *Baeometra* — 1 species in the Cape.

(25) *Dipidax* — 2 species in the Cape.

---

*) Other Liliaceae aberrant in the details of the perianth occur in Southern Africa, such as *Chlorophyton andongense*. See Schlitter in Ber. Schweiz. Bot. Gesell. 53: 491. 1943.

(26) *Wurmbea* — 8 species; 4 in Africa (Natal to Cape, Nyasaland, Cameroon and Fernando Po), 4 in West Australia.

(27) *Neodregea* — 1 species in the Southeastern Cape.

(28) *Ornithoglossum* — 3 species in South Africa.

(29) *Anguillaria* — 3 species in Tasmania and West Australia.

(30) *Iphigenia* — 10 species; *I. indica* in India, Southern China, Philippines, Northern Australia and Queensland, New Guinea; *I. Oliveri* in East Africa; *I. Schlechteri* in Natal; *I. guineensis* in Angola; *I. Ledermannii* in the Cameroon; *I. robusta* in Madagascar; *I. novaezelandiae* in New Zealand; the balance in South and Southwest Africa.

This leaves nothing to imagination. The track of *Androcymbium* is redolent of the migrations of *Erica* and *Orobanche*. *Wurmbea*, to judge by the date now in our hands, is a classic instance of a close generic *genorheitron* evolving simultaneously both in West Australia and the *Kalaharian Center*. Only *Iphigenia* seems to have one species in Madagascar, and this is the species that reaches the Far East in approved *Gondwanic* pattern, while retaining a marked tie Natal—New Zealand of pure *Afroantarctic* flavor.

A monotypic subfamily of the Liliaceae is worth last mention here, as follows,

A') Herrerioideae

(31) *Herreria* — 5/6 species in Northern Chile, Peru, Northern Argentine, Paraguay, Uruguay, Southern and Central Brazil (Bahia, Minas).

Unlike the Uvularieae and Tofieldieae which come to the New World by the line Guinea—Guianas—Southeastern United States, the Herrerioideae follow the path that takes in "Andean" America from the Southeastern Pacific. No marvel in this, of course, for all these lines and paths are standard throughout.

In conclusion, these plants originated in a first time in the *Afroantarctic* and *Gondwanic Triangles*. They left these cradles to disperse in an absolutely orthodox manner over channels of migration used over and over again. Their "cold" and "warm" components staggered themselves anywhere between Alaska and Tasmania in trite patterns of distribution.

The query: *How does a "cold" element "find a "Bridge" to migrate northward?* is certainly not difficult to answer. It can be answered curtly and efficiently in this manner; *Stop theorizing, and look at what is done by Nietneria, Uvularia, Androcymbium, Erica, Vaccinium, Cobresia, Primula, Anemone, Salix, Alnus, Drosera, Utricularia, Gaylussacia, Schoenocaulon, Potentilla, Carex, Juncus, Tilia, etc., etc.* True enough, there exist strictly "warm" groups like Cochlospermaceae and Ochnaceae which originated, and always lived, in tropical ranges of Gondwana and continents adjacent. Together with these genuine, absolutely "warm" groups coexist countless other aggregates which are not strictly "warm", witness the Ericaceae, large part of which originated together with the Empetraceae in the *Afroantarctic Triangle*.

Finally to return to *Schoenocaulon:* In this genus stand species such as these by habitat,

(1) *S. calcicola* — Oaxaca (calcareous banks, alt. 1830 m.).

(2) *S. caricifolium* — Hidalgo (calcareous mountains).

(3) *S. dubium* — Florida (Volusia Co., by sea-level, dry pine barrens).

(4) *S. megarhiza* — Sinaloa (open grassy slope, alt. 1200—1350 m.).

(5) *S. officinale* — Veracruz (dry calcareous hills); Guatemala (open wooded limestone hills; damp wooded gulch, bushy slope, alt. 1800 m.).

(6) *S. Pringlei* — Mexico D. F. (lava fields alt. 3050 m.); Puebla (Mt. Orizaba).

(7) *S. texanum* — Texas (Bexar Co., hard limestone hills).

It seems plain that this genus in tendency; (a) Avoids constantly moist or inundable habitats; (b) Is markedly calciphilous; (c) Does not stand repeated killing frosts (i.e., ca. — 2/3° Cent.; 27/28° Fahr.); (d) Is indifferent to altitude up to ca. 1500 m. (approximately 4500 feet).

In this understanding it seems again plain that, (a) *S. Pringlei* is essentially an ecotype answering decomposed or recent lava-soils, and extreme height; (b) *S. officinale* owes its wide spread to its being tolerant of extremes. Near Caracas this species lives in heavy clays by decomposed schists not at all calcareous. Other species are undoubtedly tied to calciphily.

Curious is the fact that the Cactaceae, which share many localities with *Schoenocaulon*, exhibit remarkably high percentual figures of lime in their tissues (this may well exceed 70 % in dry weight), while often occurring endemic in ranges where limestone is not to be found. Contrariwise, some of their numbers are relics indeed with marked calciphilous preferences (e.g., *Mammillaria Tamayonis:* Venezuela, Edo. Lara: calcareous hills of La Cantera near El Tocuyo — the lone known station so far).

We have no sufficient data at hand at present to press these observations farther. We record them here, nevertheless, as an indication that the *phytogeographic element* of dispersal (the element that is, which is interested in tracks purely as palaeogeographic and geographic lines conveying distribution) merges most intimately with the *ecologic element* (the element, that is, concerned with ambiental factors along the track).

It is clear that in the same genus or group we may meet species that retain ancestral preferences for certain types of climate and soil, while other of their allies readily adapt themselves to changed conditions. The latter may stand out in the end as marked species or ecotypes in their own right (e.g., *S. Pringlei*) usually of restricted distribution in "aberrant" habitats, or secure on the contrary (e.g., *S. officinale*) wide distribution.

XII. Suaeda

The maps attached (Fig. 73, 74) are bodily drawn from the work of sharp student *) of the Chenopodiaceae and the flora of arid Russia in general. It illustrates the distribution of the chenopodiaceous genus *Suaeda* apportioned by sections.

We here intend to interpret this *purely as it looks,* without any reference whatsoever to the text that originally elucidates it. As by so doing we shall not have before us an account of the probable interrelations among the sections of *Suaeda,* we will freely moot alternatives whenever necessary. We leave it to reader to verify which one of the alternatives we suggest is correct in the end.

We observe,

(a) Section *Heterosperma* — The map shows it running within a continuous belt between New Zealand and North America, not only, but extending further to South America by an outlier. We are inclined to break this dispersal in two main massings, as follows, (i) *Massing Australia—Europe.* This massing is fed from the Australian domain northward and westward over a track homologous of that of *Carex pyrenaica, Notelaea* and *Coriaria,* merely to mention obvious examples; (ii) *Massing Chile—Argentina—North America.* This is due to the influence of the Magellanian gate of angiospermy. In conclusion, the base-line of the distribution of this group squarely rests in the Western and Southern Pacific. The Western Polynesian and Magellanian gates are active. The track runs in the East Pacific in disconnection between Chile and Southern Mexico which is quite by rote. Cuba is invested, but not the Lesser Antilles, which is to be expected.

(b) Section *Lachnostigma* — The origin of this group is by the *Afroant-arctic Triangle,* as none of its species interests Madagascar or runs eastward from Beluchistan. The dispersal is by two shore-lines, (i) *Palaeogeographic* — This line agrees with the ancient inland seas or lakes once covering much of Western Africa. It runs from Angola southward to the basin of Orange River in South Africa, then veers northward by Natal and the Eastern Transvaal (these being the actual regions of entrance of the *genorheitron* into Africa). The track next follows in approved fashion through Rhodesia, Tanganyka, Ethiopia to eastern French West Africa; takes in Arabia and much of Lybia, Egypt ultimately to end by the Caspian Sea; (ii) *Geographic* — This shore corresponds with the modern one of Southwest Africa, the Cape, Natal, Mozambique, Tanganyka, Somaliland, the Red Sea generally, Arabia and Southern Persia to Beluchistan. Worthy of notice is the fact that the dispersal of this section is reminiscent of that of the cycadaceous genus *Encephalartos* (see p. 479, Fig. 93), and that there was a continuity of palaeogeographic shore leading all the way to Northern Angola. Possibly significant is the fact that the dispersals stop approximately in Uruguay and Northern Angola,

---

*) ILJIN, M. M. — Contribution à la Systématique du Genre *Suaeda* et de la Tribu Suaedeae; in Sovietsk. Bot. No. 5: 39—49 Fig. 2. 1936.

which may be taken as an indication that the connections once binding Western Africa to eastern South America had a coast stretching between these two regions.

(c) Section *Platystigma* — This group might have originated independently either from a point in the Indian Ocean by the *Gondwanic Triangle*, or have issued from the Magellanian gate in disconnection. Alternatively, it might have been derived either from sect. *Heterosperma* or sect. *Lachnostigma*. Section *Platystigma* follows in the main the shore once connecting Northern Africa with the Caribbeans by a channel of migration driving from Macaronesia to the vicinity of Guadeloupe, Cuba, Yucatán, ultimately Western Mexico. It stands to reason that if this section stems from the Caribbeans rather than from the Mediterranean the sequence of localities in the track is to be read from Western Mexico to the Mediterranean. The outlier in Central Peru might have been established in either one of two manners, (i) Directly from Cuba in the manner of *Purdiaea, Dacryodes*, etc.; (ii) Immediately from Southern Mexico in disconnection through the Eastern Pacific. All the tracks stressed in this discussion are standard.

(d) Section *Limbogermen* — May be an independent development which reached Mexico and the adjacent U.S.A. from the Eastern Pacific, or, conversely, an offshoot of either sect. *Heterosperma* or sect. *Platystigma*.

(e) Sections *Schanginia, Conosperma, Physophora* — These groups may have originated independently from the *Gondwanic* and *Afroantarctic Triangles*, or from the *Neocaledonian Center;* they may conversely stem from sect. *Heterosperma* or *Lachnostigma*, which all is to be settled with reference to their phylogeny and morphology, not by any theory. These three sections are significant in that they are rooted by the northern shore of Gondwana and were in time confined within characteristic roughly circular or oval patterns of distribution when the formerly free open sea north of Gondwana came to be surrounded by lands in emersion in a northerly direction. Patterns of distribution of the same nature are to be seen in *Menodora* (see p. 358, Fig. 68) and *Frankenia* *), which all stems from the same order of factor.

The student of *Suaeda* whom we follow is of the opinion that *Suaeda* was the modern genus not later than by the inception of the Cretaceous. This is a sound opinion.

XIII. Brunelliaceae, Cunoniaceae and Protieae

These two families, and this tribe, are of fundamental significance as regards certain aspects of New World dispersal. The reader is soon to learn why this is so.

---

*) See, for instance, SUMMERHAYES, V. S. — A revision of the Australian species of *Frankenia;* in Jour. Linn. Soc. Bot., 48: 337—387, Fig. 6 *etc.* 1930

The Brunelliaceae consist (29) of the lone genus *Brunellia*, which is distributed in the manner exemplified by the following species,

(1) *B. Oliveri* — Bolivia (Yungas).

(2) *B. hexasepala* — Peru (Puno).

(3) *B. Weberbaueri* — Peru (Huánuco, Junín).

(4) *B. ovalifolia* — Ecuador (Loja).

(5) *B. Steubelii* — Colombia (Popayán, Cauca).

(6) *B. Goudotii* — Central Colombia.

(7) *B. Funckiana* — Colombia (Santander), Venezuela (Mérida, Caracas).

(8) *B. comocladiifolia* — Colombia (Popayán, Boyacá, Santa Marta), Venezuela, Costarica, Mexico, Jamaica, Santo Domingo, Cuba, Guadeloupe, Portorico.

These eight representative species out of the fourteen or fifteen now constituting the genus describe a perfect track of the type conventionally understood as "Andean." In reality, this track is not "Andean" because of two reasons, (a) *Brunellia* was beyond doubt in the New World long before the Andes began to rise. It took to the *cordilleras*, consequently, in a second time only of its dispersal; (b) This genus breaks with the Andes in Colombia to cross over to Central Venezuela, not only, but to run a track from the boundary of Colombia and Venezuela northward to the Greater Antilles. This track is absolutely standard, and was followed by groups (e.g., *Mammillaria;* see p. 369) which are beyond doubt not "Andean."

Rather than "Andean," then, the dispersal of *Brunellia* is of a pattern essentially wedded to a basic channel of migration running from Bolivia (or Peru) to the Greater Antilles. This channel frays eastward to Venezuela; westward to Central America and Mexico, in the majority of the case.

The Brunelliaceae are altogether close to the Cunoniaceae, both these families rating in reality as offshoots out of the saxifragoid *genorheitron*. Unlike the petty Brunelliaceae the Cunoniaceae are a massive, truly formidable aggregate, with tribal groups so formed, for instance (30),

A) Belangereae

(1) *Belangera* — 8—9 species in Paraguay and Brazil (Matto Grosso, Rio Grande do Sul, São Paulo).

(2) *Gessois* — 4 species in New Caledonia, 2 in Fiji, 1 in the New Hebrides, 2 in Queensland.

B) Cunonieae

(3) *Stollaea* — 1 species in Northeast New Guinea.

(4) *Ackama* — 3 species in Australia (Queensland, New South Wales) and New Zealand.

(5) *Spiraeopsis* — 2 species in New Guinea, 1 in Celebes, 2 in the Philippines (Mindanao).

(6) *Platylophus* — 1 species in the Cape (Tulbagh, Knysna, Paarl Divisions).

(7) *Caldcluvia* — 1 species in Chile (Chiloe Island to Concepción).

(8) *Acrophyllum* — 1 species in Australia (New South Wales).

(9) *Schizomeria* — 5 species in New Guinea, 1 in North Australia, 1 in Amboina.

(10) *Ceratopetalum* — 3 species in Eastern Australia.

(11) *Adenopetalum* — 1 species in Tasmania.

(12) *Opocunonia* — 9 species in Northeastern New Guinea.

(13) *Pseudoweinmannia* — 2 species in Eastern Australia.

(14) *Wesselowskya* — 2 species in New Caledonia and Eastern Australia (New South Wales).

(15) *Cunonia* — 1 species in the Cape to Natal, 11 species in New Caledonia.

(16) *Weinmannia* — 126 species next discussed.

(17) *Pancheria* — 18 species in New Caledonia.

(18) *Callicoma* — 1 species in Eastern Australia.

(19) *Codia* — 10 species in New Caledonia.

This leaves nothing to imagination. Like the Ericaceae, the Cunoniaceae originated in the *Afroantarctic* and *Gondwanic Triangle* alike, not only, but established from the very start a strong center of evolution (see *Cunonia*) in the immediate vicinity of New Caledonia. This family, then, is both *Gondwanic* and *Afroantarctic* in origins, and it is not surprising that with this as a background it ran all manners of conventional tracks out of one or the other angiospermous gates. For instance; *Spiraeopsis* is rooted in the "Papua Coign" (see p. 111 fn.) which it reaches from the south (cf. migrations of *Ackama, Schizomeria*, etc.); *Caldcluvia* turns up in Chile; *Belangera* is contained within a track which conventionally extends that of *Caldcluvia*; etc.

As to *Weinmannia*; Its dispersal is massive in the New World; beginning right there, where *Caldcluvia* first appears (i.e., Concepción and Maule in Chile) it runs in a wholly conventional pattern to Eastern Brazil, at one hand, Bolivia, Peru, Ecuador, Colombia, Venezuela, the Guianas, Central America and Southern Mexico at the other. The West Indies are but lightly touched, which is a worthwhile suggestion to the effect that the landconnections between these islands and the South American mainland were less favorable to a plant of the ecology of *Weinmannia* (mostly a dweller of highgrounds, in the cloud-forest) than to such cactaceous genera as *Opuntia* and *Mammillaria*. No doubt, a day is to come when phytogeography — freed at least from subservience to academy and meaningless definitions — will be competent to work out the outlines of lands long vanished by reference to plants still living such as *Weinmannia* and *Opuntia*.

Outside of the New World, *Weinmannia* occurs in Madagascar, the Mascarenes and Comoros; New Guinea, Fiji, New Caledonia, New Hebrides,

Rarotonga, Samoa, Tahiti, New Zealand, Stewart Island; Moluccas, Celebes, Borneo, Philippines. Some of these stations (e.g., Tahiti, Borneo) round up the score made by the balance of the cunoniaceous dispersal, but add nothing truly important to its frame. This imposing rooster of localities is less important to our mind than the presence of a species endemic to Cerro Santa Ana in the Venezuelan state of Falcón. This "cerro" is a wholly isolated peak lifting its head to little less than 1000 meters altitude (ca. 3000 feet) out of the xerophilous plains of the Peninsula of Paraguaná. It is certainly not part of the modern Andean System, and could be reached by *Weinmannia* only over paths that have nothing to do with the Andes of today. These paths, let it be noticed, are right by the heart of the major channel of dispersal that cuts straight from Bolivia and Peru to the Greater Antilles. Accordingly, the presence of a species of *Weinmannia* on Cerro Santa Ana has precisely the same significance as the occurrence of a species of *Mammillaria (M. Tamayonis)* endemic to the heights of Lara; of a species of *Malacocarpus* and *Frailea* in Colombia, etc. *All these events are byproducts of there standing right in these regions the abutment of the major channel of dispersal alluded to above.* Were we prone to dismiss the presence of a species of *Weinmannia* on Cerro Santa Ana as insignificant, the mere byproduct of winds, birds, etc. having carried there some seeds from the modern Andes, we might be called upon to explain why also occurs on Cerro Santa Ana the showy *Utricularia alpina.* This Bladderworth belongs to a group (Subg. *Orchyllium*) known from the islands of the Gulf of Guinea quite as much as from the Pacaraima System, the modern Andes, the cordilleras of coastal Venezuela and Bolivia. Patently, we can go nowhere by postulations of dispersal through "casual agencies."

To return to the Brunelliaceae; Considering that this family is hardly better than a branch off the Cunoniaceae, we may have no fear in taking for granted that *Brunellia* entered the New World by a point located between Peru and Bolivia. Were we inclined to dispute this statement, and to assert on the contrary, that *Brunellia* came to Mexico first, we would bring nothing constructive in the discussion. The tracks of *Schoenocaulon* (see p. 375) and *Brunellia* are wholly comparable on the American mailand, and it matters little indeed that these tracks run south to north or north to south, *when the sum total of the migration remains unchanged within the same range.* In brief, the question whether a certain group entered Mexico or Bolivia first is one with the question whether a certain other group reached Eastern India or Java first. Questions of the kind lend themselves to no preconceived argument; they are to be settled on the merits of the case each time. As *Foetidia* shows (see p. 414), Eastern Bengal and Malacca could be simultaneously reached, and it should be difficult indeed to decide whether certain groups came to Mexico or Western South America first.

In conclusion, the Brunelliaceae are "American" quite as are the Cyrillaceae. However, *these two families are "American" in a very different measure, and over different approaches of dispersal. Manifestly, "American" is a meaning-*

less adjective, one of the all too many misleading and meaningless adjectives rife today in phytogeography.

It is transparent that Brunelliaceae and Cunoniaceae reached the New World before the Andes began actively to rise. There is creditable information on record (31) that by the end of the Cretaceous, or very Early Eocene, the only heights existing in what are now the tablelands of Peru and Bolivia, were the worn out stumps of an ancient Eastern Cordillera. From this beginning, and with the onset of "Andean" movements that lasted to the Miocene, a Western Cordillera began to take shape, while the ancient Eastern was rejuvenated and made higher. A time of peneplanation followed during which a lake or elevated arm of the sea were probably filled in. Only in the Miocene began the "Incaic" phase of earth-movement with concomitant block-faulting responsible for the formation of the tablelands we still know. There is evidence that by then part of this tableland stood at about 650 meters above sealevel and harbored a rich vegetation with many large animals, the whole of which was wiped out by the subsequent Pliocene uplift. Perhaps to the end of the Pliocene survived the last mastodons, remnants of one of which seem to have been found in the inter-andean regions of Ecuador together with traces of fire and man-made weapons. The Pliocene probably was the age of greater Andean uplift, which might have totalled over 1500 meters (a mile) for this single period.

It is transparent that geology and dispersal alike agree in denying that the Andes can dominate the South American biological landscape, important as their active uplift might have been in molding out secondary phases of dispersal. The Cochlospermaceae certainly were in Southern Argentina before the most active period of Andean uplift (see p. 261). Fossil Tertiary tracks of *Humiria* are so closely parallel of those of the living species of the genus as to be readily made integral part of the latter, witness the following (fossil species marked by *) (32),

(1) *H. floribunda* — Brazil (Rio Janeiro, Bahia, Minas Geraës, Para, Matto Grosso, Amazonas), Guyanas, Venezuela (Amazonas), Peru (San Martín, Loreto).

(2) *H. balsamifera* — Brazil (Amazonas), Venezuela (Amazonas), Guianas.

(3) *H. bahiensis* * — Brazil (Bahia).

(4) *H. subcrenata* — French Guyana.

(5) *H. Cassiquiari* — Venezuela (Amazonas).

(6) *H. savannarum* — Venezuela (Amazonas; Cerro Duida).

(7) *H. crassifolia* — Colombia (Caquetá: Cerro Aracoara).

(8) *H. cipaconensis* * — Colombia (Cundinamarca).

(9) *H. peruviana* * — Peru (Piura).

Humiria is treated as a monotypic family, Humiriaceae, or as a member of the Linaceae, which latter we have reviewed elsewhere in this manual. The modern and Tertiary range of *Humiria* integrate to perfection. The genus is in the Amazonian basin all the way from Eastern Brazil to the foothills of the

Andes (San Martín and Loreto of Peru) in an absolutely conventional pattern of dispersal. That it reached the New World in the first upsurge of angiospermous distribution brooks no doubt, witness the sum total of the range, not only but the occurrence at such heights as the foothills of Mount Duida and Cerro Aracoara. It is further plain that *Humiria* belongs to a *genorheitron* (linoid in general) of Gondwanic origin, so that *Humiria* reached the Americas from the east, i.e., across the modern Atlantic, not as did *Brunellia,* from the direction of the modern Pacific.

In brief, if there is anything interesting in fossil *Humiria,* this is not in the fact that the petrifacts "prove" anything which the living dispersal leaves unsettled. The contrary is true, for the fossil range, meaningless by itself, is elucidated by the living. The one point of interest is that the petrifacts tend to indicate that *Humiria* died out at the periphery of the modern optimum of dispersal (Amazonian Basin), and in two regions at least (Cundinamarca, Piura) deeply disturbed by the Andean uplift, either directly or indirectly. In conclusion, we may without fear assert that *Humiria, Brunellia, Cochlospermum, Weinmannia* etc., etc., were in the New World long before the Tertiary, for this assertion is amply supported, (a) By all the evidence from living dispersal, which shows that the Angiosperms were in active migrations by the end of the Jurassic and the very Early Cretaceous; (b) By fossil findings such as the Patuxent Beds; the discovery of *Cochlospermum, Humiria* etc. *Naturally, by far the stronger is the evidence from dispersal of living plants, because this evidence is beyond comparison richer and meatier than that from petrifacts.*

Brunellia and *Weinmannia* showed us a pattern of pseudo-Andean dispersal originating from the Eastern Pacific. The Burseraceae Protieae will now acquaint us with a pattern of conventionally "Amazonian" flavor, but somewhat different in origin from homologous patterns drawn, for instance, by *Humiria,* fossil and living, the Cochlospermaceae (see p. 257), Oxalidaceae (see p. 262), etc.

We fortunately have of the Protieae a reliable monograph (33) which is to ease our task considerably. This affinity consists of six genera, of which only two, *Protium* in part and *Garuga,* whole, belong to the Far Eastern flora. *Protium* occurs in addition in the *Gondwanic Triangle* as we are soon to see.

Considering that the dispersal of the Protieae has called for some interesting but confuse interpretations, let us begin our study in the simplest possible manner. Let us draw on the map a line which connects Madagascar with Java, and take position, figuratively speaking, right in center of this line.

In this position, we will have *Protium* on two sides as follows,

A) Sect. *Euprotium*

(1) *P. javanicum* — Java, Soenda (Bawean, Madoera, Kangean Archipelago, Soembawa, Bali).

(2) *P. serratum* — Indochina (Island of Phu-quoc; Cochinchina), Siam, Upper Burma, Assam, Eastern Bengal.

(3) *P. connarifolium* — Philippines (Palawan).

B) Sect. *Marignia*

(4) *P. obtusifolium* — Mascarenes (Mauritius).

(5) *P. Chapelieri* — Madagascar (East coast, Ile Sainte-Marie).

(6) *P. madagascariense* — Madagascar.

(7) *P. Beandou* — Madagascar (Ile Saint-Marie).

Let us take careful notice how cleancut and standard is this dispersal. It squarely rests in the Far East and Malaysia upon the "Sunda Coign," and runs the classic channel Phu-quoc—Cochinchina—Siam—Burma—Assam—Bengal with one species. A second species is in Palawan, which has oftentimes flora unlike that of the rest of the Philippines; this indicates that this flora may stem from the "Sunda Coign" and indeed belong, as it does in this case, to the first age of angiospermous dispersal. A third species stresses the Sunda insular arc, Java included. Neither Sumatra nor Malacca are interested in this distribution, which is quite illuminating in view of the tracks run by · *Triumfetta* (see p. 419) for example. The indications are here to the effect that the archetypes of *Protium* reached the vicinity of Java first from the Gondwanic cradle, and were next sucked into Malaysia and the Far East, as it were, over a line *Java to Bali—Phuquoc*, very much in the manner of *Triumfetta repens*. The triangle Malacca—Sumatra—Java, is, as usual, a cornerstone of Malaysian distribution. *Lumnitzera, Triumfetta, Protium* all stress its significance though by different patterns of dispersal.

Protium sect. *Euprotium* squarely rests upon the "Sunda Coign," which — unlike the Dipterocarpaceae, *Lumnitzera, Barringtonia*, etc. — *Protium* did not leave to migrate further east in a second epoch of angiospermous dispersal. *Protium* sect. *Marignia* on its part squarely stands by the Mascarenes and Eastern Madagascar. Significant are the station at Ile Sainte-Marie (see note, p. 348).

Clearly, *Protium* originated, as we have already stated, by the *Gondwanic Triangle*. As a group of this origin, *Protium* might be expected to have reached Continental Africa over a large front. Nothing like that happened, however. *Protium* gave Africa a wide berth to go straight to the New World by the southern path connecting the *Gondwanic Triangle* with the Magellanian gate of angiospermy, as we are next immediately to see. *Protium*, then, ran migrations very much in the style of the Cactaceae. Observe how well Continental Africa is held this time — *without being for this touched at all* — in the triangle Mascarenes—Java—Bolivia. The Cactaceae also hold Continental Africa within the triangle Mascarenes—Ceylon—Bolivia, but *Rhipsalis* touches the

Dark Continent all along its coasts. *Continental Africa could not be held within triangles of the kind and be touched, or not, if Africa were not right by the north (in the main) of the major angiospermous centers.*

It is the same order of factors that accounts for the fact, for instance, that such genus as *Croton* is massive in Madagascar, but sparingly represented on the whole in Continental Africa, and again massive in the Americas; not only, but for the circumstance that forms of *Croton* related in Madagascar and the New World are not to be seen in Africa at all.

Protium turns up in the New World with two sections, *Icicopsis* and *Euicica*. Species from the former typical of the generic dispersal in the Americas read in this pattern of distribution,

(1) *P. insigne* — Bolivia (Rio Beni, in "Amazonian" Bolivia); Brazil (Acre, Amazonas); Colombia (Northwestern Meta: Villavicencio); Dutch Guiana; Trinidad.

(2) *P. tenuifolium* — Bolivia (Rio Beni, in "Amazonian" Bolivia); Peru (Loreto, San Martín), Colombia (Magdalena, Santa Marta).

(3) *P. subserratum* — Peru (Loreto); Brazil (Amazonas, Acre); Colombia (Vaupés); British Guiana.

(4) *P. neglectum* (and varieties) — Brazil (Acre, Amazonas, Rio Janeiro (Corcovado)); Colombia (Magdalena: Santa Marta — here in mountain forest to 1100 met. alt.); Venezuela (Zulia: Perijá, Guarico, Aragua (Coastal cordillera), Anzoátegui); Trinidad; Guianas; Panamá; Costarica.

(5) *P. Warmingianum* — Brazil (Bahia, Minas Geräes).

(6) *P. fragrans* — Cuba (Oriente: Sierra de Nipe, Moa, Baracoa).

It is obvious by the face of this dispersal (Fig. 75) that *Protium* runs in the New World precisely the channels of *Brunellia*. However, *Brunellia* is kept by its ecological requirements close to the Andes, while *Protium* freely roams in the direction of the Amazonian plains and beyond. Accordingly, *Brunellia* is "Andean", *Protium* "Amazonian" by reason of a different ecology. From the phytogeographical standpoint, that is to say, based on the pure geographic approach to dispersal, these two genera are identically the same on *American soil*. Both enter the New World from a point situated between Bolivia and Peru, and both run a northward track, in the main, which ultimately leads them to the West Indies through Colombia and Venezuela.

These two genera, again, are not the same *outside American soil* from the standpoint of phytogeography, because *Brunellia* comes to the New World from a center situated in Western Polynesia and the Southern Pacific generally, whilst *Protium* invades South America from the direction of Africa. The latter genus originated seemingly in both the *Afroantarctic* and *Gondwanic Triangles*, for it occurs in the Mascarenes and Madagascar, but runs at the same time an occult "antarctic" track between the Mascarenes and Bolivia; *Brunellia*, on its part, certainly did not originated there, where *Protium* first saw the light of the day, rather in the Southern Pacific.

Protium could not enter the New World from West Africa because *Protium* is unknown on the Dark Continent *). There may be question whether it first reached Bolivia from the Eastern Pacific, or Eastern Brazil immediately from the modern South Atlantic, which question cannot be answered in detail just now. To answer it, we should enter minute considerations of intersectional affinity and local distribution which are out of our means. Whatever be the case, it is plain that *Protium* ran the channel Bolivia—Eastern Peru—Colombia, Venezuela—Cuba as did *Brunellia*, dispatching however outliers eastward which *Brunellia* did not do. Noteworthy is the fact that *P. neglectum* adapted itself to living at a comparatively high altitudinal level in Eastern Colombia, and that right in Central Colombia *Humiria* died out, as we have seen. Disturbances took place in this region which *Protium* could manage, but *Humiria* found too hard to meet.

Section *Euicica* is by far the largest of *Protium* having 60 of the 78 species now reckoned in the whole genus. Here is a list of some of the species of this section noteworthy from the standpoint of dispersal.

(7) *P. attenuatum* — Lesser Antilles (Guadeloupe, Dominica, Martinica, Santa Lucia, Saint-Vincent).

(8) *P. cubense* — Cuba (Pinar del Rio, Oriente, Isle of Pines).

(9) *P. heptaphyllum* — *Type-form:* Brazil (Amazonas, Pará, Goyaz, Pernambuco, Bahia, Minas Geraës); Colombia (Bolívar); Venezuela (Zulia: Sierra de Perijá; Portuguesa; Miranda; Carabobo; Anzoátegui, Bolívar, Nueva Esparta: Island of Margarita); Guianas; Trinidad; Saint-Vincent. — Var. *angustifolium:* Paraguay; Brazil (Pará, Bahia, São Paulo); Venezuela (Zulia); Guianas, Trinidad, Saint-Vincent. — Var. *brasiliense:* Bolivia: Santa Cruz; Paraguay (Cordillera de Peribebuy, Upper Río Yacá); Brazil (Amazonas, Pará, Matto Grosso, Pernambuco, Bahia, Minas Geräes); Venezuela (Trujillo, Sucre, Bolívar); Guianas. — Var. *floribundum:* Paraguay (Sierra de Amambay); Brazil (Ceará, Matto Grosso). — Var. *surinamense:* Brazil (Pará, Bahia); Venezuela (Bolívar); Guianas; Trinidad.

(10) *P. laxiflorum* — Brazil (Amazonas).

(11) *P. tovariense* — Venezuela (Carabobo, Caracas (D.F.), Aragua).

(12) *P. panamense* — Panamá.

(13) *P. glaucum* — Peru (Loreto).

(14) *P. Almecega* — Brazil (Amazonas, Matto Grosso, Minas Geraës, São Paulo).

(15) *P. Copal* — Panamá; British Honduras; Mexico (Tabasco, Yucatán, Veracruz, Puebla, San Luis Potosí).

---

*) This statement is based upon the face of the living dispersal, and agrees accordingly with the facts *of today*. We are far from certain, however, that *Protium* reached South America exclusively by an antarctic channel (continuous line *r* in Fig. 75), for it could use a track to the north (dotted broken line *s*). This track would manifestly cross the *Kalaharian Center* (black triangle in Fig. 75). We are inclined to think *Protium* would not fare the antarctic route because it is a genus of outspoken "warm" nature (notice stations in Madagascar). If, then, *Protium* reached Bolivia, and South America generally, by track *s*, *Protium* lost out somewhere in Africa, where it could be seen in former ages. This is not impossible, because the *Kalaharian Center* has been much modified by desiccation, even of recent date.

(16) *P. octandrum* — Colombia (Magdalena), Venezuela (Trujillo), Dutch Guiana.

(17) *P. Llewelynii* — Bolivia (La Paz: Mapiri), Peru (Loreto), Brazil (Acre, Amazonas), Venezuela (Amazonas).

(18) *P. Pittieri* — Costarica; Mexico (Oaxaca).

It will be observed, (a) *Protium heptaphyllum* var. *surinamense* yields a perfect track Eastern Brazil—Amazonian Venezuela and Guianas—Trinidad, which tallies with similar channels in *Melocactus* and, most likely, the *Cereus* affinity (see p. 369); (b) The track of *P. attenuatum* normally extends northward that of *P. heptaphyllum* var. *surinamense*, which once more suggests the run of certain species in the *Cereus* affinity (see *Pilocereus barbadensis*, etc.; p. 369) and *Melocactus;* (c) *Protium attenuatum* is closely related with *P. cubense*. Characteristically, the latter is not distributed eastward from Cuba, not only, but takes in the whole of Cuba including the Isle of Pines. As we know (see *Opuntia, Mammillaria*, p. 368) a track running from Colombia—Venezuela to Santo Domingo may fray off, giving a branch that runs eventually to Texas touching Cuba, Eastern Cuba in particular. As it is also generally known, the Isle of Pine has certain peculiarities of flora that set is apart from the rest of Cuba, Eastern Cuba particularly. It is then within reason to infer the existence of a front *Guadeloupe—Isle of Pines—Eastern Cuba—Eastern Texas* parallel with the front (next to be seen from *Tetragastris*) *Guianas—British Honduras*, not only, but with the fronts *Central Venezuela—Panamá* (see next, *P. panamense*) and *Curaçao—Los Testigos* this facing as a last front the *Venezuelan coasts* themselves. In brief, the modern Caribbean Sea was formed by a series of faultings roughly parallel with the latitudes (see Fig. 71), to judge from dispersal. Also to judge from dispersal, these faultings were concomitant with, or perhaps posterior, to the opening of two major faultings in a different direction this time (Fig. 71, dotted broken lines), one running from the vicinity of Antigua to the Venezuelan island of Tortuga; the other from the Windward Channel to the Gulf of Honduras; (d) *Protium panamense* and *P. tovariense* are close, which establishes a noteworthy connection between Central America and the coasts of Central Venezuela (Carabobo, Aragua, Caracas). We have seen previous examples (p. 321 fn.) of connections — rather, disconnections, for these connections can be *now* bridged only oversea — of the kind, and justifiably inferred that they stand for the former presence of lands standing between Central America and Venezuela; (e) Panamá, Costarica, Mexico are brought together by *P. Pittieri* and *P. Copal* which are near-allies; (f) *Protium glaucum* and *P. Almecega* also are near-allies, which binds the Peruvian Amazonas (Loreto) together with Eastern Brazil by a standard tie throughout; (g) *Protium octandrum* runs a classic channel Magdalena—Trujillo—Dutch Guiana. This channel is nearly repeated in the distribution of the type-form of *P. heptaphyllum* — Eastern Brazil to Eastern Venezuela (Bolívar—Anzoátegui—Nueva Esparta); next at one hand, always from Eastern Brazil, to the Guianas, Trinidad and St. Vincent; at the other hand to

Western Venezuela (Portuguesa, Miranda, Carabobo, Zulia) and Colombia (Magdalena). The track thus run by *P. heptaphyllum* is homologous of tracks in *Erythroxylon* (see p. 195), and is fundamental for much of the dispersal in these quarters. It accounts, for instance, for the presence of typically "Amazonian" elements in the island of Margarita (Nueva Esparta) such as *Croton Milleri*, a very close allied of *C. nervosus* of the Rio Negro; it shows that the triangle Margarita—Trinidad—St. Vincent is a door of access from the Guianas and Amazonas toward the Lesser West Indies; etc.

A genus is the immediate kinship of *Protium*, *Tetragastris*, is distributed in this manner,

(1) *T. mucronata* — Colombia (Magdalena: Santa Marta).

(2) *T. Hostmannii* — Guianas.

(3) *T. balsamifera* — Santo Domingo, Haiti, Tortuga; Portorico; Sainte-Croix — Var. *lanceifolia:* Santo Domingo, Portorico, French West Indies?

(4) *T. breviacuminata* — Brazil (Rio de Janeiro).

(5) *T. altissima* — Brazil (Acre, Amazonas), Guianas.

(6) *T. panamensis* — *Western massing:* Panamá, Nicaragua, Honduras, British Honduras; *Eastern Massing:* Guianas.

We observe, (a) Like *Melocactus*, *Tetragastris* originates by a line Eastern Peru—Eastern Brazil, and runs a main track northward from this line. This line crosses Eastern Colombia to fare straight to the West Indies; (b) *Tetragastris balsamifera* breaks into two forms, one of which stops eastward at Ste. Croix; the other, on the contrary goes beyond as far as the French West Indies. This confirms what we have learned from the Cactaceae (see p. 372). It seems altogether probable that this behavior is due to the fact that the dispersal of this species was effected prior to the opening of a line of faulting running southward from Ste. Croix at one hand, the Guadeloupe at the other; (c) *Tetragastris panamensis* exhibits a range that caved in between British Honduras/Panamá and the Guianas. This caving-in is not at all odd in view of similar occurrences reviewed throughout the pages of this manual.

The ecological requirements of *Protium* would seem to be exactly the same in Malaysia as in the New World, and to agree generally with those of the rest of the Burseraceae. Though relatively tolerant, this genus was at home primarily in habitats other than the tropical rain-forest. It stood forever ready to invade lowlands recently open to colonization on account of geological events such as re-emersions, etc. Later on it evolved, adapting itself to such conditions as now prevail in the lowlands of the Amazonian basin, for instance. We venture to suggest that *Protium* is correctly viewed as a dweller throughout of the "Monsoon" belt of tropical climate, and that in its *most distant* origins it was a plant of the immediate marine hinterland in a region of comparatively temperate climate.

XIV. Podostemonaceae

The Podostemonaceae easily rank as one of the most curious angiospermous families. They are very variable in their floral and somatic details, but so rudimentary is their gross morphology as a rule that some of them can be mistaken for Algae whenever not in flower.

This family has been the favored subject of investigation of some of the keenest minds of botany, because it is almost impossible to view it as a normal part of the Angiospermae, and nothing in orthodox thinking avails to account for their origins, morphology and evolution. On the whole, these investigations have yielded little tangible, and we are still wondering today how and why plants of the kind are unquestionably angiospermous in sexual characters.

How plants of this nature secure their dispersal is a problem which has never been solved, and one of their students concludes (34) with the laconic comment that the distribution of the Podostemonaceae is a "Complete mystery." Indeed, it is hard to account for the fact that rivers separated by an interval of few miles may harbor entirely different genera, and that cataracts in the same river may be tenanted by well marked distinct species. The habitat preferred by certain species may be so narrowly circumscribed that once taken out of it, these species promptly die. Contrariwise, some of the Podostemonaceae are widely distributed, as we shall soon see.

Most Podostemonaceae are dwellers of swift running waters throughout the tropics, but some of them reach the temperate and even cold north. As they are easily overlooked by collectors, they are probably more numerous than the present record suggests.

This family is beyond doubt one of the most primitive in existence, and its affinities are poorly understood *).

In our opinion, the dispersal of the Podostemonaceae is not so much "mysterious" as obscured by contradictory tendencies. At one hand, there occurs a massive genus-making in certain centers, the genera involved being oligotypic or monotypic; at the other certain forms remain fairly constant and run accordingly very long tracks. These tendencies will readily be understood in the light of the record of the podostemonaceous dispersal (35) as follows,

---

*) The pages of this manual are not open as a rule to considerations of morphology. It may not be useless to notice, however, that the called *haptera* (or holdfasts) of the Podostemonaceae may repay scrutiny by comparison with homologous structures in the seedling of certain parasites (see, for instance, *Orobanche* by CASPARY in Flora 37: 593. 1854; *Phrygilanthus* by REICHE in Flora 93: 271. 1904). Suggestive also are the similarities between these bodies and the "roots" of certain Lentibulariaceae (*Utricularia Hertzogii;* see LUETZELBURG in Flora 100: 200 Fig. 37. 1910). Lastly, challenging resemblances exist between Podostemonaceae and certain minute Orchids (see PFITZER in Ber. Deutsche Bot. Gesell. 2: 472. 1884), and one of the former, *Dalziella (Lawia) ramosissima* (see BAILLON, Hist. Pltes. 9: 256. 1888) is strangely close to the juncaceous *Distichia (Goudotia) tolimensis* (see DECAISNE in Ann. Sc. Nat., Sér. 3 Bot., 4: 83. 1845). The peculiar embryology of the Podostemonaceae is of course (see SCHNARF, Vergleich. Embryol. Angiosp., 109. 1931) thoroughly well known. This family is a classic instance of inhibited development carried into the adult stage, and has accordingly great phylogenetic and morphologic significance. The literature here referred to is by no means complete, merely intended to convey a hint that the Podostemonaceae require critical study in absolute freedom f rom preconceived notions as to morphology and phylogeny.

A) Intercontinental genera

(1) *Podostemon* — 14 species in Madagascar, Ceylon, India; Southern Brazil, Uruguay; 1 species in the United States and Canada (Georgia and Alabama to Kentucky, Ontario and Massachusetts).

(2) *Tristicha* — 1 species in Madagascar and the Congo; 1 species in Madagascar, Mascarenes, tropical Africa throughout, Uruguay, Eastern Brazil, Cuba, Central America and Mexico.

(3) *Dicrea* — 2 species in Madagascar, 4 or more species in Southern India.

B) African genera

(4) *Sphaerothylax* — 1 species in Natal.

(5) *Angolea* — 1 species in Angola.

(6) *Inversodicraea* — 13 species in Tropical Africa.

(7) *Leiothylax* — 1 species in East Africa, 4 species in West Africa.

(8) *Dichraeanthus* — 1 species in the Cameroon.

(9) *Saxicolella* — 1 species in the Cameroon.

(10) *Monandriella* — 1 species in the Cameroon.

(11) *Pohliella* — 1 species in the Cameroon.

(12) *Ledermanniella* — 1 species in the Cameroon.

(13) *Winklerella* — 1 species in the Cameroon.

(14) *Macropodiella* — 1 species in the Cameroon.

(15) *Anastrophea* — 1 species in the Congo and Ethiopia.

C) American genera

(16) *Apinangia* — 25 or more species in Brazil (Amazonas, Goyaz), Peru, Venezuela, Costarica.

(17) *Marathrum* — 9 species in Brazil (Amazonas), Colombia, Central America, Mexico.

(18) *Oserya* — 4 species in Brazil and Mexico.

(19) *Weddelina* — 2 species in the Amazonas and Guiana.

(20) *Mourera* — 4 species in. the Amazonas and Guiana.

(21) *Rhyncolacis* — 9 species in the Amazonas and Guiana.

(22) *Oenone* — 6 or more species in the Amazonas.

(23) *Lacis* — 1 species in the Amazonas.

(24) *Ceratolacis* — 1 species in the Amazonas.

(25) *Mniopsis* — 5 species in Brazil.

(26) *Castelnavia* — 7 species in Brazil.

(27) *Lonchostephanus* — 1 species in Brazil (Pará).

(28) *Lophogyne* — 3 species in Brazil (Rio Janeiro).

(29) *Jenmanniella* — 4 species in the Guianas.

D) Asiatic genera

(30) *Zeylanidium* — 3 species in Ceylon.

(31) *Lawia* — 1 species in Ceylon and Southern India.

(32) *Farmeria* — 2 species in Ceylon and Southern India.

(33) *Dalziella* — 1 species in Southern India.

(34) *Hydrobryopsis* — 1 species in Southern India.

(35) *Terniola* — 3 species in India and Indochina.

(36) *Griffithella* — 1 species in India and Indochina.

(37) *Hydrobryum* — 1 species in India and Burma.

(38) *Polypleurella* — 1 species in Siam (Koh-Chang Island).

(39) *Cladopus* — 1 species in Java, Southern Celebes; 1 species in Japan .

E) Australian genera

(40) *Torrentincola* — 1 species in Queensland.

The three intercontinental genera cannot be said to be difficult interpretation to the exception, perhaps, of *Podostemon*. *Tristicha* is dispersed in a typical African pattern which preludes to the further distribution in the West Indies, Mexico and Eastern South America. *Dicrea* is definitely Gondwanic. *Podostemon* is Gondwanic in part, but gives every indication of having followed the track of the Protieae (that is, from Madagascar to South America bypassing Africa in the south) for the rest of its course. It is plain that this sector of the Podostemonaceae was in active migration in the very first upsurge of angiospermous distribution, and it is possible that the range of *Podostemon* will in time be more closely knit together by new reports. The North American species is in no sense more "mysterious" than *Dionaea muscipula* or certain species of *Drosera*, witness *D. filiformis*, which are unique in the flora of the northern New World *).

The African genera would make good sense if only they were a single one, for then this single entity would be distributed as follows, Madagascar, East Africa, Ethiopia, Natal, Angola, Congo, Cameroon, which would cause no surprise. The same is true of the American genera, which might have invaded the New World in part from Africa, in part from the devious route leading to Bolivia and Mexico from the East Pacific. In sum, the backbone of this dispersal is standard, but genus-making is run with a prodigal abandon, the genus

---

*) The keen reader will notice, of course, that the migrations of *Podostemon* are in certain essentials reminiscent of those of *Plantago major* and its forms (see p. 288), and of the Liliaceae Uvularieae (see p. 377).

taking among the podostemonaceous hosts the place which in other groups is held by the species or the section.

The Asiatic genera are properly and well located in Ceylon, India, Burma, Indochina, Siam, which is to say they are wedded to the major western channel running in the Far East and Western Malaysia. Only two of their number deserve a note.

Polypleurella is so located as to suggest that it might have Australian ties *). *Cladopus* is consistent in running a short track Southern Celebes—Java. This track, short as it may be, is suggestive of two alternatives, (a) The *genorheitron* is from the Indian Ocean and the dispersal, beyond doubt incomplete at present, will seemingly be implemented in due course of time by additional records in the Philippines, Formosa and Southern China; (b) The *genorheitron* is from the Pacific, and only part of its track westward remains. Japan was reached immediately from the Celebes through the Philippines, or which is just as probable (see next, p. 438) by a disconnected track (see also p. 246) beginning in New Guinea, Queensland or New Caledonia and streaming northward over the modern Western Pacific east of the Philippines.

In conclusion, the Podostemonaceae run channels of dispersal (Fig. 76) that are anything but "mysterious." The record is manifestly incomplete on account of the difficulty of collecting these plants; huge disconnections are to be expected in a group such as this, which is exceedingly strict as to ecological requirements and environment. However, the generic massings are keyed up to certain *standard* centers of angiospermy (Western Africa, Ceylon and Southern India, Cochinchina, Madagascar and vicinity, the Guianas), so that the history of the podostemonaceous migrations is but part of a vaster and in no sense unfathomable pale of distribution.

---

*) We regret it is not possible to effect in these pages a comparison between the dispersal of the Podostemonaceae and that of *Utricularia*. A glance at Lecomte's "Flore Générale de l'Indochine", 4: 467 *et seq.* 1930, will readily reveal that certain species of the latter genus have ranges as follows, whole or in part, (1) *U. bosminifera* — Ko-chang; (2) *U. siamensis* — Ko-chang; (3) *U. hirta* — Phu-quoc; (4) *U. brevilabris* — Phu-quoc. The huge majority of the Indochinese forms of *Utricularia* turn up in Cochinchina, many of them following next trough Laos and Siam to Burma and India. In brief, the lentibulariaceous major channel of dispersal in this sector matches that of *Stylidium*, Restionaceae, Centrolepidaceae and Podostemonaceae, which is quite according to norm. Strange to say, many are the botanists who credit to miraculous intervention the fact that "delicate herbs" like *Utricularia* happen to turn up in islands like Phu-quoc and Ko-chang, when islands of the sort lie straight within major channels of dispersal that precisely fit the ecologic requirements of the Bladderworts.

BIBLIOGRAPHY

Chapter VI

(1) PERRIER DE LA BATHIE, H. — Les Brexiées de Madagascar; Bull. Soc. Bot. France 80: 205. 1933.

(2) LOESENER, TH. — *Celastraceae;* E. & P. Nat. Pflanzenf. 20b: 87. 1942.

(3) PERRIER DE LA BATHIE, H. — Révision des Celastracées de Madagascar et des Comores; (Lecomte) Not. Syst. 10: 173. 1942.

(4) BROWN, F. H. B. — see (31) Chapt. III.

(5) LOESENER, TH. — *Aquifoliaceae;* E. & P. Nat. Pflanzenf. 20b: 36. 1942.

(6) HARMS, H. — *Passifloraceae;* E. & P. Nat. Pflanzenf. 21: 470. 1925.

(7) HARMS, H. — *Achariaceae;* E. & P. Nat. Pflanzenf. 21: 507. 1925.

(8) HARMS, H. — *Caricaceae;* En & P. Nat. Pflanzenf. 21: 507. 1925.

(9) GILG, E. — *Flacourtiaceae;* E. & P. Nat. Pflanzenf. 21: 377. 1925.

(10) PERRIER DE LA BATHIE, H. — Les Passifloracées de Madagascar; (Lecomte) Not. Syst. 9: 57. 1940.

(11) BAKHUIZEN VAN DER BRINK, R. C. — Revisio Bombacacearum; Bull. Jard. Bot. Buitenzorg, III, 6: 161. 1924.

(12) BAKER, J. G. — Catostemma fragrans; Hook. Ic. Pl. 18: Pl. 1793. 1888; see further, OLIVER, D. — Catostemma fragrans; ibid. 20: Pl. 1896. 1891.

(13) MERRILL, E. D. — An Enumeration of Philippine Flowering Plants 2: 234. 1923.

(14) VAN STEENIS, C. G. G. J. — The Styracaceae of Netherlands India; Bull. Jard. Bot. Buitenzorg, III, 12: 212. 1932.

(15) KOEHNE, E. — Zur Kenntnis von Prunus Grex Calycopadus und Grex Gymnopadus Sect. Laurocerasus; Engl. Bot. Jahrb. 52: 279. 1915.

(16) VON WETTSTEIN, R. — Über die Auffindung von Rhododendron ponticum L. in der Balkanhalbinsel; Oest. Bot. Zeitschr. 67: 301. 1918.

(17) SLEUMER, H. — *Salvadoraceae;* E. & P. Nat. Pflanzenf. 20b: 232. 1942.

(18) MATTFELD, J. — *Stackhousiaceae;* E. & P. Nat. Pflanzenf. 20b: 240. 1942.

(19) UPHOF, J. C. TH. — *Cyrillaceae;* E. & P. Nat. Pflanzenf. 20b: 1. 1942.

(20) NEWBERRY, P. E. — On some African species of the genus Olea and the original home of the cultivated Olive-tree; Proceed. Linn. Soc. (London) 150: 3. 1937.

(21) CAMPBELL, C. — Sull'infiorescenza terminale nell'Olea europaea L.; Nuovo Giorn. Bot. Ital., N. S., 14: 670. 1907.

(22) CROIZAT, L. — The concept of inflorescence; Bull. Torrey Cl. 70; 496. 1943.

(23) STEYERMARK, J. — A revision of the genus Menodora; Ann. Missouri Bot. Gard. 19: 87. 1932.

(24) GILG, C. — Restionaceae; Centrolepidaceae; E. & P. Nat. Pflanzenf. 15a: 8, 27. 1930.

(25) SKOTTSBERG, C. — see (12), Chapt. I.

(26) BRITTON, N. L. & ROSE, J. N. — The Cactaceae 1—4. 1919—1923.

(27) BRINKER, R. E. — Monograph of Schoenocaulon; Ann. Missouri Bot. Gard. 29: 287. 1942.

(28) KRAUSE, K. — Liliaceae (Melanthioideae); E. & P. Nat. Pflanzenf. 15a: 254. 1930.

(29) ENGLER, A. — Brunelliaceae; E. & P. Nat. Pflanzenf. 18a: 226. 1930.

(30) ENGLER, A. — Cunoniaceae; E. & P. Nat. Pflanzenf. 18a: 229. 1930.

(31) MOON, H. P. — The Geology and Physiography of the Altipiano of Peru and Bolivia; Trans. Linn. Soc. London. 3rd ser., 1 (III). 1939.

(32) SELLING, O. H. — Fossil remains of the Genus Humiria; Svensk. Bot. Tidskr. 39: 257. 1945.

(33) SWART, J. J. — Novitates Bursacearum and A Monograph of the Genus Protium and some Allied Genera (Burseraceae); Rec. Trav. Bot. Néerland. 39: 189. 1942.

(34) VAN STEENIS, C. G. G. J. — The Podostemonaceae of the Netherlands Indies; Bull. Jard. Bot. Buitenzorg, III, 13: 530. 1936.

(35) ENGLER, A. — Podostemonaceae; E. & P. Nat. Pflanzenf. 18a: 3. 1930.

CHAPTER VII

EPOCHS IN ANGIOSPERMOUS DISPERSAL

"Primary" and "secondary" dispersal have frequently been spoken of through the preceding pages. We will endeavor in this chapter to illustrate concrete cases of one and the other.

It is generally held that mangroves are typically distributed by casual agencies of dissemination, marine currents, winds and the like. The thought that forms of the kind obey the very same requirements of plants inland as regards dispersal may seem to most altogether unlikely, when not absurd.

Let us, then, take as a preliminary critical notice of the distribution of some of these "whimsical" daughters of the sea. We are fortunate that their behavior has been studied by an author *) thoroughly well informed of their taxonomy and current distribution.

This author submits a map illustrative of the range of three species of *Lumnitzera*, a genus of the Combretaceae (Fig. 77) and comments as follows, (i) The distributions of *L. racemosa* and *L. littorea* largely overlap; (ii) *Lumnitzera racemosa* does not occur to the north of the coast of Tanganyka; (iii) This same species is restricted to the western coast of Madagascar; (iv) *Lumnitzera littorea* occurs in the Andamans to the exception of the northern islands of this archipelago, wherein *L. racemosa* thrives; (v) In straight contrast with *L. racemosa*, *L. littorea* does not reach the coasts of Southern China, Hainan and Formosa; (vi) *Lumnitzera littorea* is found on the southern coasts of Java, not on the northern; (vii) Contrariwise, *L. racemosa* is found on the northern coast of Java, not on the southern; (viii) *Lumnitzera racemosa* overruns all the coasts of Northern Australia and Queensland; *L. littorea* is restricted to Queensland; (ix) *Lumnitzera lutea* is narrowly restricted to Timor and the adjacent Alor Islands.

Faced by these facts, the author whom we quote tries to rationalize them in relation to marine currents, winds, climate, etc., without success. He implies, perhaps admits, that the problem is baffling and solving it by conventional approaches impossible.

---

*) We take for granted that the reader fully awake by now to the meaning of certain sensitive spots of the map will notice that *Lumnitzera* leaves Africa by Tanganyka, while it is precisely by Tanganyka that *arrive* to Africa *Erica arborea*, *Primula*, *Berberis*, *Populus*, etc., etc. Tanganyka, then, (not unlike Somaliland) is a door *out* of which stream certain forms, while other forms come *in*. The point of departure and arrival is always the same, *Gondwana*, that is, a now crumbled continent in the Indian Ocean.

We will tackle this very same problem in a simple manner, attacking it, that is, in the light of the same trusted generalities of method that stood us in good stead throughout the preceding pages devoted to a study of the distribution of land-plants.

By a critical analysis of factual distribution extending to many unrelated groups we have learned that the modern Indian Ocean was once the seat of a mighty landmass. This landmass crumbled throughout the Cretaceous, possibly part of the Tertiary, in eras, that is, when the Angiosperms had already widely migrated.

It stands to reason that forms, actually or potentially of the mangrove type, endemic to the landmass once standing in the Indian Ocean, would react to the shattering of this landmass in the measure of their own potential of migration and the opportunities offered to them to gain new grounds ecologically suitable. We already know that opportunities of the kind often depended from the geographic position occupied by a certain form at a certain time. To illustrate: Being rooted in the *Gondwanic Triangle* (see p. 349), the ericaceous genus *Agauria* migrated in a manner altogether at variance from the dispersal peculiar of groups like the Restionaceae, which never tenanted this triangle, but dwelt on the contrary by an antarctic shore possibly answering the outlines of the *Afroantarctic Triangle*.

A glance at the map showing the distribution of *Lumnitzera* (Fig. 77) establishes as a fact that this genus dwelt in Gondwana by the start.

It is known that Madagascar broke off the modern coast of Africa quite early. A marine transgression thus took place between this island and continental African which a mangrove narrowly localized in the vicinity of this line of faulting in origin would use for its own ends. This mangrove would doubtless tend to colonize the shores opened by this transgression, not only, *but further migrate as these shores became more extensive in any given direction.* These migrations would not be haphazard, but strictly conditioned by the terms of palaeogeography.

Lumnitzera racemosa — indeed a strange coincidence, if a coincidence at all — is narrowly wedged in the Mozambique Channel. It ranges southward to a point not far off the coast of Natal; northward to a point closely agreeing with Zanzibar, right by the northern limit of Tanganyka. These points are quite significant in dispersal, as we well know. The former stands as the door of entrance into Africa of uncounted swarms of angiospermous forms; the latter as the locus of appearance of a multitude of plants which seem to reach Tanganyka as if from nowhere, but actually from the modern Indian Ocean.

In turning up here, *L. racemosa* does not behave erratically by any means. Another of the mangroves, *Avicennia marina* var. *typica* (2) is narrowly localized among the Seychelles, the Mascarenes and the African shores adjacent to these insular groups.

Faring out of the Mozambique Channel, *L. racemosa* keeps a westerly course by the Seychelles, turns up along the Indian coasts only as far north as Goa, and further migrating in an easterly direction reaches its easternmost limits of dispersal roughly by a line Formosa—New Caledonia. While managing this stupenduous spread, this mangrove is somehow debarred from gaining a foothold on the southern coasts of Java.

Lumnitzera littorea comes to the fore by another mighty line of faulting running approximately from the Andamans to the eastern tip of Java. Quite as *L. racemosa* is born by a line of faulting investing the modern Mozambique Channel, so is *L. littorea* called to the light of the day by a similar line. Nothing is here haphazard; everything is here precise. Whether *L. racemosa* was segregated out of *L. littorea* or stands out as a "strong" species in its own right we do not know, but either alternative can be met without difficulty. A new shore came into existence while Gondwana was breaking up, running from the Andamans to Java. Right by this shore, *L. littorea* could gradually evolve as a daughter of *L. racemosa*, or conversely take to the sea on its own, as a preexisting mangrove or potential mangrove already tenanting the vicinity. Beginning active migrations in different times, and from different regions of the map, *L. racemosa* was debarred from the southern coast of Java which *L. littorea* could already occupy. We do not know the precise reason why this was so in every detail, but only palaeogeographic barriers could be responsible for events of the kind, in the very first place, as we soon are to learn.

Lumnitzera lutea is beyond doubt narrowly ranging, even if we like to suppose that part of its range is as yet unknown. This mangrove holds the immediate vicinity of Timor.

This mangrove holds this vicinity because of thoroughly understandable reason. Timor, or is vicinity, are the southernmost limit of the distribution (1) of a host of narrow-ranging mangroves, such as *Aegialitis annulata, Aegiceras floridum, Camptostemon philippinense, C. Schultzii* and *Osbornia octodonta*. In brief, *Lumnitzera lutea*, which is an exception as regards *Lumnitzera* is wholly the norm as regards numerous other mangroves. *Lumnitzera lutea* may not at all behave like *L. racemosa* and *L. littorea* do; however, *L. lutea* behaves quite as do *Aegialitis annulata, Aegiceras floridum, Camptostemon philippinense, C. Schultzii* and *Osbornia octodonta*.

Why does this happen? The answer is simple and precise. The range of all these narrow-ranging mangroves is contained within still another line of faulting (Fig. 78), that very same line of faulting which rents Malaysia as if into two phytogeographic domains, *Sundaland* to the west, *Papualand* to the east. This line of faulting coincides with the classic Malaysian "unstable region" conventionally known as *Wallacea*. To this line, or center, of faulting corresponds another, and lesser one, which roughly answers the quadrangle of modern marine waters contained among Malacca, Sumatra, Java and Eastern Borneo. Here turn up other narrow-ranging mangroves such as *Sonneratia ovata* and *Kandelia Candel*. The former, as the map shows (Fig. 78) infiltrates *Wallacea*, in addition.

In brief, the history of *Lumnitzera* is part and parcel of the general history of the mangroves. *These plants uniformly occur, or originate, by geological lines of faulting and marine transgression.*

Lumnitzera racemosa and *Avicennia marina* var. *typica* are first-born daughters of a line of faulting and marine transgression which roughly corresponds with the modern Mozambique Channel and its northern extension; *Lumnitzera littorea* owes its birth to a line of faulting and marine transgression investing the front Andamans—Java. Like *L. racemosa, Sonneratia ovata* is endemic to the northern, not the southern coast of Java, which is sound evidence to the effect that the behavior of *L. racemosa* in Javanese waters is not without reason. This behavior is in some manner or other connected with the occurrence of a marine transgression and line of faulting of minor proportions taking place among Malacca, Eastern Borneo, Sumatra and Java. *Lumnitzera lutea*, lastly, is the sister of *Aegialitis annulata, Aegiceras floridum, Camptostemon philippinense, C. Schultzii, Osbornia octodonta.* Obviously, the further history of *L. racemosa* and *L. littorea* eastward is the history of the crumbling of Malaysia, and lands beyond as far as the islands of Micronesia and Western Polynesia.

In conclusion: There, where new shores opened during the process of crumbling which invested Gondwana mangroves came to the fore. If these shores had long existence, there mangroves speciated most abundantly. As a corollary of the principle that migrations can only then take place when and where ecologic and geographic conditions are suitable, species which had from the first gained comparatively large dispersal (e.g., *L. racemosa, L. littorea*) increased this dispersal in time.

This is all so simple, and so fully in agreement with the fundamentals of ecology, phytogeography, geology and geophysics that we marvel it has not been seen before.

We claim that the "unstable region" of Malaysia, *Wallacea*, was the seat of faulting and marine transgressions in an epoch which corresponds with the active period of the breaking up of Gondwana. We claim further that before Gondwana began active disruption the Angiosperms had widely migrated. It follows, this being the case, that there must be species, or other groups, the distribution of which is disconnected in *Wallacea*. This is bound to take place because the dispersal which took place prior to disturbances in *Wallacea* was affected of necessity by the later on coming of these disturbances.

As a matter of fact, species or groups disconnected in *Wallacea* are by no means rare. We will restrict ourselves to few examples here, in the assurance the reader will without difficulty round up the score to his entire satisfaction by a further search through the data of Malaysian and extra-Malaysian distribution,

A) Nepenthaceae (*Nepenthes*, the lone genus) (3)

(1) *N. ampullaria* — Eastern massing: Eastern New Guinea; Western massing: Borneo, Malacca, Sumatra.

(2) *N. Treubiana* — Eastern massing: Eastern New Guinea; Western massing: Northern Sumatra.

In these two species the *Wallacean* disconnection is precise. Less obvious, but marked nevertheless is the same disconnection taking place in the species of *Nepenthes* Group *Regiae* as follows,

(3) Eastern massing: *N. Klossii, N. oblanceolata* in Eastern New Guinea; Western massing: *N. Boschiana, N. Burbridgeae, N. clipeata, N. ephippiata, N. fusca, N. mollis, N. pilosa, N. Rajah, N. stenophylla, N. Veitchii* in Borneo — Connecting the two massings only *N. maxima* in New Guinea, Ceram, Boeroe, Halmahera, Northern Celebes — A stray species, *N. truncata* in the Southern Philippines (Mindanao).

Most interesting variants of this disconnection occur in *Nepenthes* Group *Insignes*, of which we have already mentioned *N. Treubiana*. This group tabulates as follows,

(4) Eastern massing: *N. insignis, N. Treubiana* in New Guinea; Western massing: *N. Treubiana* in Northern Sumatra, *N. decurrens, N. Northiana, N. villosa* in Borneo, *N. Rafflesiana* in Borneo, Sumatra, Malacca.

In this group further occur,

(5) *N. Merilliana* in Northern Celebes and the Southern Philippines; *N. petiolata* in the Southern Philippines (Mindanao); *N. Burkei* in the Northern Philippines (Panay, Mindoro); *N. ventricosa* in the Northern Philippines (Luzon).

The species of this group follow two tracks, (i) *New Guinea, Sumatra, Malacca*; (ii) *Celebes, Philippines*. Both these tracks are standard, and both belong to the first age of angiospermous migrations. The former caved in in *Wallacea* throughout; the latter also caved in *Wallacea*, only the rump *Northern Celebes—Mindanao* surviving. *Nepenthes villosa* is endemic on Mt. Kinabalu; the presence of Philippine species in the same group native to Panay and Mindoro suggests that a classic, short track further ran between the heights of Mindoro and Mt. Kinabalu, which adds weight to the evidence indicating that this group indeed migrated, to begin with, in the first angiospermous age (i.e. Late Jurassic to Early Cretaceous). The ranges secured during the first migrations were later disturbed by the events taking place in *Wallacea* during the epoch of Malaysian crumbling.

B) *Deplanchea* (Bignoniaceae) (4)

(1) Eastern massing: *D. sessilifolia, D. novocaledonica, D. speciosa, D. glabra, D. tubulosa* in New Caledonia; *D. tetraphylla* in Queensland, New Guinea, Thursday Island. Western massing: *D. bancana* in Malacca, Sumatra, Bangka, Western Borneo — Connecting the two massing: *Deplanchea* sp. in Celebes *).

---

*) See VAN STEENIS in Jour. Arnold Arb. 20: 221. 1939.

It will be readily observed that the track of this genus crosses over from New Caledonia, Queensland and New Guinea to Western Borneo, Sumatra, Bangka, Malacca, and caved in in *Wallacea* but for a lone station in Celebes.

C) *Diospyros* Sect. *Hasseltia* (Ebenaceae) (5)

(1) *D. fasciculata* — New Caledonia, Queensland, Java (Besoeki).

(2) *D. hermaphroditica* — Western and Central Java, Sumatra, Bangka, Northern Borneo, Billiton, Anambas and Natoena Islands, Malacca.

(3) *D. confertiflora* — Malacca, Bangka.

(4) *D. Holttumii* — Malacca.

No comments are needed. Eastern Java (Besoeki) is immediately connected with New Caledonia and Queensland; the range has caved in in *Wallacea*.

Peculiarly, the zone where the caving-in of the dispersal takes place in the groups mentioned is precisely the range of the monotypic *Astenochloa*. This genus is allied with *Sorghum*, and stands out as the lone member of its affinity in Malaysia. It is distributed as follows (6),

(1) *A. tenera* — Java (massive in Besoeki: Mt. Idjen), Timor, Salayer, South Celebes (Mt. Bonthain), Philippines (Luzon: Benguet, Panay).

Patently, what killed off one group, favored another. *Deplanchea, Diospyros, Nepenthes* lost there, where *Aegialitis, Aegiceras, Camptostemon, Osbornia* gained. This is precisely what should be, for ecological conditions unfavorable to certain forms are by definition favorable to certain others.

We may then conclude this first part of our study as follows, (a) It is certain that dispersal effected in the first age (Late Jurassic to Early Cretaceous) was disturbed in the second age (Middle Cretaceous to Early Tertiary) of angiospermous migrations; (b) There is no doubt but that the conditions which effected the later disturbance also favored the evolution of new forms in the range disturbed; (c) It is positive that mangrove and near-mangrove do not migrate haphazardly. *Their dispersal obeys the same laws as does that of landplants, and is amenable to palaeogeographic conditions in the very first place.*

We are then in the right, in principle when not in every detail, in drawing a line of distinction between "primary" and "secondary" dispersal.

We have studied thus far the dispersal of mangroves that migrate *eastward*. We must now turn to mangroves that migrate *westward*. One of the latter is *Cassipourea* a large genus of the Rhizophoraceae, an immediate ally of such blue-blooded mangroves as *Rhizophora*. *Cassipourea* contains numerous species in distinct subgenera typically ranging in this manner (7),

A) Subg. *Weihea*

(1) *C. ceylanica* — Ceylon, Southern India.

(2) *C. Thomassetii* — Aldabra Island.

(3) *C. Salvago-Raggei* — Northeastern Ethiopia (Eritrea).

(4) *C. euryoides* — Kenya.

(5) *C. malosana* — Tanganyka, Nyasaland (Mt. Malosa).

(6) *C. microphylla* — Madagascar.

(7) *C. congoensis* — Belgian Congo, Cameroon, Nigeria to the Senegal.

(8) *C. obovata* — Mozambique.

(9) *C. Gerrardii* — Transvaal, Natal.

(10) *C. Flanaganii* — Southeastern Cape (Komgha Division).

(11) *C. trichosticta* — Angola.

B) Subg. *Dactylopetalum*

(12) *C. paradoxa* — Seychelles.

(13) *C. Redslobii* — Tanganyka, Mozambique.

(14) *C. gummiflua* — Madagascar.

(15) *C. Rogersii* — Congo.

(16) *C. verticillata* — Natal.

(17) *C. Zenkeri* — Cameroon.

(18) *C. glabra* — Southern Nigeria (Oban).

(19) *C. Barteri* — Southern Nigeria, Gaboon, Angola.

(20) *C. annobonensis* — Island of Annobon.

(21) *C. Mannii* — Island of São Thomé.

(22) *C. nodosa* — Angola.

C) Subg. *Lasiosepalum*

(23) *C. sericea* — Gaboon.

(24) *C. schizocalyx* — Gaboon.

(25) *C. kamerounensis* — Cameroon.

D) Subg. *Cassipourea*

(26) *C. plumosa* — Gaboon.

(27) *C. Afzelii* — Sierra Leone.

(28) *C. eketensis* — Southern Nigeria (Eket).

(29) *C. guianensis* — Trinidad, Guianas, Brazil (Pará, Amazonas).

(30) *C. elliptica* — Lesser Antilles (Tobago to Martinica), Portorico, Jamaica, Cuba; Panama, Honduras, Ceiba Island.

(31) *C. peruviana* — Peru (San Martín: Tarapoto).

Had mangroves migrated whimsically, and in obedience to the winds and currents obtaining today, *Cassipourea* should certainly not disperse exactly in the contrary direction of *Lumnitzera*. As things stand, on the contrary, *a line connecting Madagascar with Ceylon* (Fig. 79) *is a powerful divide in regard of mangrove dispersal.* Some cross this line, others do not. More interesting still, the mangroves that cannot cross this line eastward are forced to become largely land-plants.

Cassipourea is to all appearances a genus of *Gondwanic* origin, as is *Lumnitzera*. Out of the *Gondwanic Triangle* (Madagascar—Mascarenes—Seychelles) Subg. *Weihea* reached in a first time all the continental coasts of Africa southward to Natal, keeping clear, however, of the Mascarenes. Comparatively few of its species ranged westward through the Belgian Congo to the Cameroon, Nigeria, Senegal; only very few reached Ceylon and Southern India, and Angola.

The course of this subgenus is very nearly an exact duplicate of that of the huge "Candelabra" Euphorbias typical of the "Great Rift" of Africa *). These plants are distributed most massively between the Southeastern Transvaal and the tablelands of Ethiopia and Eritrea, but outlying forms occur as far as the Cameroon, India and New Guinea. Noteworthy is the circumstance, that these large Euphorbias are wholly unknown in Madagascar, the Seychelles and the Mascarenes, which otherwise contain a noteworthy endemic euphorbiaceous flora. Accordingly, a parallel drawn between *Cassipourea* subg. *Weihea* and the "Candelabra" *Euphorbia* — though satisfactory as to the dispersal taking place in Africa and most particularly in the vicinity of the "Great Rift," as we shall better see — must be qualified by the note that the ultimate origin of these two groups is geographically not the same. The dispersal of the "Candelabra" *Euphorbia* is such, together with the lack of these plants in Madagascar, Seychelles and Mascarenes, as to indicate that they originated close by the "Great Rift" itself, between Tanganyka and Ethiopia. *Cassipourea* subg. *Weihea* originated on the contrary along a line Madagascar—Ceylon.

Unlike Subg. *Weihea* which mostly spread along the axis Ceylon—Natal, Subg. *Dactylopetalum* ran a track westward between the Seychelles and Angola through Tanganyka, the Congo and West Africa. Subgenus *Lasiosepalum* originated in the *Nigerian Center* outright; subg. *Cassipourea*, lastly, abandoned this center to enter the New World in a conventional manner. Most interesting is the track of *C. elliptica* which, rooted in the Lesser Antilles between Tobago and the Martinica sent a two-pronged track to the Greater Antilles at one hand, Panamá and Central America at the other. Track of the kind are rare, though altogether understandable in consideration of the fact that the West Indies were once massive land, as we saw when discussing the dispersal of the Brunelliaceae, Protieae and Cactaceae.

---

*) The „Great Rift" of Africa consists in reality of a series of longitudinal rifts which broke up the East African plateau, and gave rise to considerable volcanicity during the Late Cretaceous and, particularly the Tertiary. The depressed area of these rifts (the „Great Rift," strictly speaking) is occupied by the long, narrow, precipitously walled lakes of East Africa, and, in the north, by the Red Sea.

Cassipourea is a classic mangrove, but, as its dispersal reveals, many are its species that have lost touch with the sea. This is to say that as *Lumnitzera* gained ground *eastward* in the course of the disruption overtaking Gondwana (including Malaysia), always *by the sea*, so did *Cassipourea* lose contact with the sea *westward* in the course of geological events concomitant with the disruption of Gondwana. In other words: The sea came in *eastward* and largely disappeared *westward* from a line roughly running between Natal and Ceylon. Those of the mangroves which, like *Lumnitzera*, were in advantageous geographic position to migrate *eastward* had practically unlimited spread in this direction, *for Gondwana shattered precisely in this direction.* Those of the mangroves, on the contrary, which were not in this position, and, on the contrary, had tenanted seas or great lakes covering Africa were left high and dry *westward* when these seas or great lakes came to an end. *The parting line between these two great movements, the sinking of Gondwana and the rising up of Africa lies along the axis Natal—Ceylon.*

Proof that Gondwana was shattering while Africa was rising and drying is in the fact that not few are the species of *Cassipourea* that are now "alpine." The fact is not without precedent, for the ericaceous genus *Philippia* which is "alpine" in Africa has one species, *P. mafiensis* that still thrives by the seashore in Zanzibar and the adjacent island of Mafia, under conditions so extreme that *Cocos* itself hardly can stand.

Alpine are the following species of *Cassipourea*, at least;

(1) *C. ugandensis* — Uganda, 1600 m. alt.; Congo (Mt. Ruwenzori) 2100 m. alt.

(2) *C. Salvago-Raggei* – Northeastern Ethiopia (Eritrea), 1800–2900 m. alt.

(3) *C. rotundifolia* — Tanganyka, 2300 m. alt.

(4) *C. ruwensorensis* — Uganda, 1200—2500 m. alt.; Belgian Congo (Mt. Ruwenzori, 2500 m. alt.).

(5) *C. malosana* —Tanganyka (Mt. Mfimbawa, 2300 m. alt.); Nyasaland (Mt. Malosa) 1300—1950 m. alt.

It will be observed that these "alpine" mangroves are wedded to the "Great Rift," which it proves necessary to explain.

In our opinion, the most probable explanation is the following; As revealed by the tracks of *Encephalartos;* the origin of *Genlisea* and the Rapateaceae; the presence of massive groups of aquatic or semi-aquatic forms in the Cape and north of it (Restionaceae, Centrolepidaceae); the massive development in the Karroo and adjacent regions of such plants as *Mesembryanthemum*, etc.; very nearly the whole of modern Africa was occupied by shallow seas or lakes in an area delimited by a line running from the boundary between Nigeria and the Cameroon through the French and Belgian Congo to Uganda, Rhodesia, the Western Transvaal and the Northern Cape. North of the line Nigeria— Uganda shallow seas or lakes obtained which had as their southern limits the north, respectively, of French Guinea, Nigeria, French West Africa; the central sector of the Sudan and Eritrea to the hills of Nubia.

We are not uninformed that our postulates agree with certain opinions by students of geology and geophysics. We stress these postulates, however, as something which stands quite by itself, purely on the strength of dispersal. We stress them because the limits we outline here are pretty much the same of the range of *Encephalartos* (see Fig. 93), which would be little, but correspond massively this time with two of the fundamental tracks of African dispersal. These tracks run from the *Gondwanic Triangle* in the main to the Cameroon at one hand, Natal at the other. These tracks lead to massive centers of speciation and genus making in the Cape proper; to the *Kalaharian* and *Nigerian* centers; to the New World.

Knowing this — and knowing it on a factual basis — we can draw in good conscience the conclusions which the facts justify. This is what happened. From a point leading to the waters far away occupying part of Africa (which point was located among the Seychelles, Madagascar and Uganda) *Cassipourea* infiltrated the Dark Continent prior to the epoch of active disruption of Gondwana. Having thus infiltrated the Dark Continent, *Cassipourea* proceeded to colonize the foot of the East African plateau right there, where waters from the west came to lap this plateau. Meantime, *Cassipourea* also marched on in the direction of the Cameroon, still following the geologic shores. So doing, *Cassipourea* behaved as a mangrove, or at least as a plant of the moist shore, and ran tracks leading to Natal, Angola and the Cameroon in the normal course of its spread.

This took place in a first time, and it will be observed that, unlike *Genlisea*, *Notobuxus*, etc., *Cassipourea* did not mature primarily in the *Kalaharian Center*. *Cassipourea* reached the foot of the East African Plateau, Natal and the Cameroon in the form of the modern genus which it still is. Its primary evolution, consequently, was made outside the *Kalaharian* and *Nigerian* Centers, and right in the *Gondwanic*, more precisely in the west of his last (line Madagascar—Seychelles—Ceylon).

In this first time, *Cassipourea* also extended to the New World, which was by then intimately connected with Western Africa.

In a second time, when the bodies of waters on African soil began to dry, *Cassipourea* managed to retain much of the spread it had achieved. However, caught in the general uplift taking place at the „Great Rift," several of its species were gradually turned into "alpines." It is probable, as a matter of fact, that most when not all these species developed in answer to this uplift.

As for *Lumnitzera*. This genus, represented by then by combretaceous archetypes readily amenable to aquatic, probably brackish, habitats, was rooted within the triangle Madagascar, Mascarenes, Malacca (see dispersal of *Foetidia*, p. 38) when Gondwana entered active disruption. During this disruption two huge lines of faulting, involving marine transgressions developed in what is now the Mozambique Channel and the Andaman—Java insular arc. *Lumnitzera* established itself by these shores. As these shores multiplied by the ever progressing breaking up of Gondwana and Malaysia, *Lumnitzera* gained ground always by sea. *Lumnitzera lutea* speciated — whether out of archetypes already dwelling in the vicinity of what was later to become the modern island of Timor, or by gradual segregation out of *L. racemosa* and *L. littorea,* we have not learned as yet — by another chasm in Gondwana, a chasm which rent Malaysia asunder this time *(Wallacea)*.

It is obvious that we face in these cases a pale of evidence which is beyond comparison larger than that immediately concerning *Lumnitzera* and *Cassipourea.* The evidence in question covers the whole of the geologic history of Africa and the Indian Ocean, which is to say, *the geologic history of the New World, Africa, Europe, Asia, Oceania, Antarctica as one. Genera like Lumnitzera, Cassipourea, Philippia, Aegialitis, Camptostemon, not only, but uncounted hundreds of thousands of plants of all kinds are all involved by this geologic maelstrom, whereby the entrails of the earth were shaken to the depths, and new worlds brought into being.*

It is further obvious that we go absolutely nowhere when approaching the *"Lumnitzera problem"* on the strength of considerations based upon casual agencies of dispersal, modern geography, academic notions of "speciation," etc.

We go absolutely nowhere, because by so behaving we act like mediaeval men trying to rationalize the phases of the moon in the light of the assumption that the earth is flat, not only, but the center of the cosmos *).

One of the most peculiar forms of *Cassipourea* is *C. paradoxa* endemic to the Seychelles. As the Seychelles are a cardinal outpost of dispersal, and we are fortunate in having good accounts of their flora and fauna (2) we may open a parenthesis to consider their *bios,* though most cursorily. This parenthesis will aptly close, doubtless, our comments on the dispersal of *Lumnitzera* and *Cassipourea.*

The flora of the Seychelles consists of a total of 480 species out of which some 233 are endemic. Despite the Seychelles being much closer to Africa than

---

*) As a fitting sequence to a discussion of mangroves, we ought to continue with a review of straight marine plants. A relevant article on the subject is by SETCHELL, W. A. — "Marine Plants and Pacific Palaeogeography" (in Proceed. Fifth Pacific Sc. Congr. 3117. 1930). This author writes (op. cit., 3130), for instance, "We may think of the marine flowering plants as Eur-Afro-Asian in origin, since they are overwhelmingly preponderant in the Mediterranean and Indo-Pacific, rather than Atlantic, the last being largely free from them except in boreal situations and in the Caribbean region." This opinion refers to the Hydrocharitaceae, Potamogetonaceae, Zosteraceae which are Angiosperms. "Eur-Afro-Asian" in Setchell's terminology is the equivalent of *Gondwanic* in ours. The reader may study for himself the maps which accompany Setchell's paper, in the anticipation that the principles of interpretation advanced throughout the pages of this manual will not fail him in the task of rationalizing the distributions of *Enallus, Thalassia, Halophila, Cymodocea, Zostera, Posidonia.* He will also understand what stands behind such statement as this (op. cit. 3127), "There are in the Caribbean area two members of this family *(Potamogetonaceae)* vicarious with two similar members in the Indo-Pacific area, and, although there are two members of the marine Potamogetonaceae in the Mediterranean, they are vicarious with Indo-Pacific or Australian members rather than with those of the Caribbean (or west Atlantic) area." Clearly, the whole of the dispersal is keyed up in standard fashion to a center in the Indian Ocean, and it is altogether probable that the Caribbeans and Mediterranean were reached by independent streams of migrations, both stemming, however, from a common *Gondwanic* center. Cf. also *Suaeda,* p. 382.

to India by mileage, this flora — as our authority tells us — is massively "Eastern" and mascarenean.

We will remark right here that this flora is neither "Eastern", nor "African" in this or that percentual figure. *This flora is Gondwanic through and through, and so is Gondwanic throughout the world all the flora that stems from an original center in the modern Indian Ocean, a center corresponding with Gondwana of yore. We shall never understand the origins of the modern floras until and unless we shall prove competent to bring these floras back to their ultimate centers of origin as a preliminary to study.*

Accordingly, the statement that the flora of the Seychelles is mostly "Eastern" merely means — in correct scientific language — that the Seychelles lie along tracks mostly faring eastward from Gondwana, which tracks happen to interest the Seychelles, too.

Proof of this is in the following. According to our authority, of the Seychellean strand-plants totalling 54 species, 50 further occur in India and Indochina, 52 in Malaysia, 49 in the Pacific, only 43 in the Mascarenes. The tracks of these plants, then, prevailingly run in the direction of the dispersal of *Lumnitzera*, eastward that is, India being quite as "near" the Seychelles as is Western Polynesia, which is rank absurdity in geography, but sober truth in phytogeography.

However, *Avicennia marina* var. *typica* is restricted to the Seychelles, Mascarenes and the African coasts nearby; *Cassipourea paradoxa* is part and parcel of a group which leaves the Seychelles to fare most immediately to the Cameroon and West Africa generally; certain common shore plants of the Mascarenes (witness *Triumfetta procumbens, Entada scandens, Sesuvium portulacastrum, Scaevola Plumieri* and *Pandanus odoratissimus)* are unknown in the Seychelles.

It is then clear that the Seychelles are primarily "Eastern" because they lie clear in the teeth of channels of migrations leading away from Gondwana in an easterly direction. They are less deeply "African" and "Mascarenean" because of homologous understandable reasons.

The Seychellean flora includes at least seven species which, always by our authority, have no relatives anywhere else on earth. These waifs are, (i) *Medusagyne oppositifolia*, type of the monotypic family Medusagynaceae; (ii) *Rieseleya Griffithii*, an euphorbiaceous plant of cryptic affinities; (iii) *Mimusops decipiens* of the Sapotaceae; (iv) *Protarum sechellarum*, of the Araceae; (v) *Psychotria sechellarum, P. pallida*, of the Rubiaceae; (vi) *Toxocarpus Schimperianus* (Asclepiadaceae). To us, these waifs may not have relatives anywhere else on earth *today*, but clearly had relatives somewhere else on earth *in the past*. These relatives lived in Gondwana, and went under, in better than a figurative sense, when Gondwana crumbled. In sum, these seven species are relics out of Gondwana; relics to tell us what stupenduous amount of flora must have been lost when a new world was reborn from the old. In a very definite sense, our living Angiosperms are relics, survivors out of a lost past;

survivors, however, which could in many cases avail themselves of the geological events spelling the doom of uncounted congeners to spawn hosts of new forms.

It is a matter of great regret that the general neglect of phytogeography current in our midst contributes to obscure the clear outlines of floras such as the Seychelles by half-baked generalities and inadequate assumptions. These floras will have eventually to be restudied in order that the meaning of their component plants be made clear throughout.

I. Barringtoniaceae and Triumfetta

The *genorheitron* of which the Barringtoniaceae are part is of the largest, and the families it includes are with few exceptions (e.g., Asteranthaceae) large and well known. These families are in a first place Lecythidaceae, Combretaceae, Myrtaceae, Rhizophoraceae and Punicaceae.

The range of these families is complementary, because the Lecythidaceae are the New World counterpart of the Barringtoniaceae, both strictly tropical groups. The Rhizophoraceae of which we have reviewed *Cassipourea* very nearly belong to the world at large out of a starting point which closely agrees with the Seychelles of our maps. The Myrtaceae prevailingly mass in the Southern Hemisphere, and the Australian domain most particularly, though large numbers of theirs are scattered throughout the tropical zones *). The Punicaceae consists of but two or three species of *Punica* which occupy a narrow range between Iran and Socotra — an interesting distribution in view of the importance of the Seychelles as regards plants of this alliance. It strike

---

*) BERRY, E. W., an author whose writings have done much to throw phytogeography into confusion because of fantastic postulations of "origin" and an utterly wrong chronology, is of the opinion (in Bot. Gaz. 59: 484. 1915) that the Myrtaceae had America for their original home, because in the *Lower Eocene* flora of the Mississipi there are six well marked species of *Myrcia*, four nearly equally well marked species of *Eugenia* and one species of *Calyptranthes*. Also in America are, Berry adds, Combretaceae and Melastomataceae. Early developments of the family, claims the same author, reached Europe either by the way of Asia or the North Atlantic plateau early in the Upper Cretaceous, and became cosmopolitan before the close of the Cretaceous. The types peculiar to the Australian region represent, in Berry's opinion, the relic of the "Cretaceous radiation" with numerous new forms evolved on that continent at a comparatively recent date geologically. The article in question contains the following gem (op. cit. 485), "The Leptospermae have a single monotypic genus in Chile, and the distribution of the other members of the tribe suggests the probability that the South American genus should be placed in some other alliance, since with the exception of *Metrosideros* Banks, which is represented in Africa, and the genus *Baeckea* Linn., which reaches the Asiatic mainland, all of the genera are confined to Australia or the surrounding southeast of Asia."

Most of this is, like the majority of the writings from the same pen, sheer guess. For the best part of a quarter century, this author never wavered in the opinion that everything not agreeing with his notions meant a "Taxonomic faux pas" (see p. 28, etc.). Faced with issues, such as the origin of the Fagaceae, which he could not settle in the usual vein, he roundly contradicted himself within the same article. In brief, it is fair to those who may chance to take stock in these writings here to affirm unequivocally that these writings are worthless. We may lament the necessity of being blunt, but this is a duty to our subject.

us as not impossible that the genorheitron under discussion further involves two families, Styracaceae and Cactaceae *).

In most case, when not all, the Barringtoniaceae are arrant mangroves as are the Combretaceae. Like the balance of their affinity they further occur in dry thickets, and scrub or "parkland" not far removed from the shore; the wet tropical forest; stony arid hills, and, generally, speaking, the same habitats acceptable to the Dipterocarpoids. In brief, these are plants which easily pass from the very moist to the very wet, relatively or absolutely, and are for this reason rewarded with mastery of large sectors of the earth.

The Barringtoniaceae consist (8) of five tribes, three of which Barringtonieae, Combretodendreae and Foetidieae are common to Asia and Africa, extending to Polynesia and Hawaii with the first. The remaining two, Craterantheae and Napoleoneae are restricted to Western Africa **). The massings are therefore consistent throughout for a *genorheitron* out of the African gate. Lecythidaceae and Asteranthaceae are in Tropical America, and form one of the notable constituent of the typic Amazonian forest ***). Of the Barringtoniaceae, Craterantheae and Napoleoneae (the latter oftentimes treated as a separate family, Belvisiaceae or Napoleonaceae, which is not altogether wanting in justification) are in Western Africa; Barringtonieae, Combretodendreae and Foetidieae in Asia from Afghanistan to Japan, Malaysia, Northern and Western Australia, Western Polynesia and Hawaii.

Parallel massings in different geographic ranges are taken for granted in the case of genera (e.g., *Erythroxylon*), or of groups below the rank of family, but the broad affinity under present discussion reveals that massings of the kind may also have place within a large *genorheitron*, groups of families thus replacing as "ecotypes" or "morphologic complexes" the species, or tribes, of smaller taxonomic and systematic units. This may be consistently construed

---

*) The Cactaceae are a very old family, whose genus *Rhipsalis*, as said (see p. 363), is scattered all along the coasts of Africa, Madagascar, the Mascarenes, Seychelles and Ceylon, the lone genus to reach extra-american territory. A well authenticated fossil of *Opuntia* (see CHANEY in Amer. Jour. Bot. 31: 507. 1944) is known from rocks of Eocene age in Eastern Utah, which pollen-studies show to have belonged to a flora including *Abies, Ailanthus, Alnus, Phoenix, Betula, Viburnum, Carpinus, Carya, Cedrus, Cunninghamia, Cycas* (or *Zamia*), *Dioon, Engelhardtia, Ephedra, Erica, Juglans, Liriodendron, Momisia (Celtis), Myrica, Myriophyllum, Peltandra* (Araceae), *Picea, Pinus, Potamogeton, Rhus, Salix, Smilax, Talisia* (Sapindaceae), *Taxodium, Tilia, Tsuga, Vitis*. Though unusual by modern standards, the composition of this flora — to the exception of "*Talisia*," which may perhaps be another closely allied sapindaceous plant — is quite homogeneous from the standpoint of dispersal. It includes a plain mixture of "*warm*" elements (*Phoenix* [possibly an ally of *Washingtonia*, on the contrary]; *Cycas* [most likely *Zamia* or a genus other than *Cycas*, anyhow]; *Dioon*); "*temperate*" elements (such as *Cunninghamia, Engelhardtia, Taxodium, Cedrus*); "*cold elements*" (*Alnus, Betula, Picea, Pinus, Salix*). This mixture is by itself evidence that we are in the wrong if we stress too much the nature ("warm," "temperate," or "cold") of the genera in question. In reality, the history of later epochs reveals that *Ailanthus, Alnus, Viburnum, Carpinus, Carya, Ephedra, Erica, Celtis, Myrica, Tilia, Vitis* are both "cold" and "warm." Trusting determinations made by pollen as usually reliable, we are disinclined on the contrary to accept in the flora just mentioned other plants, identified, seemingly, only by leaf-characters, such as *Banksia, Dodonaea, Ficus, Lomatia, Ochroma, Sterculia*. For these we would indeed like to have confirmation by refined "peel-method" studies.

In our opinion, there is not the slighest possibility that Cactaceae, Aizoaceae, Phytolaccaceae are related even only by *genorheitron*.

**) These two groups are of very great significance in phylogeny.

***) Notice that these plants are in the Amazonas precisely from a point of entry (West Africa—Pará) unlike the one suggested for the Protieae (Eastern Pacific—Bolivia).

to indicate that the problems involved in the origin of a variety, or species, do not fundamentally differ in the end from the issues called into play by the evolution of a whole family or order. We are bound to view the microcosm with a sense of humor in consideration of the fact that, right in back of it, looms the macrocosm, which does not bear on its part being trifled with in the name of academy.

Beginning our review with the four lesser groups of the Barringtoniaceae, we have before us the following (see Fig. 80),

A) Foetidieae

(1) *Foetidia* — 3 species in Madagascar; 2 species distributed in the following manner,

(a) *F. mauritiana* — Mascarenes (Mauritius); Var. *elongata:* Eastern Bengal.

(b) *F. ophirensis* — Malacca (Mt. Ophir).

B) Combretodendreae

(2) *Combretodendron* — 2 species ranging thus,

(a) *C. africanum* — Lower Congo, Cameroon, Ivory Coast, Angola.

(b) *C. quadrialatum* — Philippines (Luzon, Masbate, Samar, Mindanao).

C) Craterantheae

(3) *Crateranthus* — 3 species in Nigeria and the French Congo.

D) Napoleoneae

(4) *Napoleona* — 15 species in the Belgian Congo, Gaboon, Cameroon, Liberia, Fernando Po, Angola.

As to the Foetidieae we observe, (a) *Foetidia* is a genus of utmost phytogeographic significance because, (i) It range and affinity are such as to prove that the vicinity of the Seychelles is basic as regards the evolution and dispersal of most all mangroves and near-mangroves; (ii) It sends from the triangle Seychelles—Madagascar—Mascarenes *(Gondwanic Triangle)* one variety to Eastern Bengal, one species to Malacca (Mt. Ophir). This is proof that the "Sunda Coign" (see p. 389) can be freely entered either at the northern (Bengal) or the southern (Malacca) end by a form, or archetype, out of the

Gondwanic Triangle; (iii) It involves Mt. Ophir with the *Gondwanic Triangle* most directly. This very same peak is also connected with alpine systems in Europe (Alps, mountains of Transsylvania); with the Sumatran Gajolands; with the Bornean Mt. Kinabalu; with Lord Howe and Norfolk Islands. Mediately through the *Gondwanic Triangle,* Mt. Ophir is further involved with Mt. Mlanje in Nyasaland, the heights of the Cameroon and Fernando Po, the Pacaraima System of the Guianas. In brief, Mt. Ophir though insignificant as regards absolute altitude is one of the landmarks in the orogenies that were, long before the Andes, Alps and Himalayas achieved preeminence.

As to the Combretodendreae we notice, (a) The disconnection Ivory Coast—Philippines is but function of the disconnection Cameroon—Iraq—Burma illustrated by the following two tribes of the Sterculiaceae

a) Tribe *Mansonieae* (Fig. 81)

(1) *Cistanthera* — 10 species in Africa (Zanzibar, Congo, Gabun, Cameroon).

(2) *Triplochiton* — 6 species in Africa (Rhodesia, Cameroon, Nigeria, Gold Coast).

(3) *Mansonia* — 2 species in Western Africa (Cameroon); 1 species in the Far East (Burma).

b) Tribe *Theobromeae*

(4) *Glossostemon* — 1 species in Iraq and Iran (Persia).

(5) *Leptonychia* — 14 species in West Africa (Congo, Fernando Po, Cameroon, Nigeria); 1 or 2 species in East Africa (Tanganyka).

(6) *Scaphopetalum* — 6 species in West Africa (3 in Southern Nigeria by Oban; balance in the Cameroon and Ivory Coast).

(7) *Leptonychiopsis* — 1 species in Malacca.

(8) *Peniculifera* — 1 species in Malacca.

(9) *Abroma* — 10 species in Malaysia and Northern Australia.

(10) *Theobroma* — Over 20 species in Tropical America, massive in the Amazonas and adjacent regions.

(11) *Guazuma* — 10 or more species in Tropical America (Brazil to Bolivia, Cuba, Mexico).

(12) *Herrania* — 7 or more species in Amazonian Brazil, Colombia and Tropical South America generally.

It will be seen that all these disconnections come to a head along a line running as follows, French Guinea—Nigeria—Central French Equatorial Africa—Ethiopia—Iraq—Iran—Afghanistan—Northern India—Burma. This line runs in Africa by the southern limits of the present Sahara; in the tract Ethiopia—Afghanistan by a range of profound geologic disturbance. This line further rests upon the northern end of the "Sunda Coign" (Northern India—

Burma, the latter in particular, as revealed by dispersal), and may enter the Philippines directly from Burma through Tonkin and Southern China in disconnection. The incidence along this entire line of a zone of intense desiccation (Sahara, Arabian deserts) and widespread geologic disturbance (Iraq to India) readily accounts for the breaks in dispersal taking place along its course *).

It will further be remarked that *Combretodendron* invades the Philippines in the manner typical of mangroves moving in from the west that is, taking in the whole archipelago by storm. *Combretodendron* might be looked forward in the "Sunda Coign" all the way from Burma from Malacca.

As to the Craterantheae and Napoleoneae we observe, (i) These two groups are rooted in the range Congo—Liberia—Angola, with a heavy center of speciation approximately located by the Nigeria—Cameroon range (Eket, Oban, etc.). We interpret the facts in play accepting as most probable that in the course of the desiccation of huge lakes or marine transgressions once holding away over much of Central and Western Africa, extensive aquatic habitats developed which by gradual alterations of level and subsidiary geological changes furnished favorable grounds for the genus- and species-making of mangroves and near-mangroves. We see no other explanation to account for the happenings taking place in these quarters as regards Connaraceae, Bombacaceae, Erythroxylaceae, Barringtoniaceae, etc. (ii) The range Congo—Liberia—Angola was fed from two directions. From the east, immediately by the *Gondwanic Triangle* (*Cassipourea*, Connaraceae, etc., etc.); from the south, immediately from the *Kalaharian Center*. In its turn, the *Kalaharian Center* (see p. 351) was fed either from the *Gondwanic* or the *Afroantarctic Triangle*. Still in its turn, the range Congo—Liberia—Angola (*Nigerian Center;* see p. 351), fed the New World; (iii) It stands to reason that the heavy genus- and species-making taking place in the *Nigerian Center* is both secondary and primary. Primary is that part of it to be credited to an immediate influx of *genorheitra* faring to West Africa in the early stages of angiospermous dispersal; secondary, on the contrary, such species-making as exemplified for instance in the Connaraceae and Napoleoneae. We regret we cannot press the subject further here, lest we are forced to enter into extensive considerations of systematic botany and local dispersal, here out of place.

Barringtonia is a comparatively large genus with abundance of narrow-ranging species or similar forms. One of its Sections, *Agasta*, is monotypic by a species distributed in this manner,

---

*) If we are not mistaken, the species that has one of the widest range among Parrots is the one known to ornithologists of the beginning of this century as „*Palaeornis torquatus*". The range runs from the mouth of the Gambia (West Africa) through Africa to the Red Sea, India, Ceylon, Burma and Tenasserim. The African form of this Parrot is sometimes treated as a separate species, „*P. docilis*", only because of minor characters of the wings and bill. This range may be compared with that of *Combretodendron, Mansonia*, etc. It is curious that another Parrot, „*Conurus xantholaemus*", is narrowly localized, on the contrary, in St.-Thomas of the Virgin Island. This region is one of the sensitive hinges of Antillean dispersal. Accordingly, Parrots, Barringtoniaceae, Sterculiaceae, Cactaceae, etc., all speak the same language in biogeography. Their voices are doubtless louder than the confuse noise of authors who dispute the existence of "Landbridges" in the Caribbeans without having a precise idea of palaeogeography and pertinent chronology. Bird-dispersal will be amply discussed in a coming book.

(1) B. asiatica — Madagascar, Comoros, Mascarenes, Seychelles; Andaman Islands, Siam, Cambodja, Sumatra, Java, Borneo, Philippines (Luzon, Guimaras), New Guinea, Solomon Islands; Australia (Thursday Island, Queensland); Marianne and Marshall Islands, Fiji, Samoa, Society Islands, Hawaii.

Not unlike *Lumnitzera*, this mangrove is rooted in the vicinity of Madagascar, but at variance with that species it occurs in the Mascarenes and Seychelles. Peculiarly, it seems to avoid Ceylon and India, but firmly holds a second center of range in this vicinity, covering the Andaman Islands, Siam, Cambodja, Sumatra, Java, Borneo and the Philippines. Beyond the Philippines, or near them, the track forks going at one hand to Queensland, Fiji, Samoa and the Societies, at the other to the Mariannes, Marshall Islands and Hawaii. Though seemingly wild, the migration is consistent. The Andamans are beyond question an important center in the evolution of mangroves, possibly a line of ancient shore established in a first time of the crumbling of Gondwana. The branch of the track leading to New Guinea and the Solomons parallels the route of *Vatica papuana*, though Guimaras (near Panay) rather than Tawitawi is immediately involved this time. It is not to be excluded that Thursday Islands was reached directly from Java. We may not go farther in this analysis, for it is not sure that we have all the range before us. However, its main lines are not hard to read, as it appears.

The balance of *Barringtonia* consists of about a hundred species very few of which are in the Moluccas. As stated, most of these mangroves are local or narrow-ranging, but some widespread, witness,

(2) B. racemosa — Madagascar, Comoros, Seychelles, Natal, Mozambique, Tanganyka; Ceylon, Southern India; Andaman and Nicobar Islands, Malacca, Sumatra, Java, Borneo, Timor, Flores, Celebes, Ceram, Amboina; Philippines (Luzon, Basilan, Sibuyan), Formosa, Riu-kiu Islands, Palau, Mariannes, Carolines; Northern Australia, New Guinea, Bismarck and Solomon Islands, New Caledonia, New Hebrides.

The distribution is most massive and pinpoints everyone of the important lines of faulting in the Indian Ocean and its vicinity, Mozambique Channel, Andaman and Nicobar Arc, the Banda Sea, "Wallacea," etc. The Philippines are invaded in the north, south and center alike, which is typical of migrations of the age of disruption much sooner than of the first epoch of angiospermous dispersal; they are the hub of tracks fraying northward to the Riu-Kius, eastward to the Carolines and Mariannes, southeastward to the Solomons, New Caledonia and New Hebrides. This species, and the one analyzed before, are typically plants of the sea with an original range that included a sector, Madagascar—Seychelles—India and Ceylon—Malacca, beyond comparison smaller than the sum total of the dispersal achieved in a second time when the breaking up of Gondwana and the concomitant crumbling of the Western Pacific shore multiplied habitats to the mangroves' liking. The distribution, however, is never haphazard. We regret we cannot give time to a study of the consectional affinities of these two species which would be useful in the task of elucidating the most important centers of secondary speciation and distribution called into being by the final disruption of an ancient geography.

Two mangroves of this genus with unusual interesting dispersal are,

(3) *B. acutangula* — Seychelles, Afghanistan, Ceylon, India, Burma, Siam, Indochina (Laos, Annam, Cochinchina), Celebes, Queensland.

(4) *B. coccinea* — Siam, Indochina (Cochinchina) Australia (North Australia; West Australia: West Kimberley; South Australia).

The distribution of the first highlights a connection Seychelles—Afghanistan not at all surprising, though unusual. This connection is only part of a much more important channel leading northward from the Indian Ocean. *Barringtonia acutangula* is stranded in Afghanistan together with certain Palms, on account of the sea having receded without these plants either following or dying out. The rest of the range, Ceylon, India, Burma, Siam, Indochina is wholly orthodox, and certainly not reached in the second age of angiospermous migrations. Pieces may be missing in the puzzle, and the distribution we have still be far from complete, but any number of channels could lead this mangrove to Celebes and Queensland or to Queensland first and Celebes next. It is interesting that crumbling of the ancient range is suggested to have taken place between Indochina and Queensland. The dispersal of this plant will be worthy of renewed careful study once we shall have it complete, which may not yet be the case.

Barringtonia coccinea runs a short but exceedingly perspicuous route. This route, which lies along part of the western main channel of distribution faring through Malaysia (Siam—Cochinchina), further invests Northern and Western Australia, ending in South Australia. This indicates not only the existence of some ancient body of water in what is now the heart of Australia, thus confirming the indications from *Frankenia* *), but — which is far more important — shows that Western Australia could receive flora from the north, and that this flora could indeed stream all the way to South Australia.

The three remaining genera of the Barringtoniaceae are much smaller than *Barringtonia*, and range in this manner,

(1) *Careya* — 4 species in Afghanistan, India throughout, Burma, Andaman Islands, Siam, Malacca.

(2) *Planchonia* — 14 species in the Andaman Islands, Malacca, Sumatra, Java, Timor, Celebes, Borneo, Philippines, Northern Australia, New Guinea.

(3) *Chydenanthus* — 2 species in Java and Sumatra.

We observe, (a) *Careya* pinpoints the classic track India—Burma—Siam—Malacca, and adds to it a valuable record in Afghanistan, thus confirming the

---

*) A study of certain plants of the shore, actual or geological, would be most interesting. Among those well worthy of investigation are *Menodora*, *Croton* ssp. in the affinity of *C. punctatus*, *Frankenia* and *Suaeda*. See for the last two, SUMMERHAYES, V. S., in Jour. Linn. Soc., Bot., 48: 337. 1930, dealing with *Frankenia* in Australia; ILJIN, M. M., in Sovietsk. Bot. No. 5: 39. 1936, for *Suaeda*. Naturally, a detailed investigation of all mangroves would yield light upon phases of palaeogeography which could hardly be known in other ways.

significance of the record there established by *B. acutangula*. While the Afghan speaks of a sea that withdrew, the Andaman record speaks most likely of a sea that came in; (b) *Planchonia* moves parallel with the Lesser Soenda Islands to reach the Andamans, Philippines, and New Guinea, which might be read to suggest that the Andaman original faulting and marine transgression ran all the way south of the Lesser Soenda to the approaches of New Guinea, being coaeval with the marine transgression yawning in "Wallacea;" (c) *Chydenanthus* reveals that Sumatra and Java were possibly part of the primitive barringtoniaceous dispersal.

Triumfetta is a large tiliaceous genus which has about 50 species (9) in and around Africa. Its distribution is interesting (Fig. 82) because of factors that will soon transpire. The generic range is fairly well covered by this selection of species,

A) Sect. *Lepidocalyx*

(1) *T. lepidota* — Upper Sudan to French Equatorial Africa (Chari) and Northern Nigeria.

(2) *T. Amuletum* — Nyasaland, Rhodesia.

B) Sect. *Porpa*

(3) *T. repens* — Seychelles, Cocos (Keeling) Islands, islands in the Gulf of Siam and off Cambodja and Borneo, Philippines, Howick, Frankland and Northumberland Islands off the coast of Queensland.

(4) *T. procumbens* — Agalega Island, Amirante Islands, Chagos Islands, Cocos (Keeling) Islands, islands by Queensland and New Guinea (Fitzroy and Purdy Islands), Samoa, Tahiti, Cook Islands.

C) Sect. *Lasiothrix*

c') Subsect. *Graciles*

(5) *T. Kirkii* — Tanganyka.

c'') Subsect. *Digitatae*

(6) *T. digitata* — Northern Nyasaland.

(7) *T. macrocoma* — Angola (Huilla).

(8) *T. trifida* — Belgian Congo.

D) Sect. *Actinocarpae*

(9) *T. actinocarpa* — Somaliland.

(10) *T. pleiacantha* — Somaliland.

(11) *T. Welwitschii* — Rhodesia, Angola.

E) Sect. *Lappula*

(12) *T. obtusicornis* — Transvaal.

(13) *T. heterocarpa* — Mascarenes (Rodriguez), Somaliland, Eastern Ethiopia.

(14) *T. pentandra* — Nyasaland, Bechuanaland, Southwest Africa, Upper Sudan, Ethiopia, Senegal, Cabo Verde Islands.

(15) *T. annua* — Natal, Madagascar, Transvaal, Southwest Africa, Angola, Tanganyka, Ethiopia.

(16) *T. macrophylla* — Tanganyka, Kenya, Uganda; Var. *Rothii:* Ethiopia; Var. *ruwenzoriensis:* Tanganyka, Kenya, Uganda (Mt. Ruwenzori; 2100 meters alt.).

(17) *T. pilosa* — Madagascar, India, China, Malaysia; Var. *tomentosa:* Natal, Transvaal; Var. *nyasana:* Nyasaland; Var. *glabrescens:* Nyasaland, Rhodesia, Tanganyka, Eastern Ethiopia.

(18) *T. Dekindtiana* — Rhodesia, Angola, Tanganyka.

(19) *T. Antunesii* — Angola (Huilla).

(20) *T. angolensis* — Angola (Huilla).

(21) *T. delicatula* — Angola (Huilla).

(22) *T. orthacantha* — Angola (Golungo Alto).

(23) *T. glechomoides* — Angola (Huilla).

(24) *T. benguelensis* — Angola (Benguela).

Forms that migrate in the pattern of *T. repens* and *T. procumbens* are seldom, if indeed ever, considered in phytogeographic work on the ground that their distribution obeys no law. This is false, of course. To begin with, both these species turn up in the vicinity of the Seychelles, a key-center in the dispersal of plants of the sea and the sea-shore and its hinterland; both leave this vicinity faring eastward. This is all standard of a certainty.

Triumfetta repens streams in a beeline from the Seychelles to the Cocos Islands, which is consistent with a Gondwanic form faring toward Malaysia; *Foetidia* — let us not forget — also runs a beeline from the Mascarenes and Madagascar to Eastern Bengal and Malacca.

It may seem unaccountable that leaving the Cocos Islands *T. repens* next turns up in certain petty archipelagoes off Borneo and Southern Indochina. However, this is quite according to rote. While streaming eastward from the Cocos Islands, *T. repens* got caught *by the very same channel* which brought northward from Northern Australia such plants as *Stylidium, Drosera,* Podostemonaceae, Restionaceae, Centrolepidaceae, etc.; *the very same channel* that lead *Barringtonia coccinea* southward. This channel, as we know, is intimately related with a line of faulting in Western Malaysia subsidiary to *Wallacea* (see p. 402) and most likely responsible for certain vagaries in the course run by *Lumnitzera* around Java. In brief, *T. repens* acts conventionally right when it seems to obey no law. If there is a difference between it and the other groups mentioned above, this is in the fact that these groups all reach the channel in question from different directions, *T. repens* from the west; *Stylidium* from the south; *Barringtonia coccinea* most likely from the north.

It is not strange that caught by the channel in question near Malacca, *T. repens* should further turn up in certain petty islands off the Queensland coast. *Korthalsella Opuntia* is on Mt. Ophir, in Malacca, and some of these islands, too (Lord Howe, Norfolk Islands), not only, but it is right in this general vicinity that runs a track whereby *Notelaea* went from New Caledonia to Madera and *Carex pyrenaica* from New Zealand to Spain.

Triumfetta procumbens took its start very much from the general vicinity involved in the migrations of *T. repens*, and like this species fared eastward. However, for reasons of its own — which we believe to consist in the main in a more southerly run of the track — went straight to Queensland and Polynesia, without ever getting caught in the channel that sent *T. repens* to Indochina.

The rest of the dispersal of *Triumfetta* is not wanting in interest. If *T. procumbens* and *T. repens* followed more or less in the wake of *Lumnitzera*, *T. macrophylla* went on its part the way of *Cassipourea* and became alpine with a variety at least. The *Kalaharian Center*, the vicinity of the tablelands of Huilla in particular — which tablelands appear to have been in emersion when much of the rest of Western Africa was under water — is responsible for a heavy speciation, which is not surprising. Most striking is the distribution of Sect. *Actinocarpae* and *T. heterocarpa*, which suggests that Somaliland had direct connections of its own with the *Gondwanic Triangle;* in other words, Somaliland could be freely entered from archetypes immediately emerging from what is now the Indian Ocean, archetypes, that is, which never reached Somaliland from either Tanganyka or Ethiopia while faring to Angola at the other hand. It is probable that to its intimate, direct connections with lands lost in the Indian Ocean, Northern Somaliland owes its peculiar *bios.*

In conclusion, the Barringtoniaceae and *Triumfetta* add their mite of evidence to the long, heavy mass of facts accruing to these pages by the review of a quantity of unrelated groups. Everything in their dispersal is orderly, precise, clock-like. There is room, perhaps, to argue about certain details, but the fundamentals stand beyond attack. It is upon fundamentals such as these that phytogeography will have to be rebuilt in order to take its place among the exact sciences. The scarcity of marine plants in the Central and Southern Atlantic (see p. 410 fn.) harks back, for example, to palaeogeographic factors of the same order as those ruling the dispersal of *Lumnitzera, Cassipourea,* etc.

II. Dipterocarpaceae

In the classification of Engler & Prantl the Dipterocarpaceae are subdivided and distributed in this manner,

A) Monotoideae

(1) *Monotes* — 31 species in Madagascar, Rhodesia, Angola, Mozambique, Tanganyka, Cameroon, Nigeria.

(2) *Marquesia* — 3 species in Rhodesia, Angola, Gabun, Nigeria.

B) Dipterocarpoideae

a) Vaterieae

(3) *Vateria (Vateriopsis)* — 5 species in the Seychelles, Ceylon, India.

(4) *Monoporandra* — 2 species in Ceylon.

(5) *Stemonurus* — 14 species in Ceylon.

a') Vaticeae

(6) *Cotylelobium* — 6 species in Ceylon, Malacca, Borneo.

(7) *Vatica* — 65 species in Ceylon, India, Burma, Southern China (Hainan), Indochina, Malacca, Sumatra, Borneo, Philippines, New Guinea.

(8) *Pachynocarpus* — 5 species in Malacca, Borneo.

a'') Shoreae

(9) *Doona* — 12 species in Ceylon.

(10) *Hopea* — 75 species in Ceylon, India, Burma, Siam, Indochina, Southern China (Kwantung, Hainan), Java, Sumatra, Philippines, Borneo, Celebes, New Guinea.

(11) *Pentacme* — 7 species in India, Indochina, Siam, Burma, Malacca, Philippines.

(12) *Shorea* — 131 species in Ceylon, India, Burma, Indochina, Siam, Malacca, Borneo, Java, Sumatra, Philippines.

(13) *Parashorea* — 8 species in Burma, Indochina, Malacca, Sumatra, Philippines.

(14) *Isoptera* — 1 species in Malacca, Borneo, Philippines.

(15) *Balanocarpus* — 16 species in Ceylon, India, Malacca, Borneo, Philippines.

(16) *Dioticarpus* — 1 species in India.

a''') Dryobalanopseae

(17) *Dryobalanops* — 9 species in Malacca, Borneo.

C) Affinity unsettled

(18) *Upuna* — 1 species in Borneo.

The Dipterocarpoideae (or, "true Dipterocarps" in current language) are ruled out of Africa by this classification. It has been learned, however *) that this Subfamily sparingly occurs in the Congo, Gaboon and Nigeria. Fossil Dipterocarpoid wood seemingly of late Tertiary age has been reported moreover from Italian Somaliland and Mt. Elgon (Kenya, Uganda).

Most authors are agreed that the Dipterocarpaceae are of "Malaysian origin" and one of them (10; p. 351) writes as follows, "It seems that Sumatra, the Malay Peninsula, and the Philippines Islands have had former land connections with Borneo and that this land was the region where the family originated and from which it spread out."

We observe that, on the contrary, this family is most certainly of Gondwanic origin, because the face of the dispersal proves that it evolved in, and migrated from, the continental mass that once occupied the Indian Ocean of our maps. The Dipterocarpaceae behave conventionally to the very extent that they break up in two major groups, Monotoideae restricted to Africa, and Dipterocarpoideae mostly confined to Asia and Malaysia. The former — and this in the first age of angiospermous dispersal — migrated by rote to East, Central and South Africa. The latter spread on the contrary eastward and northward from the Seychelles to Ceylon, India, certain sectors of the Far East and Malaysia. Some strayers out of the true Dipterocarps reached the Dark Continent in addition through a door located between the Seychelles and Somaliland **), spreading from this point conventionally to the Congo and parts of West Africa.

The migrations of the Dipterocarpaceae, consequently, are patterned after those of *Cassipourea* and *Lumnitzera* which the reader may easily verify. It would be pointless to argue that a family which includes Monotoideae

*) See, BURTT—DAVY in Nature 136: 991. 1935; FOXWORTHY in Jour. Arnold Arb. 27: 349. 1946.
*) We cannot here devote time and attention to purely local problems of dispersal, which we often regret. Somaliland is a peculiarly interesting region, which (see, for instance, CIFERRI in Nuovo Giorn. Bot. Ital., N. S., 46: 344. 1939) falls into different, oftentimes well marked, biological provinces. The flora and fauna of Somaliland is the ultimate offshoot of a triangular play, in which Somaliland stands at the receiving end of migrational currents of continental „African" origin, or is conversely fed by channels faring to it (and simultaneously the East), immediately out of a lost center in the Indian Ocean. The phytogeography of Somaliland, consequently, cannot be understood unless in function of palaeogeographic factors, not only of Tertiary, but much earlier date. Somaliland is as to biogeography integral part both of Africa and India, with a position which is ultimately not unlike that of the Seychelles and Socotra.

together with Dipterocarpoideae would originated in Malaysia. As a matter of fact, the hinge of the distribution of these two groups is properly situated by the Seychelles, which — as we know — connect Madagascar with Ceylon in general dispersal. Though not mangroves in the strict sense of the term, many of the Dipterocarps thrive in habitats not incompatible with the lowlands by the tropical shore *). The Monotoideae, on the contrary, usually prefer the dry savanna with acidic soil.

This chapter is devoted to a consideration of epochs in angiospermous dispersal. We need not be further concerned with the Monotoideae, according-ly, which do not appear to have migrated extensively following an initial settlement in Africa. Their range varied but little in fundamentals after an initial upsurge of dispersal in the Late Jurassic and Early Cretaceous.

Different is the case with the Dipterocarpoideae. We know that many are the groups that took advantage of the breaking up of Malaysia to gain addi-tional distribution, and there is sound reason to assume that this subfamily further dispersed and evolved during the Malaysian dislocation. We will, then, study the Dipterocarpoideae in some detail.

It stands to reason that in order to verify, or disprove, the surmise that part at least of the dispersal of the Dipterocarpoideae took place in a second period of angiospermous migrations (i.e., from the Middle Cretaceous to the Early Tertiary), we should try as a preliminary to identify the probable primary range secured by this subfamily. Secondary distribution presupposes the existence of bases of migrations won in a prior time.

Doubtless, Ceylon and Southern India were part of the primary range of the Dipterocarpoideae. Ceylon, as a matter of fact, is a massive center of diptero-carpoid distribution with a whole tribe, nine genera and about fifty species in the local record. The Seychelles and parts of continental Northern India are properly viewed as appendages of Ceylon in a dispersal of this nature.

As to the rest of the primary range in play, we will be on safe grounds no doubt, if we turn to geophysical work **). Two sectors of Malaysia are: One, so called "Sunda Coign" consisting of the lands stretching westward and northward from Java. To this "coign" belong consequently, Java, Sumatra, Malacca, Indochina, Siam, Burma. The other "coign" is understood as

---

*) See, for instance, brief notes on the subject by TARDIEU—BLOT in Boissiera 7: 303. 1943. The true Dipterocarps are scarce in Indochinese regions periodically inundated, but dominant in the forest not liable to submersion. Says the author cited of these forests, "Ce sont des forêts claires, sur les sols pauvres, rocheux ou épuisés par les rays *(repeated fires)*. Entre les arbres le sol est nu ou couvert de "tranh" ou d'autres Graminés et de nombreux bambous nains. Les "terres rouges", les "terres brunes" basaltiques, les terres alluvionnaires sont recouvertes de peuplements denses et élevés." To discuss these habitats in full we should open a lengthy parenthesis, which is here impossible. Habitats of the kind, however, seems to us to constitute the "immature" backbone ("immature," that is, as to soil, and ulti-mately unsettled as to final destination between the open parkland and the tropophilous dense forest) of most tropical lands. It is lands of the kind which the forms we understand as "near-mangroves" most frequently colonize: in endless degrees, these lands stand intermediate between the dense rain-forest (or its derivatives) and the subdesertic savanna. Though oftentimes occupied by what we need consider as climaxes in adherence to current tenets of ecology and soil-science, still these lands are *essentially "seral"* in our understanding.

**) See, for instance, the short, meaty study of HOBBS, in Proceed. Amer. Phil. Soc. 88: 254, Figs. 26, 27. 1944.

"Papua Coign" and includes in the first place New Guinea. In the geophysicists' opinion, these "coigns" are sectors of fundamental importance, in the sense that it is around them that are ultimately built the continental outlines of past and present maps. We might accept them, consequently, as primaeval cores of land about the world's surface.

Between the "Soenda" and "Papua" cores, or "coigns", just outlined, stand as a more or less neutral zone Borneo *) and the very same unstable *Wallacea* we have discussed in former pages.

If, as the geophysicists claim, the "coigns" are keystones in the earth's make-up; if, as we claim, dispersal is primarily effected in obedience to ancient outlines of land and sea, it follows that the "coigns" are bound to have basic significance in dispersal. It is these cores that should retain, in a measure at least, forms of flora that are peculiarly ancient, and it is from them that migration should run its secondary course to take advantage of lands unsettled during new emersions.

As a matter of fact, the ideas entertained by geophysicists regarding the Malaysian land-cores can without the slightest difficulty be squared up with fundamental premises of Malaysian distribution **). It is a fact, that the two major channels of dispersal running in and around Malaysia are rooted each in one of the "coigns."

The range of the magnoliaceous *Schisandra*, merely to mention an example ***) squarely rests upon the "Soenda coign." We stress *Schisandra* but the reader will find without the slightest trouble rather numerous groups distributed in the *Schisandra* manner ****), mentioned in this manual and current taxonomic work. By the "Papua coign," on the contrary, are rooted the very numerous groups (e.g., *Drimys;* see p. 109) which reach northward from the *Neocaledonian* center (see p. 362) in the direction of the Northern Philippines, Northeastern Borneo (Mt. Kinabalu, in particular), Formosa and, occasionally, the southern Chinese provinces.

These two basic nodes of dispersal are served by two main channels. The *Sunda Channel* — as we shall understand it — is fed primarily from the direction of the modern Indian Ocean, which is to say from Gondwana. The *Papua Channel*, on the contrary, from the Western Pacific, which is to say,

---

*) By leaving Borneo unsettled between the "Soenda" and "Papua" coigns we depart from the understanding of the author whom we have cited in the preceding footnote. HOBBS does include Borneo in the "Soenda Coign." A full discussion of the matter cannot have place here. It is a fact that Northern Borneo is part of some "coign", for this range is as important as any other one in Malaysian dispersal. This "coign" however freely communicates both with Sumatra and Java, and Luzon and Mindoro, so that it is intermediate between the two in *dispersal*. A detailed account of the phytogeography involved in this issue cannot have place in the pages of this manual, though admittedly most interesting as such.

**) The fundamentals of distribution can further be squared up with the understanding that geophysicists entertain of other "coigns." The "Australian Coign," figured on Map 83 (after HOBBS), is an appendage of Malaysia in dispersal, indeed a necessary one.

***) See the taxonomic treatment of this genus by A. C. SMITH, who unknowingly stresses this "coign" in Sargentia 7: 92 Fig. 15. 1947.

****) See for instance the Polygonaceae monographed by DANSER in Bull. Jard. Bot. Buitenzorg. Sér. 3, 8: Maps I, II, III. 1927.

from that crumbled shore which once ran in this ocean much beyond the present coasts (see p. 69, Fig. 22).

It must be clear that the distinctions here drawn are not absolute, because numerous are the cases in which a group originating in Gondwana ranges as far eastward and southward as the Fiji Islands, for example. Contrariwise, groups that draw their being from the Western Pacific may freely migrate as far west as India, as do the Magnoliaceae, for example. However, if not absolute in detail, these distinctions answer fundamental premises, wherein phytogeography and geophysics agree throughout.

It will further be observed that the channel Java—India can be run either way, that is, from Java northward and westward to India, or in the opposite manner, depending as to whether the main track comes in from Gondwana first pointing to Java or Ceylon and India generally. We have discussed alternatives of the kind in previous pages (see for example considerations regarding the immediate geographic origin of *Schoenoxiphium* in the Sumatran Gajolands, p. 239), showing that these alternatives call into play nothing which is basic, and can be settled accordingly on the merits of individual cases and problems.

We may, then, conclude as follows; There run in Malaysia and lands immediately connected with Malaysia, two channels of distribution which on account of their fundamental nature are properly understood as primary. One of these, *Soenda Channel*, invests the sector India—Burma—Siam—Indochina—Malacca—Sumatra—Java; the other, *Papua Channel*, New Caledonia or Queensland—New Guinea—Philippines. These two channels become connected, often when not always, within a belt running in the main between Java and Northern Queensland, which belt may immediately interest in addition Northern Australia and Celebes. When not reached through Celebes from the south, the Philippines (the Northern and Central Philippines particularly) can be reached always from the same direction by a track in disconnection faring through the Western Pacific east of New Guinea and Morotai.

The belt connecting the two channels may be broken and disfigured more or less extensively by two lines of faulting, (a) One running roughly parallel with the Lesser Soenda Islands as far as the Strait of Torres; (b) A second (*Wallace's Line*, or derivatives) which takes its inception at the Lombok Strait, or, more generally, between Eastern Java and Timor, next running almost due north in the direction of the Philippines.

Borneo lies clear between the course of the two channels, and just north of their intersection in the south of Malaysia. Borneo is accordingly invested in turn by offshoots of the one and/or other of these channels. To this, and to the fact that Borneo seems to have been subjected in various epochs to gradual alterations of level, are to be credited in the main the peculiarities of the Bornean flora. Such flora may be "autochtonous" by gradual speciation and genus-making within the Bornese geographic limits, but cannot be considered to have arisen by a primary center of origins ever.

The Philippines can be further reached from the northwest by a direct prolongation of the *Soenda Channel* running from Burma through Southern China generally to Formosa and Luzon.

These outlines of the primary channels of dispersal running in Malaysia and lands adjacent can easily be visualized on the modern map (Fig. 83), and these outlines can quite as easily be grafted upon other standard tracks which reach any point we may like to fancy of the modern world.

With this clear before us, and safe already in the knowledge that the primary range of the Dipterocarpoideae covered Ceylon and India, we may conclude that this primary range additionally took in India, Indochina, Siam, Burma, Malacca and Sumatra at least. It will be observed that this primary range is wedded to the "Soenda Coign," and is further basic of the dispersal of very numerous other groups discussed throughout the pages of this manual.

A strong hint that the Dipterocarpoideae crossed over to the "Papua Coign" only in a second epoch of migration is to be had by the actual distribution of certain of their members.

For example, *Shorea* is endemic to Ceylon and India, but further occurs in Malaysia (11) with three species among others, as follows,

(1) *S. platyclados* — Malacca, Sumatra (Gajolands: Mt. Agosan, 1800 m. alt.), Borneo nearly throughout, and on Mt. Kinabalu at 1800 m. alt.

(2) *S. sumatrana* — Malacca, Sumatra.

(3) *S. seminis* — Borneo throughout, Philippines (Luzon: Albay, Camarines; Samar; Mindanao: Zamboanga).

The first of these species occupies two classic stations of primary dispersal, the Sumatran Gajolands and Mt. Kinabalu *) and these two stations are reached by a straight route Northern Sumatra—Northeastern Borneo. This

---

*) As shown by STAPF's account of the flora of Mt. Kinabalu (in Trans. Linn. Soc., Bot., 2, 4. 69. 1894), this Bornese peak is reached by streams of migration issuing from the Soenda and Papua channels alike. To a carriage of archetypes by the former are due the "Ceylonese" and "Himalayan" affinities massive in the Kinabaluese species of *Symplocos* and *Eugenia (Syzygium)*, and the following in the rubiaceous genus *Lasianthus*, (i) *L. kinabaluensis:* Mt. Kinabalu; near *L. lucidus* (Khasia, Upper Tenasserim, Java); (ii) *L. membranaceus:* Mt. Kinabalu; near *L. oliganthus* (Ceylon); (iii) *L. euneurus:* Mt. Kinabalu; near *L. strigillosus* (Southern Deccan; Travancore) and *L. Wightianus* (Malacca: Mt. Ophir). To a direct influence of the Papua channel are to be adscribed the following relationships, (i) *Vaccinium cordifolium:* Mt. Kinabalu; near *V. cereum* (Tahiti and Eastern Polynesia); (ii) "Antarctic-Austral" elements colonizing the summit, or very nearly the summit of Mt. Kinabalu, with forms of *Coprosma, Nertera, Pratia, Gaultheria* etc.

In certain cases the affinities of the flora of Mt. Kinabalu suggest the run of very long tracks, witness, (i) *Myrtus flavida:* Mt. Kinabalu; near *M. rufopunctata* of New Caledonia and further allied with the South American *M. myricoides* and *M. microphylla;* or highlight channels of dispersal of unusual interest, witness, (ii) *Clethra canescens:* Mt. Kinabalu also Borneo generally, Java, Lombok, Celebes; very close to *C. lancifolia* (Luzon) and *C. Faberi* (Southern China: Kwantung).

Striking is the position of the insignificant Mt. Ophir (insignificant, that is, from the standpoint of modern geography) as regards phytogeography. This low peak of Malacca is intimately bound, as shown in these pages (see further p. 38) with the Carpathians, Madagascar and the Mascarenes, Norfolk and Lord Howe Islands, the Gajolands and Mt. Kinabalu.

We need not insist that if certain of the opinions of STAPF regarding the immediate affinities of this or that species of Mt. Kinabalu are perhaps less than absolutely correct as to taxonomy, the whole of his conclusions is doubtless sound phytogeography.

route, be it parenthetically observed, is of great relevance because it consists of a segments of a much wider line connecting the following points, Philippines (Luzon, Mindoro), Northeastern Borneo (Mt. Kinabalu), Northern Sumatra and Malacca (Gajolands, Mt. Ophir), the *Gondwanic Triangle* (see *Foetidia*, p. 414), the *Neocaledonian Center* (see discussion of *Notelaea*, p. 355).

It seems then clear that *Shorea platyclados* runs a distribution that, insofar at least as the Gajolands and Mt. Kinabalu are concerned, belongs to the first age of angiospermous migrations. *Shorea sumatrana* and *S. seminis*, are on their part, vicariants, and the latter stands for eastern populations which in the west are identified by the epithet of the former. Be it as it may regarding taxonomic details, which every botanist is likely to manage in his own lights, these two species yield a rational pattern to the phytogeographer. The dispersal is rooted by Malacca and Sumatra, in lands, that is, won by the Dipterocarpoideae in the first upsurge of migration; leaving these lands in a second time it invades Borneo throughout and the Philippines from end to end. Well worthy of notice is the manner in which the Philippines are invested. Migrations of the first age that fare through Eastern Malaysia usually reach the Central and Northern Philippines from the direction of Celebes or New Guinea. In the second age of dispersal, on the contrary, such age as we believe to be typified by the dispersal of the species of *Shorea* under discussion the Philippines are taken by storm along their full length *from the direction of Borneo.*

In Malaysia *Vatica* has 33 species (12), spearhead of a long genorheitral line begun in the Seychelles. Not unexpectedly, these thirtythree species are massive in Malacca, Sumatra and Borneo. Two of them, however, are distributed in this manner,

(1) *V. Mangachapoi* — Northern Borneo, Philippines throughout (Luzon, Babuyan Islands, Leyte, Samar, Mindanao).

(2) *V. papuana* — Northern and Southern Borneo, Philippines (Tawitawi), Ternate, Aroe Islands, Batjan, Morotai, New Guinea, Louisiades Archipelago.

These patterns of dispersal are as transparent as we may like to have them. The two species in play originated in Borneo in a first time. The former migrated eastward only to crash into the Philippines headlong in the standard manner of forms leaving Borneo for the immediate east in the course of secondary distribution. Not so the latter; *Vatica papuana*, which of all Dipterocarpaceae reaches farthest east, barely hits the Philippines at the island of Tawitawi, in the westernmost end of the Sulu Archipelago; it is next immediately deflected to New Guinea, and New Guinea's eastern insular "tail" over a channel running through the Moluccas almost due southeast from Tawitawi. Tawitawi (and the not too distant region of Tawao in Northeastern Borneo) are standard stations in the secondary dispersal that currently follows this track. We do not exactly know just now why Tawitawi is a landmark of distribution, but feel that there is a strong probability that right here began a line of faulting leading to New Guinea rather than Mindanao or the Central Philip-

pines. It is obvious, at any rate, that migrations by occasional means of distribution do not account for the different behavior of these two species of *Vatica*.
Against the facile assertions of numerous authors who stress these means
stands the mature judgement of better informed students. One of these *)
reporting the original discovery of *Monotes* in Madagascar emphasizes the fact
that such is the bulk of the reproductive means of this plant that carriage by
occasional agencies from continental Africa to Madagascar is entirely out of the
question. The same student remarks that the most recent connection overland
between Madagascar and Africa is of Miocene age, thus inferring that *Monotes*
was in Madagascar prior to the Tertiary and by definite landconnections of
which modern maps exhibit no trace. This is the correct approach to dispersal, not that of authors who venture to insist that *Vatica*, and the Dipterocarpaceae in general, just floated aimlessly about, now hitting Tawitawi and
the Louisiades, then again running headlong into the Philippines.

Examples could be multiplied always to reach the final conclusion that
species like *Shorea sumatrana, S. seminis, Vatica Mangachapoi, V. papuana* do
not belong to the primary Dipterocarpoid distribution. Some of these plants
originated in Borneo, of course, for Borneo offered ideal grounds to the
speciation of these plants. However, even so, Malacca is close second to Borneo
in number of Dipterocarpoid species, 153 occurring in the latter, 141 in the
former. Java fares poorly in the score with but 10 species, but it is known (13)
that in the Tertiary — the heyday of the Dipterocarpaceae — this island
was richer in plants of the kind than it is today, not only, but had in its flora
genera like *Dryobalanops* which have since died out. These modifications of
floristic spectra are due to local conditions, in part, and also to the general
factor that Dipterocarpaceae multiplied most actively in an epoch of wide
submersions taking place in the Late Cretaceous and Early Tertiary. As this
factor waned, so did the Dipterocarps lose out, which is quite as true of Java
as it is of Africa.

In conclusion, it should be difficult to find a family which migrates in so
conventional and clear a manner as do the Dipterocarpaceae. The living and
fossil record both speak to the same effect, *at a glance*. The dispersal of these
plants was certainly effected in two times. The original migration laid down the
lines of the distribution of *Monotes*, which varied very little if indeed at all
past the Early Cretaceous. By this same migration the Dipterocarpoids —
not unlike *Cassipourea* — gained certain African ranges which, being less
plastic than *Cassipourea*, they could not maintain. In Africa, the Dipterocarps
have only a tiny foothold in Nigeria, which is consistent with the fact that right
by Nigeria stood a center of massive speciation and genus-making most likely
connected with the age-long persistence of conditions not unlike those responsible for the successful evolution of mangroves in "Wallacea." Always during
the epoch of primary migrations the Dipterocarps possessed themselves of a

---

*) See HUMBERT in C.-Rs. Acad. Sc. Paris 219: 341. 1944.

solid range in the "Soenda Coign," which they used as a secondary base of migrations when Malaysia came to be broken up in the manner we have seen. Everything in the tale of dipterocarpaceous dispersal is keyed up with everything else taking place inside and outside the family's limits, in so clear and simple a manner as to cause us to wonder why so much was pointlessly written of the Dipterocarpaceae and their migrations in time and space.

BIBLIOGRAPHY

Chapter VII

(1) VAN SLOOTEN, D. F. — Die Verbreitung von Lumnitzera und einiger anderen Mangrovegewaechsen; Blumea, Suppl., 1: 162. 1937.

(2) SUMMERHAYES, V. S. — An Enumeration of the Angiosperms of the Seychelles Archipelago; Trans. Linn. Soc. (London) 19: 261. 1931.

(3) DANSER, B. H. — The Nepenthaceae of the Netherlands Indies; Bull. Jard. Bot. Buitenzorg, III, 9: 249. 1928.

(4) VAN STEENIS, C. G. G. J. — see (7) Chapt. I; see further same author — The Bignoniaceae of the Netherlands Indies; Bull. Jard. Bot. Buitenzorg, III, 10: 218. 1928.

(5) BAKHUIZEN VAN DEN BRINK, R. C. — Revisio Ebenacearum Malayensium; Bull. Jard. Bot. Buitenzorg, III, 15: 78. 1937.

(6) VAN STEENIS, C. G. G. J. — The interpretation of the locality "Andor"; Bull. Jard. Bot. Buitenzorg, III, 13: 114. 1933.

(7) ALSTON, A. H. G. — Revision of the Genus Cassipourea; Kew Bull. 1925: 241. 1925. See further JONKER, F. P. — Remarks on the South-American species of the genus Cassipourea (Rhizophoraceae); Rec. Trav. Bot. Néerland. 38: 373. 1942.

(8) KNUTH, R. — *Barringtoniaceae;* E. & P. Pflanzenr. 105 (iv. 219—219b). 1939.

(9) SPRAGUE, T. A. & HUTCHINSON, J. — The Triumfettas of Africa; Jour. Linn. Soc. (London), Bot., 39: 231. 1909.

(10) FOXWORTHY, F. W. — Distribution of the Dipterocarpaceae; Jour. Arnold Arb. 27: 347. 1946.

(11) VAN SLOOTEN, D. F. — Sertulum Dipterocarpacearum Malayensium II; Bull. Jard. Bot. Buitenzorg, III, 17: 110. 1941.

(12) VAN SLOOTEN, D. F. — The Dipterocarpaceae of the Dutch East Indies. IV. The Genus Vatica; Bull. Jard. Bot. Buitenzorg 9: 67. 1927.

(13) VAN SLOOTEN, D. F. — The Dipterocarpaceae of the Dutch East Indies. VI. The genus Dryobalanops; Bull. Jard. Bot. Buitenzorg 12: 1. 1932.

CHAPTER VIII

THE SIGNIFICANE OF "LOCAL" DISPERSAL

Much has been written on plant-distribution by authors who hopefully deal with "local" problems. These authors have often selected a single species, genus or minor group as their subject, or have toiled on what they understood as the "phytogeography" of a narrow territorial compass, ultimately to reach conclusions of general significance, how plants migrate; where plants originated, etc.

It is patent that the efforts of these authors fell short of the goal sought wherever these authors attempted to formulate general ideas. It could not be otherwise, because "local" problems do not exist in phytogeography wherein everything interlocks with everything else. By artificially setting out a "local" issue, these authors need reach artificial conclusions. Problems in dispersal which parochially viewed seem to involve a few square miles or a bare handful of forms take on major importance the moment they are correctly seen against a fitting background of generalities. Conversely, issues that based upon "specialized" work seem to be important, witness "Bipolarism," are nugatory in the end.

Let it be taken for certain that dispersal is never a "local" problem. Sound phytogeographical work, whether initially conceived by a broad plan of effort, or narrowed down to single issues, is always to be approached on the strength of the same preliminaries, and to be treated by the same essentials of concept and method.

In order that these preliminaries be made clear, and these essentials stand out with the needed lucidity, we will approach in this chapter certain "local" issues. It will readily be learned that these issues are indeed anything but "local."

I. Stylidiaceae

This family occurs in Malaysia with a handful of species that suggest (1) an insignificant outlier, as follows,

A) *Stylidium* Subg. *Andersonia*

(1)   *S. tenellum* — Sumatra, Malacca, Southern Burma, Indochina.

B) *Stylidium* Subg. *Alsinoidei*

(2)   *S. alsinoides* — Celebes, Philippines, Queensland, North Australia.

C) *Affinity unsettled*

(3)   *S. inconspicuum* — Java.
(4)   *S. javanicum* — Java, Soemba.

In order to have the matter clear before us, let us complete the record of the two subgenera in question, as follows (2),

A) *Stylidium* subg. *Andersonia*

(1)   *S. tenellum* — See above.
(2)   *S. Kunthii* — Burma, Assam, Bengal.
(3)   *S. uliginosum* — Queensland, Southern China (Kwantung), Ceylon.
(4)   *Additional species* — 3 species in Queensland, 5 species in North Australia, 1 species common to North Australia and Queensland.

B) *Stylidium* subg. *Alsinoidei*

(1)   *S. alsinoides* — See above.
(2)   *S. tenerrimum* — North Australia.

The Stylidiaceae themselves tabulate in this manner (2),

A) Subf. Donatioideae

(1)   *Donatia* — 2 species in Tasmania, Stewart Island, New Zealand. Fuegia to Central Chile.

B) Subf. Stylidioideae

(2)   *Stylidium* — About 150 species throughout Australia, massive in West Australia; stray species in Malaysia and the Tropical Far East.
(3)   *Phyllacne* — 3 species in New Zealand and Fuegia.

(4) *Forstera* — 4 species in Tasmania, New Zealand and Stewart Island.

(5) *Oreostylidium* — 1 species in New Zealand.

(6) *Levenhoekia* — 6 species in Victoria, South Australia and West Australia, more numerous in the last.

This, then, is the record (Fig. 84), and it suggests the following, (a) The Stylidiaceae are doubtless antarctic, not less so than the Restionaceae and the Centrolepidaceae; (b) The Malaysian and Asiatic outlier of all these families is of the same nature, and obeys the same general factors; (c) The Malaysian species of Subg. *Andersonia* run a channel rooted by the "Soenda Coign"; those of Subg. *Alsinoidei* by the "Papua Coign"; (d) These two "Coigns" are connected in the south by a base-line which invests the whole of Northern Australia, and the region Java—Soemba in addition. This region contains species of unsettled affinity, which indicates that it mostly likely rates as a secondary important center of speciation and origin *).

Additional observations are suggested in this manner, (e) Western Australia is entered in this case by the south much sooner than by the north (see *Levenhoekia* and compare with *Barringtonia coccinea*, p. 418); (f) The range of Subg. *Alsinoidei* is a fair match of that of *Astenochloa* (see p. 405).

It is transparent that, far from being a "local" issue of Malaysian dispersal, the outlier dispatched by *Stylidium* northward is a standard contingency of migration. This outlier calls into play homologous developments in the Restionaceae and Centrolepidaceae, not only, but the fundamental division of Malaysia into two sectors, each based upon a "Coign." This outlier, too, is typical of migrations in and around Malaysia taking place in the first epoch of angiospermous dispersal, and invites attention once again upon the' region Java—Soemba. The two species of *Stylidium* in this area have unsettled status which constitutes a neat problem in systematic botany and phytogeography alike.

II. *Drosera*

The genus *Drosera* is represented in Malaysia (3) by five species, which run in this domain the following course of dispersal,

(1) *D. Burmanii* — Boeroe, Southern Celebes, Philippines (Luzon: Bontoc), Soemba, Northern Borneo, Bangka, Billiton, Karimata Islands, Malacca; *outside Malaysia* in Northwestern and Eastern Australia, Northwestern New Guinea, Indochina, India, Ceylon, Southern China, Formosa, Japan.

---

*) The matter is one we cannot consider here at length. We invite the attention of the reader, however, to the fact that centers of secondary origins often arise there, where lines of geologic faulting have place. In this regard, the Dutch West Indies (Curaçao, Bonaire, Aruba) and the Dutch East Indies (to extent of the Lesser Soenda and Java at least) are on a par. See further, speciation taking place at the Andaman Arc, the Comoros and the Mozambique Channel, etc.

(2) *D. indica* — Celebes, Philippines (Luzon, Culion) *), Borneo, Madoera, Sumatra, Malacca; *outside Malaysia* wide in the Australian domain (New Zealand and Tasmania excepted), Southeastern New Guinea, Indochina, India, Ceylon, Africa (Sudan to Nigeria across Central Africa) **); Southern China, Formosa.

(3) *D. peltata* — Timor, Lombok, Celebes, Philippines (Luzon: Bontoc, Benguet, Zambales; Mindoro: Mt. Halcon), Eastern Java; *outside Malaysia* in Tasmania, South and Eastern Australia, Indochina, Siam, India, Ceylon, Southern China, Formosa.

(4) *D. petiolaris* — New Guinea (considered here as Malaysia in a broad sense); *outside Malaysia* in the islands of the Gulf of Carpentaria, Northern Australia.

(5) *D. spathulata* — Philippines (Luzon: Albay; Mindoro: Mt. Halcon), Borneo (Mt. Kinabalu); *outside Malaysia* in Stewart Island, New Zealand, Tasmania, Eastern Australia, Formosa, Southern China, Riu-Kiu Islands, Southern Japan.

Let us ascertain the position of these five species under *Drosera* as a preliminary to reconstructing their affinities and identifying the main channels of their migrations. This is the score (4, 5),

(1) *D. Burmannii* — Belongs to *Drosera* sect. *Thelocalyx* consisting of two species only, (i) *D. Burmannii*, as described; (ii) *D. sessilifolia* in Brazil (Minas Geräes, Ceará, Piahuy, Goyaz) and the Guianas.

(2) *D. indica* — Belongs to *Drosera* sect. *Arachnopus*. Only two species go with this group, (i) *D. Adelae* in Northwestern Australia; (ii) *D. schizandra* in Northern Queensland.

(3) *D. peltata* — Belongs to *Drosera* sect. *Polypeltes*. This is a fairly large affinity of some 20 species, most of them restricted to northern and southern West Australia, *D. peltata* being the lone extra-australian member of the Section.

(4) *D. petiolaris* — Belongs to *Drosera* sect. *Rossolis* subsect. *Lasiocephala*. Only another species falls here, *D. caledonica* (New Caledonia).

(5) *D. spathulata* — Belongs to *Drosera* sect. *Rossolis* subsect. *Eurossolis*, a large group with nearly pandemic range. Typic of its distribution are, together with *D. spathulata*, the following two species, (i) *D. madagascariensis* — Madagascar, Transvaal, Tanganyka, Angola, Nigeria; (ii) *D. capillaris* var. *typica* — British Guiana, Trinidad, Cuba, British Honduras, United States (Florida and Texas to Southern Virginia); var. *brasiliensis* — Brazil (Santa Catharina, Rio Janeiro, Bahia, Goyaz).

The records in question may not be complete, and they may be freely discounted to allow for discrepancies in taxonomic interpretations, errors,

*) The reader may recall that a circuit Culion—Luzon is also part of the tracks of *Buchnera, Stackhousia*, etc. As already stated, we regret we cannot give more time and attention to the phytogeography of the Philippines in particular, and Malaysia in general.
**) A track *Nigeria—Sudan—India—Ceylon—New Guinea*, or the other way around, is commonplace. See the Celastraceae, Sterculiaceae Mansonieae, Erythroxylaceae, *Notelaea*, etc.

etc. As they read, nevertheless, they may be confidently taken to be representative.

Three of the Malaysian species, D. *indica*, D. *peltata*, and D. *petiolaris* stem from the Western Polynesian angiospermous gate. They may be strictly "antarctic" or not (see notes p. 362), which does not matter here. It is likely, at any rate, that they enter Malaysia in the manner of *Stylidium*, that is, *from the south*. We may then reconstruct their tracks (Fig. 85, 86) across Malaysia, and beyond, from this direction, as follows,

(a)   D. *indica* — The track takes in the whole front *New Guinea—Celebes— Madoera—Sumatra—Malacca—Borneo*. With this front well assured, the track follows the two standard channels of Malaysia, (i) Through Celebes to the Philippines, next reaching Formosa and Southern China; (ii) Through Malacca to Indochina, India, Ceylon, Africa to Nigeria.

(b)   D. *peltata* — The track invests the front *Timor—Lombok—Celebes— Eastern Java*, and next proceeds through the standard eastern channel to Luzon, Formosa, Southern China; through the standard western to Indochina, Siam, India, Ceylon.

(c)   D. *petiolaris* — The track stops with New Guinea. Were it possible to work out the affinities of this species outside Subsect. *Lasiocephala*, we could extend the track in standard manners much outside New Guinea.

It is manifest that these three species run standard routes. These routes belong doubtless to the first age of angiospermous migration. Worthy of notice is the fact that the dispersal of D. *peltata* does not seem to touch the Southern Philippines (e.g., Mindanao) but to enter this archipelago in the north at two points (Mindoro, Zambales) next following through Bontoc and Benguet to Formosa. *This is a common occurrence in plant migrations of angiospermous forms that appertain to the first stage of wandering (Late Jurassic to Early Cretaceous), whether the line in question is run from the south (as in the present case) or from the north.* It seems probable that the tie Mindoro—Formosa was not worked over the marine strait between Luzon and Formosa, but over the stretch Zambales—Mindoro and its approaches *). Notice how different is the behavior of mangroves and near-mangroves migrating in the second stage of angiospermy (from the Mid-Cretaceous to the Early Tertiary). These plants are stopped by the strait between Luzon and Formosa (Dipterocarpaceae), and come into the Philippines prevailingly in center from the direction of Borneo. They may be deviated southward at a front Guimaras—Tawitawi (see *V. papuana*, p. 428).

Manifestly, the dispersal of these three species of *Drosera* in Malaysia is not a Malaysian problem. *It is not a Malaysian problem because this dispersal makes no sense until and unless it is approached on a world-wide scale.*

---

*) See the peculiar coincidence in the *western* run of the track of D. *peltata* around Luzon, and the same course in the tracks of *Stackhousia* (see p. 342). The phytogeography of the Philippines and lands adjacent could easily fill a book of its own. We regret we cannot give to it detailed attention beyond what is already written in these pages. We strongly commend the subject to other workers.

The classification we follow tells us that *D. Burmannii* is part of a petty affinity reaching the New World with *D. sessilifolia*. To study this affinity throughout we ought to open here a lengthy parenthesis, which is inexpedient. Accordingly, we will moot alternatives leaving it to the reader to bring the matter to a conclusion if he should feel so inclined.

The alternatives before us are the following, (i) The affinity in question is rooted in the Indian Ocean of our maps; or (ii) It is on the contrary grounded by a center in the Western Pacific. To put this in other words, (i') The *genorheitron* of *D. Burmannii* and *D. sessilifolia* is from the *Gondwanic Triangle;* or (ii') This *genorheitron* is from the *Neocaledonian Center*.

Under either one of these alternatives we may freely begin to run the tracks from a base-line set in Northern Australia, in the knowledge beforehand that Northern Australia could be reached by standard channels of migrations either from the west or the east. Taking our start from this base-line we can state with some assurance that *D. Burmannii* espouses both "Coigns" of Malaysia. Part of the track leaves Northern Australia and New Guinea to take in Boeroe and Southern Celebes, next the north of the Philippines, Formosa and Southern China. The balance of the track enters Soemba always from Northern Australia, follows on to Bangka, Billiton and the Karimata Islands, Malacca, Indochina, India and Ceylon; Borneo might have been entered both from the Philippines or Malacca, or by either one of these lands.

As a third possibility, we may offer the following route, Ceylon—India—Indochina (from here to Southern China, Formosa, Philippines and Japan)—Malacca—Bangka—Billiton—Karimata Islands (from here to Borneo and perhaps again the Philippines)—Soemba—Northern Australia. It is further suggested that Celebes, Boeroe and New Guinea might have been entered from Northern Borneo in a manner homologous of the track of *Vatica papuana* (see p. 428) *).

It will readily be seen that the third possibility we outline in the above paragraph is predicated on the assumption that *D. Burmannii* and its *genorheitron* are strictly of Gondwanic origin, which is to say that the track comes in from the west, hitting Ceylon first and foremost. While not to be ruled out altogether, this possibility is rather more remote. Everything indicates that this group of *Drosera* enters Malaysia from the direction of Australia, precisely as does *Stylidium*, and that, though wider, the migrations of *D. Burmanii* are ultimately homologous of those of *Stylidium*, Centrolepidaceae, Restionaceae etc.

Manifestly, the dispersal of *D. Burmannii* is a not a "local" Malaysian issue. It cannot be such ever, because this dispersal makes no sense until and unless it is worked back to its ultimate origins in time and space. The mere fact that we have studied it here without working it back *fully* had brought us to consider alternatives rather than to reach finality. This is once again proof that there are no "local" issues in phytogeography.

---

*) These same alternatives may of course apply to *D. indica* and *D. peltata.*

There remains before us *D. spathulata* member of a very broad affinity (*Drosera* subsect. *Eurossolis*). We have no reason to run the dispersal in any other way but from south northward, because *D. spathulata* takes us first to Stewart Island, Tasmania and New Zealand. From these centers on the track runs through Eastern Australia then invades in disconnection the southeastern tip of Luzon (Albay), Mindoro and Northeastern Borneo (Mt. Kinabalu). This is one of the most ancient cores of Malaysia, and characteristically, this core is approached from the south in disconnection, as if the rest of the Philippines did not exist at all. Mangroves and near-mangroves behave differently as we know. Once in Luzon, the track follows northward to Formosa, Southern China, the Riu-Kius and Southern Japan. *Worthy of careful notice is the fact that the channel of migration involved, though parallel with the eastern standard one of Malaysia, is not exactly identical with it.* The channel revealed by the course of *D. spathulata* skirts the lands on the Western Pacific in the first place, barely entering Luzon, Mindoro and Mt. Kinabalu.

Drosera spathulata has the longest track of all the Malaysian species. This track takes us straight from Stewart Island and New Zealand to Central Japan (Honshu) all along the now crumbled shore (see further p. 69) of the Western Pacific and its hinterland. This species belongs to the wide-ranging Subsect. *Eurossolis*, and as it comes to a stop in Central Japan, it leaves place to other species in the same affinity which run this time tracks of this nature,

(1) *D. rotundifolia* — Yakushima Island and Japan throughout to the Kuriles and Sakhalin, Kamchatka, Northern China, Manchuria, Corea, Siberia throughout to the Caucasus and Syria (Lebanon); Bulgaria, Northern Balkans, Transsylvania, Central and Western Europe to Iceland, Spain, Northern Portugal; North America from Alaska to Idaho and Central California; Ontario to Newfoundland and Labrador, Alabama and South Carolina, Cuba and Santo Domingo.

(2) *D. anglica* — Central Japan, Kuriles, Kamchatka, Sakhalin, Hawaii; Siberia throughout to Central and Northwestern Europe (Scotland and Ireland); North America (Southern Alaska to Northern California, Idaho, Montana, the Great Lakes region, Eastern Canada and Newfoundland).

The ranges of *D. spathulata* and *D. rotundifolia* overlap (Fig. 86) between the island of Yakushima and Central Japan. It is in this region, consequently, that *Drosera* sect. *Rossolis* swaps horses, as it were, dropping the "warm" *D. spathulata* to ride the "cold" *D. rotundifolia* and *D. anglica*. This is not to surprise us, because (see p. 247) it is right by this vicinity that *Carex* also behaves like *Drosera*. In other words: An important landmark of dispersal is set between the northernmost islands of the Riu-kius proper, Amamioshima, and the southernmost insular appendage of Japan, Yakushima. In this vicinity lies an important secondary center of "cold" speciation both of *Drosera* and *Carex*.

Amamioshima stands as the northernmost limit (6) of forms such as *Panda-*

nus tectorius, Alchornea liukiuensis, Croton Cumingii, Osteomeles anthyllidi-folia and *Sporolobus virginicus*. It is understandable that some of these species do not range beyond Amamioshima, because their ecology is essentially tropical. It is strange, on the other hand, that *Osteomeles anthyllidifolia* and *Sporolobus virginicus* stop here. The latter, particularly, is by no means tropical, and ought to find congenial habitats throughout the whole of Japan. Moreover, it is a pronounced halophyte widespread in the warm and temperate regions of both hemispheres. It is difficult to find reasons why its seeds should not have been carried farther north than Amamioshima by winds, birds, marine currents, etc., etc.

Faced by this puzzle, let us investigate whether the record contributes anything toward its solution. *Osteomeles*, for instance, is distributed as follows,

(1) *O. Schwerinae* — Burma, China (Yunnan, Szechuan, Kansu).

(2) *O. anthyllidifolia* var. *subrotunda* — China (Kwantung), Formosa, Amamioshima.

(3) *O. boninensis* — Bonin (Ogasawara) Islands.

(4) *O. lanata* — Bonin (Ogasawara) Islands.

(5) *O. anthyllidifolia* — Hawaii.

This dispersal speaks for itself. Instead of leaving Amamioshima to reach Yakushima, *Osteomeles* migrates straight to Hawaii through the Bonin Islands. This can only mean that the boundary between Amamioshima and Yakushima is more important than it appears to be. This boundary is not a "Japanese" landmark, but something else, which interests a large segment of the Pacific and profoundly influences unrelated groups, like *Osteomeles, Drosera* and *Carex*.

Proof greets us here once more that dispersal does not bear being studied piecemeal, and even less can it be understood if we credit it to the ministration of haphazard agencies of dissemination. Some of the plants in this vicinity fail to cross to Japan and go to Hawaii *(Osteomeles)*, while others cross to Japan, but do not reach Hawaii *(Drosera rotundifolia)*, and others still reach both Japan and Hawaii *(D. anglica)*.

We may dismiss all this as coincidences, and conclude that the plants in question behave, after all, "mysteriously" as do many others all over our maps. So facile a view of the matter, however, cannot satisfy a cogent inquirer. If we consider that these vagaries take place there, where appears to be located a secondary center of "cold" flora, we must believe as most probable that right in this vicinity there once stood some now crumbled range. Whatever be its nature in detail, this range was significant for dispersal, and it is likely that there were different ecological conditions within its boundaries. Ranges of this nature would account for the circumstance that it is at this spot of the modern map that certain species were re-distributed to this or to that point of the compass.

Drosera tellingly illustrates the nature and significance of this lost land in the Northwestern Pacific. *Drosera spathulata*, a "warm" element, comes to a dead stop in this vicinity, but is replaced by two "cold" species, one of which *(D. rotundifolia)* turns up at Yakushima and next reaches Japan, while the other *(D. anglica)* also reaches Hawaii. *Osteomeles*, lastly, fails to reach Japan and fares straight to Hawaii.

Stopped by a mighty, if now invisible, boundary between Amamioshima and Yakushima and deflected in its course toward Hawaii, the *genorheitron* of *Osteomeles* fared in the end all the way to the New World across the Pacific. Taxonomists of a former generation used to classify as species of *Osteomeles* certain American plants which modern botanists now classify under other genera. It stands to reason that a phytogeographer is not primarily concerned in changes in classification of this nature. The *genorheitron* that centers around *Osteomeles* becomes neither looser nor more closely knit, if we decide that it consists of one instead of ten genera, or the other way around. Affinities are not for man to change with words.

The genera segregated out of *Osteomeles* in the New World tabulate at any rate as follows,

A) *Malacomeles*

(1) *M. denticulata* — Mexico (Chihuahua, Nuevo León), Guatemala.
(2) *M. nervosa* — Mexico (Nuevo León), Guatemala.

B) *Hesperomeles*

(1) *H. obovata* — Costarica (Mt. Irazú).
(2) *H. chiriquiensis* — Panama.
(3) *H. oblongifolia* — Colombia (Magdalena).
(4) *H. obtusifolia* — Colombia (Santander, Norte de Santander), Peru (Huánuco).
(5) *H. pernettyoides* — Colombia (Norte de Santander), Venezuela (Mérida), Peru (Junín), Bolivia (La Paz).
(6) *H. glabrata* — Colombia (Norte de Santander), Ecuador.
(7) *H. nitida* — Colombia (Santander).
(8) *H. latifolia* — Ecuador.
(9) *H. cuneata* — Peru (Ayacucho).

Strange to say, the affinity of *Osteomeles* — an important group in the Rosaceae — could fare at ease between Burma and Bolivia, and take the Pacific in its strides without the slightest apparent difficulty. This same affinity, however, was unable to negotiate a few miles of sea between Amamioshima and Yakushima.

Let us take notice how far ranging are, at the other hand, the migrations of the segregates of *Osteomeles* in the New World. *Malacomeles* turns up at the classic door of entrance of transpacific migrations (Mexico: Chihuahua, Nuevo León) but at the same time is also in Guatemala much to the south.

Hesperomeles staggers itself farther south still, landing by one of the classic heights of Costarica (Mt. Irazú) and in Panamá. We might anticipate that this genus follows in standard fashion along the western sector of Colombia (Antioquia, Chocó, etc.) but this anticipation comes to naught by the facts. *Hesperomeles* abandons the Pacific and moves over eastward instead, so that we find it next in Magdalena, Santander and Norte de Santander in Colombia; Mérida in Venezuela. From this point southward the track becomes speculative, but with reference to *Purdiaea, Brunellia, Protium*, etc., we have no reason not to draw it through Ecuador to Central Peru (Huánuco) and finally Bolivia (La Paz). As an alternative, we might suggest that Ecuador, or Bolivia, is directly entered from the Eastern Pacific, which does not invite speculation, because we know from a number of other genera (*Lepechinia, Cuscuta, Nama*, etc.) that there were lands leading to and from Hawaii in this vicinity.

Plainly, the Andes are not the "Bridge" which modern geography appears to offer for the comfort of our phytogeographic dreams. *Hesperomeles* does not follow the Andes, *but plays around them.* Observing the routes taken by *Malacomeles* and *Hesperomeles* we are forthwith reminded of the never explained observation by Focke, that the populations of *Rubus* in Mexico, Ecuador and Peru could be connected with no other group but certain species endemic to the Himalayas. *Rubus, Osteomeles, Malacomeles, Hesperomeles* are one in dispersal.

Sporolobus virginicus is a polymorphus species. It occurs in South Africa with two forms ranging from the Cape to Namaqualand and Natal; a variety endemic to the Mediterranean is so close to the form of Natal that standard manuals (7) identify the whole as a single species by the name *S. pungens*.

Sporolobus virginicus (or *S. pungens*, whichever be the better epithet) is reported from Australia and the Philippines (Luzon: Cagayan, Rizal). It ranges westward across Malaysia to Ceylon, northward to Amamioshima. It fails to cross over to Yakushima, but fares straight to Hawaii, where it is represented by a variety, *phleoides*, also known from Polynesia (Guam, New Caledonia, Loyalty Islands). Migrations of this lilt are not unusual. As a matter of fact, they parallel in the sector Australia—Philippines—Yakushima the dispersal of *Drosera spathulata*.

Contrary to expectations, *S. virginicus* does not occur in the Western United States or Mexico, but in the east of the New World from Brazil northward through the Caribbeans to the United States.

It is quite possible that the populations of the Caribbeans and the United States do not stem from a transpacific migration, rather from a track running northwestward out of South Africa, or still another track due westward from the Mediterranean. We cannot consequently prove that *S. virginicus* crossed the Pacific in the wake of the affinity of *Osteomeles*, entering the Caribbeans in the manner of *Washingtonia* next discussed in this manual but without leaving trace of its passage in California. *Sporolobus*, then, yields nothing final to our quest; its migrations are too broad, and admit of too many alternatives to let us know whether it did, or did not, cross the Pacific to reach the Caribbeans from the Japanese island of Yakushima.

However, *it is the law in phytogeography that what a plant does not tell another plants does reveal.* For example, the roxburghiaceous genus *Croomia* is distributed as follows,

(1) *C. kiusiana* — Japan (Amamioshima, Yakushima, Southern Kyushu).

(2) *C. pauciflora* — United States (Florida, Georgia, Alabama).

Croomia, of course, is but one of scores of migrants across the Pacific. We refer to it merely because it connects Yakushima with the Ozarkian—Appalachian center of the United States in a cleancut manner. It is clear that if the break between Yakushima and Amamioshima is not absolute — witness *Croomia* faring across it to Southern Japan — still this break is important to determine the fate of swarms of transpacific migrants. Latitude 30° North is indeed significant in the tale of transpacific migrations. Manifestly, Alaska was not *the* "Landbridge" between the Old and New World which the outlines of current maps may induce us to take for granted. *Sporolobus,* then, *probably* also crossed the Pacific in the wake of *Croomia,* only that it settled in the Caribbeans rather than in the Southeastern United States. Let us observe how foolish is the notion to insist, right or wrong, upon "proof" of this and that, as if only "proof" counts in science. This foolish notion could arise solely in the mind of men who know nothing whatever of well rounded, critical generalities. *Croomia* yields *proof; Sporolobus* furnishes only *probabllity,* but as we have as regards the one as well as the other sound generalities of interpretation the essences behind *proof* and *probability* are in this case of the same order. The line that separates the two becomes indeed so thin as to count for precious little. We should feel like smiling, if somebody were to try to convince us that this line, on the contrary, is vitally important, because only "facts" counts in "science." *)

III. Washingtonia

Washingtonia is a genus of Palms consisting of no more than three species narrowly ranging in California, Arizona and adjacent Mexico. The best known of these species is *W. filifera* (California: West and north side of the Colorado Desert; Arizona: Kofa Mts. in Yuma County; Mexico: Northern Baja California). This stately plant thrives at or above the old beach-line of the one-time interior sea that covered part of Southern California.

---

*) A curious psychological quirk is revealed by authors (see, for instance, MANGELSDORF as discussed by CROIZAT in Rev. Argent. Agron. 15: 160. 1948) who, while taking for granted definitions and methods verging on the purest twaddle, insist throughout that only "proofs" and "facts" are needed in "science". The essence of culture is in a critical approach to all values of the mind, whether these be couched as well digested definitions or contained by implication within facts and probabilities.

It should be difficult to find elsewhere on earth a "problem" quite as "local" as *Washingtonia*. This genus is patently a relic from ages past, and we could write at length, perhaps most beautifully, upon the "Concept of species-senescence" applied to the "interpretation" of the modern range and peculiarities, generally, of this Palm. We could further contribute a work upon the strange endemics of the deserts of Southern California, and "prove" that these endemics stem in reality from "Tertiary" local conditions. In short, *Washingtonia* could be handled any odd way we like, always without leaving the boundaries of Arizona and California.

We will approach the "*Washingtonia* problem" in a slightly different spirit. We will not worry, whether the problem in question is "local" or not, merely find out first what is the distributional and genorheitral background of this Palm, next reaching any odd conclusion the facts warrant.

The classification of Palms is very confuse, but compiling from the works of different authors (8, 9, 10) we may hope, perhaps, to secure a workable conspectus of the affinity including *Washingtonia*, as follows,

A) New World genera

(1) *Sabal* — 24 species in Mexico (Nayarit, Sinaloa, Sonora, Tamaulipas, Zacatecas, Oaxaca, Yucatán), United States (Western Texas, Louisiana, Florida, Georgia to the Carolinas), West Indies (Bermuda, Cuba, Jamaica, Santo Domingo, Haiti, Portorico, Trinidad), Central America (British Honduras, Guatemala), South America (Colombia; Venezuela: Zulia, Trujillo, Carabobo, Yaracuy).

(2) *Serenoa* — 1 species in the United States (Florida, Mississipi, South Carolina).

(3) *Acoeloraphe* — 1 species in Honduras, Cuba, Florida.

(4) *Brahea* — 4 species in Mexico, (Michoacán, Sinaloa, Veracruz, Nuevo León), United States (Western Texas).

(5) *Erythea* — 7 species in Mexico (Baja California, Sonora, Sinaloa).

(6) *Copernicia* — 30 species in Cuba, Santo Domingo, Dutch West Indies, Colombia, Venezuela (Guárico, Bolívar, Anzoátegui, Monagas), Brazil (Matto Grosso, Bahia, Piahuí), Paraguay, Argentina (Orán, Formosa), Uruguay.

(7) *Washingtonia* — 3 species in the United States (California, Arizona), Mexico (Baja California).

(8) *Rhaphidophyllum* — 1 species in the United States (Florida, South Carolina).

(9) *Trithrinax* — 5 species in Brazil (Matto Grosso, Rio Grande do Sul), Uruguay, Paraguay, Argentina, Eastern Bolivia.

(10) *Acanthorhiza* — 5 species in Mexico (Sinaloa, Guerrero, Oaxaca, Yucatán), Panamá, Costarica, Colombia.

(11) *Hemithrinax* — 1 species in Cuba.

(12) *Thrinax* — 11 species in Cuba, Santo Domingo, United States (Florida), Bahama, Anguilla, Honduras.

(13) *Coccothrinax* — 17 species in Cuba, Santo Domingo, Portorico, Martinica, Guadeloupe, Bahamas, United States (Florida).

(14) *Chryosophila* — 6 species in Mexico (Guerrero), Guatemala, Costarica, Panamá, Colombia.

B) Genus common to Polynesia and the New World

(15) *Pritchardia* — Over 15 species in the Tuamotus, Samoa, Fiji, Hawaii; 1 species in Cuba and the Isle of Pines.

C) Old World genera

(16) *Pritchardiopsis* — 1 species in New Caledonia.

(17) *Corypha* — 8 species in Northern Australia, Celebes, Philippines, Java, Cochinchina, Burma, Andaman Islands, India, Ceylon.

(18) *Livistona* — 24 species in Central, Eastern, Western and Northern Australia, New Guinea and the Solomon Islands, Celebes, Moluccas, Philippines, Java, Malacca, Cochinchina, Burma, India, Southern China (Kwantung), Formosa, Riu-kius, Bonin (Ogasawara) Islands, Japan.

(19) *Licuala* — About 100 species in Queensland, New Guinea, Solomon Islands, Moluccas, Celebes, Philippines, Borneo, Java, Sumatra, Malacca, Indochina, Burma, Southern China (Kwantung).

(20) *Trachycarpus* — 6 species in the Himalayas, Burma, China (Yunnan to Shantung).

(21) *Teysmannia* — 1 species in Borneo, Sumatra, Malacca.

(22) *Rhaphis* — 6 species in Indochina and Southern China (Kwantung).

(23) *Nannorhops* — 3 species in India, Beluchistan, Afghanistan.

(24) *Pholidocarpus* — Several species in the Moluccas, Celebes, Borneo, Sumatra, Malacca.

(25) *Chamaerops* — 1 species in Iran (Persia), Palestine, Italy (Sicily, Sardinia, Tuscany: Mt. Argentaro), Southern Spain and Portugal, Algeria, Morocco.

It is obvious that *Washingtonia* is not, nor could it ever be, a "local" problem. Its affinity is large, and migrates (Fig. 87) most extensively. *Washingtonia* may be a relic, crippled by a sorry case of "Species senescence," but this disease is function of a dispersal that knows the Southwestern United States only at second hand.

Washingtonia is a well marked genus in the opinion of taxonomists specialized in the classification of Palms, but so closely allied with *Pritchardia* nevertheless that it was formely currently treated as a subgenus of the latter. Considering these two genera as a single distributional and phylogenetical unit, as we must, we have before us a single perfect track leading from Polynesia to Hawaii; from Hawaii to the Western United States and Mexico; from the Western United States and Mexico to Cuba. This track is as cleancut as that of *Perrottetia*, not only, but perspicuous in every respect. *Pritchardiopsis*, the connecting-link between *Pritchardia* and *Licuala*, is endemic to New Caledonia, which all is as it ought to be according to rote. It is by New Caledonia that is set the hinge of the dispersal of the *Pritchardia—Licuala* affinity; the former conventionally streams out of this hinge ultimately to reach the Caribbeans, the latter on the contrary fares to Malaysia and Southern China always according to rule *).

Viewed against the proper background of time and space, the "*Washingtonia* problem" is certainly minor part of the issues involved in the migrations of the *Washingtonia* affinity. Its species are "senescent," because they are relics lingering on under conditions that, though still livable, are no longer as widespread and perfect as they were, let us say, *in the Early Cretaceous*. Duly to appreciate the age in question, we can certainly not restrict ourselves to study "Californian conditions."

Most interesting are some of the ties and ranges exemplified by the Palms in *Washingtonia's* kinship, witness, (a) Connections between Honduras and Cuba, Yucatán and Cuba, are revealed by the dispersal, respectively, of *Thrinax Wendlandiana, Acoeloraphe Wrightii, Sabal Japa, Sabal mayarum*. Connections of the kind are standard, but it is useful to have them stressed once again by four Palms the ecology of which is well known. We can thus visualize with fair accuracy the ecology of lands that long millions of years ago stretched between Southern Mexico, Central America and Cuba. That the ecology in question was varied is revealed by plants other than Palms, witness for example (i) *Drosera capillaris* var. *typica* — Cuba, British Honduras; Trinidad, British Guiana, United States (Texas and Florida to Virginia). In addition, we have renewed evidence to the key position held by Cuba in inter-caribbean dispersal, because this island pointedly brings together the United States as far as Virginia, and the approaches of Southern Mexico and Guatemala. Most interesting still is the fact that connections, which in certain cases call Cuba into play most directly (see *Purdiaea*, p. 343), stress at other times lands in Cuba's vicinity. Classic in this respect is the distribution of the burseraceous genus *Dacryodes*, whose Sect. *Archidacryodes* is of two species with the following range, (i) *D. excelsa* — Lesser Antilles (Martinica to the Virgin Islands), Portorico; (ii) *D. peruviana* — Peru (Huánuco). In this close interplay of tracks, some disconnected only shortly and over shallow seas, others, on the contrary, widely interrupted over deep waters, we read proof that the Caribbeans were entirely different in the past from what they are today, and that, accordingly, it is futile to argue about nugatory "Landbridges" of Tertiary age in order to elucidate the dispersal taking place in these quarters; (b) In *Sabal* there is manifest a classic disconnection between British Honduras (or Guatemala) and Colombia (or Venezuela).

---

*) Well worthy of notice is the disconnection Guadeloupe-Trinidad in the ranks of the *Washingtonia* aggregate.

There is good indication that the distribution of *Sabal mauritiaeformis* in Venezuela stops more or less in Carabobo and Yaracuy, which hints at the outlines of a crumbled front extending from these states westward to Guatemala or its immediate approaches. *Sabal, Tetragastris, Hesperomeles, Gyranthera, Gitara* *) all rhyme to the same insofar as the phytogeography of the triangle Southern Mexico (Oaxaca)—Panamá—Venezuela; (c) As regards the Old World, the affinity of *Washingtonia* brings to us a number of interesting distributions. One such is, (i) *Nannorhops'*, the range of which stresses India, Beluchistan, Afghanistan. This range is a vital one of phytogeography; it is in it that are stranded mangroves; passes a major channel of migration of cold genorheitra leading northward; come to an abrupt stop the threads of "European" flora; (ii) *Chamaerops'*, which follows the track of *Nannorhops* in a standard pattern to the Western Mediterranean; (iii) *Trachycarpus'*, that illustrates a classic track between the Himalayas and Northeastern China; (iv) *Corypha's* which departs from Northern Australia to fare the western major channel of Malaysia (Java, Cochinchina, Burma, India, Ceylon), putting up meantime an appearance in the Philippines *via* Celebes, and occupying the Andaman Islands.

Inasmuch as *Sabal* has induced us to cast a glance at Venezuela, and to formulate, or restate, certain observations of general interest in regard of the flora of the Caribbeans, we may, perhaps, tarry a little longer in these quarters.

For example, in the genus *Myrica* is known a Sect. *Faya* with three species that range in this manner,

(1) *M. Faya* — Azores, Madera, Canary Islands, Southern Portugal.

(2) *M. inodora* — United States (Florida, Alabama, Mississipi).

(3) *M. californica* — United States (Washington to California).

The tie, Portugal—Northwestern United States (Washington), is not forged by a direct line, but by an approach leading to the Southeastern United States (Ozarkian—Appalachian Node) and California first, next swerving northward. Insofar as the living allies of *Washingtonia* are concerned, the tie works exactly in the opposite. *Washingtonia* reaches the New World (as do its allies), from the direction of the Pacific, lands in California but there stops, giving free passage to its next-of-kins which follow, they, to the Carolinas.

The Sarraceniaceae are distributed in this manner,

(1) *Heliamphora* — 6 species in Venezuela and British Guiana, all confined within the Pacaraima System.

(2) *Sarracenia* — 8 species in the Eastern United States and Canada.

(3) *Darlingtonia (Chrysamphora)* — 1 species in Northern California and Southern Oregon.

These three genera are absolutely disconnected, but significantly located

---

*) *Gyranthera* is a bombacaceous genus known so far by two species, one in Panamá, the other in Venezuela (Aragua, Miranda, Carabobo); *Gitara* is a member of the Euphorbiaceae, also known from two species, one in Panamá, the other in Venezuela (Carabobo). The coastal cordillera of Western and Central Venezuela is thus immediately involved with Panamá, and beyond with Central America and Southern Mexico. The Venezuelan state of Carabobo is a landmark of potentially considerable significance.

all the while. *Darlingtonia* has its southernmost record of distribution in California (Plumas County) there precisely, where also occur *Utricularia intermedia* and *U. minor*. These two Bladderworths reached this spot from the north *), *Darlingtonia* doubtless from the opposite direction. *Heliamphora* colonized the Pacaraima region together with the first Angiosperms to reach the spot when this region was in all probability a riverine delta. *Sarracenia* dwells to this date in an ecological ambient which has something at least in common with the ambient that first greeted *Heliamphora* at the Pacaraima **). We may figure the distribution of these three genera by a line *Venezuela—Eastern United States—Western United States,* but we might do better, we believe, if we conceive of the migrations being keyed up to the Eastern Pacific in the genorheitral stage along the lines shown by the map attached (Fig. 88). We have a precedent for this understanding of the sarraceniaceous dispersal in the distribution of the vacciniaceous genus *Gaylussacia* which colonizes the New World, approximately as follows,

A) North American massing

(1) *G. dumosa* — Florida to Louisiana and New York.

(2) *G. baccata* — Georgia to Newfoundland, Nova Scotia, Saskatchewan.

B) Western South-American massing

(3) *G. buxifolia* — Colombia (Santander), Venezuela (Mérida; vicinity of Caracas).

(4) *G. loxensis* — Ecuador.

(5) *G. peruviana* — Peru (Amazonas).

C) Brazilian massing

(6) *G. cacinis* — Venezuela (Pacaraima Mountains).

(7) *G. insignis* — Brazil (Minas Geraës: Serra do Cipó).

---

*) We regret we cannot deal with *Utricularia* in the pages of this manual. *Utricularia intermedia* and *U. minor* are in one and the same case with *Drosera rotundifolia* and *D. anglica* (see p. 438) They are, in other words, "northern" species out of a definitely non-northern *genorheitron*.

**) The certain, though somewhat cryptic, homology that exists throughout between the habitats of *Heliamphora* and *Sarracenia* is bared most clearly this time by the species of the theaceous genus *Bonnettia*. In this group are included over six species endemic in the Pacaraima range, which thrive there at altitudes of between 1000 and 2200 meters (ca. 3000—7000 feet) above sealevel. Other species still occur today (see, for instance, *B. anceps*) in sandy habitats by the seashore or in arenaceous swampy grounds of Eastern Brazil (Bahia, Rio de Janeiro). The forms that eventually evolved into the modern species of the Pacaraima found in this center *in the Early Cretaceous (when not earlier)* habitats of a type which their congeners still haunt *at this hour* much farther away in Eastern Brazil. We regret we cannot expatiate on this phase of geologic ecology. See further (p. 128) for the persistency of geologic associations *Stewartia, Magnolia* and *Illicium* in a relic-station of Florida.

(8) G. *octosperma* — Brazil (Rio de Janeiro: Serra dos Orgãos).

(9) G. *brasiliensis* — Paraguay (Sierra de Maracayú), Brazil (Paraná, Minas Geraës).

(10) G. *pseudogaultheria* — Brazil (Paraná).

(11) G. *amazonica* — Brazil (Pará).

We make no claims for the taxonomic accuracy of this record, knowing full well that no two botanists agree, how to classify the Ericaceae Vaccinioideae in particular. The issue, however, is not here. The issue is with the dispersal of a genus which ranges between Saskatchewan and Paraguay, but does not show up in the Caribbeans, or if so — to say the very last — does not teem in that region.

Affinities have been hinted between *Gaylussacia* and the monotypic genus *Malea* (Guatemala, Southern Mexico; (11)) which, whether absolutely clear or not, are suggestive. The fact is that the Vaccinioideae may be for very small part in the New World from the east (i.e., along the track exemplified by *Myrica Faya* and its allies, see p. 446), but massively indeed are there from the west. *Gaylussacia* was certainly not conveyed to the Americas either from the Mediterranean or Western Africa.

The main track, then, is from the Pacific; in thoroughly standard fashion it streams to the Southeastern United States, next spreading northward and again westward. We believe Saskatchewan to have been reached from the east (i.e., along the line Florida—New York leading at one hand to Newfoundland, at the other to Saskatchewan), too, but we should not rule out as absolutely impossible a disconnected range between California, Saskatchewan and the Southeastern United States, though this strikes us as most remote. This part of the migrations, then, gives us the North American massing.

As to the Western South-American, we have before us the track Santander—Mérida—Caracas classic of such groups as stream in from Guatemala and Southern Mexico. If *Gaylussacia* is seemingly unknown there, *Malea* is right on the spot, a lucky coincidence when nothing better. The Pacaraima can be reached in no time out of the triangle Santander—Mérida—Caracas, as can Eastern Peru (compare *Purdiaea* and *Dacryodes*, which in addition invest the West Indies). Here, then, falls everything in *Gaylussacia* that happens to be located within the triangle Santander—Pacaraima Mts.—Amazonian Peru.

The distribution of the Protieae (see p. 388) leaves us in no doubt that many are the paths leading from Amazonian Peru to Paraguay and Eastern Brazil, paths to accomodate primary as well as secondary dispersal. If we so like, we may figure only migrations for *Gaylussacia*, drawing the tracks from Santander to Amazonian Peru, thencefrom to Paraguay and Brazil. Contrariwise, we may postulate a straight entrance for this genus from the Eastern Pacific to a point near Peru and Bolivia, which matters little anyway we choose. This closes the tale, for it covers the Paraguayan and Brazilian end of *Gaylussacia*.

In conclusion, the interplay of three points, Saskatchewan, Mexico (or

Colombia), Southeastern Brazil is nothing unusual *), and this interplay covers everything that interests the Sarraceniaceae and *Gaylussacia*. Like the latter, *Alnus* avoids the Caribbeans, while entering Western Venezuela (Mérida) and the United States simultaneously. *The true issue, then, is never with details, but major tracks throughout. Once these are identified, details follow to solution without difficulty. If, on the contrary, the main tracks are obscure, no amount of details can avail us to make light upon the substances in play.*

To close this parenthesis, we may take notice of two aspects of the migrations of *Vaccinium* having some bearing upon the interpretation we gave of the dispersal of *Gaylussacia*.

There is a report (12) that *Vaccinium macrocarpum* (Newfoundland to the Great Lakes, southward to Virginia and Arkansas) is endemic in the Dutch island of Terschelling. This report is credible as such, whether the track is drawn from the New World to the Old or the other way around does not matter. *Cneorum* reaches the Mediterranean from Cuba, and *Erica* and *Calluna* enter the New World from the Mediterranean. It matters not whether *V. macrocarpum* is in play *sensu strictissimo* or *sensu lato*, because phytogeography is much less interested in formal classification than in understanding higher substances that underlie it.

Vaccinium cylindraceum and *V. maderense* endemic, respectively, to the Azores and Madera are currently adscribed (13) to Sect. *Batodendron*. This Section contains four species endemic to the Eastern United States *(V. stamineum, V. arboreum, V. melanocarpum, V. neglectum)* together with three other, as follows,

(1) *V. leucanthum* — Mexico (Guerrero, Michoacán, Puebla, Veracruz, Tamaulipas, Oaxaca, Chiapas).

(2) *V. stenophyllum* — Mexico (Sinaloa, Nayarit, Jalisco, Veracruz).

(3) *V. cubense* — Cuba (Pinar del Rio, Oriente).

Manifestly, there are channels of migration directly leading from the approaches of the Eastern Pacific to Macaronesia, which channels further immediately interest the Eastern United States, and there is some indication this time that *Vaccinium* rode them from the west going east. *Gaylussacia* is not a miracle, far from it.

Assured by now that *Washingtonia* is not a local issue in dispersal, rather a typical instance of migration to the New World from the Pacific's side, we could close this part of our work. We will keep it open, however, to study some other aspects of the distribution of the Palms, judging this family to be of great value for phytogeographical studies.

---

*) This statement may seem rash, though it is conservative. We may illustrate it using a few species of one of the Sections of the so called *Junci Septati* (see (25) Chapt. II), as follows, (1) *J. canadensis* (Venezuela, United States: Louisiana to the Great Lakes, Canada: Newfoundland westward); (2) *J. guadalupensis* (Guadeloupe); (3) *J. Fauriensis* (Kamchatka, Japan). We need not argue against possible retorts that these species are too "broadly understood," or the group in question is not "sectionally perfect," for the issue is already amply covered throughout the pages of this manual. The Guadeloupe record is noteworthy.

The Old World dispersal of these plants brings before us no longer thin, more or less trite channels of migrations, but a panorama of centers of migration, some very ancient and undoubtedly "senescent," other perhaps less hoary with age. *Licuala*, for example, masses with some 16 species in Malacca alone, the massiveness of the speciation here being in harmony with the importance of Malacca as an early angiospermous center. Fifteen other species of the same genus are in Northern Borneo (Sarawak) and 21 in New Guinea. The distribution of *Livistona* is so effected in Northern Australia that, without danger of wishful thinking, we may visualize the progenitors of these Palms at the shore of the geological sea, or lakes, once covering the center of Australia. *Livistona Mariae* is native to Central Australia in a so called "Glen of Palms," an interesting match of the New World *Washingtonia* still standing watch in glens of Palms over a lost Californian sea.

Doubtless, the Palms have an important center of secondary origin in the Western Pacific, and great many are the regions of the Old and New World (for example, Andean Amazonia) responsible for a considerable amount of local species- and genus-making. However, no region of the earth is so striking as regards these plants as the immediate vicinity of Madagascar. The Seychelles and Mascarenes are tenanted by about ten endemic genera *(Roscheria, Nephrosperma, Verschaffeltia, Phoenicophorium, Lodoicea, Acanthophoenix, Deckenia, Latana, Dictyosperma, Hyophorbe)*. The Mascarenes rank moreover as a cardinal point in the distribution of *Pandanus* the migrations of which we regret not to able to follow in this manual *).

The Phoeniceae are a well marked group covering the lone genus *Phoenix*. This genus consists of about a dozen species ranging to Africa, the Canaries and Mediterranean, India and Southern China. Certain of the species of *Phoenix* are distributed in this manner,

(1) *P. reclinata* — South Africa, Mozambique, Angola, Congo, Uganda, Sudan, Senegal, Ethiopia; Var. *madagascariensis:* Madagascar.

(2) *P. paludosa* — Eastern India, Burma, Andaman Islands, Siam, Indochina, Malacca.

(3) *P. Hanceana* — Southern China (Kwantung, Hainan); Var. *philippinensis:* Philippines (Batan Islands); Var. *formosana:* Formosa.

---

*) The endemic vegetation of most small islands has been sadly depleted in the course of centuries of colonization, and the Mascarenes are not an exception. However, on certain petty islets off the coast of Mauritius (see VAUGHAN & WIEHE in Journ. Ecol. 25: 333. 1937), a typical association is formed by *Pandanus Vandermeschii* and *Latania Loddigesii*, both endemic to the spot. One of these islets, Round Island, covers but 417 acres, and though by now largely bare of vegetation, still exhibits remnants of palm-thickets in which appear endemic *Hyophorbe amaricaulis, Latania Loddigesii, Dictyospermum album* and *Pandanus Vandermeschii*. All these Palms and Screwpines are now sadly "senescent," because of a breaking up of Gondwana which began in the Late Jurassic. If things continue as of today they will become dead in the foreseable future, that is, reach one stage beyond "senescence." If, however, the petty islets in question were to be the center of active re-emersion, the "senescent" forms may then well "rejuvenate." Laden with compressed genetic powers, when plants originating from distant points were, perhaps, all concentrated in a few acres, the "rejuvenated" stock would no doubt enter a new active period of "speciation," to take advantage of wider ranges. We may imagine how difficult of classification would be the "new species" originating in the process. Difficulties seemingly of the kind are not unknown in the taxonomy of Hawaiian Palms and of the "Galápagos Finches".

It seems clear that these Palms gained rather than lost ground during the second age of angiospermous dispersal, and that *P. reclinata* — not less than *Washingtonia filifera* — stands guard upon lost African large lakes or seas. *Phoenix paludosa* runs the classic Malaysian and Asiatic channel known to us by *Stylidium, Barringtonia,* Restionaceae and Centrolepidaceae, and seeing it also occurs at the Andaman Islands we assume it fully deserves its specific epithet *paludosa.* The channel in question must have included well watered ranges, and, generally speaking, a well diversified ecology. Arresting is the coincidence that, like *Styrax* (see p. 328), *P. Hanceana* neatly hits the bull's-eye of the petty Batan Islands. The possibility that this Palm falls indeed into two genuine varieties, *philippinensis* and *formosana,* indicates that, small as they are today, the Batan Islands may have known better days in the distant geologic past, not so much as part of the Philippines as of larger landmasses in the east and west alike.

The Elaeideae are dispersed as follows,

(1) *Elaeis* — Madagascar, Nyasaland, Angola, Congo, Uganda, Zanzibar, Sudan, West Africa, Cabo Verde Islands.

(2) *Barcella* — Brazil (Amazonas and adjacent regions).

This group migrates in a manner too clear to require elucidation. It is a classic instance of the track Madagascar—Amazonas, run in the past — and surviving in part in the present — across ranges alternatively liable to heavy rains and long drought.

The Lepydocarieae are believed to be primitive Palms, and the opinion is on record (14) that *Eugeyssona* is the connecting link between this Tribe and the affinity centering around the Coconut. The Lepidocarieae consist of three subordinate units, as follows,

A) Mauritieae

(1) *Mauritia* — 16 species in Brazil (Pernambuco, Pará, Amazonas, Goyaz), Venezuela (Amazonas, etc.), Guianas, Trinidad, Colombia, Peru.

(2) *Lepidocaryum* — 8 species in Brazil (Amazonas, Pará), Venezuela (Amazonas), Peru (Loreto).

B) Raphieae

(3) *Raphia* — 8 species in Madagascar, Angola, Nyasaland, Congo, Tanganyka, Uganda, Zanzibar, Fernando Po, Sierra Leone; 1 species *(R. taedigera)* in Brazil (Paraná, Pará), Costarica, Nicaragua.

(4) *Oncocalamus* — 2 species in West Africa (Gabun).

(5) *Ancistrophyllum* — 3 species in British East Africa, Congo, Upper Sudan, Gabun, Cameroon, Fernando Po, Gold Coast, Sierra Leone.

(6) *Eremospatha* — 3 species in Congo, Gabun, Nigeria, Sierra Leone.

C) Calameae

(7) *Eugeyssona* — 10 species from Malacca to New Guinea.

(8) *Metroxylon* — 10 species from Java to New Guinea, Fiji and Society Islands.

(9) *Pigafetta* — 6 species in Celebes and New Guinea.

(10) *Zalacca* — 15 species in India, Burma and Malacca.

(11) *Korthalsia* — 25 species in Burma, Sumatra, Java, Borneo, New Guinea.

(12) *Ceratolobus* — 3 species in the Soenda Islands.

(13) *Plectocomia* — 7 species in India, Java, Borneo, Malacca.

(14) *Calamus* — Over 300 species; 4 in Africa (Gabun, Cameroon, Sierra Leone), the balance throughout tropical Asia to Southern China (Kwantung, Yunnan, Fukien), Formosa, Malaysia throughout, Northern Australia, Polynesia.

This affinity is distributed from the African gate of angiospermy to the Old and New World alike. Two genera — both of them currently accepted as primitive within the group — *Eugeyssonia* and *Mauritia*, are endemic, respectively, to Malaysia and northern South America, which matches certain aspects (see p. 319) of the distribution of the Bombacaceae. The *genorheitron* that ultimately evolved the modern genera is manifestly of most remote antiquity. Attempts to describe the Palms as newcomers in the ranks of angiospermous plant-life because their fossil record is scanty and mostly recent in the palaeobotanical sense, are patently wide of the mark. Modern distribution is, as a rule, a far safer means of gauging the antiquity of a group than sundry petrifacts, as mentioned already in these pages.

The Attaleae shape up thus, *)

---

*) The reader is doubtless well aware of the difficulty involved by an exact determination of the *immediate geographic origin* of a form which turns up in Central Chile and Southern Bolivia, whether from the Atlantic or the Southeastern Pacific. This difficulty is far from insoluble, as we have pointed out (see note on *Oxalis magellanica*, p. 262, for instance), but cannot be properly approached unless in a full fledged study of the connections between the antarctic baseline itself (i.e., such channel run, for example, by the Restionaceae and Centrolepidaceae; see p. 359) and lands in the north. Considering that the "African gate" is one of the most important, it has been possible for us to elaborate on this gate even within the limits of this manual. We have thus learned that this "gate" breaks down in reality into two ranges, one "cold" *(Afroantarctic Triangle)*, the other "warm" *(Gondwanic Triangle)*; we have further seen that the former may release tracks independently from the latter, but that both *Triangles* had very intimate connections in the early angiospermous age. In the same manner, we have learned that "cold" forms may reach Tasmania and New Zealand, respectively, faring east and west of the Tasman Sea from a southerly approach; and that there is a center of "warm" flora just north of this Sea among Norfolk and Lord Howe Islands, Fiji, Tonga, Samoa and New Caledonia, which center may be fed from the "African gate" itself, or act independently.

It is plain that relationships of the same nature must exist among the lands in the immediate vicinity of the "Magellanian gate." Given the secondary nature of this gate, however, it proves necessary to study it by investigations mostly concerned with American flora. It should be necessary, even more, to have precise knowledge of a great deal of the ecology and minute taxonomy of this flora. None of this is possible, and accessible, to us here, and the solutions we give as to *details* in these quarters of the map are open to scrutiny as a matter of course. As we are above all anxious to leave the subject in our manual clearer than we found it ourselves, we warn the reader that we make it a rule of stating problems and outlining tentative solutions when we cannot do better in the immediate present, without feeling an undue concern with the result from future investigations, ours or other authors', does not matter.

As a transparent example of species of the same genus reaching Argentina from different directions altogether we may refer here to *Drosera* (see DAWSON in Rev. Argent. Agron. 5: 231. 1938). *Drosera uniflora* is a classic antarctic element — quite as antarctic as *Stylidium*, the Restionaceae, Centrolepi-

(1) *Beccariophoenix* — Madagascar.

(2) *Jubaeopsis* — Natal.

(3) *Jubaea* — Central Chile.

(4) *Diplothenium* — Eastern and Central Brazil, Paraguay.

(5) *Orbignya* — Brazil (Piauhy, Amazonas, Goyaz, Matto Grosso), Eastern Bolivia, Eastern Peru, Guianas, Venezuela, Honduras.

(6) *Attalea* — Brazil (Rio Grande do Sul, Minas Geräes, Amazonas) Eastern Peru, Guianas, Colombia (Chocó), Haiti.

(7) *Maximiliania* — Brazil (Amazonas) Guiana, Venezuela.

(8) *Cocos* — Pantropical.

This affinity is connected with the Lepydocarieae through *Eugeyssona*, and migrates to America from the African gate in a wholly conventional manner. *Cocos* is the lone genus which ranges outside the New World.

The origin of the Coconut *(Cocos nucifera)*, whether from the Old or New World, has been debated many a time (see, for example, 14, 15), without the partecipants in the discussion ever reaching an agreement. If not inconclusive, the evidence on the score is beyond doubt confusing, which foretells controversy yet to come among the learned if nothing is done meantime to restate the issue.

Debates of this scope necessarily call into play at least three kinds of considerations, (a) *Phytogeographical;* (b) *Taxonomic;* (c) *Historical.* It is obvious that nothing useful is done when subjects involving dispersal most directly are approached without an understanding of the background of the migrations in play.

In order to deal with the issue whether the Coconut is, or is not, native to this or that part of the world — and to deal with this issue constructively from the start, leaving details for later study if need be — let us consider the dispersal of the Coconut affinity, Palmae Trib. Attaleae, according to the authority we here follow.

Doubtless, this Tribe migrates from the African gate (Gondwana) — the major gate of all Palm dispersal — to the New World in the first place. The track seemingly comes into the New World from the south *(Jubaea, Jubaeopsis)*, meaning from a channel that reaches the Eastern Pacific at a latitude intermediate between Central Chile and Bolivia. *Diplothenium, Attalea* (notice

---

daceae, etc. — allied with species in the Cape *(D. regia)*, New Zealand, Tasmania, Australia. *Drosera uniflora* reaches the southernmost tip of Argentina, accordingly, straight from the "Magellanian gate," and the south. *Drosera brevifolia* and *D. communis*, on the contrary, are from the "African gate" in the last resort, and their *exact* point of entrance into the New World is speculative. However, the ranges of these two groups of *Drosera*, the "cold" *D. uniflora* and the "warm". *D. brevifolia* and *D. communis* are sharply different in Argentina; the former is confined to the south, the latter to the northeast of that republic. It is thus easy to discriminate among them on the face of the dispersal.

Altogether different would be the case if *D. uniflora* had marched northward enough to cross the paths of, and to mingle with, *D. brevifolia* and *D. communis*, and chanced to be in addition scarcely separable by cleancut sectional bonds from the latter two species. The tangled skeins of the dispersal, and the *detailed* course of the Argentine tracks of these forms, all together, could then be unravelled, and read, only at the price of considerable work involving, for example, a careful study of climaxes, altitudinal levels, taxonomy, morphology, phylogeny etc. Ranges common to forms which offer difficulties such as those here visualized are frequently met with in Central Chile and adjacent Argentina.

the tie Eastern Peru—Haiti, and the record in the Colombian Chocó) possibly fall in with *Jubaea* underscoring this channel. *Orbignya* and *Maximiliania*, though not incompatible with it, leave it open to question whether they might not have reached the New World directly from South Africa across the Atlantic of our maps. Whatever be the details of the case, the dispersal tends to emphasize the New World first and foremost in the manner illustrated by genera like *Menodora*, and families like the Turneraceae.

Cocos nucifera, then, as integral part of the Attaleae, ought to be, in theory at least, genuinely native to the New World, the West Indies included, where it is rumored the early Spanish navigators found it at their first landing. Taking a narrow view of the matter, *Cocos nucifera* might not be native at all to the lands on the Indian and Pacific Oceans; the Coconut should be exclusively native to the Americas.

This is manifestly not the case, because it is certain that the Coconut grew in profusion in the Indian Ocean at a very early date. The issue, then, boils down purely to the question, whether *Cocos nucifera* was introduced, and when, from the Indian Ocean to Tropical America.

It might be possible, perhaps, to produce irrefutable documents of this introduction; to identify the place of original cultivation, in a word to prove that the Coconut was definitely carried by man to the New World. This is a matter beyond us. Purely as phytogeographers we would say, however, that in a Tribe migrating like the Attaleae it is perfectly possible to meet a genus, like *Cocos*, distributed from the very first both to the Old and New World. A track spreading from Africa both to the Far East and the Americas is so conventional, so usual, that by following this track *Cocos* would be acting most normally, whether the rest of its affinity, the Attaleae, does the same or not matters hardly at all. Plentiful are the case in which a genus, even a single species, runs the whole net of tracks which its immediate affinity follows in one direction only. The Turneraceae are a case in point, to mention an obvious example.

Concluding strictly from the phytogeographic viewpoint, we would say, (a) It is quite likely that *Cocos nucifera* was native to the New and Old World from the start, because its dispersal to these centers in primary or secondary dispersal (that is, when not in the Late Jurassic in the Early Tertiary) is a standard contingency; (b) Naturally, confirmatory evidence of this probability could be sought by a careful study of the migrations of all species of *Cocos* and the intertribal affinities of the Attaleae.

This would end our work. If additional positive evidence could be brought forward to the effect that *Cocos nucifera* was introduced to the New World we would consider it as a matter of course, in a purely factual spirit. However, it is pertinent to notice that evidence, for instance, that *Cocos nucifera* was actually introduced to Haiti from elsewhere would not stand as final proof that the Coconut was also brought by man to the islands at the mouth of the Amazonas or the shores of South America on the Caribbeans.

All in all, we are skeptical that positive evidence can be found that *Cocos nucifera* never was native to the New World. We are inclined to believe that the Coconut reached both Ceylon and the West Indies in the course of its normal migrations. The reader may refer to migrations such as those of the Bombacaceae (see p. 319), and gather reasons therefrom, why we are inclined to be skeptical.

The investigations we would conduct as regards *Cocos nucifera* would be substantially the same as those we would undertake concerning all other cultivated or useful plants (see also notes under *Santalum*, p. 145). We are prevented by cogent necessity from giving more time and space to these issues, which we regret. A study of the dispersal of the Malvaceae and *Gossypium*, for instance, would show that cotton was independently used in the East of the Old World and the highlands of South America. This could be proved by reference made to the course of the dispersal of *Gossypium* without making the slightest appeal to historical factors *).

A Palm is endemic to the archipelago of Juan Fernandez off the Chilean coast, *Juania*, which was classified (8) under the Iriarteae by an old monographer. In more recent times, an author (16) rejected this disposition, claiming that *Juania* would better be treated under a monotypic tribe of its own in the affinity of another group, Morenieae.

The controversy thus outlined, where *Juania* belongs in classification, offers us a good opportunity of reaching at least a beginning of decision on the strength of phytogeographic considerations. By a critical comparison of these groups we may learn which one author has the weight of probabilities in his favor, and perhaps attain a deeper insight of the relations between systematic botany and dispersal.

The Morenieae are composed in this manner,

(1) *Hyophorbe* — 3 species in the Mascarenes.

(2) *Juania* — 1 species in Juan Fernandez.

(3) *Chamaedorea* — 90 species in Brazil, Bolivia, Peru, Ecuador, Colombia, Venezuela, West Indies, Central America, Mexico.

(4) *Morenia* — 4 species in Brazil, Peru, Colombia.

(5) *Synechanthus* — 3 species in Central America.

(6) *Reinhardtia* — 7 species in Central America, Mexico.

(7) *Pseudophoenix* — 4 species in the West Indies.

(8) *Gaussia* — 2 species in the West Indies.

---

*) In one of his classic papers (Linneesvki Vid kak Sistema. 1931), VAVILOV stressed the principle of parallel morphological development. This principle can be best investigated by making reference to sound generalities of dispersal together with an application of the principle of the *genorheitron*. Though a great deal has been done to study the "Center of origin" of the cultivated plants many of the works in question are not altogether satisfactory. It is mildly curious that the author of a text on the "foundations" of phytogeography should close a review (in Torreya 43: 151. 1943) of the "Criteria for the indication of Center of Origin in Plant Geographical Studies" with the complaint that, "Assumptions arising from deductive reasoning have so thoroughly permeated the science of geography *(sic)* and have so long been a part of its warp and woof that students in the field can only with difficulty distinguish fact from fiction." Patently, the trouble is deepseated, and the current "foundations" of phytogeography stand additional props. We hope to deal with *Gossypium* in a separate study.

(9) *Kunthia (Collina)* — 1 species in Mexico.

Were *Juania* to be left under the Iriarteae, it would be part of a group of this kind,

(1) *Juania* — Juan Fernandez.

(2) *Wettinia* — 5 species in Peru (Puno, Junín).

(3) *Ceroxylon* — 14 species in Peru (Junín, Huánuco, Amazonas) Ecuador, Colombia (Cundinamarca, Magdalena), Venezuela (Mérida, Táchira, Aragua, Caracas).

(4) *Catoblastus* — 6 species in Colombia (Santander, Antioquia, Meta), Venezuela (Mérida, Carabobo, Aragua).

Classified as integral part of the Morenieae, *Juania* stands next *Hyophorbe*. We thus have before us a tie Mascarenes—Juan Fernandez which is altogether normal in the light of the evidence brouhgt to the preceding pages. As regards the Palms in particular, an homologous tie (Madagascar and Natal—Central Chile) is established outside the Morenieae by three genera of the Attaleae, *Beccariophoenix*, *Jubaeopsis* and *Jubaea*. When know full well, then, the grounds whereon we stand.

The balance of the genera of the Morenieae are distributed in a manner which is certainly not incompatible with an entrance effected into the New World from the East Pacific or the South Atlantic. We are short of detailed localities, but ranges in Mexico, Peru, Bolivia, Ecuador are in agreement, generally speaking, with such a manner of migration.

Transferred to the Iriarteae, *Juania* no longer consorts with a genus, or genera, from Africa and-the Mascarenes, but becomes part and parcel of a seemingly pure American affinity. However, the dispersal is so perspicuous in itself as to leave us in no doubt that the channel in play is by the Eastern Pacific. The migrations stress the west of South America in cleancut fashion. Localities such as Puno, Junín, Huánuco, Antioquia, Santander, Mérida are reliable indices of an invasion effected by the *genorheitron* from the East Pacific. *Ceroxylon* is in the west of Venezuela (Mérida, Táchira) and the range it holds farther east (Aragua, coastal range by Caracas) is a normal extension, witness the distribution, for example of the Protieae. *Catoblastus* fares eastward to Carabobo, a center usually in close touch with Panamá and Central America (see p. 446 fn.) not to mention coastal Colombia. The migrations of the Iriarteae, then, are of "Andean" type, as are those of *Brunellia* (see p. 384) and the Protieae (see p. 388).

Still another author (17) proposes a Tribe Ceroxyleae to which he credits but two genera, *Juania* and *Ceroxylon*. This may be better taxonomy, we do not certainly know, but does not change substances in the least. The ingredients that make up the Morenieae, Iriarteae and Ceroxyleae are all from the same kitchen. In one case *Hyophorbe* speaks, in another so does the face of the dispersal. Everything in the end rhymes to the very same. With or without *Hyophorbe*, seasoned with this or that sauce, the stew does not change. It is conventionally "Andean," which — to use language fitting the present case in science — means that the genorheitral migrations are immediately from the East Pacific or the South Atlantic. The old author who put together *Juania* with *Hyophorbe* did nothing to compound heresy from the phytogeographer's standpoint, for he stated by implication that the East Pacific was reached in this case from the direction of the Mascarenes. Nothing surprising in this, of course, for the Mascarenes stand out as one of the major centers of Palm life and origins, and a tie Juan Fernandez—Mascarenes rates merely as one of the aspects of a general tie Mascarenes—Southern South America, or Indian Ocean—Southeast Pacific.

The question next follows, is *Juania* better classified in the Morenieae, Iriarteae, Ceroxyleae of, finally, in a group as yet undescribed?

We will answer this question referring to previous comments (see p. 112) made on the score of *Geranium ardjunense* and its classification. *Juania* is one of the genera that, as yet not hardened out of the hot genorheitral matrix, can be everything to everybody, depending as to whether we like to stress a character that associates it with *Hyophorbe* or a peculiarity which places it next *Ceroxylon*. In other words, the classification of such a genus as this is bound forever to remain the plaything of opinion. Forever, that is, until the score is settled by a complete study of some fifteen or twenty genera with which this genus may be related, concluding in the end *only upon sums of probabilities*. Until then, author "A" keeping his eyes glued to the Mascarenes and *Hyophorbe* may utter oracles which author "B" looking at *Gaussia* in the West Indies will dismiss as idle vaporings. It is sufficient for us to know here that *Juania* cannot be properly classified because of certain well definite reasons which are the same why *Geranium ardjunense* is a "difficult" species, and *Phrygilanthus* a "hard" genus. If there is anything satisfactory in all this for us, this is the fact that by a knowledge of dispersal, and its interplay with *genorheitron*, we can reach at least a beginning of solution in issues that remain obdurately shut before our peering eyes in other manners. *We are at least in the safe position of discriminating between facts and fiction.*

If we further observe the dispersal of the tribes of Palms we have so far studied, we must reach the conclusion that, once on American soil their *genorheitra* tended to mass in two distinct centers, western and eastern. The eastern center rested in the main upon the "serras" of coastal Brazil, which, not unlike the Appalachians of North America, are one of the oldest lands in past and present emersion. The western center, on the contrary, is keyed up to the Andes of our maps, and for this very reason if often spoken of as "Andean."

In reality this center is not "Andean" because it became operative in a geological era long antedating the rise of the modern Andes. It might be called "Andean" in the purely descriptive sense making reference to the geography of the present. It further might be connotated "Andean" to mean, presumably, that the majority of the genera and species which tenant it today had their immediate origins from the geologic and climatic revolution attending the peak of the "Andean" uplift in the comparatively late Tertiary. Be this as it may, it seems certain that the indiscriminate use of terms such as "Andean," "Cordilleran," etc. has done great mischief because it has put basic issues under a false light from the start.

IV. Tracks in the high north and the "Ice Age."

Dispersal in the high north is commonly considered to be a strictly local problem. So far as known, no author specialized in polar floras ever made consistent attempts directed to rationalizing this type of distribution in function of a world-wide pattern of migrational channels. This statement is not to be construed as meaning that authors of this bent have neglected to study actual distribution. Indeed, some among them have done most painstaking work in this field and contributed a stupendous wealth of data. The point is, rather, that all this work has remained focussed throughout in the direction of the North or South Pole to the neglect of less frigid distributional landscapes.

The attitude just described toward the issues of boreal or strictly antarctic dispersal is largely understandable. Were a student of distribution in the high north curious to learn something of the ultimate end of the migrations of immediate interest to him, he would find nothing in print fit to satisfy his desire. Moreover, the high north is not rich as regards genera and families, and the species there becomes one of the major units of immediate work. Specific tracks then, are an object of paramount investigation up north, and tracks of the kind usually do not lead very far. Investigators of dispersal in the tropics do not labor under similar limitations, because they are constantly called upon to deal with masses of different genera and families, and forced as a rule to take immediate interest in three to four continents.

We take the opportunity which offers here to underscore the psychological factor in research, because we are satisfied this factor has vital importance. We believe that, for example, botany has not always gained at the hands of authors who laid down the fundamentals of ecology and phytogeography on the strength of what was known to them in the northern hemisphere only. Many of the concepts and definitions current work badly, if at all, as regards the floras of the tropics. These concepts and these definitions may not be erroneous throughout, be this clear, but they seldom are completely correct, on the other hand, and prove for this reason objectionable in the end.

Divorced from a full understanding of migration, wherever migration may lead, many of the works written on the floras of the north, the high north in particular, are, we take, faulty as to chronology. For example; seeing that a plant "A" is native to Greenland, and the Alps as well, many authors are tempted forthwith to assume that this plant "originated" during the "Ice Age," at a time when this "Age" was knocking at the doors of the Alps, and was left "stranded" in "Alpine" position by subsequent recessions of the "Ice Sheets." Other authors devise strangest migrations around the "Margin of the Ice Sheets" in order to account for the dispersal of this or that species in, let us say, Greenland and Siberia. Hardly credible statements are on record regarding the origin of certain "Alpine" species, as if these species were necessarily born yesterday.

In our opinion, the problems of dispersal in the high north do not differ in the least as to prime movers from the problems of distribution in other regions of the world. True enough, these problems involve a stringent reckoning of comparatively recent glaciations, and invite attention to issues of speciation and distribution of a peculiarly narrow compass. However, dispersal was effected to the high north long before the advent of comparatively recent glaciations, and the effect of these glaciations as to distribution cannot be fully understood if proper perspectives of time and space are lacking.

We intend to review in the coming pages the distribution of certain species of the north and high north, investigating so far as possible their genorheitral ties, wherever these may be found. While we anticipate that nothing of the utmost importance will greet us in the course of this work, still we believe that we can only gain in depth of achievement and general information by bringing the events of the high north in direct touch with events elsewhere in the world. This is a necessary undertaking on our part, at any rate, for the purpose of our work is to contribute generalities of method and treatment regardless of latitudes.

Alchemilla group *Vulgares* consists of about 21 species. Five of these species (18) belong to the north and the high north. *Alchemilla Muerbeckiana* is native to Central Asia (Altais, Dzungaria, Tarbagatai) and Siberia (region of Tiumen); *A. subcrenata* occurs in Siberia (region of the Obi). Leaving these regions behind, the migration next streams westward investing Central and Northern Russia *(A. subcrenata)*, possibly Eastern Germany *(A. Wichurae)*, Scotland, the Alps and Pyrenees *(A. glomerulans)*. Greenland is either barely touched at the east coast *(A. Wichurae)*, or massively entered *(A. glomerulans, A. filicaulis)* to about 70°. Also invaded are Labrador and Newfoundland *(A. glomerulans, A. filicaulis)* seemingly as a direct consequence of the general westward sweep of the tracks.

The caryophyllaceous genus *Stellaria* has at least two species in the high north. *Stellaria longipes* (19) departs from Central Siberia *eastward*, and crossing into the New World eventually finds its way into Western Greenland, here ranging to about 64° Latitude. Its ally, *S. crassipes*, fares in the high north only, reaching the Spitzbergen, Novaia Zemlia, the Polar Urals, Northern Scandinavia, Greenland northward from 70° Lat. and Ellesmereland.

The migrations of *Alchemilla* and *Stellaria* to the high north begins, *eastward or westward*, from a center which closely corresponds with the approaches of the Altais. This center, as we know, is a capital node of dispersal for the boreal lands throughout, because the tracks coming to this region, either from the African (see, for example, Ericaceae Andromedeae, p. 171) or the Western Polynesian gate (see, for example, Ericaceae Rhododendroideae), are next re-routed here to their ultimate destination. A knowledge of this fundamental of distribution, consequently, materially lightens our investigations of tracks and speciation throughout the high north. Doubtless, not all species in the high north *originated* at the Altai Node, which is readily demonstrable with reference, for example, to the range of *Stellaria crassipes*. Reference to this center, however, is necessary in order generally to understand what goes on eastward, westward, northward and southward as well.

Could we affirm, then, that *Alchemilla glomerulans*, which occurs in the Alps and Pyrenees quite as much as in Greenland, Labrador and Newfoundland was "telescoped" into the Alps only as the immediate result of the "Ice Age"? Indeed, not at all. Its pattern of migration is by far too broad and too involved to support so easy an explanation. The weight of probability is, on the contrary, that *A. glomerulans* had reached all, or very nearly all, its present stations before the Pleistocenic Glaciations, and survived these glaciations more or less there, where we find it now.

Naturally, its range might have been more extensive than it is now, but that this species was born during the Pleistocene does not seem probable. In our opinion, it is a fair guess that great many, perhaps most, of the primary species of the high north originated *before the Ice Age*. The supposition that this may well be the case is a prerequiste toward the solution of "local" issues of dispersal in the high north that have so far baffled our efforts. In other words, as we see it, *the issue is fundamentally one of correct general chronology*, and if chronology is vitiated by assumption that the forms of the north are necessarily of the Pleistocene, or the like, we may not hope to approach the dispersal in the north with promise of success.

Our viewpoint will be made clearer, perhaps, if we refer to one of the "alpine" Spurges of the Western Alps. *Euphorbia Valliniana* is endemic to the slopes of Mt. Viso, and has been reported from the French Alps to the west. We are under the impression, though we have not so far investigated the matter in full, that *E. Valliniana* is allied with *E. pauciflora* of the Pyrenees and the mountains of Aragón, and may further have kinship with plants known from the Sinai in Arabia. Be it as it may, *E. Valliniana* is beyond doubt most closely allied with *E. Potaninii* narrowly localized in Eastern Mongolia. It may be added that also in the direction of Central Asia, and beyond, are to be looked for the relatives of the Edelweiss, *Leontopodium alpinum*.

To deal with *E. Valliniana* as a plant *of the Alps* in the first place is downright mistaken. There is no *E. Valliniana* for a phytogeographer, but a complex of forms ranging from the Pyrenees, possibly the Sinai, to the Alps and Eastern Mongolia alike. The distributional problem involved calls for a consideration of all these points as one, and it is to be taken for granted — until and unless the contrary is proved true — that *Euphorbia Valliniana*, and perhaps as well *Leontopodium alpinum*, never were "Alpine" in the first place.

Forms very much like these, if not identically the same, belonged doubtless to a very ancient flora on the northern shores of an Early Cretaceous continent, the outlines of which clearly transpire through numerous examples of dispersal introduced to the pages of this manual. Whether these forms were by then plants strictly confined to indifferent grounds by the sea, or, already indeed, dwellers of hills and mountains antedating the Alps and Pyrenees we do not know. The fact stands that the quickest way of stultifying our understanding of the origin of these plants, and their dispersal, is in making of them "Problems of Alpine distribution" within the narrow compass of the Tertiary.

As it is not within our immediate possibilities to give much space to local problems of dispersal, we summarily cast on a map (Fig. 89) the tracks of some forms of *Alchemilla* and *Stellaria* revised in previous paragraphs. We bring these tracks uniformly to a head at the Altaian node, as the evidence advises. It is in this node that ultimately originated the affinity of *Alchemilla Vulgares*, and it is from this node southward that dispersal and genorheitral ties are to be worked out in order that we may understand the ultimate origin of the affinity in question. If the Altaian node is not in play as regards other plants in the north (witness *Erica* and *Calluna*, p. 160, 166), still the procedure we outline cannot change as to substance. The starting point of the tracks northward is to be sought *somewhere* in the first place, for only at this price we can understand as an harmonious whole the progress of dispersal and speciation as we move away from the equator toward boreal lands.

Alchemilla is a large genus of confuse classification. To work out the main threads of its dispersal would require more time than the result would be worth at this stage of our investigations. Doubtless, *Alchemilla* is well represented in the flora of Africa. It numbers about a dozen species in Natal, Transvaal, Basutoland and the Cape. Three species at least are native to Usambara and the Kilimajaro, one of which is an "Alpine" that significantly associates (20) with the ericaceous *Ericinella*. Still another species, *A. javanica*, is known from India and Ceylon, not only, but Java, where it occurs restricted to relic-ranges on Mt. Papandajan and Mt. Idjen. These two heights are the home of species having spectacular disconnections *(Potentilla Mooniana* and *Alisma natans)*, which proves that their flora is not altogether modern in origin. If it is true that *Alchemilla* more abundantly speciated in the north than in Africa and Tropical Malaysia, still the few records here mentioned are adequate to prove that the dispersal of this genus is, in part at least, such as to match that of the *Schoenoxiphium—Cobresia* (see p. 237) affinity.

In the New World *Alchemilla* is represented — whether wholly or in part we do not presently know — by a group, *Lachemilla*, variously treated in taxonomy as a separate genus or a section of *Alchemilla*. Typical of the dispersal of this group (21) are the following species,

(1) *A. aphanoides* — Bolivia (Santa Cruz, La Paz), Peru (Cuzco, Ayacucho, Huánuco, Amazonas), Ecuador, Colombia (Cauca, Tolima, Huilla, Caldas, Cundinamarca, Santander, Magdalena), Venezuela (Mérida, Caracas);

Var. *subalpestris:* Costarica, Guatemala, Mexico (Oaxaca, Michoacán, More-los, Hidalgo, Mexico, San Luís Potosí, Chihuahua, Baja California).

(2) *A. frigida* — Argentina (Córdoba, Tucumán), Bolivia (La Paz).

(3) *A. domingensis* — Santo Domingo.

(4) *A. Moritziana* — Colombia (Magdalena: Sierra Nevada de Santa Marta).

The *genorheitron* of this affinity might have reached the New World from several directions, which by now the reader can figure out for himself without difficulty. We would guess it came into South America precisely as did *Ilex* subg. *Yrbonia, Brunellia, Protium,* that is, from the Pacific, first hitting Bolivia and Peru, next migrating in part southward (Argentina), but mostly north-ward. The var. *subalpestris* of *A. aphanoides* is native to a consistent range well known to us from the distribution of part of *Zamia* (see p. 477) and Baja California of our maps was doubtless in emersion when the first Angiosperms came to the New World. The track of *A. aphanoides* (typical form) is consis-tent, too, for it veers eastward from the Colombian province of Cundina-marca ultimately to reach the region of Caracas, Venezuela. This deviation is commonly observed in dispersal, and is to be credited, we believe, to some feature of early American palaeogeography anterior to the Andean uprising. The occurrence of two species in Santo Domingo and the Sierra Nevada de Santa Marta, respectively, also indicates a dispersal of very great age. The former is one of the lands in the Caribbeans that stood in emersion when the Angiosperms first knocked at the doors of the New World; the latter a moun-tain in part recent, but as to remote origins quite hoary with geological age. It will readily be observed that the track run by this group of *Alchemilla* is a classic of American dispersal, which leaves the continent at the usual center, between Eastern Colombia and Western Venezuela to hit Santo Domingo. This channel is used by all manners of *genorheitra,* "cold" and "warm" alike, and *Cardamine* found it useful (see p. 93) as a connection between the mountains of East Africa and certain heights which, on their part, must have antedated the modern Andes by ages.

It is manifest that a genus which migrated like *Alchemilla* both in the Old and New World was active in the very first age of angiospermous dispersal, and had no more difficulty in reaching the Altai Node than it did in finding Mt. Idjen or the Sierra Nevada de Santa Marta. Patently, reference to the Altai Node as the point of secondary origin of *Alchemilla* Group *Vulgares* — be this reference only tentative at the outset — is bound to clarify the ultimate relationship of this aggregate with the balance of the genus; the nature of the boreal forms, whether primary or secondary; the course of the all the tracks running in the high north. Without reference to this Node, the issue of the dispersal of *Alchemilla* Group *Vulgares* must be handled within an hyper-boreal vacuum.

Stellaria is a large and, taxonomically speaking, obscure genus. The dis-persal of certain of its main subdivisions (22) reads in this manner,

(1) Sect. *Eustellaria* — Subsect. *Petiolares:* Chile to Mexico and North America, Africa, Islands of the Gulf of Guinea; Subsect. *Insignes:* Himalayas to Mongolia, China, Japan, Pacific North America, Russia and the Alps; Subsect. *Holsteae:* Europe; Subsect. *Larbreae:* Northern Hemisphere generally to the Mediterranean, Mexico and the Himalayas; Subsect. *Spinescentes:* Tasmania, Southeastern Australia.

(2) Sect. *Schizothecium* — Ceylon, Himalayas.

(3) Sect. *Adenonema* — Himalayas, Tibet, Southwestern China, Eastern Siberia, Argentina, Peru.

(4) Sect. *Pseudalsine* — Iran (Persia).

(5) Sect. *Leucostemma* — Himalayas.

(6) Sect. *Oligosperma* — Western India, Asia Minor.

The distribution of the last five Sections in this tabulation removes every doubt we may entertain as to the possibility of *Stellaria* having reached the high north from the Altai Node. It is manifest that this genus had an important center of secondary origins within the triangle, Near East—Eastern Siberia—Southwestern China. The South American records in Sect. *Adenonema* are repeated in Sect. *Eustellaria* Subsect. *Petiolares,* and may be referred to the elucidations we next supply of the dispersal of *Carex incurva.* Section *Eustellaria* Subsect. *Spinescentes* could easily be delivered by tracks homologous of those of *Pelargonium* Sect. *Peristera* or the Centrolepidaceae. The range Africa, Islands in the Gulf of Guinea in Sect. *Eustellaria* Subsect. *Petiolares* speaks for itself. Finally, Sect. *Eustellaria* Subsect. *Larbreae* is of a conventional "northern" type, such as originates from a stream of migration hitting the Himalayas from the Indian Ocean.

It is sure, at any rate, that *Stellaria* had a marked center of evolution in Eurasia, as stated, and it is also well established that this center was fed immediately from the south. Genera like *Stellaria* would be excellent material for studies in dispersal if reliable detailed monographic work were available *).

A taxonomist whose name is not wholly unknown to phytogeographers contributes (23) a map of the dispersal of *Carex incurva* var. *setina* (Fig. 90), which is so perfectly circumpolar as to yield no clue at first glance, whether this tiny Sedge reached the high north from the Altais or any other secondary center of distribution, west or east. The best known monographer of *Carex* (29) tells us, on his part, the following,

(1) *C. incurva* — *South America:* Fuegia, Argentina (Mendoza), Chile (Aconcagua); *North America:* Colorado to Alaska and Greenland; *Asia:* Afghanistan, Northwestern Himalayas (Kashmir and Karakorum), Transbaikalia, Kamchatka, shores of the Arctic Ocean in Siberia; *Europe:* Caucasus and the Alps northward to Scotland and the Faer Oes, Scandinavia, Novaya Zemlia, Spitzbergen, Jean Mayen.

---

*) The reader has doubtless noticed a peculiar coincidence. "Cold" groups like *Potentilla, Alchemilla, Stellaria,* etc., may occur throughout Africa, in Ceylon, Java, the Himalayas, and generally all manners of "tropical" stations. Groups of the kind, however, do not turn up as a rule in Madagascar and the Mascarenes. These islands are center of origin of a flora which is *genuinely* "warm."

(a) Var. *setina* — As per map (Fig. 90).

(b) Var. *melanocystis* — Argentina (Mendoza), Chile (Atacama).

(c) Var. *misera* — Argentina (Mendoza), Chile (Atacama).

(d) Var. *chartacea* — United States (Colorado).

    (a') Forma *pallens* — Northern Greenland.

    (b') Forma *erecta* — Strait of Magellan, Scotland, Norway, Western Greenland.

This species is a perfect instance of "bipolarism", for it occurs in Fuegia, as well as in Northern Greenland. "Bipolarism", as we know, is a bugaboo of academic phytogeography, which is usually hard put to work out "Bridges" to carry "bipolar" species to their ultimate destinations in the very cold north and south, or to invent birds dispersing seeds between the intervals of a "Glacial Age,", etc., etc.

We have shown in previous pages (see p. 21) that "bipolarism" in one continent can be accounted for by making reference to dispersal taking place in another. For example, American "bipolarism" may be the byproduct of distribution from a base-line in the Southern Pacific which principally interest Western Polynesia, Malaysia and the Pacific Far East, not the New World.

The distribution of *Carex incurva* is no exception to the rule we have thus outlined.

We know that a plant of "cold" nature can reach without difficulty the whole of the north and high north once it has gained access to the Altai Node. We further know that this Node can be reached by tracks coming in from the general direction of the Indian or Pacific Ocean. This established, it follows that we can easily solve any problem in "bipolarism" centering at the Altais, or vicinity, the moment we can correlate dispersal in this center with dispersal in the Indian and Pacific Ocean. A correlation achieved between these two dispersals forthwith connects the deep south with the deep north of the map.

This is a basic theorem of phytogeography, which it is necessary to discuss.

In the classification we follow, *Carex incurva* is isolated under Sect. *Incurvae*, which as its nearest affinity another monotypic group, Sect. *Physodes* (24), as follows,

(1) *C. physodes* — Afghanistan, Turkestan, Iran, Lower Volga, Urals.

Taxonomists may debate among themselves, whether the monographer we follow is to be accepted as the final authority on *Carex*, or is, on the contrary, an indifferent worker, which can interest us but mildly. As phytogeographers we observe, (a) *Carex incurva* and *C. physodes* are properly distributed. On the strength of their range, they indeed represent divergent parallel branches of the same immediate *genorheitron*, one *(C. incurva)* definitely boreal, the other *(C. physodes)* not quite so. We know nothing just now of the ecology of these species, but are inclined to suspect that they may not favor identically the same habitants, either. In brief, these two species are seemingly both primary, hence they exhibit sectional characters or (if this be more acceptable as to terminology), very strong specific individuality. They are, in addition, well and properly distributed from the standpoint of pure phyto-geography; (b) The *genorheitron* of these two species reached Asia, and the rest of the range, primarily from a center in the modern Indian Ocean corres-ponding with a now crumbled continent, Gondwana; (c) Were it ever so, that *C. incurva* and *C. physodes* are wholly unrelated from the standpoint of system-atic botany, still their origin, severally when no longer singly, might remain set by the center of the Indian Ocean we have just characterized.

So much laid down as introduction, let us proceed to consider the dispersal of *C. incurva*.

In this dispersal are evident two centers, (a) *Eurasian*, beginning at the Al-tai Node (line Afghanistan—Transbaikalia); (b) *South American*, Fuegia to Northern Chile. Out of these two ranges are left, (c) The High North generally; part of the Western United States (Colorado).

As we know (see *Lepechinia, Juania, Jubaea, Menodora*, etc., etc.), the African gate (i.e., the heart of the modern Indian Ocean, meaning Gondwana) has intimate connections with the south of South America. A tie between Fue-gia and a point in the Indian Ocean is thoroughly standard and usual. The two massings of *C. incurva* in Eurasia and South America are accordingly thoroughly well connected, and are so by a standard tie. This tie, and its further normal extension northward, is sufficient to account for the "bipola-rism" of *C. incurva*, or any other species homologously dispersed. We refer the readers to tracks in the Pacific (see p. 247) which can give further "bipo-larism" along a different channel of migration, always standard however.

The Western United States (Colorado) could be reached in any one of two ways, (a) From the High North, by a track originating at the Altai Node, cros-sing over into North America of our map through the Northern Pacific; (b) From Magellania in disconnection by a crumbled shore in the Eastern Pacific.

It is possible that a majority of the readers will accept the former alternative as the more likely. They may assume that this alternative, after all, agrees with the outline of modern maps, and is "safer." We take both alternatives to be equally possible, and in the abstract, perhaps might incline to favor the latter. *The outlines of modern maps have nothing whatever to do with "modern" dis-persal, which is an axiom of phytogeography.* We definitely know that discon-nections in the Eastern Pacific are normal (see, for instance, *Menodora, Cus-cuta*, possibly *Phyrgilanthus*, etc., etc.), and if it be true that these disconnec-tions are mostly shorter than the one Northern Chile—Colorado now facing us, it is also true that Colorado can be reached from California, very easily, witness the distribution of *Carex* Group *Foetidae* (25), as follows,

(1) *C. perglobosa (C. incurva* var. *chartacea)* — Colorado.
(2) *C. vernacula* — California, Washington, Wyoming, Colorado.
(3) *C. incurviformis* — Alberta.

We stress California, because California is a standard station of disconnect-ions reaching southward to Northern Chile and Bolivia. In sum, a disconnection Northern Chile, even Southern Chile—Colorado, is in no sense more extraordinary than a disconnection Bolivia—California.

Concluding, the "bipolarism" of *C. incurva* in the New World cannot be fully understood without making preliminary reference to dispersal from the Altai Node in Asia. However, the strictly hyperboreal range of this sedge is both so complete and so fragmentary that we incline to the opinion that this range was fed at both ends, namely, (a) From the Altai Node; (b) From Colorado, this outpost having first been reached in disconnection from the direction of Northern Chile. It will be observed that, were our opinion erroneous, and the range of *C. incurva* all from the Altaian node, or all from a Colorado secondary center of speciation, nothing much would change of the fundamentals in the dispersal. Were we to study here the matter to an end, we would certainly consider everyone of the possible hypotheses and alternatives suggested by the face of the dispersal. Hypotheses and alternatives are fundamental in science, and, most definitely, tools of refined phytogeographic work. It is transparent, we believe that, even when solely confined in the high north, groups like *Alchemilla, Stellaria, Carex* do not automatically become a fare reserved for specialists in boreal distribution. There is hardly a pattern of boreal distribution that is without root elsewhere than in the north.

We will observe that everything in the evidence before us, the broken range of *C. incurva* in Kamchatka quite as much as the heavy speciation of *Bombax* in certain centers of West Africa, depends from the dominant influence of certain few essentials. We believe that the monotonousness of our elucidations is, perhaps, the best proof that our method of approach to distribution and its problems is, when not absolutely correct in detail, at least satisfactory as an introduction to scientific phytogeography.

Although the analysis of the distribution of *C. incurva* just given may be allowed to stand upon its own merits, we can hardly resist the temptation of introducing two more examples relating to dispersal in the High North, or the north.

The cruciferous genus *Subularia* contains but two species within present notice, distributed in this manner,

(a) *S. monticola* — Mountains of Eastern Africa and Ethiopia.

(b) *S. aquatica* — Altai Mts., *westward* to Southeastern (Orenburg) and Central Russia, Europe generally to Northern Scandinavia, Iceland, Ireland with scattered stations reaching southward to the Balkans, Alps, Pyrenees; *eastward* to Kamchatka *), Alaska and Southern Greenland, Eastern California, Northern Nevada, North Dakota, New Hampshire and Newfoundland.

This dispersal leaves nothing to imagination, and reaches all the way from East Africa to Greenland in less time than it takes to scan the map. It is in no sense more extraordinary than the distribution of *Schoenoxiphium* and *Co-*

---

*) We owe these data largely to the work of HULTÉN, Lunds Univ. Årsskr., N. F., Avd. 2, 41 (1) 813. 1945, an outstanding example of painstaking reconstruction of modern ranges.

bresia, which it rigorously parallels. Fittingly to account for one of these patterns of dispersal we are forced to devise methods and concepts fit to explain *all as one.* The mountains of Eastern Africa and Ethiopia *) stand in intimate contact with a center in the Indian Ocean — a center rich from the start in "cold" flora — which has vanished from our maps, *but is quite as tangible and material today as if still were solid land.* This center fed both these mountains by a track faring westward, and the Western Himalayas and Altais by a track going directly northward. Tanganyka and Ethiopia are classic regions of breaks, not only, but there is evidence that Somaliland had ties with South Africa, Madagascar and the Mascarenes over this same channel in the modern Indian Ocean. *It is to this channel that Northern Somaliland in particular seemingly owes most of its faunistic and floristic peculiarities.*

A simple diagram will show why this track runs in disconnection in the modern Indian Ocean. For the purpose, let us spread before us a map of the world in Mercator's Projection, and draw three lines, one exactly to match 60° Latitude South, two to agree, respectively, with 60° Longitude West and East. This done, let us observe, and remark, (a) The line that matches the latitude is consistent with a line connecting the Magellanian with the African gate of angiospermy. This line contains the whole of the "antarctic" connections between these two gates, not only, but invests primary regions of angiospermous origins *(Afroantarctic* and *Gondwanic Triangles).* This, too, is the line that rationalizes such types of distribution as *Lepechinia, Juania,* etc. Obviously, this line is standard; (b) The intersection of 60° Lat. South with 60° Long. West (Magellanian gate) is situated in close proximity by South America. Contrariwise, the intersection of the same latitude with 60° Long. East (approaches of the African gate) is very distant from South Africa; (c) A *genorheitron* of "cold" affinity issued from the former intersection has abundant suitable ranges in near southern South America. A *genorheitron* of the same nature issued from the latter has few footholds nearby (Kerguelen, Amsterdam, Saint Paul Islands) and, if it misses them, can find no suitable ranges until it reaches the mountains of East Africa or, better still, the Western Himalayas—Altais **).

These data are culled out of a diagram (Fig. 91) which is to warn the reader that they are not mathematically exact in detail. However, they well answer the substances in play, revealing that *Magellania and the Altais lie exactly on the same "latitude" from the standpoint of phytogeography. In other words, a genorheitron can reach the Altais with the very same ease it reaches Magellania, always starting from a point in the deep south.*

---

*) We may repeat that geophysicists who believe that the vicinity of the Kilimajaro is a "horst" straight from the crumbled Gondwana are amply supported by plant-dispersal.

**) Though the matter does not bear being made the subject of precise homology, plants that fare to Asia along Long. 60° East are usually rewarded for their longer track by reaching centers (Himalayas, Altais, etc.) which generally secure them in the end a much wider range, possibly circumboreal throught. Plants faring the Long. 60° West channel, on the contrary — if truly "cold" — have for their own but Magellania and the Andes as a rule. Exceptions, however, are not unknown which go to Mexico and beyond along this route.

Applied to the dispersal of *Carex incurva*, the diagram just offered works tellingly. This diagram further shows that the whole of the modern Indian Ocean was once land. The testimony of this diagram is borne out by the entire evidence of dispersal, which is to conclude that we do not dream in accepting its outlines as a sound elucidation of the main substances in play.

One of the finest problems of "Post-glacial" distribution within our immediate notice is in the modern range of *Sarracenia purpurea*. This pitcherplant has a form, subsp. *venosa*, thinly spread along the coast of the Southeastern United States as far north as New Jersey. Northward from New Jersey, this subspecies yields to another, and this time truly wideranging entity, subsp. *gibbosa*. The map illustrating this dispersal (Fig. 92), speaks for itself.

We believe it to be well established (26), that subsp. *venosa* holds on to a range which was established when the ancestors of *S. purpurea* first reached what are now the Southeastern United States. Doubtless, this range was first occupied quite before the inception of the Cretaceous. It was modified in latter ages, but we cannot question its original antiquity. Obviously, this subspecies is a relic.

As to subsp. *gibbosa*, we face two hypotheses, (a) This form was in existence long before the Pleistocene glaciation, and might have begun to evolve on its own, as a matter of fact, in the Cretaceous or Early Tertiary. With the oncoming of the Pleistocene glaciations, subsp. *gibbosa* lost heavily, possibly surviving for a time only in some *nunatak* anywhere in the north (it might have reached Greenland in the first flush of migration), or some well protected strip of the North American coast north of New Jersey. At the end of the Ice Age, *Sarracenia purpurea* subsp. *gibbosa* was gradually freed, and reoccupied large areas in the wake of the retreating ice-sheets; (b) This subspecies, on the contrary, is a form which originated during the glaciations in answer to special glacial conditions. It is not pre-glacial (at least in a measure), but wholly post-glacial.

Let us observe at the start that, whatever the case be, *S. purpurea* subsp. *gibbosa* is a remarkably good example, how a track is brought into being. Hemmed in by glaciations, this form starts at the end of these glaciations a new migration, and a sizeable one at that. Had a migration of this kind taken inception, let us say, at the latitude of Madagascar ages ago, it could well reach very far, for the potential of dispersal revealed by this pitcherplant is remarkable, purely as the post-glacial record reads.

Returning now to the alternatives suggested above, the facts warrant the conclusion that the history of *Sarracenia purpurea* has two phases, (a) *Preglacial*, which dates from the close of the Jurassic, when the sarraceniaceous ancestors came to North America, adhering in the main to a region of "Pine Barrens" and sandy plains; (b) *Glacial and Post-glacial* which has its immediate inception with the colder eras of the Pleistocene. It is obvious that a study of this pitcherplant which intends to be critical is bound to take strict account of these two phases. We do not know, whether *S. purpurea* subsp. *gibbosa* antedates, or not, the Pleistocene, but our guess is that it might. At any rate, we would investigate the matter using all the tools of modern research, cytology, taxonomy, ecology, etc., in close reference to the two hypotheses outlined above. Pure phytogeography, consequently, would be a preliminary to whatever we may do.

Were we bent upon rounding up our investigations, and give ourselves the fullest possible reason of what *Sarracenia purpurea* stands for in the world of botany, we could not overlook a third phase in the history of this plant. This is the phase which opens when the Sarraceniaceae still were in the genorheitral stage *).

Regarding the *genorheitron*, we believe that the Sarraceniaceae, and the droseraceous genus *Dionaea* in particular have affinities toward *Caltha* Sect. *Psychrophila* and *Parnassia*. *Caltha* Sect. *Psychrophila* is conventionally antarctic (30), because it ranges from Australia to Ecuador. Though not strictly "insectivorous", its foliage is strikingly reminiscent of *Dionaea* **). Contrary to current belief, "digestive" glands are certainly not the fundamental character of the insectivorous plants, because an endless variety of absorbing "hairs" are known from a multitude of plants that bear to the carnivorous group no immediate kinship. Contrariwise, *Roridula*, a form which has the entire trappings of carnivory, with seemingly perfect "insectivorous" glands, is no longer capable of digesting animal prey. This loss of function (27) may perhaps be due to the mycorrhizae in the modern root-system of this genus (28) which supply essential nitrogenous elements ***).

So far for *Dionaea*, and the circumstance that *Caltha* sect. *Psychrophila* is not immediately "carnivorous." *Drosera*, on its part, is beyond question intimately related with the Pittosporaceae and the Ochnaceae (*Sauvagesia*

---

*) The full extent of our ignorance regarding the dispersal and origin of the Sarraceniaceae is documented by the following statement by an author (GLEASON in Bull. Torrey Cl. 58: 366. 1931) whose devotion to ecology and "phytogeography" belongs also to the record, "The sole genus of pitcher-plants in South America, so far as known, is *Heliamphora*.... Since the two other genera (*Sarracenia* and *Darlingtonia*) of the family are both North American, the presence of this isolated genus on Roraima *raises phytogeographical questions of great interest for which we have at present no answer whatever*" (*Italics ours*).

**) An incredible amount of loose, if seemingly "orthodox" morphology has been written on the nature of the pitchers of the Sarraceniaceae, and homologous and analogous structures, whether "leaf" or "stem," etc. We cannot discuss the matter here, but call to the attention of the reader the following, (a) The pitcher of *Utricularia Menziesii*, an Australian plant (see LLOYD, F. E., The Carnivorous Plants, pl. 35 figs. 10, 11. 1942) is homologous of the standard pitcher of *Nepenthes* to the point of looking very much like it; (b) This pitcher passes by degrees into very much simplified forms of "traps," such as are typical — supposedly — of *Utricularia* (see loc. cit., figs. 1, 2, 3, etc.); (c) There exists, on the other hand, a form in *Utricularia*, *U. tubulata* (see op. cit., pl. 36, figs. 10, 11), suggesting transitions in foliage between conventional *Utricularia* and *Aldovrandra* (also the myriophyllaceous clan, generally); (d) Structural and morphological premises competent to furnish fullest transition between the *Aldovrandra* type of "trap" and the conventional nepenthaceous pitcher are exhibited by the former genus (see op. cit., pl. 19, fig. 6, a to h). It is worthy of notice that *Utricularia Menziesii* and *U. tubulata* are Australian plants, and that the track of *Aldovrandra* is a good match of that of *Carex pyrenaica* in essentials.

***) An exceedingly fine study in biochemistry is suggested by this plant, which seems to contain a relatively high percentage (28) of rubber. The problem of latex, fundamental in many respects, is immediately called into play by *Roridula*. Investigations assuming that latex is integral part of the present make-up of plants may be vitiated at the start. We incline to suspect that latex, and all the laticiferous system, are in reality *relic-structures long antedating in phylogeny the present morphology and physiology*.

affinity) through the Byblidaceae (Roridulaceae), opinions to the contrary notwithstanding (29), which mistakenly dismiss the affinities between Droseraceae and Byblidaceae as "superficial." We are further inclined to include within the immediate genorheitral affinities of the Droseraceae, as already stated, the small family Tremandraceae and the whole of the affinity centering around *Linum*.

In conclusion, the parentage of *Sarracenia* and the Sarraceniaceae is to be sought in a *genorheitron* having ties in the direction of, (a) Ranunculaceae *(Caltha);* (b) Saxifragaceae *(Parnassia);* (c) Pittosporaceae; (d) Ochnaceae *(Sauvagesia);* (e) Droseraceae; (f) Tremandraceae; (g) Linaceae Lineae. Nepenthaceae and Cephalotaceae further belong in this list. This *genorheitron* is proteiform from the standpoint of morphology, but not at all so obscure as to hide manifest ties of consanguineity, which indeed reach beyond the families mentioned toward other groups which it would be too long to discuss be it so very briefly *).

The migrations of this *genorheitron* stem from a primary center located in and around modern Australia (see for part of it, p. 357) which still easily ranks as the region of the modern world where carnivorous plants are most common and best diversified. Leaving this center, the Sarraceniaceae fared to the New World over a path which took them neither to Magellania, nor to other parts of South America which we may surely indicate, but seemingly directly to the riverine plains later to become the "mesas" of Guiana (e.g., Mt. Roraima). It is possible, as a matter of fact, that this path is the very same one taken by the ancestors of *Gaylussacia* (see p. 447), a genus which today still consorts with the Sarraceniaceae in Venezuela and the Southeastern United States, but, like the Sarraceniaceae, is not known from the West Indies. It is further possible that the ancestors of *Darlingtonia*, a monotype now strictly localized in Oregon and California, took their start from a point in the Pacific, wherein *Heliamphora* and *Sarracenia* also broke company. In brief, we would look for this point at a center in the Pacific of today very nearly equidistant from Oregon, New Jersey and Venezuela, as shown by Fig. 92 and Fig. 88.

Some of the readers may reject as fantastic the idea that a study of *Sarracenia*, purposely intended to cover only part of the United States, can be extended to take in the whole field we have outlined. We feel otherwise, taking for granted that a day is to come, when neither *Sarracenia* nor any other genus will be studied as a "Local problem." We do not believe that "Science" has become too much "specialized" in our time and era. Science can no more become too "specialized" than can common sense be parcelled out to cover marital relations as something entirely divorced from correct behaviour in the halls of a learned society. It will be expedient, of course, to

---

*) Readers interested in the morphology of the "carnivorous clan," ought to turn their attention to the problem offered by *Isoetes* "roots." These „roots" have been shown (STEWART, W. N., in Amer. Jour. Bot. 34: 315. 1947) to be homologous of Stigmarian "appendages." We think time will not be lost by workers effecting critical comparison, *in full freedom from preconceived definitions*, between these "roots" and "appendages", and the "pitcher" of such carnivorous genus as *Genlisea*.

draw in the future a much sharper line than it is done today between the technician and the thinker, who can use many technicians, though perhaps he cannot perform well the more or less exalted tasks of a single one.

Botanical thought has become stilted precisely in the measure that it has become "specialized" and "technical." Uncapable, or unwilling, to consider as one the three basic factors of evolution, Time, Space and Form, this thought now takes pride in announcing from the halls of formidably equipped universities that a certain wood has been found to be "vesselless", or such other similar thing. We think more and better possibilities are open to students of nature, one of the vastest fields of knowledge on earth.

BIBLIOGRAPHY

Chapter VIII

(1) VAN SLOOTEN, D. F. — The Stylidiaceae of the Netherlands Indies; Bull. Jard. Bot. Buitenzorg, III, 14: 169. 1937.

(2) MILDBREAD, J. — *Stylidiaceae;* E. & P. Pflanzenr. 35 (iv. 278). 1908.

(3) VAN STEENIS, C. G. G. J. — Droseraceae; Bull. Jard. Bot. Buitenzorg, III, 13: 106. 1933.

(4) DIELS, L. — *Droseraceae;* E. & P. Pflanzenr. 26 (iv. 112). 1906.

(5) DIELS, L. — *Droseraceae;* E. & P. Nat. Pflanzenf. 17b: 766. 1936.

(6) MASAMUNE, G. — Floristic and Geobotanical Studies on the Island of Yakusima, Province Osumi; Mem. Fac. Sc. & Agr. Taihoku Univ. 11 (Bot.) No. 4. 1934.

(7) STAPF, O. — Graminaceae; Dyer's Fl. Cap. 7: 587. 1900.

(8) DRUDE, O. — *Palmae;* E. & P. Nat. Pflanzenf. 2 (3): 1. 1889.

(9) DAHLGREEN, B. E. — Index of American Palms; Field Mus. Nat. Hist., Bot. Ser., Publ. 355, 14. 1936.

(10) BECCARI, O. — Le Palme Americane della Tribú Corypheae; Webbia 2: 1. 1907; see further same author — Recensione delle Palme del Vecchio Mondo appartenenti alla Tribú delle Corypheae; ibid. 5 (1): 5. 1921.

(11) LUNDELL, C. L. — New vascular plants from Texas, Mexico and Central America; Amer. Midl. Nat. 29: 484. 1943.

(12) DRUDE, O. — see (16) Chapt. IV (see here p. 51).

(13) HEGI, G. — Vaccinium; Ill. Fl. Mittel-Europa; 5 (3): 1667. 1926.

(14) CHIOVENDA, E. — La culla del Cocco; Webbia 5 (1): 199. 1921.

(15) COOK, F. O. — History of the Coconut Palm in America; Contr. U. S. National Herb. 14 (2): 270. 1910.

(16) SKOTTSBERG, C. — see (10) Chapt, I.

(17) HUTCHINSON, J. — The Families of Flowering Plants, II. Monocotyledons. 1934.

(18) SAMUELSSON, G. — Die Verbreitung der *Alchemilla*-Arten aus der *vulgaris*-Gruppe in Nord Europa (Fennoskandien und Daenemark); Act. Phytog. Suec. 16: 1. 1943.

(19) HULTEN, H. — Stellaria longipes Goldie and its allies; Bot. Notiser 1943: 251. 1943.

(20) ENGLER, A. — Über die Vegetationsformationen Ost-Afrikas; Zeitschr. Gesell. Erdkunde Berlin No. 4/6: 254. 1903.

(21) PERRY, L. M. — A tentative revision of *Alchemilla* sect. *Lachemilla;* Contr. Gray Herb. 84: 1. 1929.

(22) PAX, F. & HOFFMANN, K. — *Caryophyllaceae;* E. & P. Nat. Pflanzenf. 16c: 274. 1934.

(23) FERNALD, M. L. — Recent discoveries in the Newfoundland flora; Rhodora 35: 56. 1933.

(24) KUEKENTHAL, G. — see (1) Chapt. V.

(25) RYDBERG, P. A. — Flora of the Rocky Mountains and adjacent Plains. 1917.

(26) WHERRY, E. T. — The geographic relations of *Sarracenia purpurea;* Bartonia No. 15: 1, 1933.

(27) LLOYD, F. E. — Is *Roridula* a carnivorous plant?; Canad. Jour. Res. 10: 780. 1934.

(28) MARLOTH, R. — Flora of South Africa 2: 26. 1925.

(29) HUTCHINSON, J. — The Families of Flowering Plants, I. Dicotyledons. 1926 (see here p. 158).

(30) HILL, A. W. — The genus *Caltha* in the Southern Hemisphere; Ann. Bot. 32: 421. 1918.

CHAPTER IX

NON-ANGIOSPERMOUS DISPERSAL

Non-angiospermous dispersal ought to be studied in a special manual because of its importance and vastness in time and space. We review here, nevertheless, certain Cycads, Conifers and Ferns in order to prove that non-angiospermous plant-life migrates in a manner consistent with the dispersal of the Angiosperms throughout. So much cannot be strange. There never was, in the dim past and the immediate present as well, a channel of migration on earth, reserved to the inroads of a single plant or phylum. *Pinus* and *Podocarpus* never could be made recipient by nature of favors denied *Erythroxylon* or *Sarracenia*, because no plant ever migrated alone.

As we have seen, the origin of a group can be known whenever it proves feasible to effect critical comparison between *genorheitron* and migration. This is because plant-life evolves in the measure it migrates. Formal evolution and dispersal only then make sense when both are seen to proceed along parallel lines. We are fortunate in that these lines are, on the whole, fairly discernible as regards the Angiosperms, for this group appears in the south of the modern maps in an epoch anterior to the end of the Jurassic, and from then on can be followed in migration to the farthest corners of the earth that still is ours. This is not to say that the Angiosperms actually originated as late as the end of the Jurassic, but to affirm that their migrations, and the progress of their evolution as well, can be mapped out with definite reference to certain epochs and certain lines of distribution.

The case is quite other as regards Cycads, Conifers, Ferns, Liverworths, etc., for the origins of these plants is lost, so far as presently known, in the dimmest night of the ages. We can follow their migrations of course, so long as we compare them as such with those of the Angiosperms, but we cannot safely orient them. In other words, finding a certain Angiosperm in Mexico we can trace its affinities up to the point when we are reasonably sure by the facts on record that this plant reached Mexico faring in the main from the south to the north of our maps. Not so as regard such a genus as *Zamia*, for example. *Zamia* might have originated for all we know in a land which is today the United States, and migrated to Bolivia southward from a starting point in the north. In brief, we understand the starting points of angiospermous tracks, but we do not know the inception of the cycadaceous channels of migration. We can to a point orient the former, but we cannot orient the latter.

This is a fundamental difference, but it is not so important as to prevent our effecting fruitful comparison between angiospermous and non-angiospermous migrations. If *Zamia* ranges from Mexico to Northern Brazil and Bolivia, it stands to reason that these lands were open to *Zamia's* inroads, and it is always possible for us to compare the behaviour of *Zamia* in these lands with the behaviour in them of the Angiosperms. In the end, we cannot receive indications from the non-angiospermous plants of trends and fundamentals in migration which basically conflict with the indications we secure from the Angiosperms.

We might add that not even in the Angiosperms identity of migration always bespeaks identity of origin. To illustrate: *Drosera spathulata* and *D. rotundifolia* (see p. 438) share a common track between the approaches of Southern Japan and Central Japan. This notwithstanding, the origin of the former species can be traced back to the neighbourhood of New Zealand, much to the south of Japan, while that of the latter is set within a secondary center of speciation located among Japan, Formosa and Hawaii.

In conclusion, we can at present study the tracks of non-angiospermous plants only as comparative material. We can hardly hope to handle them by themselves. However, certain common factors stemming from the geological vicissitudes of the earth, have in time molded all these distributions, non-angiospermous and angiospermous alike, in a similar manner. The paramount effect of these factors upon general dispersal has been detected by authors other than ourselves (1, 2; cf. with 3), who, aware that unrelated plants migrate alike, have correctly inferred that dispersal is keyed out to a few essentials.

If we consider that life interlocks throughout, we may further advance the sober conclusions that animal dispersal bears being compared with plant distribution. Comparisons between the two dispersals, animals and plants, are entirely out of our present purposes, however, because too large a mass of evidence cannot be considered within the pages of a manual intended to remain true to the essentials factually revealed by plant-life. We are certain, nevertheless, that the task we cannot fulfill today will be achieved in the not too distant future by biogeographers who will find useful we here the tracks supply in regard of plant-dispersal. These tracks cannot materially differ from those of animal life coaeval with the Angiosperms, or of later origin, that is to say in the main, of animal life of Late Jurassic or posterior age. The outline of these tracks, no doubt, will contribute to rid the field of biogeography of the mysterious and the unusual which have absolutely no place in science.

I. Cycadaceae

It is not to be doubted that this family was already old in the latter part of the Jurassic, and by then a conspicuous element in the pre-angiospermous floras. It origins are obscure, but it is possible that the ultimate cycadaceous origins are to be sought in some pteridospermous stock.

To orient the tracks of the Cycadaceae is impossible at this hour, because we are entirely in the dark as to the centers of their origin. This is no bar against using their distribution in phytogeographic work, however; it will soon be learned that this falls in very well indeed with the main lines of angiospermous dispersal.

The Cycadaceae consist (4) of ten genera in two subfamilies, Cycadoideae and Zamioideae.

The former contains but a single living genus, *Cycas*, of about 8 species. As the map reveals (Fig. 93), *Cycas* is a transparent example of pure Gondwanic dispersal. It ranges throughout the tropical Far East, Malaysia and Western Polynesia as far as Fijis, Carolines and Bismarck Archipelago. Its westernmost limits fall in with the Himalayas of Nepal, where the widespread *C. circinalis* occurs most frequent in pine-woods. On the African side of the dispersal, *C. circinalis* is well represented by a subspecies, *madagascariensis*, distributed to Madagascar, the Mascarenes, Comoros, Zanzibar and part of the African coast in the immediate hinterland of the last named island. It should be indeed difficult to imagine a more conventional dispersal.

We often mentioned Indochina and some the islets nearby as among the most ancient lands of the Far East. Right here occurs *Cycas* sect. *Indochinenses* with two species, (a) *C. siamensis* (Pulo Condor, Cochinchina, Siam, Burma, Southern China); (b) *C. Micholitzii* (Annam); which is quite according to anticipations. These two species do not invest Java, Sumatra or Malacca but like *Protium*, *Triumfetta repens*, etc. stress a region father east and north. This region is well known to us, and beyond doubt a very important one in dispersal.

Cycas Rumphii is none the less distributed within a classic range, Ceylon—Nicobar Islands—Java—Celebes—Amboina—Northern Queensland—New Caledonia reminiscent of the eastern spread of mangroves or near-mangroves. Also classic, though this time directed toward other territorial goals, is the dispersal of *C. revoluta* by a channel Eastern Bengal—Tonkin—Southern China—Formosa—Riukius—Southern Japan.

The wide dispersal still retained by *Cycas* in the modern world can be explained without difficulty in consideration that *Cycas* is most cases a plant of the marine shore or its immediate hinterland. As a near-mangrove, *Cycas* was favored by the very same factors which assisted *Lumnitzera*, the Dipterocarpaceae, Barringtoniaceae, etc. Though very old, this genus never became "senescent," because it always found abundance of habitats to its liking, not only, but these habitats became more diffusive and numerous by the breaking up of Gondwana, Malaysia and the Western Pacific shore generally. Considering that *Cycas* made no dent in Continental Africa to the exception of an insignificant lodgement in Tanganyka *), we must conclude that *Cycas* would

---

*) Geophysicists who believe that the Kilimajaro and some of the adjacent heights are part of a "horst" representing a surviving segment of Gondwana — we repeat — have much in their favor by the facts of dispersal.

be today very poorly represented had it not partaken from the very start of the nature and preferences characteristic of near-mangroves. *Cycas* would have survived as a rare relic there only, where the ranges it espoused *in the Late Jurassic* lasted practically intact to this day. We may safely draw the inference from this, that the pre-angiospermous floras were killed off wholesale not on account of an inherent biological weakness of all their components but because of the near-total subversion of the ranges in which they had established their climaxes. In the last analysis, a climax is a source of strength quite as much as a mark of "senescence" and a danger-line. However, even plants of very ancient vintage, like *Cycas*, can weather any storm so long as they constantly find ranges to establish climaxes or semi-permanent seres to their convenience. In sum, *Cycas* is pretty much of one cloth with *Lumnitzera* and the Dipterocarpoideae which explains its amazing survival and large modern range despite the relative paucity of its species [*]).

The Cycadaceae Zamioideae contain a number of petty genera, none of which is wanting in interest. *Bowenia*, a monotype typified by *B. spectabilis* is narrowly confined to Northern Queensland, which is one of the oldest lands in continuous emersion and a cornerstone of angiospermous dispersal both from the Indian Ocean and the Western Pacific. *Dioon* is in three species, as follows, (a) *D. Purpusii* — Mexico (Oaxaca); (b) *D. spinulosum* — Mexico (Oaxaca, Yucatán, Veracruz); (c) *D. edule* — Mexico (Veracruz, San Luís Potosí, Tamaulipas, Nuevo León). The range tenanted by *Dioon* is recognized without difficulty as one of the most ancient of North America, the home of a host of odd cactoid monotypes or oligotypes and, oftentimes, also of a unique angiospermous flora. It might not be impossible that the survival of *Dioon* in the range indicated is due to the fact that Western Mexico had in ages bygone connections with Eastern Mexico only of such a nature which *Dioon* could not use (see, e.g., p. 358 fn.).

Zamia, the largest cycadaceous genus living is distributed in the following pattern,

A) Sect. *Caribeae*

(1)   *Z. floridana* — Florida.
(2)   *Z. pygmaea* — Western Cuba.
(3)   *Z. angustifolia* — Eastern Cuba.
(4)   *Z. Chamberlainii* — Cuba.
(5)   *Z. media* — Bahamas, Cuba, Santo Domingo, Jamaica, Portorico.

B) Sect. *Mexicano-meridionales*

(6)   *Z. Fischeri* — Mexico.
(7)   *Z. Wielandii* — Mexico.

---

[*]) *Gnetum* (see p. 271) is of one cloth, too, with *Cycas*.

(8) *Z. Loddigesii* — Mexico (Veracruz) British Honduras, Guatemala, Salvador, Venezuela (Caracas, Aragua).

(9) *Z. muricata* — Mexico (Oaxaca), Guatemala, Colombia, Venezuela (Falcón, Carabobo, Aragua).

(10) *Z. furfuracea* — Mexico (Veracruz), Colombia (Antioquia).

(11) *Z. Lawsoniana* — Mexico (Oaxaca), Brazil.

(12) *Z. obidensis* — Brazil (Pará).

(13) *Z. Lecontei* — Brazil (Pará, Amazonas).

(14) *Z. boliviana* — Brazil (Matto Grosso, Eastern Brazil), Bolivia (Santa Cruz).

C) Sect. *Centrali-meridionales*

(15) *Z. Verschaffeltii* — Mexico.

(16) *Z. Turckeimii* — Guatemala.

(17) *Z. acuminata* — Nicaragua.

(18) *Z. Skinneri* — Panamá, Costarica.

(19) *Z. obliqua* — Colombia (Chocó).

(20) *Z. Chigua* — Colombia (Chocó).

(21) *Z. cupatiensis* — Colombia (Cerro de Cupaty).

(22) *Z. montana* — Colombia (Antioquia).

(23) *Z. Wallisii* — Colombia (Antioquia).

(24) *S. Lindenii* — Ecuador (Guayas).

(25) *Z. Poeppigiana* — Colombia (Nariño), Peru (Amazonas).

(26) *Z. pseudoparasitica* — Colombia (Valle del Cauca), Ecuador, Peru.

(27) *Z. Ulei* — Peru (San Martín), Brazil (Amazonas).

We cannot orient the tracks of *Zamia*, because we have no means of knowing whether this genus originated in Bolivia rather than in Mexico, though the latter is the more probable hypothesis in the light of the massiveness of the cycadaceous flora endemic to North America in the Jurassic. If we cannot orient the tracks of *Zamia*, we can understand its distribution well indeed, however. Section *Caribeae* is an harmonious whole that brings together lands such as Florida and Cuba which it is difficult indeed to separate with finality in dispersal. The Bahamas are the standard approach to Florida from the direction of Santo Domingo, Portorico and Jamaica, not to mention Cuba.

Section *Mexicano-meridionales* is firmly rooted in Mexico, and from Mexico runs channels of migration some of which are striking, but none of which is without precedent. What we have learned of the distribution of *Tetragastris* (see p. 393) and other genera readily accounts for the ranges of such species as *Z. Loddigesii*, *Z. muricata*, *Z. furfuracea* and *Z. Lawsoniana*. This group moreover use the standard channel leading from Colombia and Venezuela to the southwestern Amazonian regions and, finally, Bolivia.

Section *Centrali-meridionales* is more firmly anchored to Western America than the section discussed immediately above. A dualism in distribution of the sort is evident in groups other than the Cycadaceae (see, for example, in the Cactaceae the *Mammillaria-Malacocarpus* affinity as compared with the group *Trichocereus-Echinocereus*), and answers fundamental realities of American palaeogeography. Section *Mexicano-meridionales* mostly used crumbled ranges in the modern Caribbeans; Sect. *Centrali-meridionales*, on the contrary, mostly went by crumbled ranges in the modern Pacific.

In sum, it should be difficult to unearth an *angiospermous* group running quite as true to the *angiospermous* norm as does the *non-angiospermous* genus *Zamia*. The massing and course of the species of *Zamia* is contrived by the book, as it were; we could hardly hope to do better, if we proposed to arrange a synthetic example, how angiospermous tracks ought to run in all the *Zamia* range. Whether this cycad originated in Mexico, Florida or Bolivia matters not at all. Indeed, the Angiosperms themselves run through the *Zamia* range one way or the other, south to north or north to south, witness *Brunellia* and *Schoenocaulon*. Patently, to orient the tracks of *Zamia* is a matter of secondary importance once we have of these tracks a precise understanding in the light of an imposing array of angiospermous distribution. *Certain proof is here reached that the tracks of dispersal we can draw on modern maps are not subjective and fictitious, rather stand as the genuine expression of an ancient geography underlying the geography of the present.*

Encephalartos is a genus in the affinity of *Zamia* endemic to Africa. About two dozens of its species are reported (5) from South Africa alone. The Transvaal accounts for little less than a third of this total. Four of the Transvaalian species seem to be confined within the alpine arc running from the Zoutpansberg (Northern Transvaal) to about Barberton (Eastern Transvaal). This is one more reminder of the preeminency that this small town holds in things phytogeographical. Other South African forms of *Encephalartos* are endemic to the Western Cape, Zululand, Natal and Mozambique.

Outside of South Africa, as delimited, are found,

(1) *E. gratus* — Nyasaland (Mt. Zomba and Mt. Mlanje).

(2) *E. villosus* — Cape, Natal, Pondoland, Uganda.

(3) *E. Hillebrandtii* — Zanzibar, Tanganyka, Uganda.

(4) *E. Laurentianus* — Uganda, Congo.

(5) *E. septentrionalis* — Southern Sudan, Northwestern French Equatorial Africa.

(6) *E. Barteri* — Nigeria, Dahomey, Togo, Gold Coast between 7° and 9° Lat. North.

We may not contend that *Encephalartos* originated within the *Afroantarctic Triangle*, but we are reasonably sure that it came to Africa from this center, and not from the direction of Gondwana, because, (i) *Encephalartos* is wholly unknown in Madagascar and the Mascarenes; (ii) A massive concentration of its species is present in Natal and the Transvaal; (iii) *Stangeria*, a monotype of the Zamioideae, is also endemic to Natal. Natal and the immediately adjacent part of Transvaal, then, stand as the door through which *Encephalartos* and *Stangeria* both filtered into Africa from the *Afroantarctic Triangle*.

Out of Natal and the Transvaal, *Encephalartos* runs a conventional course, keeping clear altogether of the region once occupied in Central and Western Africa by large lakes or inland seas. This course is fairly closely paralleled by *Suaeda* sect. *Lachnostigma* (see p. 382) and is generally standard of uncounted Angiosperms.

The last remaining cycadaceous genera, all of them of the Zamiodeae, are the monotypic *Microcycas*, endemic to Eastern Cuba; the oligotypic *Ceratozamia*, restricted to Mexico with two species; and *Macrozamia*.

Macrozamia consists of about ten species, most of which are in Eastern Australia (New South Wales and Queensland). Only one species, *M. Preissii*, is endemic to Central and Southwestern Australia. The distribution of this species is of great interest because it immediately connects the MacDonnell Range of Central Australia with the Swan River of Southwestern Australia. The latter is a significant landmark of distribution because it stands as the northernmost limit of the massive dispersal of forms that belong to the genuine Southwestern Australian flora. In other words, New South Wales and Queensland, Northern and Central Australia and Western Australia form a block as far south and west as the Swan River. South of this river begins the domain of another flora, part of which reached Western Australia by purely antarctic paths. This "antarctic" flora is the one commonly accepted as the "genuine" Western Australian plant-world. We regret we may not discuss the matter further.

In conclusion, the Cycadaceae, though not angiospermous, migrate very much in the angiospermous manner. As a matter of the record, *Cycas* yields perfect Gondwanic patterns of distribution; *Dioon* stresses significant Mexican ranges; *Zamia* illustrates most tellingly vital phases of American dispersal; *Encephalartos* is typical of the course in Africa of non-Gondwanic elements; and confirms that the region Natal—Transvaal is of fundamental significance; *Macrozamia* is most revealing of certain important connections between Eastern and Western Australia; *Microcycas, Ceratozamia, Stangeria, Bowenia* uniformly pinpoint centers of utmost significance in general dispersal.

Things could not be otherwise indeed. The earth was not made to take care of the Angiosperms in one way, the non-Angiosperms in another. As we have ly understood angiospermous dispersal perhaps correctly in the main, so we have no difficulty in bringing this dispersal into harmony with non-angiospermous distribution.

II. Coniferae

The dispersal of plants of the gymnospermous description is passingly mentioned in various chapters of this manual. To round up the record rather than to add much to the weight of what we have stated regarding dispersal in general, we will additionally revise here certain Conifers.

Araucaria has a long history which goes back to the Carboniferous and the Permian. Fossils of this genus, or probably to be credited to its immediate ancestors, are reported from Europe, South America, New Zealand, Tasmania, Java, Kerguelens, Madagascar and the Nile Valley, North America. Antarctica. In short, *Araucaria* seems to have avoided in its age-long wanderings only one continent, Asia, and it is probable that here, too, it will sooner or later be discovered fossil.

The modern genus is distributed (6) as follows,

A) Sect. *Colymbea*

(1) *A. araucana* — Chile, Southwestern Argentina.

(2) *A. angustifolia* — Brazil (Rio Grande do Sul, São Paulo, Minas Geräes).

(3) *A. Bidwillii* — Queensland.

(4—6) *A. Hunsteinii, A. Schumanniana, A. Klinkii* — New Guinea.

B) Sect. *Eutacta*

(7) *A. Cunninghamii* — New South Wales, Queensland.

(8) *A. excelsa* — Norfolk Island.

(9—13) *A. columnaris, A. Rulei, A. Muelleri, A. Balansae, A. montana* — New Caledonia *).

The dispersal of *Araucaria* (Fig. 94) is conventionally "antarctic," and the Araucariaceae are identified as an "antarctic" element for this reason. It is patent that, for example, this genus followed the shore of a geologic austral continent to reach both Chile and Norfolk Island. Its migrations in these quarters bear being compared with those of numerous Angiosperms, and yield the same indications in detail. Norfolk Island is a key-point, and it is not surprising that it harbors an endemic species of *Araucaria* together with endemic forms of various angiospermous plants.

However, we are by no means certain that *Araucaria* reached Norfolk Island from the south or the southeast. This genus might have reached this island, and the antarctic shore generally, from the north or the west. We do not know its *genorheitron* well, and even less are we prepared to identify its

---

*) In a letter (November, 1948), Prof. J. T. BUCHHOLZ reported tot the auhor of this manual the discovery of new species of *Araucaria* and other Conifers in New Caledonia. The massive speciation of these plants in New Caledonia, then, is not due to "Taxonomic faux pas," but to the circumstance that New Caledonia has retained much of a flora nearly extinct elsewhere. This flora includes, too, a world of significant Angiosperms of every description. Believers in the "Holarctic" origins of the Angiosperms have nothing to contribute to the pointed question, *why New Caledonia, the Mascarenes, etc. rate as high as they do in the annals of dispersal.* That New Caledonia emerged only in the Tertiary is of course impossible. This island is one of the oldest pieces of land afloat insofar as we are concerned.

geographic origin *). The contrary is true of the Angiosperms, for we are fairly well informed of their *genorheitra*, and a critical comparison of the tracks of some threehundred of their families always makes it possible for us to get bearings with some finality.

If we cannot be clear, where and when *Araucaria* actually originated, we may not doubt on the other hand that the lands wherein *Araucaria* dwells in the immediate present, (1) Were once suitably connected overland; (b) Have remained in state of undisturbed, or scarcely disturbed, emersion for long ages; (c) Retained throughout conditions of climate and soil agreable to this genus, or at least well within its genetic powers to meet.

It will also be noticed, (a) The dispersal of *Araucaria* and *Libocedrus* broadly follows the outlines of the geologic shores of the Pacific; (b) *Araucaria* sect. *Eutacta* is rooted very much by the center, New Caledonia—Norfolk Island—Fiji—Samoa—Tonga, which is important for the Western Polynesian Violaceae (see p. 181) in the affinity of *Melicytus* and *Hymenathera*, and the bird-life, generally, of the Pacific.

Had anyone of these conditions failed, *Araucaria* would have become extinct in Brazil, Norfolk Island or Queensland as it became in the Kerguelen archipelago or Madagascar, and could never have reached both Argentina and New South Wales. It is accordingly clear that this genus bears testimony to the fact that migrations were possible in the distant past between Magellania and the Western Pacific, and that these lands contain cores of the earth the origin of which is lost in the very night of the ages. Let us observe that if we may not claim that *Araucaria* originated in Gondwana or Antarctica, we may on the other hand use the present dispersal of this genus as evidence that what the Angiosperms reveal is true and correct. The occurrence of fossils of *Araucaria* in the Kerguelen Islands (see p. 253) is certain proof in itself that these outposts were once connected with the "antarctic" shore.

Genus *Libocedrus* is dispersed as follows (Fig. 94; see also Fig. 11),

A) Subg. *Eulibocedrus*

(1) *L. uvifera* — Magellania (Fuegia to Southern Chile).
(2) *L. chilensis* — Magellania (Southern to Central Chile).
(3) *L. Bidwillii* — New Zealand.
(4) *L. plumosa* — New Zealand.
(5) *L. torricellensis* — New Guinea.
(6) *L. arfakensis* — New Guinea.

---

*) It is claimed (HOLLICK, A., & JEFFREY, E. C., in Amer. Natur. 40: 189. 1906) that the Cretaceous *(Raritan Formation) Protodammara* probably was the "Last survivor of an ancient Araucarian line of descent joined near its base with the primitive stock of the Abietineous and Cupressaceous series." This might be satisfactory as to phylogeny, but altogether too late as to age. The line in question was by then thoroughly well differentiated, and fossils of this sort throw little, if any, light on the problem of origin in time and space.

(7) *L. papuana* — New Guinea, Moluccas?
(8) *L. austrocaledonica* — New Caledonia.

B) Subg. *Heyderia*

(9) *L. macrolepis* — Formosa, China (Yunnan), Indochina (Annam).
(10) *L. decurrens* — United States (California to Oregon).

Insofar as the Pacific is concerned, this type of dispersal (Fig. 94) matches in every essential that of the Coriariacee, Magnoliaceae and Fagaceae. *Libocedrus* disappears between New Guinea and Formosa as does *Coriaria*, and migrates from Formosa to Southwestern China and Annam in a manner characteristic of the main trends of the dispersal of *Magnolia* and its group. Like the Magnoliaceae and Fagaceae, *Libocedrus* has one subgeneric group ranging in the Pacific from New Guinea to Chile, another in the northern waters of the same ocean.

It would seem that climate has been the agent that molded *Libocedrus*'s dispersal throughout. We do not know what once stood in the open Pacific of today east of the Philippines, but, as we have suggested, it is altogether likely that land, possibly high chains of mountains were in these quarters. A "cold" track runs here as certainly as it does in the Indian Ocean and the Atlantic.

The path taken by *Libocedrus* to North America is less speculative than live and fossil distribution appears to suggest. We see no reason to believe that in ages past Greenland was any colder than Fuegia is today. Knowing that the maps of the present are in no way a reliable index of the geography of the past, we are not averse to believing that Greenland was actually "floated" northward somewhat to occupy the position which it holds today. Far as we are from accepting the Wegenerian cosmology (see p. 512), we admit that moderate shifts of certain lands northward is a possibility. All in all, we see no reason whatever to reject as contrary to the trends of the evidence in our hands the probability that *Libocedrus* reached its northern stations, fossil and live alike, from a point of original landing in the New World roughly agreeing with modern California *). The tracks of *Libocedrus* are comparable to those

---

*) The whole of the angiospermous and non-angiospermous dispersal to North America will have to be reinvestigated in the light of the fact that modern Alaska is certainly not *the* "Landbridge" between Eurasia (with the Far East) and the northern New World. Alaska, on the contrary, might have been reached by angiospermous life *moving northward by secondary migrations only*. Beyond doubt, two of the major landbridges feeding angiospermous genorheitra to the United States, and the Caribbeans generally, were located (a) In the vicinity of modern California; (b) In the vicinity of modern Florida; and operated, respectively, from the Pacific and the Atlantic. The Southern Appalachians and the Ozarks appear indeed to be one of the foremost phytogeographic nodes of the United States, for they were easily reached both from the Pacific and the Atlantic. We have discussed Alaska in a previous chapter (see p. 33) as an illustration of what we understood as a "Landbridge," but we warn again the reader that we never intended *modern* Alaska to be *the* "Landbridge" between the Old and the New World.

of *Euphrasia*, as we have shown in a previous chapter (see p. 29), not only, but those as well of the Magnoliaceae, Fagaceae, Coriariaceae. The geologic valley now sunk under the South China Sea (see p. 123) was doubtless inhabited by *Libocedrus* in association with *Magnolia, Quercus, Prunus, Buxus*, etc., when the Cretaceous was as yet very young, and perhaps millions of years before then.

Libocedrus belongs to the Cupressaceae, which, according to our authority, are further distributed in this manner,

(1)   *Fitzroya (Diselma)* — 2 species in Chile and Tasmania.

(2)   *Callitris* — 20 species in Tasmania, Australia, New Caledonia.

(3)   *Callitropsis* — 1 species in New Caledonia.

(4)   *Actinostrobus* — 2 species in West Australia.

(5)   *Widdringtonia* — 5 species in Southern and Eastern Africa.

(6)   *Tetraclinis* — 1 species in Morocco, Southern Spain, Algeria, Tunisia*).

(7)   *Fokienia* — 2 species in China (Yunnan, Kwangsi, Kweichow, Fukien, Chekiang) Indochina (Tonkin, Annam).

(8)   *Thujopsis* — 1 species in Japan.

(9)   *Thuja* — 6 species in Eastern Asia and North America.

(10)   *Cupressus* — 12 species in the Mediterranean, Africa (Central Sahara), Asia and Western North America **).

(11)   *Chamaecyparis* — 6 species in Formosa, Eastern Asia, North America.

(12)   *Arceuthos* — 1 species in Central Greece.

(13)   *Juniperus* — 60 species in the northern hemisphere, southward to the West Indies, Nyasaland ***), India, Formosa.

---

*) *Tetraclinis articulata* is said (MARKGRAF in Ber. Deutsch. Bot. Gesell. 52: 68. 1934) to have lived in the Pliocene near Frankfurt a/M, which is a standard possibility. The author cited credits it to the "Altmediterrane Element," and adscribes further *Caralluma europaea* to the "Palaeotropichen Altareal"; *Pinus Peuce, Prunus lusitanica, Forsythia* to the "Arktotertiaere Element." To still another, this time unnamed „Element," he credits *Cneorum, Liquidambar, Geum*, etc. This last "Element", we believe, is a mixture from African and Eastern *standard* tracks to Europe, which the reader may easily ascertain by himself, (e.g. *Geum coccineum* — Bosnia, Hercegovina, Serbia (Old), Bulgaria, Thrace, Macedonia, Albania, with 4 allies in the Mediterranean, 1 in the Cape, 4 in South America, 1 in North America). The author cited concludes, however, "Trotzdem bleibt diese Beziehung die Engler für Makaronesien ebenfalls über Asien auflösen wollte, noch etwas ungeklärt." We do not think so, naturally, after we know how to read elementary tracks, and stop using such terms as "Altmediterrane," "Palaeotropisch Altareal," etc.

**) *Cupressus Dupreziana* is one of the most striking relics discovered not so long ago in the Central Sahara (Hoggar, Tassili-n-Ajjer, Djanet, Ghat). Its range connects, at least in the geographical sense, that of *Widdringtonia* with that of *Tetraclinis*, and immediately preludes to the Mediterranean and Western Asiatic distribution of *Cupressus sempervirens*. The balance of *Cupressus* belongs to the East (Himalayas, China) and North America (Mexico and the United States). *Cupressus* has typic "Holarctic" distribution, accordingly, but this distribution is matched by the "Antarctic" ranges of *Fitzroya, Callitris, Callitropsis, Actinostrobus*, which shows once again that the "Holarctic" hypothesis of dispersal does not hold firm even as regards the Conifers. Like *Washingtonia, Cupressus* suffers from a bad case of "Species senescence," (see WOLF & WAGENER, The New World Cypresses (El Aliso 1): 9—10. 1948) which is due to the same causes throughout. We regret we could not learn more of the affinities of *C. Dupreziana*, particularly as regards the controversy still raging about the origin and range of *C. lusitanica*. This species is credited by WOLF & WAGENER to a range Central Mexico to Guatemala and Costarica, which is not the opinion of MARTINEZ (see Anal. Inst. Biol. México 18: 89—90 1947).

***) *Juniperus* is one more of the "cold" groups that turn up between the Transvaal and Ethiopia, and more frequently so between Tanganyka and Ethiopia. See further *Vaccinium, Primula, Salix, Populus*. This evidence is to be read in conjunction with the fact that a genorheitral stream of "cold" forms

The Taxodiaceae are a most ancient group of Conifers and suggest affinities both toward the Pinaceae and Cupressaceae. They consist of but few genera, thus,

(1) *Sciadopytis* — 1 species in Japan.

(2) *Sequoia* — 2 species in California.

(3) *Taxodium* — 3 species in Mexico and Southeastern North America.

(4) *Glyptostrobus* — 1 species in China (Kwantung, Fukien).

(5) *Cryptomeria* — 1 species in Japan (Shikoku, Kyushu, Honshu, Yakushima), China (Anhwei, Fukien, Kweichow, Szechuan, Yunnan), Northern Burma, Eastern Himalayas (Sikkim).

(6) *Taiwania* — 1 species in Formosa, China (Yunnan), Southeastern Tibet, Northern Burma.

(7) *Cunninghamia* — 2 species, 1 *(C. Konishii)* in Formosa, 1 *(C. lanceolata)* in China (Kiangsu, Chekiang, Fukien, Kwantung, Hupeh, Hunan, Kweichow, Kwangsi).

(8) *Arthrotaxis* — 3 species in Tasmania.

Once more, we have no information worth the name as regards the *genorheitra* of these genera and tribes. Doubtless, some of their number (e.g., *Sciadopitys*) are of capital importance for the phylogeny of the "pinaceous" leaf. Some of their number, too, surely evolved to their present forms (that is to say, *originated*) in southernmost Western Polynesia or Antarctica, witness *Fitzroya* and *Arthrotaxis*. It seems safe in addition to conclude that *Callitris*, *Callitropsis*, *Actinostrobus* and *Widdringtonia* belong to a closely knit *genorheitron* which, matching in this numerous Angiosperms (e.g., *Geniostoma* and *Vaccinium;* see p. 169), streamed into the modern world out of the African and Western Polynesian gates. *Tetraclinis* falls within this *genorheitron*, too, for the range it haunts has ties in the Southern Atlantic, not only, but Eastern Africa as well.

Characteristically, *Callitris*, *Widdringtonia* and *Tetraclinis* are within reach of three widely scattered but otherwise narrow-ranging insectivorous genera, *Byblis* (Northern Australia) *Roridula* (Cape), *Drosophyllum* (Morocco, Portugal, Southern Spain), and the last is reminiscent of the most peculiar of the North American species of *Drosera* (*D. filiformis:* Var. *typica:* Delaware to Massachussets; Var. *Tracyi:* Florida to Louisiana, Georgia and South Carolina). Droseraceae and Conifers seem to agree, whether it be in the Philippines, the "Pine-barrens" of the North American Atlantic coast, or the five continents.

---

running into the several hundreds, possibly more, follows a track in the waters of the modern Indian Ocean more or less parallel with the eastern coasts of modern Africa. It is not unlikely that the influence of this track is felt also in the respect of fauna. It is known (HEPTNER in Bull. Soc. Nat. Moscou, Sect. Biol., 50: 36. 1945) that the animal world of Turkestan is permeated by certain "African" elements most definitely associated with psammophily. The general rigidity of animal dispersal would seem to be quite as marked as plant-life's. The author cited affords of it good examples (see, e.g., op. cit. footnote p. 38). To return to plants: see further the distribution of the Sapindaceae Koelreuterieae (p. 61); the *Cobresia—Schoenoxiphium* affinity (p. 237); *Potentilla* and *Carex* in the affinity of *C. incurva* and *C. physodes* (p. 463, 464); the report of a fossil wood *Aceroxylon* with the characters of *Acer campestre* in the Cretaceous of Madagascar (p. 214), etc., etc.

If we may not speak of tracks in the conventional sense as regards the Conifers generally, we are free to affirm, on the other hand, that these plants migrated using all the channels open to angiospermous inroads, which the genera briefly discussed make clear. The distribution of the Chinese and Japanese Conifers makes it clear furthermore, (a) That Southern and Central China are fragments of the Jurassic "Angaraland"; (b) That the climate and conditions generally of this region underwent comparatively insignificant changes from the Late Jurassic — at least — to the immediate present. Shifts of the poles in the Wegenerian manner (see also p. 513) are accordingly wholly out of question, all the more so that the persistence of *Araucaria* in Norfolk Island, New South Wales and Brazil rhymes to the same as does the dispersal of the Chinese and Japanese Conifers. These plants live today in the very same lands which were theirs in the Late Jurassic, and the fact we speak of them under the names Szechuan or Yunnan does not make these lands any more modern. We are all too prone to imagine that "Angaraland" and "Tethys" belong to the mythology of the earth, while, in reality, fragments out of these odd sounding territories may be right under our feet biologically speaking.

The Conifers clearly outline two lines of penetration and approach to Japan, both of which have deep significance, witness,

(1) *Cryptomeria japonica* — Eastern Himalayas, Burma, China (Yunnan, Szechuan, Kweichow, Fukien, Anhwei), Japan (Yakushima, Kyushu, Shikoku, Honshu).

(2) *Cunninghamia lanceolata* — China (Kweichow, Kwangsi, Hunan, Hupeh, Kiangsu, Chekiang, Fukien).

(3) *Cunninghamia Konishii* — Formosa.

(4) *Taiwania cryptomerioides* — Burma, Southeastern Tibet, China (Yunnan), Formosa.

One of the lines in question *(Cryptomeria)* leaves the Himalayas to cross China and reach Japan at, or beyond, the island of Yakushima; the other *(Cunninghamia, Taiwania)* departs from the Himalayas straight to Formosa, there being seemingly no known species of the Taxodiaceae common to this island and Japan. Let us notice that sound evidence suggests that the "bridge" departing from the Far East to reach the New World started precisely between Central China and Yakushima *) (see p. 438).

---

*) The wealth of plant-life in this islands, including an extraordinarily rich cryptogamic flora, is tellingly described by WILSON, E. H., in the "Conifers and Taxads of Japan" (Publ. Arnold Arb. No. 8, 1916), p. 67. The flora of Yakushima numbers families or genera having classic transpacific migrations such as *Croomia, Hydrangea, Distylium, Illicium, Stuartia, Rhododendron, Astilbe, Rhus, Castanopsis, Abies, Pinus, Chamaecyparis, Torreya,* Lardizabalaceae, Chloranthaceae, etc.; relic-forms like *Trochodendron, Daphniphyllum,* etc. QUOTH WILSON, "The forest floor (of Mt. Miyanoura on Yakushima) and tree trunks support an extraordinary rich Cryptogamic flora; nowhere else, not even on famed Mt. Omei, in western China, have I seen such a wealth of this vegetation.... The flora is a wonderful Cryptogamic kingdom with a few low shrubs under a vast evergreen canopy. To me the most interesting and remarkable forest in all Japan is this on Yaku-shima, where the Cryptomeria has its southern home." This sounds very much like the description of a left-over from the days when Cryptogams and Conifers (making up 60 % of the vegetation described by WILSON) still were dominants. Manifestly, Yakushima means something both by itself, and as a weighty phytogeographic landmark (see discussion p. 438).

These two lines are standard. *Juniperus* adds to these two channels a third, for certain of its species range in this manner,

(1) *J. tsukusiensis* — Formosa, Yakushima, Japan (Kyushu).

(2) *J. formosana* — Formosa, China (Fukien, Chekiang, Anhwei, Kiangsu, Kweichow, Szechuan, Yunnan, Hupeh, Shensi, Shansi, Hopei).

(3) *J. lutchuensis* — Riu-kiu.

(4) *J. taxifolia* — Bonin (Ogasawara) Islands.

These four species are not of the same value from the taxonomic standpoint, for one *(J. formosana)* is undoubtedly "good," while the specific status of the balance may be controversial. Their distribution, on the other hand, is transparent, and taken in conjunction with that of *Cryptomeria, Cunninghamia* and *Taiwania* previously revised, tells us that a three-cornered exchange of flora took place in the dim Jurassic or very early Cretaceous past among three points, the Eastern Himalayas, Formosa and Japan. One of the sides of the triangle so formed (Eastern Himalayas to Kwantung and/or Formosa) is steady nearly throughout, which is reason why this very same side also stands as one of the major alleys of angiospermous dispersal both coming and going from the Pacific to the Indian Ocean. This side, as a matter of fact, is part of the hoary Angaraland, and never was in complete submersion. A second side is steady only for the major part of its course (the Eastern Himalayas to Kiangsu). The third side which runs between Formosa and Japan, and roughly parallels the modern Chinese coast is known to have undergone vast crumblings and to have been unstable for aeons of time. This is the side tenanted most frequently by petty segregates and controversial forms, and it is also here that phytogeographic landmarks appear (e.g., the break between Amamioshima and Yakushima) that have in their power to dispatch a genus or species to Japan or Hawaii and the New World. At the northern end of this unstable side appears a secondary important center of "cold" flora (see *Carex* and *Drosera;* p. 438). In short, vegetation and the geological past are in unison throughout, and Conifers forever confirm the same as does the dispersal of the Angiosperms.

The Pinaceae are constantly being spoken of as "Holarctic," and it is not our immediate intention to challenge this belief. *Cupressus, Thuja* and *Sequoia* are certainly "Holarctic" as regards their present distribution, but abundant fossil evidence vouches for the last of these forms at least having wandered all over the earth. It well may be that, sooner or later authentic, pinaceous fossils are dug out in the southern hemisphere.

An immediate discussion of the origin and possible migrations of the Pinaceae strikes us as all the more untimely in that, as repeatedly noticed (see p. 93), pollen of allegedly cleancut abietineous character is on record from a prevailingly angiospermous flora of the Cretaceous of Southwest Africa. It is manifest that origins cannot be competently discussed until and unless we are at ease in our mind as to the meaning of the term in rigorous language. We had better work out a rational outline of coniferous phylogeny before venturing to deal with the coniferous tracks by themselves.

In our deliberate opinion, *origin* refers in precise thinking to a point in space, and a moment in time, where and when a living form begins to be recognizable as the form which it still is today. The meaning of this term cannot be arbitrarily stretched beyond these limits, lest it fizzles out into eternity to embrace the cell primaeval. This meaning cannot fall short of these limits, either, lest, mislead by its improper application, we fail to see phylogeny against the background which essentially belongs to it.

Pollen is not a final character of identification by any means *), but rates as a significant one, and if abietinous pollen indeed occurs in a fossil flora of Southwestern Africa not later in age than the Cretaceous, we cannot fail to accept a presumption that the Pinaceae (at least as to the immediate affinity of *Abies*) *originated* in part at least elsewhere than in the north. Too much remains to be known of the fossil floras of Southern Africa in particular for us to accept as well founded the current belief, or notion, that the Pinaceae are necessarily "Holarctic."

The *modern* Angiosperms are an exceptional group because they can as a rule be traced to common centers, and dated with sufficient accuracy as regards their most active period of migrations. The case is other with the Conifers, for these plants have a long geologic past, and their fossil record is comparatively obscure.

It strikes us as desirable that, in the very first place, a broad and rational scheme of phylogeny be worked out for the coniferous clan. This scheme requires — *in the very first place* — a correct understanding of the "coniferous" leaf, for the cone, male and female, is derived from, and compound of leaves and their cladodial derivatives. There is no evidence, in our opinion, that this scheme is at hand.

Once this scheme is prepared — which we believe is feasible — it may be possible to interpret the fossil evidence with close reference to the maps of the geologic past. This will put us in the position of reaching at least tentative generalizations as to the origin in time and space of the major coniferous group.

Broad generalizations and discussions are consequently premature at this hour. We do not know enough of the Conifers as yet, but we may, at the other hand, already assert that there is nothing whatever in the dispersal of these plants that contradicts the conclusions to be derived from the dispersal of the Angiosperms. This belief is not personal with us, for authors, whose conclusions we do not share in many and vital respects (see p. 475) agree with us that all plants migrate along comparable tracks. Indeed, it could not be otherwise, for all channels of migrations are conditioned by geography, and it never was in the design of creation that two sets of maps of the Jurassic and Cretaceous, for instance, be prepared, one to suit the Conifers, the other the Angiosperms.

---

*) See POLLOCK, J. B., in Amer. Natural. 40: 253. 1906, for instability in the pollen-grain of *Picea*, and interesting general notes on the variability of the male gametophyte in conifers.

III. Ferns and Liverworts

A student of Ferns who has of these plants throughout the world good knowledge believes (7; p. 267) that well over three-fourths of those now living are of "Clearly austral origin," and acknowledges surprise (8; p. 188) at the preponderance of the austral over the boreal element.

This ratio is surprising, doubtless, if we keep only Ferns in mind. These plants are among the oldest denizens of the earth, and the fossil record is positive that they dwelt in both hemispheres for uncounted millions years. As this records reads, little is in it that may by itself explain, why only Ferns of "antarctic origins" seemingly predominate today.

If, however, we consider as one the dispersal of Ferns and Angiosperms, we no longer have reason to marvel. The Angiosperms gained ascendancy, and were confirmed in it, by at least two well marked geologic revolutions. A palaeobotanist writes, for example (9; p. 340) as follows, "In Maryland and Virginia, there are beds of clays, sands, and sandstones, that make up what has been called the Potomac Formations, so named from the Potomac River, which separates these two states. These beds are divided into three distinct divisions, which together represent practically the whole of lower Cretaceous time in this region. The lowest or oldest of these three divisions, called the Patuxent, contains a flora of about 100 species of plants, of which about thirty-five are ferns, twenty-five Cycads, and over forty are Conifers .... Curiously enough there appeared suddenly, following the maximum inundation of the land, a rich Angiospermous flora."

We see nothing curious in this, because "Maximum inundations of the land" are precisely the condition that (a) Would alter climatic and edaphic factors ruling in, and near, the regions where these inundations were felt; (b) Would for this reason break up ancient and well established climaxes, replacing them with seres of transitory significance; (c) Would favor the evolution of new forms in agreement with changing conditions; (d) Would prompt the creation of new tracks of migrations moving away from the lands stocked with plant-life toward lands recently out of the waters and untenanted, or poorly tenanted, by vegetation. "Maximum inundations of the land" repeated at different intervals would, doubtless, wipe out the immense majority of plants depending for their living and reproduction upon definite ambiental requirements, and replace these plants with aggressive vegetation of "weedy" behavior. Were we boldly to generalize, we would say that the triumph of the Angiosperms over a more ancient type of flora is in the last resort the triumph of the mangrove and the "steppe" back of it over plants that were unfit for anything but life within steady climaxes under conditions peculiarly of their own choosing *). It stands to reason that the plant-world eventually wiped out by the Angiosperms in the late Jurassic had itself replaced an older plant-world in an epoch

---

*) Weeds widespread as pests in the Tropics often occur wild, and quite scarce, in the heliophilous formations on river-banks *(personal observations)*.

long antedating the Jurassic. This stands to reason because each well marked epoch of geologic revolution begot its own flora and fauna *).

It further stands to reason that if the Angiosperms could take good advantage in the earliest Cretaceous of "Maximum inundations of the land" to conquer the region of the Potomac, this they did because they were in the Potomac's vicinity — and thoroughly modern at that — prior to the inception of the Cretaceous. The Pacaraima Mountains, as a matter of fact, bring before us a living analogy of the angiospermous fossil "Potomac Flora." The Pacaraiman flora, too, took advantage of "Maximum inundations of the lands" to sinks its roots into certain riverine deposits which came to emersion in the Late Jurassic or Early Cretaceous. The evidence interlocks at all sides, and palaeobotany, phylogeny, dispersal, taxonomy of the living plants all agree in the end.

The Angiosperms that duly invaded the Potomac were accompanied, doubtless, by contingent of Ferns which had fared with them from distant lands in the south, or had become intimately associated with them somewhere south of the modern Potomac. These Ferns, then, came into conflict for survival with other Ferns which they found already endemic to the Potomac's vicinity.

Let us observe — and do so carefully, for the subject demands it — that the conflict in question did not involve Ferns only. It involved in reality groups of plants, that is, associations, in which two "climaxes" can be identified, (a) One formed by non-angiospermous plant-life of conventional Jurassic type, including Conifers and Pteridophytes; (b) A second, constitued by Angiosperms of modern type, together with certain Conifers and Pteridophytes. In brief, while at one hand, modern Angiosperms faced conventional Jurassic flora (whatever this be), at the other *Conifers and Pteridophytes associated with the Angiosperms came to grip with Conifers and Pteridophytes associated with this Jurassic flora.* The struggle, then, was between groups of plants, each one of them including different phyla.

In the end, the Angiosperms gained the upper hand, and it stands to reason *that the upper hand also was simultaneously gained, in the main, by the Conifers and Pteridophytes which had a long previous history of associations with*

---

*) The thought occured to us several times that the modern Angiosperms might have been favored in the first place by the Permo-Carboniferous glaciations, and really come into original being at their close. By then, however, the Angiosperms were not in the position of pre-eminence required to effect a comparatively rapid colonization of de-glaciated lands, and to replace, in fact, the ancient non-angiospermous climaxes. They lingered on in the continents of the south, where the Permo-Carboniferous glaciations had been most acutely felt, allowing a conventional "Jurassic" flora to gain the upper hand for a geologic while. This flora was, perhaps, not very rich, though the Bennetitean stock had part in it, which, from the standpoint of morphology and general evolution, very closely approached the modern Angiosperms. Their real opportunity was offered the Angiosperms by the age-long revolution which, begun in the Late Jurassic ended in the Early Tertiary. It is worthy of notice that this opportunity primarily arose in the wake of a geological epoch of wide recastings in the earth's surface, for reproduction by seeds was by no means unknown in plants other than the modern Angiosperms which the Angiosperms eventually supplanted. The *initial geographic position* occupied by the Angiosperms, then, was one of the foremost factors in their ultimate triumph during the geological revolution in question. Had such continent as Gondwana survived, we would be well stocked with early Pre-Angiospermous fossils immediately allied with the Angiosperms living to day. These fossils, alas, are not easily found because of understandable reasons.

the Angiosperms. However, the triumph of these Conifers and Pteridophytes did not mean the entire extermination of the Conifers and Pteridophytes belonging to the opposite camp, that is, formerly associated with the non-angiospermous flora. Part of the latter switched allegiance, as it were, and aligned itself with the winning hosts. On this account, the Angiosperms (and the Conifers and Pteridophytes associated with them) eventually received in their associations certain waifs and aliens. These were the Conifers and Pteridophytes that had managed to survive the destruction of the Jurassic flora with which they had lived in "climaxes" for ages long prior to the advent of the Angiosperms and their non-angiospermous allies *).

The performance unfolded before us at the Potomac, mute testimony of which remains in certain fossil beds, is pregnant with meaning also in other directions. The "Conventional Jurassic flora" which died out at this particular spot — if representative at all in its fossil state — was already biologically poor and scarcely balanced, as we understand the term today. We might compare it with a sere still keeping up the outlines of a former climax, but already so depauperated as to insure the total elimination of the climax-elements in the near future. We cannot reconstruct with accuracy the ecological step leading to this sere, but the facts clearly seem to point in the direction we underscore. At the Potomac, then, was played the last act of a performance that must have lasted ages long. *To speak of the appearance of the Angiosperms as "sudden" is to do violence to the normal sequence of biological events, and to refer to these plants as "mysterious in origin" is to ignore a very great deal which can be plainly read in the record in our hands.*

To return now to the Pteridophytes living in our times. These plants clearly include two elements, (a) *Element that had associated with the Angiosperms from the start;* (b) *Element which had not done so, but was later absorbed within the angiospermous floras.* It is easy to understand with reference to the events that insured the dominance of this type of flora, that the "antarctic" element in the living ferns overtops the "non-antarctic." The former overtops the latter, because the former travelled with Angiosperms that were themselves "antarctic." The latter is a rump, picked up, as it were, along the northward course of the angiospermous tracks.

Our authority tells us that living Ferns are 75% of "antarctic" origins, and 25% of other derivation. This tallies well indeed with the 100% which is "antarctic" in the living Angiosperms, and the tiny residuum in our floras still constituted by the Cycadaceae.

The question next arises, are the living Ferns factually "antarctic" to the tune of 75%, the like are "antarctic" 100% the Angiosperms themselves?

Our answer is that we do not think so. It stands to reason that certain of these Ferns are thoroughly "antarctic," in the meaning we credit the adjective when speaking of such Angiosperms as the Restionaceae (see p. 359).

---

*) Such survival well might be the origin of much genus- and species-making.

In other words, these Ferns originated in the deep south or south of the map in the same sense and manner as did the Angiosperms. On the other hand, it is probable that other Ferns are "antarctic" only in the sense that they were picked up at a very early date by the Angiosperms, and made to travel tracks which can still be traced back to southern or antarctic lands. To elucidate, pteridophyte genus "A" might have originated right by the primary center of angiospermous genorheitral origins out of other prototypic ferns, and travelled all the way northward with the Angiosperms; genus "B," on the contrary, might have originated in what is now Northeastern Brazil, and have mixed with the Angiosperms, both north and south, taking advantage of landconnections used by the seed-plants.

Patently, genus "A" is "antarctic" throughout in the conventional sense of the term; genus "B" is "antarctic" not as to origin, but as to part, perhaps large part, of its migration. We discriminate most definitely between these concepts ("antarctic," that is, *by origin*, and "antarctic" *by migration*), not because such a discrimination is immediately necessary to us here, but because it will have to be effected when the dispersal of the Ferns is to be rigorously analyzed. Suffice it to say for our purposes that the fact that living Pteridophytes are 75 % "antarctic" is altogether possible, not only, but that this percent handsomely bears out our estimate of the Angiosperms, credited by us 100 % to "antarctic" ultimate origin and migration alike *).

The mere fact that 75 % of the Pteridophytes are "antarctic" destroys the validity of any claim made to the effect that the Angiosperms can be "Holarctic" to any material extent. There is a block of angiospermous groups (e.g., Restionaceae, Centrolepidaceae, *Astelia*, etc.) which is uncontroversially accepted as "antarctic," and this percentual part of Angiosperms sums itself forthwith with the Pteridophytes to run into a formidable score.

This being rockbottom, authors who still insist that the Angiosperms are "Holarctic" are forthwith drawn of necessity to figure out *two sets of opposite tracks*, one for the "Holarctic"; the other for the "Antarctic" Angiosperms, and further to complicate the record to make room for the Pteridophytes. In brief, authors of this persuasion can never hope to reach a rational and orderly understanding of dispersal. All they can do is to confuse the record, and to offset this in the end to have recourse to pious theories, claims involving "mysteries," etc. This, as a matter of fact, is precisely what authors of the sort have contributed to phytogeography. The record in print is so well known that to cite and quote it would be redundant.

Were the "Holarctic" origin of the Angiosperms a tenable doctrine, *Vaccinium* and *Erica* should have begun to run southward from the "North Pole" right at the time, when *Pernettya* would start migrating in the opposite direction from the "South Pole." To uphold this transparent nonsense we ought

---

*) We pointedly warn the reader that we use here the adjective "antarctic" in its conventional sense, meaning *of the south in general*. The reader knows that we discriminate as to what arose in the *Afro-antarctic* and *Gondwanic Triangle*, respectively; see p. 357.

to postulate "polyphiletic" origins for the Angiosperms right and left; scramble the whole of their *genorheitra;* misread their entire migrations; do violence to everything we safely know of ecology; in sum, turn phytogeography into a sorry hodgepodge. This is indeed more than the evidence demands, and common sense can tamely grant.

Always dealing with the origin of Ferns, a pteridologist whose thinking is usually sharp (7; p. 166) speaks of the Ophioglossaceae in these terms, "The Ophioglossaceae are regarded as the most primitive extant Ferns, and therefore as very old . . . . On the whole the distribution of both genera indicates the great age ascribed to them on other grounds. We may suppose that they were represented in Antarctica during the last period of fern luxuriance there, and that some emigrants thence still survive; but that another and probably large element . . . . survives from a fern flora living elsewhere at the same time."

This text is lucid. It tells us that the Ophioglossaceae consist of two groups, one of "unknown" origin, the other, on the contrary, of "antarctic" extraction. In our opinion, the group of "unknown" origin is the one which managed to be eventually absorbed into angiospermous climaxes at the end of the Jurassic, after having long lived in non-angiospermous (i.e., conventionally "Jurassic") climaxes. The "antarctic" group, on the contrary, is the group which, originally established in the deep south of the map, fared northward with angiospermous migrants, and has to this day survived with them. Admittedly, this rationalization is not free from the necessity of certain qualifications but we submit it precisely as it reads in order that there be no doubt about the thought which it expresses.

It is now convenient to review the Ophioglossaceae, particularly so that we have of them a recent monograph (10) making it easy to study their dispersal.

The Ophioglossaceae consist of three genera, *Helmintostachys, Ophioglossum* and *Botrychium*. The first consists of but one species having the following distribution,

(1) *H. zeylanica* — Ceylon, India (Assam), Indochina (Laos), China (Kwantung, Hainan), Formosa, Riu-kiu, Malacca, Sumatra, Java, Borneo (Mt. Poi, Mt. Kinabalu), Philippines (Mindanao, Luzon), New Guinea, Solomon Islands, Queensland.

This is a well-knit, and thoroughly consistent range, matched by countless Angiosperms. The dispersal of *Selaginella* also fits this pattern of distribution (11, 12), although many of the species of this genus are local or narrow-ranging. We find in *Selaginella* species such as these, for example,

(1) *S. Mayeri* — India, Malacca, Sumatra.

(2) *S. remotifolia* — Java, Philippines, New Guinea, China, Japan.

(3) *S. caulescens* — India, Sumatra, Flores, Celebes, China.

(4) *S. plana* — Malacca, Sumatra, Java, Krakatoa, Bali, Soembawa, Flores, Timor, Celebes, Boeroe, Ceram, Amboina, Banda, Kei Islands, Timor Laoet.

(5) *S. tamariscina* — India, China, Siberia; Java, Lombok, Celebes.

Selaginella remotifolia intergrades with the African *S. Kraussiana*, which repeats, and perfectly matches, certain affinities under *Cuscuta* (see p. 218), and leaves us in no doubt as to the course of the tracks. The tie is the usual one, South and Eastern Africa—Java.

Selaginella plana is a classic instance of distribution along the entire Soenda arc. Krakatoa is most likely a recent station, but the rest of the dispersal is doubtless ancient. *Selaginella tamariscina* is a xerophyte that "curls up" in times of rest, and pays no attention whether rest is induced by spells of drought in "Monsoonland," or the rigors of the Siberian winter.

Unlike *Helminthostachys*, *Ophioglossum* is a comparatively large genus of close to thirty species distributed into four subgenera. The largest of these groups is Subg. *Euophioglossum* which is distributed in this manner,

(1) *O. vulgatum* — North America (Nova Scotia to Alaska and Central Mexico), Europe (Iceland to the Azores, Italy, Russia and Sweden), Island of Madera, Asia (Syria, Persia (Iran), India, Japan, Kamchatka).

(2) *O. sarcophyllum* — Madagascar, Mascarenes, Ceylon.

(3) *O. angustatum* — India, China (Anhwei, Kiangsu), Japan, Riu-kiu, Bonin (Ogasawara) Islands.

(4) *O. reticulatum* — Northern Argentina, Brazil, Bolivia, Peru, Colombia, Venezuela, Guianas; Galápagos Islands; Panama, Costarica, Honduras, Guatemala, Mexico (Mexico, Jalisco, Veracruz, San Luís Potosí), West Indies, (Grenada, Martinica, Guadeloupe, Saint-Thomas, Hispaniola, Jamaica, Cuba); Natal, Mascarenes, Nyasaland, Cameroon, Liberia, Cabo Verde, Madera; Ceylon, India (Nilghiri Mts. to Assam), China (Hupeh), Corea (Quelpaert Island), Japan (Honshu), Philippines (Luzon: Batangas, Bataan, Benguet, Ilocos Norte), Formosa, Bonin (Ogasawara) Islands; Eastern Polynesia (Easter Island).

(5) *O. Harrisii* — Grenada, Jamaica, Hispaniola.

(6) *O. petiolatum* — Madagascar, Tropical Africa, Northern South America, West Indies (including Trinidad), Mexico, Florida, Ceylon, India, Siam, China, Japan, Sumatra, Java, Borneo, Philippines, New Guinea, New Caledonia, Fiji, Samoa, New Zealand.

(7) *O. concinnum* — Hawaii.

(8) *O. Aitchinsonii* — Ethiopia, Western India, Afghanistan.

(9) *O. Engelmanii* — Mexico (Oaxaca to Baja California, Chihuahua), United States (Louisiana to Arizona, Illinois, Virginia, Florida).

(10) *O. pedunculosum* — Cape, Southwest Africa, Ethiopia, Madagascar, Ceylon, Southern India, Sumatra.

(11) *O. ellipticum* — Bolivia, Brazil (Piauhy, Pernambuco), Guianas, Panama.

(12) *O. nudicaule* — Var. *typicum:* Cape, Madagascar, Fernando Po, Florida, India (Pulney Hills), China (Yunnan), Java, Lombok, Queensland, South Australia, New Caledonia, Samoa (Upolu); Var. *tenerum:* Angola, Argentina (Formosa, Chaco), Brazil (Rio Grande do Sul, Soã Paulo, Minas Geraës, Pernambuco, Matto Grosso, Amazonas) Bolivia, Colombia (Mt.

Tolima), Venezuela (Mérida), Guianas, Cuba, Hispaniola, Mexico (Mexico, Sinaloa), United States (Florida to Texas and Alabama), Sumatra, Philippines (Luzon); Var. *minus:* Guiana, Florida; Var. *vulcanicum:* Panama; Var. *macrorhizum:* Paraguay, Brazil (Santa Catharina, Amazonas), Guianas, India (Pulney Hills); Var. *laxum:* Brazil (Pernambuco), Colombia (Mt. Tolima), East Africa; Var. *grandifolium:* Philippines (Leyte).

(13) *O. rubellum* — Eastern Africa (Uganda).

(14) *O. Thomasii* — Eastern Africa (Uganda).

(15) *O. fernandezianum* — Juan Fernandez (Masatierra).

(16) *O. scariosum* — Bolivia (La Paz), Peru (Junín).

(17) *O. Schlechteri* — Northeastern New Guinea.

(18) *O. opacum* — Tristan da Cunha, Saint-Helena.

(19) *O. crotalophoroides* — Argentina (Buenos Aires), Uruguay, Bolivia (La Paz), Chile (Valparaiso), Peru (Cuzco), Colombia (Santander), Venezuela (Mérida), Honduras, Guatemala, Mexico (Mexico, Michoacán), United States (Texas to South Carolina, Alabama).

(20) *O. lancifolium* — Philippines (Mindanao) Madagascar, Eastern Africa (Uganda).

(21) *O. lusitanicum* — Subsp. *typicum:* Azores, Madera, Algeria, Portugal, France, Iceland, Germany, Austria, Italy, Sardinia, Dalmatia; Afghanistan, India; Subsp. *californicum:* Mexico (Central Mexico, Baja California), California; Subsp. *coriaceum:* Bolivia (Cochabamba, La Paz), Chile (Aconcagua), Easter Island, New Zealand, Tasmania, Australia (New South Wales, Victoria), New Caledonia.

(22) *O. gramineum* — Africa (Angola), India, New Guinea, Queensland.

(23) *O. lineare* — Bismarck Archipelago (New Guinea).

The balance of the genus shapes up as follows,

B) Subg. *Cheiroglossa*

(24) *O. palmatum* — Mascarenes, Indochina, America (Southeastern Brazil to Mexico, West Indies and Florida).

C) Subg. *Ophioderma*

(25) *O. pendulum* — Subsp. *typicum:* Madagascar, Mascarenes, India, Ceylon, Malacca, Sumatra, Java, Borneo (Mt. Kinabalu), Philippines (Mindoro, Luzon, Mindanao), Amboina, New Guinea, Mariannes, Carolines, New Hebrides, Fiji, Samoa, Tahiti, Hawaii, Australia (Queensland, New South Wales); Subsp. *falcatum:* Mariannes, New Hebrides, Hawaii.

(26) *O. intermedium* — Sumatra, Java, Borneo (Mt. Kinabalu), Philippines (Mindoro), New Guinea.

(27) *O. simplex* — Eastern Sumatra.

D) Subg. *Rhizoglossum*

(28) *O. Bergianum* — South Africa (Cape: Stellenbosch).

The dispersal of *Ophioglossum* (Fig. 95) is keyed to the continent (Gondwana of most authors) which, prior to the Cretaceous, brought together as one America, Africa and the Far East. Were this plant an Angiosperm, we should say that it migrated out of the African gate in the first place. The distribution of individual species of this genus is oftentime a telling match of standard angiospermous dispersal. If we may no longer speak of tracks in the conventional sense of the term, because a track must be oriented, and we cannot orient the channels of migration through which flow the Ophioglossaceae, still we can recognize well defined ranges. Taking position at a point in the heart of the Indian Ocean, we detect, for example, the following as to *Ophioglossum*, (a) The wide-ranging *Ophioglossum reticulatum* is contained at least within a double channel, (i) Natal and the Mascarenes to Cameroon and Madera, next the Guianas, Brazil and the West Indies; (ii) Natal and Mascarenes to India, China, the Bonin (Ogasawara) Islands, the New World from Mexico through Central America to South America, including the Galápagos, and, lastly Easter Island. This species crosses two modern Oceans, Atlantic and Pacific, with significant dispersal in the Eastern Pacific *); (b) The equally wide ranging *O. nudicaule* is rooted alike in the Western Pacific (Western Polynesian gate) and the Indian Ocean (African gate) and from the latter, or both, takes in the Far East and Malaysia. The American end of its dispersal is suggested to have been effected prevailingly, possibly altogether, along the main track Madagascar—Bolivia on account of certain peculiarities, witness, (a) The stop put on eastern dispersal in Mérida of Venezuela; (b) The occurrence of stations on Mt. Tolima in Colombia; (c) The occurrence of a single variety in disconnection in the Guianas and Florida (see Sarraceniaceae, *Gaylussacia*, p. 447); (d) A track run by var. *macrorhizum* from Paraguay to Southern Brazil, ultimately the Guianas, together with an outlier in India (see *main* trends of the migrations of *Carex incurva*, p. 463); (e) The presence of a variety in Panamá. *Ophioglossum petiolatum*, on the contrary, seemingly reaches the New World from the African gate (Gondwana) directly across the modern Atlantic, covering meantime the Far East in a wholly standard fashion; (c) *Ophioglossum ellipticum* may stem both from the Eastern Pacific or the Atlantic, and only a careful study of its affinities and populations may help us in deciding by sums of probabilities. *Ophioglossum lusitanicum* subsp.

---

*) The keen reader will doubtless suggest that the possibility exists that a third track is in play, quite as standard as the other two. This track (see Palms, p. 452) could lead straight from Natal and the Mascarenes to Easter Island, Bolivia, Northern Argentina, the Galápagos, Colombia, Venezuela, Central America and Mexico (see Protieae, p. 388), not only, but the West Indies as well, always from the Eastern Pacific, not from the Atlantic. We agree that a dispersal such as that of *O. reticulatum* can be legitimately read also the way the keen reader suggests. To discriminate in details, we should have a far deeper insight of the populations of this species than we possess at present. Phytogeography still babbles in its infancy, and this manual is perhaps the first published work in which an attempt is made to establish this science globally upon bases of fact. The lot of pioneers is not always enviable.

typicum runs a classic course from India through Afghanistan to North Africa and Europe, becoming "atlantic" by dispersal in the last. The subsp. *californicum* is strewn on the contrary along a none less classic domain in Mexico and Baja California, which neatly hinges upon the range of subsp. *coriaceum*. This last enters Chile and Bolivia from the Pacific, and is thus further at home in Easter Island, New Zealand, Tasmania and Australia; (d) Hardly anything need be said of the ranges of species such as *O. sarcophyllum* (Madagascar, Mascarenes, Ceylon), *O. angustatum* (India, Southeastern China, Japan, Riu-Kiu and Bonin (Ogasawara) Islands), *O. Aitchinsonii* (Ethiopia, Western India, Afghanistan), *O. opacum* (Tristan da Cunha, Saint-Helena), *O. lancifolium* (Madagascar, Uganda, Philippines), *O. crotalophoroides* (United States to Chile through Mexico, Central America and the west of South America generally, including an outlier to Western Venezuela), *O. gramineum* (Angola, India, Queensland, New Guinea), etc., because their dispersal is uniformly contained within sectors of standard angiospermous tracks.

In brief, the statement that *Ophioglossum* is "Gondwanic" in dispersal is amply supported by the facts. Even its "Holarctic" species, *O. vulgatum*, is in tune with that ancient continent. The distribution is "atlantic" on both sides of the modern Atlantic Ocean, including Madera. It further reaches eastward to Japan and Kamchatka, which is good match of certain groups of *Ilex* (p. 310), merely to mention an obvious parallel. Only one of the forms reviewed, *O. lusitanicum* subsp. *coriaceum*, is "antarctic," and this subspecies is not harder to understand than part of *Carex incurva* (see p. 463), for instance. Most revealing is the line Southern Australia, New Caledonia, Tasmania, Easter Island, Chile, Bolivia, which pinpoints a direct continuous "track" across the wide expanses of the Southern Pacific.

Botrychium is divided by the monographer we follow in three Subgenera, as follows,

A) Subg. *Osmundopteris*

a) Sect. *Lanuginosae*

(1) *B. lanuginosum* — Var. *typicum:* Ceylon, India (Punjab to Assam), Southwestern China (Yunnan), Sumatra, Java, Philippines (Luzon: Benguet); Var. *leptostachyum:* India (Himalayas), Formosa, Japan.

b) Sect. *Virginianae*

(2) *B. virginianum* — Subsp. *europaeum:* Northern Russia, Sweden, North America southward to Colorado, Wisconsin and Massachusetts; Subsp.

meridionale: Mexico (Hidalgo, Morelos, Chiapas); Subsp. *typicum:* Europe, India, China, Japan, Mexico, United States (Florida, California, Arizona), Canada (British Columbia, Prince Edward Island).

(3) *B. cicutarium* — West Indies (Cuba, Jamaica, Santo Domingo), Mexico, Central America (Guatemala, Costarica, Panamá), South America (Colombia: Antioquia, Santander, Ecuador, Peru, Bolivia, Brazil (Minas Geraës, Santa Catharina)).

(4) *B. strictum* — Southwestern China (Szechuan), Japan (Honshu, Hokkaido).

(5) *B. Chamaeconium* — Africa (Uganda, Cameroon).

B) Subg. *Eubotrychium*

a) Sect. *Lunaria*

(6) *B. Lunaria* — Var. *typicum:* Kashmir, Afghanistan, Europe (Northwestern Russia, Ireland, Faer Oes, Iceland), North America (Idaho, Mackenzie region to Greenland), South America (Patagonia), Tasmania, New Zealand, Australia; Var. *onondagense:* Kashmir, Europe (Hungary, Germany), North America (Labrador, Newfoundland, Quebec, Maine, Vermont, New York, Michigan, Montana, Washington); Var. *minganense:* North America (Quebec, Mackenzie, Hudson Bay, Alberta, Alaska, Montana, Colorado, Nevada, California, Wisconsin); Var. *Dusenii:* Patagonia.

(7) *B. simplex* — Var. *typicum:* Japan (Yezo), Europe (Corsica, Germany, Danemark, Finland, Scandinavia), North America (Newfoundland, Nova Scotia, British Colombia to California and New Mexico); Var. *compositum:* Europe in the north, North America (New England and the Western U.S.A. to California); Var. *laxifolium:* North America (Vermont, Massachusetts, Connecticut, New York, New Jersey, Pennsylvania); Var. *tenebrosum:* Austria (Tyrol), North America (Quebec, Ontario, Maine, New Hampshire, Vermont, Massachusetts, Connecticut, New York, New Jersey, Pennsylvania, Maryland, Michigan, Minnesota, Washington).

(8) *B. pumicola* — North America (Oregon).

(9) *B. boreale* — Var. *typicum:* Corea, Kamchatka, Russia, Sweden, Norway, Alaska; Var. *crassinervium:* Siberia, Sweden; Subsp. *obtusilobum:* Eastern Asia (Amur), Aleutian Islands, British Columbia, Alberta, Montana, Oregon, Washington.

(10) *B. matricariaefolium* — Subsp. *typicum:* Corea, Europe (Italy, Switzerland, Germany, England, Sweden), North America (Newfoundland and New Brunswick to New Jersey and Pennsylvania, Maryland, westward to Minnesota); Subsp. *hesperium:* Colorado; Subsp. *patagonicum:* Patagonia.

b) Sect. *Lanceolatae*

(11) *B. lanceolatum* — Subsp. *typicum:* Japan, Kamchatka, Sweden, Scotland, North America (Greenland, Labrador, Quebec, Maine to Wyoming, Washington, Colorado and Alaska); Subsp. *angustisegmentum:* North America (Newfoundland, New Brunswick, Quebec, New Jersey, Pennsylvania, Ohio, Michigan).

c) Sect. *Multifidae*

(12) *B. multifidum* — Subsp. *silaifolium:* United States and Canada (New Brunswick to British Columbia, California, Pennsylvania); Subsp. *typicum:* North America (New Brunswick to British Columbia, Michigan, New York), Europe (Northern Europe southward to Eastern France, Australia, Rumania), Asia (Western Siberia, China: Yunnan, Szechuan); Subsp. *robustum:* Alaska; Subsp. *Coulteri:* California to Washington, Montana, Colorado; Subsp. *californicum:* California.

(13) *B. australe* — Var. *typicum:* Australia (Victoria, New South Wales), New Zealand; Var. *millefolium:* New Zealand; Var. *erosum:* New Zealand; Subsp. *Negeri:* Chile (Valdivia).

(14) *B. Schaffneri* — Var. *typicum:* Argentina, Bolivia, Peru, Colombia, Guatemala, Mexico (Oaxaca, Mexico, Hidalgo, San Luís Potosí, Durango, Chihuahua); Var. *pusillum:* Mexico (Mexico, Hidalgo, San Luís Potosí).

(15) *B. ternatum* — India (Punjab, Sikkim), Indochina (Tonkin), China (Kwantung, Kweichow), Corea (Quelpaert Island), Japan (Kyushu, Shikoku, Honshu, Hokkaido).

d) Sect. *Biternatae*

(16) *B. biternatum* — Southeastern United States (Florida to South Carolina and Alabama).

(17) *B. Jenmanii* — Cuba (Oriente), Santo Domingo, Portorico, Jamaica.

(18) *B. alabamense* — Southeastern United States (Florida, Georgia, Alabama, North Carolina).

(19) *B. Underwoodianum* — Costarica, Colombia (Santander), Venezuela (Mérida), Jamaica.

e) Sect. *Elongatae*

(20) *B. dissectum* — Var. *oneidense:* North America (New Brunswick to Ontario, Minnesota, Ohio, North Carolina); Var. *obliquum:* North America (New Brunswick to Ontario, Wisconsin, Arkansas, Alabama), Jamaica; Var. *enuifolium:* North America (Maryland to Missouri, Texas, Louisiana, Flori-

da); Var. *typicum:* North America (Nova Scotia to Minnesota, Illinois, Kentucky, Florida); Subsp. *decompositum:* Guatemala, Mexico.

(21) *B. japonicum* — China (Kwantung), Japan (Kyushu, Shikoku, Honshu).

(22) *B. subbifoliolatum* — Hawaii.

(23) *B. daucifolium* — Ceylon, India (Deccan, Assam), Burma, China (Yunnan), Java, Borneo (Mt. Kinabalu), Philippines (Luzon), Fiji.

Subgenus *Osmundopteris* is described by the pteridologist we here follow as "Most Fern like." We do not exactly know what this mean, but assume that this is the group of *Botrychium* which resembles more than all others what we may understand as a conventional fern. In other words, we may not be mistaken in believing that this subgenus has some phylogenetic significance.

Peculiarly, Subg. *Osmundopteris* is so distributed as to tie in directly with the ranges of *Helminthostachys* and *Ophioglossum*. It would seem, accordingly, that these three genera bring to a single head the involved threads of their dispersal in one and the same center.

In order that we may see these threads in the proper light, and fully appreciate the meaning of the center in question, let us take position in the geographic heart of the modern Indian Ocean, which is a focus of primary importance for the Angiosperms. We know that by taking position at this point we do not stand amid the waters of the modern ocean, but on the solid ground of a continent that was. *This point might as well be called Gondwana.* If what the Angiosperms taught us is true and correct, we must from this same point easily understand the dispersal of the Ophioglossaceae. Let us repeat that two sets of maps were never made separately to accomodate Angiosperms and Ophioglossaceae.

Let us begin with the lone species of *Botrychium*, *B. Chamaeconium* endemic to Africa. No doubt, were this fern an angiospermous plant it could not be better located. The range it follows, Uganda—Cameroon, is absolutely standard, and leads from the Indian Ocean straight to the modern Atlantic in readiness for a crossing to the New World. This crossing is indeed effected by *B. cicutarium*, endemic to the West Indies, Mexico, Central America, Eastern Brazil, Colombia, Ecuador and Peru. All this is perfectly in tune with angiospermous distribution.

The Angiosperms told us that the track Uganda—Cameroon has for its counterpart tracks sallying forth from the modern Indian Ocean toward the Far East and the Himalayas generally. It is gratifying to learn that these tracks are not at all neglected by *Botrychium*. *Botrychium strictum* is strewn along a standard segment of this circuit leading to Southwestern China and Japan. *Botrychium lanuginosum*, a well marked species, further stresses the same circuit, and more completely. It occurs in Ceylon, the Himalayas, Southwestern China, Japan, Sumatra, Java, Philippines and Formosa. All these localities are standard for tracks fanning out of the Indian Ocean to the Pacific. *The agreement is accordingly perfect between the doings of Botrychium and the*

Angiosperms. This is necessarily so because these, and uncounted other groups, uniformly migrated in the direction dictated by the same palaeogeography.

The distribution of *B. virginianum* does not wholly agree as to geography with those of the species so far discussed. It is made up of sundry massings which bear being analyzed as follows (i) *Asiatic:* India, China, Japan; (ii) *European:* Europe generally to Sweden; (iii) *American:* Central Mexico to the United States and Canada. It will be observed in relation to these massings, (1) The *Asiatic* is standard for the whole of the subgenus, mediately *(B. Chamaeconium)* or immediately; (2) The *European* represents an extension northward and westward of the channels feeding in the first place the *Asiatic;* (3) The *American* and the *European* form a single transatlantic whole. This is perfectly in agreement with standard dispersals in the Angiosperms, proof again that *Botrychium* and the Angiosperms migrated together.

It is transparent that *Botrychium* Subg. *Osmundopteris* is harmoniously distributed. Nothing in it contradicts standard angiospermous channels. The Patagonian outlier of *B. Lunaria* in the different Subg. *Eubotrychium* cannot strike us as odd. This species is in Afghanistan and the Kashmir, quite as *Carex physodes* and its ally *C. incurva*, and in Patagonia again as *C. incurva*. Tasmania, New Zealand and Australia can easily be reached either from the Indian Ocean directly (see *Pelargonium* sect. *Peristera*, p. 263) or the other way around crossing the Pacific. Either way, the dispersal is standard.

The north and high north are within reach of *Botrychium* either from the Altai Node, or by transatlantic migrations next working up along both sides of the Atlantic. Other centers bearing watching are off the coast of Eastern Japan, where stands a secondary point of "cold" speciation. We need not reconstruct all these channels in detail as regards the Ophioglossaceae, and *Botrychium* in particular, because we have seen them all while studying Angiosperms. *Botrychium boreale* can be seen moving step by step away from the Altai Node, *westward* to Siberia and Sweden, *eastward* to the Amur, Kamchatka, the Aleutian Islands, Alaska, British Columbia. This is the path indicated by modern maps, but it is altogether likely that the early migrations of *Botrychium* in these quarters took place when none as yet of these regions had its modern outlines.

Botrychium australe is a perfect specimen of "antarctic" dispersal, matched by the "antarctic" distribution of *Ophioglossum lusitanicum* subsp. *coriaceum* in essentials.

These, and similar displays, of "antarctic" and "bipolar" fireworks leave us by now quite unmoved. We no longer have reason to marvel, how far-flung ranges can be secured by plants with indifferent "Means of dispersal." Everything is by now so closely intervowen before our eyes that there is no longer a point of the map that can be said to stand isolated from all others. Alaska and Tasmania, the Altais and Patagonia are all next door. Seen for what it is, even "bipolarism" is commonplace.

We now drop the Ophioglossaceae, because we have abstracted their juices for the purposes immediately at hand. What this family has shown us contains nothing which we have not seen before. Every step of the ophioglossaceous migrations can be matched by angiospermous dispersal. The very same point of vantage in the Indian Ocean which makes light on the migrations of these Pteridophytes illuminates those of the Angiosperms. There is some reason to affirm that the true center of origin of the Ophioglossaceae might be located in the modern Indian Ocean, south of modern India. Accordingly, *Helminthostachys*, *Ophioglossum* and *Botrychium* may be "antarctic" as to migration — in part, when not whole — but "non-antarctic" as to ultimate origins. Nobody can safely guess, where they originated, but the lost continent they highlight must certainly have been of colossal antiquity. Strange that even so everlasting a landmass could come to its end, while *living plants* still bear uncontrovertible testimony to the former existence of mountains and plains vanished from sight for ever. Who might think that a delicate Fern of our glades is more lasting, in a sense, than a continent? Who might guess that such a Fern vouches at this hour for events that came to pass well over fifty millions years ago?

Dicksonia is a genus of large Ferns distributed in this manner,

(1) *New World* — 1 species in Mexico, 3 in Central America, and Colombia, 1 in Ecuador and Brazil, 1 in Peru.

(2) *Old World* — 1 species in Saint-Helena, 1 in Malaysia, 4 in New Guinea, 3 in New Caledonia, 1 in Fiji and Samoa, 1 in Queensland, and New South Wales, 1 in Australia and Tasmania, 3 in New Zealand.

Allied with *Dicksonia* are two genera, as follows,

(1) *Thyrsopteris* — 1 species in Juan Fernandez.

(2) *Cibotium* — 6 species in Mexico and Central America, 4 species in Hawaii, 3 species in China and Malaysia.

European fossils were referred to *Thyrsopteris*, which the pteridologist we follow (8; p. 175) is inclined to discount. We would like better evidence than he offers in order to tag these fossils as misidentifications. We know that a direct channel of migration leads from Juan Fernandez to Bolivia and Peru, at one hand, the Mascarenes at the other. The former of these channels reaches the gates of the Mediterranean in no time (cf. *Bystropogon* and its allies, p. 93), and very little effort (see *Orobanche*, for instance, p. 286) takes us from South Africa and its purviews to this same sea. Indeed, following *Erica* and *Calluna* (see p. 160) we can fare from the Cape to Newfoundland and Norway without a serious break. *Thyrsopteris* may well have lived in geologic Europe, if *Pellaea hastata* and *Carex pyrenaica* still live in Europe today.

Cyathea is a large genus of unsettled limits. Supposedly primitive in its affinity is genus *Lophosoria*, consisting of a variable species which is most abundant in Chile and Juan Fernandez but ranges as far northward as Mexico.

Cyathea itself teems in the Pacific, and one of its species, *C. medullaris*, holds preeminence, because all the species east and south of Tahiti (13) are closely allied with it, to the exception of a single one, *C. decurrens* (Tahiti,

Rarotonga, Samoa, New Caledonia). The distribution of *C. decurrens* might well indicate that its ultimate origin is to be sought in New Caledonia or Samoa, while the *C. medullaris* affinity perhaps stems from the immediate vicinity of Tahiti. This surmise would automatically account for the fact that east and south of this island every species is closely allied with *C. medullaris*. The existence of well defined centers of secondary evolution in the Pacific is suggested by the distributional peculiarities in *Cyathea* just noticed. These peculiarities are not the result of pure coincidence, because the same occurs in *Vaccinium* as does in *Cyathea*. All the species of *Vaccinium* in Polynesia belong to a Section or Subgenus, *Macropelma*, barring (see p. 170) *V. Whitmeei* endemic to Savaii.

Cyathea medullaris is reported (14) to range in Rapa, Stewart and Chatham Islands, New Zealand, and to migrate from Three Kings Island to Northern Australia. This range is not entirely unknown to us (see p. 182), because it is a standard alley of secondary dispersal in the Pacific. There is thus a very good indication that *Cyathea* ranged in the Southern Pacific when this ocean was studded by lands of which there is no longer trace. However, we may not be sure that *Cyathea* originated in these quarters. *Lophosoria*, supposedly the primitive form in the *Cyathea* affinity, is confined to Western America, between Juan Fernandez and Mexico. We know from the dispersal of *Zamia* that Western America was in emersion by the close of the Jurassic, and the whole of the distribution of the angiospermous *genorheitra* (see *Cuscuta*, for example, p. 223) suggests that the stretch Mexico—Chile is one of the most important regions of the world for the phytogeographer. This was long before the Andean uplift, when the shores of the New World on the Eastern Pacific reached much farther west than they do today. We might summarize the matter with the statement that plant-dispersal gives precise indications to the effect that the whole of the lands around the Pacific crumbled and receded away from the modern center of this ocean. As a corollary, it seems most likely the whole of Western America moved bodily eastward away from this same center. Purely relying upon the strength of factual, repetitious evidence from distribution, we might draw a shore-line for the Pacific running from Japan through Hawaii to the Revilla Gigedos, the Galápagos, Easter Island, Tahiti, New Zealand, the Bonin (Ogasawara) Islands and again Japan. The academic questions to which "Landbridges" gave rise are childish in the light of this direct evidence. *This evidence, we repeat, is that the Pacific shores receded and crumbled all over, and that this Ocean was in the absolute much narrower in the days when the Angiosperms first reached the lands on its margins than it is now.* As we know, geophysicists are already on record (see p. 69) for a crumbling and a recession of the Pacific shores *in the west*, but hardly anything has been said to our immediate knowledge of a crumbling and a recession of these shores *in the east*. Plant-dispersal speaks for both with the same strong accents. We confess our utter ignorance as to the geophysical reasons of this massive crumbling and receding, but the hypothesis that stresses from the ocean-floor are active against the shore (15) is far from absurd in the light of the factual records of phytogeography. Something is here that bears lucid, dispassionate study, for dispersal speaks.

The Schizaeaceae are described as intermediate between the lower Osmundaceae and modern Ferns of a much less primitive type. The Schizaeaceae are for this reason a peculiarly interesting group, for, unlike the Ophioglossaceae, they yield at least some indication of having evolved, perhaps, in some center which we may hope to identify.

The typic-genus *Schizaea* is distributed in the following manner,

A) Subg. *Euschizaea*

(1) *S. fistulosa* — Falkland Islands, Fuegia, Auckland Island, Society Islands, Norfolk Island, New Guinea, New Zealand, Tasmania, New Caledonia.

(2) *S. bifida* — New Zealand, Australia.

(3) *S. tenella* — South Africa.

(4) *S. pectinata* — Madagascar, South Africa, Saint-Helena.

(5) *S. malaccana* — Burma, Malacca.

(6) *S. Hallieri* — Borneo.

(7) *S. robusta* — Hawaii.

(8) *S. incurvata* — Tropical South America.

(9) *S. pusilla* — Atlantic North America (New Jersey to Nova Scotia and Newfoundland).

B) Subg. *Lophidium*

(10) *S. dichotoma* — Madagascar, India, New Guinea, Southeastern Polynesia.

(11) *S. rupestris* — Australia.

(12) *S. elegans* — Tropical America.

(13) *S. Poeppigiana* — Tropical America.

(14) *S. fluminensis* — Tropical South America.

(15) *S. pacificans* — Brazil.

(16) *S. Sprucei* — Brazil (Amazonas).

C) Subg. *Actinostachys*

(17) *S. digitata* — Madagascar, Ceylon, Philippines, Moluccas, New Guinea, Palau, Carolines, Fiji.

(18) *S. inopinata* — Malacca, Philippines (Bohol).

(19) *S. Wagneri* — Bismarck Archipelago.

(20) *S. melanesiaca* — New Caledonia, Fiji.

(21) *S. Balansae* — New Caledonia.

(22) *S. plana* — New Caledonia.

(23) *S. tenuis* — New Caledonia.

(24) *S. intermedia* — New Caledonia.

(25) *S. laevigata* — New Caledonia.

(26) *S. Biroi* — New Guinea.

(27) *S. spirophylla* — Amboina.

(28) *S. Pennula* — Tropical America.

(29) *S. penicellata* — Tropical South America.

(30) *S. orbicularis* — Colombia.

(31) *S. Germainii* — West Indies, Florida.

The dispersal of this genus is so wide, and so disconnected in detail — at least, in the tabulations of ranges and species available to us at present — that no "track" can be identified at a glance, which should not be the case when facing *standard* types of distribution. However, certain ranges stand out reasonably well, witness, (a) *Schizaea fistulosa* is neatly "antarctic" between Fuegia and Tasmania. Considering that this species is native to New Caledonia, and that New Caledonia is beyond doubt an important secondary center of evolution of *Schizaea*, it is probable that the sum total of the range was reached with New Caledonia as its hub. We further stress the fact that conventional "Jurassic" floras shortly before the inception of the Cretaceous (see Patuxent fossil beds, p. 489) included large numbers of conifers and ferns. New Caledonia is extraordinarily rich in the former, and certainly not without the latter. It is accordingly possible that in this island the facies of the vegetation, though altered by the addition of numerous peculiar Angiosperms, may yet retain here and there a "Jurassic" habit. This possibility is worth studying, and even more, New Caledonia ought to be set up as a center of biological and botanical investigations thoroughly well organized to last. This island is well worth it *); (b) *Schizaea pectinata* highlights a classic segment of the African gate (Madagascar, South Africa, Saint-Helena) which easily might open the way to Southeastern Brazil and Fuegia alike; (c) *Schizaea digitata* migrates seemingly eastward in a "mangrove pattern."

This is about all that can be made out with reasonable assurance of the distribution of this genus. It is furthermore clear that Atlantic North America was reached at an early date, because two species, each of a different section, *S. pusilla* and *S. Germainii*, belong to the North American flora, and the latter also occurs in the West Indies. There is room to suspect, consequently, that *Schizaea* followed part at least of the channels of migrations we have explored in the case of the Ophioglossaceae. The sum total of the distribution is to the effect that in a very early stage of its wandering *Schizaea* was well rooted in what is now the Indian Ocean, and Western Polynesia quite as well. Speaking

---

*) Most interesting studies are suggested to effect certain comparisons between New Caledonia and Yakushima. See p. 438, 442, for the latter.

of this genus, our authority *) says that it contains, "Some thirty recognized species, predominantly southern and of the most evidently Antarctic ancestry." This opinion requires stringent qualifications. It is true that *Schizaea* followed angiospermous tracks leading from the south northward in the main; it is correct that the majority of the known living species are endemic south of the Tropic of Cancer; but there is certainly no warrant to assert that *Schizaea* originated in Antarctica. By insisting upon a precise application of the term *origin* we much less please a whim of ours than serve objective discussion. *Schizaea* might have originated anywhere within the net of "tracks" it runs today, and the guess it originated in Antarctica, or near what is now New Caledonia or Ceylon, is, and remains, a guess. Before passing judgement upon the origins of this genus, we will have to consider in detail its affinity and to consult with a critical mind the fossil record, which all is still of the future.

Of another Schizaeaceous genus, *Lygodium*, our authority writes (see footnote below for reference, op. cit. 24) as follows, "Thirty-nine recognized species, pantropic, south to New Zealand and South Africa, north to Japan, with one species, *L. palmatum* (Bernh.) Swartz in the Eastern United States. Antarctic origin of the extant species is most probable; but fossil *Lygodium* is reported from Europe in the Cretaceous and subsequent rocks. The genus is natural, and so homogeneous that there has been no agreement as to its division into natural groups. On whatever, basis this is attempted, the several resulting groups seem, like the genus as a whole, to be Antarctic in ancestry."

Without a precise tabulation of species, and specific localities, we may not be free to discuss these statements, but it seems to us that to them applies the very same we have stated as regards *Schizaea*. Having one species in the Eastern United States now alive, and fossil Cretaceous records in Europe, *Lygodium* may very well be one, as to dispersal, with the Ophioglossaceae. Its "tracks" are "antarctic," but its true origin is again something else.. All in all, three centers of pteridophytic evolution can be made out on the basis of existing knowledge, namely, (a) Modern Indian Ocean (Gondwana); (b) Modern Eastern Pacific; (c) Modern Southern Pacific. A form originating in (a) or (b) will be "Holarctic" and "Antarctic" depending as to whether we begin to run its "track" from the bottom of the map, or the other way around. The plain truth is we have no knowledge today how these "tracks" can be run to agree with evolution, so that orienting them in any rigorous manner is now impossible. All we can safely identify are *centers of evolution*, primary as well as secondary, and with this knowledge we must rest satisfied today. An immense work remains to be accomplished on the phylogeny and morphology of the Pteridophytes before we can approach a study of their "tracks" with some confidence that we may be able to orient them as oriented they must be to serve the aims of a rigorous study.

---

*) The statement quoted is contained in the recent *Genera Filicum* by E. B. COPELAND (Chronica Botanica ed.), 1941, and occurs p. 23.

Anemia is a third schizaeaceous genus, of which our authority writes, "Range: Tropical America, north to Florida and Texas; Africa and Madagascar, one species reaching India. A genus of about 90 species, nearly all in America, where the recent evolution of species has evidently been active."

This is interesting, because the Angiosperms were also actively "speciating" (and better) in the New World shortly after they had reached it from channels connecting as one India, Madagascar, Africa and the Americas. It seems plain that *Anemia* came to America from the modern Indian Ocean, following very much the "track" of *Botrychium*. We regret we lack just now an accurate account of the species of *Anemia*, and a reliable monograph to elucidate its intergeneric and interspecific affinities, for we think *Anemia* is first rate material for a distributional study mainly of New World interest.

The last generic group of the Schizaeaceae is *Mohria*, which, according to the usual authority, "Is peculiar in its family in bearing sporangia on its vegetative frond, and in bearing paleae. The former feature may be primitive. The latter can hardly be so. The paleae intergrade with hairs, and are to be regarded as dilated hairs, such as are found on *Lindsaea* and other genera of chaetopterid ancestry."

These statements contain a tantalizing fare, but, alas, our authority says no more, though he evidently accepts *Mohria* as monotypic. This genus, then, is endemic to South Africa, a land where anything may happen. *Cheilanthes* is characterized by our authority in the same work as "Genus arduum" of about 180 species characteristic of dry places — though not exclusively confined to them — in the tropical to warm-temperate regions of the world. We do not believe it is possible to effect a satisfactory disposition of paleae merely by curtly dismissing them as "Enlarged hairs." Were it possible to do so here, we should like to discuss them, showing that such "Enlarged hairs" occur in great many plants besides Ferns, not only, but are *vital part in the phylogenetic history of the "leaf."* The evolution of the "leaf" is fundamental, and we never will understand angiospermous (and non-angiospermous) phylogeny until and unless we give ourselves a rational account how the "leaf" came into being *).

Mohria, then, is probably the genus which, studied in full freedom from preconceived definitions, is to make light upon the ancestry and affinities of the Schizaeaceae. A very great deal in the distribution of this family matches standard angiospermous dispersal, but if we wish, at long last overdue, to understand origins — *with the rigor the importance of the subject demands* — we cannot rely upon the fact that in certain families a majority of the species may be endemic south or north of the equator. In reality, the term "antarctic" may be used in a sense quite as misleading, or devoid of precise meaning as its counterpart "holarctic." The Ophioglossaceae, for example, can be used — if misunderstood as to origins — to prove anything we may like to conclude.

---

*) Meaningful "Enlarged hairs" occur, e.g., among the Melastomataceae.

The suggestion that a broad range formerly extended far to the west from the present American coasts, deeply biting into what is now the open Pacific, is supported, as we have seen, by a massive evidence, only part of which it has proved possible to bring to the pages of this manual. Vouched for by Ophioglossaceae, this vanished range seems essential to account for certain fossil findings of *Ginkgo*, or forms allied with it *). In conclusion, we are not far from thinking that much vital work can be done by gradual steps in the following manner, (a) The angiospermous tracks, such as outlined in the pages of this manual, should be referred to the "tracks" (correctly, ranges) of non-angiospermous plants in a first place. A critical reference so made will elucidate the nature and location of the primary and secondary centers of origin of plants of diverse phyla, and identify regions of the modern maps fundamental in dispersal; (b) This done, the phylogeny of non-angiospermous groups may be worked out on the basis of range and a morphology broad enough not to take for granted that paleae are strictly and exclusively "hairs." This will correlate the ranges of non-angiospermous plants, assisting in the further identification of main centers of origin; (c) The fossil record is last to be critically scanned, so as to date the first appearance of interesting forms. Palaeogeographic and geologic work is to conspire to the same effect.

These three steps, indeed necessary steps, will ultimately make it possible for us (a) To reconstruct on the basis of dispersal outlines of lands that were, and to contribute accordingly vital evidence to geology. Geology is handicapped most definitely by the fact that, so far at least, it has been forced to stop there, where modern oceans start. This handicap is beginning to lessen by the frank recognition that a full exploration of the ocean's bottom is vital, and initial works have been done to this effect. However, phytogeography has in its power to span the modern oceans by nets of tracks, and to indicate with great accuracy outlines of land and water which no longer exist. *This is far more than other branches of lore can do in the present. We are certain that the examples contributed in these pages are a mere beginning. We do not doubt at all that by working out critically the dispersal of enough plants with due regard of their ecology and taxonomy we can secure results beyond comparison more accurate than those obtained throughout these pages. The results here secured are intended to verify certain essentials of method and principle, for, indeed, it should be impossible to include within the covers of such a manual as this detailed studies of hundreds of different genera. Studies of this scope and nature will soon be undertaken, however, and within a short span of years we shall*

---

*) *Ginkgoites eximia* discovered in Patagonia (ca. 45° Lat.) is said (Feruglio in Notas Mus. La Plata, Paleont. No. 40, 7: 93. 1942) to be close to *G. antarctica* of Queensland, and to suggest in addition *G. minor* of the Cretaceous of Alaska. Without giving these hints more value than they possess, it is clear that they point to the existence of Ginkgoid forms in former Alaska, Patagonia and Australia. These three points are easily connected within a net of dispersal homologous of part of the Ophioglossaceae's. SHAPARENKO's imaginative account of the migrations of *Ginkgo* (in Bull. Bot. Inst., Acad. Sc. U.S.S.R., Ser. I, No. 2, 3. 1937) receives scanty comfort from the Patagonia fossil. "Phytogeographical" work done along the lines of SHAPARENKO's is not at all rare in print, but we do not feel as a rule cogent obligation of referring to it.

know a very great deal of which we are today still wholly ignorant. We are certain that botany can easily be turned into a form of exact thinking and phytogeography sharpened to approach most closely pure mathematical sciences. It purely is a question of methods and approaches, and these we can hammer out to suit ourselves, and those who must learn from us.

As a contributory example of the possibilities open by a comparison of morphology and ranges we have recent work (16, 17) on the immediate affinity of *Polypodium vulgare*. This fern, depending upon the nature, presence, or absence of certain glandular bodies (Paraphyses) borne in conjunction with the sori, breaks down as follows,

(1) *Group with unbranched paraphyses present* — Eastern Siberia (Amur), Sakhalin, United States eastward from the Rocky Mountains.

(2) *Group with branched paraphyses present* — Madera, Canary Islands, Morocco, Algeria, Tunisia, Balearic Islands.

(3) *Paraphyses absent* — All the rest of the ranges, including the western United States.

These results are as yet not such to allow conducting critical exhaustive work. It seems still possible to correlate ranges and peculiarities of morphology much more closely, and perfectly, by a study of larger quantities of specimens and plants of *Polypodium* than have been investigated so far. However, the group with branched paraphyses highlights beyond doubt a consistent phytogeographic node of considerable significance. Peculiarly, unbranched paraphyses belong to plants which cross into the *eastern* United States going from the regions of the Amur through Sakhalin. It is perhaps possible that this one type of dispersal is due to the existence of palaeogeographic connections much to the north of the mighty "Landbridge" abutting to California. Whether these palaeogeographic connections have something to do — like those in their south — with phytogeographic boundaries in the Far East is at present purely speculative. It may be interesting to remark that as there is a "break" between the Japanese islands of Amamioshima and Yakushima, so there occurs in Sakhalin a certain "Schmidt Line" that interests plant-distribution. We feel confident that parallel "breaks" or "lines" exist also on the American side of the Pacific.

A glance at two diagrams (Fig. 96, Fig. 97) illustrating the distribution of certain Liverworts (2) reveals aspects of dispersal which confirm what we have learned from the Angiosperms. The fragmentary ranges of *Jaegerina* and *Rhegmatodon* can very easily be connected by "tracks" of pure angiospermous flavor. The distribution of *Anisothecium Hookeri* sharply highlights what throughout these pages we understood as the primary angiospermous center of origins, and pinpoints a door to South Africa through Natal which is of utmost relevancy in angiospermous dispersal. The same order of remark applies as regards *Psilopilum* and *Hymenoloma*.

Dicranoweisia and *Encalypta* are massive in the north, but sparingly represented in the south. It will be noticed that the "bridges" leading them northward closely correspond with the standard angiospermous gates, and would answer their purpose regardless as to whether the tracks are worked south or north. It is furthermore manifest that the ranges south of the equator are of relic-nature. This may be due in part to deficiencies in the record, but must have on the whole a basis in fact. The mute evidence from the dispersal of *Dicranoweisia* and *Encalypta* is that these two Liverworts formerly lived in certain austral landmasses, and survived only in parts of them to their own convenience after these landmasses left our maps. However, these two Liverworts had, in former days, also established contacts northward with certain centers climatically suitable to them, wherein they tarried. When these centers became intimately connected with lands gradually emerging to the north, *Dicranoweisia* and *Encalypta* followed, actively colonizing the lands in gradual emersion precisely in the same manner as *Sarracenia purpurea* subsp. *gibbosa* (see p. 468) left its "nunataks" or refuges to go forward in the wake of retreating ice-fields. In sum, the record is clear, and simple as to prime movers. Obviously, *Dicranoweisia* and *Encalypta* would not have mastered the north had their ecological requirements prevented them from so doing. As a matter of fact *Encalypta* remained, even in the north, south of a line marking the southernmost limit of the dispersal of *Dicranoweisia*. *Jaegerina* and *Rhegmatodon* which reached Central Mexico and the Himalayas northward might, in theory, have secured a vast boreal dispersal to match that of *Dicranoweisia* and *Encalypta*. They did not, however, which may be due at least to, (a) Their own ecological preference; (b) Their having been confined from the start in relic-ranges, without efficient contacts with the north.

BIBLIOGRAPHY

Chapter IX

(1) GUPPY, H. B. — see (4) Chapt. I.

(2) IRMSCHER, E. — see (22) Chapt. VIII.

(3) CROIZAT, L. — see (19) Chapt. II.

(4) SCHUSTER, J. — *Cycadaceae;* E. & P. Pflanzenr. (iv. i.) 1932.

(5) HENDERSON, M. R. — Materials for a revision of the South African species of *Encephalartos;* Jour. South Afr. Bot. 5. 1945.

(6) PILGER, R. — *Coniferae;* E. & P. Nat. Pflanzenf. 13: 121. 1926.

(7) COPELAND, E. B. — Antarctica as the source of existing Ferns; Proceed. Sixth Pacific Sc. Congr. 4: 625. 1939.

(8) COPELAND, E. B. — Fern evolution in Antarctica; Philippine Jour. Sc. 70: 157. 1939.

(9) DARRAH, W. C. — see (16) Chapt. I.

(10) CLAUSEN, R. T. — A monograph of the Ophioglossaceae; Mem. Torrey Bot. Cl. 19 (2): 5. 1938.

(11) ALSTON, A. H. C. — The Selaginellae of the Malay Islands I. Java and the Lesser Sunda Islands; Bull. Jard. Bot. Buitenzorg, III, 13: 432. 1935.

(12) ALSTON, A. H. G. — The Selaginellae of the Malay Islands II. Sumatra Bull. Jard. Bot. Buitenzorg, III, 14: 1937.

(13) COPELAND, E. B. — Ferns of Southeastern Polynesia; Bishop Mus. Occ. Paps. 14: 45. 1938.

(14) BROWN, E. D. W. & BROWN, F. H. B. — Flora of Southeastern Polynesia, II. Pteridophytes; Bishop Mus. Bull. 89. 1931.

(15) HOBBS, W. H. — Mountain growth; A study of the Southwestern Pacific region; Proceed. Amer. Phil. Soc. 88: 221. 1944.

(16) MARTENS, P. — Les organes glanduleux du *Polypodium virginianum (P. vulgare* var. *virginianum);* Bull. Jard. Bot. Bruxelles 17: 1. 1943.

(17) MARTENS, P. & PIRARD, N. — Les organes glanduleux de Polypodium virginianum. II. Structure, origine et signification; Trav. Biol. Inst. Carnoy, No. 39 (Cellule 49): 385. 1943.

CHAPTER X

GEOGRAPHY OF THE PAST:
A PHYTOGEOGRAPHIC HYPOTHESIS

It is accepted procedure in manuals dealing with plant-dispersal to discuss purely extra-botanical theories of the earth's past which have, or are supposed to have, importance as regards distribution.

It seems to us that this procedure is not entirely justified, and that its results are not encouraging. Confusion is now so great at all quarters that nothing short of trenchant and fundamental discussion based upon facts, not theories, can lessen its burden. A phytogeographer has abundant facts in his chosen field. It is but fair to the reader, and even more the evidence, that he discusses these facts as fully as possible, steering clear of compilation, and aiming, on the contrary, to present a cleancut outline of vegetable dispersal which entirely rests upon its merits and demerits.

A theory holds the field, however, which we may not entirely overlook. This theory, Koeppen & Wegener's, has enjoyed success, we believe, much beyond its merits because it strives at least to meet two essential needs, (a) A global presentation of dispersal in all its aspects, past and present; (b) A concise rationalization of the earth's past based upon the history of life, and plant-life in particular. These two needs are genuine, and accordingly keenly felt. Regardless of its intrinsic merits, an outline which strives to meet them is assured of eager readers. The world demands something along these lines, and all botanists will agree that if Koeppen & Wegener's theory is unsatisfactory, something else quite as general is, doubtless, welcome to take its place sooner or later.

In dealing with the theory of Koeppen & Wegener it proves necessary at first to underscore the all essential fact that these two authors are not the inventors of "floating" continents. Accordingly, we need not underwrite their tenets, if we intend to accept "floating" of this sort as desirable or necessary. The idea of continental shifts arose in the mind of Lamarck, well over a century ago (1), and in one form or other has remained alive ever since. Koeppen & Wegener, consequently, have nothing to do with its paternity.

The core of the Wegenerian hypothesis is elsewhere, which few seem to have realized. Insofar at least as it interests the botanist, this core is in the statement (2; p. 3) that the climate of any given region depends primarily from the position which a region occupies as regards pole and equator *(Die Klimageschichte eines Ortes is daher in erster Nährung die Geschichte seiner Lage zu Pol und Äquator)*.

It follows from this basic Wegenerian principle that a fossil flora of the Jurassic, for example, dug out today near the equator, but containing "cold" forms and, perhaps, supposed traces of alluvial deposits of "glacial" origin, must have lived in origin away from the equator. This is evidence, Koeppen & Wegener argue, that the poles "wandered" and that the equator of the Jurassic might have been located there, where the 70° Lat. North is found today.

This principle is alluring, and answers reality to the extent that it is warmer near the equator than it is near the pole, *in general*. This principle contrariwise is unmitigately false in other, practically vital respects, and leads accordingly to absurdity.

There is no possibility, in the first place, of diagnosing the nature of a plant, whether "cold" or "warm" except by actual investigation of its ultimate nature and immediate surroundings. *Empetrum* thrives in the island of Tristan da Cunha, where the climate is equable the year around, and frosts are seldom felt. This very same genus makes its home in polar Novaia Zemlia, and its fossil rests may accordingly prove anything we like. We may credit them to a catastrophic alpine landslide having taken place right at the equator, or an exceptional summer thaw on the shore of a thoroughly arctic sea. A large litterature could easily be brought forward on this subject, and examples multiplied without end. *Empetrum*, in our opinion, is adequate proof, and may accordingly stand here alone as example *).

If we follow the 0° Cent. isotherm in January we learn from the maps of the modern world that this isotherm varies of close to 40° latitude between its southernmost reach (near St. Louis, U.S.A., and Nanking, China) and its boreal limit in Northern Norway. *It is accordingly manifest that the presence of continental masses or marine currents may deflect an isotherm vital to vegetation nearly half the distance between the pole and equator.*

It is furthermore manifest that glaciations do not absolutely require polar conditions and position. As of today (3) heavy snowfalls are the rule in the east and north of the high mountains of Burma at 10.000 feet (about 3000 meters) altitude, and in certain large valleys the snow persists at this level until June or July. At 12000 feet (about 3700 meters) the snow endures to August and September, and never disappears between 13000 (about 3950 meters) and 14000 (about 4300 meters) feet. Based upon a different outline of land and sea, the nearness of mighty orogenies and other purely physiographic factors, a heavy glaciation would be possible in the present latitude of Burma. The January 0° Cent. isotherm just mentioned leaves us in no doubt as to this.

---

*) In the pages of this manual is contributed what we understand as a beginning of rationalization of the issue as to what is "cold" and "warm" in plant-life, *with factual reference to origins and dispersal*. It is obvious that a genus may be in Tanganyka, Ceylon and Java without for this being necessarily "warm." This same genus may associate in these same regions with authentic "warm" groups. Origins — *correctly understood* — are fundamental to the decision as to what is "cold" and "warm" in plant-life, witness *Erica, Linum*, etc.

The Permo-Carboniferous glaciations were very severe and long lasting in the Australian sector, less so in South Africa, Brazil and India. We have no reason to reject as unfounded — rather the contrary — the hypothesis that the continental masses of the dim past "floated" in some measure to latitudes other than those which their modern counterparts exactly occupy in the map but also see no reason whatever for building up a fanciful cartography along Wegenerian lines as the beginning of solid investigation. In other words, we are certainly not inclined to generalize, and to enter a theoretical straight-jacket at the outset. To begin with, as noticed, the Permo-Carboniferous glaciations were of uneven intensity, and to all appearances, much more intense in the regions (Australia and vicinity) which are still today nearest the pole. Moreover, it is certain that climate is also molded by factors which do not primarily depend from latitude, and we are not averse to the thought that the last Ice Age stemmed from conditions which — against Koeppen & Wegener's belief — were not immediately correlated with the position of the north pole *). The theory that nearness to the north and cold are synonymous occurs to, and impresses most deeply, investigators accustomed to the comparably steady climates of most of Southern and Western Europe, but seems much less well founded to other students. Those of them, for instance, who happen to be familiar with conditions in the Northeastern United States at the Atlantic coast well know that the position of zones of high and low pressure can radically alter the readings of the thermometer in a matter of hours, and that latitude does not necessarily rule temperature.

The absurdities which Koeppen & Wegener are ultimately induced to accept on the strength of a principle so uncautiously assumed are perhaps best proved by their maps of the Pliocene and Quaternary. Hypnotized, as it were, by the presence of huge icecaps centering in Greenland, the two authors hang their "wandering pole" straight over Hudson Bay at 70° Lat., with the result that the shores of Siberia on the Arctic Ocean fall in the Wegenerian cartography close to the latitude of modern Louisiana. As a corollary, the two authors assume that the lack of massive glaciations in Eastern Siberia at the time was, seemingly, due to a milder climate than the present in these quarters *).

---

*) We have no title to speak in the matter with authority, but do not reject as absurd the thought that the Pleistocene Glaciations stemmed, in part, from changes in the sea-currents now running by Eastern North America and north of it. The climatology of this region is so unsettled, in potential when not in fact, that a slight alteration of physiography favoring certain factors as against certain others, might radically alter the whole pale of its weather. Orogenies might in geological time have caused glaciations together with such alterations in physiography as we have just mentioned. We see no need whatever for "Wandering Poles" in all this, though, as we state in the progress of the main text, we are far indeed from rejecting as preposterous the suggestion that Greenland, in reality, was "shifted" northward to the position it occupies today, etc.

*) A good deal of quibbling, in our opinion, attends the statement that a distinction must be made between the *variable* "poles" of shifting landmasses, and the *steady* astronomic "poles" of the earth. This statement frequently recurs in the arguments of authors who believe in Koeppen & Wegener's theory. We need not enter lengthy discussions, merely notice that if we hang the north pole at Hudson Bay in the Pleistocene and Quaternary — as Koeppen & Wegener openly do — the Siberian latitudes — as Koeppen & Wegener openly admit in their maps — are no longer suitable to well known Siberian conditions.

The truth is probably other. To this day, Eastern and Northern Siberia receive comparatively little moisture in winter, because moisture-laden winds cannot arise from a icebound Arctic Ocean, nor can they reach the high north from the south across the high mountains and tablelands that fringe Siberia toward China. It is probable that Siberia was in the Pleistocene invaded by an overflow of ice from the Arctic Ocean, but it is also quite as probable that it was not buried under heavy icecaps originating on land precisely on account of the same climatic factors that rule there today.

Koeppen & Wegener have no idea worth the name of dispersal and its requirements. They are accordingly innocent of the status of the Pacific Ocean in migration, and they do not know that even ornithologists who decry as not "Serious" landbridges in and around Hawaii are willing to accept a pronounced lowering of the Pacific's floor *as a whole* (see p. 12) to get away from under the nutcracker of certain "Real quandaries." Uninformed as to the matter, the two authors bunch their continents around Africa, mistakenly supposing, we think that this continent hardly "floated, and leave to the Pacific a berth at least twice as wide as its present size.

All this is thoroughly absurd in the understanding of anybody having *a smattering* of sound phytogeography and rational general botany. It is to be rejected as a matter of course. Koeppen & Wegener's theory is not a legitimate scientific theory, but a dream without status in legitimate investigation. The Wegenerian assumptions are false as regards the nature and possibilities of what is "cold" and "warm" in plant-life; they are false moreover as to fundamentals of climatology and dispersal. All it can be said for the theory of these two authors is that it understands the needs of the hour without for this satisfying these needs in the very least. That this theory could be seriously received is proof not of its value, but of our general ignorance.

Insofar as we are concerned, the history of the modern Angiosperms begins in the Jurassic, and the ultimate migration of these plants takes its start as a rule from a range in the south of the map connected with certain geologic lands, now vanished, but once occupying large sectors of the modern oceans, Pacific, Atlantic and Indian alike. Leaving these centers, the Angiosperms streamed generally northward when the ranges now crumbled made contact with lands still extant on modern maps. Be it clear that we stress this foremost in obedience to the pedagogical requirements of our subject.

It is possible, using dispersal, to trace the rough outlines of the world that was (Fig. 98) when the Angiosperms began their wanderings *).

---

*) In submitting such a map as this to the reader, and quite aside from what is said in the main text we should add that efforts of the kind cannot be wholly successful as a reconstruction of the earth that was. For instance; It is patent that North America was entered from the direction of the Far East, and massively so — almost exclusively so, we would say — in Lat. 30° North approximately. To show this, we are bound to put on our map a continuous land-belt in this latitude running clear across the Pacific. It is sure, on the other hand, that had new grounds opened to colonization of plant-life from North America toward the Far East dispersal would have run just the opposite way it took. This considering, we are most likely in the right, if we visualize a gradual emersion of lands tending to span the Pacific *west to east*, which led the angiospermous *genorheitra* from the Far East to the New World. It is most likely that a complicate play of emersions and submersions, with connections broken and made over

This map is generalized, and based upon the total primary dispersal of the modern Angiosperms. It agrees in the main with outlines that *might* be supposed to have lasted between the Mid-Jurassic and the earlier phases of the Cretaceous. We are thoroughly aware that this map does not give indication of local factors, and is patently unsatisfactory as to important details. This map contributes, nevertheless, the *essential outlines of continental masses and oceans* required by phytogeography, if dispersal is to make sense. This map therefore disposes of the Wegenerian cartography, not only, but of assumptions that dispersal could be effected, and was effected, on maps even faintly resembling those of the present. To this extent, *this map is positive,* and its contribution to the discussion quite direct. We press no other claim.

Dispersal was further molded in a second epoch, possibly lasting from the Mid-Cretaceous to the Early Tertiary, when the ancient continental mass filling the Indian Ocean entered its last phase of crumbling, and also crumbled Malaysia and the shores of the Pacific most actively. Maps for this secondary dispersal are not within our immediate possibilities, because a vast array of plant-life is to be studied in much detail to follow constantly changing outlines of lands and sea. This will be readily apparent from the migrations of such genera as *Lumnitzera* (see p. 400), for instance.

Strict requirements of space restricted our choice of the groups that could be discussed in these pages to a minimum. We venture to state, nevertheless, that vegetation pinpoints today with great accuracy geological features no longer agreeing even with the roughest outlines of present geography, and are accordingly of the opinion that a vast field of enterprise lies open before the joint undertakings of the phytogeographer and geophysicist. Undertakings of this kind could not be thought of in the past because phytogeography had nothing tangible to contribute on its part.

The modern Angiosperms first reach the world of our maps at certain regions of the southern hemisphere which we have for this reason identified as *gates of angiospermy.* We say, *first reach,* because it is comparatively easy to drive consistent tracks of migration from these regions going northward in the main. We have also commented, that the tracks move consistently in the main direction of the north in the wake of a geologic wave of profound disturbances spreading from the south northward. As a corollary, we have stressed the circumstance that dispersal and geology cannot be studied apart.

It may be thought that these gates of angiospermy have ephemeral value, and, though important in some respects, are on the whole hardly better than illusory. We do not share this opinion, because we believe that these gates

---

various times in succession obtained at the Magellanian gate. Moreover, thoroughly to understand migration it is necessary to have some appreciation of physiography, because this is what determines ecology in the last resort, and ecology is capital in biogeography which claims to work out details critically.

On the other hand, the existence of Gondwana is an axiom; the presence of shallow seas or lakes over much of Africa very much sure, etc. In sum, such map as we present here is a mixture of truth, half-truth and opinion, and the reader should not view it unless with a sharp sense of the limitations connatural with the subject itself.

answer on the contrary positive realities of ancient geography. Good indications of the presence of mighty phytogeographical boundaries are found in tracts of the Indian and Pacific Oceans where nothing is to be seen today but an immensity of waters. Moreover, the African gate of angiospermy, merely to mention one, is used by different groups of plants in different manners. We hold to the opinion that if the gates of angiospermy were illusory, matters would be different. In other words, *what is today the ocean would not behave precisely as if it were land*. The evidence need not be stressed beyond sober reason to conclude that the gates of angiospermy were in touch with landmasses to the south and north that did not agree with the present limits of Antarctis.

A student of the natural history of New Zealand (4; p. 135) is of the opinion that these islands were of continental dimensions during the Cretaceous, and that in the Late Jurassic and Early Cretaceous lands were emerging all around the Pacific. He says, "The Antarctic continent during this period of elevation no doubt extended farther to the north, approaching perhaps within a few hundred miles of the New Zealand continent. Possibly Macquarie Island was much larger than at present . . . . It contains altered sedimentary rocks of unknown age."

This statement throws light, broadly speaking in the manner in which contacts might be established between Antarctis and the landmasses to the north. If Antarctis came to within "A few hundred miles of the New Zealand continent," no obstacle exists to the belief that these few hundred miles might have been factually erased at some time or other during this era by a northward extension of Antarctis, or by a southward peninsular prolongation of New Zealand. It is furthermore possible to conceive of a minor landmass emerging between Antarctis and New Zealand to furnish a bridge between the two. The geologic history of these regions is entirely lost to us under the waters of the southern oceans, and it is accordingly reasonable to admit that the connections between Antarctis and the lands to the north might have been established not in one definite manner only, but by a combination of all the possibilities just outlined. Physical contiguity is not even necessary, for chains of islands separated by a comparatively small mileage of sea may be equivalent to solid landbridges. It is sure in our opinion that an entire pale of geography lasting many millions years has forever vanished from sight in the deep south of our maps, and the modern oceans of today generally.

We could end our review of the earth's past at this point, because our statements already cover everything which phytogeography demands. Phytogeography is satisfied with starting points of migration well ascertained and with tracks that run consistent courses, and is not primarily interested with the building up of theories to cover much of the earth's past. Considering, however, that phytogeography and geology are closely correlated, we may have some excuse if we venture to record in print certain ones of the thoughts that occurred to us in the preparation of this manual. Having dismissed Koeppen & Wegener's theory as unworthy of serious attention, we might do well, perhaps, in stating what we accept as most probable in regard of the so called "Continental Shifts." Many are the botanists and phytogeographers accustomed to the idea that these "Shifts" are integral part of phytogeography, and we owe them a word at least of comment.

Though we are not at all well informed as to the details of such "Shifts," and are satisfied in last resort that deep familiarity with them is not a vital ingredient of plant-geography, still we, too, came to accept them in a way. We pointedly say, *came to accept them,* because it took us time to get rid of the preconception that a landmark of geography is absolutely fixed. This preconception left us gradually, perhaps reluctantly, when massively dealing with factual records of dispersal we found ourselves handling something so fluid and so monotonous at the same time that we were forced to change mind. The map, at first immovable before our eyes, gradually began to stir and move in unison with plants, and we had to follow.

A region of the modern maps struck us as peculiarly instructive. This region (Fig. 99) lies between the southern tip of South America (Fuegia) and Antarctica (Graham Land, roughly speaking). This region is connected by an arc of islands much distended eastward, and bulging farthest in this direction with South Georgia and the Sandwich Islands.

No comparable arc of islands connects South Africa, Tasmania and New Zealand with Antarctica, though scattered insular outposts (Bouvet, Prince Edward, Crozet, Kerguelen, Heard, Amsterdam, Saint-Paul, Macquarie, Campbell, Auckland, Antipodes Islands, etc.) lie to the south of these continents and domains in the general direction of Antarctica. The relation that some of these outposts bear to the landmasses in the north is peculiar. Streams of migration traceable to the Kerguelen Islands, for example (see p. 252), seem to have a tendency to avoid Africa, and to move on the contrary toward the more distant South America.

We do not know whether it is pure coincidence that South America, connected with Antarctica by a clearly delineated insular arc, is also the continent that reaches farthest south. It might be thought, on the other hand, that this arc still exists *because* South America so reaches. In other words: Had South America moved northward, this arc might have been distended beyond recognition, and its insular components be reduced to the uncertain geographic position of the islands ranging, for example, between the Indian Ocean, South Africa and Antarctica.

Always as regards the same arc, two alternatives readily suggest themselves, (a) The main stress molding this arc appears to act eastward in a course roughly parallel with latitudes. Should this stress continue without being offset by competent counterstresses at the periphery, the arc in question would be completely shattered, and its segments either left as roughly parallel insular chains, or eventually rolled up against the eastern coast of South America, and the coast of Antarctica; (b) Should the main stress be counteracted by peripheral stresses, and its trend reversed, the arc in question would shift to reestablish a fairly compact landbridge between South America and Antarctica.

These thoughts may be wholly unworthy of the attention of geophysicists, and we make no claim for them, for geophysics are not of our resort. A geophysicist, however, writes (5; p. 213) as follows, "Only Antarctica, for some obscure reason, remained anchored and more or less constantly polar; all other continental masses worked toward the equator . . . . This could be achieved only by pushing the masses in the northern hemisphere toward and over the north pole. How this was possible against the trend to the forces that should have prevailed there? This is the real problem."

Antarctica, we surmise would no doubt remain in a polar position by stresses acting along the lines of those suggested by the arc of islands connecting South America and Antarctica. These stresses work in three directions, east, north and south (Fig. 99), and Antarctica would be pressed all around its periphery in a southern push, while the lands to the north would be shifted northward and possibly eastward meantime.

It so happens that Antarctica has the highest average altitude (about 2000 meters) of all other continents on the modern earth. Is this a coincidence? Could it be possible that the *high* Antarctica of today arose by degrees out of the compression of the *broad* Antarctica of the past around a central polar core?

These thoughts would not occur to us, if dispersal did not require continental masses in the south which are no longer. Where did the earthy components, *sial* and *sima*, of this former *broad* Antarctica end?

Hardly anything is so clearly proved by dispersal as the fact that Africa and South America once formed a single landmass. The stresses acting against the arc of islands between South America and Antarctica act east, north and south. Is it a coincidence that while Antarctica appears to have been compacted all around by stresses to the south, Africa appears to have been shifted as regards America northward and eastward? Climate need not fundamentally alter because of this.

An overwhelming evidence from dispersal further tells us that Africa and Asia as far as Malaysia were once welded together by a single continental mass, which we have currently understood in these pages as Gondwana, but other authors call Lemuria. This landmass is a keystone of dispersal, and to doubt its former existence is impossible. Its presence is still so overpowering as regards dispersal that, we may well say, *Gondwana exists*. We, at least, would stoutly reject every suggestion that this continent is hypothetical. Further to clarify our thoughts in this essential matter, we might make use of a simile, and state that Gondwana is quite as factual in the immediate present as is factual the plasma of our ancestors within our own body. We could not be, had it not been for this plasma, nor could dispersal be molded in its present shape, had it not been for this continent.

Gondwana had intimate connections with lands to the south, and a thoroughly continental geography. Where did this landmass go? Parts of it might

well have been absorbed within such landmasses as form today other continents, but was it not shattered in the first place by a system of stresses mainly
arising from the south?

Here we end our questions, in the understanding that we are wholly ignorant of geophysics. All we have done in asking them, is to ventilate passing
thoughts, not to lay down premises even of tenuous hypothetical nature. As a
matter of fact, we need not ask such questions at all. As phytogeographers, we
find them at bottom immaterial. They have nothing to with the essences of
dispersal, which consists of facts from the records of taxonomy and distributions, not theories how the earth could be remolded one way or the other.

Perhaps of greater value to the readers may be our thoughts as to certain
features in the continents of today which dispersal shows to have been different in the distant geological past. While making bold to write out these
thoughts, we warn the reader that in the pages of this manual we have conveyed the barest essentials, not only, but have conveyed them by degrees, as
it were, gradually working around these essentials as we further advanced in
our examples and interpretations. We have at first barely outlined main tracks,
purposely refraining from mentioning Gondwana, lest the readers might
assume, we were theoreticians to whom the existence of Gondwana was essential as a preliminary. This, as the reader knows by now, never was the case.
We accepted Gondwana after — *but not before* — the facts in dispersal told us
that this ancient continental mass necessarily had material existence.

It stands to reason that the series of steps whereby we were led to accept
Gondwana, and to mention it openly in print, need not be cut short with the
broad realization that Gondwana indeed existed. These steps — we firmly
believe — could be carried much farther. By an appropriate critical reference
to the dispersal of ever larger groups of plants from all phyla, we could refine
our knowledge much beyond its present limits. *We could reach — we no
less firmly believe — a much more detailed understanding of the geography
that was. The aspect of dispersal which actually surprised us, and still surprises
us after years of more or less intimate acquaintance with our subject, is the manner in which the plants living at this our bear witness to events buried in the
night of remote geological ages. There is seemingly no limit to the questions we
can ask, and the questions plants will answer. Obviously, answers as to details,
secondary tracks, lesser centers of evolution cannot be exacted without correlating
phytogeography with ecology, taxonomy in all its aspects, constructive morphology, palaeobotany and geology.*

We make these declarations wishing that it be clear that our thoughts are
still far from having reached full maturity *as regards details*, perhaps important
details, details indeed so important as to approach at times essentials. Very
often, we keenly felt our inhability to go farther, because we lacked relevant data of ecology and taxonomy. Not seldom we suspected that some of the authors
we consulted in regard of issues important to us were less clear in their own
minds than could be anticipated. It frequently happened that we dimly felt,

rightly or wrongly, the importance of certain questions, only to find that the general limits of our knowledge, and the means to our disposal in general, were below the task of bringing these issues (whether in the end nugatory or capital we could not say) into light. Writing, then with a sharp, unequivocal understanding of our own limitations, we venture to outline certain of our thoughts relating to the past of the modern continents, as dispersal would have it. These thoughts are still crude as to details, and our only justification for broadcasting them is in the fact that they may be useful to the readers, whether they be proved satisfactory, or discarded as erroneous in the end, does not matter.

Starting with the Pacific, an ocean that interests today as one Oceania, Asia, America and Antarctica, we may repeat what we have stated already. Precise evidence is in dispersal that at the time of the first angiospermous migrations — corresponding roughly with the latter part of the Jurassic — this ocean was not as wide as it is today. This ocean could be reduced almost to a quadrangular body of waters held under a broad approximation within the following landmarks, (i) Mariannes; (ii) Midway Island; (iii) Galápagos Islands; (iv) Sala y Gomez Island. The range contained within this quadrangle rates with us as the original Pacific.

Always dealing for the present with the static elements in our visualization — for we will summarily consider the shifts that altered it in a second part of this discussion — we conceive the modern Caribbeans as of land throughout. This land stands connected within a single body with, (i) The Southeastern United States as far as the Ozarks and the Appalachians; (ii) The eastern and central regions of Mexico together with California, which are separated by a shallow sea from the western; (iii) Central America; (iv) The northern belt of South America to include parts of Colombia, Venezuela and the Guianas; (v) Through the lands just described in (iv), with two ranges separated again by a regions of shallows, intermittently occupied by an inland sea or great lakes, roughly stretching north to south, namely, (vi) Western South America as far south as Central Chile; (vii) Eastern South America as far south as Uruguay. These two sectors (i.e., vi and vii) are in more or less intimate communication through islands now represented by certain heights in Paraguay, and the Matto Grosso of Brazil. To complete the picture, a third region, now part of South America, but originally constituted by lands from Central Chile southward to Antarctica is separated from Western and Eastern South America by an arm of the sea.

This covers lands which we are wont to understand as America in our maps. However, these lands are not just slices of the modern New World. They are something else again. As a corollary of the much lesser size of the Pacific by then, the western coasts of California and Mexico run much farther west to include the Hawaii Islands, Revilla Gigedo, the Galápagos. The western coast of South America — which is contiguous with that of North and Central America — also reaches westward far beyond its present limits to engulf the

Juan Fernandez Archipelago, and to reach perhaps the vicinity of Sala y Gomez Islands. The Caribbeans which we have arbitrarily assumed as the hub of the visualization we describe are integral part, together with the Eastern United States, Mexico and Northern South America, of a continental mass which absorbs within its limits the southern districts of part of Europe (Spain at least) and the whole of Africa.

Farther east still modern Africa becomes immediately confluent with a massive continent, Gondwana, that fills the whole of the modern Indian Ocean, not only, but has for its limits on the Pacific an approximate line Japan—Mariannes—Western Carolines—Samoa—Tuamotus—Tonga. This line further extends eastward between the 45° and 30° latitudinal belt to make contact with the southernmost end of South America. In the north, the boundaries of Gondwana are identified by a line which runs more or less by Afghanistan—Northern India—Central China—Japan. Between the 30° and 40° Lat. North stretches a wide belt of lands leading all the way to California, Mexico and Western South America.

These are the main outlines of a world in the Late Jurassic or very Early Cretaceous fit to account for the migrations of the Angiosperms in a first time. This world, then, looks like our map (Fig. 98) indicates. It will be noticed that certain lands in the north remain out of the main continental masses, as shown by the map. Moreover, Australia and Africa are deeply indented by huge marine or lacustrine basins. It will be seen that, for example, a mangrove or near-mangrove originating by East Africa would tend to migrate *westward* in the main.

This world was later modified as follows, (a) By a general widening of the Pacific; (b) By the opening of the Indian and Atlantic Oceans; (c) By a compacting of Antarctica in strictly polar position; (d) By a general accretion of lands northward; (e) By the filling in of the lacustrine and marine depressions in Australia, South America, Africa. These changes involved the establishment of huge lines of faulting and marine transgressions some of which are indicated on the map attached. With reference to these lines, it may be noticed, (a) A mangrove or near-mangrove originating by Madagascar would migrate *eastward*, not westward; (b) A plant of the same description rooted in the first age of migrations near the "Great Rift" would in the second age become "alpine" by this shore's uplift; (c) The formation of inlands seas or lakes homologous of the modern Caspian, for example would favor the multiplication of mangrove and near-mangroves in Nigeria; (d) Likewise, Angola would become a center of secondary speciation and origin by gradual withdrawing of waters opening new grounds to colonization; (e) The filling-in of Paraguay would open the sluice to active invasion of the modern Amazonian basin and its southern extension by forms already rooted in eastern and western South America.

Considering that the New World is *uniformly* at the receiving end of angiospermous tracks throughout the first epoch of migration (Late Jurassic to

Early Cretaceous at the latest) we are drawn to conclude that none of the primary angiosperms ever originated in the Americas. The course of the migrations *constantly* points to, (a) Geologic Antarctis; (b) Gondwana; (c) The Western Pacific; as the cradles of angiospermy.

To repeat what we have already stated: *We are certain that by adequately extended studies of large masses of plants the bare bones of the ancient land-masses we have outlined can be filled with relevant orogenies, valleys and the like.* Correlations in lithology that are rash when effected in vacuo become possible when plants open the way; heights, whether rejuvenated or not, that seemingly stand wholly disconnected in the modern world, may be brought together by dispersal to form the backbones of ancient chains; etc. We trust that the work ahead, which we foresee, but cannot accomplish with the means and time now at our dispersal, will some day be successfully achieved.

We may now be asked, What about "continental shifts"? Did "shifts" of the kind take place in our opinion, or not?

Our answer is that we are ignorant of geophysics, but altogether willing to believe that the western, eastern, southern and northern shores of the Pacific alike were "shifted" in the position which these shores occupy today. So might Africa have been "shifted" away from South America, and America itself have been "compacted". Greenland might have "moved" northward somewhat. All in all, however, it is unconceivable to us that these "shifts" involved changes in the poles' position.

We are ignorant of geophysics, we repeat, but it seems to us that "shifting" and "compacting" become very much synonymous in the tale which the Angiosperms reveal. *A certain fact is that the maps radically altered by different ages; that the geologic revolution begun in the Late Cretaceous, and more or less accomplished by the end of the Tertiary, is the master-key to an understanding of the history of modern plant-life. Obviously, the reciprocal is true, so that this history is one of the most valuable adjuncts to an understanding of palaeogeography, consequently of geology in general.*

The student of dispersal faces in his works not one, but endless types of geography. As he deepens his quest, the outlines of the maps of today gradually vanish, and as if a light shone from underneath to reveal the outlines of the maps of the past, these outlines come to the fore one by one. The very though that stuff *so basic* can be handled by "casual agencies of dispersal," academic arguments, etc., is soon perceived in its true size, *as something plainly ridiculous.* It is the past and present of the earth that speak, and we must be blind altogether to reduce all that to a handful of theoretical recipes, none of which works when confronted with the facts. The question is not longer with honestly arguing the evidence, merely telling those who do not know it, *Look, and keep silent awhile ere you speak.*

We can easily draw a simile, possibly an effective simile. Let us suppose that we have before us a glass disk, and mark upon it certain sectors as the centers of origin of the angiospermous hosts. Let us suppose next that we give this disk a sharp rap with a hammer right there, where these centers are located. The disk will fly into pieces. Let us finally rearrange these pieces in such a manner that pieces where the Angiosperms did not originate come into contact with fragments wherein, on the contrary, these plants first appeared. Let us rearrange this broken stuff again more or less in a disk-like outline, and observe.

We will doubtless notice that, (a) All the centers can still be connected over lines with the positions they formerly occupied; (b) Each center may in its new positions feed angiospermous life farther afloat; (c) Every point occupied by this life stands in contact over more or less extended lines with every other point and center.

This is pretty much the very same that happened to the earth and its plant-cover. The disk formerly entire may be compared with the pre-Cretaceous world, wherein the angiospermous prototypes were few, and concentrated. The disk next shattered and rearranged represents, on the contrary, the post-Cretaceous world. *Tracks* in the shape of lines will connect forever the secondary and primary centers among themselves, not only, but all these *tracks* will come back in the end to the original starting points, the *gates of angiospermy*. These *gates* may no longer be whole glass, and be represented by tiny, disfigured fragments only (e.g., Mascarenes, New Caledonia, Juan Fernandez) out of their former outlines. These *gates*, however, will still be tangible on account of the effects they begot.

BIBLIOGRAPHY

Chapter X

(1)   LAMARCK, J. B. — Hydrogéologie (see a review, most likely by Cuvier, in Jour. Litt. France, Pluviôse An X (approximately: February 1802)).

(2)   KOEPPEN, W. & WEGENER, A. — Die Klimate der Geologische Vorzeit. 1924.

(3)   KINGDON—WARD, F. — A sketch of the geography and botany of North Burma; Jour. Bombay Nat. Hist. Soc. 44: 550. 1944; ibid. 45: 16. 1944.

(4)   OLIVER, W. R. B. — see (4) Chapt. V.

(5) VAN DER GRACHT (A. J. M. van Vaterschoot van der Gracht) *(et al.)* — Theory of Continental Drift. A symposium. 1928.

CHAPTER XI

PHYTOGEOGRAPHY: ITS PURPOSE, METHODS AND NATURE

The purpose of phytogeography is of the simplest. This branch of botany studies dispersal not only in order to learn how plants migrate, but ultimately to ascertain how they evolve while migrating. The why of this double purpose is self-evident, considering that there is no possibility of following plants in their dispersal without coming sooner or later to a region of the earth, where old affinities yield place to new. If phylogeny is not necessarily an integral part of phytogeography, still phylogeny and phytogeography work together so closely that their aims and doings cannot be dissociated.

The methods of phytogeography are of the simplest. These methods do not demand studies in earth's "Pendulations;" investigations of "Species Senescence;" speculations regarding "Extraordinary capacities for transoceanic dispersal;" and the like. Even less do these methods strive to separate "Descriptive" and "Interpretative plant-geography," because the two are one and the same. Description without interpretation is transparent cataloguing; interpretation without description a form of mysticism or propaganda. Sound phytogeographical methods require that dispersal be plotted in the first place on the map. This done, suitable tracks are drawn to connect known stations, and general principles of interest to phylogeny and dispersal alike are ultimately inferred by a critical comparison of as many forms and as many tracks as are necessary to make light upon the substances in play.

Phytogeography is accordingly a science which rests upon the facts of dispersal, purely and simply, and interpret these facts in the light of a general knowledge of botany and geography. To put this in other words: Phytogeography is common sense directed to observe and interpret the behaviour of plants in motion over the maps of the past and the present alike. If this be done, and this method be followed, phytogeography remains science. If, contrariwise, this purpose and method are replaced by preoccupations with ready-made definitions, traditional issues and the like, phytogeography is no longer science, but academy.

To introduce the principles of phytogeography to students already schooled in the essentials of botany requires no effort at all. These students are faced by a map of the world, and one of their numbers is required to read the records of he distribution of different taxonomic groups, while another pinpoints the localities in the map. Tracks are thus built right under the eyes of the class, and the concept of orderly migration and *genorheitron* introduced without difficulty at one stroke. Seeing how repetitious are the tracks, and how all of them eventually end at certain angiospermous gates, *the students forthwith grasp the fundamental that dispersal is beyond comparison greater in time and broader in space than they had probably been led to believe*. Accordingly, they will be immediately put on their guards against assumptions and theories which fail to take into account this essential.

It is not too much to hope that in less than four lessons students so taught will know more of the essentials of phytogeography than they can learn in a much longer period of time devoted to the review of sundry theories, compilations, and issues that have but little to do with real dispersal.

It is clear, moreover, that this approach to dispersal is calculated to make an appeal to a natural feeling of curiosity and wonderment which is the root of scientific thinking. *Imagination will receive its just deserts together with reason, which is of the essence of sound educational methods.* *)

It is not for us to follow the matter in detail, for the preparation of a syllabus is not within our immediate purpose. It is obvious, nevertheless, that phytogeography easily lends itself to observations as regards the entire field of botany. Problems of morphology, phylogeny, ecology and taxonomy are constantly met with in dispersal.

Brief notes on the kind of morphology and phylogeny essential to phytogeography, and, we further believe, to botany in general are contributed elsewhere in this manual. It is the place here, on the other hand, to comment on the limits which are common to ecology, taxonomy and phytogeography.

Geography itself consists of two fields. One of these fields deals with the generalities of the earth in their broadest aspects, and is concerned only with great distances and main features. This field considers the continents, their major political divisions, oceans, and the essential features of the world. The other of these fields, which is properly understood as physiography, takes an interest in more minute objects such as the nature of a definite region, studying in detail its rivers, plains, mountains, and the local interrelation of these factors.

These definitions are of necessity imperfect, because it proves impossible to draw a precise line where geography ends, and physiography narrowly understood begins. They answer, nevertheless, fundamental generalities, and for this reason are not devoid of meaning.

Phytogeography also is geography, but applied to an investigation of plants in motion. In other words: It is of the resort of phytogeography to study main

---

*) When we mention imagination, we intend *disciplined imagination*, the insight, that is, which intuitively derives from the facts inner causes, and builds therefrom a reasonable picture in the future. The art of disciplined imagination seems lost in these times and age. The less imaginative the student of science, the better he is now believed to be, on the grounds that he is fit to live only by "facts." Those who so live, alas, are oftentimes "specialists" competent to split a lone hair four ways, without for this knowing *what* to do with the four ends, and *why* the hair was split in the first place.

tracks; to consider extensive migrations; to deal with masses of plant-life in the act of their translation from an important division of the earth to another. This aspect of phytogeography naturally follows out of the necessity of considering as one the five continents and intervening oceans. Tracks are not stopped by physiographic barriers, and demand for this very reason a consideration of the earth which is essentially geographic.

Ecology, contrariwise, is vegetable physiography. It deals with the interrelations of plants already brought by phytogeographical means to definite, comparatively narrow ranges; it studies the manner in which these plants take advantage of the ground, and react in harmony with the ground. A critical consideration of vegetable associations is of the resort of ecology quite as a much as a critical study of broad channels of migration is of the essence of phytogeography.

It proves impossible to discriminate finally between geography and physiography, and very much for the same order of reasons it is not feasible to draw final boundaries between phytogeography and ecology. Geography and physiography, phytogeography and ecology are complementary in the last resort and interlocking in great many respects. In reality, no one can be well schooled in anyone of these branches of lore who does not master its correlatives.

Phytogeography invariably reaches a point where its conclusions need be supplemented by those of ecology, and ecology is forever blind without the ministrations of phytogeography. The two are essentially different *in tendency*, factually identical in the scope which they seek to achieve. It is easy for this reason to confuse the two, and hopefully to present taxonomic work but thinly venereed with ecological preoccupations as "Phytogeographical studies" (e.g. 1), which is beyond doubt incorrect, therefore misleading. The two can no longer be confused once it is clear how they differ, and how they agree.

One of the chief difficulties found in our work has been throughout the lack of reliable ecological data. We have followed tracks more than once which led us to alternatives that we proved unable to settle. A final understanding of the details of the phytogeography of Malaysia, for example, requires a knowledge of ecology which is not yet on record. The interrelation of certain regions of the Philippines, Luzon and Formosa, Palawan and Southern China, for instance, suggests problems to a phytogeographer which cannot be dealt satisfactorily without critical and extensive ecological investigations. In cases of doubt as to the origin of a certain taxonomic entity, whether it be from the African or Western Polynesian gate, a knowledge of the associations (whether *seres* or *climaxes*) in which this entity takes part is beyond doubt most desirable.

Another aspect of the interrelation of phytogeography and ecology lies in a field which interests most immediately also palaeogeography and geology. The dispersal of living plants pinpoints with accuracy — *sometimes amazing accuracy* — features of the geography of ages past of which hardly a trace is left in the maps of the present. Arrant mangroves like *Cassipourea* became "alpine" following the prior occupation and subsequent uplift of certain African geological shores. The tiliaceous *Triumfetta* speaks of the geologic past of South and Central Africa, not only, but the Indian Ocean as well, bringing before us vivid pictures out of the Cretaceous and very Early Tertiary. A small town of South Africa, Barberton, and an insignificant height of Malacca, Mt. Ophir, loom larger in phytogeography than half of Canada; etc. An enormous amount of work remains to be done in which phytogeography, ecology, geology and taxonomy must all lend a hand, if the details of these problems are ever to be filled in, as they must be sometime.

Phytogeography further appropriates and readily makes its own principles which so far have been regarded as essentially ecological. The concept of ecotype, for example, is by no means limited, the like current opinion takes for granted, to the narrow compass of ecology proper. This concept validly applies to families, for example, because Cruciferae and Capparidaceae are in realities ecotypes of the *genorheitron* which sired them both. The former consist in the main of ubiquists or psammophytes with "cool" to "cold" preferences; the latter, on the contrary, are at bottom xerophytes with "warm" to "tropical" requirements. The brief discussion of certain mangroves given elsewhere sufficient indicates how groups of this nature and tendencies may arise from a common progenitor which current taxonomy may treat as a single undifferentiated species. Physiology and destination of tissues are readily seen as the most powerful agencies of morphology regardless of the time and space required for their action. In a very definite sense, a variety and a family are equivalent in the scale of ultimate biologic realities.

None of the problems which ecology has so far considered as peculiarly its own is unknown to phytogeography. The interrelation of surroundings and life which yields petty forms and varieties in recently de-glaciated regions has been instrumental in ages past to evolve genera and families against a much larger background of time and space. There is sound indication, for example, that a study of the intergeneric relationships of *Nepenthes* in Malaysia is at bottom homologous of the study of comparatively petty populations of *Hieracium* in areas covered by the ice-fields of the Pleistocene.

In these, and similar respects, ecology and phytogeography readily interchange their fundamental concepts, and the study of the macrocosm (phytogeography) borrows the tools of that of the microcosm (ecology). This is all as it should be, for time and space, as such, are irrelevant in nature. It is man who magnifies them beyond reason, for he himself lives a short life, and only very recently has learned to understand the meaning of distance. As a matter of fact, this meaning has as yet failed to sink deep into his consciousness, and is accordingly still colored by everyday's preoccupations which tend to obscure its ultimate importance toward a more perfect understanding of nature and its works.

It strikes us that, had some of the generalities here briefly outlined been understood, Willis' famed theory of "Age and Area" (2) would have called for soberer and more constructive discussions. Willis' thoughts were colored by the study of the flora of Ceylon which many of its critics, and admirers, knew hardly at all. This flora, like that of Madagascar for example, is of very ancient composition, therefore not exactly comparable *at a glance* with that of the regions in the northern hemisphere that underwent comparatively recent glaciation.

Willis might not have been a profound philosopher, but was beyond doubt a willing thinker and a keen observer. Studying the flora of Ceylon first and foremost, Willis observed that *on the average*, the species endemic to this island occupied the smallest area; those also ranging to Peninsular India (but not beyond) areas rather larger, and those lastly that extended beyond Peninsular India the largest areas. He further remarked that the number of species in each one of these categories diminuished or increased in proportion of the area involved. These findings Willis later corroborated by an analysis of the flora of New Zealand (2), which told him that species having in these islands a range of between 881 and 1080 miles were endemic to them to the total of 112, and wideranging, on the contrary, to that of 201. Species on the other hand, having in New Zealand ranges of between 1 and 160 miles gave 296 endemics and 30 wides, most of the latter being of recent introduction.

A sagacious student of the flora of Madagascar who had with Willis' theory no concern made observations that agree with Willis' own. This student noticed, for instance, that certain genera have in this island (3) a species peculiar of the marine shore which is wide-ranging together with other species confined inland and narrowly dispersed. He further remarked that some of the best characterized species of the ericaceous genus *Philippia* are confined to a lone peak, even a few crags, while forms of the same genus morphologically unsettled, therefore "difficult" from the taxonomic standpoint, are widespread.

From these observations, this student argued that the species having narrower ranges were probably older, a conclusion of which Willis would not have approved. The fact remains that, regardless of individual interpretations, both this student and Willis were impressed by the same order of facts.

Were we to account for the facts in question, we would say that whenever a *genorheitron* comes to a large and ultimately well diversified land, this *genorheitron* makes the best of circumstances. It releases out of its folds, forms or species that narrowly espouse certain habitats and ranges, and speciate all the more strongly as they remain for a longer time indisturbed in the possession of these haunts. This very same *genorheitron* will also release, or *perpetuate*, in this same lands form having less definite requirements, which will in time achieve much larger distribution, or, conversely, lose none of the large distribution which was theirs in a first place.

A brief comment, how this may happen is offered elsewhere, and there is indeed nothing extraordinary in that it happens, once events are seen against the background of time and space which is theirs.

Willis' theory rests consequently upon certain aspects of the general problem of dispersal and speciation. It brings forth the all essential fact that speciation is not haphazard, but takes place within definite limits of time and space, and so it does with some constancy. Its greatest merit was in dealing the deathblow to the notion that dispersal is the work of casual agencies of dissemination, and furnishing *a simple and factual basis* of evaluation of the work of speciation in time and space. The denomination "Age and Area" fits Willis' effort in a first place.

What Willis did not see, in our opinion, is that the age of a species cannot be inferred from the space this species occupies in present dispersal. This cannot be inferred on account of two main reasons, (a) No ground in principle exist for assuming that a species that narrowly ranges is necessarily younger — or older, conversely — than another which is widely dispersed. Age implies derivation, evolution that is, and evolution is not a subject which bears being handled by statistics and figures. Evolution is in the last resort a philosophical concept of the interrelations among time, space and form, and its effects are not primarily gauged by miles; (b) A species which narrowly ranges in the immediate present may be a relic which has lost most of its former wide compass in consequence of adverse geologic factors.

Willis' theory consists accordingly, as we see it, of two propositions of very unequal value, one generally true of all insular domains that were colonized in epochs most remote and scarcely disturbed by geological catastrophes ever after; the other generally erroneous if erected into a principle, that range and age are definitely correlated.

Everything considered, Willis' theory rendered a signal service in that it pointed out that dispersal is not unfathomable as it was supposed to be in his days, and still it is believed today to be by all too many authors. This theory highlighted the problem of speciation or, better to say, of form-making in time and space. This theory, lastly, had a sound basis of fact in many respects, and involved as one phytogeography and ecology.

Willis' work could readily afford a basis of concrete discussion and sober criticism, if only those to whom it was addressed had been familiar with the facts and floras to which Willis referred in a first place. Some of them who had of these floras knowledge, and retained a measure of restraint, greeted this theory with plaudits or at least looked upon it as a serious effort. Others, on the contrary, attacked it virulently.

The clamors that greeted the appearance of Willis' theory have subsided, but this theory is to this day the object of conflicting opinions. It is not altogether surprising, for example, that naturalists well versed in the flora of New Zealand take it seriously. Other authors, on the contrary, are on record (5) who charge Willis with having "Naively" listed many pseudocosmopolitan species in his works, and accordingly vitiated his statistics. One of these critics remarks that Willis had credited *Montia fontana* to the native flora of New Zealand, while the true *M. fontana*, "Is characterized by having dull or opaque, plump back *(sic)* seeds conspicuously and closely covered with acutish to blunt tubercles and its is known only from Eurasia and western North America; the plant of New Zealand, as anyone can quickly see by examining New Zealand specimens, has much larger lenticular seeds which are highly lustrous and with flattened and obsolete tubercles." *)

This same critic flays Willis in the same paper, as follows, "Other features of Willis' work which at once demonstrate that no reliance can be placed upon the precision of his method and consequently upon the soundness of his deductions are numerous. One case in particular is very pathetic or very ludicrous according to the degree of compassion one may feel. Professor James Small contributes a chapter of *Age* and *Area*, and *Size and Space*, in Compositae, in which he develops his ideas of the evolution of the group. Because of the great extension of the genus *Senecio* in the northern Andean region he localizes the evolution of the Compositae through that genus in the Andes in upper Cretaceous time. But, as Berry remarks, the fact that Small "by age-and-area methods, finds that the great alliance of the Compositae originated in the mountains of northwestern South America at a time when there were neither mountains nor even land in that region, but seas, does not add to my confidence in the general method;" and Bateson politely said: "The inclusion of the chapter of the Compositae reflects more credit on Dr. Willis's candor than on his scientific judgement." The fact is, that the great elevation of the Andes, where *Senecio* how has its phenomenal development, did not occur until the Pliocene and early Pleistocene; and, although Willis finds Small's work a "remarkable verification" of Age and Area, opponents might be tempted to point out that, since it is based on a colossal error, it is *naturally* a verification of Age and Error!"

These criticisms require no comment beyond the statement that Berry is responsible for questions and notes elsewhere dealt with in these pages (see p. 261, 283), and that the author from whom we quote has the paternity of a map of the dispersal of *Erechtites*, *Senecio's* close ally, which will be discussed in a coming part of this chapter.

Constructive if not altogether successfull efforts like Willis' require something less superficial than quips aimed to amuse, but not to enlighten the public. Listeners of good will and common sense are spurred by efforts of this kind, to learn how they can be bettered. A track New Zealand—Eurasia—Western North America is standard, and so endures whether *Montia Fontana* is a single species or consists, on the contrary of two subordinate entities

---

*) We have not much to add to what we state in the main text. Naturally, the crass futility of these strictures forthwith stares a competent phytogeographer in the face. We deal here with a complex, consisting in the end most likely of better than the two forms which Fernald characterizes from the seed only. This complex, which may be expressed with the formula, (*Montia fontana* sensu stricto + Spp. vel formae aff. *Montiae fontanae*), has a global range, New Zealand—Eurasia—Western North America. A competent phytogeographer will immediately be reminded of the essentials, for example, in the dispersal of *Euphrasia* (see p. 19), and proceed accordingly to reconstruct the migrations in play by degrees on a purely factual basis. In reality, we might identify *Montia fontana*, typic, as A, and all the forms in its affinity as B, C, D, etc., without for this disturbing our studies at all.

with slightly different seeds. Willis' work was not fittingly received either by those who pointlessly derided it, or those who applauded it beyond the requirements of performance, but the former did less for phytogeography and botany than the latter.

In conclusion, Willis' theory could easily introduce thoughts fit to enlighten both the phytogeographer and the ecologist, for it contained implicit at least the rudiments of the concepts of orderly migration and *genorheitron*. This theory dealt with some of the most important aspects of the problem of evolution and migration, and dealt with them on the strength of facts rather with the usual run of academic preoccupations. Some of these facts might not be entirely correctly stated, others perhaps misapplied, and eager followers of Willis further might err on their own. Nothing of this could be as important, and welcome, as the tangible accretion brought by Willis to the lean store of thought in the bins of phytogeography and ecology. It is significant that this accretion was hardly seen at all, and that both these branches of lore preferred, in the main, to remain where they stood, laughing at the joke that "Age and Area" were, on the contrary, "Age and Error."

The interrelations of phytogeography and taxonomy are transparent. The former uses the records furnished by the latter taking for granted that classification is a critical account of affinities identified by suitable nomenclatural means. Taking this for granted, phytogeography transfers to the maps of the world the data it secures from taxonomy and next interprets them for its own aims.

It is fair to state that these data are almost invariably amenable to reason, which is proof that neither phytogeography nor classification are tainted with gross errors. Cases are on record (see, for instance, *Geranium, Phrygilanthus Juania*) in which formal taxonomy is unable to deal with forms that are immature, as it were, and retain ties with different *genorheitra*. In our opinion, formal taxonomy is thoroughly well advised if, recognizing its own limitations, strives in cases of the sort to write classification that is useful rather than classification which tries to be "natural." Diluted affinities are not the meat of nomenclature, and it is a mistake of judgement, we believe, to try to handle them by nomenclature supported by halfbaked notions of phylogeny. It seems less than necessary to fill the literature with tremenduous masses of synonyms, and constantly changing treatments when the result is bound in the end to remain the same. Two generic names, *Rhododendron* and *Azalea*, are supposed to identify lesser branches of the ericoid *genorheitron*, of which the former retains its foliage throughout, the latter, on the contrary, drops it in season. It proves impossible to reconcile the viewpoints of the monographer of both these branches who, aware to the existence of intermediates, retains them both under the generic name *Rhododendron*, and the viewpoint of the local florist who seeing among his plants only the extremes uses both *Rhododendron* and *Azalea*. The classification written by taxonomists who serve at bottom different aims is forever bound to differ, unless common sense steps in to suggest needed compromises.

Alluding to these compromises, and their desirability, we are by no means blind to the difficulty of effecting them, and to the factual impossibility of dictating, when individual judgement must, after all, be allowed some scope. It merely is our purpose — and a definite purpose at that — to underscore that phytogeography is not the handmaid of formal classification, and cannot take a hand in these controversies, standing by *Rhododendron* or *Azalea* to the bitter end. Phytogeography is not taxonomy, proof of which is in the fact that phytogeography still sees its way clear out of difficulties which formal taxonomy attempts to settle by a deluge of synonymy going in reality nowhere while so doing.

Tremendous damages were, and still are being inflicted upon phytogeography and taxonomy alike by the fallacy that only "perfect" taxonomy can make "perfect" phytogeography. It is current practice to take the work of the phytogeographer apart showing that the phytogeographer is guilty of having omitted certain records, used certain improper names, referred to certain species in a sense rather than in another, next concluding that since the phytogeographer's records are taxonomically "imperfect," his writing are worthless throughout.

This brand of criticism may be satisfactory to authors who believe they can write "perfect" taxonomy, but is definitely injurious to botany. Common sense dictates that no work is ever perfect, but that work is useful which, by suitably presenting evidence, makes it possible to approach ancient, and thus far insoluble, problems under a new light, and to dispose of obstacles that are otherwise immovable. Nature is an endless chain of riddles, and that work serves its purpose which by disposing of some of these riddles eases the way to tackle some others.

None of the fundamental works which grace the annals of thinking mankind ever endured beyond a definite lapse of time. The very fact that these works dispose of extant conundrums, and open the way to solve others, conspires to cast these works into obsolescence. Works of this exalted nature are sooner or later left behind, but their significance remains, not because they were "perfect" in the beginning, *but because without them further progress would have been impossible.*

All these works were attacked the moment they left the press, and it is fair to say that the literature on the score has its humorous side. Nothing is ever new under the sun.

It goes without saying that these thoughts have no bearing upon our thoroughly modest efforts which are evidently in need of constructive criticism and further elaboration. It is clear, on the other hand, that only constructive criticism has value. It is furthermore transparent that constructive criticism must be directed at the thought which a work represents, and that an issue of "imperfect" records can only then be properly raised, when these imperfect records can be shown to have vitiated thought to the point of impairing its usefulness. A sense of humor requires that the "perfect" records of today be viewed in perspective, and that the believers in these "perfect" records ask of themselves whether these records will still be "perfect" twenty years hence. If the answer is that they so will be, we will return no comment, for the matter is settled to the satisfaction of any candid judge. If, on the contrary, the answer is that these records will be imperfect in due course of time, there can be no argument to start with.

No plant migrates in such a manner as to stagger the record of its presence on the map without break. Disconnections either overland or oversea sooner or later crop up, which can be bridged only by critical comparison of an adequate number of tracks, and making references to generalities critically ascertained in the first place. This rule suffers of no exception, the fact notwithstanding that certain records are more important than others, therefore more desirable.

If we consider the tracks of *Perrottetia* (see p. 54), we may readily come to the conclusion that the record established by this genus in Hawaii is significant, for this record clinches the evidence of a transpacific crossing between the Far East and North America in the latitude of those islands. It would be clearly undesirable to omit this record by oversight, or to have of it no knowledge for any other reason. It is also clear that the lone species of *Schoenoxiphium* in the Gajolands of Sumatra, *S. kobresioideum*, is of greater interest to a phytogeographer than other species in Africa, and that two or three of the latter could be omitted from the record in preference to the lone former.

Supposing, however, that both these important records had escaped our attention, we would certainly not be driven to deny the reality of a transpacific track in the latitude of Hawaii, or be induced to set the origin of *Schoenoxiphium* in "Holarctis." Phytogeography does not rest its conclusions upon a lone record ever, and the presence of *Perrottetia* in Hawaii and *Schoenoxiphium* in Sumatra means in the end very little to a student who has mastery of this science. By the same token, this student would merely shrug his shoulders if a taxonomist were to belabor him, on the ground that the supposed species of *Perrottetia* in Hawaii and *Schoenoxiphium* in Sumatra are in reality the types of new "genera." The significance of the presence of *Perrottetia* and *Schoenoxiphium* in Hawaii and Sumatra is at bottom educational, for these records can be displayed for the benefit of the inquirer who has as yet of phytogeography scanty knowledge, impressing upon him the reality of migrations across the Pacific and Indian Oceans. An old hand at the game of dispersal will view these records as contributory evidence of something he has learned by other, and in the end *beyond comparison stronger means*. The presence of scores of genera of animals and plants on both shores of the Pacific is *proof* of the reality of transpacific tracks, not the existence of *Perrottetia* in Hawaii. By overstressing Hawaii in the name of *Perrottetia* we may, as a matter of fact, create in untutored minds the impression that Hawaii stood in the center of a thin "Land-bridge" leading from the Far East to the New World (or the other way around), which is certainly not the whole truth.

Minds educated to precise thinking are never brought to belief by a lone happening, striking as it may be in appearance. These minds seek out, and are impressed in the first place by, consistent generalities and sums of probabilities and strive above all to attain concepts having the value of laws. Minds of the kind are not interested in solving one puzzle at the price of allowing others to stand unsolved, for the mysterious and the unusual are abhorrent to them wherever found. Naturally, those who think along these molds do not take as an article of faith migrations in the latitude of Hawaii merely because of *Perrottetia*, because they realize all too clearly that this genus might be in Hawaii by chance. Only then they credit with significance the presence of this plant in these islands when other solid evidence has satisfied them that chance is in this particular case wholly out of order.

This being of the substances in play, it is patent that phytogeography is proof against errors and omissions of detail in the taxonomic record, and in no way bound to wait for the "perfect" taxonomy to appear ere it can begin to organize its own evidence. This taxonomy will never appear, and it is pure quibble to maintain or imply the contrary. Moreover, if there is a definite interrelation between phytogeography and taxonomy, still the former is not the helpless handmaid of the latter, for it can interpret and correct the latter, not only, but take sharp reckoning of the latter's substances and appearances.

We indulge in considerations on the score which are beyond doubt more extensive than the intelligent reader requires. In reality, these considerations would be redundant altogether, if statements were not on record which stand for the opposite. Typical of these statements is the following (6; p. 235), "I have used the term phytogeography, not because the term as often used in America signifies an accurate knowledge of plant-distribution, but because it is a term which ought to stand for a scholarly and precise branch of our science. Unfortunately, many Americans who have styled themselves phytogeographers have not hesitated to stultify the subject by the publication of the point of view that, from the phytogeographer's standpoint, the exact identity of the plants is of little consequence. So long as any "phytogeographers" hold such views they must not expect to win the commendation of those who are striving for final truth. Imagine such sentiments expressed by Linnaeus, Wahlenberg, Alphonse de Candolle, Darwin, Hooker, or Gray! In the American rush to see ourselves in print and not to trouble about precision of detail we are too apt to forget the wise saying of Dr. Holmes: "Knowledge and timber shouldn't be used till they are seasoned." As I have elsewhere had occasion to say, "Much inaccurate and unscholarly publication has seriously injured taxonomy; the same tendency intensified has cheapened ecology; and, unless we take the utmost pains to verify all compilations and to publish only what we have critically studied and digested, we shall soon cheapen and discredit phytogeography as well."

The strange assumption is transparent in this text, that a critically minded peruser of endless "compilations" (in reality, the output of taxonomy, as the text in question implies) may be forever unable to abstract out of these "compilations" anything which stands *for averages*, and to formulate its own thoughts accordingly. This strange assumption also confuses formal "perfection," and preoccupations with trifling aspects of taxonomy, with the substances underlying classification and its proper use in other sciences. In other words: A concept is here apparent of taxonomy as something that stands alone and has in its power to reach "Final truth." This concept is essentially shorn of an idea of humor, limits and values, and is the very last which Linnaeus and all masters of the natural sciences would countenance as the facts in the record prove.

As this concept propounds the impossible as the measure of "Final truth," and in reality rehashes the convenient opinion that nothing is to be done while the "Final truth" impends, this concept evidently plays false to those who hold dear to it. Those who so do are perhaps hazy in the first place about the purpose of phytogeography and its means, and pursuing strange ideals miss the obvious meantime.

In the very same work from which we have just quoted (6) a map is offered of the dispersal of the Compositae genus *Erechtites* (Fig. 100) which properly read, that is to say, viewed against a background of orderly migration, and with a mastery upon the elements of phytogeography, forthwith suggests two possibilities at least, (a) The existence of a track immediately running to South and North America at one hand, New Zealand and Australia at the other; (b) The existence, conversely, of a track running to South America, New Zealand and Australia in a prior place, next streaming northward to Asia and ultimately eastward again to reach the Americas in the approximate latitude of Hawaii across the Northern Pacific.

Tracks of this sort have been with us through the pages of this manual from beginning to end, and no reader of its pages can fail to identify them at once.

As offered by the author in question, the map is not satisfactory. A Compositae genus massive in Central Brazil and the Atlantic shore is not likely to be unknown altogether in Bolivia. It is also more than likely that a genus of this nature ranges northward from Australia. Java and Celebes may be suggested offhand as regions where it might turn up in the first place. In short, whether we call an entity having this range *Erechtites* or "A" makes no difference in the first place, for the track speaks on its own account, and immediately suggests pointed inquiries in various directions. "Perfect" taxonomy may come in, of course, and we will welcome all at all times, but there is no evidence to begin with that the tracks themselves need it. It is obviously right at this juncture that taxonomy is one thing, phytogeography another, and that when coming together, the latter does not fall helplessly into the arms of the former.

It so happens that *Erechtites hieracifolia* and *E. valerianifolia* had already been reported from Bolivia (7) when the map under discussion was in course of preparation. The former, moreover is widespread in Northern Argentina which represents its southernmost boundary. The omission of the Bolivian records may be minor, but these records definitely belong to the map on the very face of the dispersal.

Erechtites also widely ranges north of Australia, and some at least of the records established in this quarter were known when the map under discussion was being concocted. *Erechtites hieracifolia* is reported from New Guinea and the Philippines, and so is *E. valerianifolia*. A genus in the immediate affinity of *Erechtites*, *Arrhenechtites*, has five species in New Guinea, and seemingly authentic forms of *Erechtites* also are endemic to this large island, *E. arguta*, *E. bukaensis*, *E. erechtitoides*, *E. papuana*. Of these species, *E. arguta* further occurs in New Zealand, Tasmania, Australia, and is represented by var. *dissecta* in the Lesser Soenda Islands (Lombok). The Brazilian *E. petiolata* is credited to Mt. Singalang in Sumatra, and *E. pyrophila* is reported from Eastern Java, with the note (8) that it may be identically the same as *E. quadridentata* of Australia and New Zealand. Celebes is the seat of an unnamed species. So far as known, *Erechtites* does not range north of Malaysia, but it may easily be looked for in certain drier ranges of Indochina, Southern China and Siam.

All these records merge within an harmonious pattern of dispersal. The report of *E. petiolata* from Sumatra begs confirmation, but is by no means to be ruled out as a taxonomic fauxpas, for the American genus *Hedyosmum* also has a thin outlier in Hainan and Malaysia. There is no warrant, likewise, to dismiss records of *E. hieracifolia* and *E. valerianifolia* outside of North America as "accidental introductions." Some of them at least may not be so at all. We regret we have no certain knowledge of the ecology of *Erechtites* throughout its range, but the dispersal is Malaysia at least is suggestive of a form seeking open grassy slopes, or homologous habitats, in ranges having seasonal rains in the main. These indications are corroborated by the presence of the genus in Bolivia and Central Brazil at least.

It is certain, accordingly, that the map of the distribution of *Erechtites* under discussion is marred by significant omissions, and omissions in quarters which a knowledge of the rudiments of phytogeography would suggest to investigate forthwith.

May we argue that this map is worthless, and not fit to convey an idea of the dispersal of this genus?

Indeed, to argue in this sense would be unnecessary. This map is not the work of a competent phytogeographer, and its author had scanty understanding of regular channels of migrations in the first place. Moreover, and regardless of its origin, this map suggests its own remedy at a glance. In other words: This map is not useless, despite its being imperfect, for the missing ranges it tolerates can readily be filled in by anybody genuinely schooled in dispersal.

In reality, it is possible that today even we have not all the records of this genus in the Far East, as we have suggested. Moreover, the limits of *Erechtites* toward *Senecio* are not absolutely definite, so that we may anticipate taxonomic changes and additions in the future.

In conclusion, no taxonomic record ever is, or can ever hope to be perfect. For this simplest of all reason, a phytogeographer cannot be expected to chase idle fancies, and to hire a battery of taxonomists to write special monographs of all the groups which he must study, eventually to use some of them only. This phytogeographer knows thoroughly well that these taxonomists will never agree among themselves as to what is "perfect;" and may not be honestly expected to anticipate discoveries still in the future. This phytogeographer, finally, is a man of common sense, for the earth and its problems, mankind included, are to be approached with a sense of humor.

This phytogeographer, accordingly, will take the record as it is, and with a sharp understanding of its limitations. Knowing that the record is not "perfect," he will form his thought at the very first in the light of critically approached generalities. Ideas are essential to science, not standardized formulae and definitions, and even less mere tabulations of facts. The world was full of facts, and contained the raw materials of everyone of the great discoveries of mankind aeons of time before man was thought of in the scheme of creation. Man got where he is now not by piling up idle data, but by striving forthwith to organize what he knew along imaginative, broad lines of thought. The idea that a rapid successions of explosions could be harnessed to cause wheels to turn seems most strange, yet is the root of a very great deal which moves today on earth. To conceive of it, active imagination was necessary in the first place. We have forgotten this homely truth and gone back in botany at least to the cult of endless compilation, perhaps because we are now too comfortable to recall that we owe to that truth everything which is now ours.

BIBLIOGRAPHY

Chapter XI

(1) RAUP, H. M. — Phytogeographic Studies in the Peace and Upper Liard River Regions, Canada; Contr. Arnold Arb. 6. 1934.

(2) WILLIS, J. C. — Age and Area. A study in geographical distribution and origin of species. 1922.

(3) WILLIS, J. C. — The distribution of species in New Zealand; Ann. Bot. 30: 437. 1916; same author — The sources and distribution of the New Zealand flora, with a reply to criticism; ibid. 32: 339. 1918.

(4) PERRIER DE LA BATHIE, H. — see (22) Chapt. 4 and (1) Chapt. VI.

(5) FERNALD, M. L. — The antiquity and dispersal of vascular plants; Quart. Rev. Biol. 1: 212. 1926.

(6) FERNALD, M. L. — see (13) Chapt. IV (see here Fig. 110).

(7) RUSBY, H. — An enumeration of the plants collected in Bolivia by Miguel Bang, Pt. IV; Bull. N. Y. Bot. Gard. 4: 392. 1907.

(8) VAN STEENIS, C. G. G. J. — see (12) Chapt. III.

CHAPTER XII

SUMMARY

As phytogeography we understand that branch of botany which interprets plant-migrations in time and space.

As phytogeographers we need no theories. All we need is factual records of plant-distribution such as happen to be contributed in standard taxonomic and systematic work. We transfer these records to a map, connect them by lines representing tracks of actual migration, and thus secure charts of dispersal. A sufficient amount of charts of the kind yields generalities of distribution because migrations are at bottom orderly and repetitious. They need be so, because everything of nature is simple at bottom.

Once we have secured what we believe to be generalities of distribution, such as, for instance, definite points of inception of tracks; repetitious routes of distribution throughout; etc., we test these generalities by making reference to specific problems of phytogeography, seeking, for example, a rationalization of types of migrations which have so far remained "mysterious." If these generalities work — as they must because critically elaborated generalities cannot lie — we forthwith become possessors of efficient tools in all manners of phytogeographical undertakings.

By adhering to the methods just described we come to understand in the end the following,

(a) The main angiospermous forms arose *to the last* in Pre-Cretaceous continents of which but remnants are to be seen in modern maps. The largest of these remnants is possibly constituted by existing antarctic lands.

(b) Foremost among the Pre-Cretaceous continents that stood as the cradle of modern angiospermy was a landmass which occupied practically the whole of the modern Indian Ocean. This continent — *Gondwana* of usual terminology — welded as one Africa, Asia, Malaysia, Australia and additional lands in the Western Pacific.

(c) The landmasses so welded further had contiguity with South America, the Caribbeans generally, and certain sectors of North America and Europe.

(d) The existence of these landmasses brooks no refutation. *It survives in the dispersal it brought into being.* Without these landmasses of old, dispersal — animals' and plants' alike — makes no sense.

(e) Naturally, preoccupations with thin "Landbridges" of Post-Cretaceous age; reliance upon modern maps; acceptance of academic tenets and theories; etc., do not elucidate, rather confuse phytogeography. Arguments against "Landbridges" based on preconceptions, and mistaken assumptions of chronology, etc., have no standing in fact.

(f) Beliefs to the effect that the "poles" wandered, etc., have no basis in fact. Dispersal shows that changes in climate from such causes as these never took place.

(g) Migrations are orderly, precise, repetitious. *So orderly, precise, repetitious are migrations, as a straight matter of the record, that the whole of the essentials of phytogeography can be taught in a few lessons,* with nothing better at hand than a map of the world, and a handful of taxonomic records from such large genera, for instance, as *Carex, Euphorbia,* etc. The map speaks for itself, and refutes theories with the crushing weight of repeated facts which ultimately fall within an harmonious interpretative whole.

(h) It may be affirmed that biogeography is one. The methods and principles valid in an effective study of plant-dispersal are efficient as regards all manners of problems in animal-distribution.

(i) *A study of biogeography is in the last resort a critical investigation of evolution in time and space.* This investigation requires to be successful, (a) A fitting understanding of past and present geography, which is to say a pertinent idea of chronology as regards the origins of the groups taken under study; (b) An adequate concept of morphology. It is selfevident that the archetypes which originated the migrations of modern groups were not the exact counterpart of their modern offsprings. We conceive, consequently, of archetypal streams of migration departing from certain points of ancient maps, at a certain epoch in the past. As the migrations unfolded themselves, and the maps changed in time, these streams released along their path new forms of life in agreement with their genetic powers, and in answer to the requirements of the ambient. Streams of archetypes so moving and evolving are understood in this manual as *genorheitra.*

(j) Inasmuch as migration is a continuous process in time and space, the only fitting way of investigating dispersal is by following dispersal wherever it may lead. A form of plant-life that occurs in Madagascar, Ceylon, Sumatra, Borneo, the Philippines, etc., cannot be studied severally in one or the other of these domains. It must be studied throughout the range *in the beginning,* though the investigation may later be properly restricted to such details as are most immediately sought. It stands to reason that there is no warrant for concluding that such form of plant-life is a "Madagascan" element in the flora of Ceylon, or, conversely, a "Ceylonese" elements in the flora of Madagascar, *when it is not certainly known whether the migration flows from Madagascar to Ceylon or the other way around.*

(k) Considering that phytogeography has so far proved unable to identify the starting point of angiospermous tracks, phytogeography has confused — not elucidated — its record by coining endless terms, adjectives, etc., such as

"Madagascan," "Ceylonese," "Holarctic," etc., etc. Terms and adjectives of the kind have no status in science because they rest on nothing demonstrable and precise.

It follows from the principles just stated that a knowledge of the starting point of the angiospermous tracks is essential to critical phytogeography. It further follows that the geological epoch in which these tracks began to run must be known within reason.

A knowledge of the kind can be secured, as already stated, by the simple device of plotting actual patterns of migrations. If this be done, it will soon be learned,

(i) The tracks of all Angiosperms take their start in the south of modern maps. They uniformly do so in certain well defined geographic centers, which we have identified as "Gates of angiospermy." These "gates" are essentially three, (a) African; (b) Western Polynesian; (c) Magellanian.

(ii) These "gates" were connected within a geologic shore.

(iii) A final study of all the "gates of angiospermy", and their immediate connections, call into play an investigation of large masses of flora, sizeable part of which is of essentially local interest. This study proved impossible in the pages of this manual, but it was feasible at least to investigate the major aspects of dispersal at certain of the "gates."

(iv) The "African gate" is doubtless of foremost importance. This "gate" can actually be broken down (see Fig. 67) into two distinct ranges, namely, (a) Madagascar—Mascarenes—Seychelles; (b) Natal—Kerguelen Islands—Tristan da Cunha. It can be shown that certain Angiosperms originated in one or the other of these ranges; but it is certain, on the other hands, that plants in the same taxonomic family, or tribe, may have originated from one or the other. It proves impossible for this very reason to effect a final separation as between these two ranges, which we understand, respectively, as *Gondwanic* and *Afroantarctic Triangle*.

(v) Characteristically, forms of flora essentially wedded to lands of warm or tropical climates (e.g., Cochlospermaceae) usually originate in the *Gondwanic Triangle*. Contrariwise, plants that, though often endemic to tropical lands make their home also in the north or far south, and are for this reason tolerant of cold to cool (e.g., *Erica*) climates, stem as a rule from the *Afroantarctic Triangle*. It proves possible on the strength of this observation to gain a much deeper insight into the nature of what is "warm" and "cold" in plant-life, and as a corollary issues of "bipolarism," "climatic landbridges," etc. of long standing can be disposed of quickly as entirely nugatory.

(1) What is generally true of the "African gate" is also correct as regards the "Western Polynesian gate". The latter rests upon two ranges, which, formerly represented by broad landmasses, have today insular nature only. These ranges are, (a) *Neocaledonian Center*, composed of the lands among New Guinea, Queensland, Lord Howe Island, Tonga, Samoa and Fiji; (b) *Macquarian Center*, made out of Tasmania and certain petty islands south of New

Zealand (Macquarie, Campbell, Auckland, Antipodes). The *Neocaledonian Center* immediately originated certain modern angiospermous *genorheitra* (e.g., Stackhousiaceae, Magnoliaceae Magnolioideae), but mostly received them from, (a') The "African gate" by tracks crossing the Indian Ocean of our maps, and what is now the northern half of Australia (e.g., Oleaceae, Violaceae, etc.); (b') The *Macquarian Center*. This latter originated certain *genorheitra* (e.g., Stylidiaceae), but received many archetypes (e.g., Restionaceae, Centrolepidaceae) also from the "African gate" by strictly "antarctic" paths crossing the high south of the Indian Ocean and the waters meridional to modern Australia. The *Neocaledonian* and *Macquarian Center*, respectively, interplay as do the *Afroantarctic* and *Gondwanic Triangles*; they, too, release "warm" and "cold" flora northward and southward.

(vii) The "Magellanian gate" is less important than the two just mentioned, "African" and "Western Polynesian." Its operations are as a rule closely involved with those of the remaining two "gates."

(viii) In addition to the "gates," certain important centers of *secondary* evolution are apparent throughout the map. These are, (a) *Kalaharian* and *Nigerian* in Africa; (b) *Roraiman* and *Ozarkian-Appalachian* in the Americas; etc. Not to confuse our subject with definitions, we have identified by name only the most important of these centers, and purposely refrained from cataloguing them all. The reader will find, however, abundant material to perfect his own investigations, and identify by name — referring to proper tracks and starting points in each case — all the secondary, tertiary, etc., centers which may be of particular interest to him.

As the angiospermous tracks uniformly return to certain points of the map, and follow uniform channels throughout, *it proves possible to use plant-tracks to reconstruct the outlines of the world that was when the Angiosperms began to migrate, or migrated most vigorously. It proves also possible to correlate these outlines with trusted data of palaeobotany and palaeography.* If this be done, it is shown that, (I) The modern Angiosperms began active migration at the close of the Jurassic and the very Early Cretaceous; (II) These plants had a second period of active migration in an epoch of general submersions and emersions which lasted from the closing eras of the Cretaceous to the first eras of the Tertiary; (III) No ultimate angiospermous archetype ever originated in "Holarctis," or some such mythical continent in the north.

To summarize what we have brought forward so far, we will put on record the following;

The modern Angiosperms originated long before the Jurassic. They began actively to migrate at the close of this age, and went on migrating ever after, more vigorously so in epochs of geologic revolutions. Their history is accordingly wedded throughout to the process of alteration whereby the Pre-Jurassic world was changed through the Cretaceous and the Early Tertiary into the world we know. It stands to reason that angiospermous evolution is part and parcel of the

*geologic history of the earth, and that plant-life is — in reality — a geologic
layer of the earth.*

*Obviously, attempts made at interpreting plant-dispersal (and all manners
of biogeography as well) by "Occasional means of seed-distribution;" vague
generalization as to "Floating Continents," "Polar Shifts;" preconceived theories as to the existence, or non-existence of "Landbridges;" lead absolutely
nowhere. These attempts are faulty by the base, and cannot be dignified with
the name of science. Science demands precise generalities; sharp methods of
investigation and interpretation, not sheer opinion and pious generalities. No
primary modern angiospermous group ever originated elsewhere than in lands
of the south. The "Holarctic" theory of angiospermous origins has no support
in facts.*

We stress *modern* Angiosperms throughout. It is well known that forms with
characters pointing to angiospermy (see, for instance, ELIAS in Sc. Bull. Univ.
Kansas 20: 115. 1931) are known from the Lower Pennsylvanian of the U.S.A.
The origin in time and space of the earliest prototypes out of which the *modern* Angiosperms were to descend cannot be of our immediate concern. It
cannot be of our concern at present because we have no factual understanding
of the subject; it would serve no purpose to debate issues which could bring
us back to the Carboniferous and Permian without at the same time furnishing
light upon the paths in time and space whereby these earliest prototypes came
to the lands in the south from which the *modern* Angiosperms were next
fated to emerge. As we are not interested in speculations, we leave it to the
future to investigate questions of the sort. It is sufficient for us to take the *modern* Angiosperms in hand by the end of the Jurassic, in the understanding that
this is the first step necessary to a rational, ultimate understanding of all phases
of dispersal.

To continue: The reason why there exist "gates of angiospermy" is of the
simplest. The "gates" in question are part of, or stand close by, the continents
which sired the modern Angiosperms. Considering that it is *from these
crumbled continents* that the lands still on our maps received the angiospermous archetypes, it follows that the tracks of the Angiosperms are bound
to point with accuracy to the regions where contact was originally made which
brought the Angiosperms to the modern world.

Naturally, all angiospermous tracks go back to a "gate" in the end, either
directly, or by part of their *genorheitron*. The correlation between cause and
effects is here so palpable that we wonder why it was never recognized.

There are groups among the Angiosperms which originate by one "gate,"
others on the contrary which take their start by two, or, finally, three "gates."
The last, obviously, are those which achieve in the end the widest possible
dispersal (e.g., Celastraceae), because theirs are all the standard tracks of
angiospermy.

Inasmuch as the world that was when the Angiosperms began to migrate
bears to the world of today no resemblance whatsoever, *the standard tracks*

*of angiospermy run courses which wholly disregard the modern outlines of land
and sea.* It can factually be proved that dispersal is vitally affected by certain
"barriers" operative in regions where nothing at all can be seen today that may
be so effective. Contrariwise, seemingly massive land-obstacles do not have in
their power to deflect the course of migration at all. This leads to very striking
aspects of dispersal — in the light of modern maps — but these aspects become common-place once we rationalize them — *as we must* — with reference
to the competent cartography of the past.

To summarize once again;

*Biogeography in general, and phytogeography in particular, are sheer bypro-
ducts of the geological history of the earth. Nothing of these sciences can be
understood if this elementary fact is unrecognized or denied.*

The term *originate* has been much abused in biogeography, and has accordingly lost the sharp edge which is demanded by scientific terminology. To
us, *originate means to become recognizable as a modern group.* In other words,
we are not going to seek the origin of the Magnoliaceae in the cell primaeval
nor at the place where *Magnolia tripetala* is found today. We seek this origin
there on the map, where the *modern* Magnoliaceae first appear as the *modern*
Magnoliaceae. The center of the map where the Magnoliaceae can be shown to
originate, that is, *to form their ranks as the modern family*, can be easily located
by a critical study of the tracks and phylogeny of the whole group, not only,
but can further be critically compared with the center of origins of all other
Angiosperms. Analogies that are repetitious enough become homologies, and
homologies that are constantly with us have very nearly the status of authentic
scientific proof. Final scientific proof is reached, eventually, when the application of certain general principles leads without further to the satisfactory
solution of concrete cases. It is with reference to proof such as this that the
origin of the modern Magnoliaceae can be shown to take place in certain
regions of the maps, and a certain time. Naturally, the determination of "centers of origin" is thus brought down to trusted factual principles of method
and practice.

Considering that principles of the kind exist, not only, but are so transparent
that they can taught with telling effect in the classroom, a well trained phytogeographer is competent to make the most even of incomplete and imperfect
records of taxonomy and distribution. It is downright untrue that phytogeography is subservient to taxonomy, because it can be factually shown that
phytogeography can often set taxonomy aright, not only, but account for
factors which forever escape the understanding of the latter. The actual course
of a long track can as a rule be inferred by the statement of no more than three
or four localities. It further proves possible to guess where a plant may be
found, once the geographic fundamentals of its distribution are on record.

If it proves possible successfully *to guess* where a certain plant can be
found on the strength of the geographic elements of its dispersal, something
more than a knowledge of these elements is required to perfect the score. To

perfect the score, and to form a well-rounded opinion of distribution, we need take into account two factors throughout, (a) *Phytogeographic;* (b) *Ecologic*.

Phytogeography studies the run of the track on the face of the map as a line, or front, connecting a series of localities in proper sequence. *Ecology*, on the contrary, consider ambiental conditions along the track without necessarily being concerned with its geographic aspects.

To elucidate: Let us suppose we deal with plant "A" which is recorded from Tanganyka, the Congo, Cameroon, the Guianas and "Amazonian" Colombia. Based upon pure *phytogeography*, we may safely advance the conclusion that plant "A" is most likely to be found also in "Amazonian" Venezuela, because the track in play is standard, and includes this sector of Venezuela as a normal station. However, plant "A" only then will be found in "Amazonian" Venezuela, when in this range occur habitats favorable to its normal survival. Accordingly, *phytogeography* and *ecology* constantly interplay, and interplay so closely that we cannot hope to understand the operations of the former so long as we have no precise idea of the effects of the latter.

It will be observed — to continue in the example above — that if plant "A" is not to be found in "Amazonian" Venezuela because of unsuitable habitats, a form close to "A" may there occur, which is suitable by the local ecology. It is of the essence of *genorheitral* operations in time and space, as a matter of fact, that the archetypal stream *(genorheitron)* release along its course — *so far as possible* — all manners of forms adapted to different ambients. Forms so adapted, naturally, will differ by degrees as varieties or species, and if old and advanced enough, as genera and higher categories. *Thus, phylogeny, taxonomy, ecology and phytogeography all closely interplay, and their interplay can only then be fully understood when we have sound ideas of morphology, geology and palaeogeography. It stands to reason that, this being the case, we can never hope to fathom dispersal, and its byproducts, if we forget that a plant-form is necessarily part and parcel of an association and formation in the ecological sense. No plant ever travelled single throughout.*

It is possible that the principles laid down in the pages of this manual make it at last feasible to draw a cleancut line of discrimination between phytogeography and ecology. These two branches of sciences have so far been hopelessly confused.

In conclusion, we affirm that phytogeography is not ecology, though these sciences are intimately correlated. We must know — *and know critically* — how an archetypal form can reach a certain locality in order that we may next understand how this form react by the ambient it finds in this locality. Confusion in fundamentals cannot be tolerated, and nothing is worthy of the name of science, the principles of which are not lucidly stated on the strength of few tested generalities.

It is a truism stemming immediately from the principles here stated that phytogeography (and all manners of biogeography) and geology have exactly the same status in an investigation of the past of the earth. These sciences are facets of the same prism. *That which a critical study of dispersal affirms cannot be untrue in the light of geology.*

A day is soon to come in which the geologist and geophysicist will be acquainted with the fundamentals of biogeography *as a matter of course*, not only, but will use biogeography as a necessary check of their own conclusions whenever possible. Naturally, by then phytogeography — and biogeography generally — may not be the confuse hodgepodge of theory and half-baked notions which they are today still.

It stands to reason that modern dispersal elucidates palaeobotanical findings of a date later than the closing eras of the Jurassic. It cannot be otherwise; live dispersal is massive and can be studied with the necessary lucidity and perfection; petrifacts are comparatively few, and oftentimes more than dubious. *Palaeobotany of ages later than the closing days of the Jurassic, then, is the handmaid of phytogeography, not the other way around.*

If we consider that dispersal is repetitious for the very good reason that dispersal was molded by geologic events responsible in the end for modern geography, we must conclude that the major channels of migration are few.

This conclusion is correct, because few indeed are these channels. They can be outlined as follows,

A) By and around the "African Gate."

As we know, the "African gate" breaks down into two ranges, respectively, Madagascar—Mascarenes—Seychelles; Natal—Kerguelen Islands—Tristan da Cunha. These two ranges are known to us as the *Gondwanic Triangle* and the *Afroantarctic Triangle*.

The basic tracks that issue from the *Gondwanic Triangle* are (Fig. 101),

(1) *Madagascar to Peru*, through Tanganyka—Cameroon—Amazonian Brazil.

(2) *Madagascar to Fiji*, through Malaysia and the Coral Sea.

We define these tracks here referring to their maximum extension, and the main stations along their course. It stands to reason, nevertheless, that these tracks may indulge in endless vagaries of detail. To illustrate: Track (1) may stop short in Senegal while at the same time running all the way south to Transvaal; or, conversely, freely cross to the Guianas, take in the West Indies and end in Central Mexico, etc. It will readily be seen that these and similar variants are function of the main course, important as may be their territorial effects in the end. Likewise, track (b) may stop in the "Soenda Coign" (Soendalan) of Malaysia, or fare northward from the Himalayas to Southern China, eventually Japan. These variants are relevant, no doubt, but are function still of the main track itself. We abstain from considering here these variants and alternatives because (i) We intend to deal with nothing more than generalities; (ii) The reader will find ample examples throughout the pages of this manual, how the fundamental tracks vary in detail.

Both these tracks, (1) and (2), are "warm" as a rule, which is to say that they are followed by groups of strictly tropical nature.

From the *Afroantarctic Triangle* are released the following main channels,

(3) *South Africa to Europe*, possibly North America, through East Africa, Northern Africa generally and the Mediterranean.

(4) *Central Asia (with outliers to East Africa, Java, Sumatra, the Himalayas on occasion) to all circumboreal lands by the Altai Node.*

(5) *South Africa to Australia, Tasmania, New Zealand, Magellania.*

These three tracks (Fig. 102) are "cold," as a rule, which is to say that, though patronized by groups oftentimes well represented in the tropics (e.g., *Erica, Orobanche, Pimpinella, Oreolobus, Centrolepis*, etc.), they are not devoted to the carriage of genuine tropical forms (e.g., Cochlospermaceae, Linaceae exclusive of *Linum* in particular, etc.).

Certain capital tracks (Fig. 103) cannot be safely adscribed to either one of the *Triangles* in detail. These tracks are,

(6) *Kalaharian Center to Brazil, ultimately reaching the Caribbeans, parts of North and South America.*

(7) *South Africa, generally, Madagascar and the Mascarenes to Patagonia, Bolivia and points beyond in the New World.*

Turning to the Pacific, we have before us tracks that originate from the Western Polynesian and Magellanian gates, in this manner,

B) By and around the "Western Polynesian Gate" (Fig. 104).

(8) *Neocaledonian Center to the New World (North, Central or South America.*

This track takes its inception in the Coral Sea. It streams to Eastern Australia, New Guinea, Eastern Malaysia and Asia as a rule, often, however, hitting only the Northern Philippines. It next follows to the Northern Riu-Kiu Islands, then sharply veers eastward to cross the modern Pacific. It enters the New World anywhere between California and Bolivia. In the American part of its course, this track may keep to the East Pacific in the main, or freely cross over Northern or Southern Mexico into the Caribbeans, further invading South America.

(9) *Neocaledonian Center to Europe and Northern Africa.*

The origin is the same as in (7) above. However, the track crosses over invading Western Malaysia. It next follows by path *a* (Fig. 104) to India and Ceylon, and, after an eclipse usually taking place between India and Asia Minor, is again seen in the Mediterranean, following all the way to the Iberian Peninsula, possibly the Canary Islands, and Northern Africa. It may occasionally run alternative route *b* which calls for even more massive disconnections.

Tracks immediately issuing from the *Macquarian Center* (e.g., those possibly of Stylidiaceae, Centrolepidaceae, Restionaceae, certain groups in *Carex*) are shorter — as a rule — than those from the *Neocaledonian Center* *), and may easily be harmonized with the course of the latter.

Track (8) may be "warm" or "cold." Track (9) is prevailingly "cold."

C) By and around the "Magellanian Gate."

(10) *Central Chile or Bolivia northward to Mexico, the U. S. A., possibly Hawaii, and the Caribbeans, oftentimes sending outliers to Uruguay and South-eastern Brazil.*

This track (Fig. 105) is both "warm" and "cold." It may fare by the East Pacific over a disconnection involving on occasion the Galápagos and Revilla Gigedo Islands, or, conversely, invade the American continent throughout. In this case it follows the modern Andes veering eastward from Southern Colombia to a point between Colombia and Venezuela, where it ultimately leaves South America to go to the Greater Antilles, Central America, Mexico and the U.S.A. This track may send outliers to the Guianas and Eastern Brazil generally, crossing the Amazonian Basin and its southern approaches.

These are the ten fundamental tracks followed by the Angiosperms. Considering that no form ever had the monopoly of the maps of the past or the present, these very same tracks, or their segments, can be shown to have been used by all manners of plants of non-angiospermous descriptions such as Cycadaceae, Conifers, Pteridophytes, Liverworts etc.

In addition to these ten fundamental tracks are certain other channels of migrations, as follows (Fig. 105; 11—12—13),

(11) *From the Mediterranean to the Caribbeans and Mexico.*

This track may graft itself upon the channel described in (3), (4) or (12). It is worked sometimes in the reverse, from the Caribbeans to the Mediterranean.

(12) *From the South Atlantic to the North Atlantic.*

This track emerges as a variant of (5), (6) or (7).

(13) *From West Africa to Burma and the Philippines.*

This channel is a variant of track (1) which doubles back, as it were, following mostly in disconnection the route Central Africa—Arabia (or Iraq)—Iran—India.

We have mentioned morphology as the necessary ingredient of a correct appreciation of evolution, therefore as a vital adjunct to well rounded studies in dispersal.

By *morphology* we understand in the very first place a factual, rational appreciation of organ- and tissue-destination. In other words; Supposing we have

---

*) Noteworthy exceptions are in *Drosera* and *Carex*. These unusual tracks may run all the way from New Zealand to Japan and Spain, respectively.

before us a rod of organized vegetable matter (a *branch*, roughly speaking) which emits series of non-sporogenous and sporogenous bodies, it is for us to know in the first place all *possibilities of evolution* inherent in this rod and its appendages. It is for us to have definite, factual ideas how this rod may stop apical growth, and by what organ or tissue the apical inhibited meristem may be replaced. It is for us to reach at least a sound preliminary understanding how the actually sporogenous bodies borne by this rod, and the sterile or sterilized tissues alternating, following, or preceding them may combine together to yield the "flower" in the end clearly, a flower of *Ficus* is not homologous of one of *Magnolia*.

Anatomical morphology — such morphology which is mistakenly dignified by the straight appellation *morphology* in deferences to tradition — is entirely subservient to this play of possibilities. This play of possibilities is not necessarily revealed by studies in anatomical morphology to those who have of it no knowledge *in the first place.*

As an example of the type of morphology relevant to a stringent study of evolution in time and space, therefore also to phytogeography in its ultimate aspects, we may refer to the Barringtoniaceae, a family considered elsewhere (see p. 412) in the pages of this manual.

The embryo of *Barringtonia Vriesei* (1) is fusiform in shape and factually shorn of cotyledons. These are replaced (Fig. 106) by scales spirally arranged in a ratio approaching 2/5 in standard notation which contain in their axils weak buds. Good authority states (2) that the scales apical to the embryo crowd to form a structure approaching a bud which ultimately develops into the main stem of the seedling. This stem bears scales at base which are replaced higher up by normal leaves. Should the main stem be broken, a shoot from the bud in the axil of a scale replaces it. Internal to the body of the embryo or very young seedling is a jacket of tissue which differentiates as phloem and xylem independently from the leaf-traces in a first time. Later on, this jacket establishes connections with these traces.

As the seedlings grows, a ring of phellogen develops immediately outside the phloem which eventually cuts off the cortex from the central column, and causes it to be shed. The roots are endogenous, and take their origin from the jacket mentioned above.

Types of development analogous, when not downright homologous, of the ones described are reported (3) in another mangrove, *Avicennia nitida*. Certain seedlings are of slow growth, soon lose their fleshy cotyledons, have a poorly developed fibrous root-system and are essentially woody because of the early formation of abundant cork-tissue. Others, on the contrary, make much more rapid growth, retain their cotyledons for a longer time, do not become woody, and have a well developed root-system. The formation of the root appears to be influenced by the surroundings, because stout, unbranched roots are associated with growth in mobile sands; fibrous roots with mud and peat.

It takes no effort of imagination to realize that the fixation of either one of these types of growth, and their continuation throughout the life of the plant, may yield in the end different "species," perhaps "genera" *). It is possible to conceive of an avicenniaceous archetype releasing out of its *genorheitral* folds forms competent to thrive in swampy grounds quite remote from the coast, possibly in subalpine surroundings. In other words, lowlands and mud flats originally occupied by avicenniaceous prototypes may acquire by a subsequent process of uplifiting and orogeny — carried on during geological ages — "species" and "genera" which are no longer *Avicennia nitida,* even *Avicennia* to a taxonomist. Meantime, at the other hand, the same prototypes may become fixed at the seashore as absolute mangroves, or take to sands back of the coast, ultimately to yield therein subxerophytic or xerophytic forms with much reduced leaves, etc.

We need not go back to the Jurassic or the Early Cretaceous to have before us a concrete picture of possibilities in evolution and their meaning for descriptive botany and ecology. *Avicennia nitida* gives us everything needed to, (a) Populate the seashore; (b) Insure vegetation to a subxerophylous domain; (c) Fill highlands with forms adapted to the deep soil of the primary forest or the open bog. These are no idle imaginations. There is positive evidence, as a matter of fact, that the archetypes of the Angiosperms were plants quite as plastic as is *Avicennia nitida* today, and that many of them originated in the same type of habitats.

A sagacious observer of the vegetation of Madagascar comments (4) that numerous are the genera *(Terminalia, Calophyllum, Cordia, Tournefortia, Erythrina, Barringtonia,* etc.*)* that have in this island a wide-ranging species by the seashore and various endemic species inland. This is understandable. Were new land to emerge by the present haunts of *Avicennia nitida,* and soar ultimately into hills and peaks, *Avicennia nitida* — even as *Avicennia nitida* of today — could fill in time the whole. If *Avicennia nitida* is wedded to the mangrove *today* this is because it can stand the mangrove which its potential competitors are unable to invade, as plant-life goes *today.* Were conditions to change, *Avicennia nitida* would certainly not die out easily. We know, as a matter of fact, that "alpine" mangroves exist (see p. 161, 408), forms, that is, which did not die out when the geologic shores by which they once grew were uplifted.

Were the seedling of *Barringtonia Vriesei* to continue growth along the lines laid down for the embryo and the younger seedling-stage, a cactoid plant would easily result. This plant would have, (a) A well developed region of ground-tissue and cortex with comparatively weak development of vasculation. This vasculation could originate independently both by the central cylinder or at some reduced leaf, reaching more or less perfect fusion in the end; (b) Scaliform leaves (or, indeed, no leaves at all) carrying in their axils dormant buds, or systems of buds, some dormant throughout, other active or potentially active.

---

*) This is verified in the Cactaceae, to mention an obvious case.

The homology between growing-points of this nature and the cactoid areole is palpable enough not to require comment. By their nature spinescent "leaves" of this kind and "cataphylls" are beyond doubt quite close; (d) Fibrous endogenous roots; (e) A succulent habit throughout.

In conclusion, it may be left to disciplined and well informed imagination what the *genorheitron* of the Barringtoniaceae and Avicenniaceae could yield in time, and under the impact of different geological and ecological conditions. There is no doubt that were such lands to emerge again fit to bring together the coasts of Venezuela and Cuba, *Avicennia nitida* would populate all these lands. Were later on these lands to be modified by uplifts, *Avicennia* well might espouse new habitats and new surroundings with forms adapted to the new environments.

It is seemingly true that "casual dissemination" would be the instrument to scatter *Avicennia nitida* at first all along the lands we have visualized connecting Venezuela and Cuba. Doubtless, by virtue of this "dissemination" *Avicennia nitida* would follow a track leading from Venezuela to Cuba all the way, or most of the way, depending upon the nature of the land-connections in play in detail.

However, this "casual dissemination" would only apparently be "casual." This dissemination could not be "casual" in reality, because, (a) It would spend itself along a definite chain of lands in gradual emersion; (b) It would be only then possible, if the habitats were suitable; (c) It would originate from definite points of the map moving in the direction of other, and not less definite, points.

Obviously, we mouth sheer nonsense when we speak of "casual dissemination" in the abstract, and in the faith that dissemination can indeed be genuinely "casual."

Although we have repeatedly made passing mentions of homologies and identities in dispersal as between animals and plants, we have sought no support for our interpretation of dispersal in the pages of non-botanical work. This we have done on purpose, intending that this interpretation exclusively rests upon its own merits and demerits, thus opening in the end the way to constructive joint endeavors of phytogeographers and other students of nature. We have quoted, as a matter of fact, zoologists who do not agree with us, and used their works against their authors (see p. 16, for example).

In fairness to our subject, we are bound to declare, nevertheless, that there is fairly good agreement between our own basic conclusions and those of numerous zoologists. One of them concludes (5) for example as follows,

(a) The hypothesis of "Holarctic" dispersal holds good for man and most, though not all, of *Tertiary* mammals, but does not hold good for Mesozoic dispersal, nor for such groups as crawfishes and certain frogs *).

---

*) The author of this statement has at least a cleancut understanding of the fact that dispersal cannot make sense if we are in the dark about the chronology of origins. We are not sure that the hypothesis of "Holarctic" dispersal holds good for man and certain "Tertiary" mammals if the origin of man and these mammals are brought *back enough in time*. South Africa, for example, was fairly rich in hominid forms.

(b)   Australia has been isolated from the northern landmasses since the Late Mesozoic, and the portions of its fauna which had to depend upon land-migrations came in from the south.

(c)   The Antarctic continent had, during the greater part of the Mesozoic (Triassic to Cretaceous), and the Early Tertiary, a temperate climate, and a rain-forest flora, and was therefore habitable by animals.

(d)   There are geographical indications of former connections between Australia, South America, New Zealand and Africa which taken in conjunction with the evidence from faunal distribution, render it probable that land-migration has from time to time been possible along such connections. No other hypothesis is adequate to explain some cases at least of dispersal.

(e)   The evidence from parasites strengthens the faunal distribution, and minimizes the suggestion of multiple and convergent origins.

(f)   The Andes have always formed the only temperate pathway through the tropics *).

(g)   The Australian marsupials, Parastacid crawfishes, Leptodactyl and Hylid Frogs, and a host of other forms, vertebrate and invertebrate, reached Australia from South America by way of Antarctica **).

The author whose conclusions we quote pointedly and finally rejects the hypothesis (6) that "Antarctica" never was a faunal route. As "Antarctica" was *the basic* floral archetypal route, the notion that "Antarctica" never gave way to animals can only be absurd, because animals and plants are fated to travel together if the former are to survive massively *at all*. The same author (7) pertinently identifies as crucial in favor of "antarctic" migrations the evidence from the distribution of the Crawfishes. These Crustaceans are divided into two distinct groups, Potamobiidae restricted to the Northern, and Parastacidae peculiar to the Southern Hemisphere, and come together only in Central America.

The Parastacidae occur in South America, Madagascar, Mascarenes, Australia, New Zealand and New Guinea, which is a close match of the distribution of many Angiosperms. Significant is the recurrence of the Mascarenes — altogether petty islands as such — in vital pales of animal and vegetal dispersal alike. We might not say that a track *Mascarenes—Madagascar—Australia—New Zealand—New Guinea—South America* is certainly and absolutely "antarctic," because of reasons dealt with at length throughout this manual. However, it is pacific that a track of the kind, when not "antarctic," is "Gondwanic," and could never have been run if the world had not been once very different from the world of today, and well provided with land-connections in the Southern Hemisphere throughout.

---

*) With this we sharply disagree on general grounds. It *might* be "proved" correct with reference to few animals, as it can be with reference to a few plants, but partial "proof" of the kind is misleading in the end.

**) To judge from plant-dispersal it is perhaps more probable that Australia and South America were simultaneously reached from points in the modern map stretching about the southern tip of Africa, in many when not in all cases.

Another zoologist (8) is induced by a study of Frogs and their parasites to conclude in this manner,

(a)   There existed a continent, Gondwanaland, in the Triassic which brought together Africa and South America, either directly or by way of Antarctica.

(b)   Australasia was separated from Malaysia since early Cretaceous times.

(c)   There were continuous fluctuations during the Tertiary in the Malaysian islands; Malaysia had temporary connections at least with New Guinea.

(d)   A Cretaceous arm of the sea connected the western Gulf of Mexico with the Arctic Ocean.

(e)   A Cretaceous land-strip through Southern Alaska connected the Far East with the New World, running parallel to Western America and certainly reaching Central America, possibly extending southward to Ecuador and Chile.

(f)   A Pre-Cretaceous or Cretaceous connection was established between central and northern South America and New Guinea with Australia across islands in the Central Pacific.

(g)   A bar to migration between Southern Brazil and Patagonia present in the Cretaceous or a little later, but disappearing probably in Mid-Miocene.

(h)   An extended Antarctica uniting Australia, New Zealand, Patagonia the subantarctic islands, possibly Africa and South America. The South American connections probably disappeared before the Transargentine Sea was obliterated. Greater Antarctica was probably Cretaceous and persisted during the Early Tertiary.

(i)   A Triassic continent, Lemuria, bringing together Africa, Madagascar, Ceylon, Southernmost India, the islands to the southwest of Malacca, possibly some of the southwestern islands of Malaysia. This continent broke off from Africa during the Jurassic, and was fragmented into separate islands in the Tertiary.

These conclusions differ *in detail* from ours, but agree *in essentials* to an extent at least fit *to promote concrete, constructive debate on lesser points of honest divergence*. It is reasonable to believe that these points can be smoothed out in the end.

The user of this manual may readily satisfy himself by a perusal of the *factual records* upon which we have relied that we have not been dreaming throughout. The question then follows, *How could have dreamt the authors of the conclusions quoted? Where did they get these conclusions from? Why is it that we all roundly agree in fundamentals?*

If the user of this manual can reach the legitimate conclusion in the end that all of us have sired chimeric thoughts; that the facts in the record are illusory; that only coincidence is responsible for the agreement we have pointed out; we will accept his verdict as something that does not bear being discussed, be it in the least. Time never need be wasted.

We will never believe, however, that we must go on keeping faith in the miraculous, the incredible, the odd, the unusual in a branch of science such as phytogeography. Too many are the groups that may be made to yield their juices, and to tell the truth, to stand by meekly when sheer chatter is heard about "Extraordinary means of transoceanic dispersal," "Holarctis", "Cretaceous radiations," etc., etc. Scientific patience has a limit, and the advice readily comes to the lips of those who know some at least of the relevant facts, *Why do you theorize instead of looking at the facts straight in the face? Please, give them an opportunity to speak their message while you learn.*

BIBLIOGRAPHY

Chapter XII

(1) TREUB, M. — Notes sur le sac embryonnaire et l'ovule; 5. L'embryon du *Barringtonia Vriesei* T. et B.; Ann. Jard. Bot. Buitenzorg 4: 101. 1884.

(2) THOMSON, T. — On the structure of the seed of *Barringtonia* and *Careya;* Proceed. Linn. Soc. (London), Bot., 2: 47. 1858.

(3) CHAPMAN, V. J. — 1939 Cambridge Univ. Exped. Jamaica: Pt. 3. The morphology of *Avicennia nitida* Jacq. and the function of its pneumato-phores; Jour. Linn. Soc. (London), Bot., 52: 487. 1944.

(4) PERRIER DE LA BATHIE, H. — see (1) Chapt. IV (refer to footnote 1, p. 213).

(5) HARRISON, L. — The migration route of the Australian marsupial fauna; Austral. Zool. 3: 247. 1924.

(6) SIMPSON, G. G. — Antarctica as a Faunal Migration Route; Proceed. Sixth Pacific Sc. Congr. 755. 1939.

(7) HARRISON, L. — Crucial evidence for antarctic radiation; Amer. Natural. 60: 374. 1926.

(8) METCALF, M. M. — Frogs and Opalinidae; Science, N. S., 79: 213. 1934.

NAME- AND SUBJECT-INDEX

In view of massive repetitions throughout the text of taxonomic names (species, genera, etc.) and geographic designations of all kinds, in this Index,

1) Are disregarded names incidentally mentioned or irrelevant.

2) Are not recorded generic and specific names, nor localities introduced in lists of families, genera, etc. These names should be looked for under the families or other taxonomic groups to which they belong. The Chapter-Index will be found useful in this regard.

3) Mention of particular subjects is reduced to a minimum.

4) Only names of persons are set in italics.

566

574

580

582

584

445; New Caledonia—Caribbeans—
S. China 445; Madagascar—Amazonas 451; in the "High North" 458
et seq.; Natal—Mascarenes—Easter
Isld. — Mexico — Caribbeans 496;
Madagascar—Bolivia 496; S. Australia—New Caledonia—Tasmania
—Easter Isld.—Chile—Bolivia 497;
"Atlantic" in Ferns; New Zealand—
Eurasia—W. North America 532
African: V-shaped or ray-like playing around Africa 16, 26, 49, 113,
170, 317 fn., 322 fn., 332, 389;
Tanganyka—Cameroon 162; "cold"
and "warm" in the Indian Ocean
247; skipping Tanganyka and Uganda 335; within Africa 352; alternatives within Africa and south of it
391; Seychelles—Angola 407; Indian Ocean—Somaliland 421; Uganda—Cameroon 500
American: Argentina—N. America
27; California—SE. USA 77; (Far
East)—Ozarkian Node—New Brunswick—British Columbia 77; Colombia—Caracas 99, 448, 462; Chile—
Mexico 99; Galápagos—Revilla Gigedo Islds.—(Hawaii) 100; to and
from the New World in the Pacific
102; S. California—Cuba—Lesser
Antilles—Amazonas 127; Juan Fernandez—Galápagos 168; Bolivia—
Cuba—Guianas 196, 391; Brazil—
Peru — Colombia — Guatemala —
Mexico 266; Bolivia—Argentina—
Uruguay—Brazil 273; E. Brazil—
Guianas—Grenada Isld.—Santo Domingo 325 fn.; Guadeloupe—Portorico—Haiti—Jamaica—Cuba *(Prunus occidentalis)* 330, 331 fn. 2;
Cuba—E. Colombia—Peru 345, 370,
371; Cuba—Mexico—British Honduras 346; Curaçao—Ste. Croix
Isld.—Bahamas—Florida—(Texas,
Cuba) 368; Colombia — Central
America—Mexico 371; Cuba—Yucatán—Texas—N.E. Mexico—Bahamas—Florida and USA 371; in
E. and W. Mexico 369; Peru—
Colombia — Greater Antilles — C.
America — Mexico 372; Bolivia —
Ecuador—Mexico 374, 440 *et seq.*;
Colombia—Venezuela—Greater Antilles 384; E. Brazil—Venezuela—

Guianas — Trinidad 392; Tobago
Isld. — Martinica — Portorico — Jamaica—Cuba and Panama—Honduras—Ceiba (Coiba) Isld. 406,
407; Mexico—Guatemala—Costarica (Mt. Irazú)—Panama—Colombia
—Bolivia 440, 441; Martinica—Portorico 445; Saskatchewan—Florida
448; Santander—Mérida—Caracas
448 (see further, p. 99); (Japan and
Kamchatka)—Canada—USA—Guadeloupe Isld.—Venezuela 449 fn.;
*Asiatic and in the Western Pacific
generally:* in the Pacific east of the
Philippines 53, 98, 247; W. Pacific—
Hawaii 53; S. China—Indochina—
Formosa — Luzon — Sumatra —
Java—Lombok 79; Rapa Isld.—
Tahiti — Fiji — Samoa — Micronesia—Lord Howe Isld.—New Zealand — Queensland — W. Malaysia
105; along mountains on the Western
Pacific 111; Borneo—Philippines—
Celebes—Australia 111; Lombok—
Culion—Luzon 114; Annam—Palawan — Borneo — SE. China 114;
from the Pacific avoiding Africa 178;
in Polynesia 178; "cold" and
"warm" among Tahiti, New Caledonia and the islands by New
Zealand 181 *et seq.*; emerging out of
the Indian Ocean 219; "cold" and/or
"warm" in the Indian Ocean and the
Western Pacific 247, 483; China—
Japan 292; Borneo (Tawao)—Laos—
E. Himalayas 323; Malacca—N.
Borneo—Palawan 324; Sumatra—
Java—Palawan 327; centering around
the Philippines 328, 417; (Australia)
—New Guinea—Moluccas—Sumatra — Guimaras — Culion — Luzon
—Palau—Carolines 342; (Australia)
— Philippines — Borneo — Tonkin
363; Phuquoc Isld.—Cochinchina—
Burma—E. India 389; S. Celebes—
Java 397; Celebes—Philippines—
Japan 397; New Guinea—Sumatra—
Malacca 404; Celebes—Philippines
404; Mindoro—Mt. Kinabalu 404;
(New Caledonia and NE. Australia)
—New Guinea—W. Borneo—Sumatra—Bangka—Malacca 405; India—
Burma—Siam—Malacca 418; N.
Sumatra—NE. Borneo 427; Timor

MAPS
1 to 105

FIGURE
106

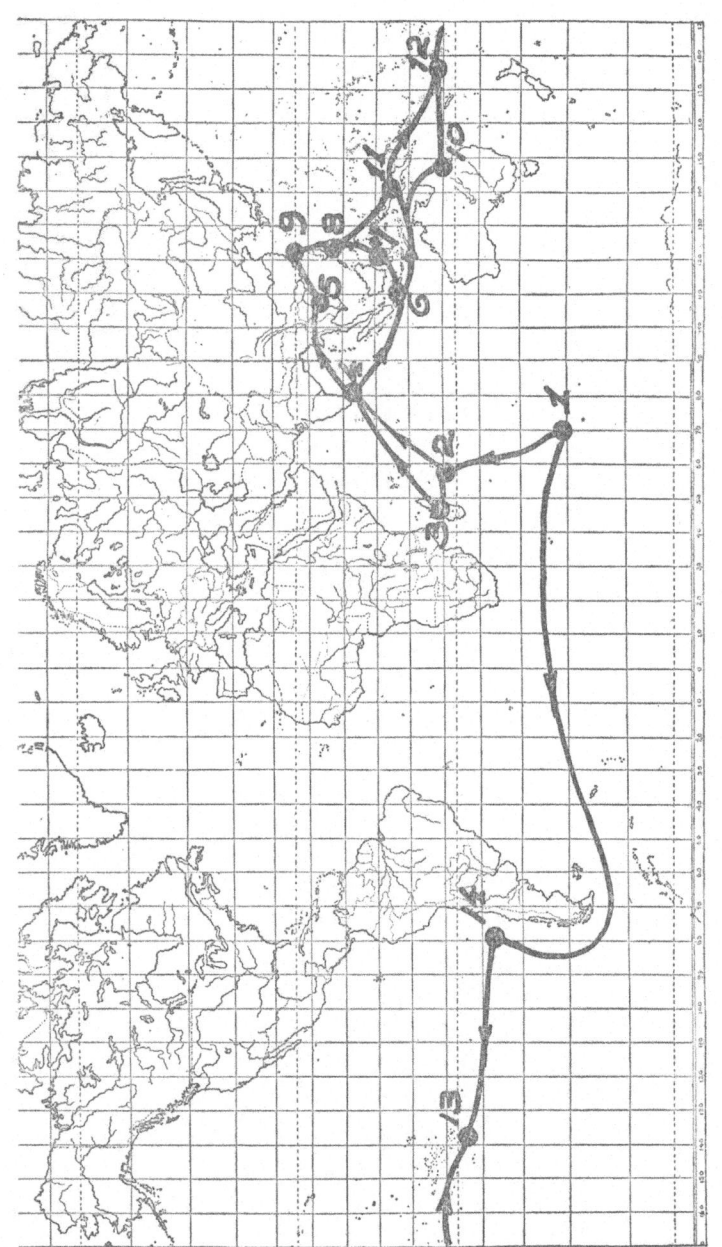

Fig. 1. Kerguelen Island 1; Mascarenes 2; Madagascar 3; Ceylon 4; Indochina 5; Java 6; Celebes 7; Philippines 8; Formosa 9; Queensland 10; New Guinea 11; Fiji 12; Rapa Island 13; Juan Fernandez Island 14. The arrows give potential orientation of the tracks. See text for elucidation, p. 9.

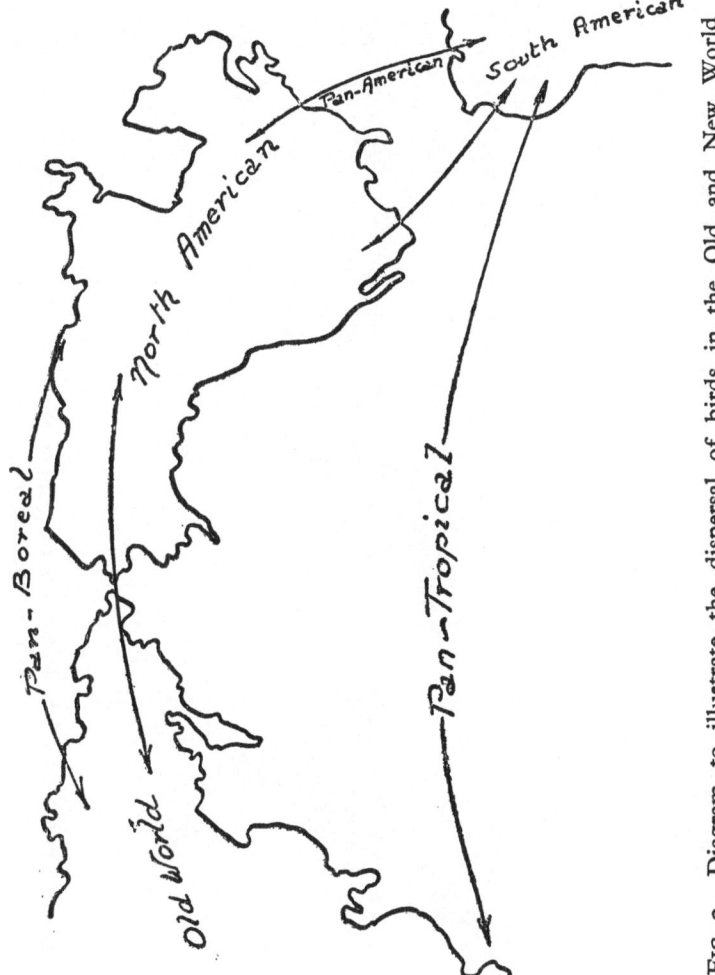

FIG. 2. Diagram to illustrate the dispersal of birds in the Old and New World (after (2)). Notice the transpacific range of the "Pan-tropical" element. See text for elucidation, p. 12.

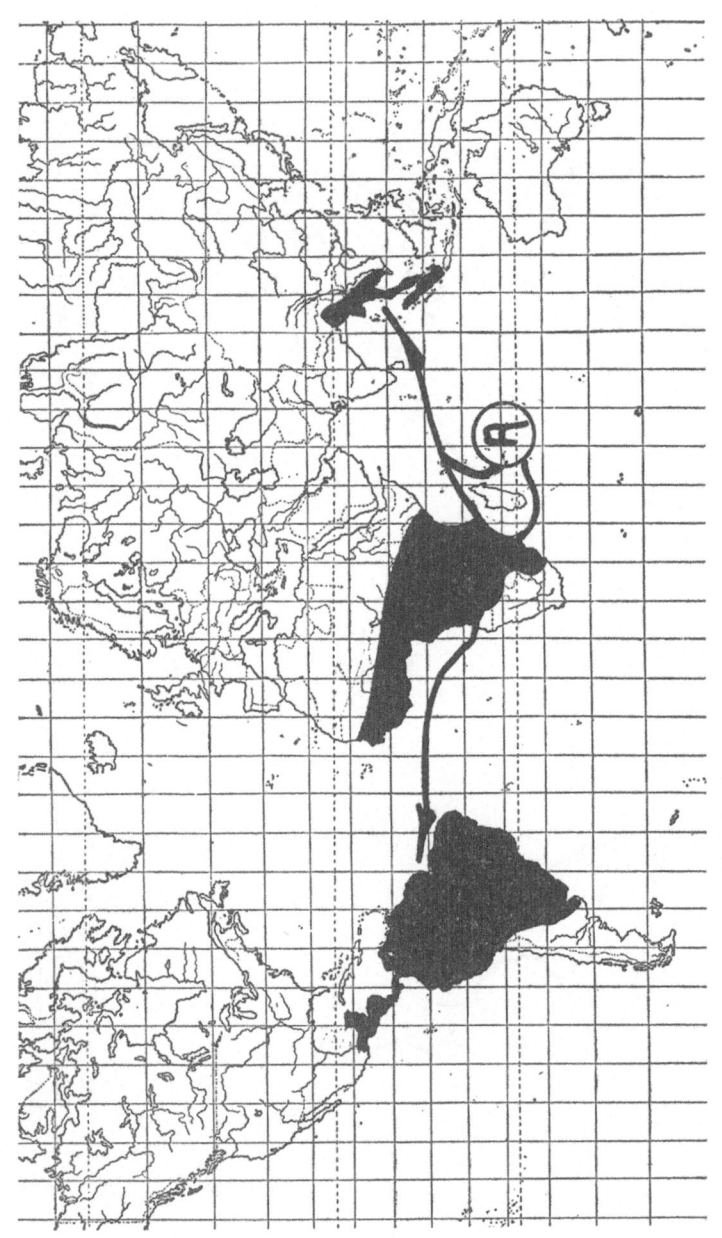

Fig. 3. Living distribution of birds *Heliornithidae* (Sungrebes) in solid black. Tracks running out of the African Gate of Angiospermy (A), such as are standard for plant-migration, added for reference and comparison. See text for elucidation, p. 16.

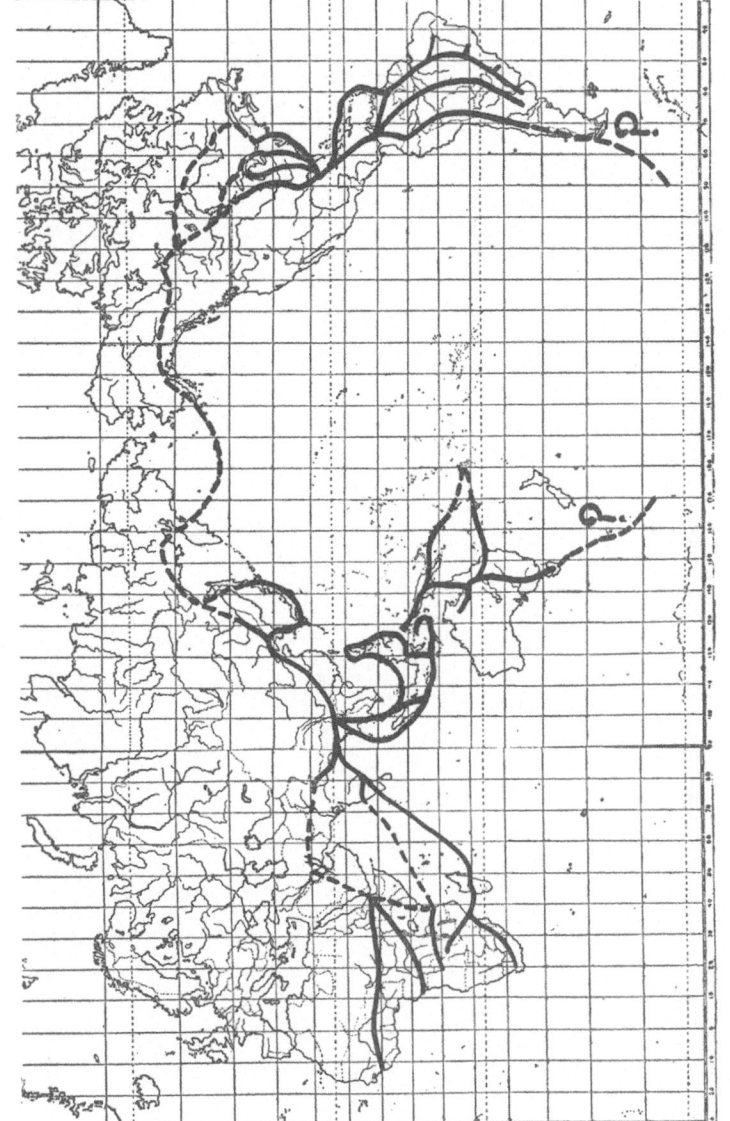

Fig. 4. Distribution of Bignoniaceae (after (7)). Actual range in unbroken, presumed connections in broken lines. See text for elucidation, p. 17.

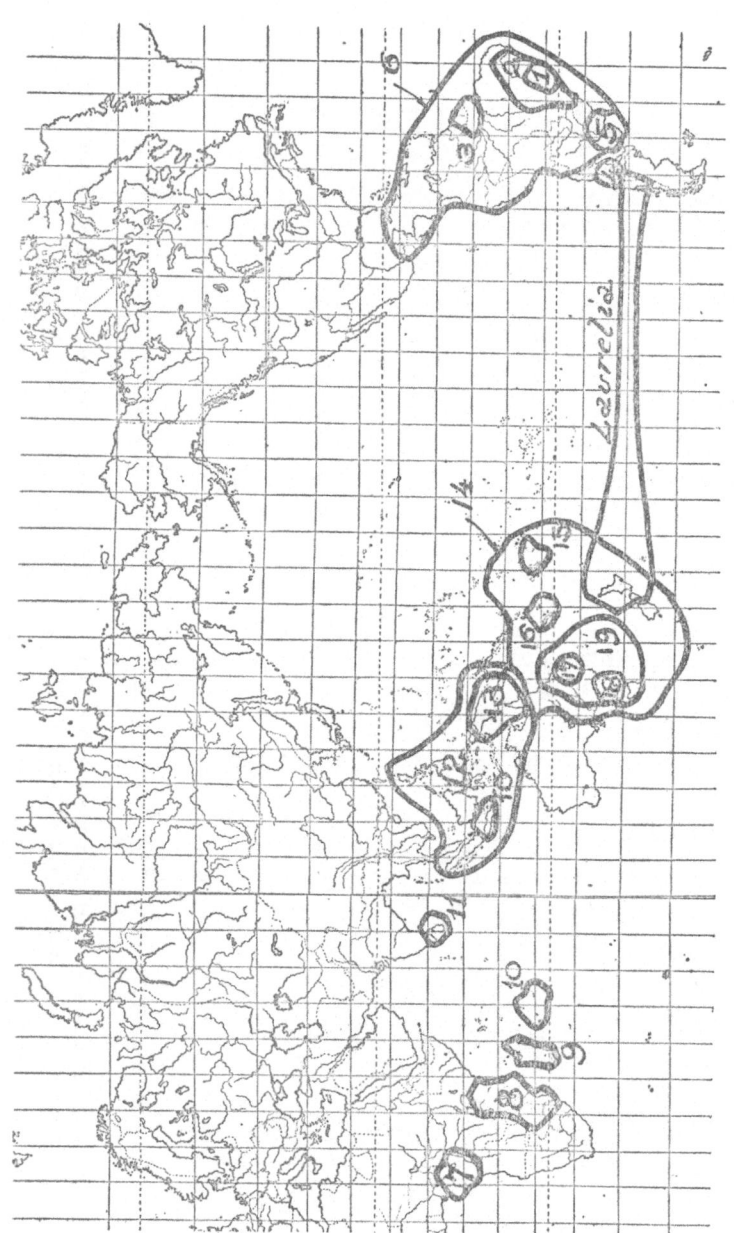

FIG. 5. The distribution of the Monimiaceae (after (9)). *Macrotorus* 1; *Macropeplus* 2; *Cornuleum* 3; *Peumus* 4; *Hennecartia* 5; *Siparuna* and *Mollinedia* 6; *Glossocalyx* 7; *Xymalos* 8; *Ephippiandra* 9; *Monimia* and *Tambourissa* 10; *Hortonia* 11; *Matthaea* and *Kibara* 12; *Leviera* and *Palmeria* 13; *Hedycarya* 14; *Trimenia* 15; *Amborella*, *Carnegiea* and *Nemuaron* 16; *Piptocalyx* and *Daphnandra* 17; *Doryphora* 18; *Atherosperma* 19; *Laurelia* unnumbered. See text for elucidation, p. 18, 279.

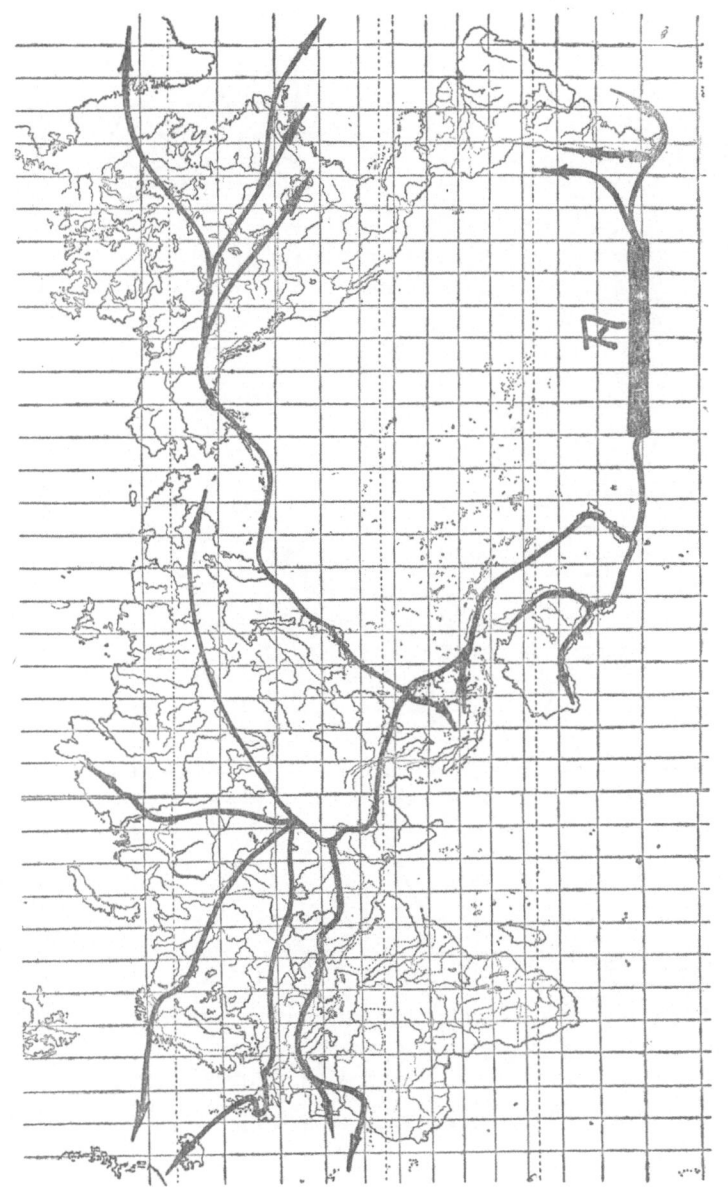

Fig. 6. The main arteries of the dispersal of *Euphrasia* (after (8)). The base-line of distribution in A, as heavy bar. See text for elucidation, p. 19

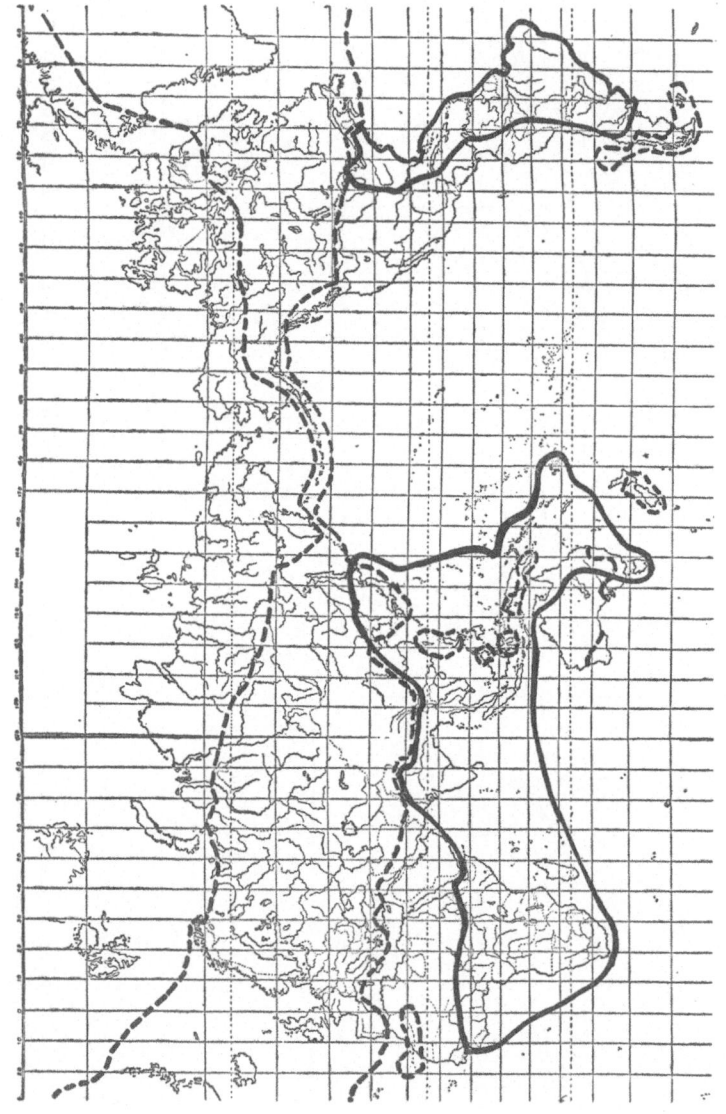

FIG. 7. Comparative ranges of Bignoniaceae (in unbroken line) and *Euphrasia* (in broken lines)
See text for elucidation, p. 19.

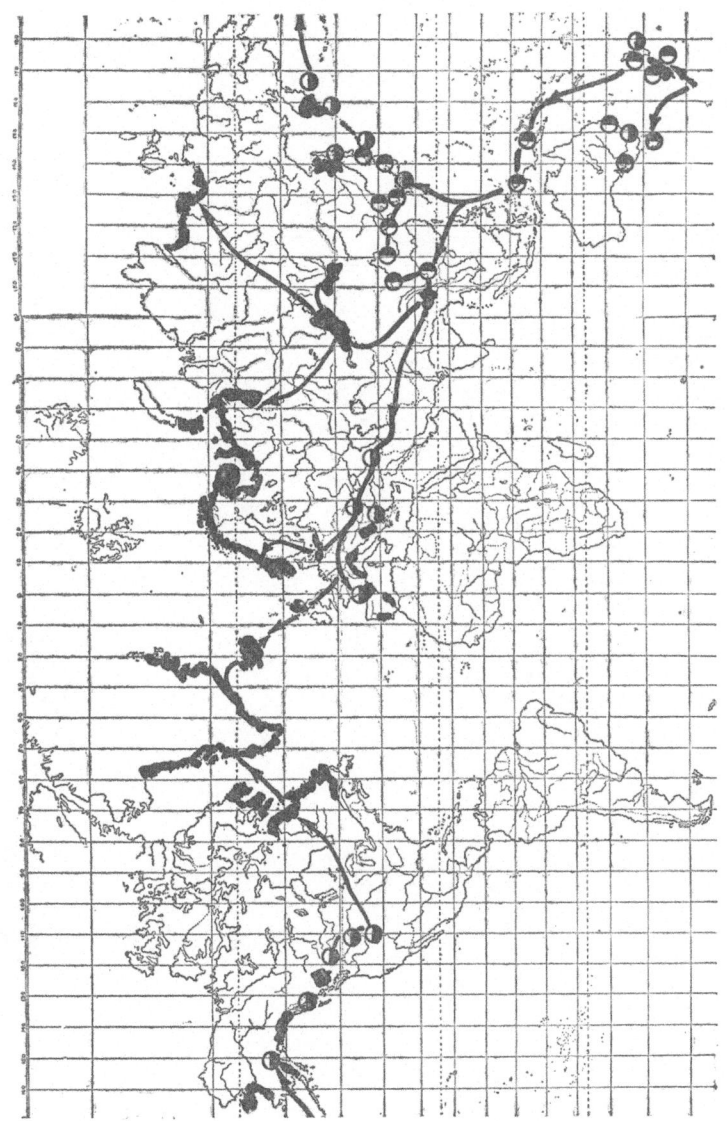

FIG. 8. Dispersal of *Carex Lachenalii* (range all in heavy black) *C. pyrenaica* (circle with black below), *C. Gaudichaudiana* (circle with black to right) after (8). Lines with arrows indicative of the trends of distribution. See text for elucidation, p. 23.

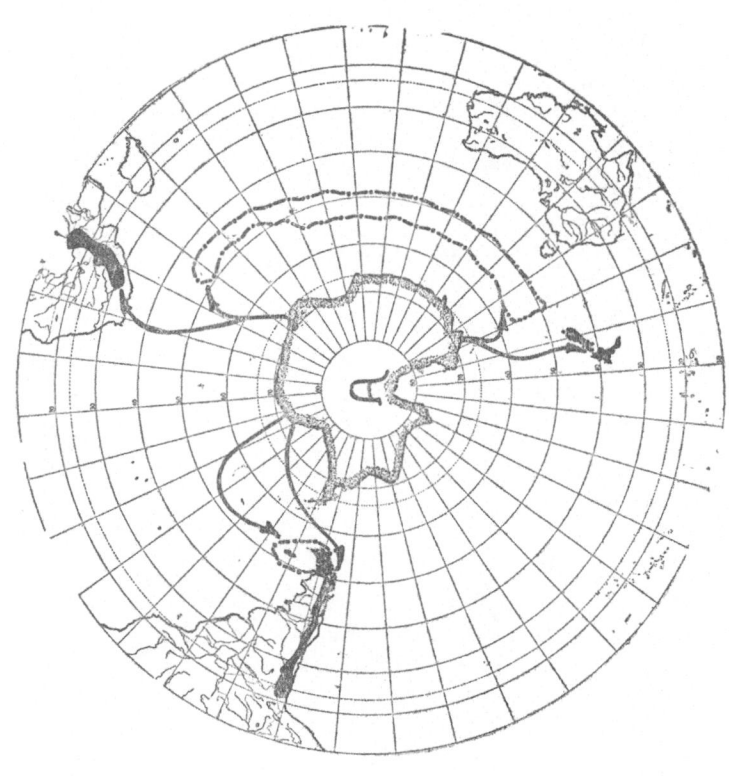

FIG. 9. Range of *Taraxacum magellanicum* (heavy black) and *Azorella Selago* (in broken dotted line). The center of dispersal in A (Antartica). Diagrammatic. See text for elucidation p. 23.

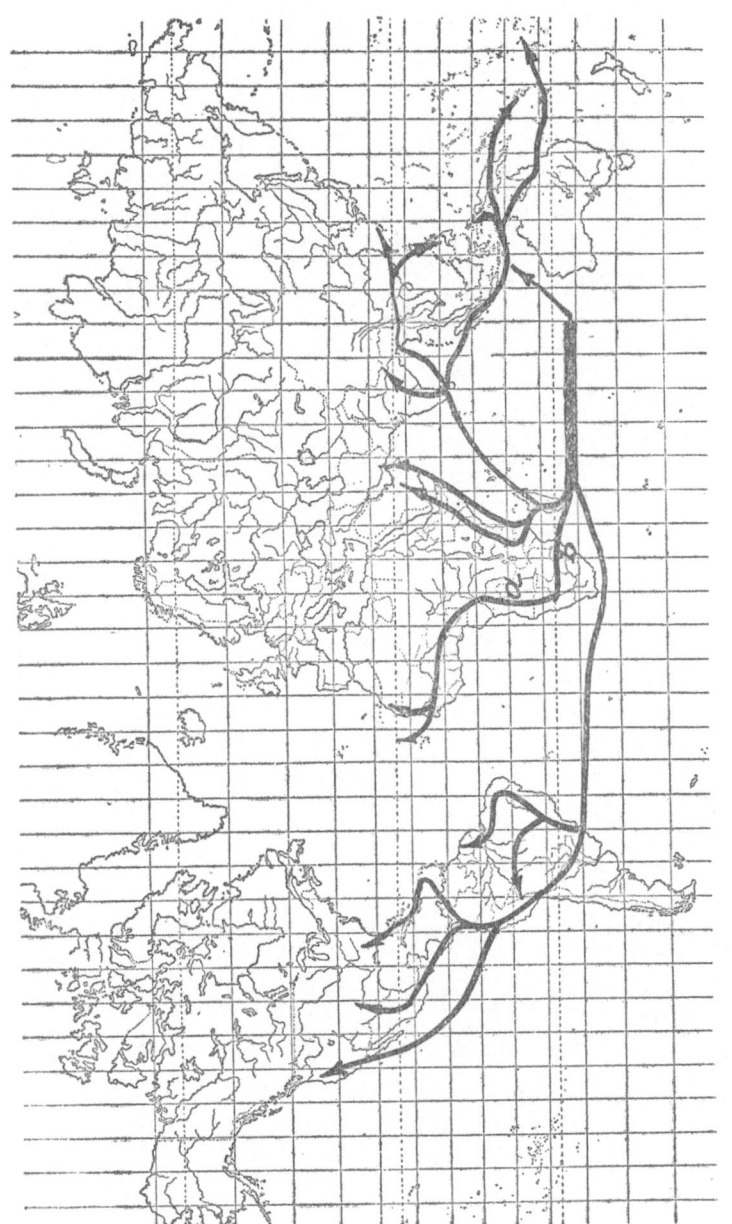

FIG. 10. The dispersal of *Evolvulus* subsect. *Pedunculati*. Angola is *a*, Transvaal is *b*, base-line of distribution as heavy black bar. See text for elucidation, p. 26.

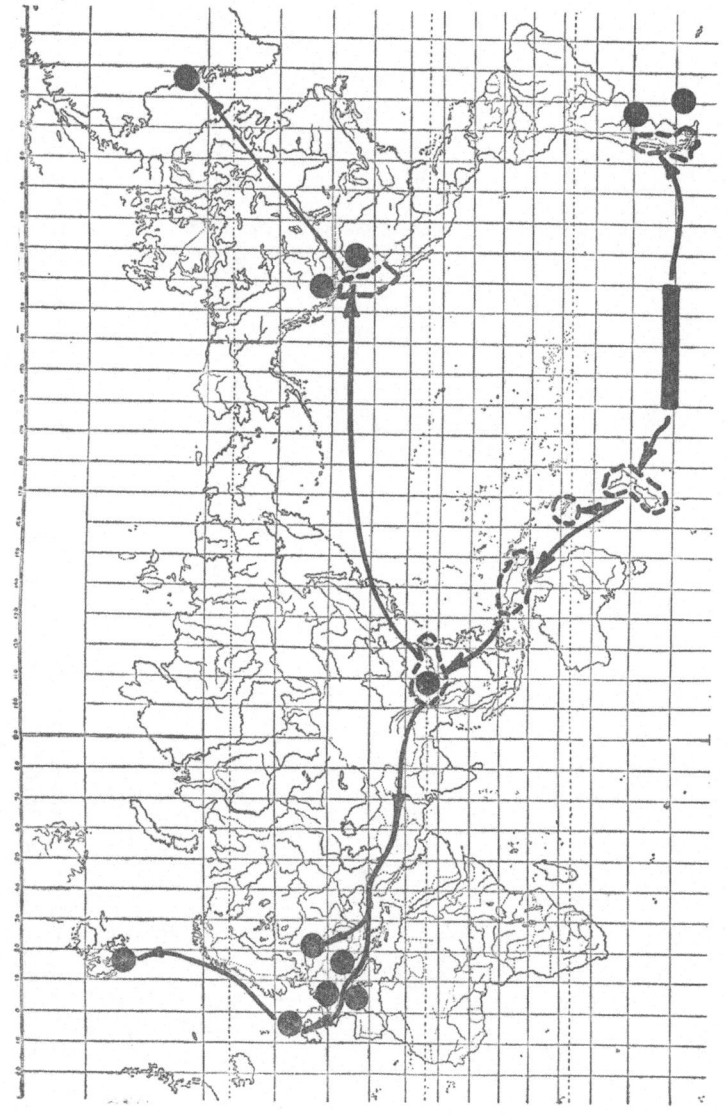

FIG. 11. Dispersal of *Euphrasia* (schematic; lines with arrows to indicate trends) compared with that of *Libocedrus* (living range within broken line; fossil stations as black dots after (9)). Baseline in heavy black. See text for full elucidation, p. 29.

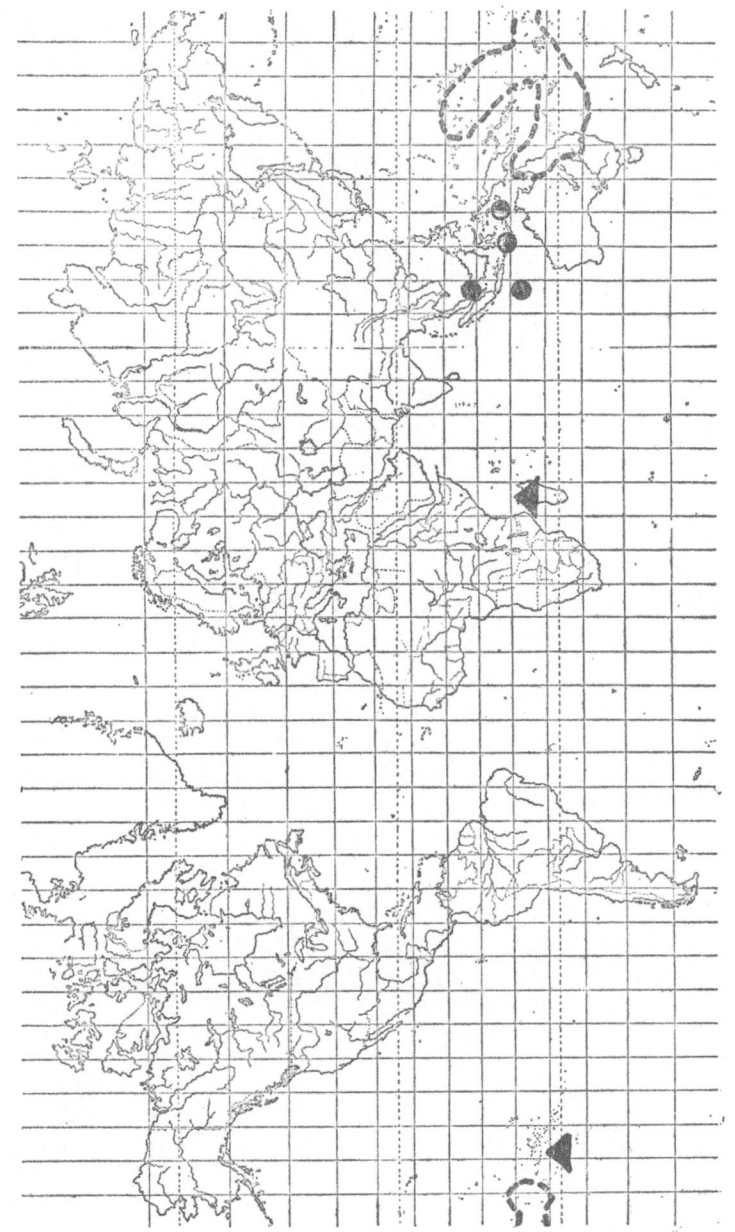

FIG. 12. Dispersal of *Balanophora*: *B. fungosa* in broken line; *B. Hildebrandti*, black triangles; *B. insularis*, black circle; *B. Micholitzii*, circle with black at right; *B. Zollingeri*, circle with black below. Ranges diagrammatic. See text for elucidation p. 40.

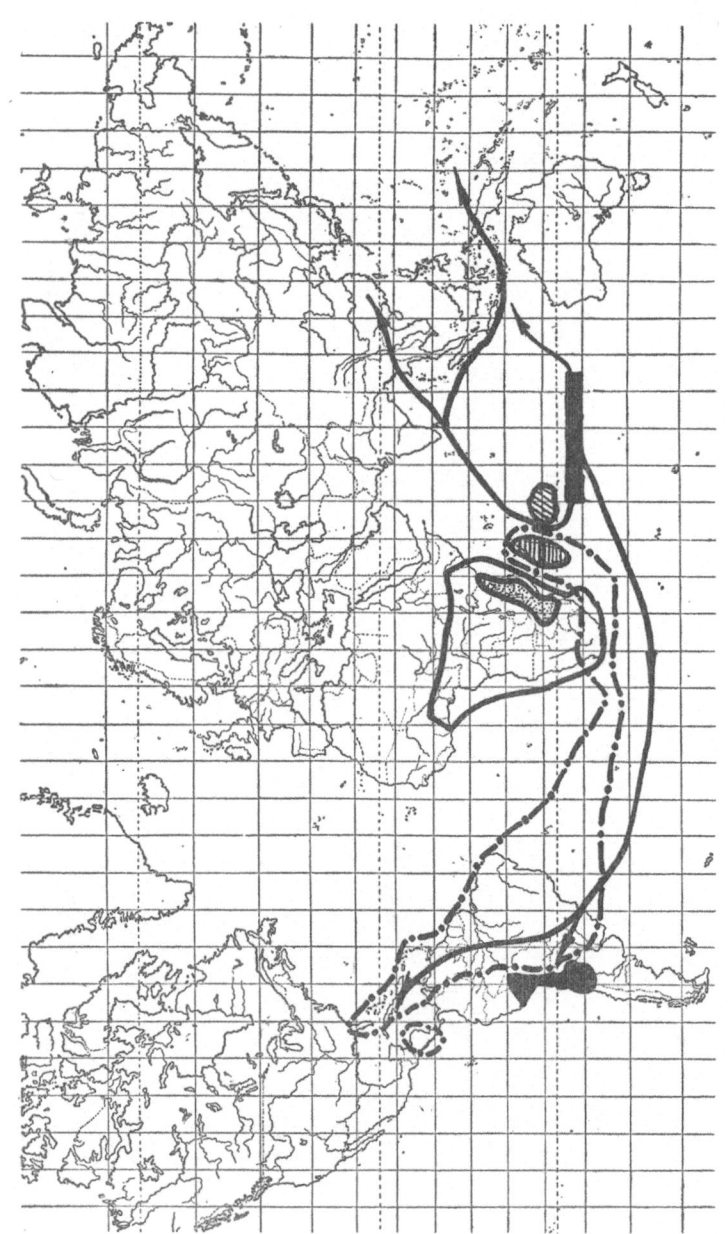

FIG. 13. Dispersal of Turneraceae and Malesherbiaceae (latter range in solid black). See text for *Loewia.* *Wormskioldia*, in unbroken line; *Streptopetalum*, stippled; *Mathurinia*, diagonally hatched; *Hyalocalyx*, horizontally hatched. Range of *Piriqueta* and *Turnera* in broken dotted line; track with arrows for *Turnera ulmifolia*: base-line of dispersal in heavy black. Ranges diagrammatic. See text for elucidation p. 45.

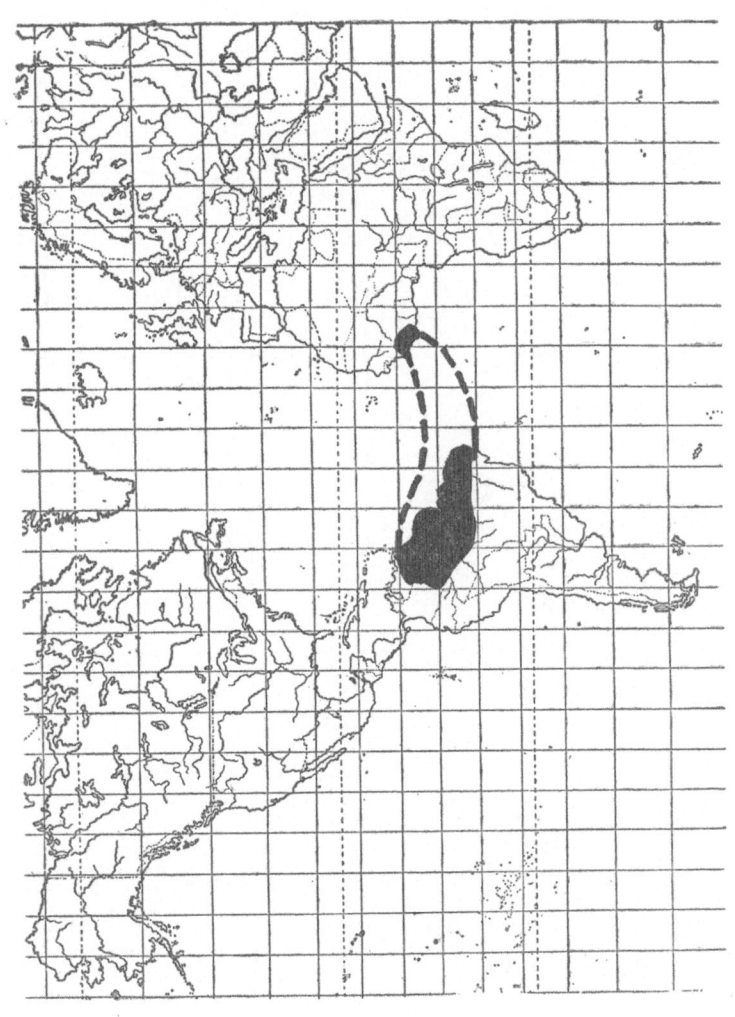

FIG. 14. Range of the Rapateaceae in solid black. (diagrammatic); See text for elucidation, p. 46.

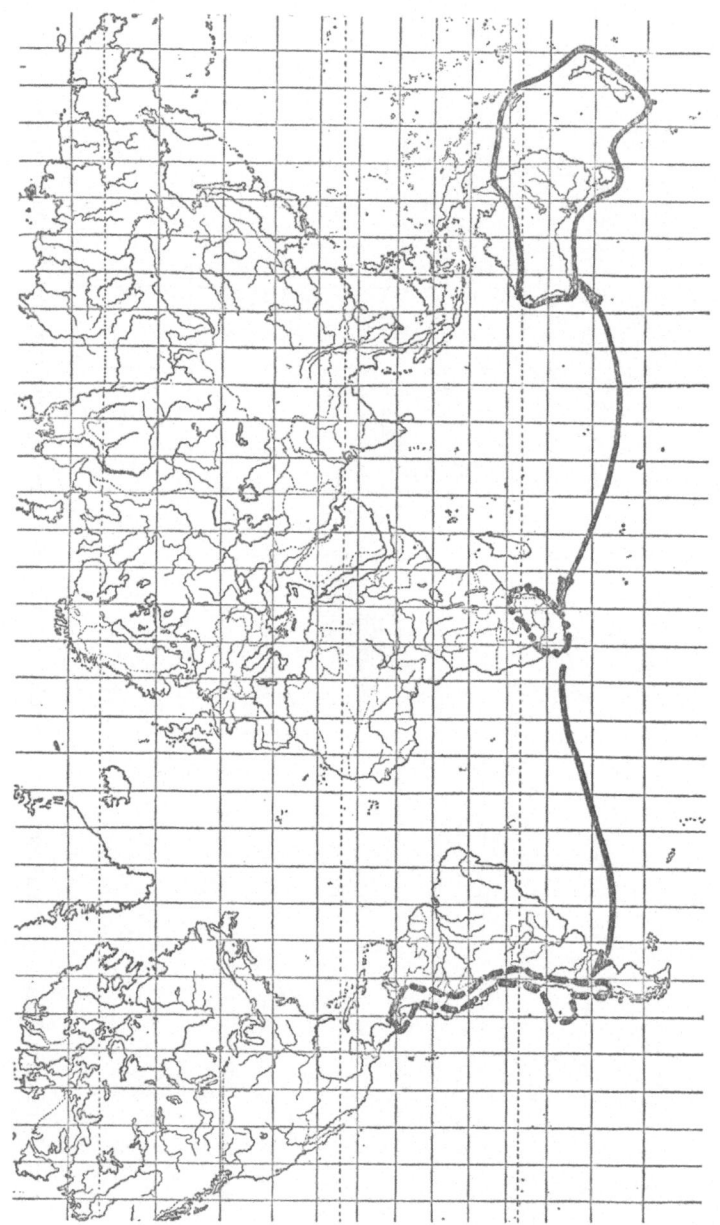

FIG. 15. Dispersal of the euphorbiaceous groups, *Dysopsis* (broken line at left of map); *Seidelia* and *Leidesia* (broken line with dots, in center); *Seidelia* and *Leidesia* (continuous line at right.) Range diagrammatic. See text for elucidation, p. 51.

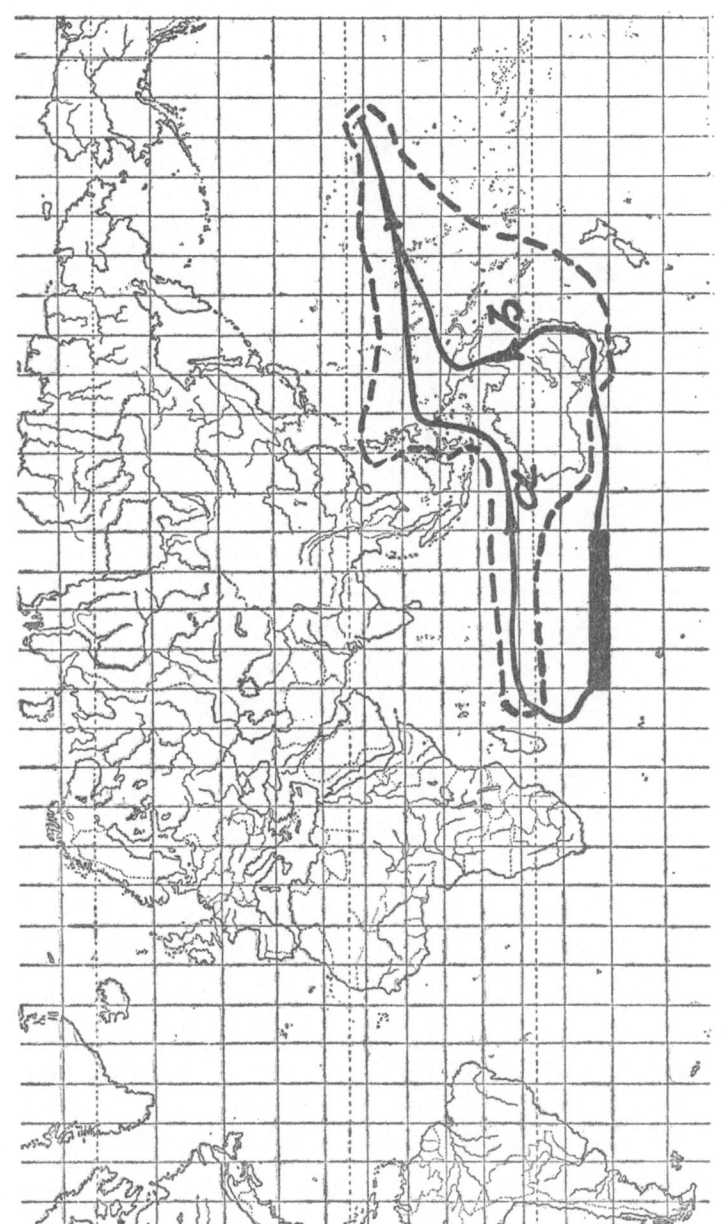

Fig. 16. Range of phyllodine *Acacia* (diagrammatic) within broken line. Tracks in *a* and *b*: base-line of dispersal heavy black. See text for elucidation p. 53.

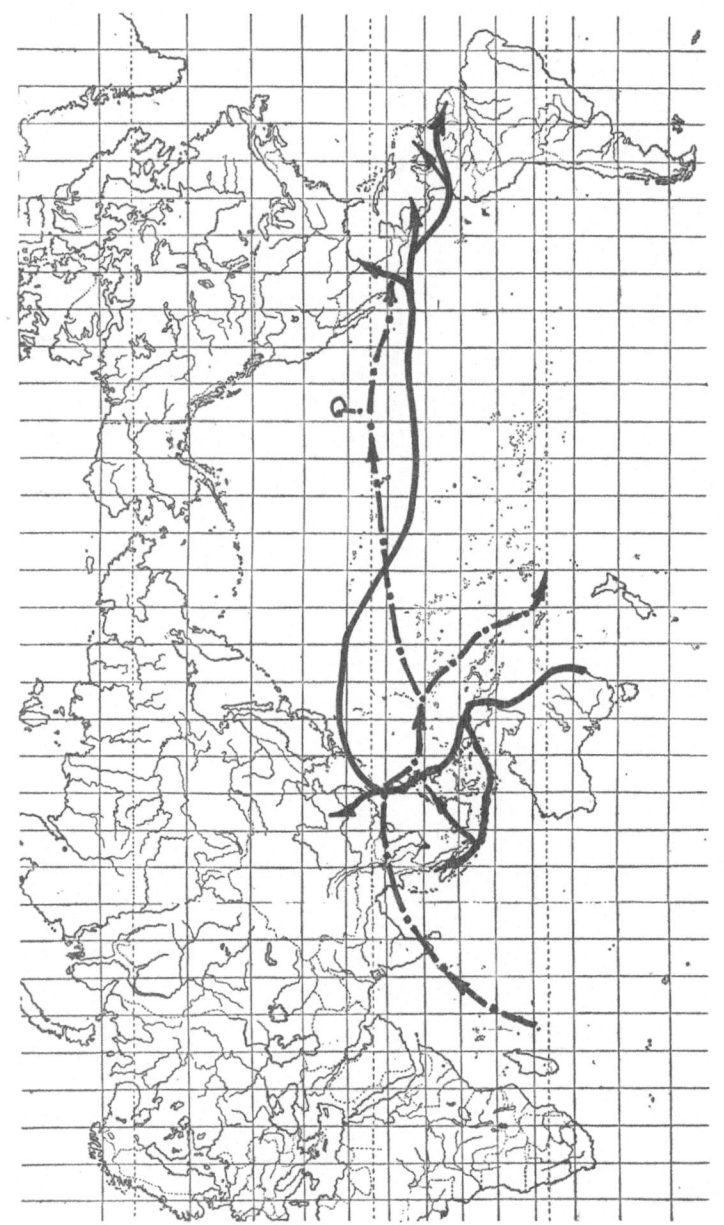

FIG. 17. The migration of *Lysimachia mauritiana* (broken line), and *Perrotetia* (continuous line). The question mark connotates the probable track of *Lysimachia glaucophylla* to Western Mexico. See text for elucidation, p. 54.

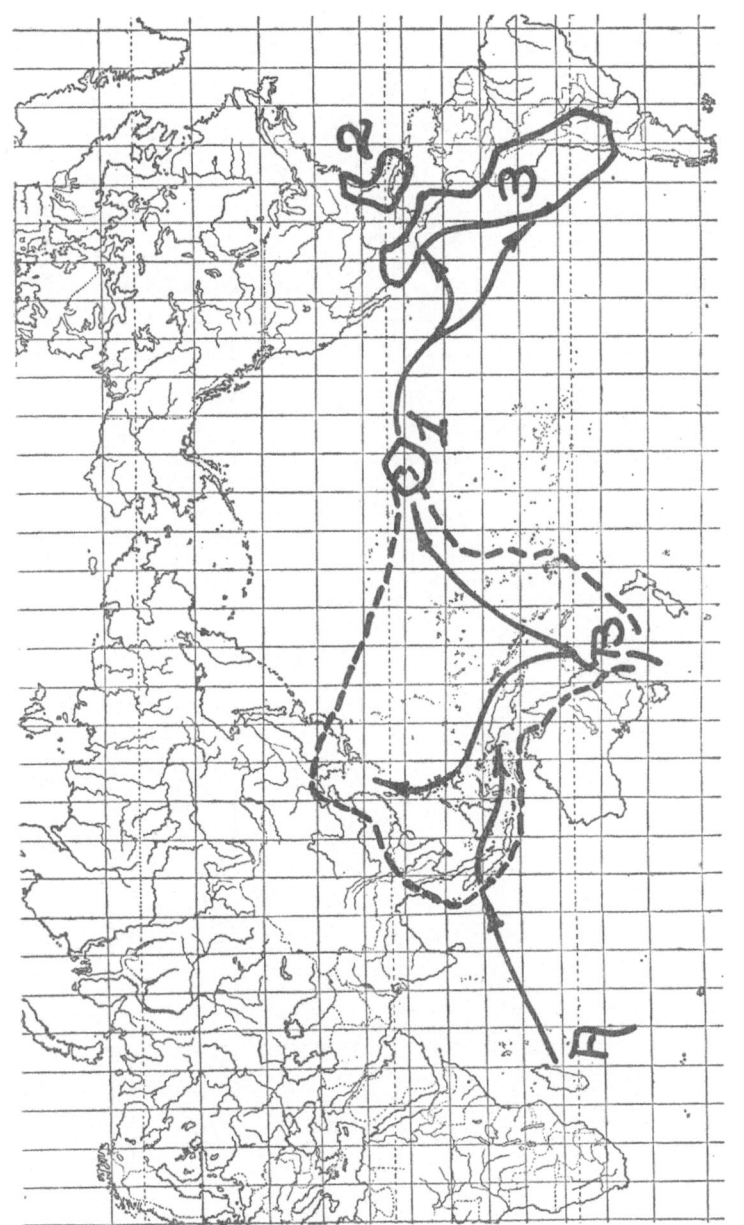

FIG. 18. Actual modern range of genus *Vallesia* in 1, 2, 3; range of *Vallesia's* affinity within broken line. African gate in A; Western Polynesian in B. See text for elucidation, p. 54.

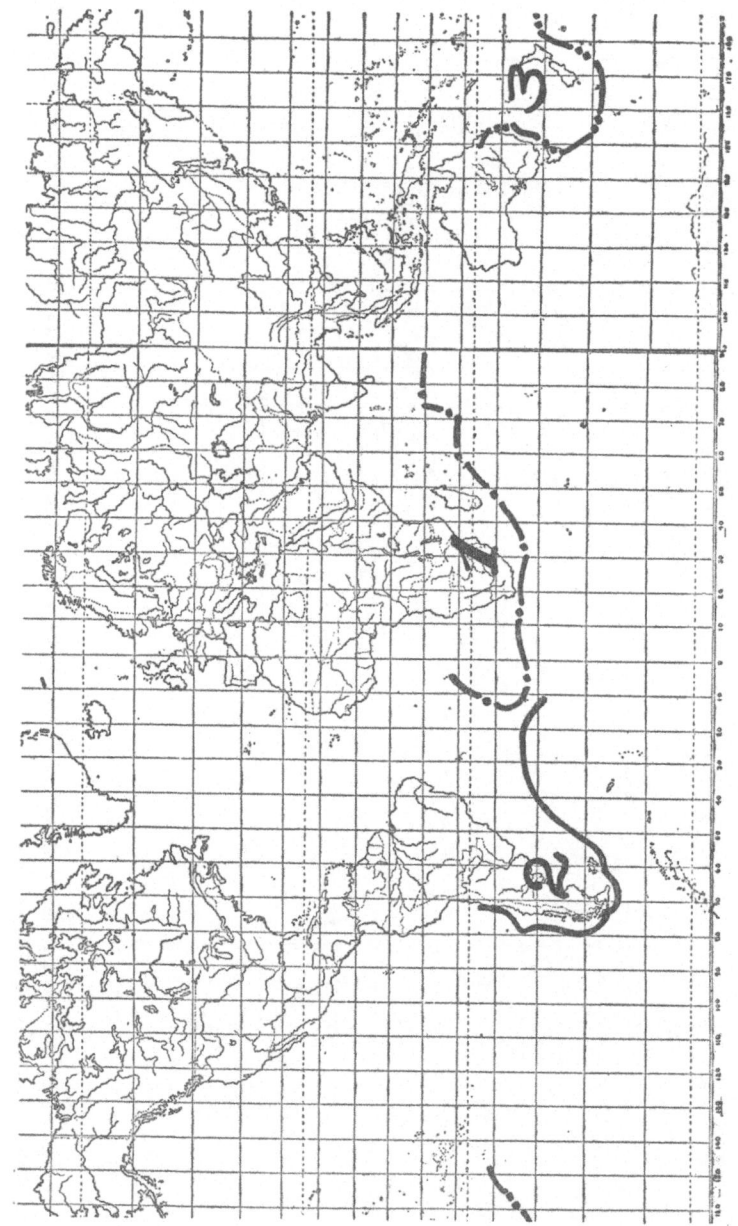

FIG. 19. The *Gates of Angiospermy*. African Gate in 1; Magellanian Gate in 2; Western Polynesian Gate in 3. Diagrammatic; see text for elucidation, p. 59.

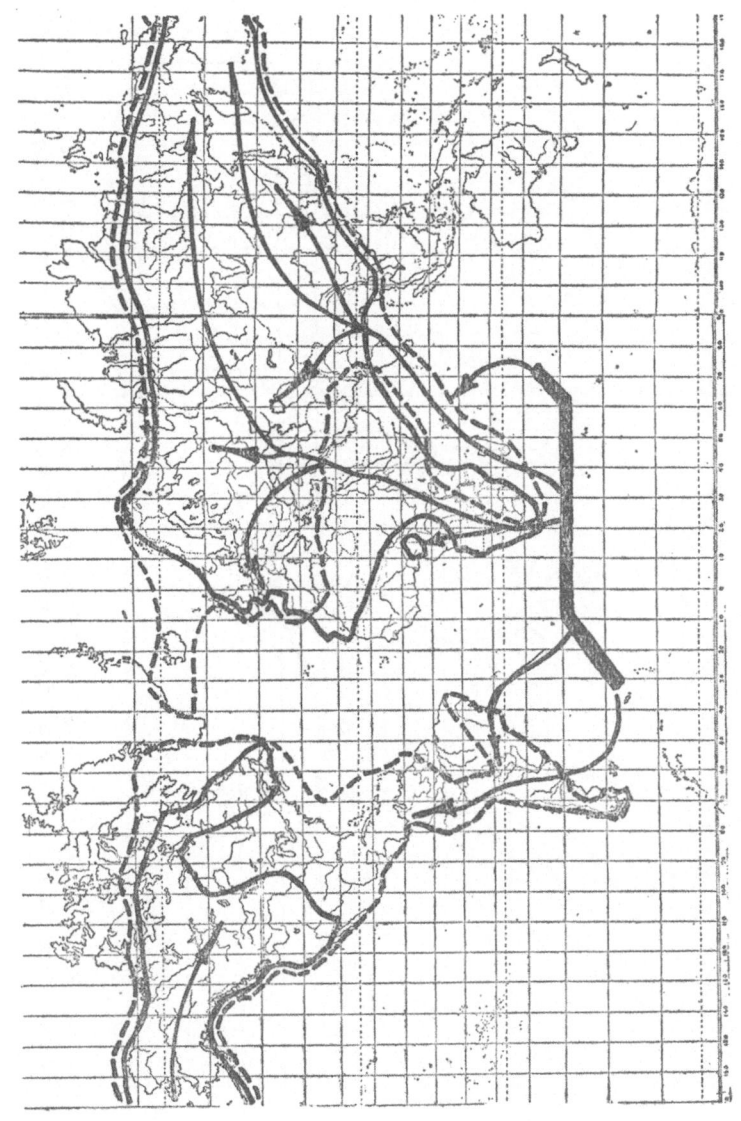

FIG. 20. Range of *Crepis* (solid line) and *Hieracium* (broken line) with arrows indicating the main trends of the dispersal (diagrammatic). Base-line of dispersal as heavy black bar. See text for elucidation, p. 63. Range after (17).

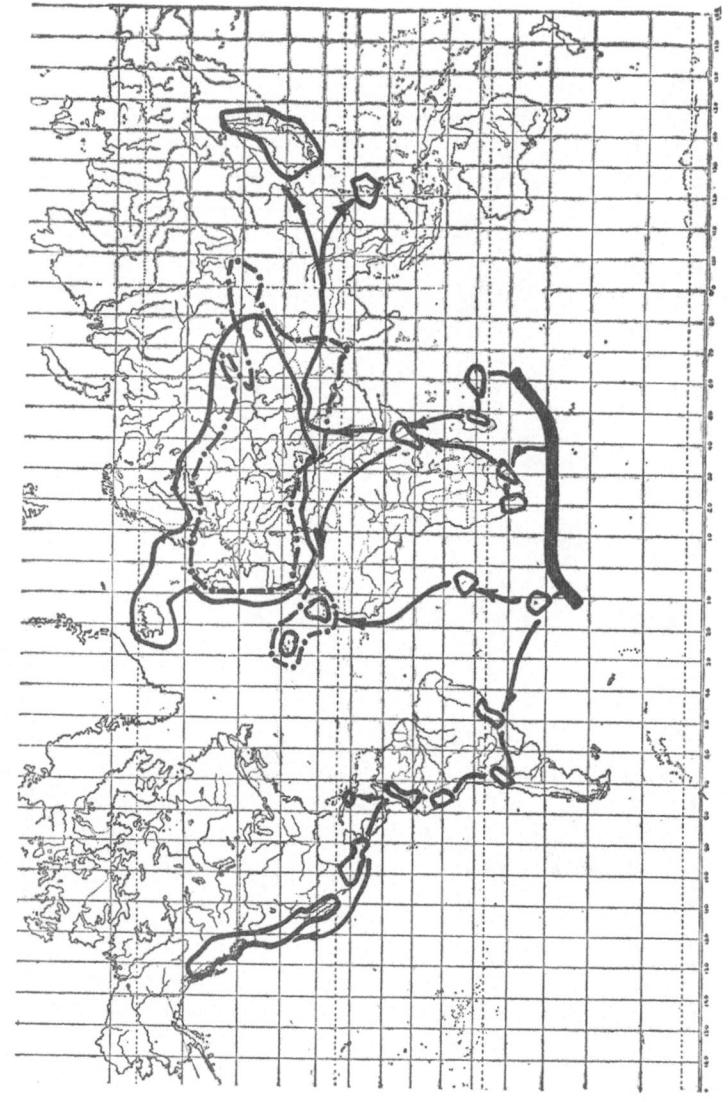

FIG. 21. Range of *Juncus effusus* (solid line) and *J. inflexus* (broken dotted line) after (26). Dispersal from the African gate of angiospermy (heavy black line). See text for elucidation, p. 64.

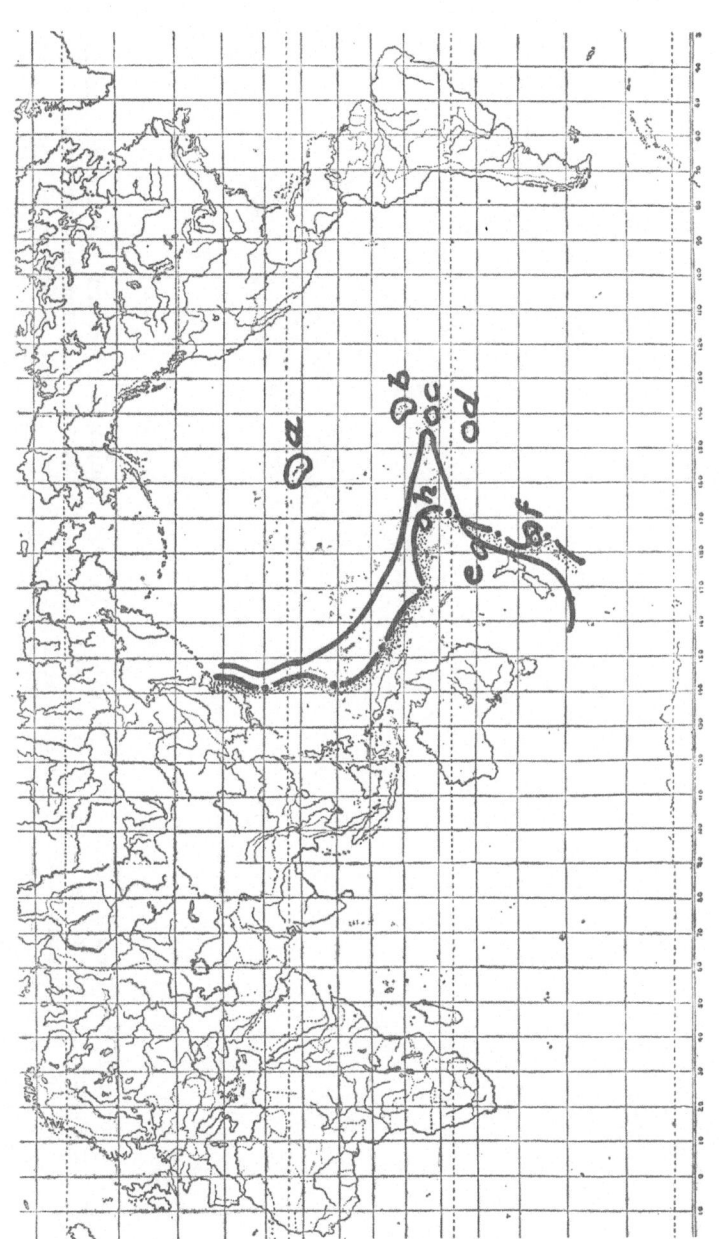

FIG. 22. Geological continental shelf of the Western Pacific (stippled broken dotted line; after (1), Chapt. III, in the main). Solid line delimits region of fairly large to large Orchid flora (after (2), Chapter III). Hawaii in *a*, Marquesas in *b*, Tuamotus in *c*, Rapa in *d*, Kermadec Islands in *e*, Chatham Island in *f*, Samoa in *h*. See text, p. 69.

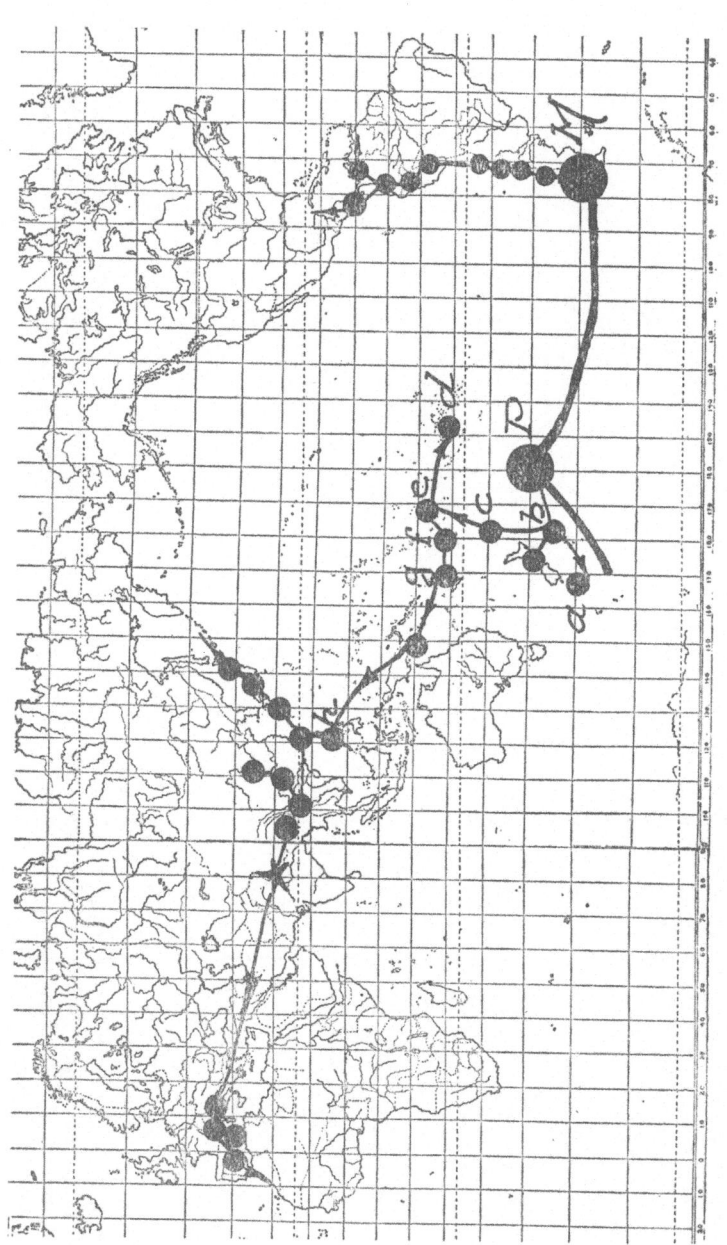

FIG. 23. Dispersal of *Coriaria*. Gates of angiospermy: *M*, Magellanian; *P*, Western Polynesian. Specific localities indicated by black circles connected with tracks (lines with arrows). Relevant geographic localities: *a*, Stewart Island; *b*, Chatham Island; *c*, Kermadec Islands; *d*, Society Islands; *e*, Samoa; *f*, Fiji; *g*, New Hebrides. Himalayan record to match European fossil indicated with star. Diagrammatic; see text for discussion and elucidation, p. 72.

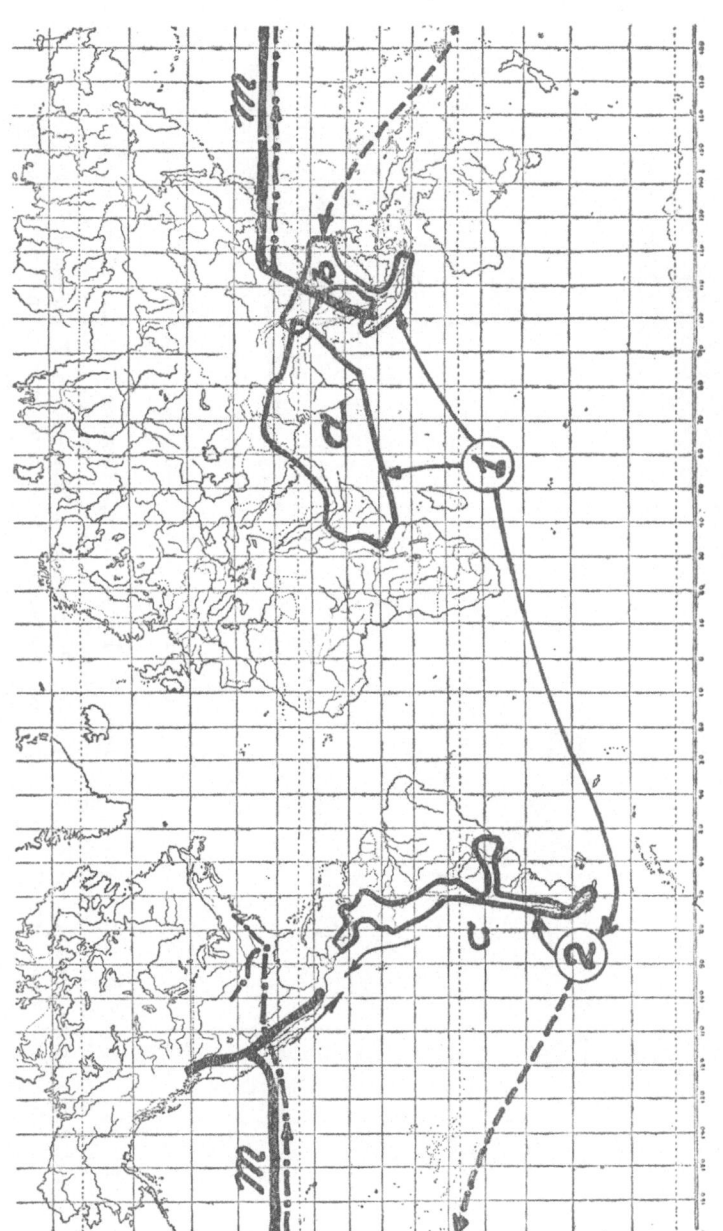

FIG. 24. Trends in the dispersal of the Berberidaceae. African gate of angiospermy, 1; Magellanian gate of angiospermy, 2. *Berberis* Sect. *Tinctoriae*, *a* within continuous line; *Berberis Wallichiana*, *b* within continuous line; *Berberis*, *c* within continuous line. Dispersal of *Mahonia* as heavy bar *m*: next to it, as broken dotted line, the dispersal of the Northern American species of *Berberis*. Main tracks from the African gate as continuous lines with arrows; alternative track from the Magellanian gate to the Far East as broken line in the Pacific Ocean. Essentially diagrammatic; see text, p. 79, for elucidation.

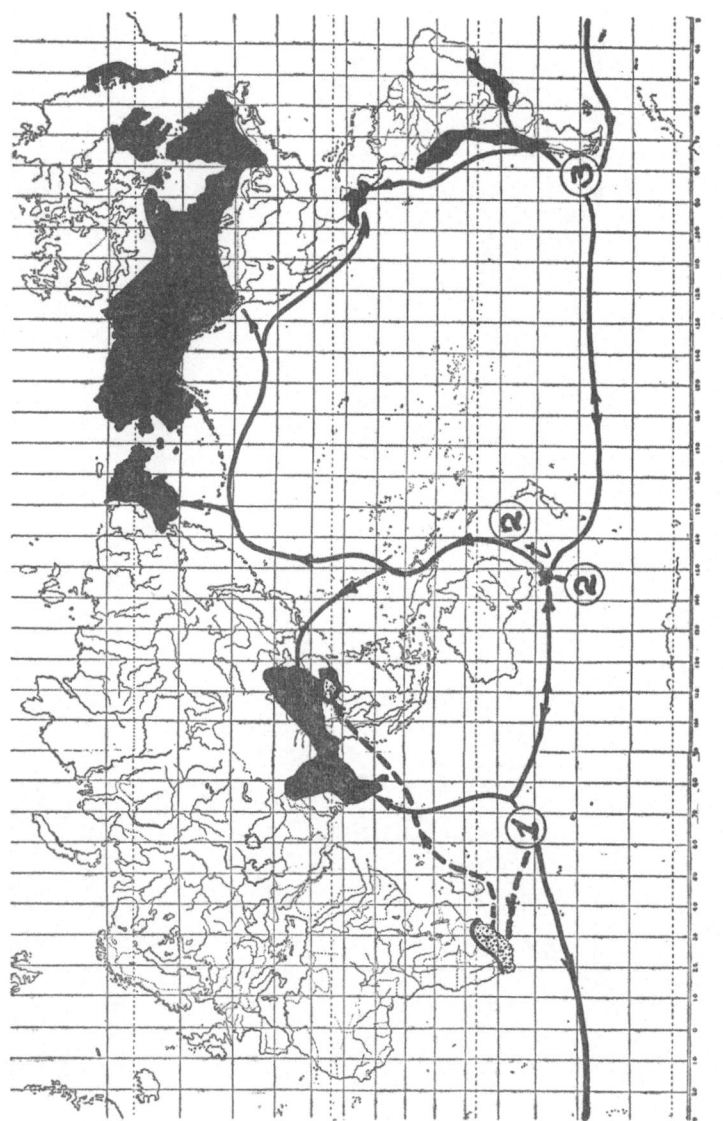

FIG. 25. Trends in the dispersal of *Anemone*. Heavy black, *Anemone* Sect. *Rivularidium*: stippled within continuous line, *Anemone* Sect. *Pulsatilloides* Subsect. *Longistylae*. Gates of angiospermy: 1 African; 2 (both centers; see Fig. 67) Western Polynesian; 3 Magellanian. Migrations in *t*. Migrations in Sect. *Rivularidium* shown by tracks as continuous lines with arrows; in Sect. *Pulsatilloides* Subsect. *Longistylae* by tracks as broken lines. Essentially diagrammatic; see text, p. 81, for elucidation.

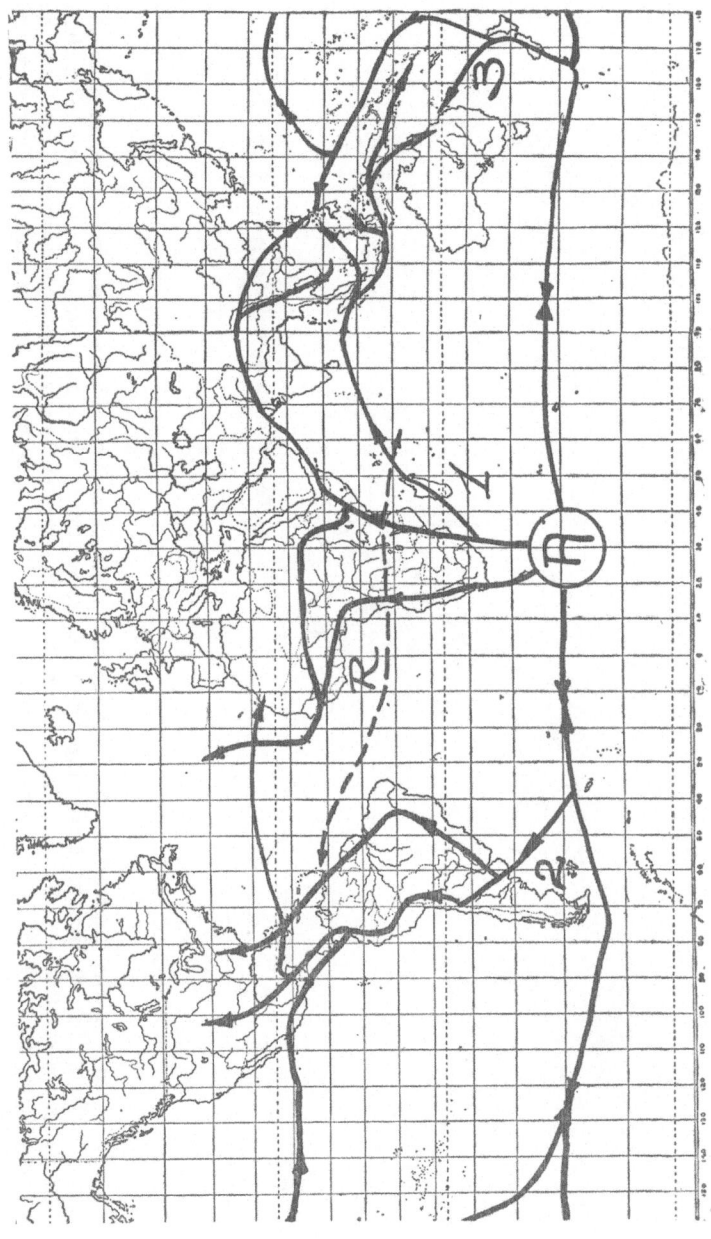

FIG. 26. Trends in the dispersal of the Myrsinaceae. Main angiospermous center, A. Gates of angiospermy: 1, African; 2, Magellanian, 3 Western Polynesian. Tracks as continuous lines with arrows for orientation. Broken line R, alternative track securing crossing between Africa and America in the Central Atlantic. Essentially diagrammatic; see text, p. 86, for further elucidation.

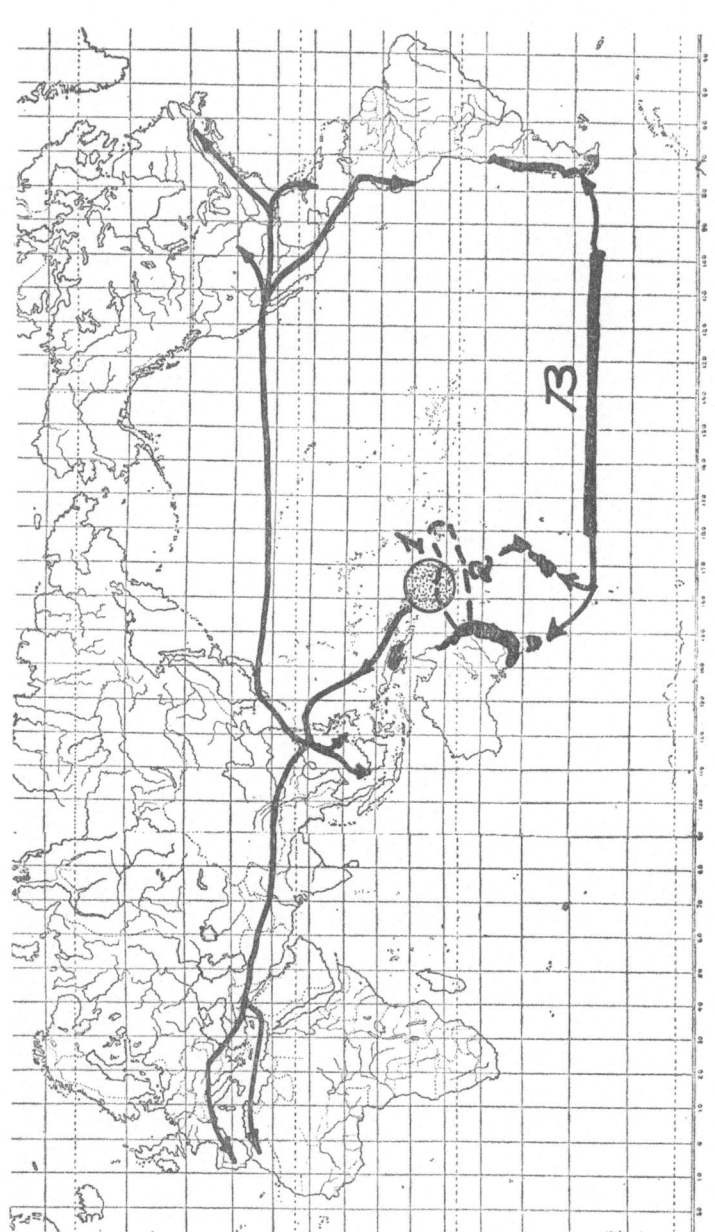

FIG. 27. Trends in the dispersal of the Fagaceae and their immediate alliance. Base-line of dispersal in the South Pacific as bar B. Heavy black, range of *Nothofagus*: dotted circle 1, center of origin of the modern Fagaceae (see *Neocaledonian Center: Fig. 67*); range within broken line 2, Balanopsidaceae. Tracks of the Fagaceae (essentially based upon *Quercus*) as continuous lines with arrows for orientation. Diagrammatic; see text, pp. 95, 108, for further elucidation.

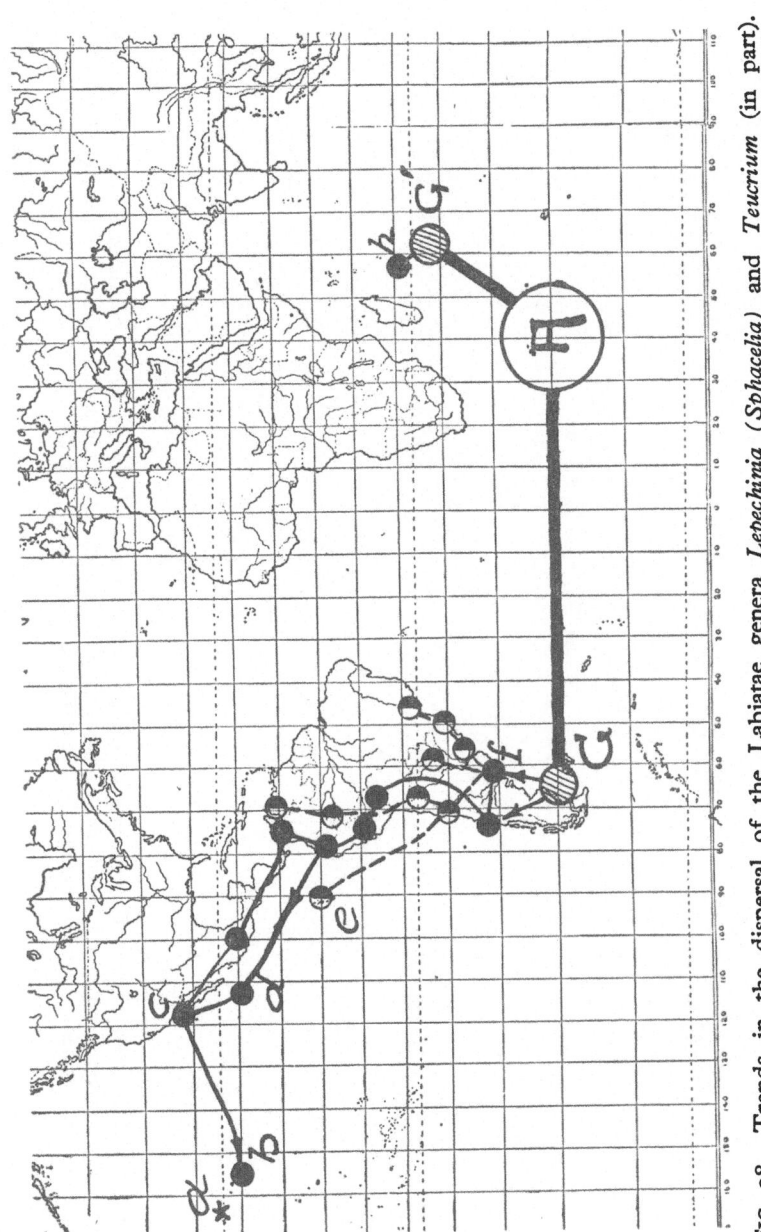

FIG. 28. Trends in the dispersal of the Labiatae genera *Lepechinia* (*Sphacelia*) and *Teucrium* (in part). Main center of angiospermous *genorheitra* in A, connected with gates (G, Magellanian; G', African) by heavy bar (standing diagrammatically for geologic "antarctic" shore). Specific records of *Lepechinia* as black circles; of *Teucrium* as circles with black at right. Relevant geographic localities: *a* with star, Laysan Island (see discussion of *Nama* in text, p. 100); *b*, Hawaii; *c*, Baja California; *d*, Revilla Gigedo Islands; *e*, Galápagos Islands; *f*, Sierra de la Ventana (in Argentina); *h*, Mascarenes (Islands of Mauritius and Réunion). Essentially diagrammatic. See text, p. 99, for elucidation.

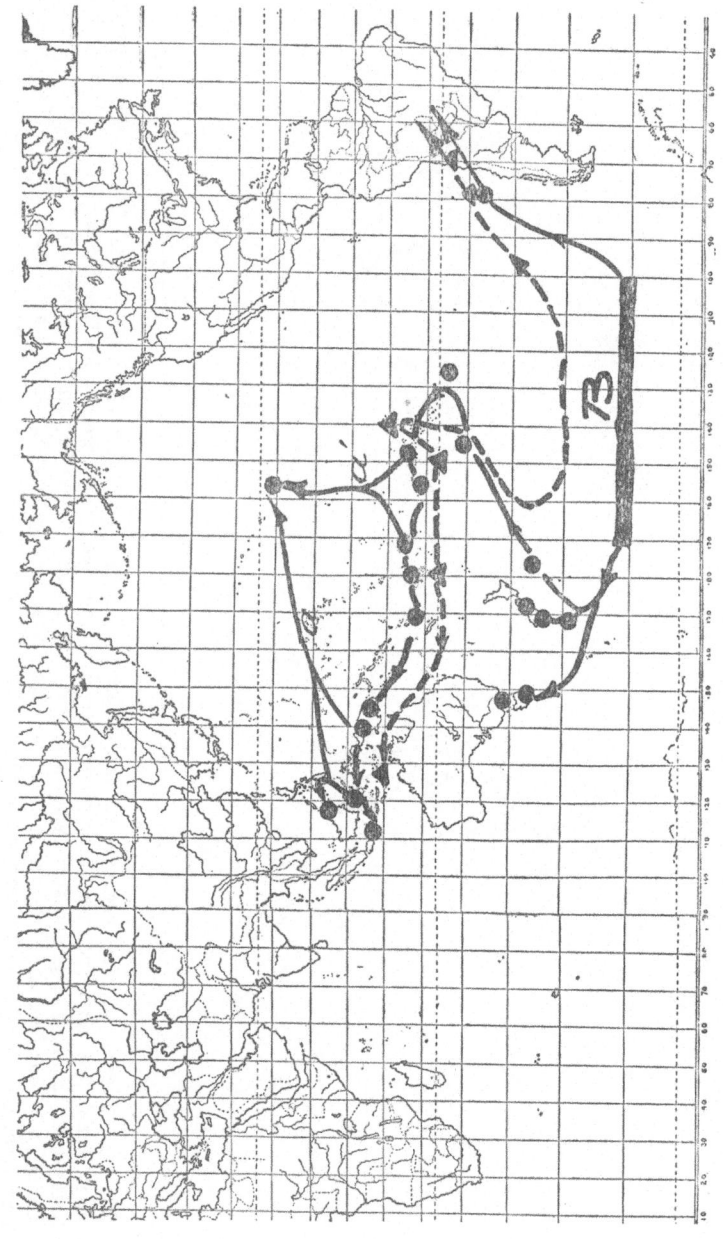

FIG. 29. Illustrative range of *Coprosma* as black circles connected by tracks as continuous lines. Tracks a, a′ are alternative. Records of *Peperomia* as black triangles similarly connected. Base-line of distribution in the South Pacific as heavy bar B. Essentially diagrammatic. See text, p. 101.

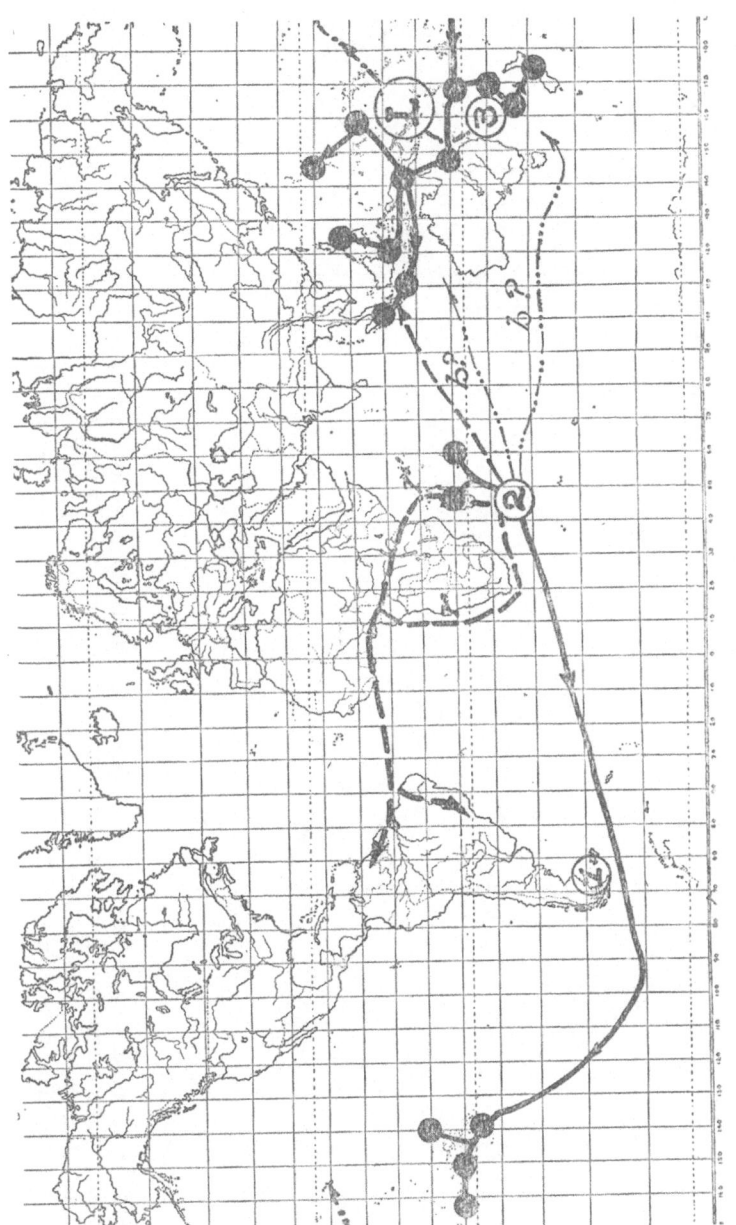

FIG. 30. Trends in the dispersal of the Loganiaceae. Range of *Geniostoma* by a selection of typical species as black circles, connected by tracks as continuous lines. Broken dotted lines *b?*, potential alternative tracks from the Indian to the Western Pacific Ocean. Distribution of the Antonieae indicated by broken lines, with alternative routes in Africa, *r*, *r′*. Center of (secondary) origin of *Labordea* as circle L with track (broken dotted line) to Hawaii. Gates of angiospermy: I Magellanian; 2 African; 3 Western Polynesian. Essentially diagrammatic. See text, p. 105, for further elucidation.

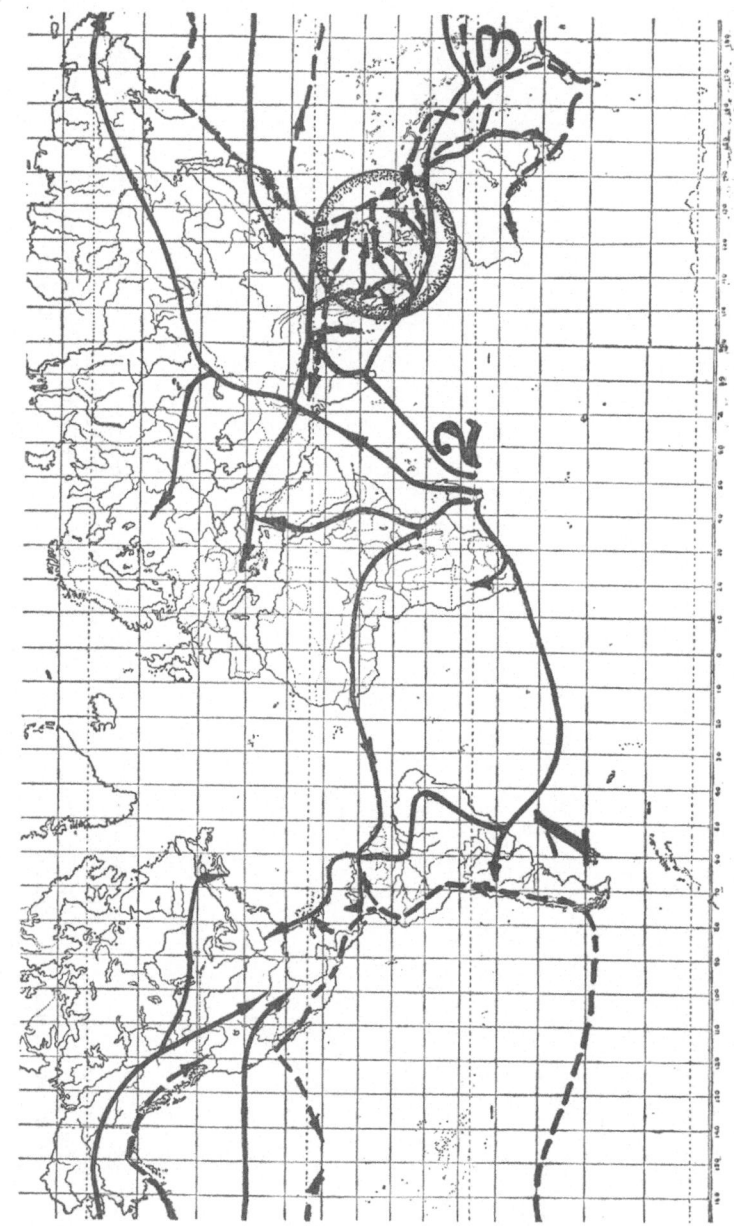

FIG. 31. Tracks in and around Malaysia, within stippled circle to illustrate impossibility of handling Malaysian phytogeography as a local problem. Tracks from the Pacific as broken lines; from Africa and the Indian Ocean as continuous lines. African gate of angiospermy in 2; Western Polynesian gate of angiospermy in 3; Magellanian gate in 1. Diagrammatic; see text for elucidation, p. 109.

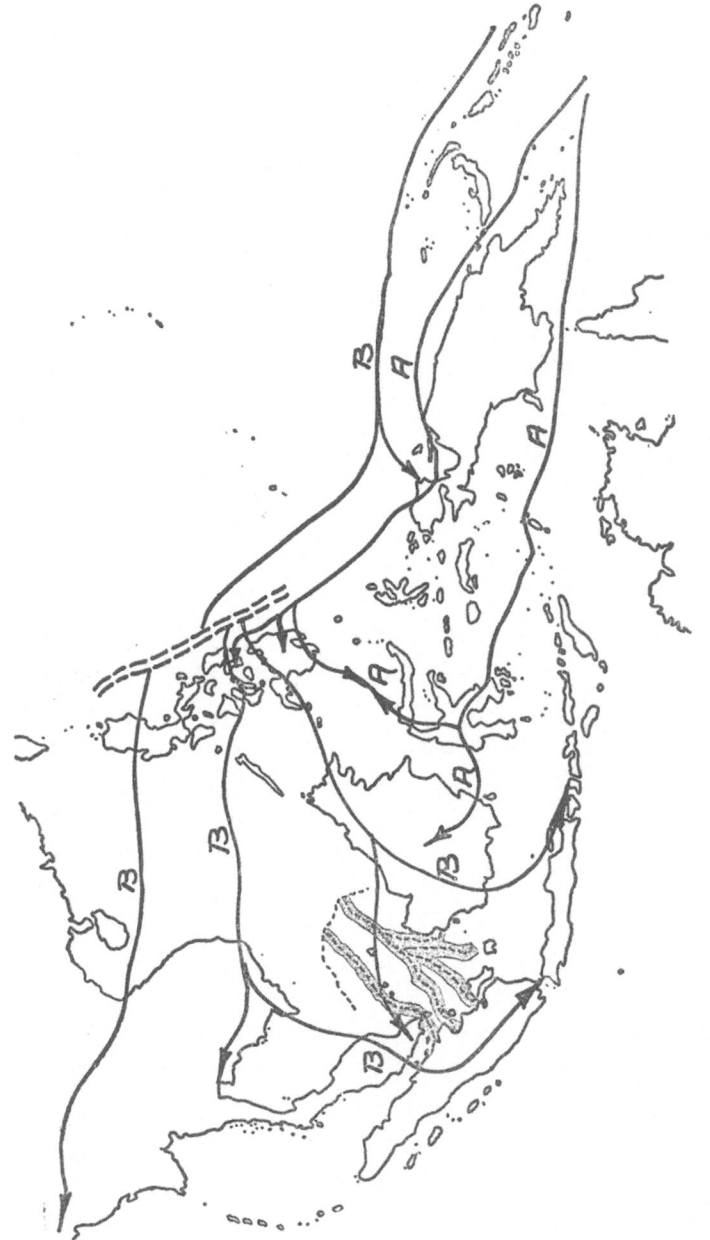

FIG. 32. Dispersal of certain Magnoliaceae around, and in, Malaysia. Tracks of *Elmerrillia* in A (the southern track possibly alternative of the northern); tracks of *Talauma* in B; base of invasion of the Philippines and the Far East by *Talauma* as heavy double broken lines. The "Molengraff River" in broken lines marginally stippled. See text, pp. 115, 117, 121, for further elucidation.

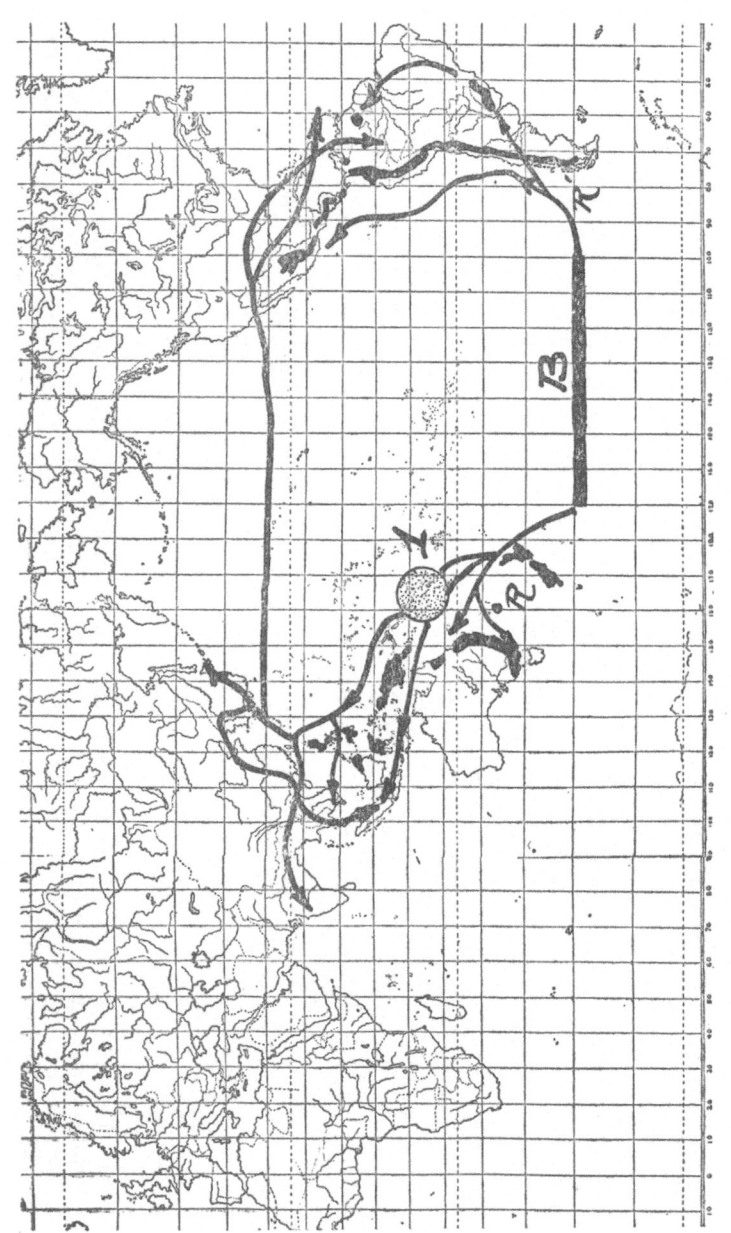

FIG. 33. Trends in the dispersal of the Magnoliaceae (and allies). Base-line of distribution in the South Pacific as bar B. Center of origin of modern family in dotted circle 1. Range of *Drimys*, heavy black. Gates of angiospermy: R Western Polynesian; R' Magellanian. Tracks as continuous lines with arrows for orientation. Essentially diagrammatic. **See text, p. 121,** for elucidation.

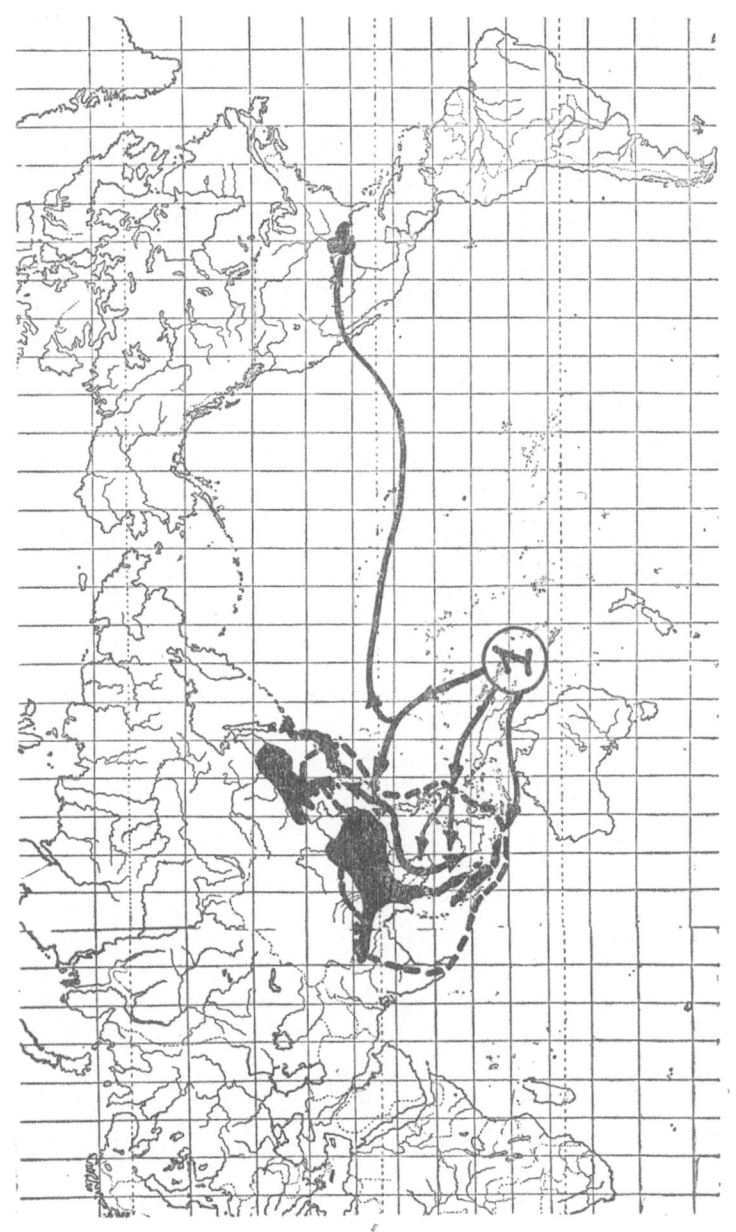

FIG. 34. Dispersal of the magnolioid *Kadsura* (range within broken line) and *Schisandra* (range heavy black). Center of origin of the modern magnolioid alliance (*Drimys* and immediately related genera perhaps excluded) in circle I. Essentially diagrammatic. See text for elucidation, p. 126 fn.

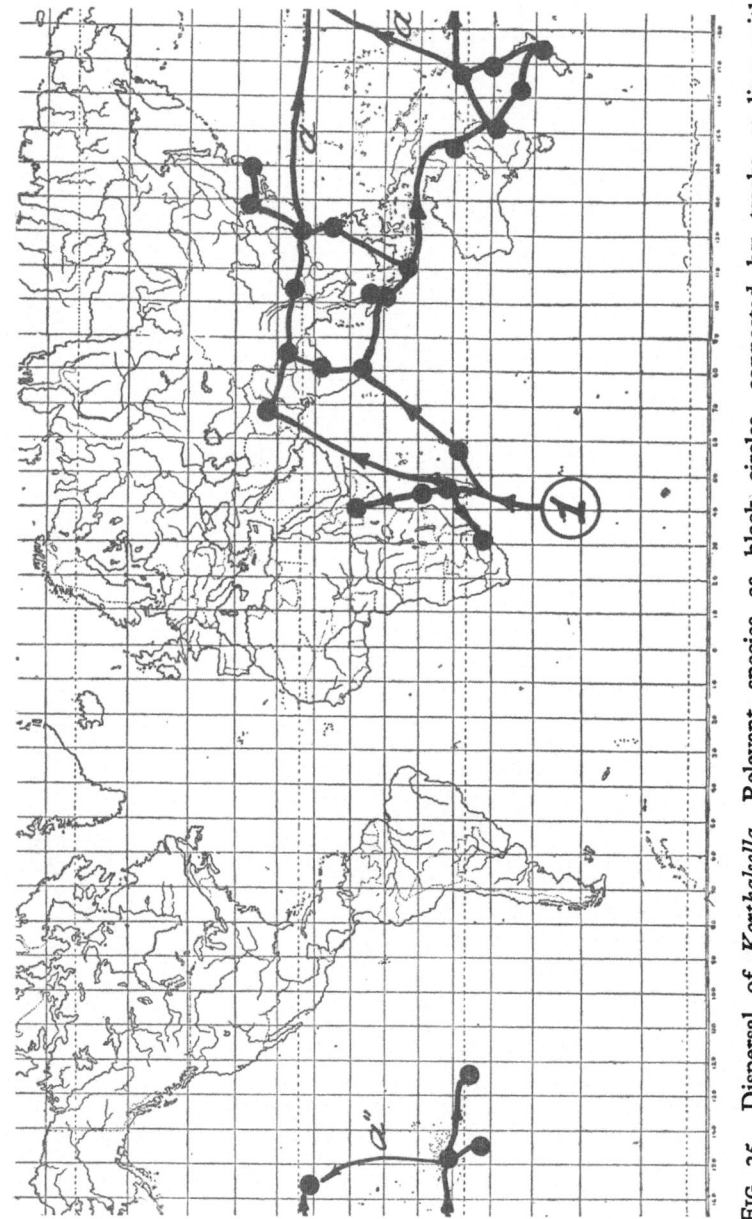

FIG. 35. Dispersal of *Korthalsella*. Relevant species as black circles, connected by tracks as lines with arrows. Main angiospermous center in circle 1. Tracks *a*, *a′*, *a″* are alternative to Hawaii. Essentially diagrammatic. See text, p. 143, for elucidation.

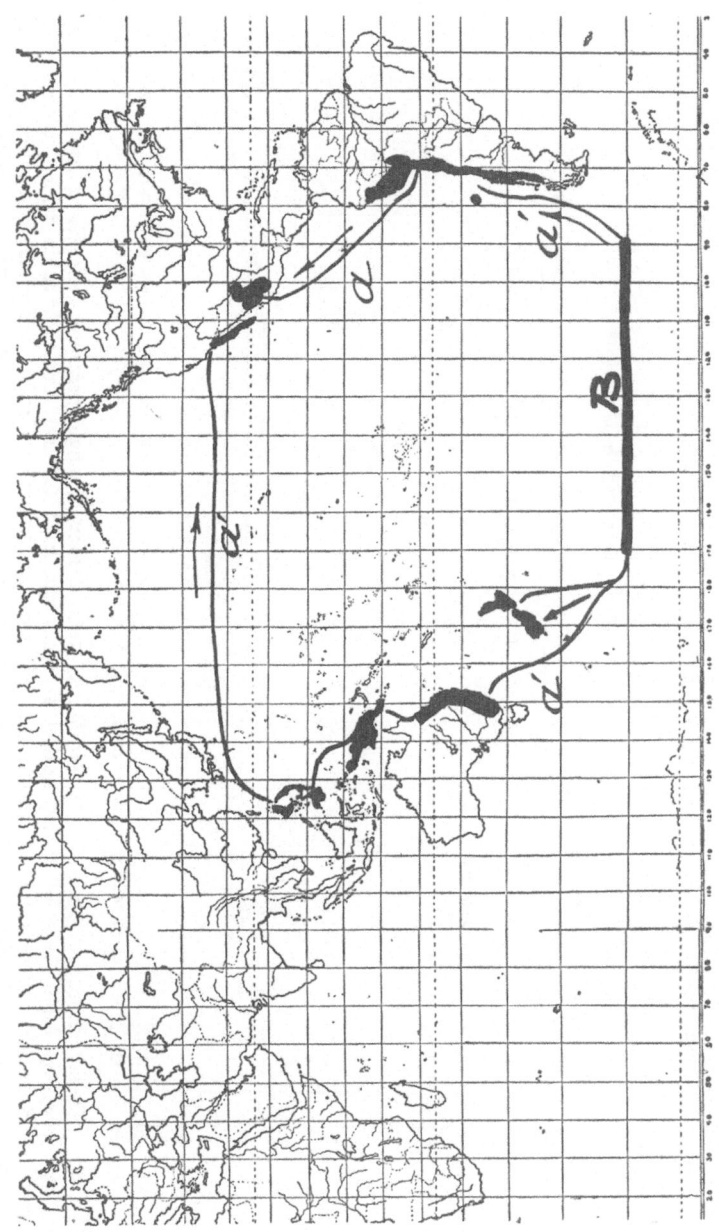

Fig. 36. Dispersal of *Phrygilanthus*. Base-line of dispersal in the South Pacific as heavy bar B. Actual range in heavy black. Essentially diagrammatic. See text, p. 146, for elucidation, including symbols.

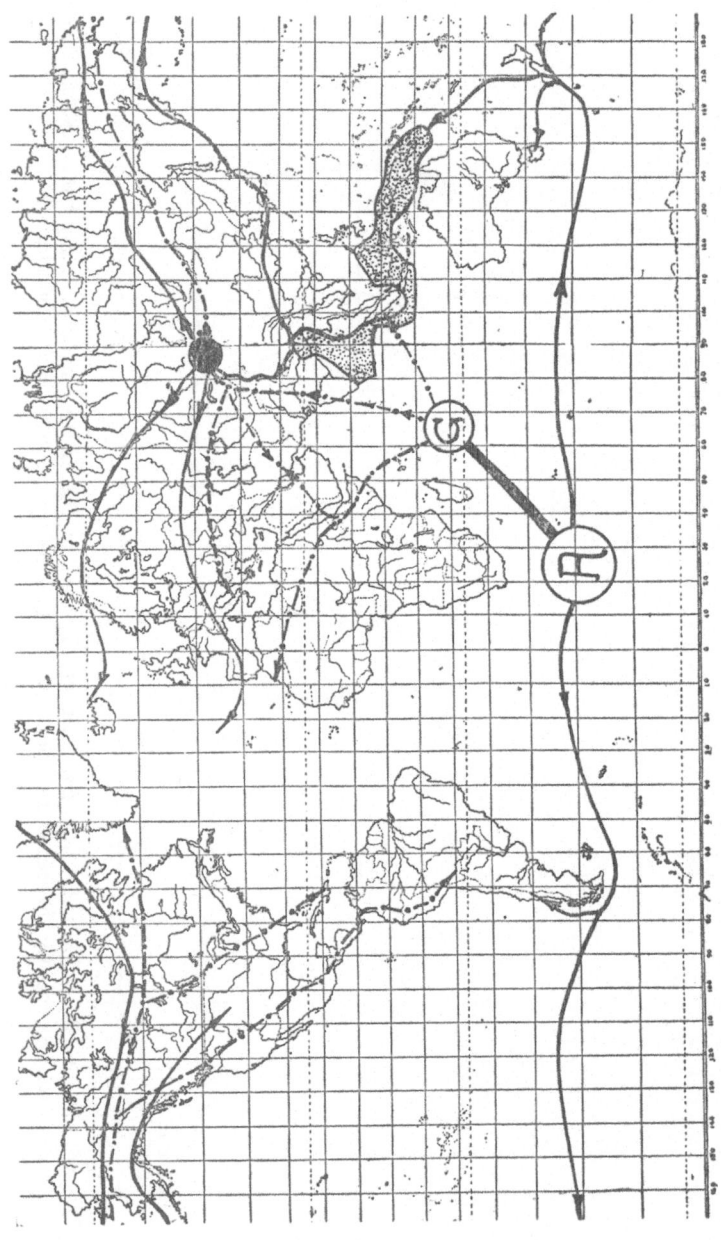

FIG. 37. Trends in the dispersal of *Potentilla*. A, main center of angiospermous *genorheitra* connected with African gate, G. Altai Node of redistribution as black circle in Central Asia. Range in which appear vicariants of *Potentilla anserina* dotted within line. Tracks of *Potentilla* Group *Anserinae* as continuous lines; of *Potentilla* Group *Sundaicae* as broken dotted line. Magellanian and Western Polynesian gates not specifically identified (see text, footnote p. 154, 217). See text, p. 154, for elucidation.

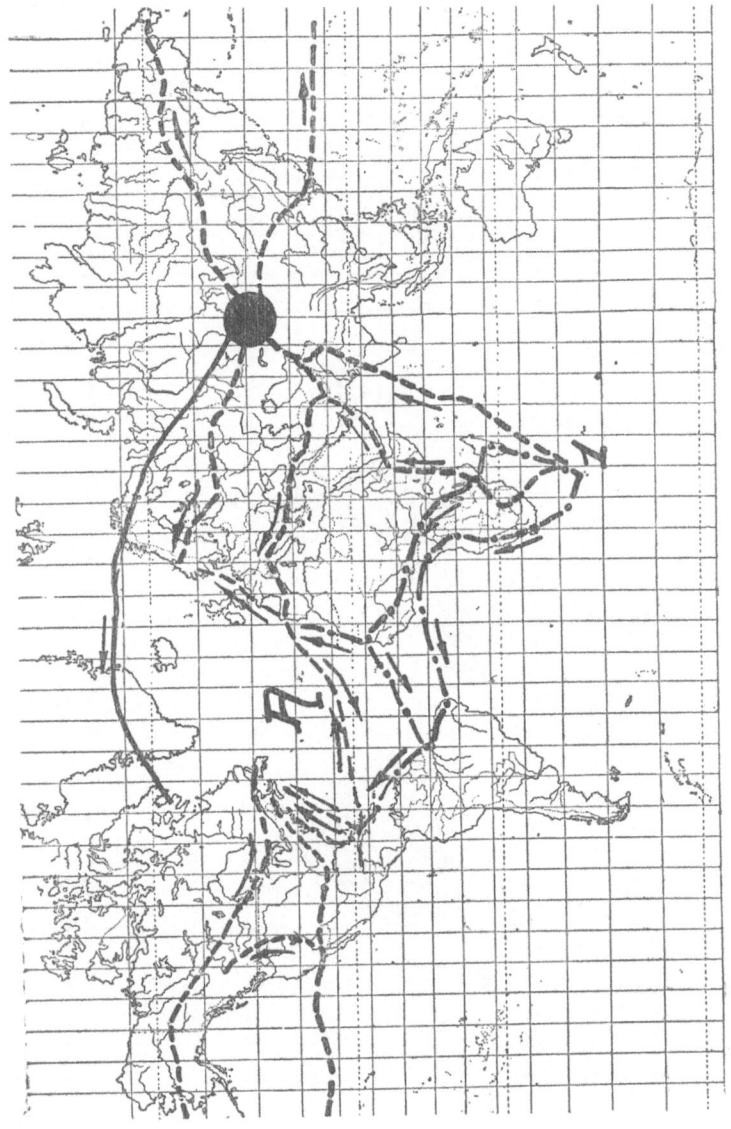

FIG. 38. Diagram to illustrate potential "Circumatlantic" distribution. Main angiospermous center, I; Atlantic Ocean, A; Altai Node, black circle. "Cold" tracks as broken lines; "warm" tracks as broken dotted lines. Strictly "polar" track in the high north as continuous line. Essentially diagrammatic. See text, p. 155, 217, for elucidation.

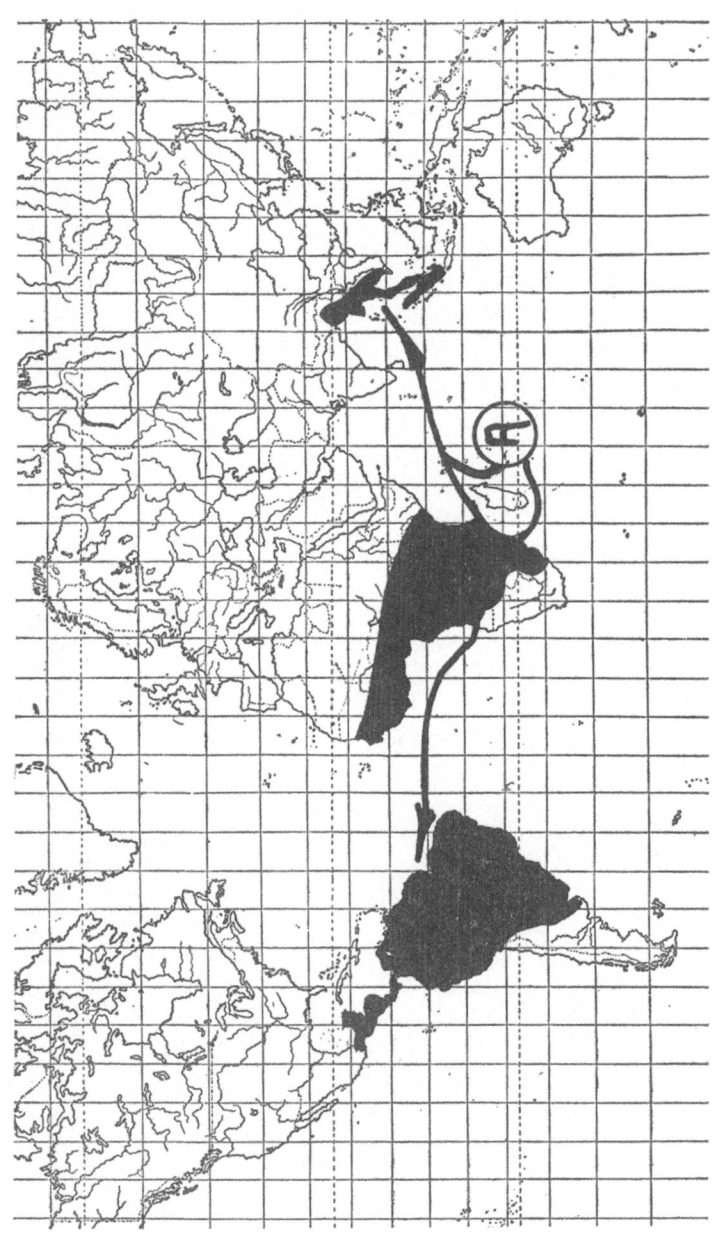

FIG. 39. Trends in the dispersal of the Primulaceae (in part). Range of *Dionysia*, 1; of *Douglasia*, 2 (in two massings in Europe and North America); of *Stimpsonia*, 3; of *Ardisiandra*, 4; of *Primula* (as to the Magellanian outlier), 5. Main tracks as dotted lines uniformly from the African angiospermous gate, G. Altai Node as black circle. Essentially diagrammatic. See text, p. 159, for further elucidation, also Fig. 17 and 40.

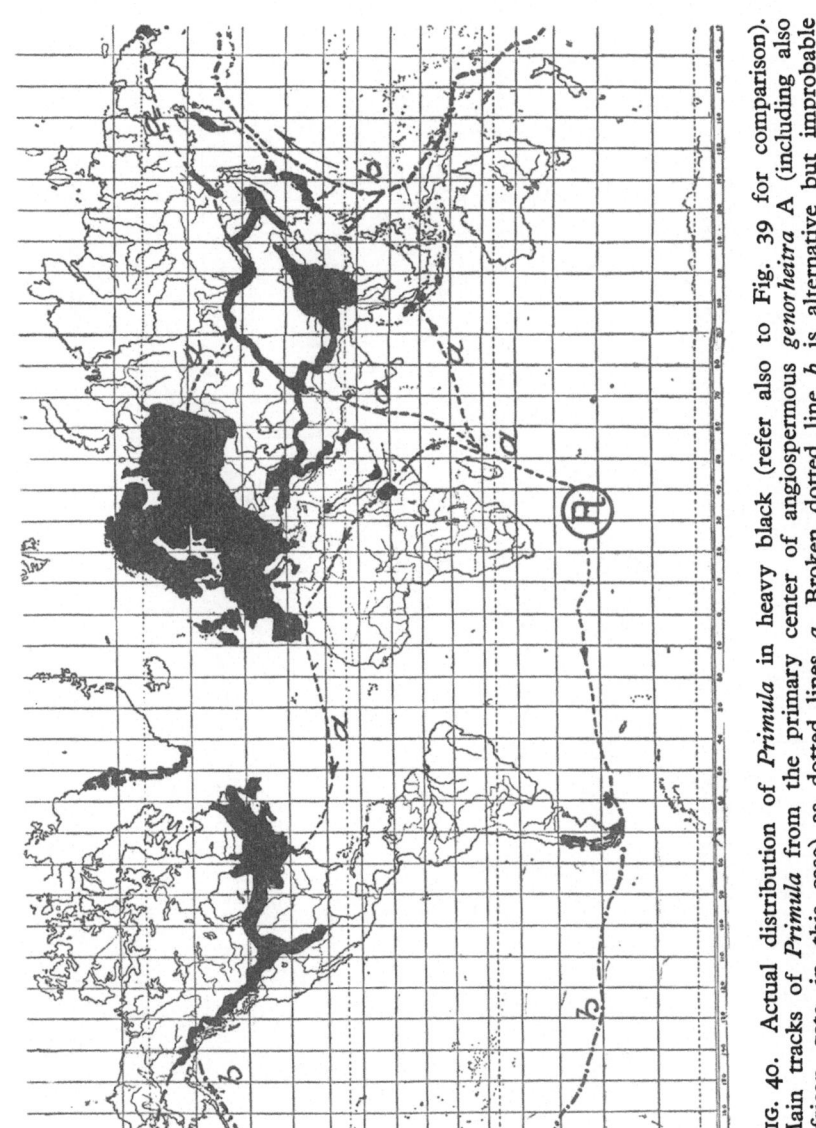

Fig. 40. Actual distribution of *Primula* in heavy black (refer also to Fig. 39 for comparison). Main tracks of *Primula* from the primary center of angiospermous *genorheitra* A (including also African gate in this case) as dotted lines *a*. Broken dotted line *b* is alternative but improbable track running in the Pacific. See text, p. 159, for elucidation.

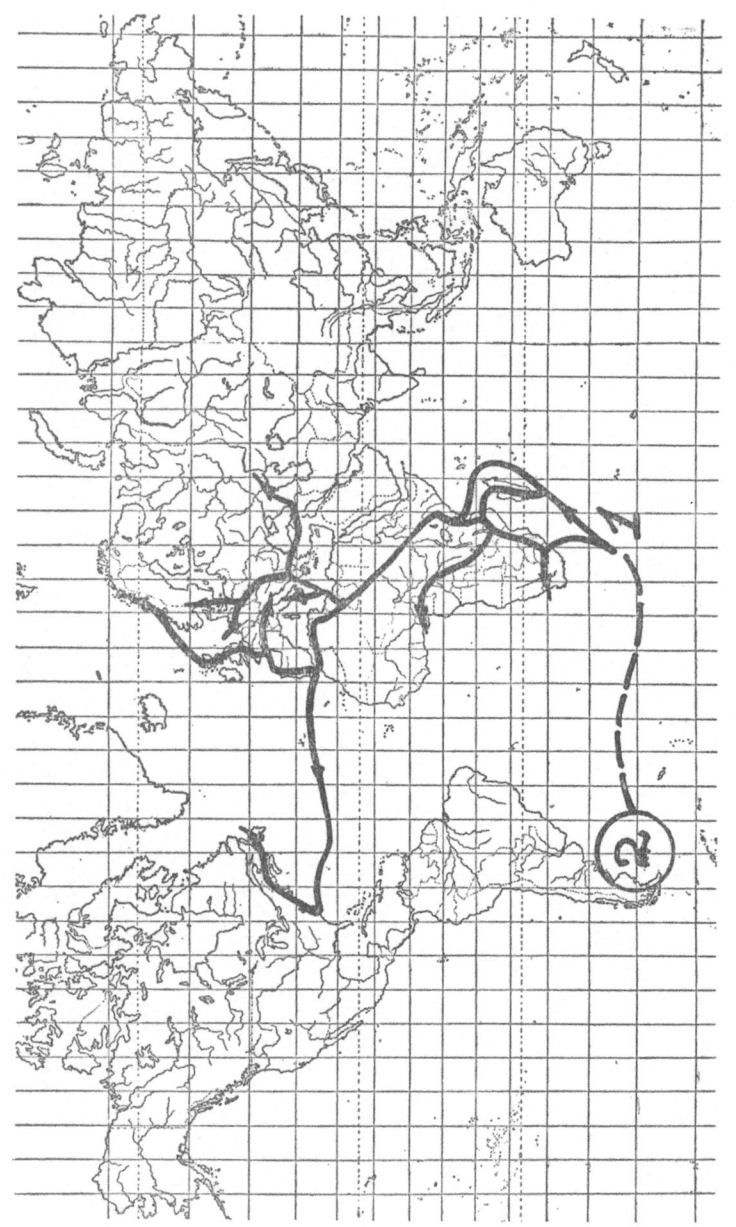

FIG. 41. Trends in the dispersal of the Ericaceae Ericoideae. Main angiospermous center, 1. Magellanian gate, 2. Broken line in the South Atlantic indicative of the connection between Ericaceae Ericoideae, in particular, and Empetraceae. Essentially diagrammatic. See text, p. 160, for elucidation.

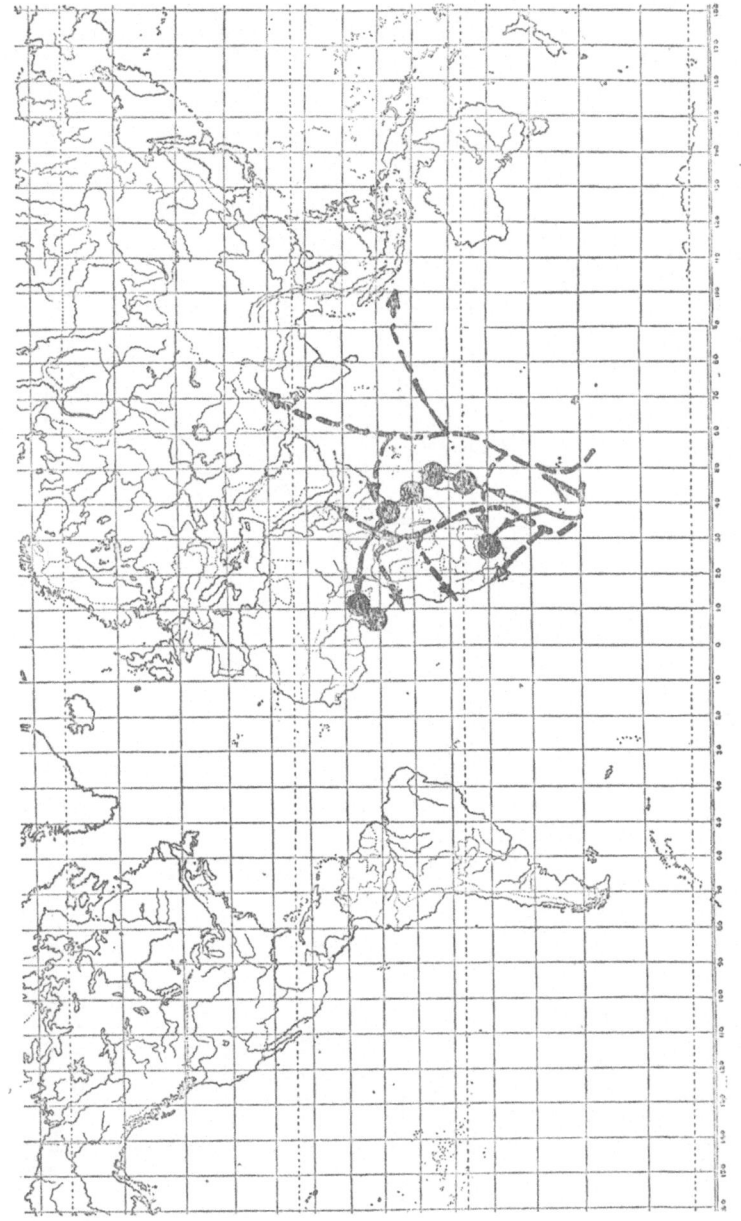

FIG. 42. Diagram to show how Africa is invaded from an easterly direction. Main angiospermous center (i.e., range approximately surrounding the *Afroantarctic* and *Gondwanic Triangles*: see Fig. 67) in 1. Black circles, species of *Philippia* in the affinity of *P. Mannii*. "Cold" tracks in broken lines; "warm" in continuous lines. Essentially diagrammatic. See text, p. 161, for elucidation.

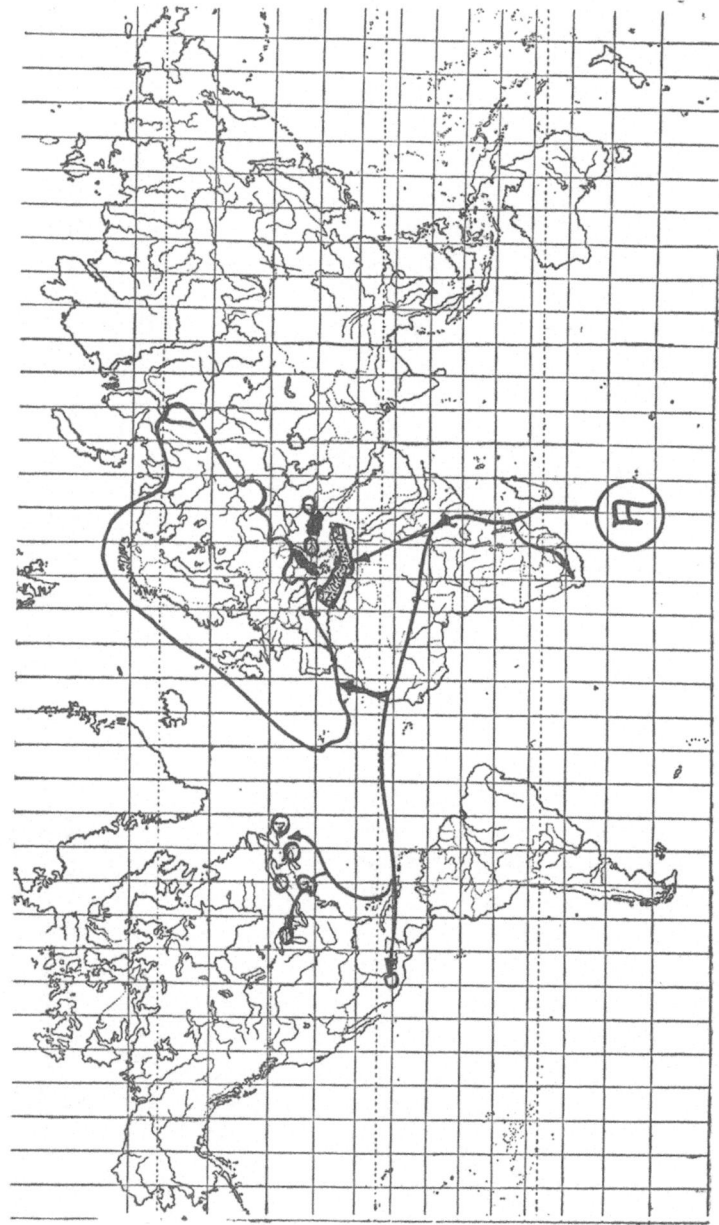

Fig. 43. Actual range of *Calluna* within black line. Ranges of *Erica* subg. *Pentaptera* dotted; of *Bruckenthalia* heavy black. Primary center of angiospermous *genorheitra* in A. Tracks as lines with arrows, indicating direction of distributional flow. See text, p. 166, for elucidation.

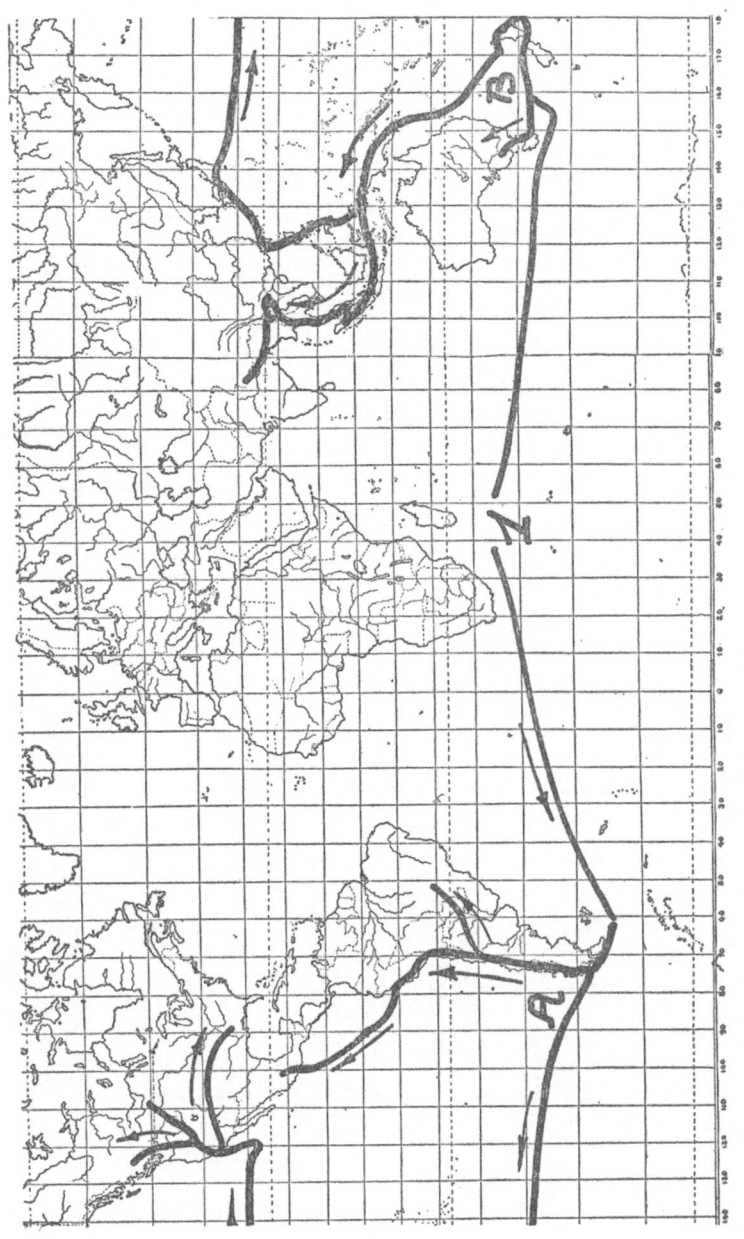

Fig. 44. Trends in the dispersal of the Ericaceae Gaultherioideae. Main angiospermous center, 1. Gates of angiospermy: A, Magellanian; B, Western Polynesian. Essentially diagrammatic. See text, p. 167, for elucidation.

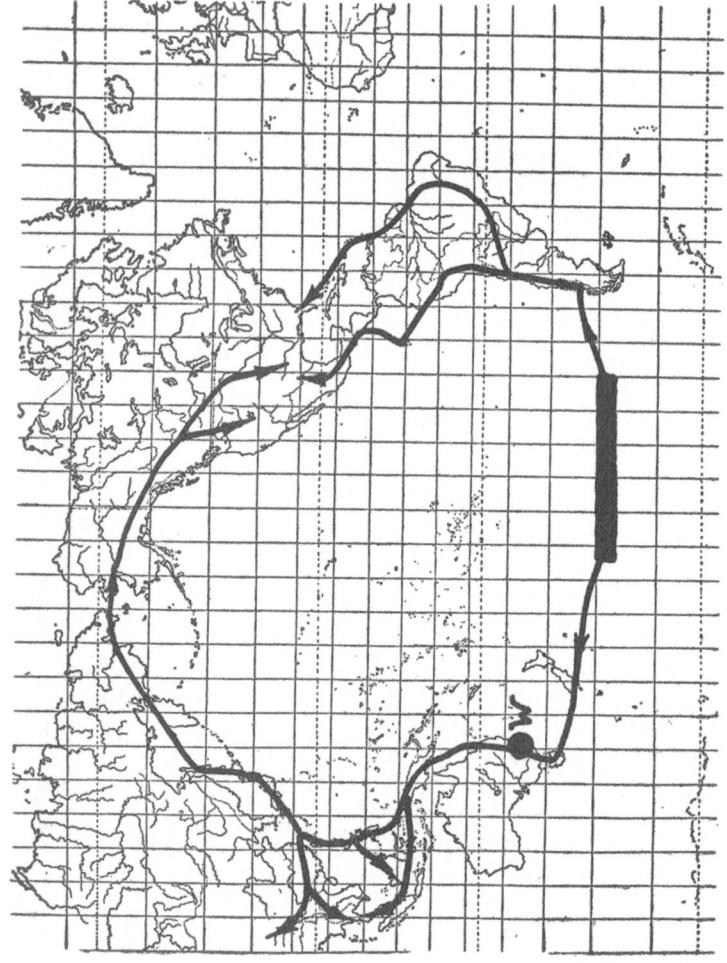

Fig. 44A. Tracks (diagrammatic) of the Ericaceae Gaultherioideae. Base-line of dispersal as heavy bar in the Southern Pacific. Range of *Wittsteinia*, black circle W.
See text for elucidation, p. 167. This map is submitted for the sake of comparison
with Fig. 44.

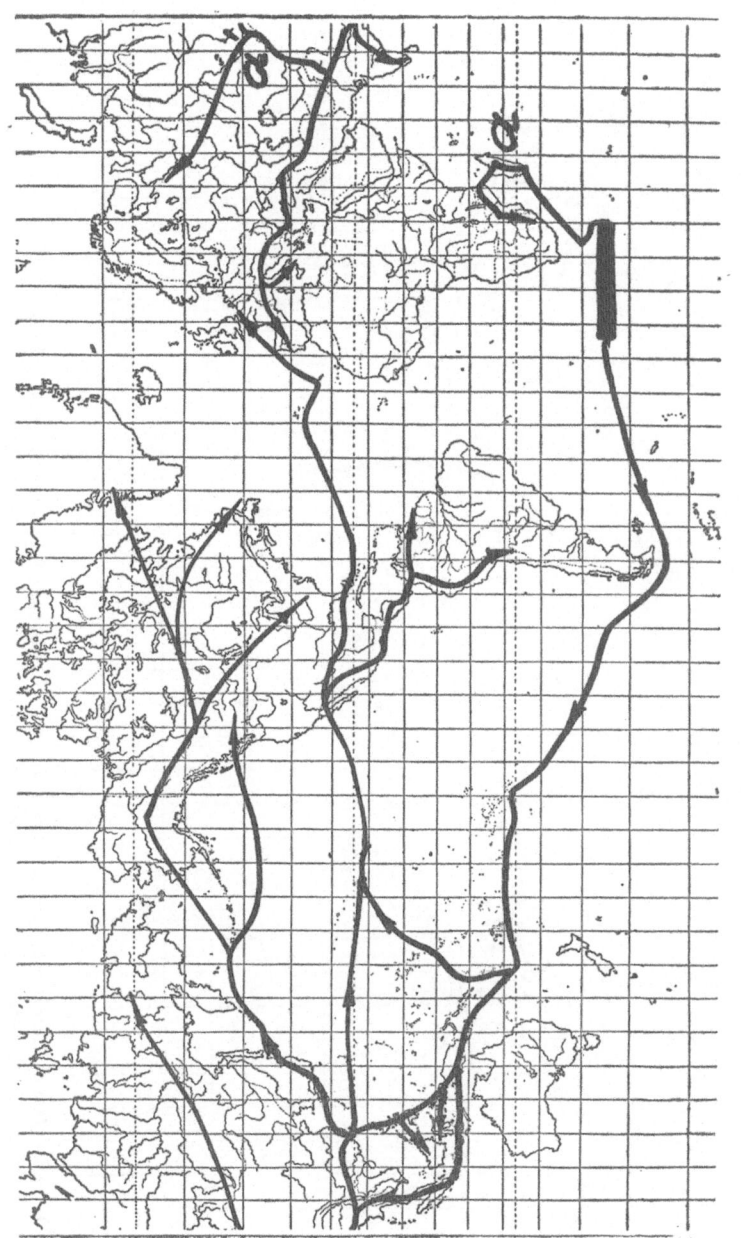

FIG. 45. Main tracks of the Ericaceae Vaccinioideae. Base-line of dispersal as heavy bar south of Africa. South African outlier in *a*: Altai center of dispersal in *a'*. See text for elucidation, p. 169.

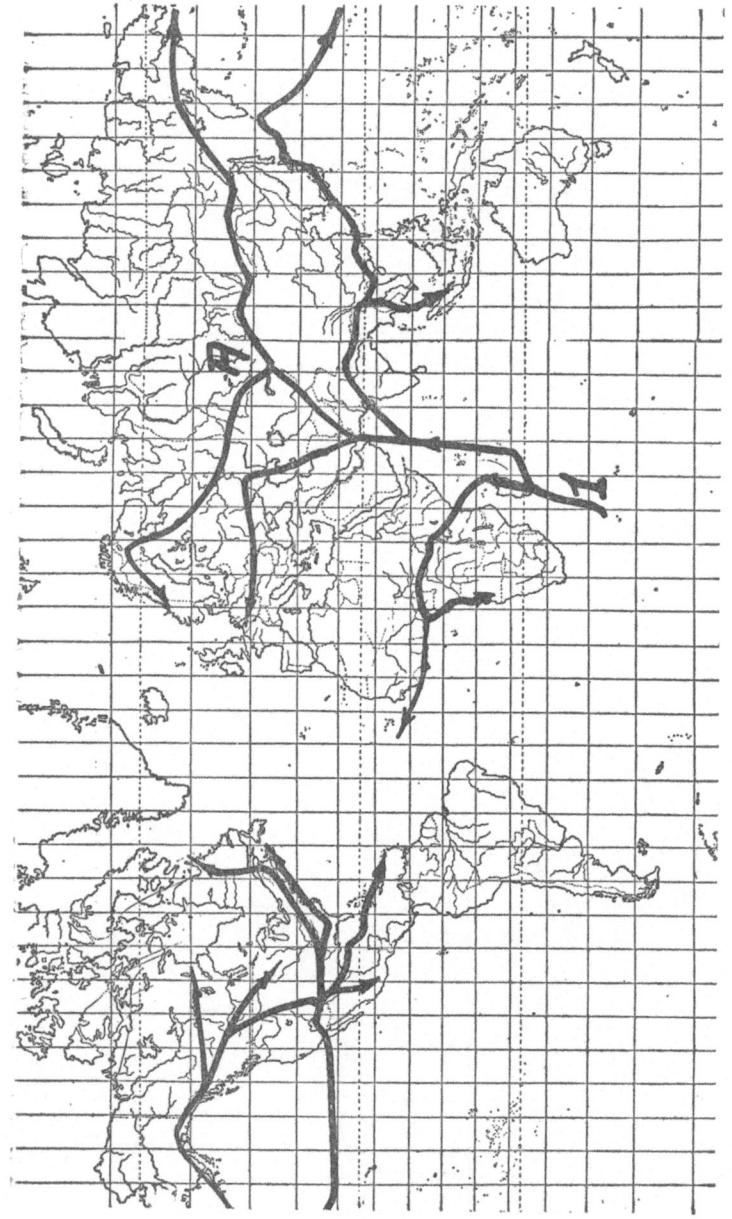

FIG. 46. Main trends in the dispersal of the Ericaceae Andromedoideae. Main angiospermous center, 1; Altai Node, A. Essentially diagrammatic. See text, p. 171, for elucidation.

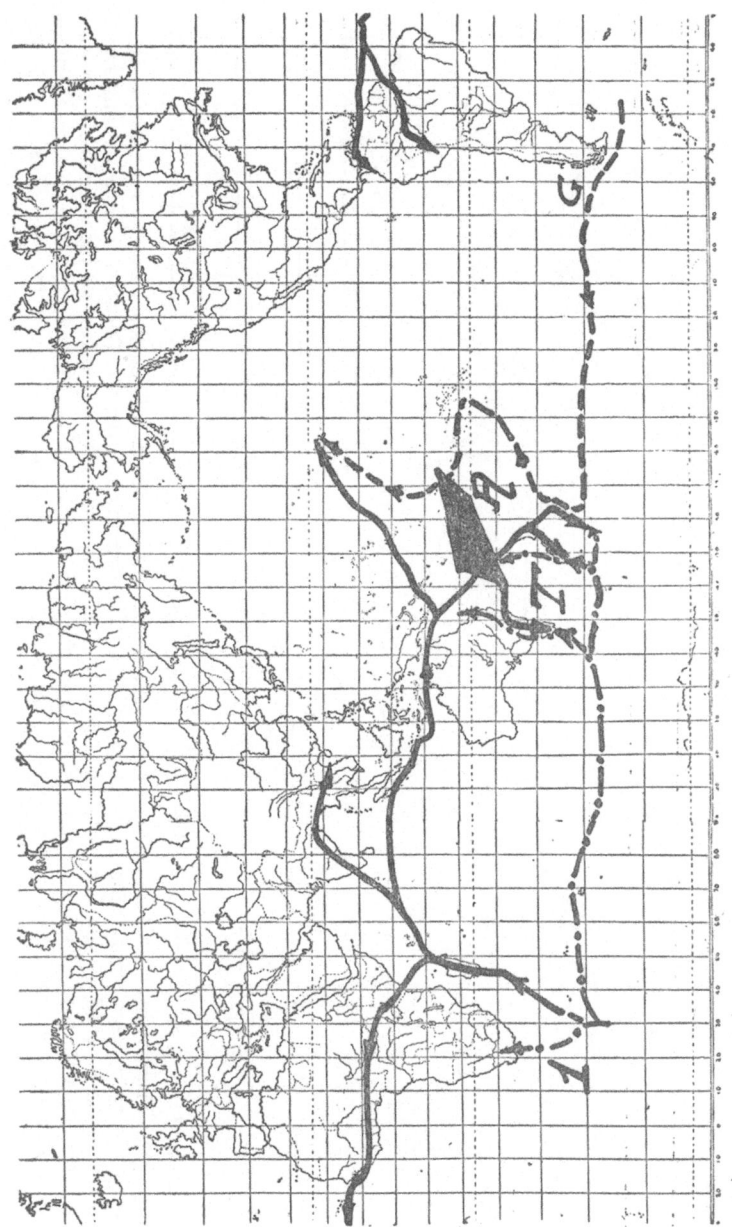

FIG. 47. Diagram to illustrate tracks and centers in the Western Pacific. Main angiospermous center, I; Magellanian gate, G. Tracks out the main angiospermous center as lines (continuous, if "warm," broken and dotted if "cold") running to a Western Polynesian center, A. Tasman Sea in T. Track in the Southeast and South Central Pacific to Eastern Polynesia and Hawaii as broken line. Essentially diagrammatic. See text, p. 184, for elucidation.

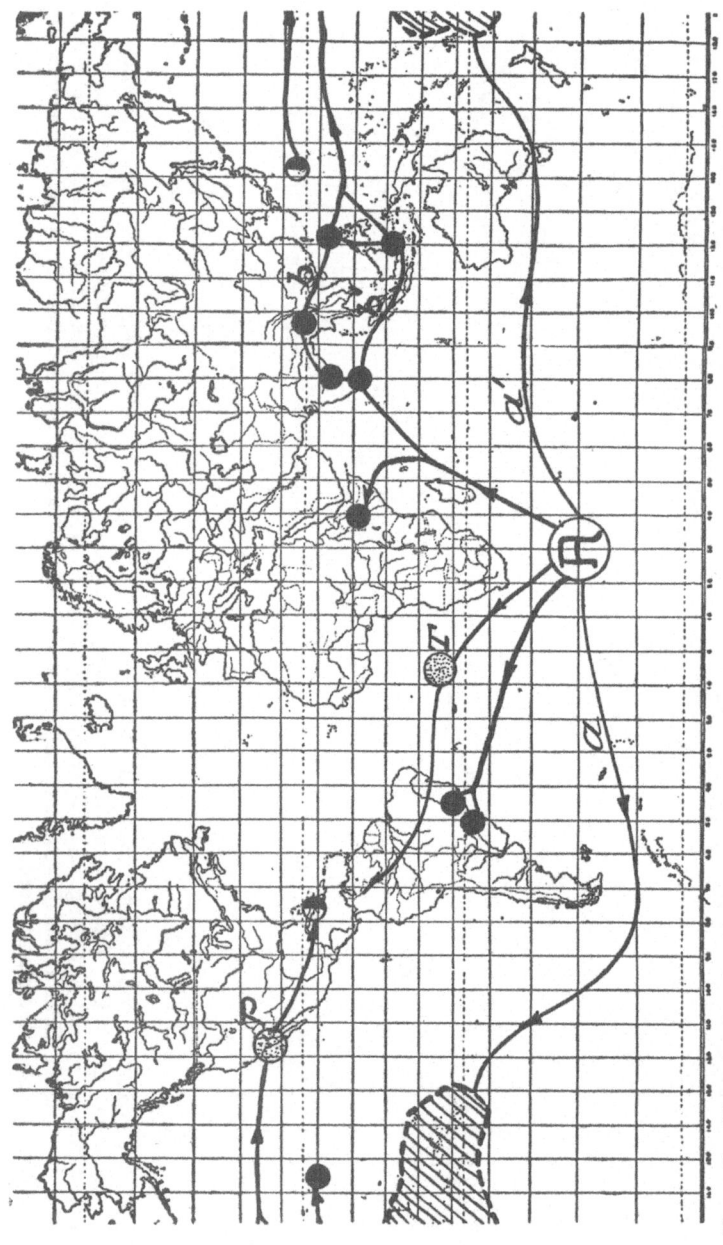

FIG. 48. Trends in the dispersal of the Lobeliaceae. Main center of angiospermous genorheitra in A, connected by conventionally antarctic tracks a, a' (highly diagrammatic) with the Polynesian range of controversial genera (cross-hatched within broken line). Species of *Lobelia* sect. *Rhyncopetalum* as black circles, with tracks as continuous lines to South America, Africa and the Far East including Hawaii; tracks b, b' are alternative or complementary, not necessarily exclusive. As dotted circles; range of *Trimeris* in Saint-Helena (T), with track pointing to South America (affinities toward *Siphocampylos*); range of *Palmerella* in Southern California (P). Species of *Lobelia* Sect. *Tylomium* as circle with black to right. See text, p. 209, for elucidation.

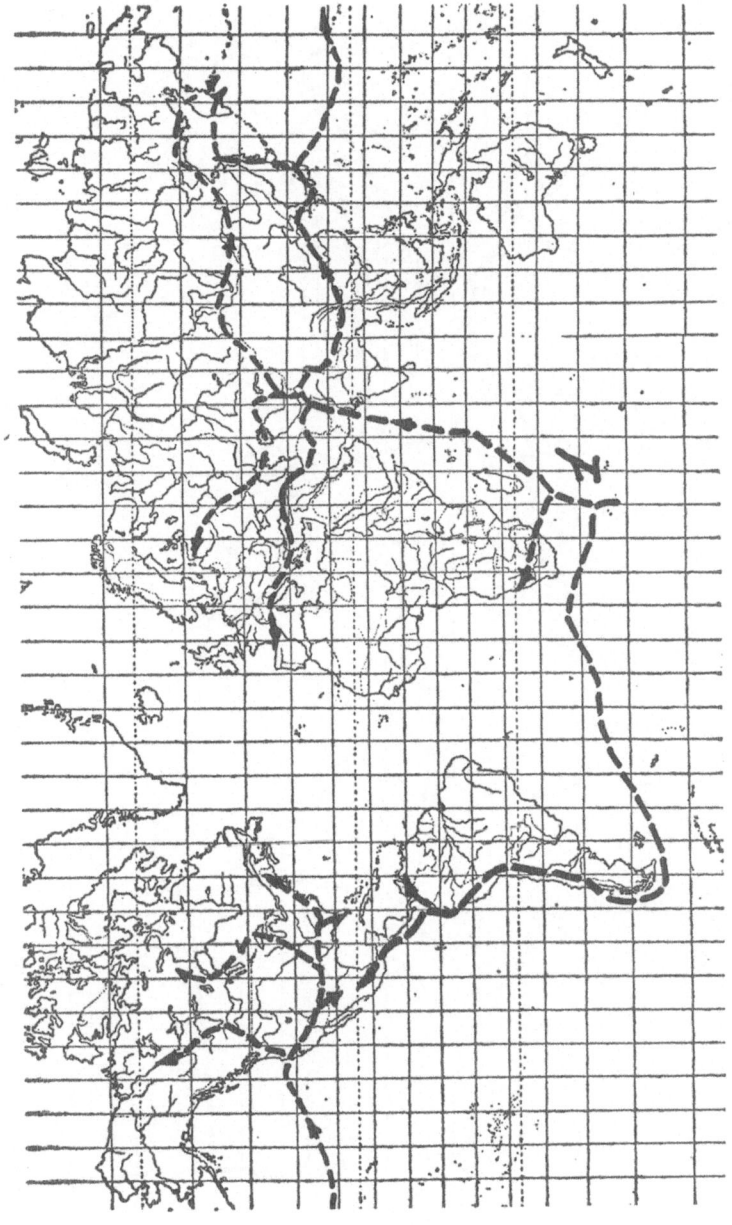

FIG. 49. Trends in the dispersal of *Alnus*. Main angiospermous center, A. Essentially diagrammatic. See text, p. 213, for elucidation.

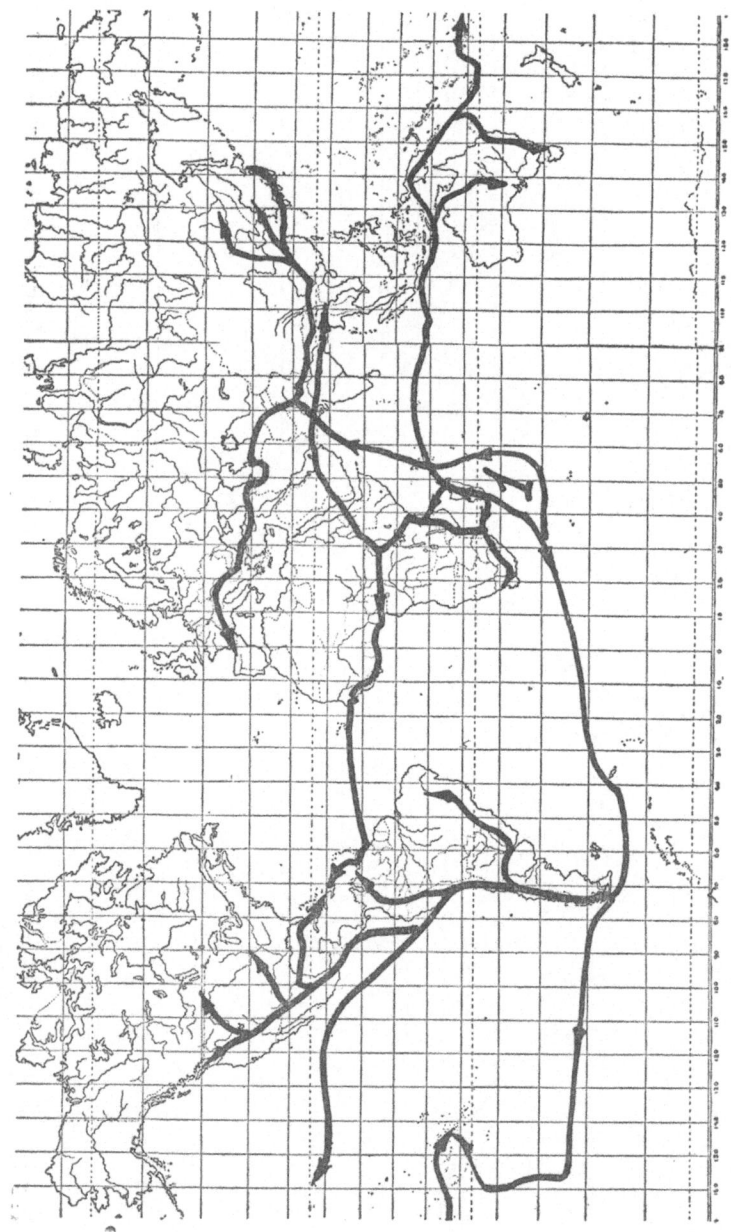

FIG. 50. Trends in the dispersal of *Cuscuta*. Main angiospermous center, 1. Essentially diagrammatic. See text, p. 223, for elucidation.

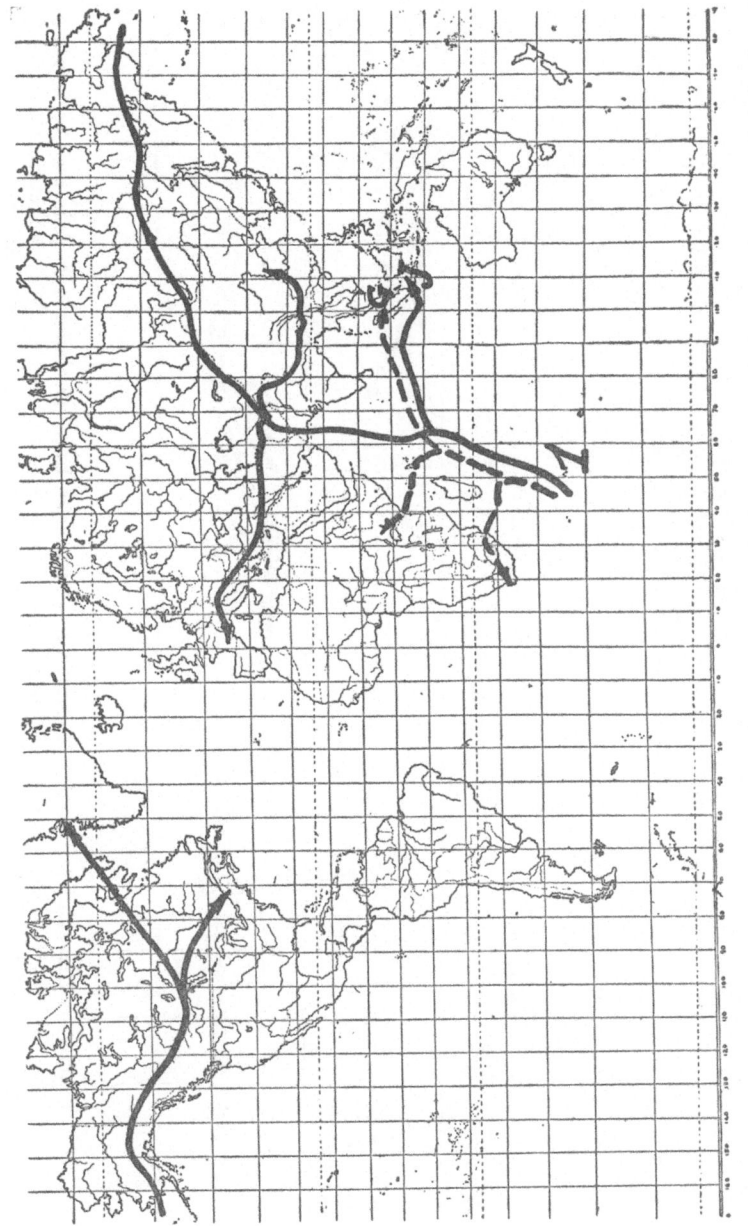

FIG. 51. Dispersal of the *Schoenoxiphium-Cobresia* affinity. Main angiospermous center, I; Gajolands of Northern Sumatra, G; Java, J. Broken line, tracks of *Schoenoxiphium*; continuous, of *Cobresia*. Essentially diagrammatic. See text, p. 241, for elucidation.

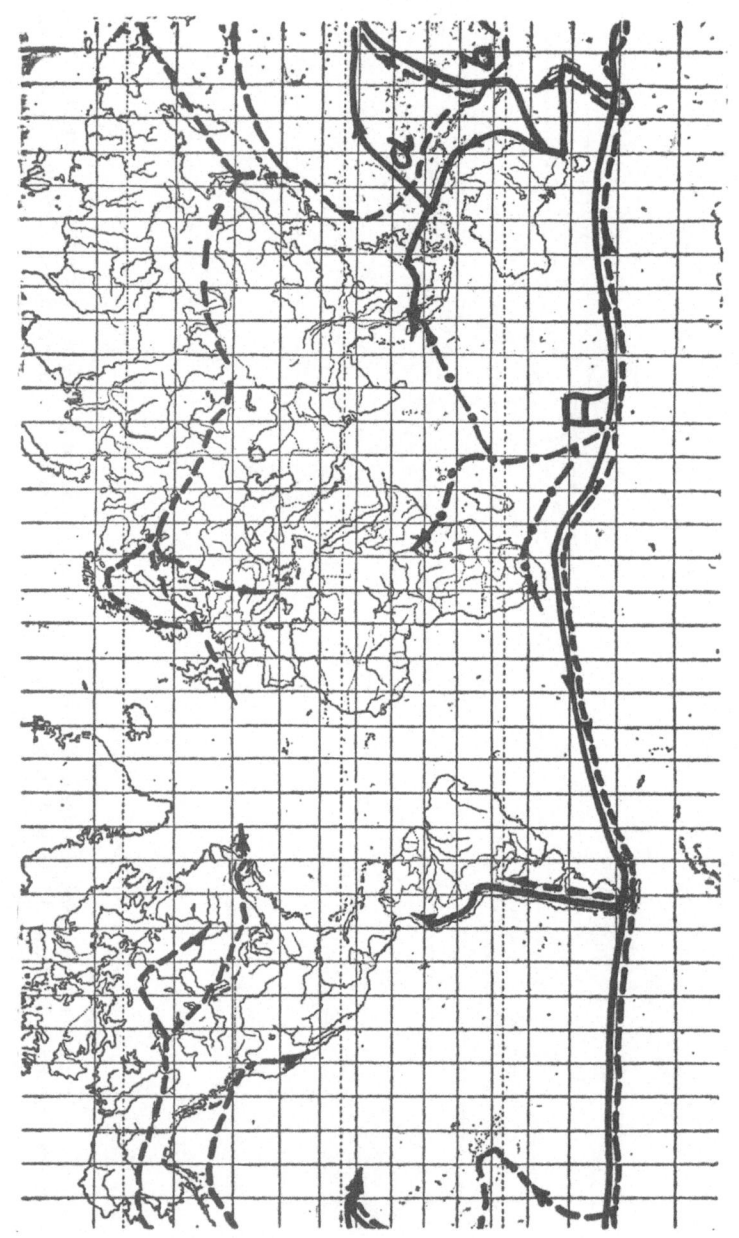

FIG. 52. Trends in the dispersal of the Cyperaceae. Continuous lines, *Oreolobus* and *Schoenus*; broken lines, *Carex* Sect. *Acutae* Subsect. *Cryptocarpae*; broken dotted lines, *Schoenoxiphium*. Main angiospermous center, A. Tracks *a, b* are alternative to Hawaii. Essentially diagrammatic. See text, p. 243, 246, 249, for elucidation.

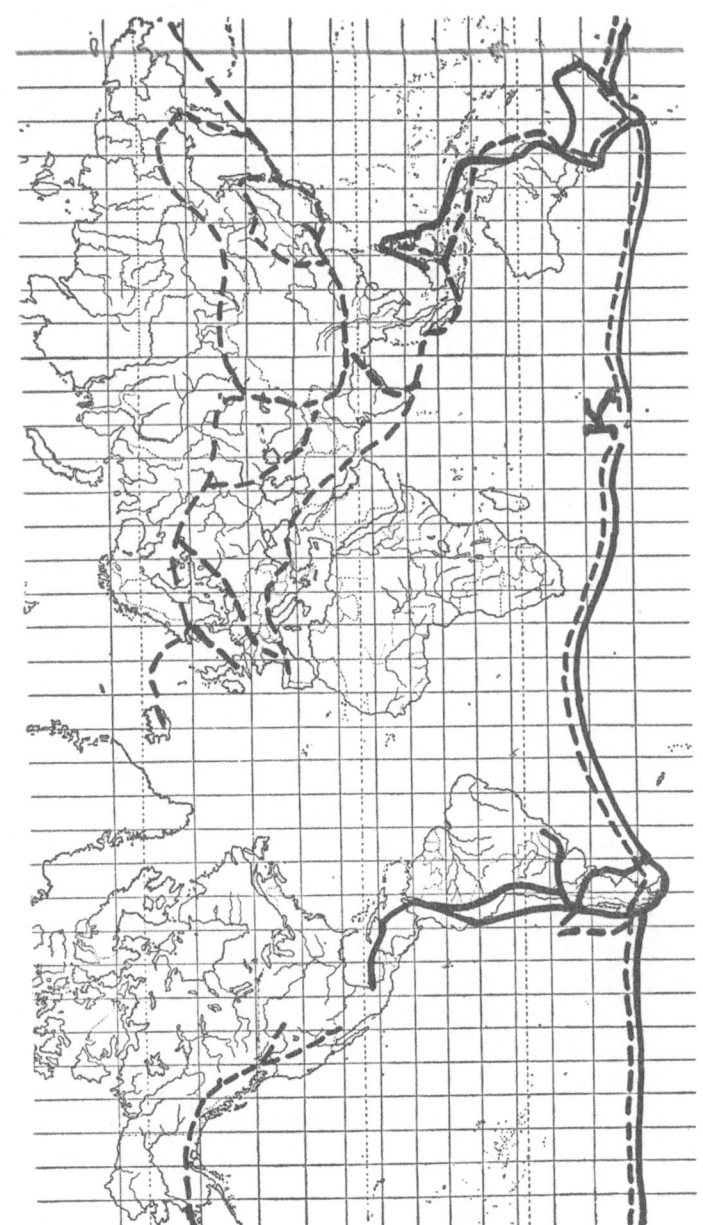

FIG. 53. Trends in the dispersal of the Cyperaceae (continuous lines, *Uncinia*; broken lines, *Carex* Sect. *Unciniaeformes*). Main center (agreeing this time fairly closely with the modern Kerguelen Islands) in K. Essentially diagrammatic. See text, p. 249, 250, for elucidation.

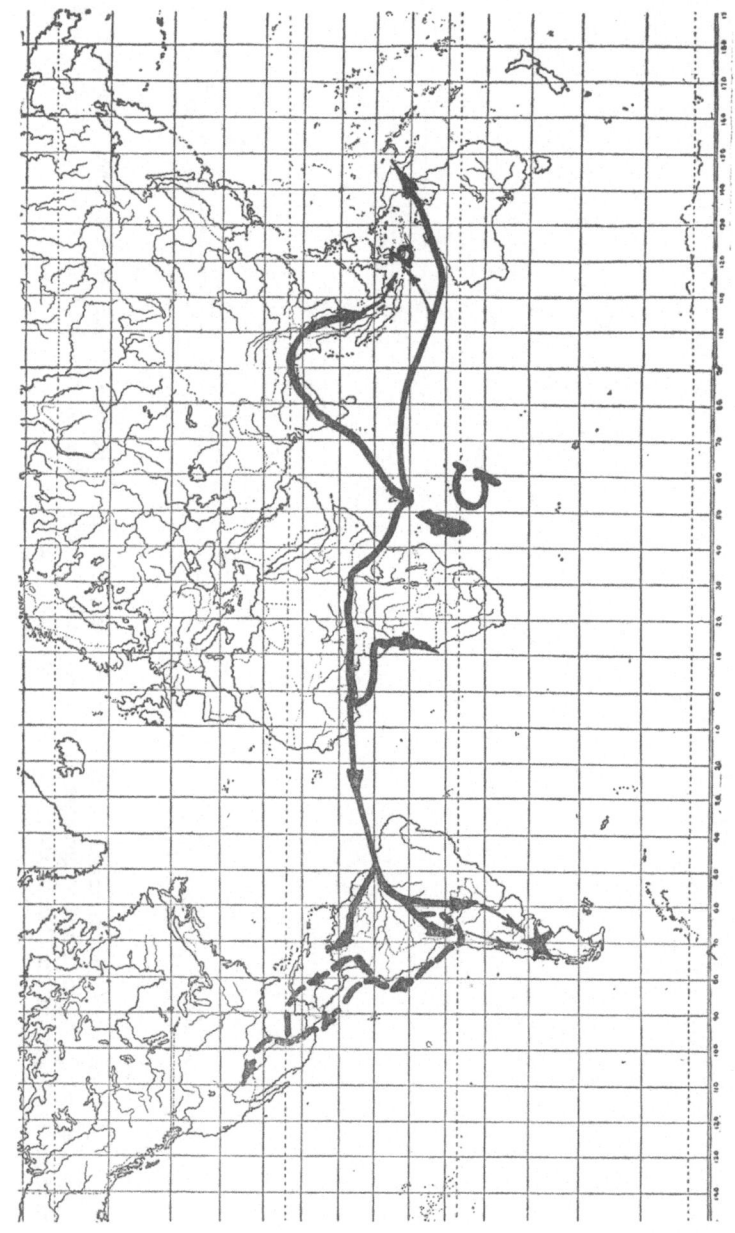

FIG. 54. Trends in the dispersal of the Cochlospermaceae. African gate of angiospermy, G. Range of *Sphaerosepalum* (Madagascar) in heavy black. Tracks of *Cochlospermum* as continuous lines; tracks of *Amoureuxia* as broken lines. Island of Bali in *b*; fossil record of "*Cochlospermum praevitifolium*" Berry, by star in South Argentine. Essentially diagrammatic. See text, p. 258, 261, for elucidation.

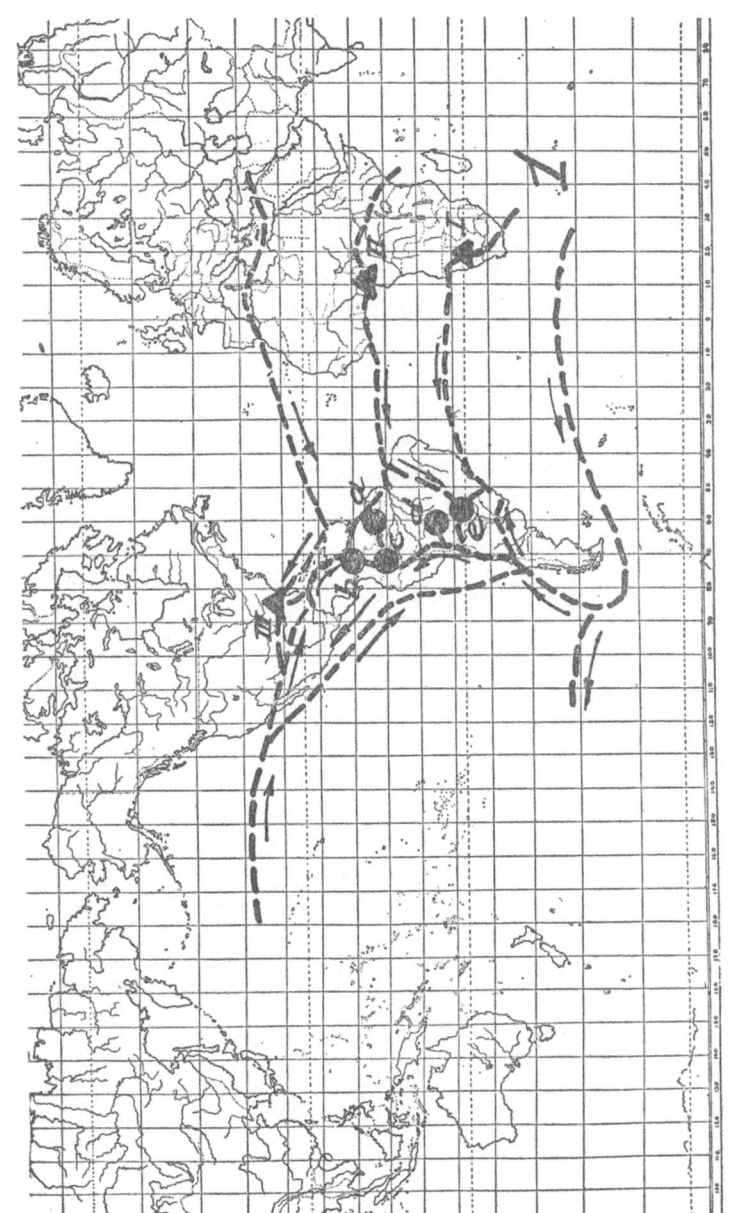

Fig. 55. Tracks (partial) running to the Americas. Main angiospermous center, 1. Secondary centers of origin as triangles, I *Kalaharian:* II *Nigerian:* III *Ozarkian-Appalachian* (see p. 357). Ancient pre-Andean orogenies and heigts as black circles, a (Mt. Duida, Mr. Auyan-Tepui, Mt. Roraima, Kaieteur Savanna, Tafelberg), b (Sierra Nevada de Santa Marta, Sierra de Perijá), c (Serrania de Chiribiquete, Cerro de Aracoara, Cerro de Cupaty), d (Sierra do Parecis, Sierra de Chiquitos), e (Sierra de Amambay). Essentially diagrammatic. See text, p. 266, 278, for elucidation.

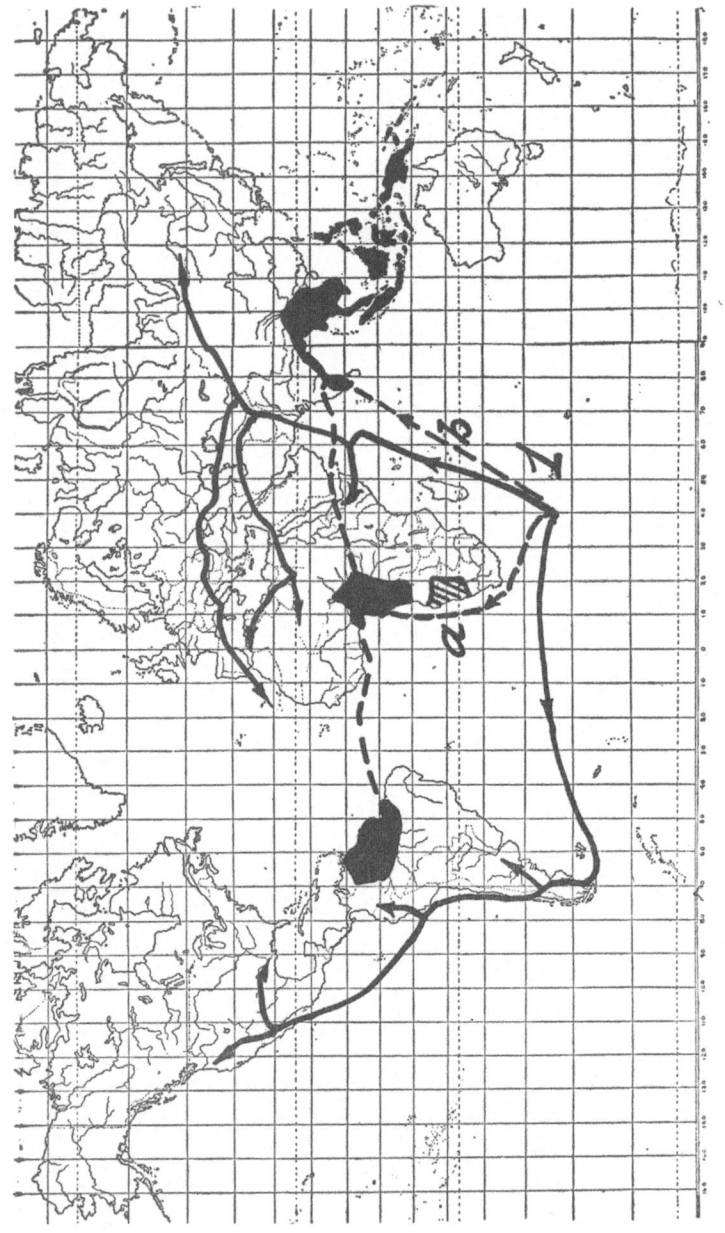

FIG. 56. Trends in the dispersal of the Ephedraceae. Range of *Gnetum*, heavy black; range of *Welwitschia*, cross-hatched. Main angiospermous center, 1. Tracks of *Ephedra* as continuous lines; tracks of *Gnetum* as broken lines (tracks *a*, *b* are complementary). Essentially diagrammatic. See text, p. 275, 279, for elucidation.

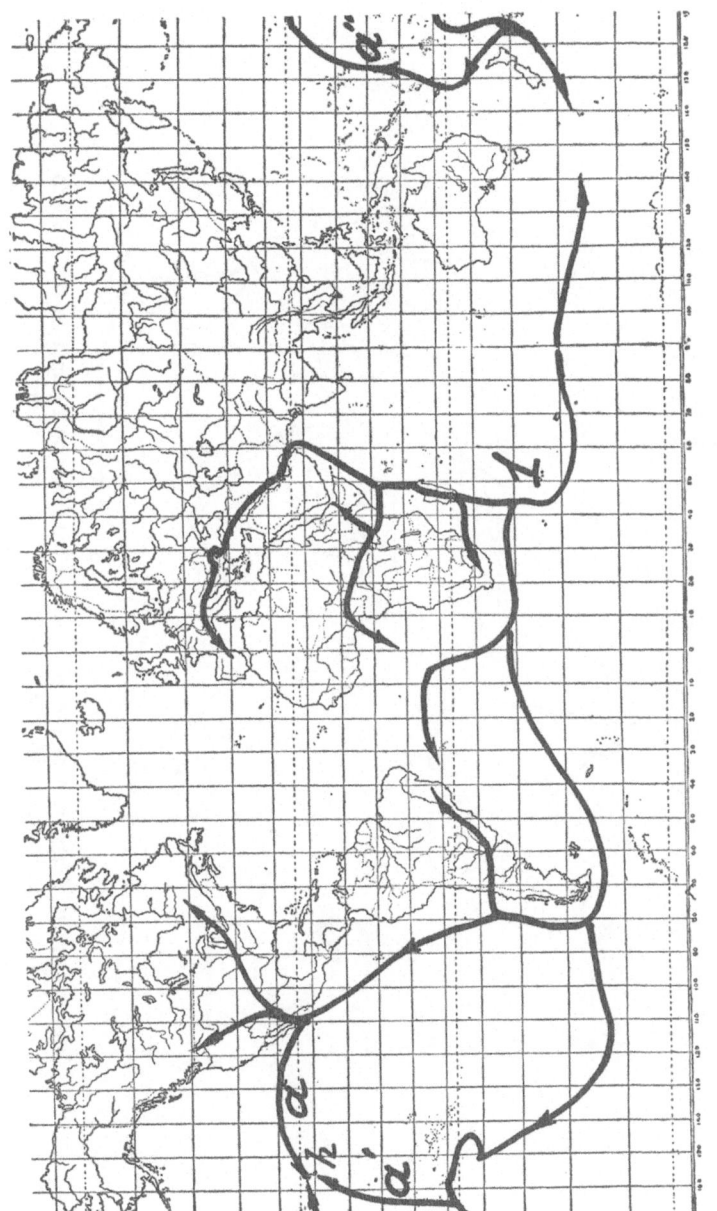

FIG. 57. Trends in the dispersal (in *h*) of *Plantago* Sect. *Palaeopsyllium*. Main angiospermous center, I. Tracks *a*, *a'*, *a"* are alternative or complementary to Hawaii. Essentially diagrammatic. See text, p. 284, for elucidation.

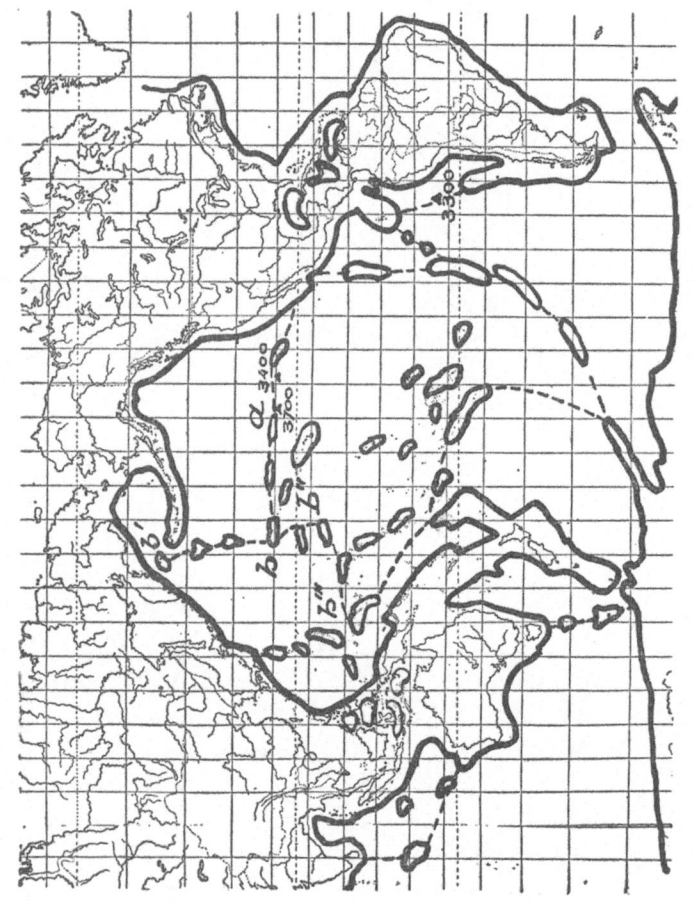

FIG. 58. Map illustrative of the *3000 meters* bathymetric reading in the modern Pacific Ocean. Diagrammatic. See text, p. 287, for discussion.

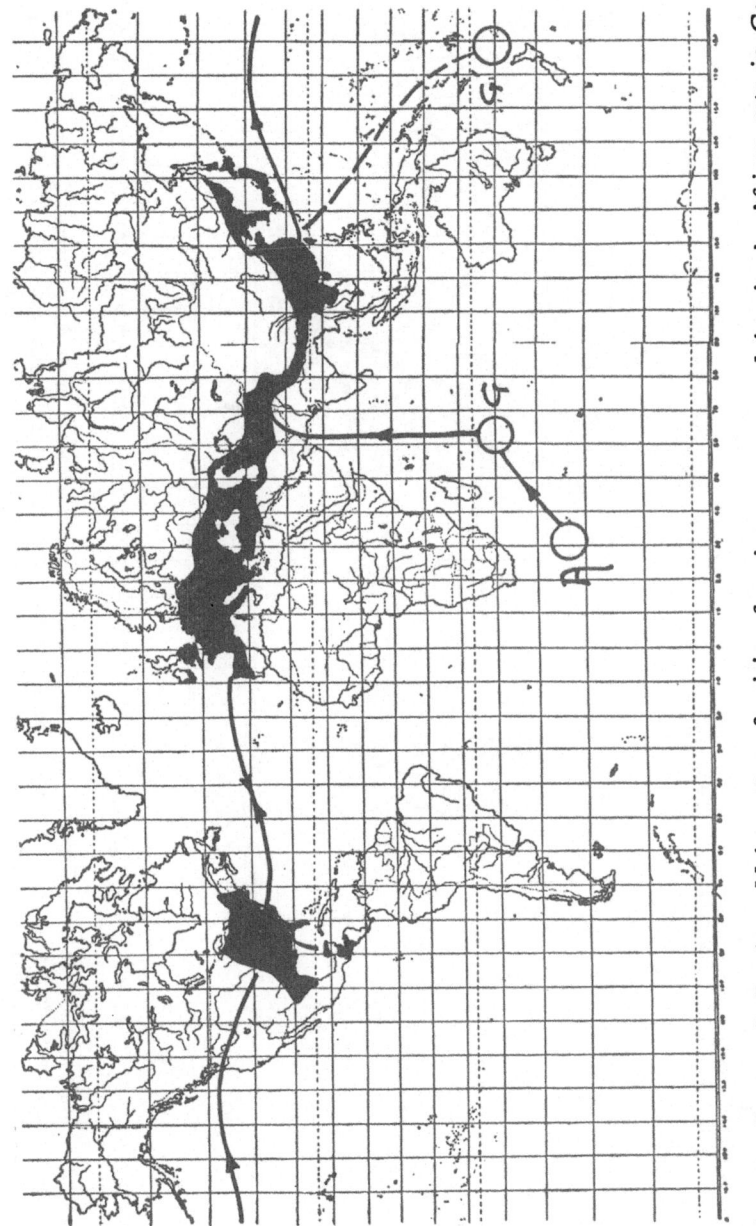

FIG. 59. Range of *Carpinus*. Main center of origin of angiospermous *genorheitra* in A; African gate in G; Western Polynesian gate in G′ with possible track northward as broken line. See text, p. 297, for elucidation.

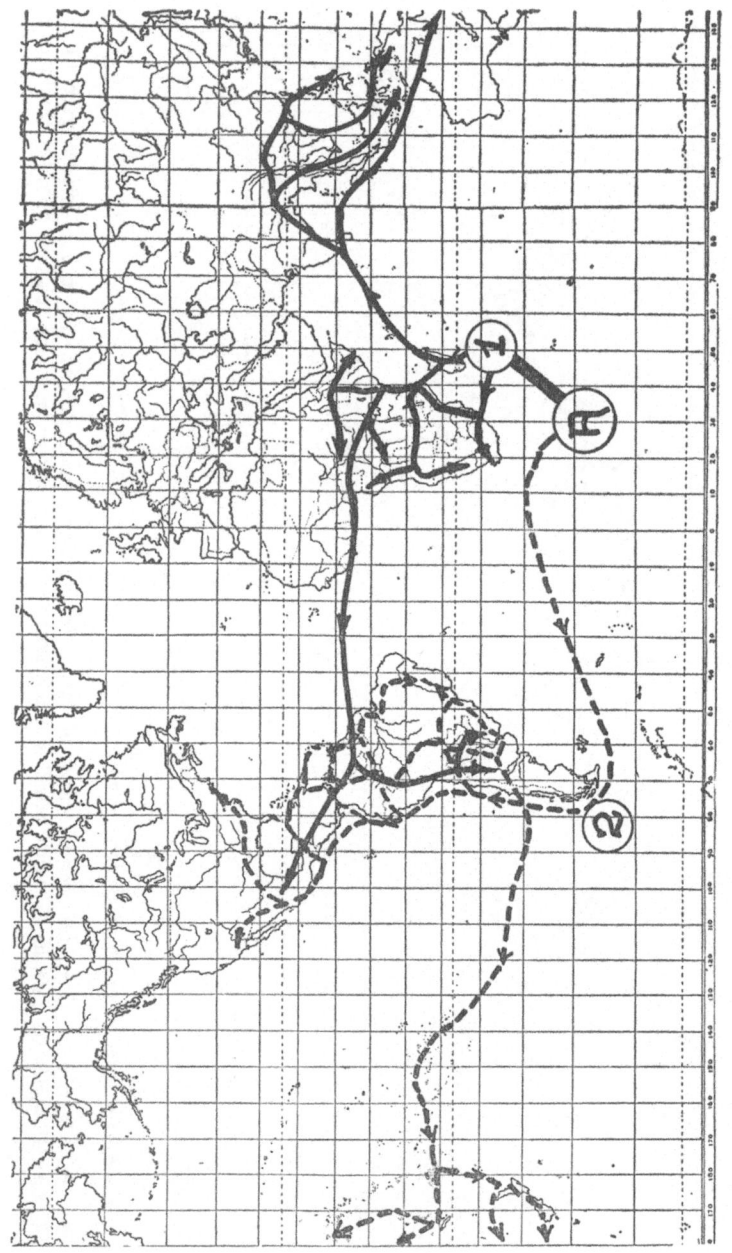

FIG. 60. Main trends in the dispersal of the Passifloraceae, Caricaceae and Achariaceae. Main center of angiospermous *genorheitra* in A, connected with African gate (1). Magellanian gate (2). Tracks *(Passiflora* and *Terrapathea* excepted) as continuous lines. Tracks of *Passiflora* and *Terrapathea* as broken lines. See text, p. 316, for discussion.

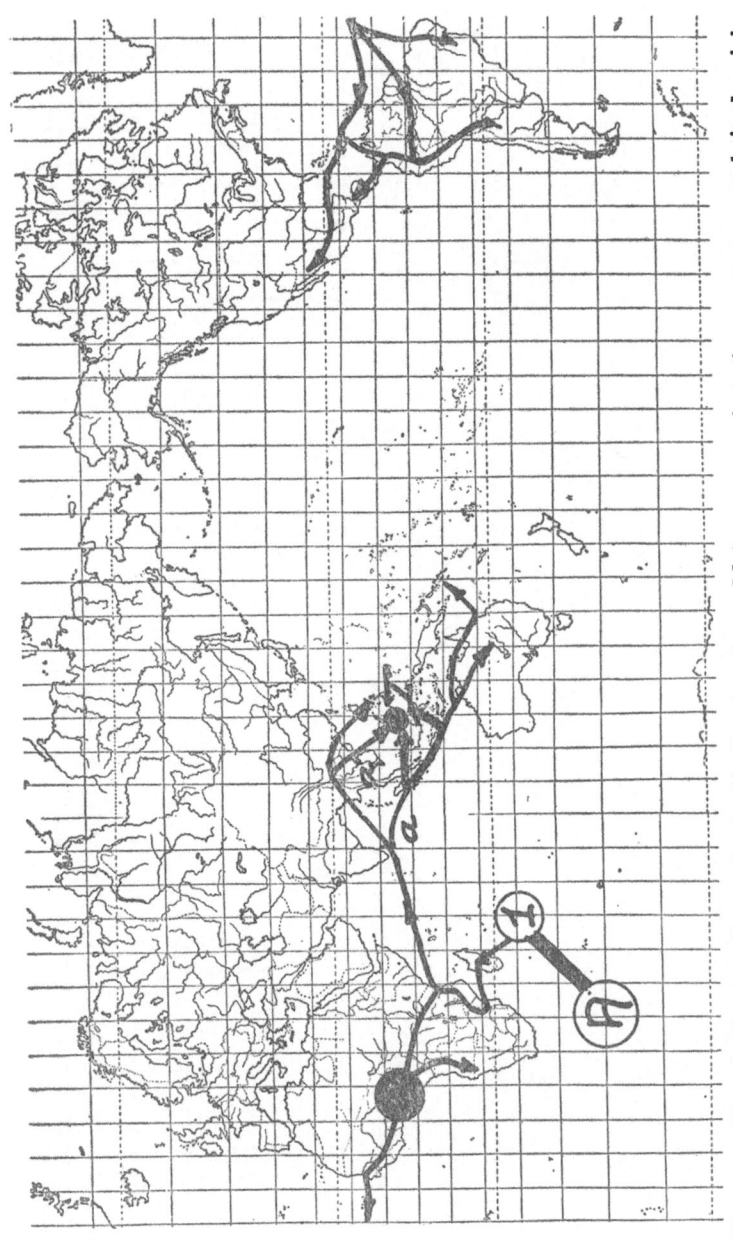

FIG. 61. Trends in the dispersal of the Bombacaceae. Main center of angiospermous genorheitral origins in A, connected with the African gate, I. Important secondary center of evolution in West Africa as black circle. Tracks *a*, *a'* are alternative to the region of Tawao (black circle T). Refer to text, p. 326, for elucidation.

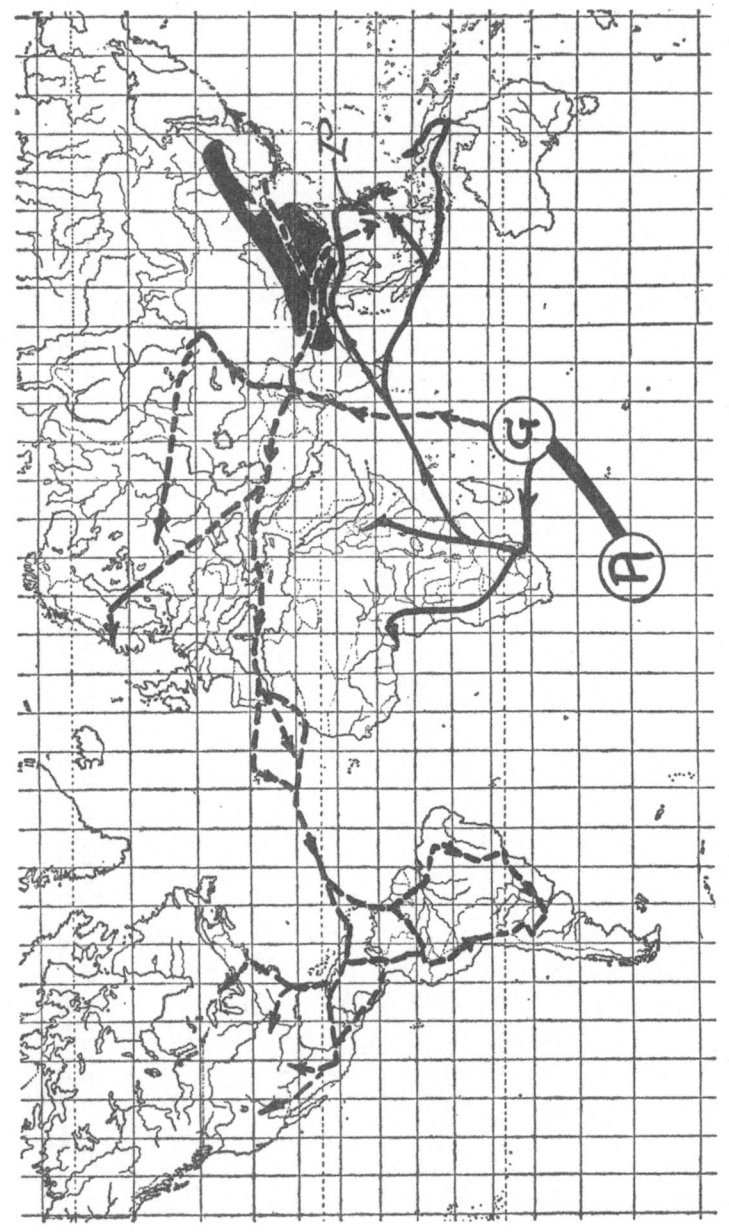

FIG. 62. Trend in the dispersal of the Rosaceae Prunoideae. Ranges of *Prinsepia, Maddenia, Dichomantes* heavy black (in the Far East). Tracks of *Pygeum*, continuous lines; tracks of *Prunus*, broken lines. Main center of origin of angiospermous *genorheitra* in A. African gate of angiosperny in G. Island of Palawan in P (indicated by small arrow). See text, p. 327, for elucidation.

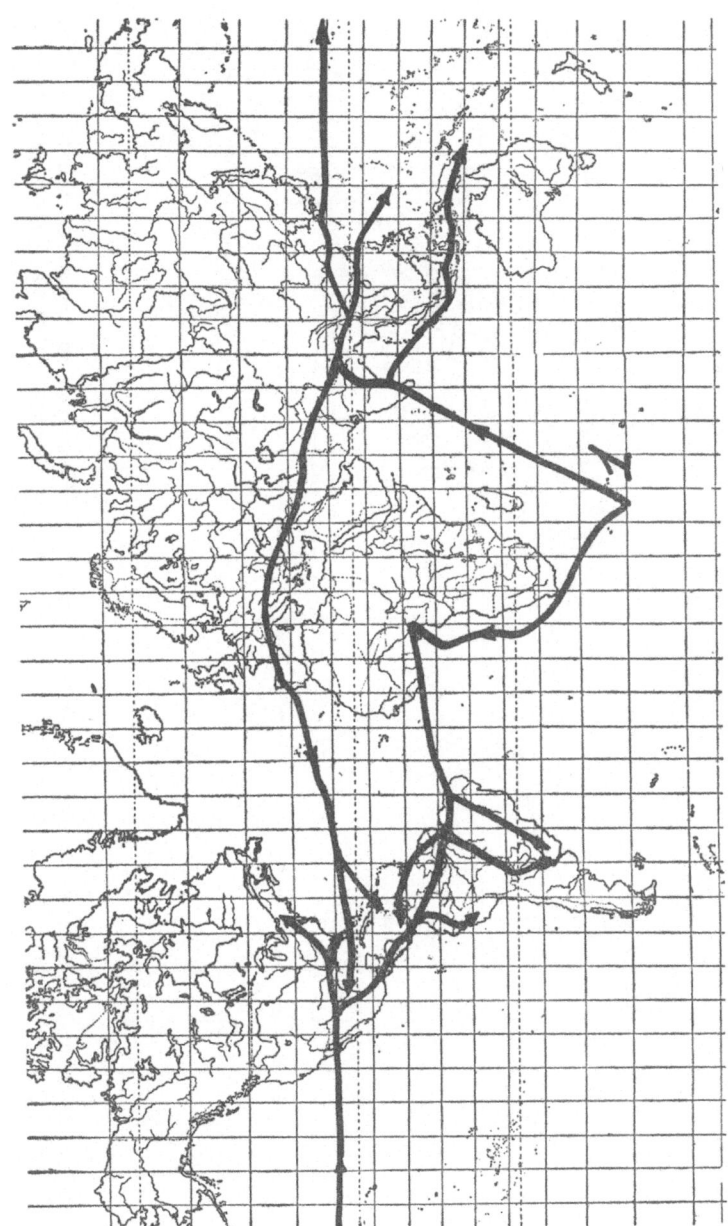

Fig. 63. Trends in the dispersal of the Styraceae (based on *Styrax*). Main angiospermous center, 1. Notice Philippines within the prongs of a track pointing eastward. Essentially diagrammatic. See text, p. 329, for elucidation.

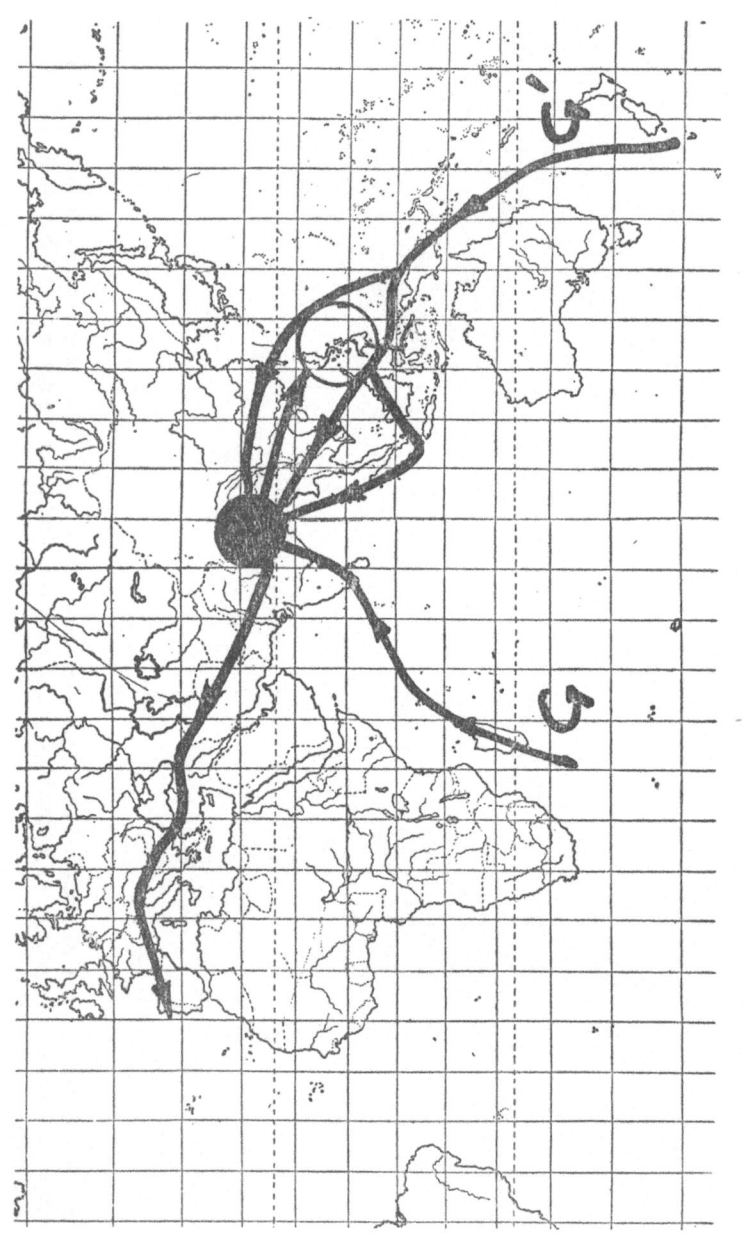

Fig. 64. Diagram to illustrate ambiguous nature of the term *Himalayan* in regard of the flora of the Philippines (within circle). Angiospermous gates, G African; G′ Western Polynesian. Himalayas as black circle. See text, p. 117, 336, for elucidation and compare with Fig. 31.

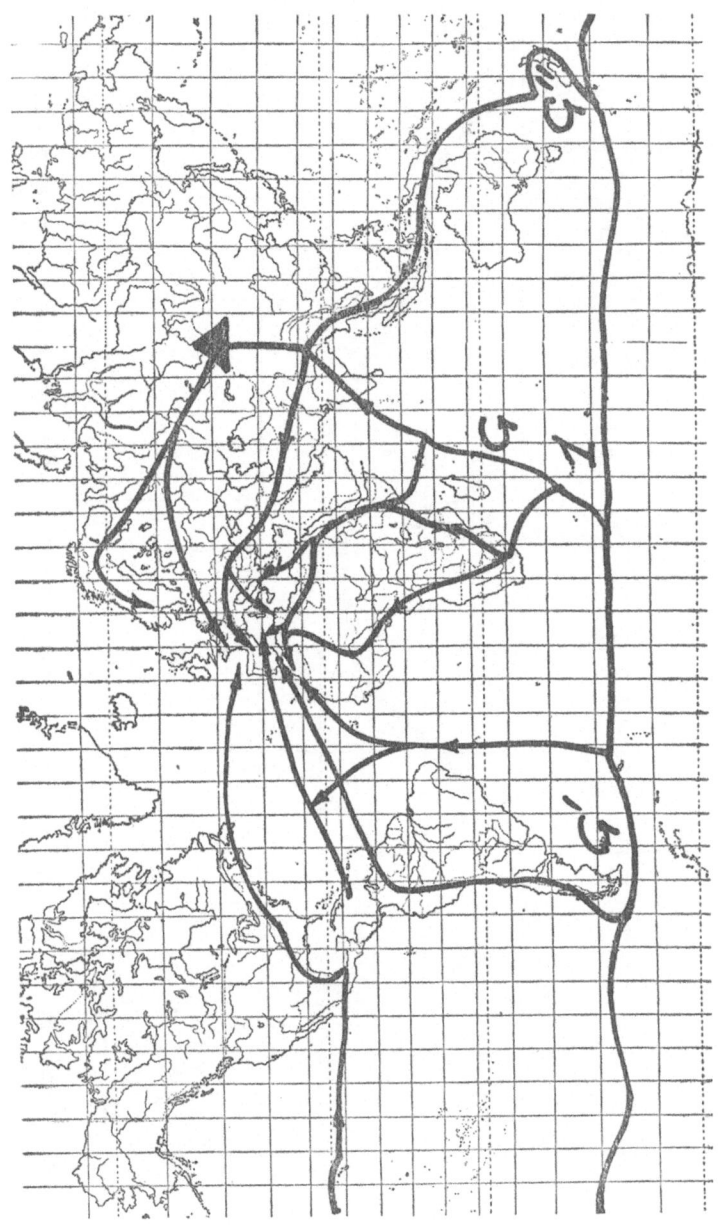

FIG. 65. Diagram to illustrate main tracks running to Europe. *Altai Node* as black triangle; gates of angiospermy, G African; G′ Magellanian, G″ Western Polynesian. Main angiospermous center, I, connected with all gates by "antarctic" tracks. Diagrammatic. See text, p. 337, for elucidation.

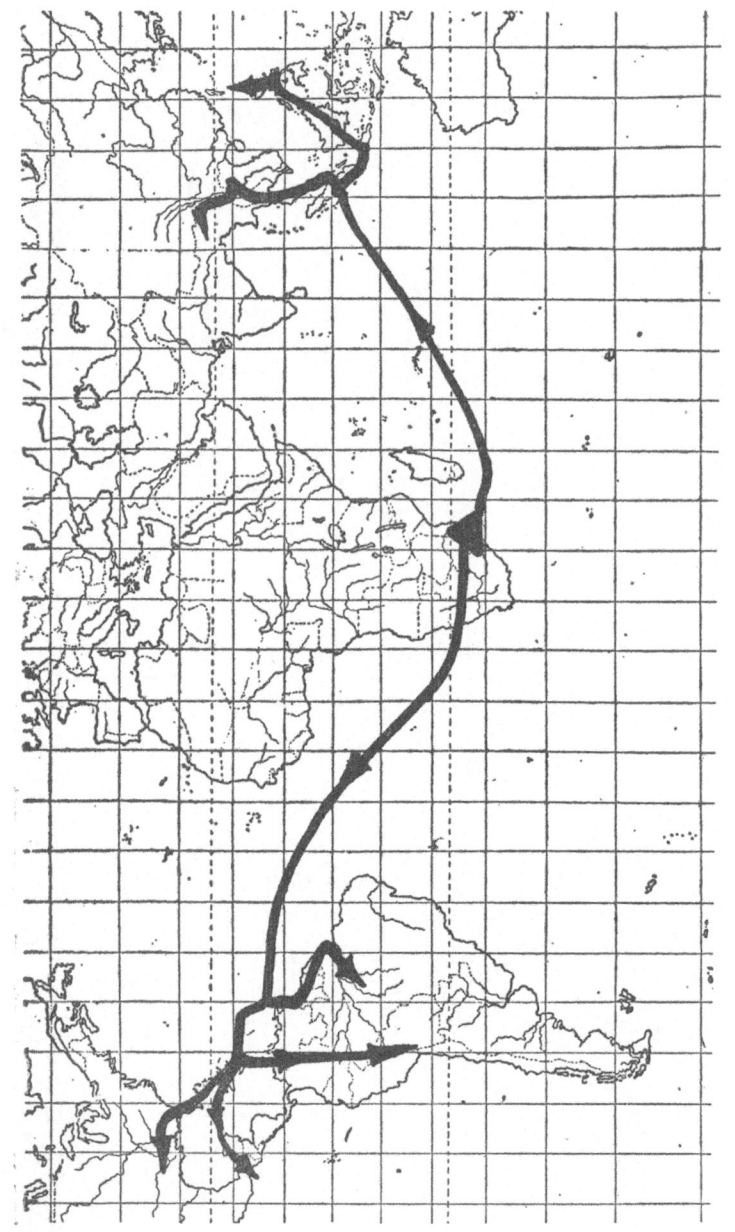

FIG. 66. Trends in the dispersal of the Cyrillaceae and their allies. Range of the Heteropyxidaceae as black triangle by South Africa. Essentially diagrammatic. See text, p. 347, for elucidation.

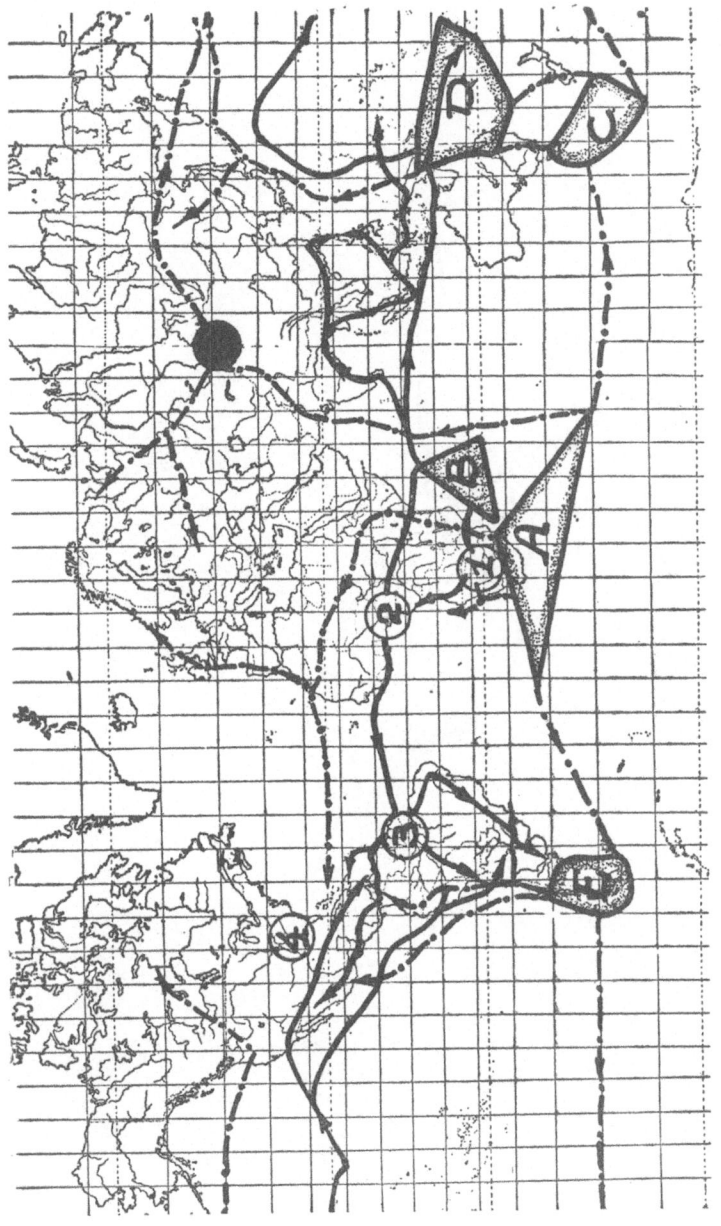

FIG. 67. Main angiospermous centers. A. *Afroantarctic Triangle*: B, *Gondwanic Triangle* (the two making the *African Gate*) — C, *Macquarian Center*: D, *Neocaledonian Center* (the two making the *Western Polynesian Gate*). E, *Magellanian Gate*. Important secondary centers, 1 *Kalaharian*: 2 *Nigerian*: 3 *Roraiman*: 4, *Ozarkian-Appalachian*. *Altai Node* of re-distribution as black circle. Main tracks: Continuous lines, *warm*: broken dotted lines, *cold*. Essentially diagrammatic. See text, p. 349, 352, 357, for elucidation.

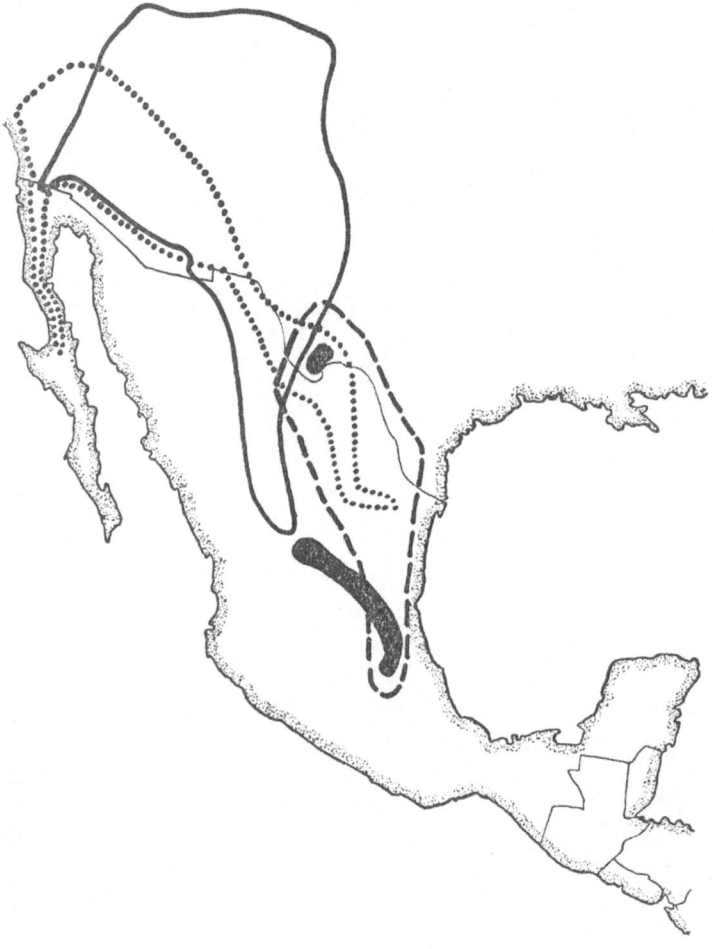

FIG. 68. Ranges of species of *Menodora*. Continuous line, *M. scabra;* dotted line, *M. scoparia;* broken line, *M. longiflora;* heavy black *M. decemfida* var. *longifolia.* See text, p. 358, for elucidation.

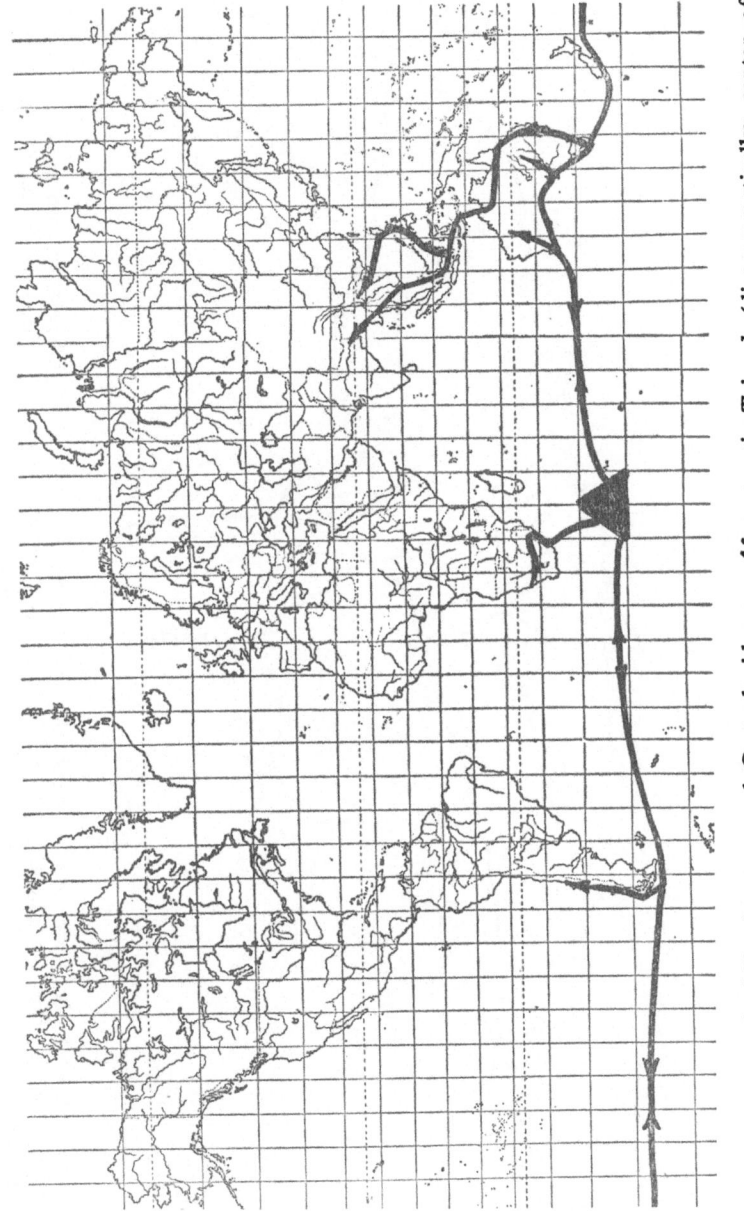

FIG. 69. Dispersal of Restionaceae and Centrolepidaceae. *Afroantarctic Triangle* (diagrammatically; center of origin of these two families) heavy black. See text, p. 361, for elucidation.

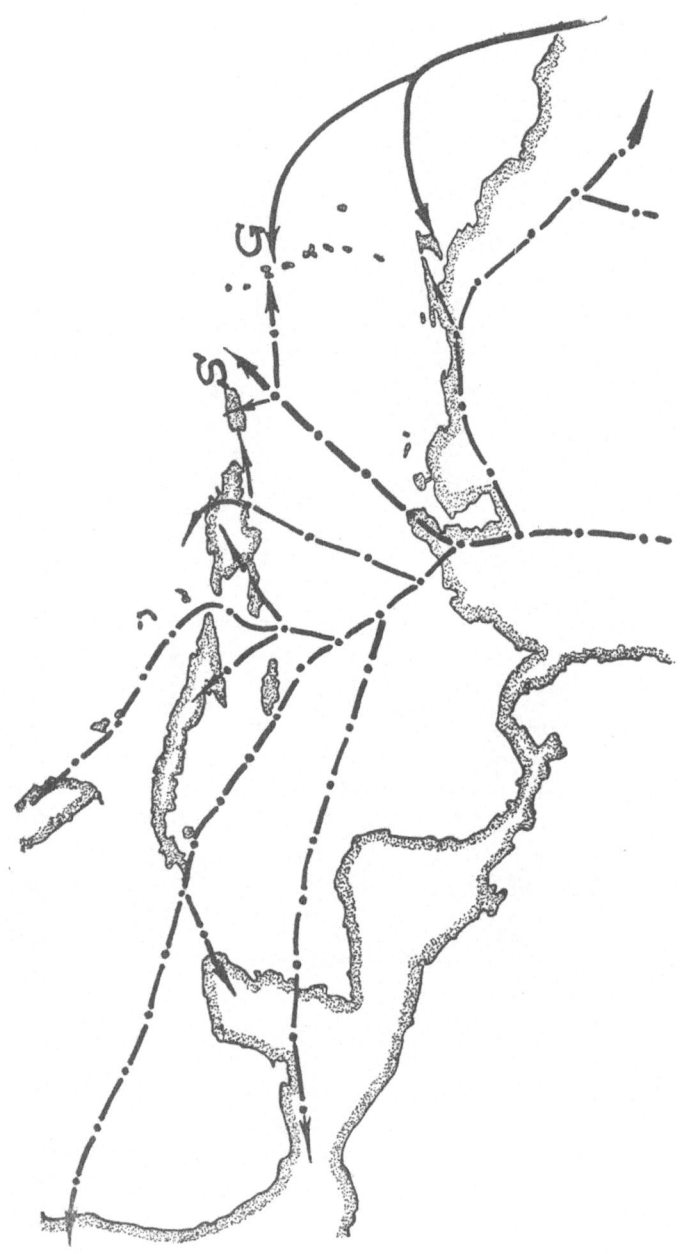

Fig. 70. One of the main sets of tracks leading to, and from, the Caribbeans, as broken dotted lines. Additional tracks from the direction of the Guianas and Brazil as continuous lines. S, island of Ste.-Croix; G, island of Guadeloupe. See text, p. 367, 369, for elucidation.

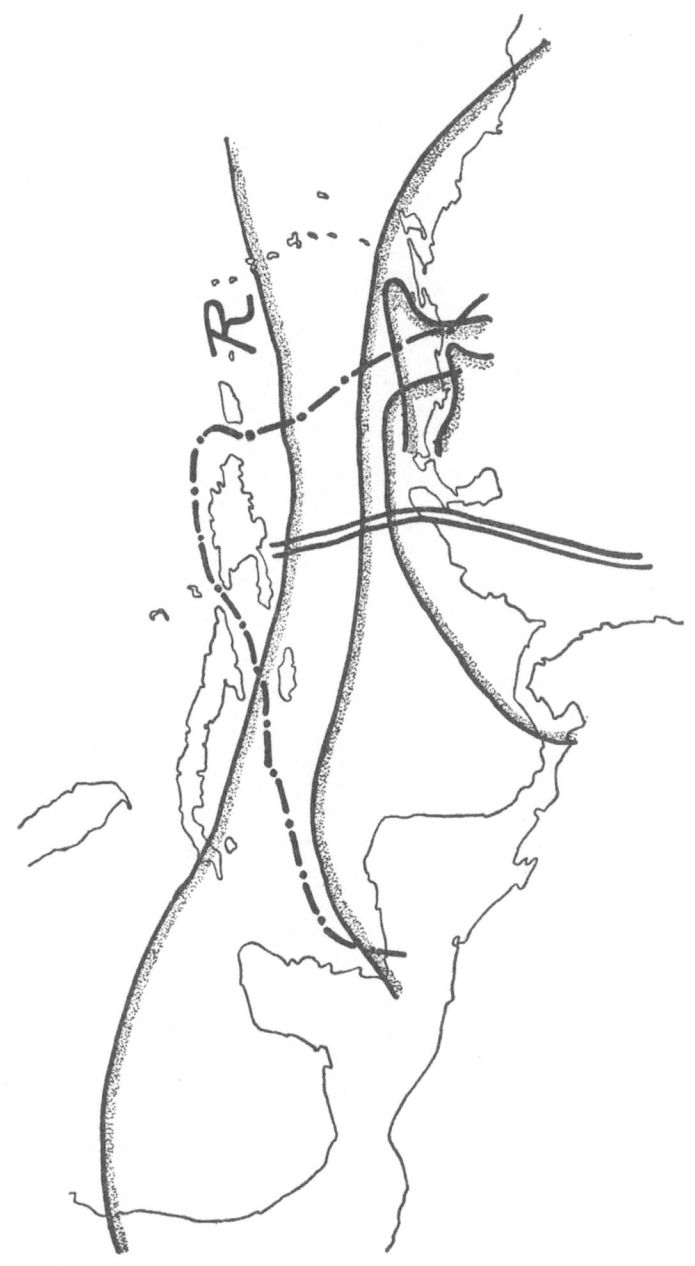

Fig. 71. Regions of disconnection in the West Indies, the "fronts" thus resulting as continuous stippled lines. Major track *West Indies-Peru* as double line. Zones of major depth in the Caribbeans as broken dotted line. Region of separation between Greater and Lesser Antilles in R. See text, p. 368, 392, for elucidation.

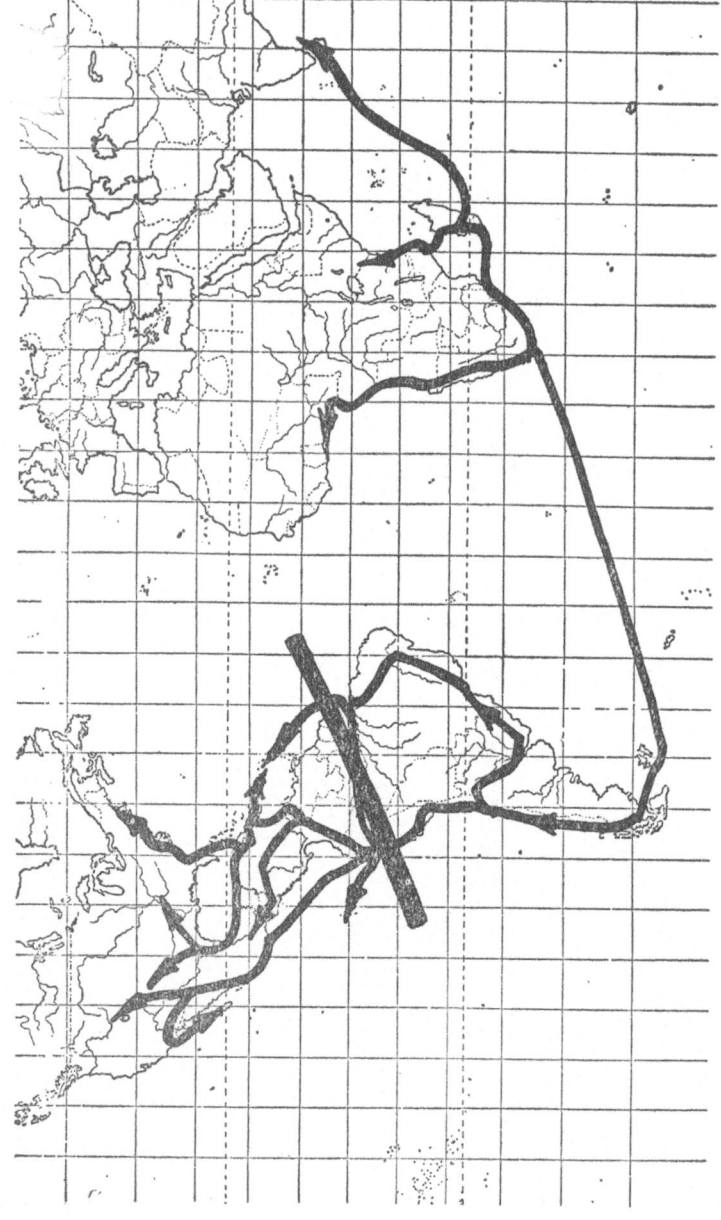

FIG. 72. Trends in the dispersal of the Cactaceae. Heavy bar marking approximate line of division between "southern" and "northern" Cactaceae in the New World. Diagrammatic. See text, p. 371, for elucidation.

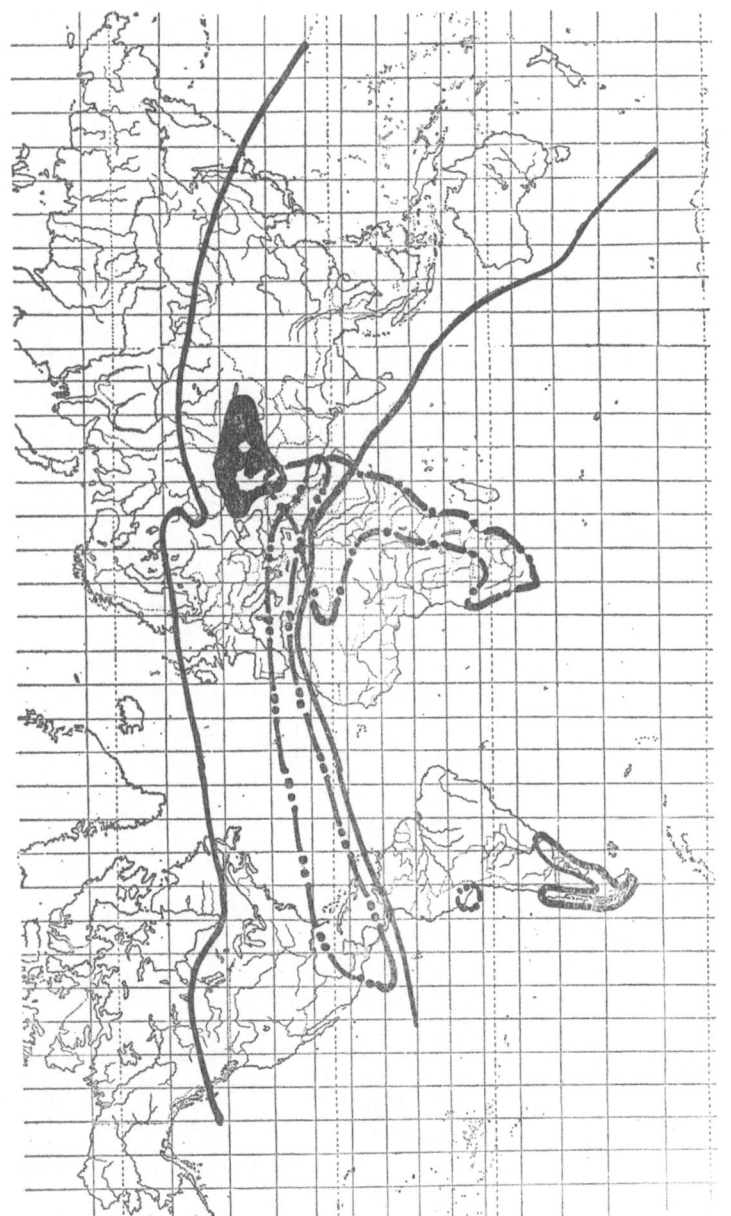

Fig. 73. Ranges of *Suaeda*. Section *Heterosperma*, continuous line; Sect. *Lachnostigma*, broken dotted line; Sect. *Platystigma*, broken double dotted line; Sect. *Physophora*, heavy black. See text, p. 382, for elucidation.

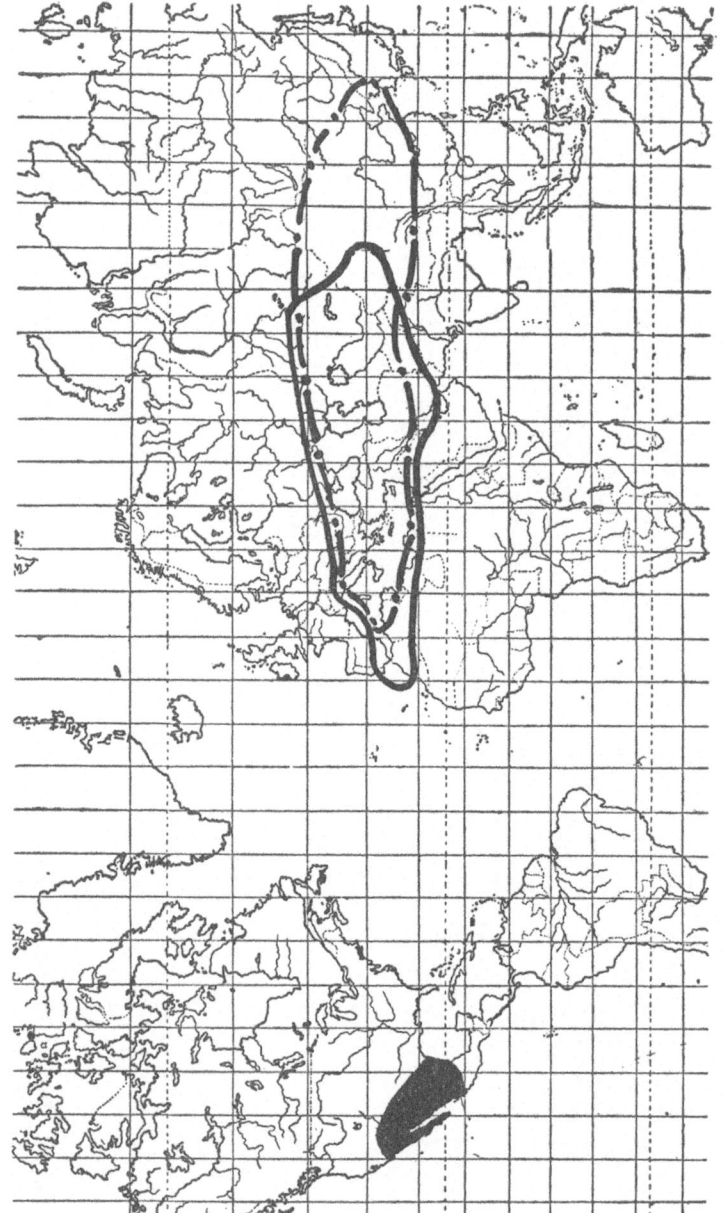

FIG. 74. Ranges of *Suaeda*. Section *Limbogermen*, heavy black; Section *Conosperma*, continuous line; Section *Schanginia*, broken dotted line. See text, p. 382, for elucidation.

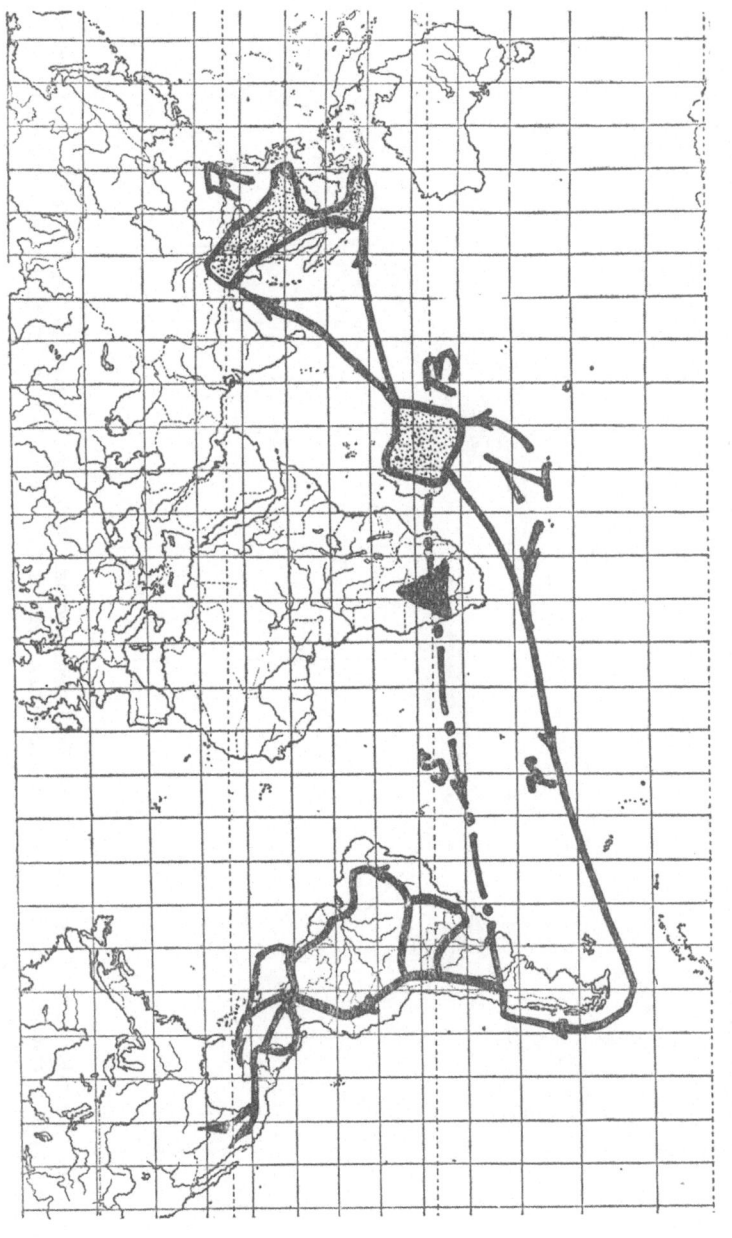

FIG. 75. Trends in the dispersal of the Proteae. Main angiospermous center, I. Massings of *Protium* in the Old World, A, B (dotted within continuous lines). *Kalaharian Center* as black triangle. Alternative tracks to the New World *r*, *s*. Essentially diagrammatic. See text, p. 390, for elucidation. Track *s* should be drawn to Northern Bolivia.

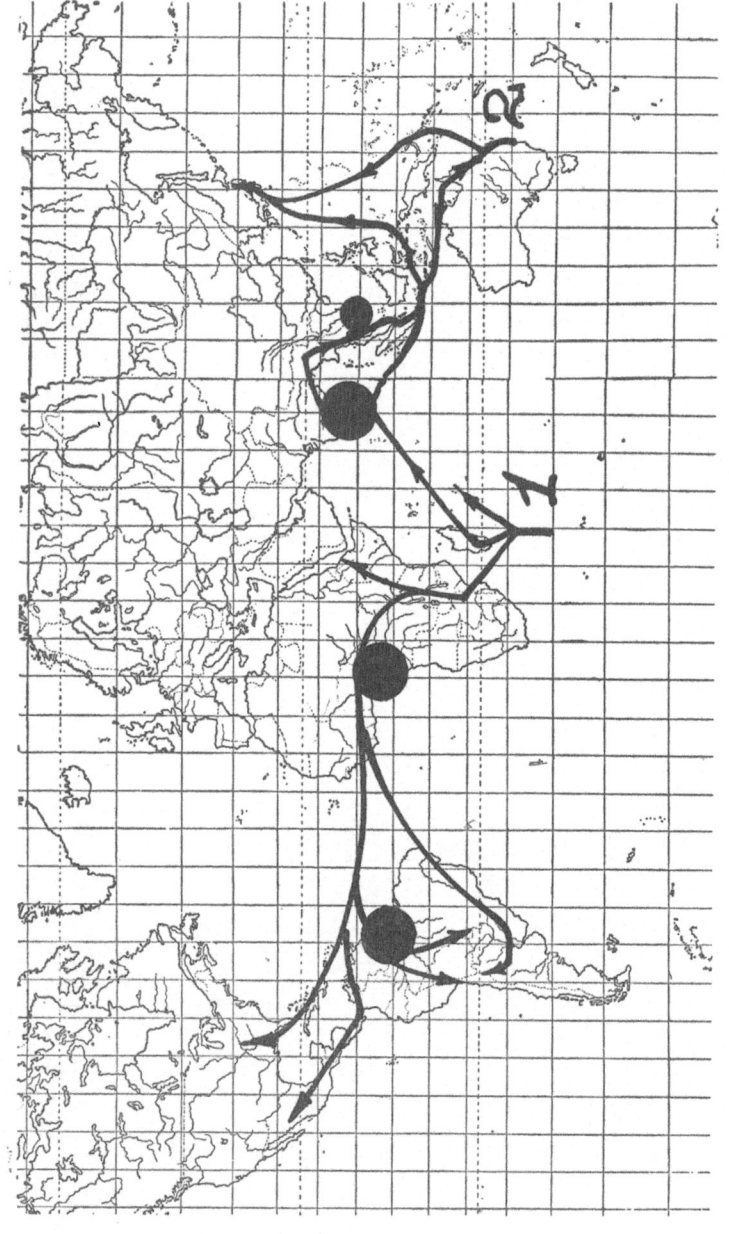

FIG. 76. Trends in the dispersal of the Podostemonaceae. Main angiospermous center, 1. Western Polynesian angiospermous gate, 2. Important secondary centers for the family as black circles. Essentially diagrammatic. See text, p. 397, for elucidation.

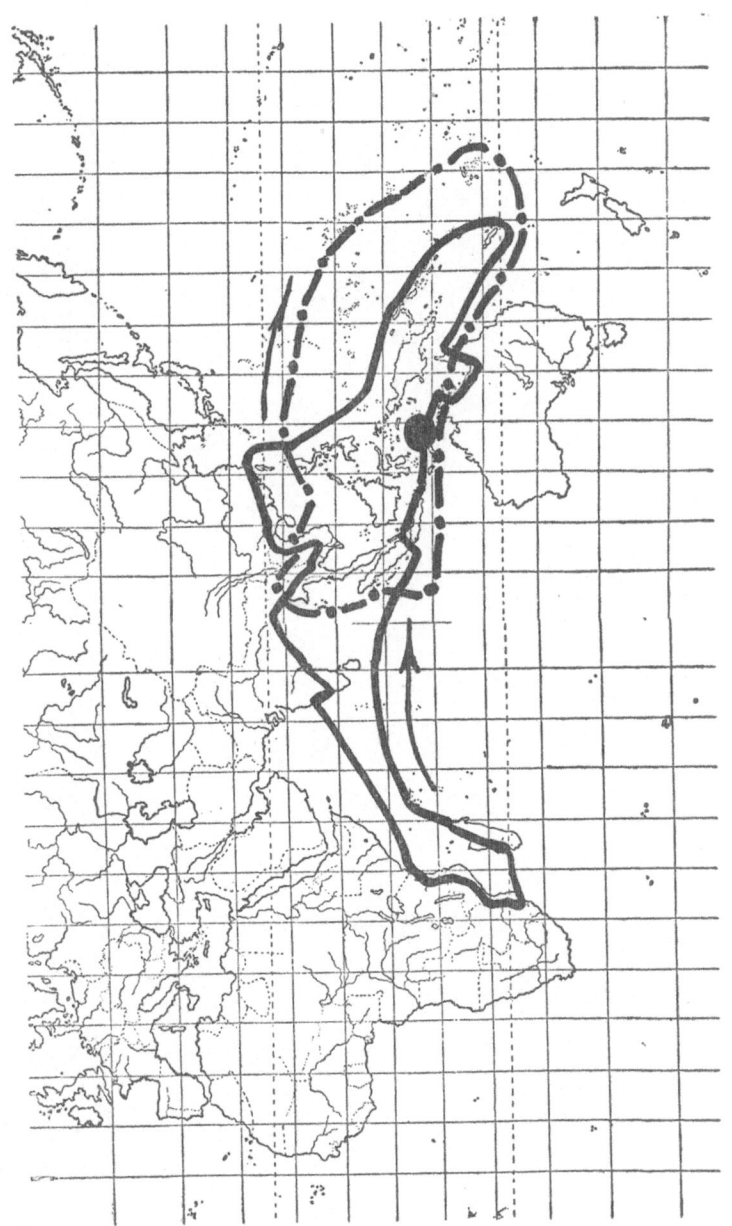

FIG. 77. Dispersal of *Lumnitzera*. Continuous line, *L. racemosa;* broken dotted line, *L. littorea:* black circlet, *L. lutea.* Direction of migration by arrows. See text, p. 400, for elucidation.

Fig. 78. The mangrove-belt in Wallacea. Wallacea in stippled outline, *Wallace Line* (as modified by various authors) in W'; the original Line cutting between the Philippines and New Guinea, Borneo and Celebes in W. *Aegiceras* black circles; *Camptostemon* circle with black at right; *Sonneratia* circles with black below; *Aegialitis* triangle within circle; *Osbornia* cross within circle; *Asthenochloa* black·squares. *Deplanchea* white triangles; the broken range indicated by line *a*. *Diospyros* sect. *Hasseltia* black triangles, the nearly broken range indicated by line *b*. The eastern extension of range of *Diospyros* and *Deplanchea* suggested by arrows ending the disconnected tracks. Diagrammatic. See text for elucidation, p. 402.

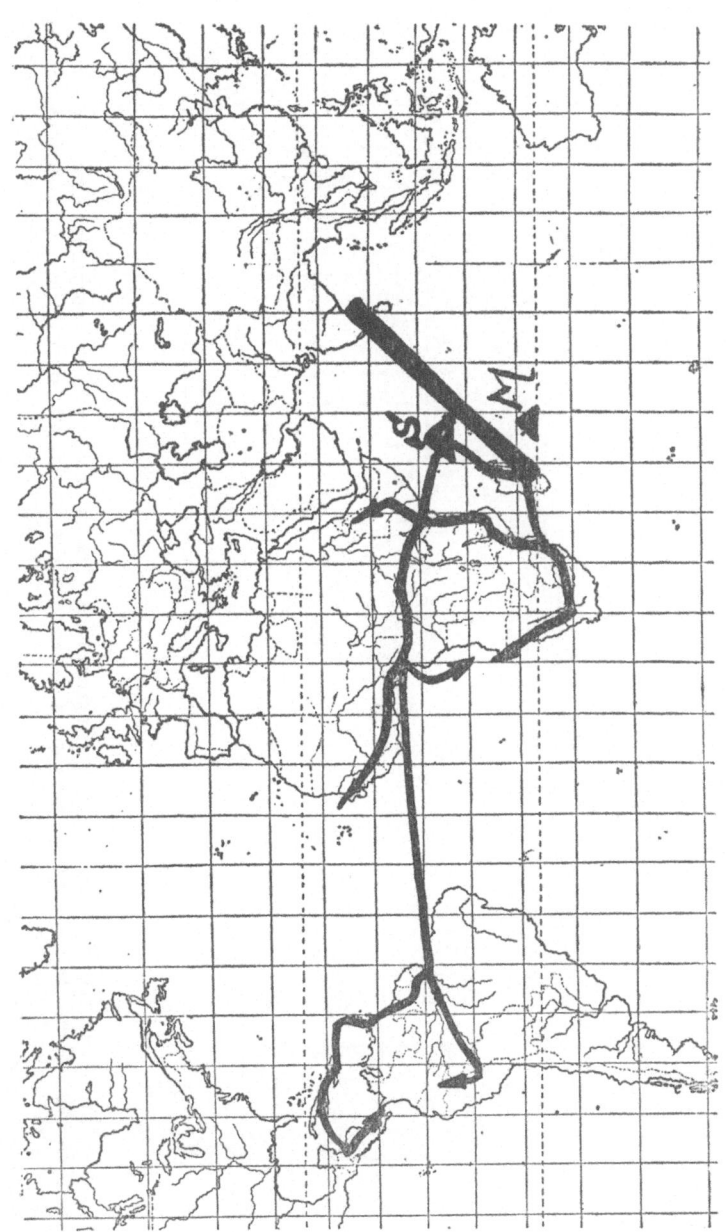

FIG. 79. Dispersal of *Cassipourea*. Heavy bar, approximate boundary between "eastern" and "western" mangrove distribution. Seychelles in S, Mascarenes in M. Essentially diagrammatic. See text, p. 407, for elucidation.

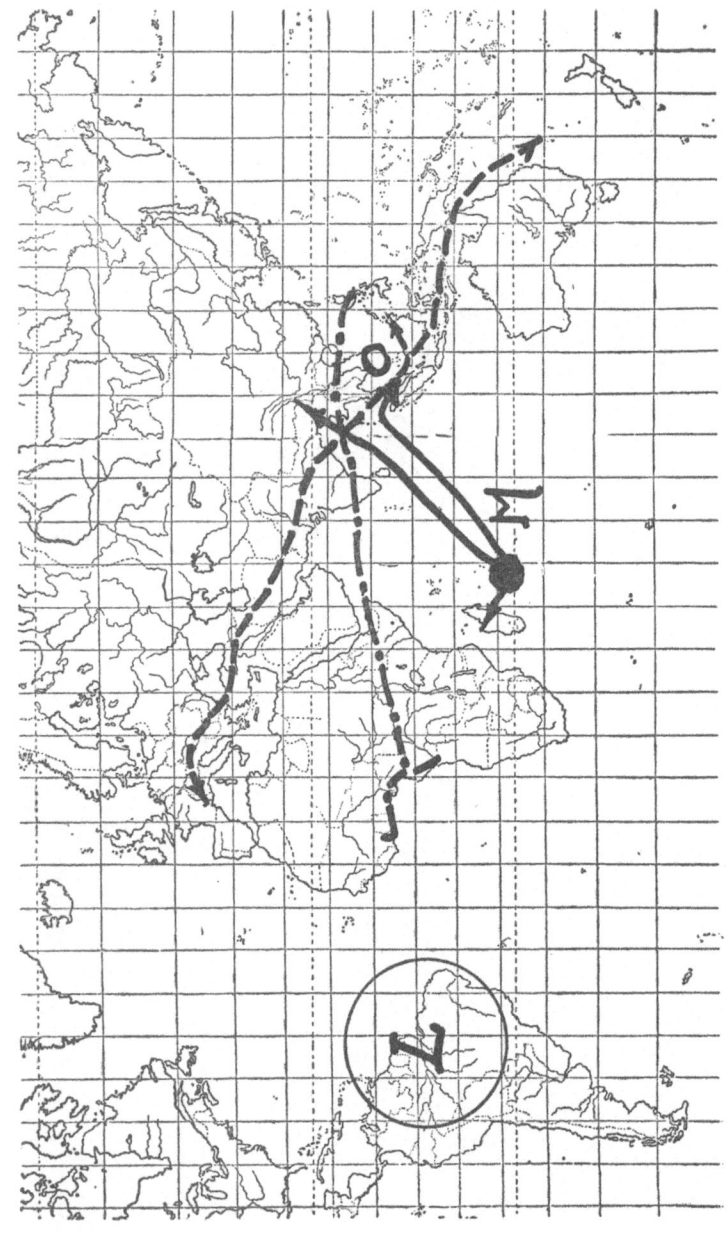

FIG. 80. Trends in the dispersal of the Barringtoniaceae. Tracks of *Foetidia* as continuous lines; M, Mascarenes; O, Mt. Ophir in Malacca. Tracks of *Combretodendron* as broken dotted line. Connections of Mt. Ophir in the west (Alps; see *Linaria alpina*, mostly) and east (approaches of New Zealand; see *Korthalsella*, mostly) as broken line. Center of the Lecythidaceae, L within circle. Essentially diagrammatic. See text, p. 414, for elucidation.

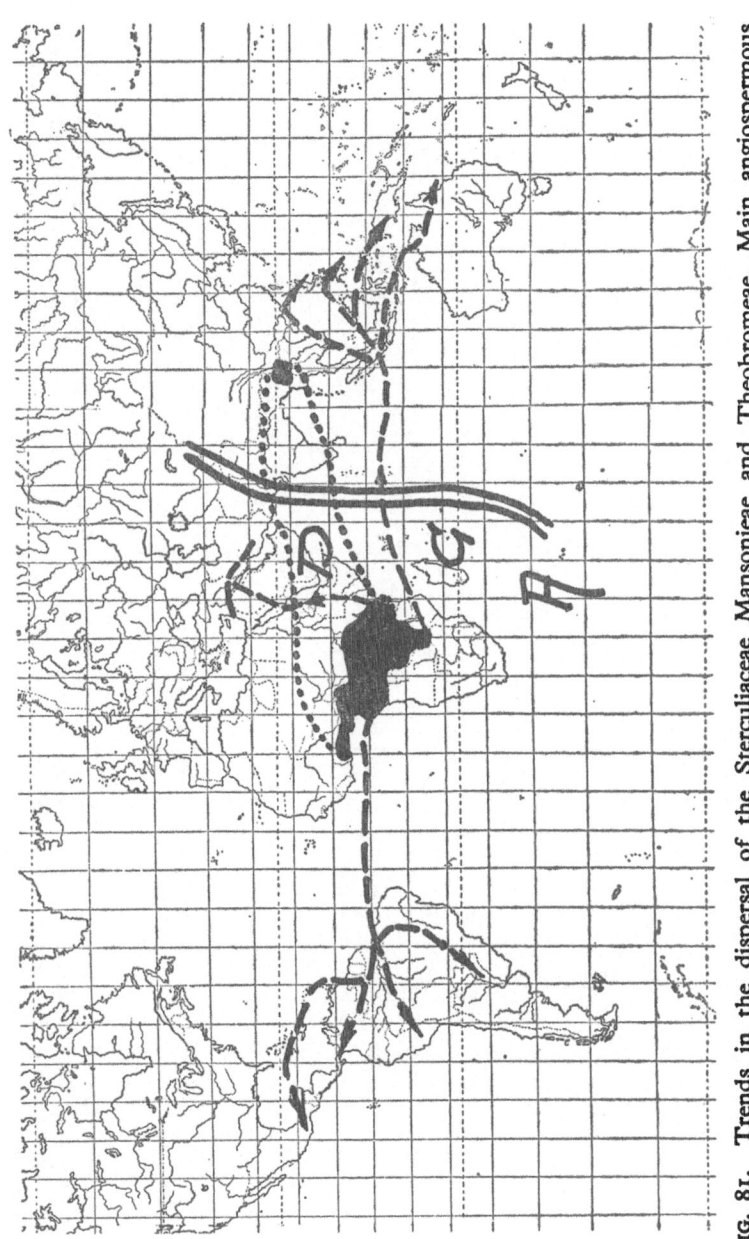

FIG. 81. Trends in the dispersal of the Sterculiaceae Mansonieae and Theobromeae. Main angiospermous center in A, with main "cold" track northward as double line. African gate of angiospermy in G. Tracks of the Theobromeae as broken lines. Range of *Mansonia* in heavy black, the disconnected ranges brought together by dotted lines. Essentially diagrammatic. See text, p. 415, for elucidation.

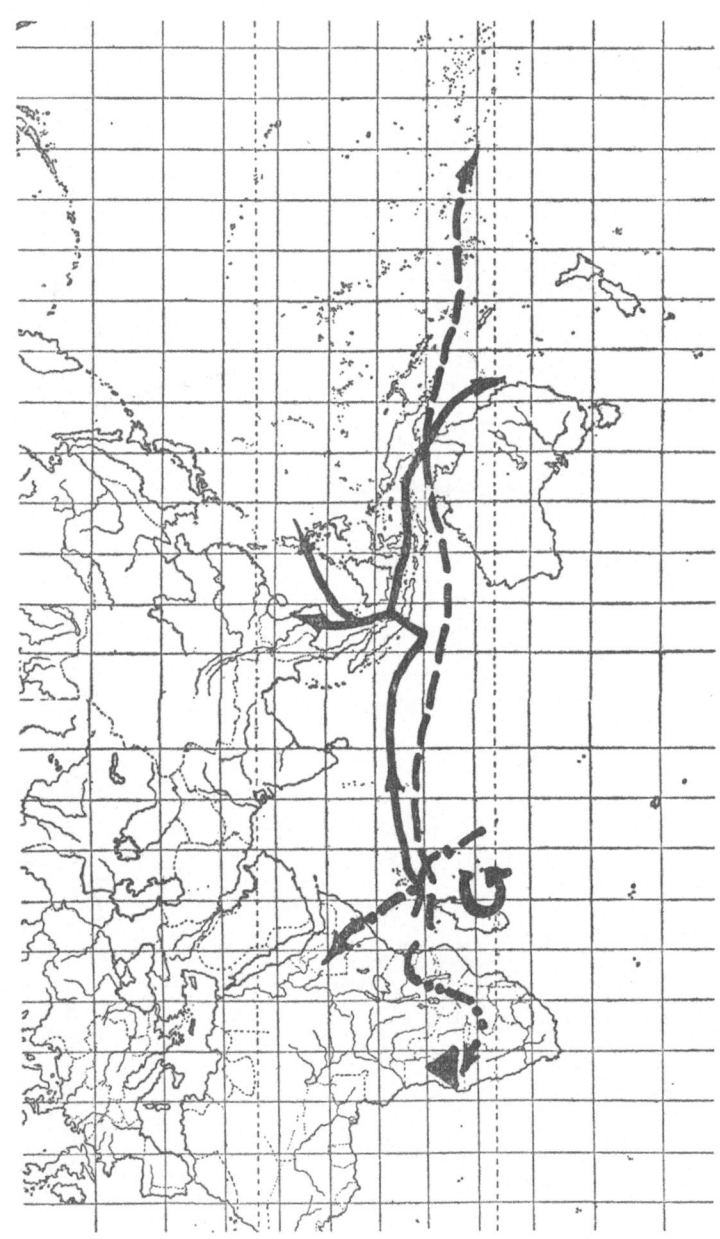

FIG. 82. Dispersal of *Triumfetta* (partial). Tracks of *Triumfetta repens* as continuous lines; *Triumfetta procumbens* as broken line; *Triumfetta heterocarpa* as broken dotted line; *Triumfetta Dekindtiana* as broken double-dotted line. African gate of angiospermy G; *Kalaharian Center* as black triangle. Essentially diagrammatic. See text, p. 419, for elucidation.

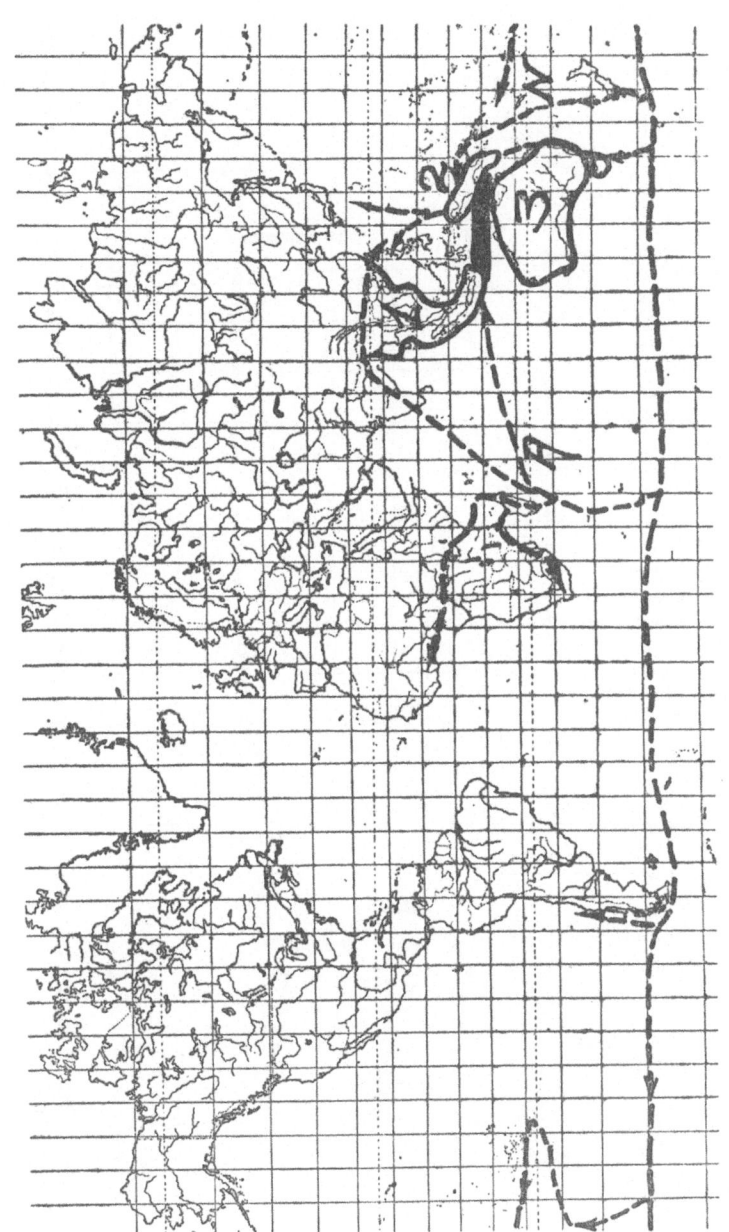

Fig. 83. Diagram to illustrate three main "Coigns" in, and immediately around, Malaysia. *Soenda Coign*, 1; *Papua Coign*, 2; *Australia Coign*, 3; all *Coigns* connected through heavy bar. A, African gate of angiospermy; N, Western Polynesian. Main tracks as broken lines. Essentially diagrammatic. See text, p. 427, for elucidation.

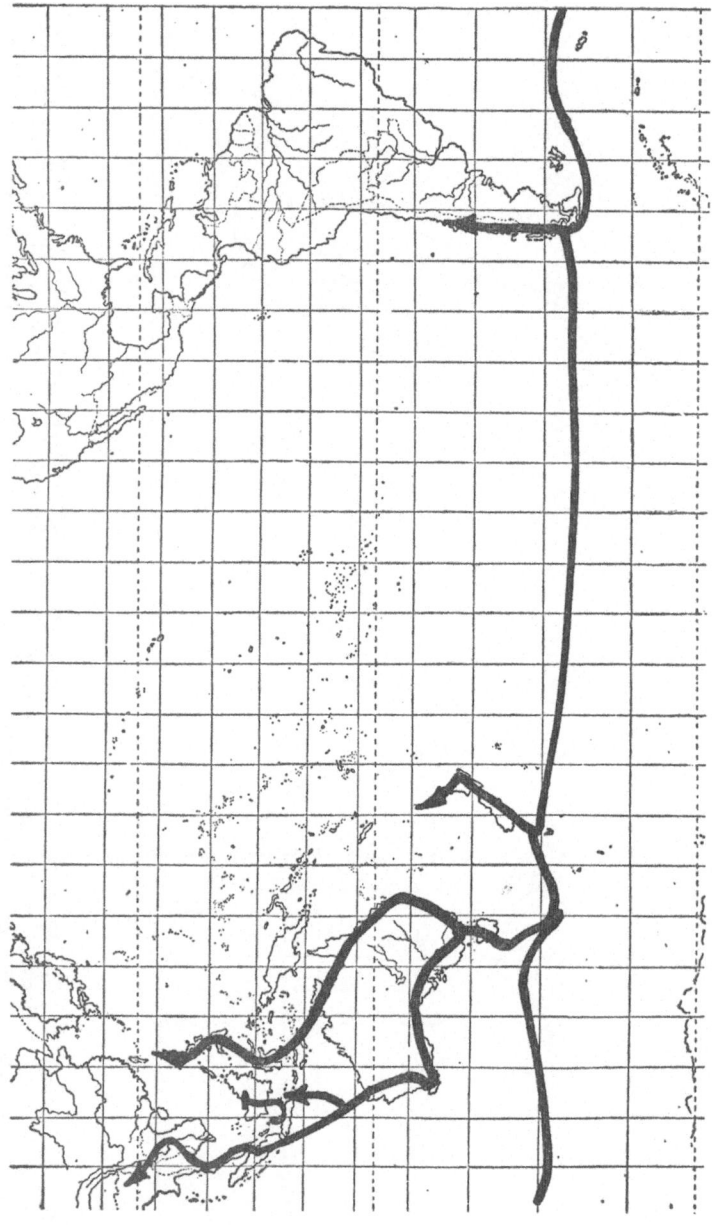

FIG. 84. Trends in the dispersal of the Stylidiaceae. Java in J. Essentially diagrammatic. See text, p. 434, for elucidation.

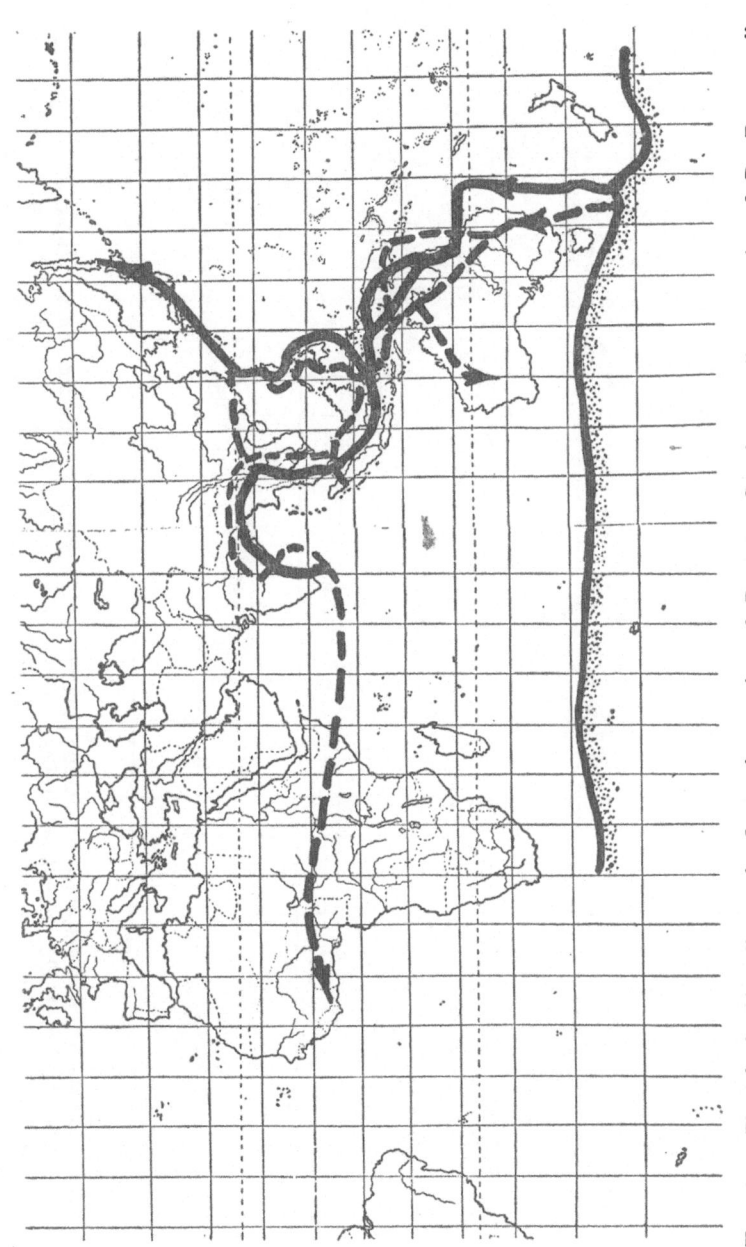

FIG. 85. Trends in the dispersal of certain species of *Drosera*. Continuous lines, tracks of *D. Burmanni*: broken lines, tracks of *D. indica*. "Antarctic" continental shore by stippled line. Essentially diagrammatic. See text, p. 436, 437, for elucidation.

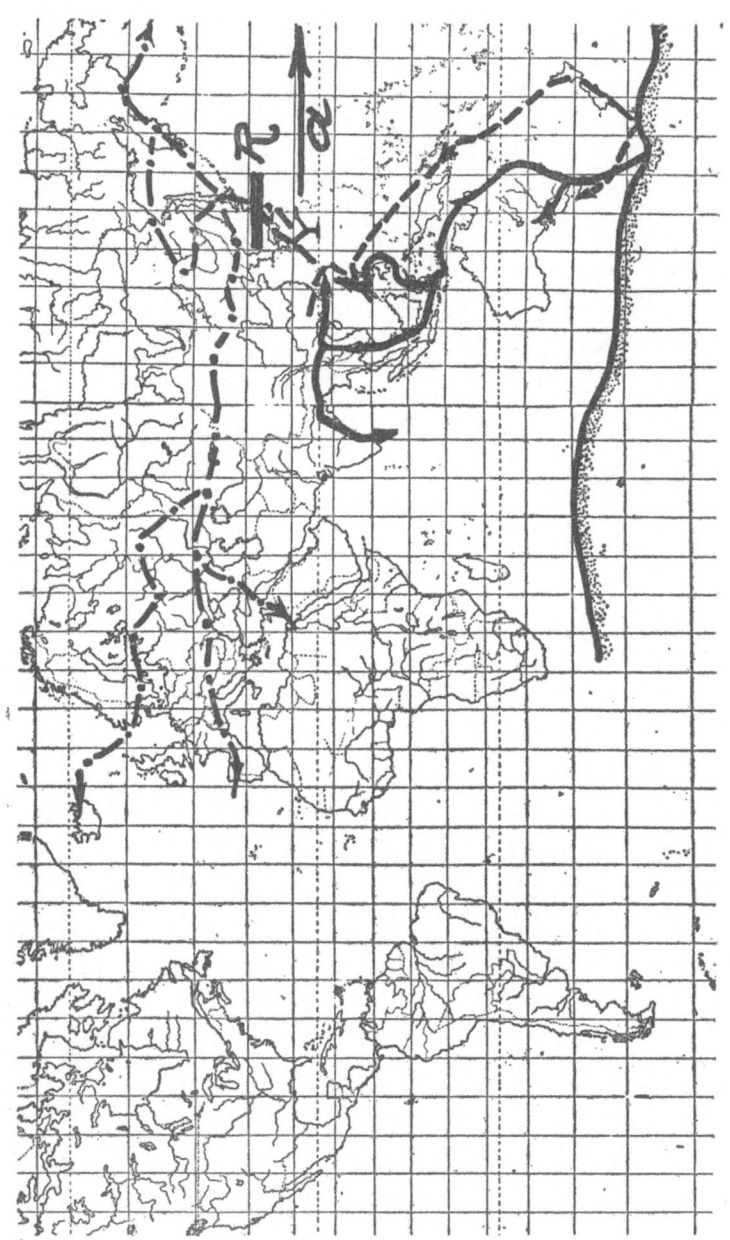

Fig. 86. Trends in the dispersal of certain species of *Drosera*. Continuous lines, tracks of *D. peltata:* broken lines, tracks of *D. spathulata.* Broken dotted lines, tracks (in part) of *D. rotundifolia.* Region of break between "cold" and "warm" forms as heavy bar R; track to the New World indicated by arrow *a.* Island of Yakushima in Y. Essentially diagrammatic. See text, p. 436, 438, for elucidation.

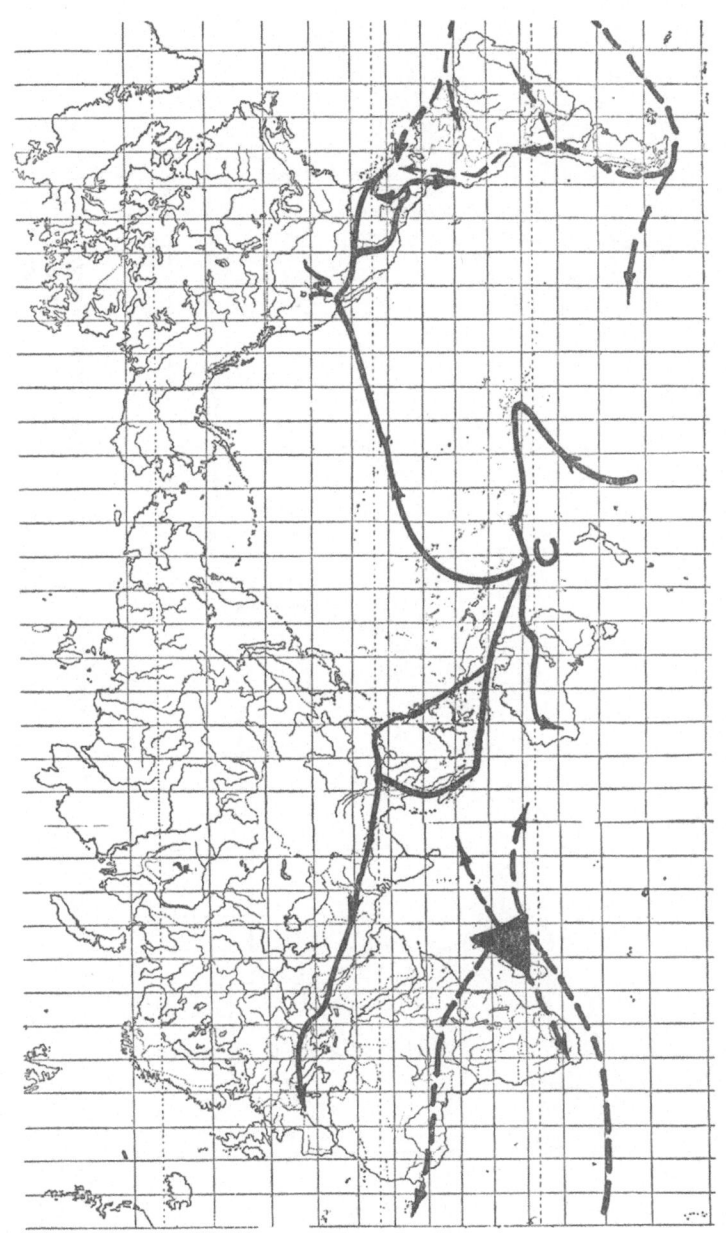

FIG. 87. Trends in the dispersal of Palms of the *Washingtonia* affinity shown by continuous lines; range of *Washingtonia* in W; center of re-routing of tracks (New Caledonia) in C. Main center of origin of Palms as black triangle, with suggestive tracks therefrom as broken lines. Essentially diagrammatic. See text, p. 444, for elucidation.

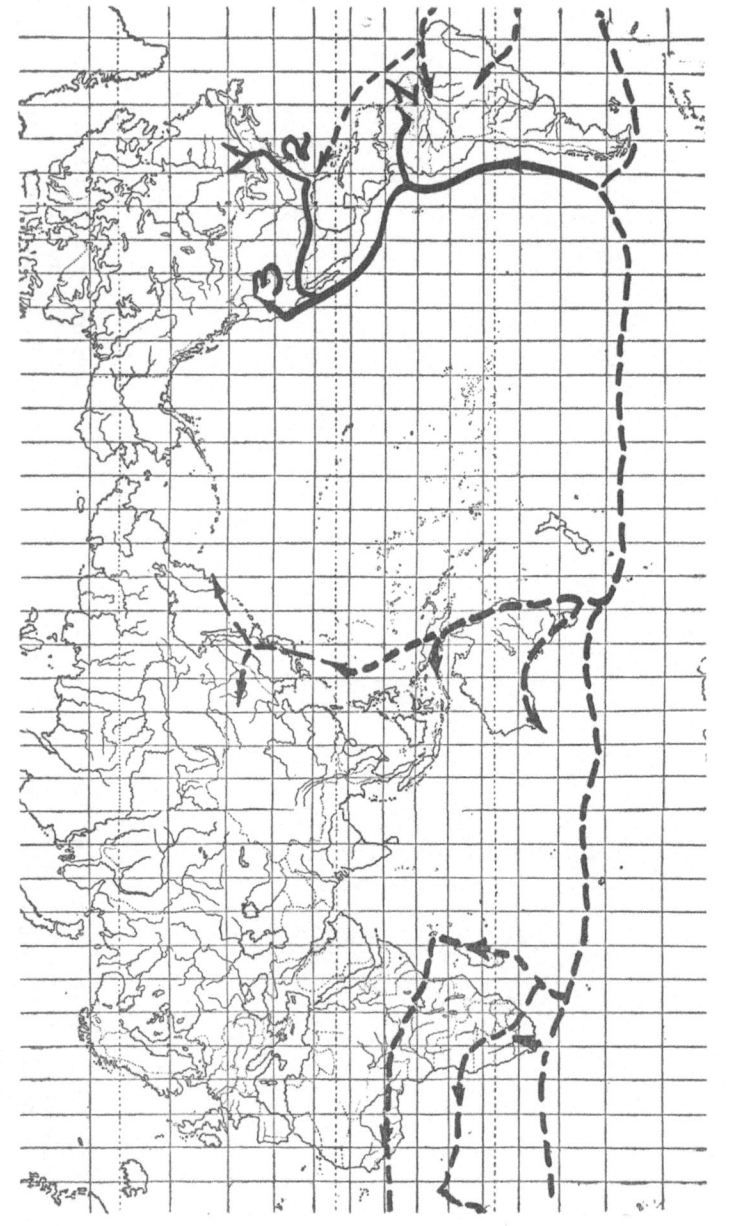

FIG. 88. Trends in the dispersal of the Sarraceniaceae as continuous lines. Range of *Heliamphora*, 1; *Sarracenia*, 2; *Darlingtonia* (*Chrysamphora*), 3. Dispersal of other members of the *genorheitron* (*Caltha*, *Utricularia*, etc.) suggested by broken lines. Essentially diagrammatic. See text, p. 447, 470, for elucidation.

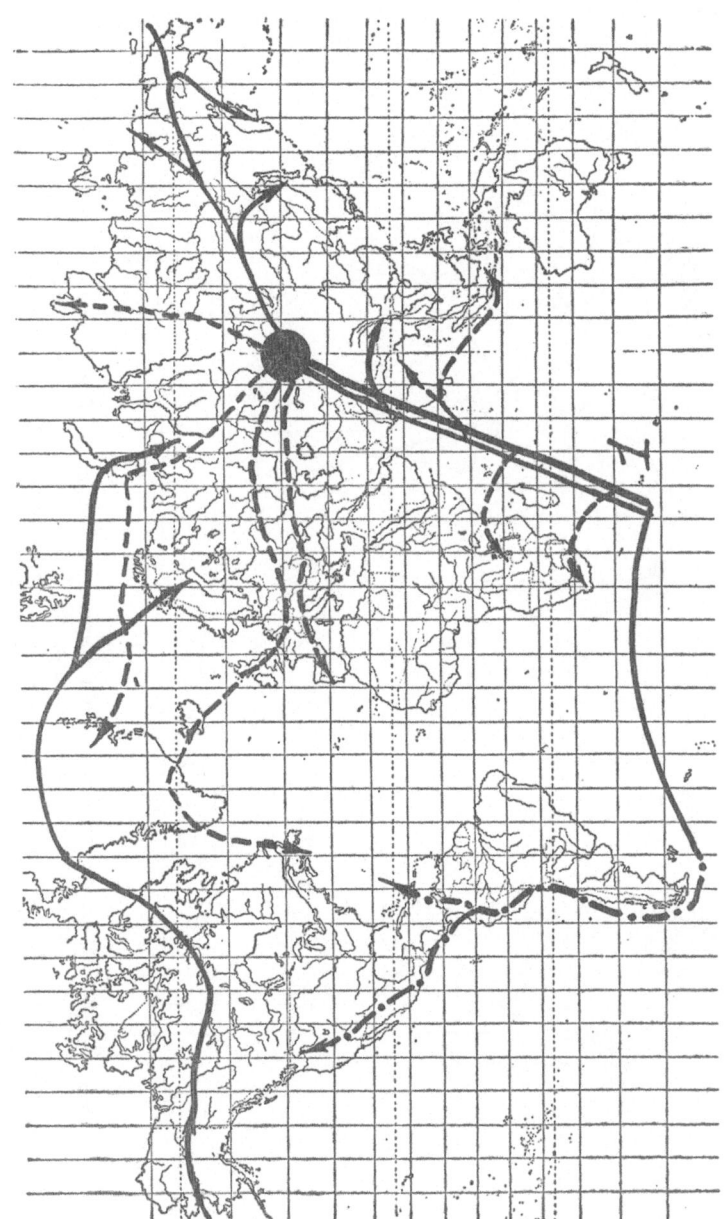

FIG. 89. Trends in the dispersal of *Stellaria* (continuous lines), *Alchemilla* Group *Vulgares* (broken lines; added tracks of African and Malaysian species of the genus), *Alchemilla* Sect. *Lachemilla* (broken dotted line). Altai Node as black circle; main track to the Altai Node as double line. Essentially diagrammatic. See text, p. 461, for elucidation.

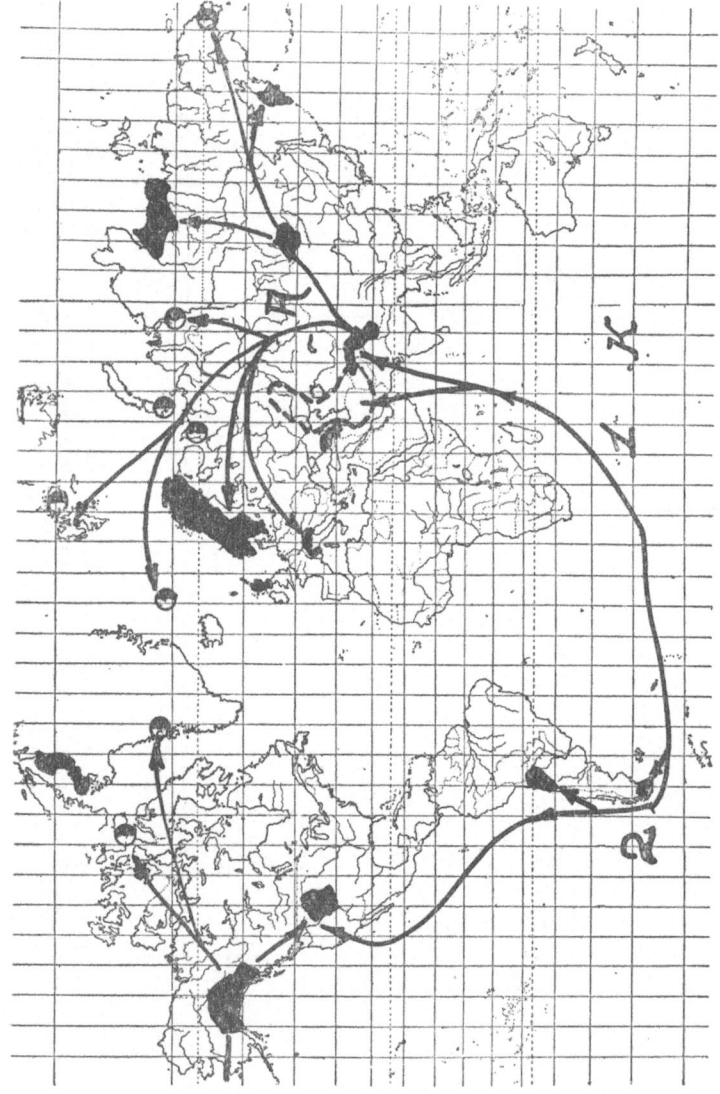

FIG. 90. Trends in the dispersal of certain Cyperaceae. Range of *Carex physodes* within broken line; range of *C. incurva*, heavy black; range of *C. incurva* var. *setina*, circles with black at right. Main course of tracks by continuous lines with arrows. Angiospermous gates, 1 African; 2 Magellanian. Kerguelen Islands in K. Essentially diagrammatic. See text, p. 463, for elucidation.

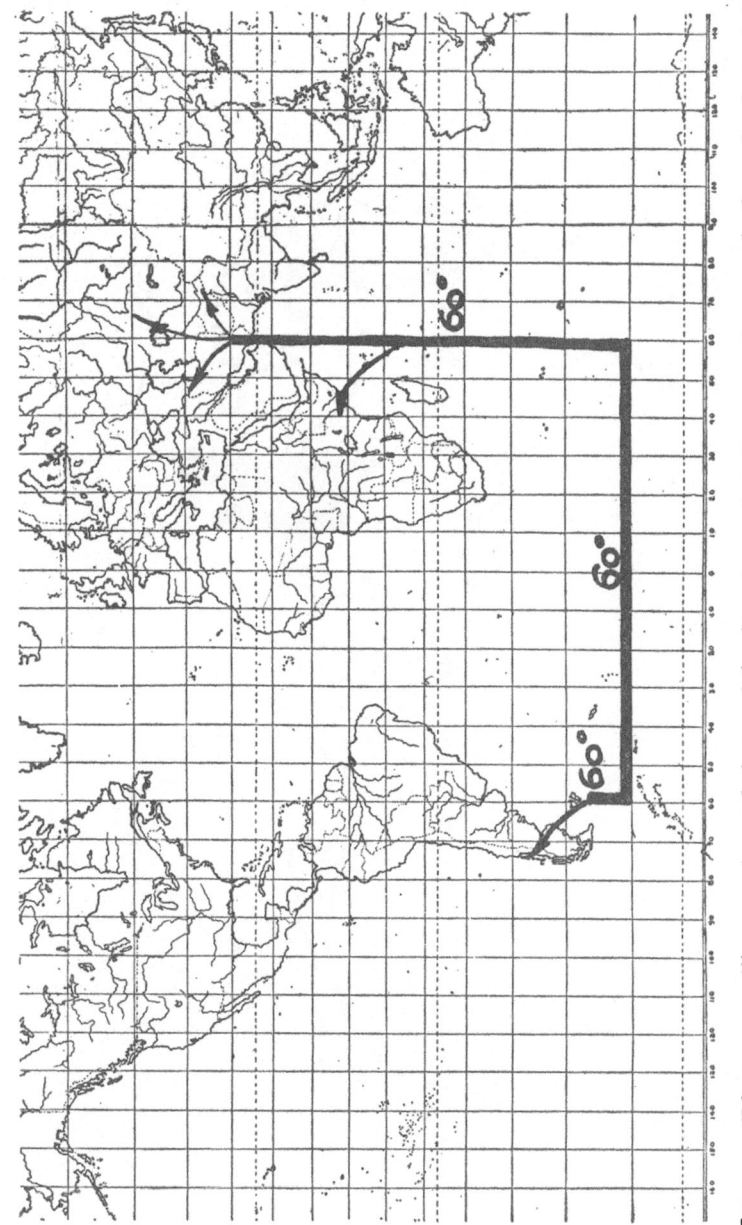

FIG. 91. Diagram to illustrate interrelations of land and sea by the run of a standard track along 60°
Longitudes. See text, p. 467, for elucidation.

FIG. 92.   Existing ranges of the Sarraceniaceae. *Sarracenia* in A; *Darlingtonia*
(*Chrysamphora*) in B; *Heliamphora* in C. Genorheitral tracks to the New
World as lines with arrows. See text for discussion, p. 468.

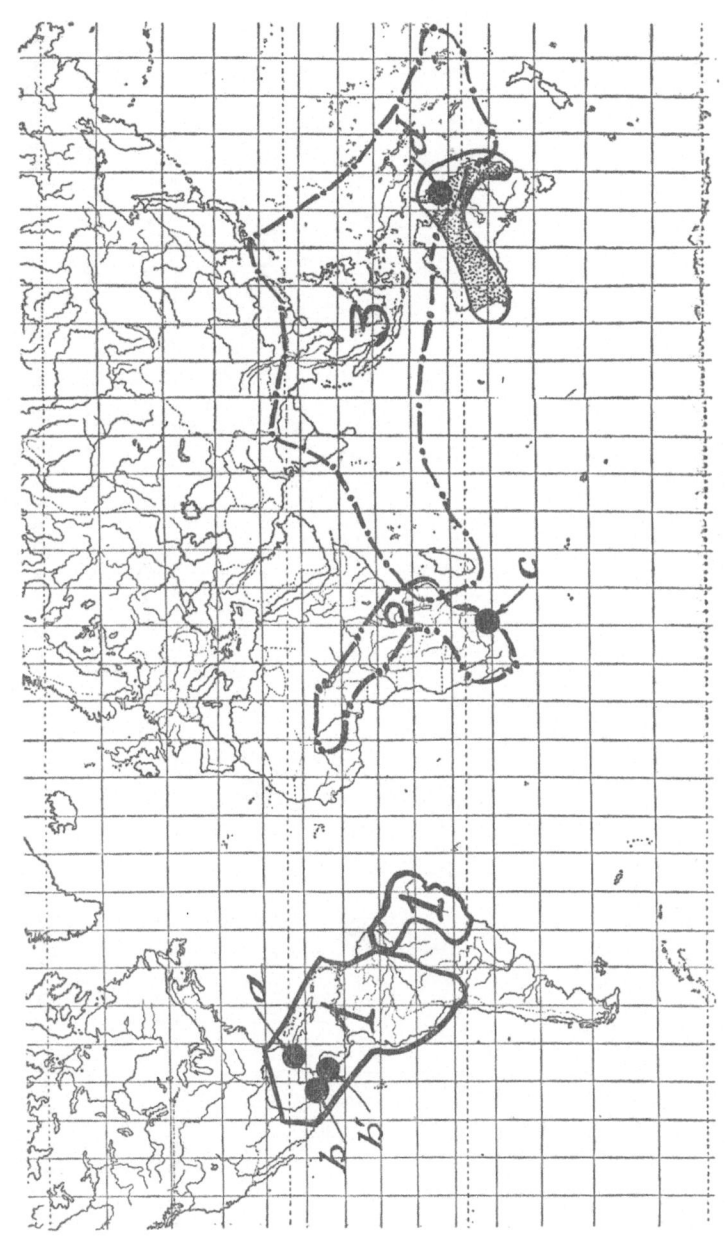

FIG. 93. Existing ranges of the Cycadaceae. *Major genera:* 1 (in two main groups), *Zamia;* 2, *Encepha-lartos;* 3, *Cycas;* dotted (Australia), *Macrozamia. Minor genera* (diagrammatic, as black small circles): *a,* *Microcycas; b, Dioon; b', Ceratozamia; c, Stangeria; d, Bowenia.* See text, p. 476, for elucidation.

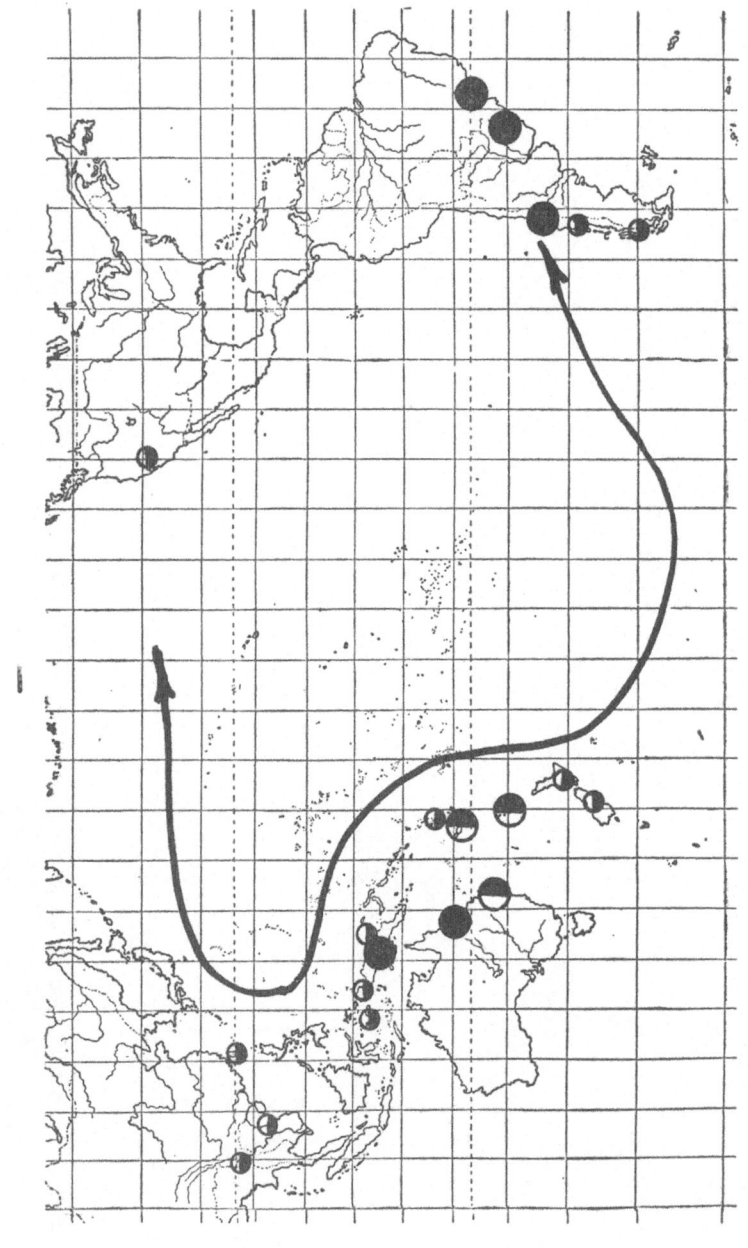

FIG. 94. Trends in the dispersal of *Araucaria* (large circles, black to right for *Eutacta*; black for *Colymbea*) and *Libocedrus* (small circle). Essentially diagrammatic. See text, p. 481, for elucidation.

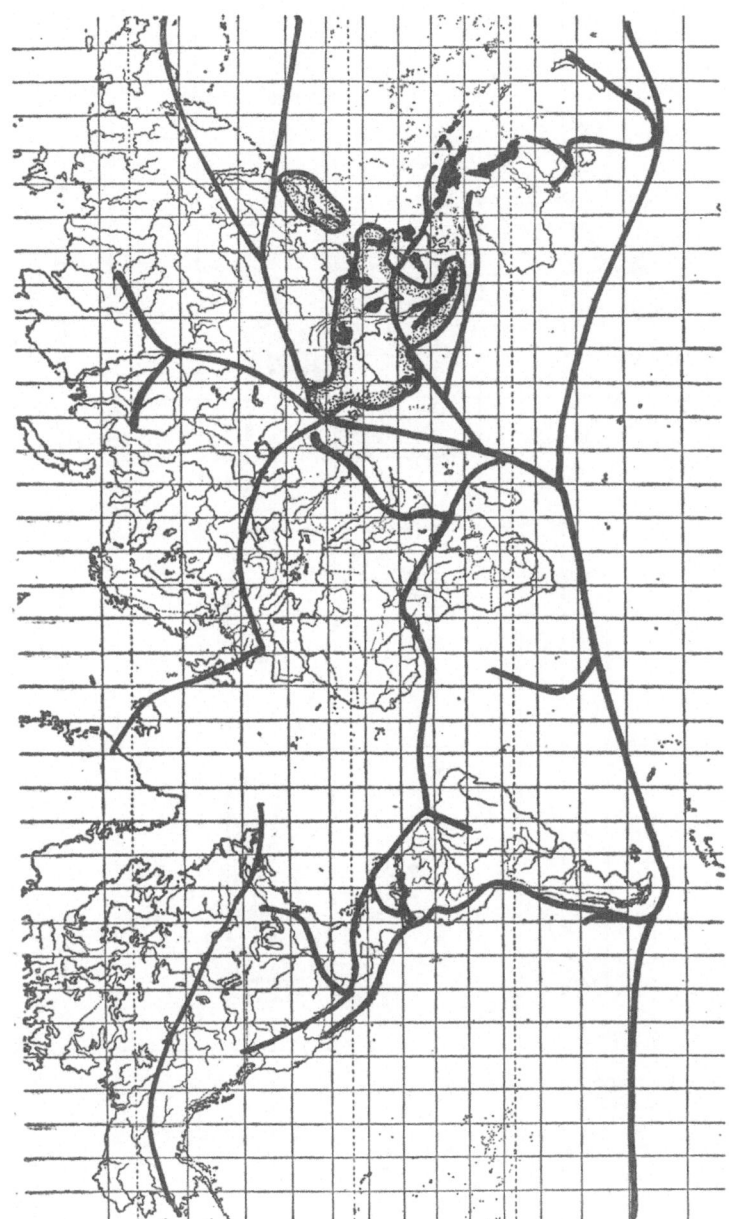

FIG. 95. Trends in the dispersal of the Ophioglossaceae. Range of *Helminthostachys*, solid black; range of *Botrychium* Sect. *Osmundopteris* within stippled lines. Essentially diagrammatic. See text, p. 496, for elucidation.

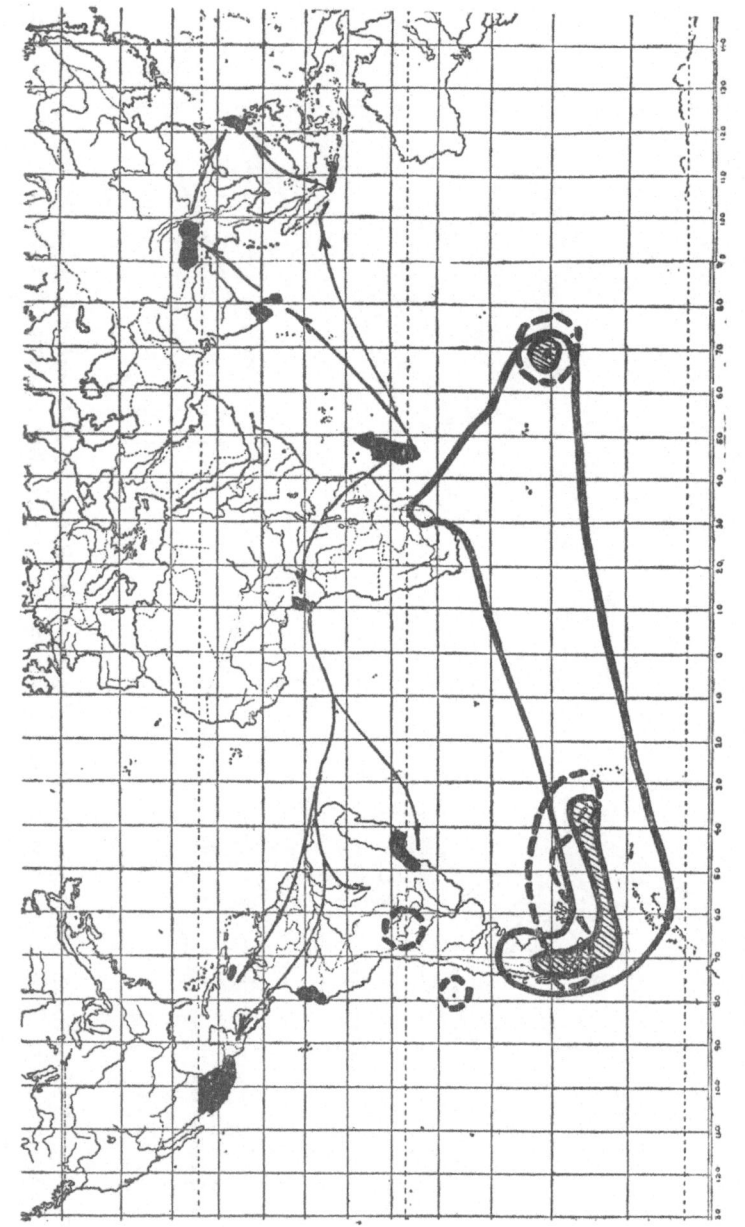

FIG. 96. Ranges of certain Liverworts (*Hepaticae*). *Jaegerina* and *Rhegmatodon* jointly, heavy black, with "tracks" indicative of potential connections: *Anisothecium Hookeri*, continuous line; *Psilopilum antarcticum* broken line; *Hymenoloma*, hatched. Essentially diagrammatic. See text, p. 509, for elucidation.

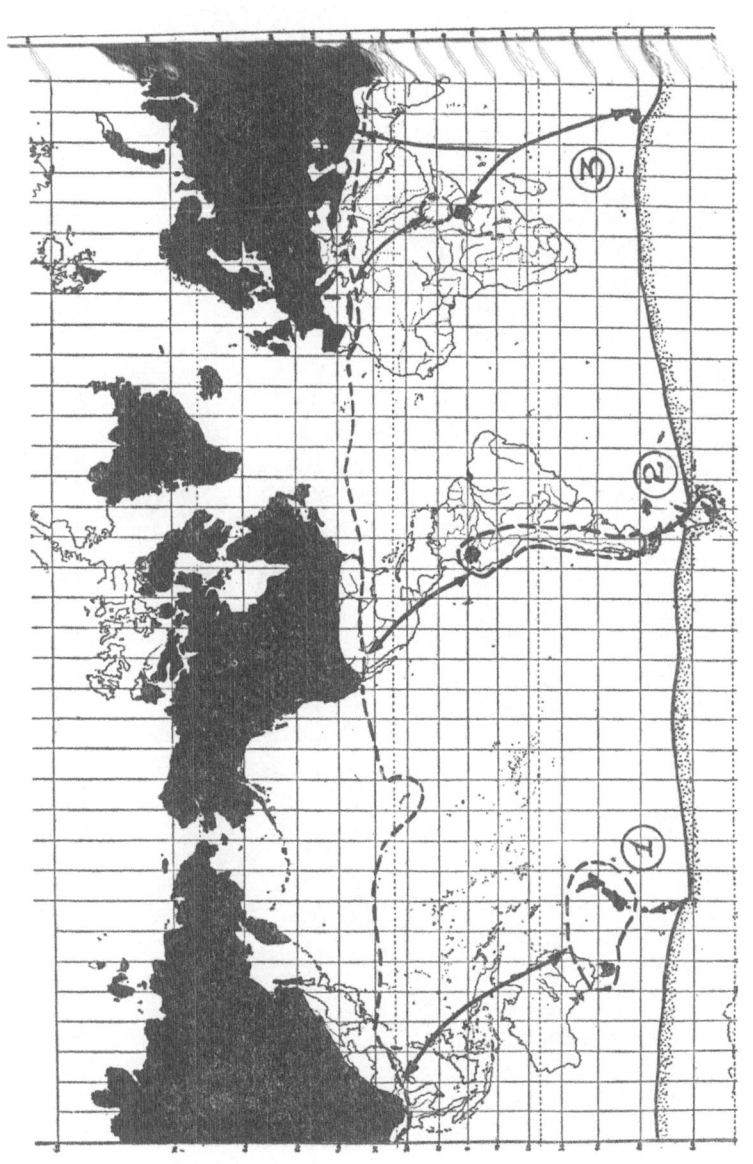

Fig. 97. Distribution of certain Liverworts (Hepaticae). *Dicranoweisia*, heavy black; *Encalypta*, within broken lines, and north of uppermost broken line. Gates of angiospermy as reference, 1 Western Polynesian; 2 Magellanian, 3 African. Essentially diagrammatic. See text, p. 509, for elucidation.

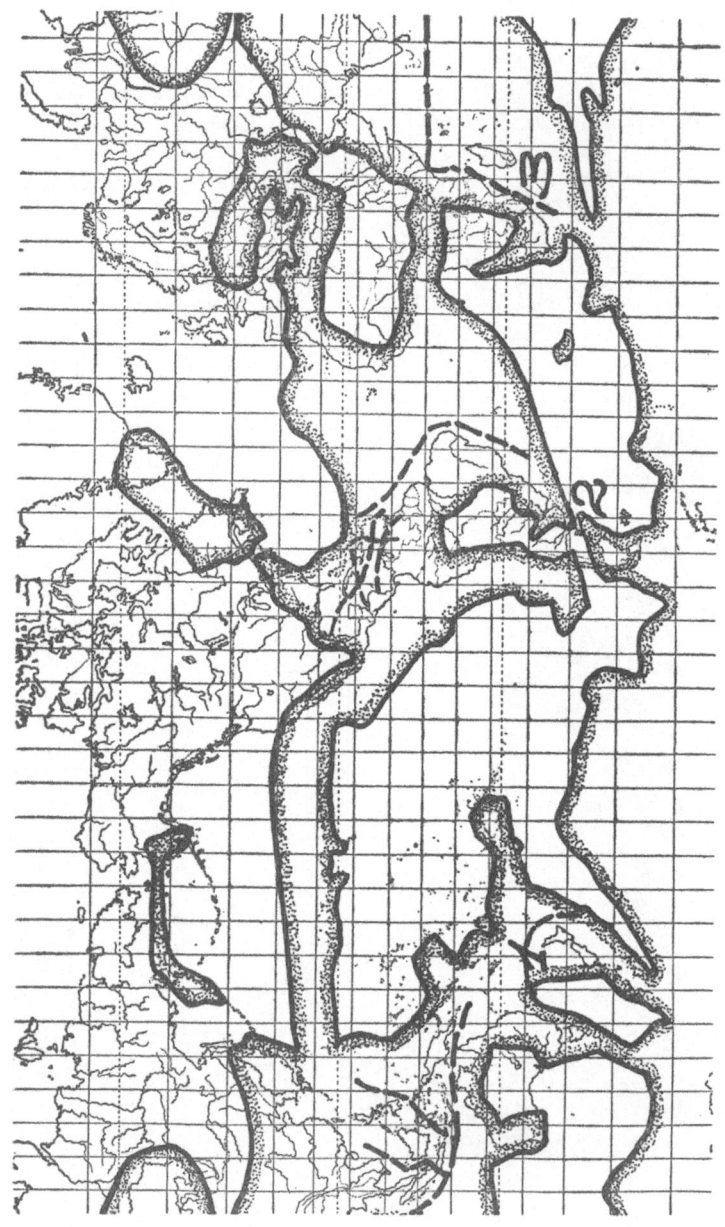

Fig. 98. Outlines illustrative of the earth at the inception of active angiospermous migrations. Lands within stippled lines; seas unstippled. Main zones of breaking, or faulting, as broken lines. Gates of angiospermy: 1, Western Polynesian; 2 Magellanian; 3 African. See text, p. 515, 522, for elucidation.

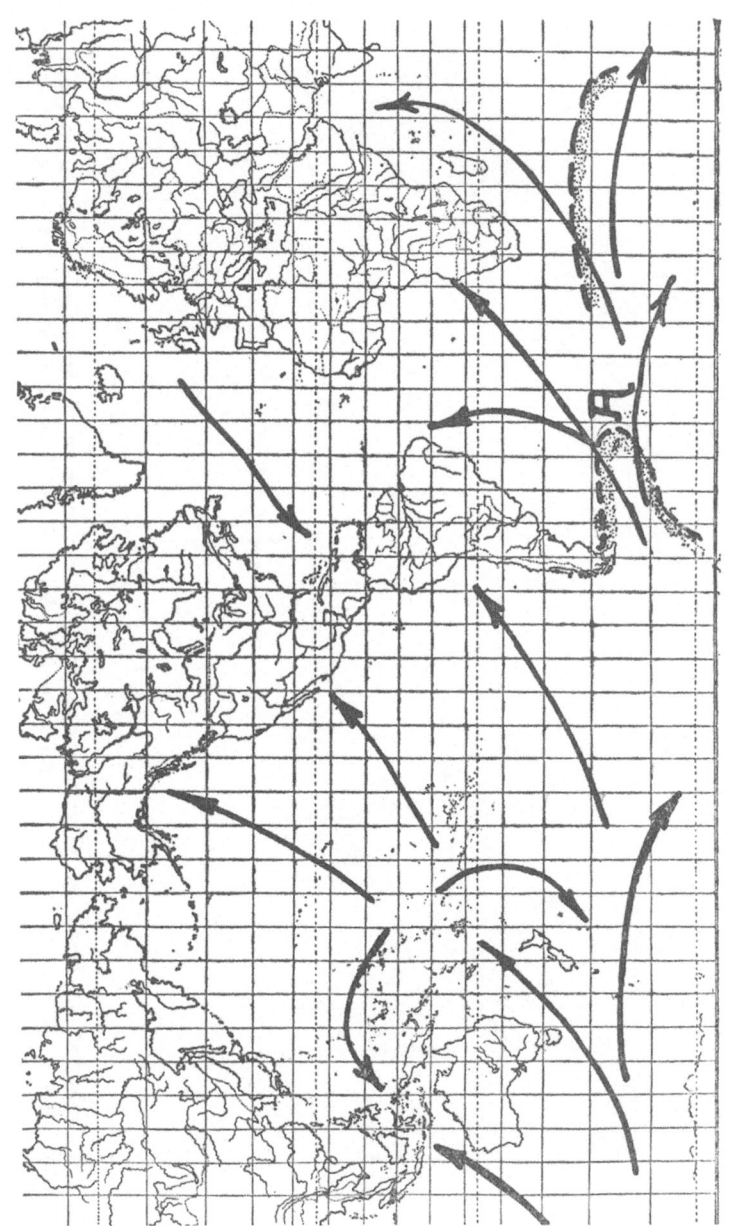

FIG. 99. Diagram to illustrate potential lines of stresses in the making of modern geography. Insular arc between South America and Antarctica in A (broken stippled line); shattered "insular arc" south of Africa as stippled line eastward. See text, p. 518, for discussion. The arrows bear no proportion to possible "floating".

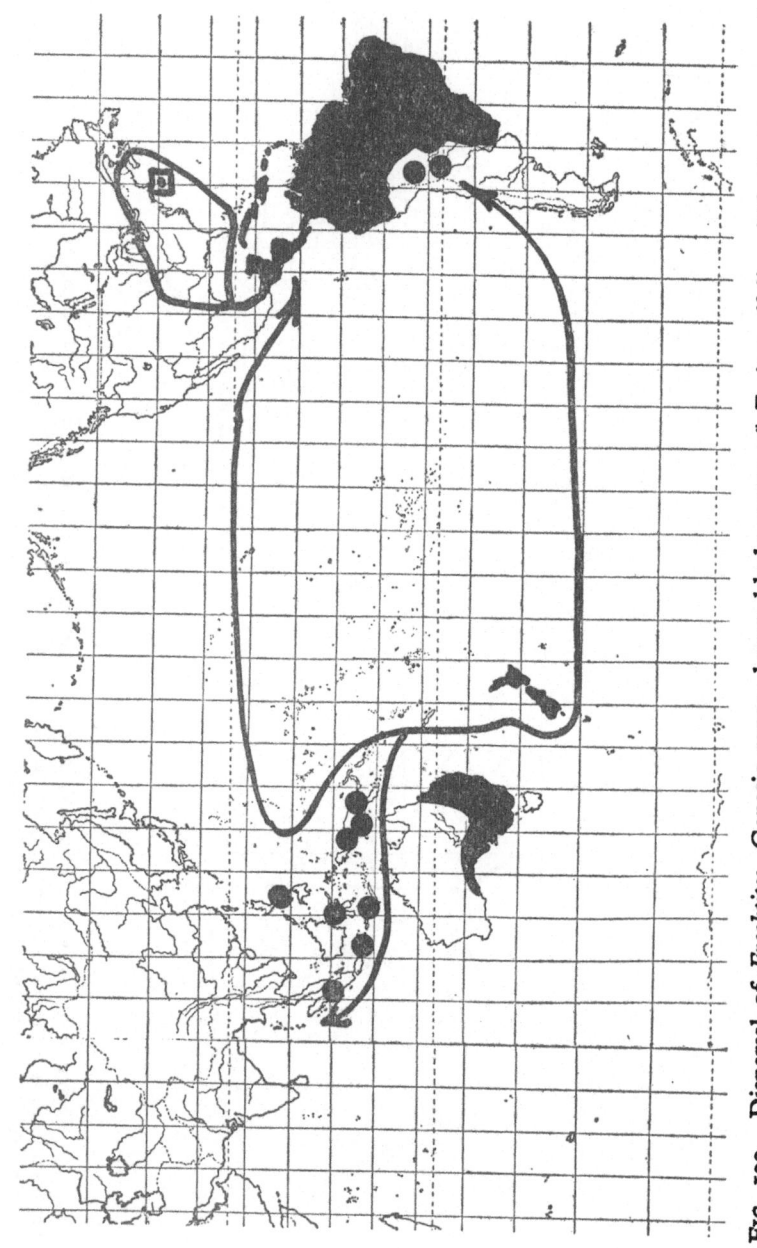

FIG. 100. Dispersal of *Erechtites*. Generic range, heavy black; range of *E. heracifolia* within line; range of *E. megalocarpa* as dotted square (all according to (6)). Missing or overlooked records as black circles. See text, p. 537, for discussion.

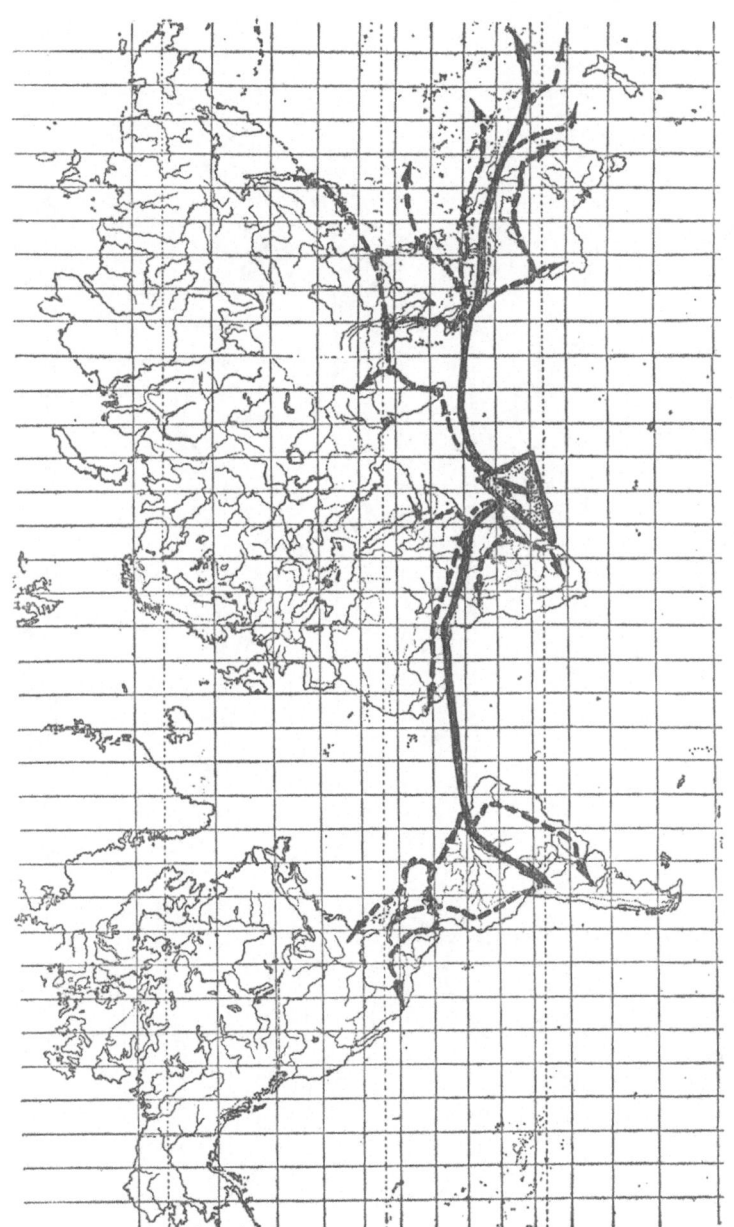

Fig. 101. The two main tracks out of the *Gondwanic Triangle* (1 within stippled outline). Main tracks as continuous lines; variants of the main tracks as broken lines. Essentially diagrammatic. See text, p. 548, for elucidation.

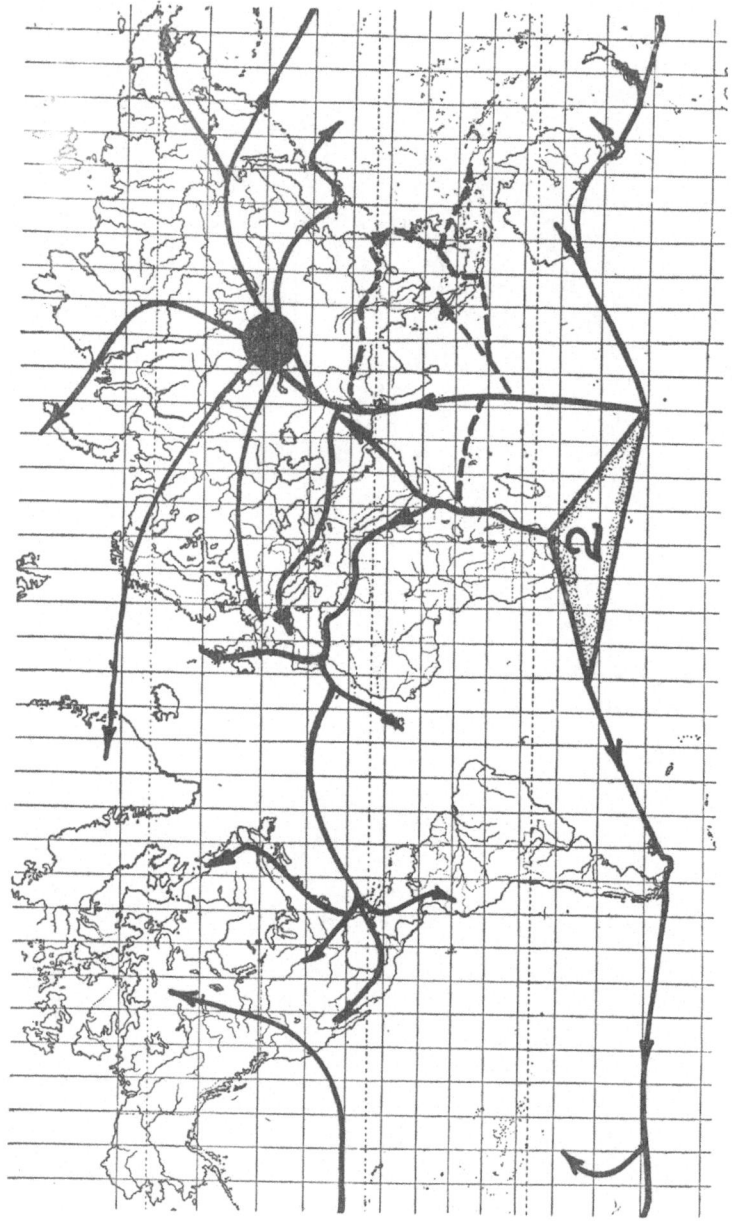

FIG. 102. Tracks out of the *Afroantarctic Triangle* (2 within stippled outline). Main tracks as continuous lines; variants and outliers as broken lines. Essentially diagrammatic. See text, p. 549, for elucidation.

Fig. 103. Tracks out of the *African Gate* (A, within stippled outline). *Magellanian Gate* in 4. Essentially diagrammatic. See text, p. 549, for elucidation.

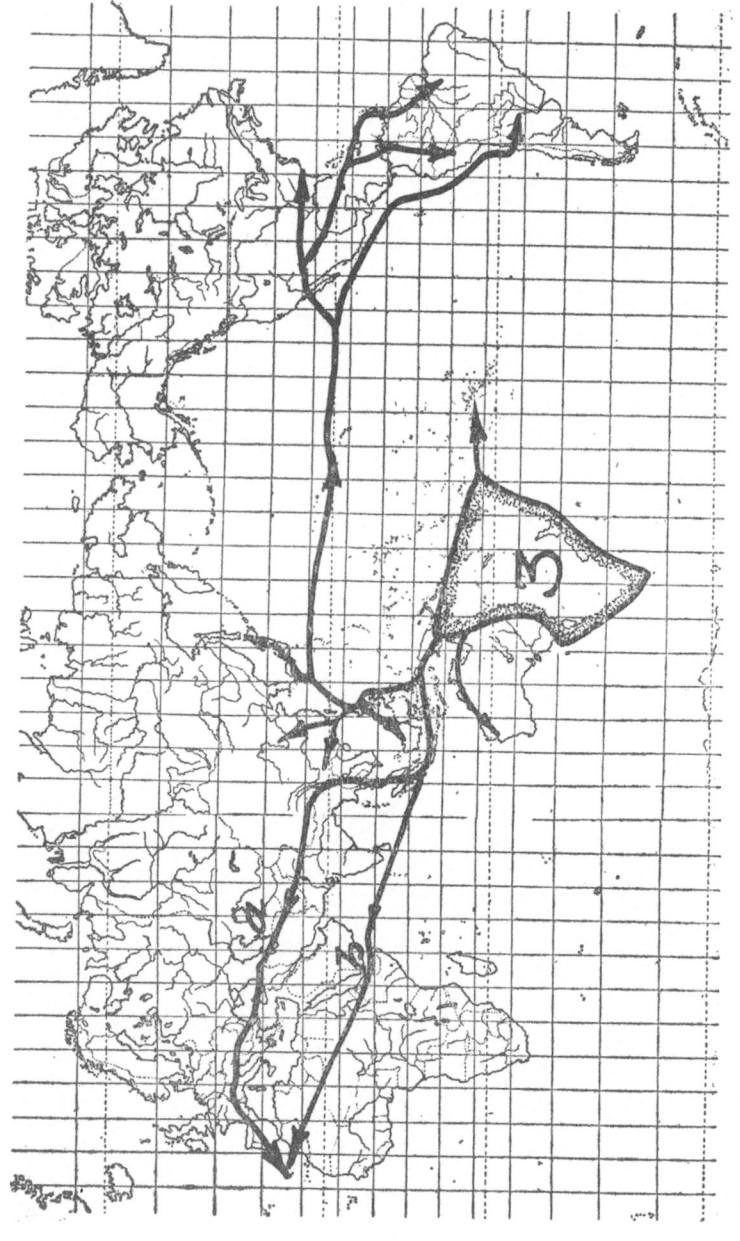

FIG. 104. Tracks out of the *Western Polynesian Gate* (3, within stippled outline). Tracks *a*, *b* are alternative. Outliers and variants as short spurs; track New Zealand—Magellania not shown. Essentially diagrammatic. See text, p. 549, for elucidation.

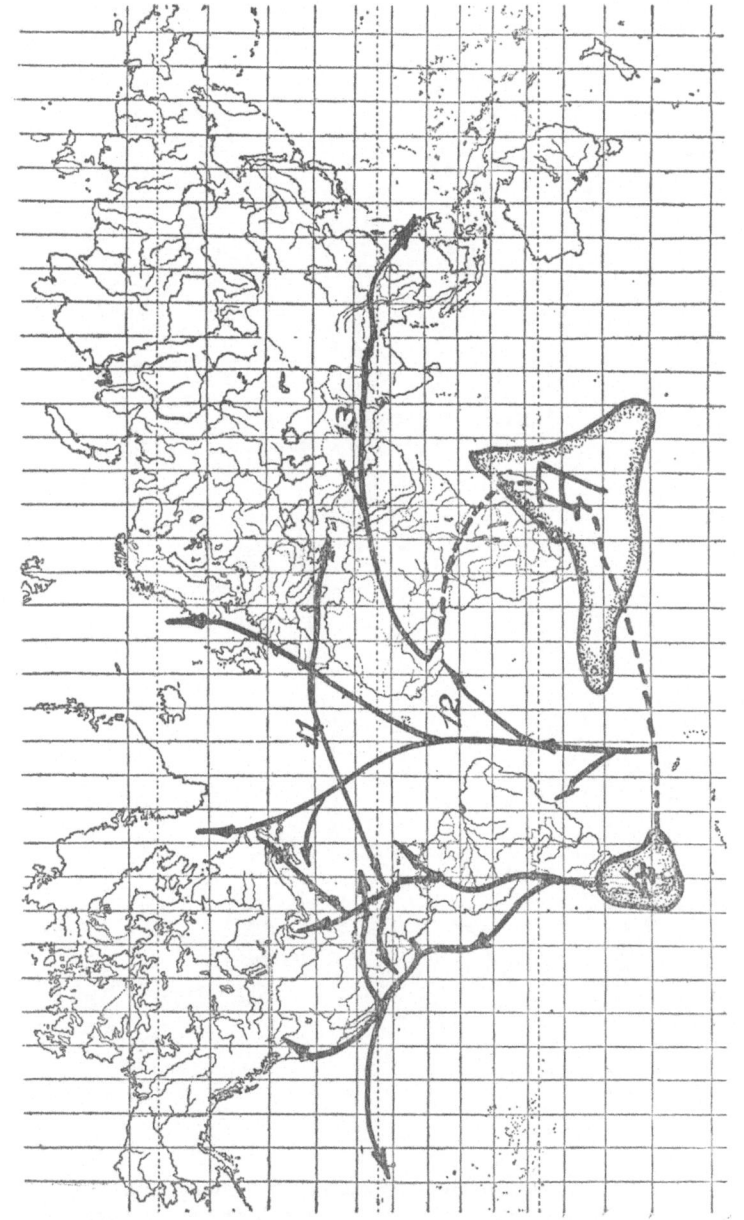

FIG. 105. Tracks out of the *Magellanian Gate* (4, within stippled outline). Added are tracks 11, 12, 13. *African Gate*, A within stippled outline. Essentially diagrammatic. See text, p. 550, for elucidation.

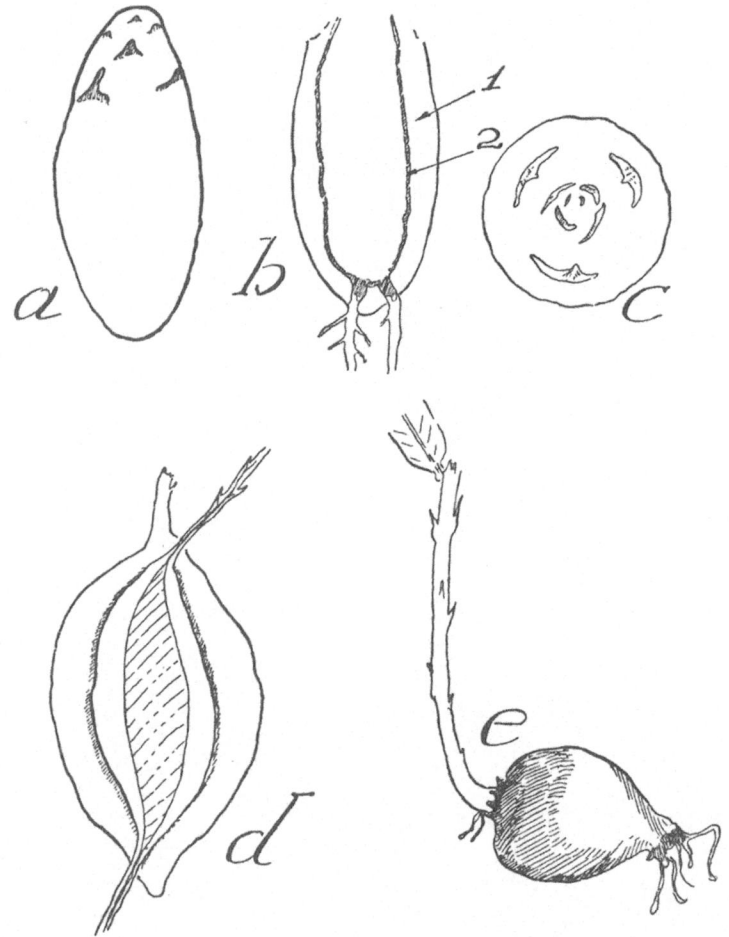

FIG. 106. Embryos of *Barringtonia Vriesei* (a, c) showing scales; seedling of same in b, revealing in longitudinal section the cortex (1) and the "inner jacket" or central cylinder (2) from which roots arise (all from [2]); Young plant of *Barringtonia racemosa*, showing regeneration from "inner jacket" by lateral bud (d); Young plant of *Careya arborea* showing habit of growth (all from [3]). See text for further elucidation, p. 551.